Benchmark Papers in Electrical Engineering and Computer Science

Series Editor: John B. Thomas
Princeton University

Volume

1 SYSTEM SENSITIVITY ANALYSIS / *José B. Cruz, Jr.*
2 RANDOM PROCESSES: Multiplicity Theory and Canonical Decompositions / *Anthony Ephremides and John B. Thomas*
3 ALGEBRAIC CODING THEORY: History and Development / *Ian F. Blake*
4 PATTERN RECOGNITION: Introduction and Foundations / *Jack Sklansky*
5 COMPUTER-AIDED CIRCUIT DESIGN: Simulation and Optimization / *S. W. Director*
6 ENVIRONMENTAL MODELING: Analysis and Management / *Douglas Daetz and Richard H. Pantell*
7 AUTOMATIC CONTROL: Classical Linear Theory / *George J. Thaler*
8 CIRCUIT THEORY: Foundations and Classical Contributions / *M. E. Van Valkenburg*
9 DATA COMMUNICATION: Fundamentals of Baseband Transmission / *L. E. Franks*
10 NONLINEAR SYSTEMS: Processing of Random Signals—Classical Analysis / *A. H. Haddad*
11 RANDOM PROCESSES, II: Poisson and Jump-Point Processes / *Anthony Ephremides*
12 DIGITAL FILTERS AND THE FAST FOURIER TRANSFORM / *Bede Liu*
13 DETECTION AND ESTIMATION: Applications to Radar / *Simon S. Haykin*
14 DATA COMPRESSION / *Lee D. Davisson and Robert M. Gray*
15 ACTIVE *RC* FILTERS: Theory and Application / *L. P. Huelsman*
16 NONLINEAR SYSTEMS: Stability Analysis / *J. K. Aggarwal and M. Vidyasagar*
17 LINEAR LEAST-SQUARES ESTIMATION / *Thomas Kailath*
18 OPTICAL COMMUNICATION THEORY / *R. O. Harger*
19 ERGODIC AND INFORMATION THEORY / *Robert M. Gray and Lee D. Davisson*
20 TWO-DIMENSIONAL DIGITAL SIGNAL PROCESSING / *Sanjit K. Mitra and Michael P. Ekstrom*
21 COMPUTER ARITHMETIC / *Earl E. Swartzlander, Jr.*
22 ARRAY PROCESSING: Applications to Radar / *Simon Haykin*
23 NONPARAMETRIC DETECTION: Theory and Applications / *Saleem A. Kassam and John B. Thomas*
24 DECONVOLUTION OF SEISMIC DATA / *V. K. Arya and J. K. Aggarwal*
25 REPRODUCING KERNEL HILBERT SPACES: Applications in Statistical Signal Processing / *Howard L. Weinert*

Benchmark Papers
in Electrical Engineering
and Computer Science / 25

A BENCHMARK® Books Series

REPRODUCING KERNEL HILBERT SPACES

Applications in Statistical Signal Processing

Edited by

HOWARD L. WEINERT

The Johns Hopkins University

Hutchinson Ross Publishing Company

Stroudsburg, Pennsylvania

Benchmark Papers in Electrical Engineering and Computer Science, Volume 25
Library of Congress Catalog Card Number: 82-9332
ISBN: 0-87933-434-7

84 83 82 1 2 3 4 5
Manufactured in the United States of America.

LIBRARY OF CONGRESS CATALOGING IN PUBLICATION DATA
Main entry under title:
Reproducing kernel Hilbert spaces.
(Benchmark papers in electrical engineering and computer science ; 25)
Includes bibliographies and indexes.
1. Signal processing—Addresses, essays, lectures. 2. Statistical communication theory—Addresses, essays, lectures. 3. Kernel functions—Addresses, essays, lectures.
4. Hilbert space—Addresses, essays, lectures.
I. Weinert, Howard L., 1946- . II. Series.
TK5102.5.R44 1982 621.38'043 82-9332
ISBN 0-87933-434-7 AACR2

Distributed worldwide by Van Nostrand Reinhold Company Inc.,
135 W. 50th Street, New York, NY 10020.

CONTENTS

SERIES EDITOR'S FOREWORD

This Benchmark Series in Electrical Engineering and Computer Science is aimed at sifting, organizing, and making readily accessible to the reader the vast literature that has accumulated. Although the series is not intended as a complete substitute for a study of this literature, it will serve at least three major critical purposes. In the first place, it provides a practical point of entry into a given area of research. Each volume offers an expert's selection of the critical papers on a given topic as well as his views on its structure, development, and present status. In the second place, the series provides a convenient and time-saving means for study in areas related to but not contiguous with one's principal interests. Last, but by no means least, the series allows the collection, in a particularly compact and convenient form, of the major works on which present research activities and interests are based.

Each volume in the series has been collected, organized, and edited by an authority in the area to which it pertains. In order to present a unified view of the area, the volume editor has prepared an introduction to the subject, has included his comments on each article, and has provided a subject index to facilitate access to the papers.

We believe that this series will provide a manageable working library of the most important technical articles in electrical engineering and computer science. We hope that it will be equally valuable to students, teachers, and researchers.

This volume, *Reproducing Kernel Hilbert Spaces,* has been edited by H.L. Weinert of The Johns Hopkins University. It contains thirty papers on the theory of reproducing kernel Hilbert space and its application to problems of detection, estimation, and approximation. This collection should be valuable to students and researchers interested in system theory, communication theory, and the structure of random processes.

JOHN B. THOMAS

PREFACE

Each previous volume in this series has focused on a single problem area. In contrast, this volume deals with a single methodology and its application to several different areas. Reproducing kernel Hilbert space (RKHS) theory provides an elegant and unified framework for a large class of problems in detection, estimation, and approximation. It allows us to see the intimate connections that may exist between problems that, at first glance, appear totally unrelated. These connections permit quantitative and qualitative results for one problem to be applied immediately to others, thus synthesizing a large body of specific results into a general framework. Just as important is the fact that in several instances RKHS methods have provided viable solutions where none has previously existed.

In editing this volume, I hope to partially fill the gap created by the absence of an English-language text on reproducing kernels, and thereby help students and researchers who might benefit from an awareness of this methodology and its existing applications. The papers I chose to include in this volume address problems of interest to the electrical engineering community and make significant use of RKHS methods in the analytical development of their results. Many of them also explore the connections discussed above. Furthermore, the reference lists at the end of each section provide a fairly complete bibliography.

I wish to thank the authors of the included papers for providing reprints, in many cases their last remaining copy. I am grateful to Professor John B. Thomas of Princeton University, the series editor, for giving me the opportunity to carry out this enterprise, and especially to Professor Thomas Kailath of Stanford University for introducing me to reproducing kernels and the beauty of connections.

HOWARD L. WEINERT

CONTENTS BY AUTHOR

REPRODUCING KERNEL HILBERT SPACES

INTRODUCTION

Reproducing kernel Hilbert space (RKHS) methods have been in use for at least seventy-five years, although the terminology and general theory were established only in the 1940s. Applications of significant interest to electrical engineers and statisticians date back some twenty-five years. Since then, RKHS techniques have provided solutions and a unified theoretical framework for many problems in detection, estimation, and approximation. Many of the significant papers dealing with these applications are reprinted in this volume.

A reproducing kernel Hilbert space is a Hilbert function space characterized by the fact that it contains a kernel that reproduces (via an inner product) every function in the space, or, equivalently, by the fact that every point evaluation functional is bounded.

RKHS methods are useful in certain detection and estimation problems because every covariance function is also a reproducing kernel for some RKHS. As a consequence, there is a close connection (an isometric isomorphism, in fact) between the closed linear span of a random process and the RKHS determined by its covariance function. These detection and estimation problems can then be solved by evaluating a certain RKHS inner product. Thus it is necessary to be able to determine the form of the inner product corresponding to a given reproducing kernel.

In optimal curve fitting problems, in which one is reconstructing an unknown function based on its samples, it is necessary that point evaluation functionals be bounded, so one is forced to formulate the problem in a RKHS whose inner product is determined by the quadratic cost functional that must be minimized. To solve these problems one must find a basis for the range of a certain projection operator, an easy task if one can determine the reproducing kernel corresponding to the given inner product.

This volume is organized as follows. The papers reprinted in Part I present the relevant aspects of RKHS theory. Those in Part II survey a large number of statistical signal processing applications and address the problem of evaluating RKHS inner products. Part III contains papers that deal with particular applications in Gaussian and non-Gaussian detection and in parameter estimation. Part IV deals with applications to structural problems of stochastic processes, including representation of nonlinear functionals, generalized innovations representations, and Markovian properties of random fields. The papers in Part V present applications to optimal-curve and surface fitting using splines. Each part contains a supplementary reference list.

There are some other interesting RKHS applications that, for reasons of space, could not be included in this volume. Hitsuda and Ihara [1] derive expressions for mutual information in Gaussian feedback channels. Messerschmitt [2] uses RKHS methods in the study of intersymbol interference. Csibi [3] develops iterative algorithms for adaptation and learning. Applications in system identification and quantum optics are discussed by Mosca [4] and Klauder and Sudarshan [5], respectively. Relations between RKHS and orthogonal polynomials are studied and exploited by Vieira and Kailath [6], Kailath et al. [7], and Dym and Gohberg [8].

REFERENCES

1. M. Hitsuda and S. Ihara, Gaussian Channels and Optimal Coding, *J. Multivar. Anal.* **5:**106–118 (1975).
2. D. G. Messerschmitt, A Geometric Theory of Intersymbol Interference, *Bell Syst. Tech. J.* **52:**1483–1539 (1973).
3. S. Csibi, *Stochastic Processes with Learning Properties,* Springer-Verlag, Berlin, 1975, 150p.
4. E. Mosca, A Deterministic Approach to a Class of Nonparametric System Identification Problems, *IEEE Trans. Inform. Theory* **IT-17:**686–696 (1971).
5. J. R. Klauder and E. C. G. Sudarshan, *Fundamentals of Quantum Optics,* Benjamin, New York, 1968, 279p.
6. A. Vieira and T. Kailath, On Another Approach to the Schur-Cohn Criterion, *IEEE Trans. Circuits and Syst.* **CAS-24:**218–220 (1977).
7. T. Kailath, A. Vieira, and M. Morf, Inverses of Toeplitz Operators, Innovations, and Orthogonal Polynomials, *Soc. Ind. Appl. Math. Rev.* **20:**106–119 (1978).
8. H. Dym and I. Gohberg, On an Extension Problem, Generalized Fourier Analysis, and an Entropy Formula, *Integral Equations and Oper. Theory* **3:**143–215 (1980).

Part I

THEORETICAL BACKGROUND

Editor's Comments on Papers 1 and 2

1 **ARONSZAJN**
Theory of Reproducing Kernels

2 **STEWART**
Positive Definite Functions and Generalizations, An Historical Survey

Reproducing kernels were used as early as the first decade of this century, but it was not until 1943 that Aronszajn [1] wrote the first systematic development of reproducing kernel Hilbert space theory. It was he, in fact, who coined the term "reproducing kernel." Paper 1 is an expanded version of this early work and is the standard reference for RKHS theory. The introduction to Paper 1 gives the history of reproducing kernels and their applications up to 1948. At that time, some independent work was being carried out in the Soviet Union by Povzner, who derived many of the basic properties of RKHS in "On a Class of Hilbert Function Spaces" [2], and gave several examples in "On Some Applications of a Class of Hilbert Function Spaces" [3]. Krein [4] also derived some RKHS properties in his extensive study of kernels satisfying certain invariance conditions.

Other works dealing with RKHS theory are Meschkowski [5], Hille [6], Davis [7, Ch. 12], and Shapiro [8, Ch. 6]. Some particular reproducing kernel Hilbert spaces are treated at length in the books by Bergman [9], deBranges [10], and Duren [11]. A more abstract development of the theory appears in a paper by Schwartz [12]. With few exceptions, however, Paper 1 contains all the theoretical results needed for applications. The reader can find brief treatments of RKHS theory in some of the other papers included in this volume.

As discussed in the introduction to Paper 1, RKHS theory and the theory of positive definite functions are two sides of the same coin. In Paper 2 Stewart gives a concise historical survey of positive definite functions and their principal generalizations as well as a useful bibliography. Sections 1, 3, 8, 9 of Paper 2 are of particular interest for the applications discussed in this volume.

REFERENCES

1. N. Aronszajn, The Theory of Reproducing Kernels and Their Applications (in French), *Cambridge Philos. Soc. Proc.* **39:**133–153 (1943).
2. A. Povzner, On a Class of Hilbert Function Spaces (in Russian), *Dokl. Akad. Nauk. SSSR* **68:**817–820 (1949).
3. A. Povzner, On Some Applications of a Class of Hilbert Function Spaces (in Russian), *Dokl. Akad. Nauk. SSSR* **74:**13–16 (1950).
4. M. G. Krein, Hermitian-Positive Kernels on Homogeneous Spaces (in Russian), *Ukrainian Math. J.* **1:**64–98 (1949) and **2:**10–59 (1950); English translation in *Am. Math. Soc. Transl.*, ser., 2, **34:**69–164 (1963).
5. H. Meschkowski, *Hilbert Spaces with Kernel Functions* (in German), Springer-Verlag, Berlin, 1962, 256p.
6. E. Hille, Introduction to General Theory of Reproducing Kernels, *Rocky Mt. J. Math.* **2:**321–368 (1972).
7. P. J. Davis, *Interpolation and Approximation,* Dover, New York, 1975, 393p.
8. H. S. Shapiro, *Topics in Approximation Theory,* Springer-Verlag, Berlin, 1971, 275p.
9. S. Bergman, *The Kernel Function and Conformal Mapping,* American Mathematical Society, New York, 1950, 161p.
10. L. deBranges, *Hilbert Spaces of Entire Functions,* Prentice-Hall, Englewood Cliffs, N. J., 1968, 326p.
11. P. L. Duren, *Theory of* H^p *Spaces,* Academic Press, New York, 1970, 258p.
12. L. Schwartz, Hilbert Subspaces of Topological Vector Spaces and Associated Kernels (in French), *J. d'Analyse Math.* **13:**115–256 (1964).

1

Reprinted by permission from *Am. Math. Soc. Trans.* **68**:337–404 (1950)

THEORY OF REPRODUCING KERNELS[1]

BY

N. ARONSZAJN

PREFACE

The present paper may be considered as a sequel to our previous paper in the Proceedings of the Cambridge Philosophical Society, *Théorie générale de noyaux reproduisants—Première partie* (vol. 39 (1944)) which was written in 1942–1943. In the introduction to this paper we outlined the plan of papers which were to follow. In the meantime, however, the general theory has been developed in many directions, and our original plans have had to be changed.

Due to wartime conditions we were not able, at the time of writing the first paper, to take into account all the earlier investigations which, although sometimes of quite a different character, were, nevertheless, related to our subject.

Our investigation is concerned with kernels of a special type which have been used under different names and in different ways in many domains of mathematical research. We shall therefore begin our present paper with a short historical introduction in which we shall attempt to indicate the different manners in which these kernels have been used by various investigators, and to clarify the terminology. We shall also discuss the more important trends of the application of these kernels without attempting, however, a complete bibliography of the subject matter.

In Part I, we shall discuss briefly the essential notions and results of our previous paper and give a further development of the theory in an abstract form. In Part II, we shall illustrate the results obtained in the first part by a series of examples which will give new developments of already known applications of the theory, as well as some new applications.

TABLE OF CONTENTS

Presented to the Society, December 28, 1948, under the title *Hilbert spaces and conformal mappings*; received by the editors November 26, 1948.

(1) Paper done under contract with the Office of Naval Research, N5ori 76-16-NR 043-046, Division of Engineering Science, Harvard University.

HISTORICAL INTRODUCTION

Examples of kernels of the type in which we are interested have been known for a long time, since all the Green's functions of self-adjoint ordinary differential equations (as also some Green's functions—the bounded ones—of partial differential equations) belong to this type. But the characteristic properties of these kernels as we now understand them have only been stressed and applied since the beginning of the century.

There have been and continue to be two trends in the consideration of these kernels. To explain them we should mention that such a kernel $K(x, y)$ may be characterized as a function of two points, by a property discovered by J. Mercer [1](²) in 1909. To the kernel K there corresponds a well determined class F of functions $f(x)$, in respect to which K possesses the "reproducing" property (E. H. Moore [2]). On the other hand, to a class of functions F, there may correspond a kernel K with "reproducing" property (N. Aronszajn [4]).

Those following the first trend consider a given kernel K and study it in itself, or eventually apply it in various domains (as integral equations, theory of groups, general metric geometry, and so on). The class F corresponding to K may be used as a tool of research, but is introduced a posteriori (as in the work of E. H. Moore [2], and more recently of A. Weil [1], I. Gelfand and D. Raikoff [1], and R. Godement [1, 2]). In the second trend, one is interested primarily in a class of functions F, and the corresponding kernel K is used essentially as a tool in the study of the functions of this class. One of the basic problems in this kind of investigation is the explicit construction and computation of the kernel for a given class F.

(²) Numbers in brackets refer to the bibliography at the end of the paper.

The first of these trends originated in the theory of integral equations as developed by Hilbert. The kernels considered then were continuous kernels of positive definite integral operators. This theory was developed by J. Mercer [1, 2] under the name of "positive definite kernels" and on occasion has been used by many others interested in integral equations, especially during the second decade of this century. Mercer discovered the property

$$(1) \qquad \sum_{i,j=1}^{n} K(y_i, y_j)\xi_i\bar{\xi}_j \geqq 0, \qquad y_i \textit{ any points}, \ \xi_i \textit{ any complex numbers}(^3),$$

characterizing his kernels, among all the continuous kernels of integral equations. To this same trend belong the investigations of E. H. Moore [1, 2] who, during the second and third decades of the century, introduced these kernels in the general analysis under the name of "positive hermitian matrices" with a view to applications in a kind of generalization of integral equations. Moore considered kernels $K(x, y)$ defined on an abstract set E and characterized by the property (1). He discovered the theorem now serving as one of the links between the two trends, proving that to each positive hermitian matrix there corresponds a class of functions forming what we now call a Hilbert space with a scalar product (f, g) and in which the kernel has the reproducing property.

$$(2) \qquad f(y) = (f(x), K(x, y)).$$

Also to the same trend (though seemingly without any connection to previous investigations) belongs the notion introduced by S. Bochner [2] during the third decade of the century under the name of "positive definite functions." Bochner considered continuous functions $\phi(x)$ of real variable x such that the kernels $K(x, y) = \phi(x-y)$ conformed to condition (1). He introduced these functions with a view to application in the theory of Fourier transforms. The notion was later generalized by A. Weil [1] and applied by I. Gelfand and D. Raikoff [1], R. Godement [1, 2], and others to the investigation of topological groups under the name of positive definite functions or functions of positive type. These functions were also applied to general metric geometry (the Hilbert distances) by I. J. Schoenberg [1, 2], J. v. Neumann and I. J. Schoenberg [1], and S. Bochner [3].

The second trend was initiated during the first decade of the century in the work of S. Zaremba [1, 2] on boundary value problems for harmonic and biharmonic functions. Zaremba was the first to introduce, in a particular case, the kernel corresponding to a class of functions, and to state its reproducing property (2). However, he did not develop any general theory, nor did he give any particular name to the kernels he introduced. It appears that nothing was done in this direction until the third decade when S. Berg-

(3) Mercer used only real numbers ξ_i as he considered only real kernels K.

man [1] introduced kernels corresponding to classes of harmonic functions and analytic functions in one or several variables. He called these "kernel functions." They were introduced as kernels of orthogonal systems in these classes for an adequate metric. The reproducing property of these kernels was noticed by Bergman [1] (also by N. Aronszajn [1]), but it was not used as their basic characteristic property as is done at present.

In the third and fourth decades most of the work was done with kernels which we shall call Bergman's kernels, that is, kernels of classes of analytic functions f of one or several complex variables, regular in a domain D with the quadratic metric

$$\int_D |f|^2 d\tau.$$

A quantity of important results were achieved by the use of these kernels in the theory of functions of one and several complex variables (Bergman [4, 6, 7], Bochner [1]), in conformal mapping of simply- and multiply-connected domains (Bergman [11, 12], Zarankiewicz [1] and others), in pseudo-conformal mappings (Bergman [4, 5, 8, 9], Welke [1], Aronszajn [1], and others), in the study of invariant Riemannian metrics (Bergman [11, 14], Fuchs [1, 2]), and in other subjects.

The original idea of Zaremba to apply the kernels to the solution of boundary value problems was represented in these two decades by only a few papers of Bergman [1, 2, 3, 10]. Only since the last war has this idea been put into the foreground by a series of papers by Bergman [13], and Bergman and Schiffer [1, 2, 3]. In these investigations, the kernel was proved to be a powerful tool for solving boundary value problems of partial differential equations of elliptic type. By the use of variational methods going back to Hadamard, relations were established between the kernels corresponding to classes of solutions of different equations and for different domains (Bergman and Schiffer [1, 3]). For a partial differential equation, the kernel of the class of solutions in a domain was proved to be the difference of the corresponding Neumann's and Green's functions (Bergman and Schiffer [1, 3]) (in the special case of the biharmonic equation a relation of this kind was already noticed by Zaremba). Parallel to this revival of the application of kernels to partial differential equations there is developing a study of the relationship between these kernels and Bergman's kernels of analytic functions (Bergman [12], M. Schiffer [1, 2]). Also the application of kernels to conformal mapping of multiply-connected domains has made great progress as all the important mapping functions were proved to be simply expressible by the Bergman's kernel (Bergman [11, 12], P. Garabedian and M. Schiffer [1], Garabedian [1], and Z. Nehari [1, 2]). Quite recently, the connection was found between the Bergman's kernel and the kernel introduced by G. Szegö (P. Garabedian [1]).

In 1943, the author ([4] also [6]) developed the general theory of reproducing kernels which contains, as particular cases, the Bergman kernel-functions. This theory gives a general basis for the study of each particular case and allows great simplification of many of the proofs involved. In this theory a central role is played by the reproducing property of the kernel in respect to the class to which it belongs. The kernel is defined by this property. The simple fact was stressed that a reproducing kernel always possesses property (1) characteristic of positive hermitian matrices (in the sense of E. H. Moore). This forms the second link between the two trends in the kernel theory (the theorem of E. H. Moore forming the first link was mentioned above).

The mathematicians working in the two trends seem not to have noticed the essential connections between the general notions they were using. At present the two concepts of the kernel, as a positive hermitian matrix and as a reproducing kernel, are known to be equivalent and methods elaborated in the investigations belonging to one trend prove to be of importance in the other.

We should like to elaborate here briefly on the matter of the terminology which has been used by various investigators. As we have seen above, different names have been given to the kernels in which we are interested. When the kernels were used in themselves, without special or previous consideration of the class to which they belonged, they were called "positive definite kernels," "positive hermitian matrices," "positive definite functions," or "functions of positive type." In cases where they were considered as determined, and in connection with a class of functions, they were called "kernel functions" or "reproducing kernels." It is not our intention to settle here the question of terminology. Our purpose is rather to state our choice and to give our reasons for it together with a comparison of the terminology we have chosen with that used by other authors.

It would seem advisable to keep two names for our kernels, the function of each name being to indicate immediately in what context the kernel under consideration is to be taken. Thus, when we consider the kernel in itself we shall call it (after E. H. Moore) a *positive matrix*(4), in abbreviation, *p. matrix*, or *p. m.* When we wish to indicate the kernel corresponding to a class of functions we shall call it *the reproducing kernel of the class*, in abbreviation, *r. kernel* or *r.k.*

As compared to other terminology, we believe that the name "positive definite function" or perhaps better "function of positive type" will probably continue to be used in the particular case when the kernel is of the form $\phi(x-y)$, x, y belonging to an additive group. This term has been used in a few instances for some more general kernels, but we believe that it would prove to be more convenient if it were restricted to the particular case

(4) We drop here the adjective "hermitian" since the condition that the quadratic form (1) be positive implies the hermitian symmetry of the matrix.

mentioned above.

Although the name of "positive definite kernels" would seem, somehow, more adequate than "positive matrices," especially since it was introduced first, we have chosen rather the term used by E. H. Moore. This is because we wish to reserve the notion of positive definite kernel for more general kernels which would include the positive definite matrices as well as some other non-bounded kernels (such as the kernels of general positive definite integral operators and also the recently introduced pseudo-reproducing kernels [Aronszajn 5, 6]).

On the other hand, when we have in mind the kernel corresponding to a given class of functions, the simplest terminology is to call it "the kernel of the class" or "the kernel belonging to the class." But when some ambiguity is to be feared, or when we wish to stress its characteristic property, we use the adjective "reproducing."

PART I. GENERAL THEORY

1. Definition of reproducing kernels. Consider a linear class F of functions $f(x)$ defined in a set E. We shall suppose that F is a *complex* class, that is, that it admits of multiplication by complex constants.

Suppose further that for $f \in F$ is defined a norm $\|f\|$ (that is, a real number satisfying: $\|f\| \geqq 0$, $\|f\| = 0$ only for $f = 0$, $\|cf\| = |c|\,\|f\|$, $\|f+g\| \leqq \|f\| + \|g\|$) given by a quadratic hermitian form $Q(f)$

$$\|f\|^2 = Q(f).$$

Here, a functional $Q(f)$ is called quadratic hermitian if for any constants ξ_1, ξ_2 and functions f_1, f_2 of F

$$Q(\xi_1 f_1 + \xi_2 f_2) = |\xi_1|^2 Q(f_1) + \xi_1 \bar{\xi}_2 Q(f_1, f_2) + \bar{\xi}_1 \xi_2 Q(f_2, f_1) + |\xi_2|^2 Q(f_2).$$

$Q(f_1, f_2) = \overline{Q(f_2, f_1)}$ is the uniquely determined bilinear hermitian form corresponding to the quadratic form $Q(f)$. This bilinear form will be denoted by $(f_1, f_2) \equiv Q(f_1, f_2)$ and called the scalar product corresponding to the norm $\|f\|$ (or the quadratic metric $\|f\|^2$). We have

$$\|f\|^2 = (f, f).$$

The class F with the norm, $\|\ \|$, forms a normed complex vector space. If this space is complete it is a Hilbert space.

If F is a class of real-valued functions forming a *real* vector space (that is, admitting of multiplication with only real constants), if the norm, $\|\ \|$, in F is given by $\|f\|^2 = Q(f)$ with an ordinary quadratic form Q (that is, for real ξ_1, ξ_2, $Q(\xi_1 f_1 + \xi_2 f_2) = \xi_1^2 Q(f_1) + 2\xi_1\xi_2 Q(f_1, f_2) + \xi_2^2 Q(f_2)$, where $Q(f_1, f_2)$ is the corresponding bilinear symmetric form), and if F is a complete space, it is a real Hilbert space. The scalar product is given there by $(f_1, f_2) \equiv Q(f_1, f_2)$.

Every class F of real functions forming a real Hilbert space determines a complex Hilbert space in the following way: consider all functions f_1+if_2 with f_1 and f_2 in F. They form a complex vector space F_c in which we define the norm by $\|f_1+if_2\|^2=\|f_1\|^2+\|f_2\|^2$. F_c is a complex Hilbert space.

The complex Hilbert spaces F determined in this way by real Hilbert spaces are characterized by the two properties:

(1) if $f\in F$, $\bar{f}\in F$ ($\bar{f}$ is the conjugate complex function of f),

(2) $\|f\|=\bar{f}$.

Let F be a class of functions defined in E, forming a Hilbert space (complex or real). The function $K(x, y)$ of x and y in E is called a *reproducing kernel* (*r.k.*) of F if

1. For every y, $K(x, y)$ as function of x belongs to F.
2. *The reproducing property*: for every $y\in E$ and every $f\in F$,

$$f(y) = (f(x), K(x, y))_x.$$

The subscript x by the scalar product indicates that the scalar product applies to functions of x.

If a real class F possesses a r.k. $K(x, y)$ then it is immediately verified that the corresponding complex space F_c possesses the same kernel (which is real-valued):

From now on (unless otherwise stated) we shall consider only complex Hilbert spaces. As we have seen there is no essential limitation in this assumption.

It will be useful to introduce a distinction between the terms subclass and subspace. When F_1 and F_2 are two classes of functions defined in the same set E, F_1 is a subclass of F_2 if every f of F_1 belongs to F_2. F_1 is a subspace of F_2 if it is a subclass of F_2 and if for every $f\in F_1$, $\|f\|_1=\|f\|_2$ ($\|\ \|_1$ and $\|\ \|_2$ are the norms in F_1 and F_2 respectively). $F_1\subset F_2$ means that F_1 is a subclass of F_2.

2. Résumé of basic properties of reproducing kernels. In the following, F denotes a class of functions $f(x)$ defined in E, forming a Hilbert space with the norm $\|f\|$ and scalar product (f_1, f_2). $K(x, y)$ will denote the corresponding reproducing kernel.

The detailed proofs of the properties listed below may be found in [Aronszajn, 4].

(1) *If a r.k. K exists it is unique.* In fact, if another $K'(x, y)$ existed we would have for some y

$$\begin{aligned} 0 < \|K(x, y) - K'(x, y)\|^2 &= (K - K', K - K') \\ &= (K - K', K) - (K - K', K') = 0 \end{aligned}$$

because of the reproducing property of K and K'.

(2) *Existence.* For the existence of a r.k. $K(x, y)$ it is necessary and sufficient that for every y of the set E, $f(y)$ be a continuous functional of f running through the Hilbert space F.

In fact, if K exists, then $|f(y)| \leq \|f\|(K(x, y), K(x, y))^{1/2}=K(y, y)^{1/2}\|f\|$.

On the other hand if $f(y)$ is a continuous functional, then by the general theory of the Hilbert space there exists a function $g_y(x)$ belonging to F such that $f(y) = (f(x), g_y(x))$, and then if we put $K(x, y) = g_y(x)$ it will be a reproducing kernel.

(3) $K(x, y)$ is *a positive matrix* in the sense of E. H. Moore, that is, the quadratic form in $\xi_1, \cdots, \xi_n$

$$\sum_{i,j=1}^{n} K(y_i, y_j)\bar{\xi}_i\xi_j \tag{1}$$

is non-negative for all $y_1, \cdots, y_n$ in E. This is clear since expression (1) equals $\| \sum_1^n K(x, y_i)\xi_i\|^2$, following the reproducing property. In particular it follows that

$$K(x, x) \geqq 0, \quad K(x, y) = \overline{K(y, x)}, \quad |K(x, y)|^2 \leqq K(x, x)\, K(y, y).$$

(4) The theorem in (3) admits a converse due essentially to E. H. Moore: *to every positive matrix $K(x, y)$ there corresponds one and only one class of functions with a uniquely determined quadratic form in it, forming a Hilbert space and admitting $K(x, y)$ as a reproducing kernel.*

This class of functions is generated by all the functions of the form $\sum \alpha_k K(x, y)$. The norm of this function is defined by the quadratic form $\| \sum \alpha_k K(x, y_k)\|^2 = \sum\sum K(y_i, y_j)\bar{\xi}_i\xi_j$. Functions with this norm do not as yet form a complete Hilbert space, but it can be easily seen that they may be completed by the adjunction of functions to form such a complete Hilbert space. This follows from the fact that every Cauchy sequence of these functions (relative to the above norm) will converge at every point x towards a limit function whose adjunction to the class will complete the space.

(5) If the class F possesses a r.k. $K(x, y)$, every sequence of functions $\{f_n\}$ which converges strongly to a function f in the Hilbert space F, converges also at every point in the ordinary sense, $\lim f_n(x) = f(x)$, this convergence being uniform in every subset of E in which $K(x, x)$ is uniformly bounded. This follows from

$$\begin{aligned} |f(y) - f_n(y)| &= |(f(x) - f_n(x), K(x, y))| \leqq \|f - f_n\|\,\|K(x, y)\| \\ &= \|f - f_n\|(K(y, y))^{1/2}. \end{aligned}$$

If f_n converges weakly to f, we have again $f_n(y) \to f(y)$ for every y (since, by the definition of the weak convergence, $(f_n(x), K(x, y)) \to (f(x), K(x, y))$. There is in general no increasing sequence of sets $E_1 \subset E_2 \subset \cdots \to E$ in each of which f_n converges *uniformly* to f.

If a topology (a notion of limit) is defined in E and if the correspondence $y \leftrightarrow K(x, y)$ transforms E in a continuous manner into a subset of the space F, then the weakly convergent sequence $\{f_n\}$ converges uniformly in every compact set $E_1 \subset E$. In fact E_1 is transformed into a compact subset of the space F. For every $\epsilon>0$ we can then choose a finite set $(y_1, \cdots, y_l) \subset E_1$ so that for every $y \in E_1$ there exists at least one y_k with $\|K(x, y)-K(x, y_k)\| \leqq \epsilon/4M$, where $M=\text{l.u.b.}_n \|f_n\|$.

Further, if we choose n_0 so that for $n>n_0$, $|f(y_k)-f_n(y_k)|<\epsilon/4$, we shall obtain for $y \in E_1$.

$$\begin{aligned} | f(y) - f_n(y) | &= | (f(y) - f(y_k)) + (f(y_k) - f_n(y_k)) + (f_n(y_k) - f_n(y)) | \\ &\leqq | f(y_k) - f_n(y_k) | + | (f(x) - f_n(x), K(x, y) - K(x, y_k)) | \\ &\leqq \frac{\epsilon}{4} + \|f - f_n\| \, \|K(x, y) - K(x, y_k)\| \leqq \frac{\epsilon}{4} + 2M \frac{\epsilon}{4M} < \epsilon. \end{aligned}$$

The continuity of the correspondence $y \leftrightarrow K(x, y)$ is equivalent to equicontinuity of all functions of F with $\|f\|<M$ for any $M>0$. This property is satisfied by most of the classes with reproducing kernels which are usually considered (such as classes of analytic functions, harmonic functions, solutions of partial differential equations, and so on).

(6) If the class F with the r.k. K is a subspace of a larger Hilbert space $\mathfrak{H}$, then the formula

$$f(y) = (h, K(x, y))_x$$

gives the projection f of the element h of $\mathfrak{H}$ on F.

In fact $h=f+g$, where g is orthogonal to the class F. $K(x, y)$ as a function of x belongs to F and so we have $(h, K(x, y))=(f+g, K(x, y))=(f, K(x, y)) =f(y)$ by the reproducing property.

(7) If F possesses a r.k. K, then the same is true of all closed linear subspaces of F, because if $f(y)$ is a continuous functional of f running through F, it is so much the more so if f runs through a subclass of F. If F' and F'' are complementary subspaces of F, then their reproducing kernels satisfy the equation $K'+K''=K$.

(8) If F possesses a r.k. K and if $\{g_n\}$ is an orthonormal system in F, then for every sequence $\{\alpha_n\}$ of numbers satisfying

$$\sum_1^\infty | \alpha_n |^2 < \infty,$$

we have

$$\sum_1^\infty | \alpha_n | \, | g_n(x) | \leqq K(x, x)^{1/2} \left(\sum_1^\infty | \alpha_n |^2 \right)^{1/2}.$$

In fact, for a fixed y, the Fourier coefficients of $K(x, y)$ for the system $\{g_n\}$ are

$$(K(x, y), g_n(x)) = \overline{(g_n(x), K(x, y))} = \overline{g_n(y)}.$$

Consequently

$$\sum_1^\infty g_n(y)^2 \leqq (K(x, y), K(x, y))_x = K(y, y).$$

Therefore

$$\sum_1^\infty |\alpha_n| |g_n(x)| \leqq \left(\sum_1^\infty |\alpha_n|^2\right)^{1/2} \left(\sum_1^\infty |g_n(x)|^2\right)^{1/2}$$
$$\leqq K(x, x)^{1/2}\left(\sum_1^\infty |\alpha_n|^2\right)^{1/2}.$$

3. **Reproducing kernels of finite-dimensional classes.** If F is of finite dimension n, let $w_1(x), \cdots, w_n(x)$ be n linearly independent functions of F. All functions $f(x)$ of F are representable in a unique manner as

(1) $$f(x) = \sum_1 \zeta_k w_k(x), \qquad \zeta_k \textit{ complex constants.}$$

The most general quadratic metric in F will be given by a positive definite hermitian form

(2) $$\|f\|^2 = \sum_{i,j=1}^n \alpha_{ij}\zeta_i\bar{\zeta}_j.$$

The scalar product has the form

(3) $$(f, g) = \sum_{i,j} \alpha_{ij}\zeta_i\bar{\eta}_j, \qquad \textit{where } g = \sum \eta_k w_k.$$

It is clear that

(4) $$\alpha_{ij} = (w_i, w_j).$$

Therefore the matrix $\{\alpha_{ij}\}$ is the Gramm's matrix of the system $\{w_k\}$. This matrix always possesses an inverse. Denote by $\{\beta_{ij}\}$ the conjugate of this inverse matrix. We have then

(5) $$\sum_j \alpha_{ij}\bar{\beta}_{jk} = 0 \text{ or } 1 \quad \textit{following as } i \neq k \text{ or } i = k.$$

It is immediately verified that the function

(6) $$K(x, y) = \sum_{i,j=1}^n \beta_{ij} w_i(x)\overline{w_j(y)}$$

is the reproducing kernel of the class F with the metric (2).

The matrix $\{\beta_{ij}\}$ is hermitian positive definite. From the preceding developments we get, clearly, the following theorem.

THEOREM. *A function $K(x, y)$ is the reproducing kernel of a finite-dimensional class of functions if and only if it is of the form* (6) *with a positive definite matrix $\{\beta_{ij}\}$ and linearly independent functions $w_k(x)$. The corresponding class F is then generated by the functions $w_k(x)$, the functions $f \in F$ given by* (1) *and the corresponding norm $\|f\|$ given by* (2), *where $\{\alpha_{ij}\}$ is the inverse matrix of $\{\overline{\beta_{ij}}\}$.*

4. Completion of incomplete Hilbert spaces. In applications, we often meet classes of functions forming incomplete Hilbert spaces, that is, linear classes, with a scalar product, satisfying all the conditions for a Hilbert space with the exception of the completeness. For such classes, two problems present themselves. Firstly, the problem of completing the class so as to obtain a class of functions forming a complete Hilbert space and secondly, to decide (before effecting the completion of the class) if the complete class will possess a reproducing kernel.

A few remarks should be added here about the problem of the completion of a class of functions forming an incomplete Hilbert space. Consider such a class F. It is well known that to this class we can adjoin ideal elements which will be considered as the limits of Cauchy sequences in F, when such a limit is not available in F, and in such a way we obtain an abstract Hilbert space containing the class F as a dense subset. This space, however, will not form a class of functions. In quite an arbitrary way we could realize the ideal elements to be adjoined to F as functions so as to obtain a complete space formed by a linear class of functions, but, in general, this arbitrary manner of completion will destroy all the continuity properties between the values of the functions and the convergence in the space.

In this paper when we speak about the functional completion of an incomplete class of functions F, we mean a completion by adjunction of functions such that the value of a function f of the completed class at a given point y depends continuously on f (as belonging to the Hilbert space)(5). From the existence theorem of reproducing kernels we deduce the fact that a completed class has a reproducing kernel. In this way the problem of functional completion and of the existence of a reproducing kernel in the complete class is merged into one problem. We shall prove here the following theorem:

THEOREM. *Consider a class of functions F forming an incomplete Hilbert space. In order that there exist a functional completion of the class it is necessary and sufficient that* $1°$ *for every fixed $y \in E$ the linear functional $f(y)$ defined in F be bounded;* $2°$ *for a Cauchy sequence $\{f_m\} \subset F$, the condition $f_m(y) \to 0$ for every y implies $\|f_m\| \to 0$. If the functional completion is possible, it is unique.*

Proof. That the first condition is necessary is immediately seen from the

(5) A more general functional completion was introduced in connection with the theory of pseudo-reproducing kernels (N. Aronszajn [5]).

existence theorem of reproducing kernels, since the complete class would necessarily have such a kernel. The necessity of the second condition follows from the fact that a Cauchy sequence in F is strongly convergent in the complete space to a function f, and the function f is the limit of f_m at every point y of E. Consequently, $f=0$ and the norms of f_m have to converge to the norm of f which is equal to zero. To prove the sufficiency we proceed as follows: consider any Cauchy sequence $\{f_n\}\subset F$. For every fixed y denote by M_y the bound of the functional $f(y)$ so that

$$f(y) \mid \leqq M_y\|f\|. \tag{1}$$

Consequently,

$$| f_m(y) - f_n(y) | \leqq M_y\|f_m - f_n\|.$$

It follows that $\{f_n(y)\}$ is a Cauchy sequence of complex numbers, that is, it converges to a number which we shall denote by $f(y)$. In this way the Cauchy sequence $\{f_n\}$ defines a function f to which it is convergent at every point of E.

Consider the class of all the functions f, limits of Cauchy sequences $\{f_n\}\subset F$. It is immediately seen that it is a linear class of functions, and that it contains the class F (since the Cauchy sequence $\{f_n\}$ with $f_n=f\in F$ is obviously convergent to f). Consider, then, in the so-defined class $\overline{F}$, the norm

$$\|f\|_1 = \lim \|f_n\| \tag{2}$$

for any Cauchy sequence $\{f_n\}\subset F$ converging to f at every point y. This norm does not depend on the choice of the Cauchy sequence: in fact, if another sequence, $\{f_n'\}$ converges to f at every point y, then $f_n' - f_n$ will be a Cauchy sequence converging to zero, and by the second condition the norms $\|f_n' - f_n\|$ converge to zero.

Consequently,

$$| \lim \|f_n'\| - \lim \|f_n\| | = \lim | \|f_n'\| - \|f_n\| | \leqq \lim \|f_n' - f_n\| = 0.$$

On the other hand, it is readily seen that $\|f\|_1^2$ is a quadratic positive form in the class $\overline{F}$; it is obviously 0 for $f=0$, and it is positive for $f\neq 0$ because of (1). This norm defines a scalar product in $\overline{F}$ satisfying all the required properties. It remains to be shown that $\overline{F}$ is complete and contains F as a dense subspace.

The second assertion is immediately proved because $F\subset\overline{F}$. For elements of F, the norms $\| \quad \|$, $\| \quad \|_1$ coincide, and every function $f\in\overline{F}$ is, by definition, the limit of a Cauchy sequence $\{f_n\}\subset F$ everywhere in E. It follows that f is a strong limit of f_n in $\overline{F}$ since by (2)

$$\lim_{n=\infty} \|f - f_n\|_1 = \lim_{n=\infty} \lim_{m=\infty} \|f_m - f_n\| = 0.$$

To prove the first assertion, that is, the completeness of $\overline{F}$, we shall consider any Cauchy sequence $\{f_n\} \subset \overline{F}$. Since F is dense in $\overline{F}$, we can find a Cauchy sequence $\{f_n'\} \subset F$ such that

$$\lim \|f_n' - f_n\|_1 = 0.$$

The Cauchy sequence $\{f_n'\}$ converges to a function $f \in \overline{F}$. This convergence is meant at first as ordinary convergence everywhere in E, but the argument used above shows that the f_n' also converge strongly to f in the space $\overline{F}$. It follows immediately that f_n converges strongly to f. The uniqueness of the complete class is seen from the fact that in the completed class a function f must necessarily be a strong limit of a Cauchy sequence $\{f_n\} \subset F$. Since a reproducing kernel must exist for the completed class, this implies that f is a limit everywhere of the Cauchy sequence $\{f_n\}$ which means that it belongs to the above class $\overline{F}$. As the norm of f has to be the limit of $\|f_n\|$ it necessarily coincides with $\|f\|_1$. It is also clear that every function f of $\overline{F}$ must belong to the completed class. In summing up the above arguments we see that any functional completion of F must coincide with $\overline{F}$ and have the same norm and scalar product as $\overline{F}$.

It should be stressed that the second condition cannot be excluded from our theorem. We shall demonstrate this by the following example:

Consider the unit circle $|z| < 1$. Take there an infinite sequence of points $\{z_n\}$ such that

$$\sum (1 - |z_n|) < \infty.$$

We shall denote by E the set of all points z_n, and in E we shall consider the class F of all polynomials in z. It is obvious that the values of a polynomial cannot vanish everywhere in the set E if the polynomial is not identically zero. Consequently the values of a polynomial on the set E determine completely the polynomial. We define the norm for a function f of the class F by the formula

$$\|f\|^2 = \iint_{|z|<1} |p(z)|^2 dxdy, \qquad z = x + iy,$$

where p denotes the polynomial whose values on the set E are given by the function f. We see that F satisfies all requirements for a Hilbert space with the exception of completeness. The first condition of our theorem is satisfied but the second is not. To prove the last assertion we take the Blaschke function $\phi(z)$ corresponding to $\{z_n\}$. This function has the following properties

$$\phi(z_n) = 0, \qquad n = 1, 2, 3, \cdots,$$
$$|\phi(z)| < 1 \qquad \text{for } |z| < 1.$$

The function $\phi(z)$ is a strong limit of polynomials $p_k(z)$ in the sense

$$\lim \iint_{|z|<1} |\phi(z) - p_k(z)|^2 dxdy = 0.$$

Consequently the sequence $\{p_k\}$ is a Cauchy sequence in our class F and the polynomials $p_k(z)$ converge at each point z_n to $\phi(z_n)=0$, in spite of the fact that the norms $\|p_k\|$ converge to $\|\phi\|>0$.

This example also shows us the significance of condition 2°. We can say that if condition 2° is not satisfied it means that the incomplete class is defined in too small a set. If we added a sequence of points $\{z_n'\}$ of the unit circle with a limit point inside the circle to the set E of our example, then the class of polynomials, considered on this enlarged set E_1 with the same norm as above, would satisfy condition 2°. There are infinitely many ways of enlarging the set E where an incomplete class F is defined so as to insure the fulfillment of condition 2°.

In the general case we can always proceed as follows. We can consider the abstract completion of the class F by adjunction of ideal elements. This completion leads to an abstract Hilbert space $\mathfrak{H}$. To every element of $\mathfrak{H}$ there corresponds a well determined function $f(x)$ defined on the set (the limit of Cauchy sequences in F converging to this element). But to different elements of $\mathfrak{H}$ there may correspond the same function f. The correspondence is linear and the functionals $f(y)$ are continuous in the whole $\mathfrak{H}$. To every point $y\in E$ there corresponds an element $h_y\in\mathfrak{H}$ such that $f(y)=(\bar{f}, h_y)$, where $\bar{f}$ is any element of $\mathfrak{H}$ corresponding to the function $f(x)$. As there are elements of $\mathfrak{H}$ different from the zero element and which correspond to the function identically zero on E, it is clear that the set of elements h_y is not complete in the space $\mathfrak{H}$. (This is characteristic of the fact that condition 2° is not satisfied.) To the set of elements h_y we can then add an additional set of elements so as to obtain a complete set in $\mathfrak{H}$. This additional set will be denoted by E'. We can then extend the functions of our class F in the set $E+E'$ by defining, for any element $h\in E'$, $f(h)=(f, h)$. This class of functions, so extended in $E+E'$, will then satisfy the second condition.

We shall complete this section by the following remark: It often happens that for the incomplete class F a kernel $K(x, y)$ is known such that for every y, $K(x, y)$ as function of x belongs to F (or, even more generally, belongs to a Hilbert space containing F as a subspace), this kernel having the reproducing property

$$f(y) = (f(x), K(x, y)) \qquad \textit{for every } f\in F.$$

It is immediately seen that the first condition of our theorem follows from this reproducing property so that it suffices to verify the second condition in order to be able to apply our theorem.

5. **The restriction of a reproducing kernel.** Consider a linear class F of

functions defined in the set E, forming a Hilbert space and possessing a r.k. $K(x, y)$. K is a positive matrix. If we restrict the points x, y to a subset $E_1 \subset E$, K will still be a positive matrix. This means that K will correspond to a class F_1 of functions defined in E_1 with an adequate norm $\| \ \|_1$. We shall now determine this class F_1 and the corresponding norm.

THEOREM. *If K is the reproducing kernel of the class F of functions defined in the set E with the norm $\| \ \|$, then K restricted to a subset $E_1 \subset E$ is the reproducing kernel of the class F_1 of all restrictions of functions of F to the subset E_1. For any such restriction, $f_1 \in F_1$, the norm $\|f_1\|_1$ is the minimum of $\|f\|$ for all $f \in F$ whose restriction to E_1 is f_1.*

Proof. Consider the closed linear subspace $F_0 \subset F$ formed by all functions which vanish at every point of E_1. Take then the complementary subspace $F' = F \ominus F_0$. Both F_0 and F' are closed linear subspaces of F and possess reproducing kernels K_0 and K' such that

$$(1) \qquad K = K_0 + K'.$$

Since $K_0(x, y)$, for every fixed y, belongs to F_0, it is vanishing for $x \in E_1$. Consequently,

$$(2) \qquad K(x, y) = K'(x, y), \qquad \textit{for } x \in E_1.$$

Consider now the class F_1 of all restrictions of F to the set E_1. If two functions f and g from F have the same restriction f_1 in E_1, $f-g$ vanishes on E_1 and so belongs to F_0. Conversely, if the difference belongs to F_0, f and g have the same restriction f_1 in E_1. It is then clear that all the functions $f \in F$ which have the same restriction f_1 in E have a common projection f_1' on F' and that the restriction of f_1' in E_1 is equal also to f_1. It is also clear that among all these functions, f, f_1' is the one which has the smallest norm. Consequently by the definition of the theorem, we can write

$$(3) \qquad \|f_1\|_1 = \|f_1'\|.$$

The correspondence between $f_1 \in F_1$ and $f_1' \in F'$ obviously establishes a one-to-one isometric correspondence between the space F_1 with the norm $\| \ \|_1$ and the space F' with the norm $\| \ \|$.

In order to prove that for the class F_1 with the norm $\| \ \|_1$ the reproducing kernel is given by K restricted to E_1, we take any function $f_1 \in F_1$ and consider the corresponding function $f_1' \in F'$. Then, for $y \in E_1$, $f_1(y) = f_1'(y) = (f_1'(x), K'(x, y))$.

Since $K'(x, y)$, for every y belongs to F', we may now write $f_1(y) = (f_1'(x), K'(x, y)) = (f_1(x), K_1(x, y))_1$, where $K_1(x, y)$ is the restriction of $K'(x, y)$ (considered as function of x) to the set E_1.

By hermitian symmetry we obtain from (2)

$$K(x, y) = K'(x, y), \qquad \text{for every } y \in E_1.$$

This shows that the restriction $K_1(x, y)$ of $K'(x, y)$ coincides with the restriction of K to the set E_1, which completes our proof.

The norm $\| \ \|_1$ is especially simple when the subclass $F_0 \subset F$ is reduced to the zero function. In this case $F' = F$, each function $f_1 \in F_1$ is a restriction of one and only one function $f \in F$ (since every function of F vanishing in E_1 vanishes identically everywhere in E), and therefore $\|f_1\|_1 = \|f\|$ for the function f having the restriction f_1.

6. **Sum of reproducing kernels.** Let $K_1(x, y)$ and $K_2(x, y)$ be reproducing kernels corresponding to the classes F_1 and F_2 of functions defined in the same set E with the norms $\| \ \|_1$ and $\| \ \|_2$ respectively. K_1 and K_2 are positive matrices, and obviously $K = K_1 + K_2$ is also a positive matrix.

We shall now find the class F and the norm $\| \ \|$ corresponding to K. Consider the Hilbert space $\mathfrak{H}$ formed by all couples $\{f_1, f_2\}$ with $f_i \in F_i$. The metric in this space will be given by the equation

$$\|\{f_1, f_2\}\|^2 = \|f_1\|_1^2 + \|f_2\|_2^2.$$

Consider the class F_0 of functions f belonging at the same time to F_1 and F_2 (F_0 may be reduced to the zero function). Denote by $\mathfrak{H}_0$ the set of all couples $\{f, -f\}$ for $f \in F_0$. It is clear that $\mathfrak{H}_0$ is a linear subspace of $\mathfrak{H}$. It is a closed subspace. In fact if $\{f_n, -f_n\} \to \{f', f''\}$ then f_n converges strongly to f' in F_1, and $-f_n$ converges strongly to f'' in F_2. Consequently, f_n converges in the ordinary sense to f' and $-f_n$ to f'', which means that $f'' = -f'$ and f' and f'' belong to F_0. $\mathfrak{H}_0$ being a closed linear subspace of $\mathfrak{H}$ we can consider the complementary subspace $\mathfrak{H}'$ so that $\mathfrak{H} = \mathfrak{H}_0 \oplus \mathfrak{H}'$. To every element of $\mathfrak{H}$: $\{f', f''\}$ there corresponds the function $f(x) = f'(x) + f''(x)$. This is obviously a linear correspondence transforming the space $\mathfrak{H}$ into a linear class of functions F. The elements of $\mathfrak{H}$ which are transformed by this correspondence into the zero function are clearly the elements of $\mathfrak{H}_0$. Consequently, this correspondence transforms $\mathfrak{H}'$ in a one-to-one way into F. The inverse correspondence transforms every function $f \in F$ into an element $\{g'(f), g''(f)\}$ of $\mathfrak{H}'$. We define the metric in F by the equation

$$\|f\|^2 = \|\{g'(f), g''(f)\}\|^2 = \|g'(f)\|_1^2 + \|g''(f)\|_2^2.$$

Our assertion will be that to the class F with the above-defined norm there corresponds the reproducing kernel $K = K_1 + K_2$. To prove this assertion we remark:

(1) $K(x, y)$ as a function of x, for y fixed, belongs to F. Namely, it corresponds to the element $\{K_1(x, y), K_2(x, y)\} \in \mathfrak{H}$.

(2) Denote for fixed y, $K'(x, y) = g'(K(x, y))$ and $K''(x, y) = g''(K(x, y))$. For a function $f \in F$, we write $f' = g'(f)$, $f'' = g''(f)$. Consequently, $f(y) = f'(y) + f''(y)$, $K'(x, y) + K''(x, y) = K(x, y) = K_1(x, y) + K_2(x, y)$, and thus $K''(x, y) - K_2(x, y) = -[K'(x, y) - K_1(x, y)]$, so that the element $\{K_1(x, y) - K'(x, y),$

$K_2(x, y) - K''(x, y)\} \in \mathfrak{H}_0$. Hence

$$\begin{aligned} f(y) &= f'(y) + f''(y) = (f'(x), K_1(x, y))_1 + (f''(x), K_2(x, y))_2 \\ &= (\{f', f''\}, \{K_1(x, y), K_2(x, y)\}) \\ &= (\{f', f''\}, \{K'(x, y), K''(x, y)\}) \\ &\quad + (\{f', f''\}, \{K_1(x, y) - K'(x, y), K_2(x, y) - K''(x, y)\}). \end{aligned}$$

The last scalar product equals zero since the element $\{f', f''\} \in \mathfrak{H}'$ and the element $\{K_1(x, y) - K'(x, y), K_2(x, y) - K''(x, y)\} \in \mathfrak{H}_0$. The first scalar product in the last member is, by our definition, equal to $(f(x), K(x, y))$ which proves the reproducing property of the kernel $K(x, y)$.

We can characterize the class F as the class of all functions $f(x) = f_1(x) + f_2(x)$ with $f_i \in F_i$. In order to define the corresponding norm without passage through the auxiliary space, we consider for every $f \in F$ all possible decompositions $f = f_1 + f_2$ with $f_i \in F_i$. For each such decomposition we consider the sum $\|f_1\|_1^2 + \|f_2\|_2^2$. $\|f\|$ will then be defined by

$$\|f\|^2 = \min\ [\|f_1\|_1^2 + \|f_2\|_2^2]$$

for all the decompositions of f. To prove the equivalence of this definition and the previous one we have only to remember that $f(x)$ corresponds to the element $\{f_1, f_2\} \in \mathfrak{H}$ and also that f corresponds to $\{g'(f), g''(f)\} \in \mathfrak{H}'$, that is, $f = f_1 + f_2 = g'(f) + g''(f)$. Consequently,

$$f_2 - g''(f) = -\ [f_1 - g'(f)]$$

so that $\{f_1 - g'(f), f_2 - g''(f)\} \in \mathfrak{H}_0$ and

$$\|f_1\|_1^2 + \|f_2\|_2^2 = \|\{f_1, f_2\}\|^2 = \|\{g'(f), g''(f)\}\|^2 + \|\{f_1 - g'(f), f_2 - g''(f)\}\|^2,$$

and this expression will obviously get the minimal value if and only if $f_1 = g'(f)$, $f_2 = g''(f)$ and its value is then given by

$$\|\{g'(f), g''(f)\}\|^2,$$

which by our previous definition is $\|f\|^2$.

Summing up, we may write the following theorem:

THEOREM. *If $K_i(x, y)$ is the reproducing kernel of the class F_i with the norm $\|\ \|_i$, then $K(x, y) = K_1(x, y) + K_2(x, y)$ is the reproducing kernel of the class F of all functions $f = f_1 + f_2$ with $f_i \in F_i$, and with the norm defined by*

$$\|f\|^2 = \min\ [\|f_1\|_1^2 + \|f_2\|_2^2],$$

the minimum taken for all the decompositions $f = f_1 + f_2$ with $f_i \in F_i$(6).

It is easy to see how this theorem can be extended to the case where

(6) This theorem was found by R. Godement [1] in the case of a positive definite function in a group.

$K(x, y) = \sum_{i=1}^{n} K_i(x, y)$. A particularly simple case presents itself when the classes F_1 and F_2 have no function besides zero in common. The norm in F is then given simply by $\|f\|^2 = \|f_1\|_1^2 + \|f_2\|_2^2$.

In this case (and only in this case) F_1 and F_2 are complementary closed subspaces of F.

If we denote by $\bar{F}$ the class of all conjugate functions of functions of a class F, then the kernel of $\bar{F}$ is clearly $K_1(x, y) = \overline{K(x, y)} = K(y, x)$ (K is the kernel of F and in $\bar{F}$ the scalar product $(\bar{f}, \bar{g})_1$ is given by (g, f), the norm $\|\bar{f}\|_1$ by $\|f\|$). Consequently Re $K(x, y) = 2^{-1}(K(x, y) + K(y, x))$ is the reproducing kernel of the class F_0 of all sums $f + \bar{g}$, for f and g in F with the norm given by

$$(1) \qquad \|\phi\|_0^2 = 2 \min [\|f\|^2 + \|g\|^2],$$

the minimum taken for all decompositions $\phi = f + \bar{g}$, f and g in F.

If F is a complex space corresponding to a real space, that is, if $F = \bar{F}$ and $\|f\| = \|\bar{f}\|$ (see §1), it is clear that $F_0 = F$ and $\|f\|_0 = \|f\|$. Consequently the kernel $K = \text{Re } K$ is real and this property characterizes the kernal K corresponding to a real space.

7. **Difference of reproducing kernels.** For two positive matrices, $K_1(x, y)$ and $K(x, y)$, we shall write

$$1) \qquad K_1 \ll K$$

if $K(x, y) - K_1(x, y)$ is also a p. matrix.

From $K_1 \ll K_2 \ll K_3$, it follows clearly that $K_1 \ll K_3$. On the other hand, if $K_1 \ll K_2$ and $K_2 \ll K_1$, it follows that $K_1 = K_2$. In fact, we then have $K_2(x, x) - K_1(x, x) \geqq 0$ and also $K_1(x, x) - K_2(x, x) \geqq 0$, which means $K_2(x, x) - K_1(x, x) = 0$. Further, by the property of positive matrices,

$$|K_2(x, y) - K_1(x, y)|^2 \leqq [K_2(x, x) - K_1(x, x)][K_2(y, y) - K_1(y, y)] = 0,$$

so that $K_2(x, y) = K_1(x, y)$ for every x, y in E. Thus we see that the symbol $\ll$ establishes a partial ordering in the class of all positive matrices.

THEOREM I. *If K and K_1 are the r.k.'s of the classes F and F_1 with the norms $\| \;\|$, $\| \;\|_1$, and if $K_1 \ll K$, then $F_1 \subset F$, and $\|f_1\|_1 \geqq \|f_1\|$ for every $f_1 \in F_1$.*

Proof. $K_1 \ll K$ means that $K_2(x, y) = K(x, y) - K_1(x, y)$ is a positive matrix. Consider the class of functions F_2 and the norm $\| \;\|_2$ corresponding to K_2. As $K = K_1 + K_2$ we know by the theorem of §6 that F is the class of all functions of the form $f_1(x) + f_2(x)$ with $f_1 \in F_1$ and $f_2 \in F_2$. In particular, when $f_2 = 0$, the class F contains all functions $f_1 \in F_1$ so that $F_1 \subset F$. On the other hand, in F we have, by the same theorem,

$$\|f_1\|^2 = \min [\|f_1'\|_1^2 + \|f_2'\|_2^2]$$

for all decompositions $f_1 = f_1' + f_2'$, with $f_1' \in F_1$ and $f_2' \in F_2$. In particular, for the decomposition $f_1 = f_1 + 0$, we obtain

$$\|f_1\|^2 \leqq \|f_1\|_1^2$$

which achieves the proof.

THEOREM II. *If K is the r.k. of the class F with the norm $\| \ \|$, and if the linear class $F_1 \subset F$ forms a Hilbert space with the norm $\| \ \|_1$, such that $\|f_1\|_1 \geqq \|f_1\|$ for every $f_1 \in F_1$, then the class F_1 possesses a reproducing kernel K_1 satisfying $K_1 \ll K$.*

Proof. The existence of the kernel for F involves the existence of constants M_y such that $|f(y)| \leqq M_y\|f\|$ for all $f \in F$. In particular, for $f_1 \in F_1 \subset F$, we get

$$|f_1(y)| \leqq M_y\|f_1\| \leqq M_y\|f_1\|_1$$

which proves the existence of the kernel K_1 of F_1.

Let us now introduce an operator L in the space F, transforming the space F into the space F_1 and satisfying the equation

$$(f_1, f) = (f_1, Lf)_1, \qquad \text{for every } f_1 \in F_1.$$

The existence and unicity of Lf is proved in the following way: (f_1, f), for fixed f, is a linear continuous functional of f_1 in the space F. A fortiori, it is a similar functional in the space F_1. As such, it is representable as a scalar product in F_1 of f_1 with a uniquely determined element Lf of F_1.

The operator L is everywhere defined in F, linear, symmetric, positive, and bounded with a bound not greater than 1. It is clear that L is everywhere defined and linear. It is symmetric because for any two functions f, f' of F, $(Lf, f') = (Lf, Lf')_1 = \overline{(Lf', Lf)_1} = \overline{(Lf', f)} = (f, Lf')$. It is positive because $(Lf, f) = (Lf, Lf)_1 \geqq 0$. It is bounded with a bound not greater than 1 because $(Lf, f) = (Lf, Lf)_1 = \|Lf\|_1^2 \geqq \|Lf\|^2$. Consequently, $\|Lf\|^2 \leqq (Lf, f) \leqq \|Lf\| \cdot \|f\|$, and thus $\|Lf\| \leqq \|f\|$.

Consider now the operator $I - L$ (I being the identical operator). This operator clearly possesses the same properties as those enumerated above for L. Therefore there exists a symmetric, bounded square root L' of this operator. (In general there will be infinitely many L' available and we choose any one of them.) Hence

$$L'^2 = I - L.$$

We define F_2 as the class of all functions $f_2 = L'f$ for $f \in F$. Denote by F_0 the closed linear subspace of F transformed by L' into 0, and by F' the complementary space $F \ominus F_0$. The functions of F_0 are also characterized by the fact that $L'^2 f = 0$ (from $L'^2 f = 0$ it follows that $(L'^2 f, f) = (L'f, L'f) = \|L'f\|^2 = 0$) which is equivalent to $f = Lf$.

Now denote by P' the projection on F'. Every two functions f, g of F transformed by L' into the same function f_2 of F_2 differ by a function belonging to F_0. Consequently, they have the same projection on F'. It is also clear that the class F' is transformed by L' on F_2 in a one-to-one way. We remark further that

$$F_2 \subset F'$$

since for $f_0 \in F_0$, $(f_0, L'f) = (L'f_0, f) = (0, f) = 0$. In the class F_2 we introduce the norm $\| \quad \|_2$ by the equation

$$\|f_2\|_2 = \|P'f\|, \qquad \text{for any } f \text{ with } f_2 = L'f.$$

With this metric the class F_2 is isometric to the subspace $F' \subset F$ (the isometry being given by the transformation L'). It follows that F_2 is a complete Hilbert space and we shall show that the class F_2 with the norm $\| \quad \|_2$ admits as reproducing kernel the difference $K(x, y) - K_1(x, y)$. To this effect we remark firstly that the operator L is given by the formula

$$f_1(y) \equiv Lf = (f, K_1(x, y)).$$

In fact, since $Lf \in F_1$, $f_1(y) = (f_1(x), K_1(x, y))_1 = (f(x), K_1(x, y))$. Consequently for any fixed z, the operator L applied to $K(x, z)$ gives the function $LK(x, z) = K_1(y, z)$. If follows that the operator $I - L = L'^2$ transforms $K(x, z)$ into $L'^2K(x, z) = K(y, z) - K_1(y, z)$. This formula proves, firstly, that $K(x, z) - K_1(x, z)$ as a function of x belongs to F_2, since it is the transform of $L'K(x, z)$ by L'. Secondly, we prove the reproducing property for any $f_2 \in F_2$:

$$\begin{aligned} f_2(y) &= (f_2(x), K(x, y)) = (L'f, K(x, y)) = (f, L'K(x, y)) \\ &= (P'f, P'L'K(x, y)) = (L'f, L'L'K(x, y))_2 \\ &= (f_2(x), K(x, y) - K_1(x, y))_2. \end{aligned}$$

In these transformations we took f as any function of F such that $f_2 = L'f$ and we used the property $L'K(x, y) \in F_2 \subset F'$. This finishes the proof of our theorem.

The proof of Theorem II has established even more than the theorem announced: namely, it gives us the construction of the class F_2 and the metric $\| \quad \|_2$ for which the difference $K - K_1$ is the reproducing kernel. Let us summarize this in a separate theorem.

THEOREM III. *Under the hypotheses of Theorem* II, *the class* F_2 *and the norm* $\| \quad \|_2$ *corresponding to the kernel* $K_2 = K - K_1$ *are defined as follows: the equation* $f_1(y) \equiv Lf = (f(x), K_1(x, y))_x$ *defines in* F *a positive operator with bound not greater than* 1, *transforming* F *into* $F_1 \subset F$. *We take any symmetric square root* $L' = (I - L)^{1/2}$. F_2 *is the class of all transforms* $L'f$ *for* $f \in F$. *Let* F_0 *be the closed linear subspace of all functions* $f \in F$ *with* $f = Lf$ *and let* F' *be* $F \ominus F_0$. L' *establishes a one-to-one correspondence between* F' *and* F_2. *The norm* $\|f_2\|_2$ *for* $f_2 = L'f'$, $f' \in F'$, *is then given by*

$$\|f_2\|_2 = \|f'\|.$$

The construction of the class F_2 is somehow complicated because we need the square root L' of the operator $I-L$. However, there exists a much easier way of constructing an everywhere dense linear subclass of F_2 and the norm $\|\ \|_2$ in this subclass. Namely, we can consider the class F_2' of all transforms $(I-L)f=f-Lf=L'^2f$, of $f\in F$, by the operator $I-L$.

This clearly is a linear subclass of F_2. The norm $\|\ \|_2$ in F_2' can be defined as follows:

$$\|f_2'\|_2^2 = \|L'f\|^2 = (L'f, L'f) = (f, L'^2 f) = (f, f - Lf) = (f, f) - (f, Lf)$$
$$= (f, f) - (Lf, Lf)_1.$$

In these transformations we considered $f_2' = f - Lf = L'^2 f$.

F_2' is everywhere dense in F_2 (in respect to the norm $\|\ \|_2$), otherwise we would have a function $f_2 = L'f' \neq 0$ with $(L'f', L'^2g)_2 = 0$ for all $g\in F$. We can suppose here $f'\in F'$ so that $0=(L'f', L'^2g)_2=(f', L'g)=(L'f', g)$ whence $L'f'=0$, $f'\in F_0$, which is impossible.

The simplest case is the one where $F_2' = F_2$. By using the spectral decomposition, we can easily show that this case presents itself if, and only if, the zero is not a limit point of the spectrum of L', which is the equivalent of saying that 1 is not a limit point of the spectrum of L. In this case $F_2 = F'$. If L has a bound <1, the spectrum does not contain 1 and the subspace F' coincides with the whole space F so that $F_2 = F$.

When L is completely continuous, the only limit point of the spectrum of L is zero, so that 1 is certainly not a limit point, and $F_2' = F_2 = F'$.

We shall add still another theorem which results immediately from Theorem III.

THEOREM IV. *Let K be the r.k. of class F. To every decomposition $K = K_1 + K_2$ in two p. matrices K_1 and K_2, there corresponds a decomposition of the identity operator I in F in two positive operators L_1 and L_2, $I = L_1 + L_2$, given by*

$$L_1f(y) = (f(x), K_1(x, y)), \qquad L_2f(y) = (f(x), K_2(x, y)),$$

such that if $L_1^{1/2}$ and $L_2^{1/2}$ denote any symmetric square roots of L_1 and L_2, the classes F_1 and F_2 of all transforms $L_1^{1/2}f$ and $L_2^{1/2}f$ respectively, $f\in F$, correspond to the kernels K_1 and K_2. If F_{i0}, $i=1, 2$, is the class of all $f\in F$ with $L_if=0$, and if $F_i'' = F\ominus F_{i0}$, then $L_i^{1/2}$ establishes a one-to-one correspondence between F_i'' and F_i and the norm $\|\ \|_i$ in F_i is given by $\|L_i^{1/2}f\|_i = \|f\|$ for every $f\in F_i''$.

Conversely, to each decomposition $I = L_1 + L_2$ in two positive operators there correspond classes F_i with norms $\|\ \|_i$ defined as above. The corresponding r.k.'s K_i are defined by $K_i(x, y) = L_iK(x, y)$ and satisfy the equation $K = K_1 + K_2$.

8. **Product of reproducing kernels.** Consider two positive matrices K_1

and K_2 defined in a set E. Using a classical result of I. Schur concerning finite matrices, it is easy to prove that their product $K_1 \cdot K_2$ is also a positive matrix. We shall prove this theorem, constructing at the same time the class F and the norm $\| \ \|$ corresponding to the matrix $K = K_1 \cdot K_2$(7). To this effect, we consider the class F_i and the norm $\| \ \|_i$ corresponding to K_i and form the direct product F' of the Hilbert spaces F_1 and F_2, $F' = F_1 \otimes F_2$(8). We construct this direct product in the following manner: We form the product set $E' = E \times E$ of all couples of points $\{x_1, x_2\}$, $x_i \in E$. In the set E' consider the class of all functions $f'(x_1, x_2)$ representable in the form:

$$f'(x_1, x_2) = \sum_{k=1}^{n} f_1^{(k)}(x_1) f_2^{(k)}(x_2), \tag{1}$$

with $f_1^{(k)} \in F_1$ and $f_2^{(k)} \in F_2$. As scalar product of two such functions we define:

$$(f', g')' = \sum_{k=1}^{n} \sum_{l=1}^{m} (f_1^{(k)}, g_1^{(l)})_1 (f_2^{(k)}, g_2^{(l)})_2, \tag{2}$$

where m is the number of terms in the representation of g'. The same function f' may admit of many different representations of type (1). The scalar product, $(f', g')'$, is independent of the particular representation chosen for f' and g'. In fact, we see immediately from (2) that

$$(f', g')' = \sum_{l=1}^{m} ((f'(x_1, x_2), g_1^{(l)}(x_1))_1, g_2^{(l)}(x))_2 \tag{3}$$

which proves that (f', g') is independent of the particular representation of f'. In a similar way we prove that it is independent of the particular representation of g'. We still have to prove that $(f', g')'$ satisfies all the requirements for the scalar product. It is clearly seen that it is a bilinear hermitian form in f', g' and it remains to be proved that $(f', f')' \geqq 0$ and is equal to zero only when $f' = 0$. In order to prove this, we take any representation of f' of type (1) and orthonormalize the sequences $\{f_1^{(k)}\}$ and $\{f_2^{(k)}\}$ in the spaces F_1 and F_2 respectively. Denote by $\{f_1^{(k,1)}\}$ and $\{f_2^{(l,1)}\}$ the orthonormalized sequences, where $k = 1, 2, \cdots, n_1$, $l = 1, 2, \cdots, n_2$.

Every function $f_i^{(k)}$ is then a linear combination of the orthonormal functions $f_i^{(k',1)}$ so that we obtain a representation for f' as a double series

$$f'(x_1, x_2) = \sum_{k=1}^{n_1} \sum_{l=1}^{n_2} \alpha_{k,l} f_1^{(k,1)} f_2^{(l,1)}. \tag{4}$$

We then obtain for $(f', f')'$ the following expression

(7) The idea of the proof was arrived at independently by R. Godement and the author. Godement applied it only to positive definite functions.

(8) For the notion of direct product of abstract Hilbert spaces see J. v. Neumann and F. J. Murray [1].

$$(5)\qquad \begin{aligned}(f', f') &= \sum_{k=1}^{n_1}\sum_{l=1}^{n_2}\sum_{k'=1}^{n_1}\sum_{l'=1}^{n_2} \alpha_{k,l}\bar{\alpha}_{k',l'}(f_1^{(k,1)}, f_1^{(k',1)})_1 (f_2^{(l,1)}, f_2^{(l',1)})_2 \\ &= \sum_{k=1}^{n_1}\sum_{l=1}^{n_2} | \alpha_{k,l}|^2.\end{aligned}$$

It is clear from this representation that $(f', f') \geqq 0$ and equals zero only when all the $\alpha_{k,l}=0$, that is, when $f'=0$. The class of all functions f' of type (1) does not yet form in general a Hilbert space because it may not be complete. To complete this class with respect to the norm $\| \quad \|'$, we consider a complete orthonormal sequence $\{g_i^{(k)}\}$ in the space F_i, $i=1, 2$. It is obvious that the double sequence $\{g_1^{(k)}(x_i) \cdot g_2^{(l)}(x_2)\}$ is composed of functions of type (1) and is orthonormal in respect to the norm $\| \quad \|'$. Consider then, all the functions g' of the form

$$(6)\qquad g'(x_1, x_2) = \sum_{k=1}^{\infty}\sum_{l=1}^{\infty} \alpha_{k,l} g_1^{(k)}(x_1) g_2^{(l)}(x_2)$$

with

$$(7)\qquad (g', g')' = \sum_{k=1}^{\infty}\sum_{l=1}^{\infty} | \alpha_{k,l}|^2 < \infty.$$

It is clear that any finite sum of type (6) is also of type (1) and that the norm $\| \quad \|'$ for such finite sums coincides with the norm introduced in (7). We prove firstly that every sum of type (6) is absolutely convergent for every x_1, x_2. In fact, as the class F_1 possesses the reproducing kernel K_1, in view of (7) we have

$$\sum_{k=1}^{\infty} | \alpha_{k,l}| \, | g_1^{(k)}(x_1) | \leqq [K_1(x_1, x_1)]^{1/2}\left[\sum_{k=1}^{\infty} | \alpha_{k,l}|^2\right]^{1/2}.$$

Then,

$$(8)\qquad \begin{aligned}&\sum_{k=1}^{\infty}\sum_{l=1}^{\infty} | \alpha_{k,l}| \, | g_1^{(k)}(x_1) | \, | g_2^{(l)}(x_2) | \\ &\qquad \leqq \sum_{l=1}^{\infty} | g_2^{(l)}(x_2) | \, [K_1(x_1, x_1)]^{1/2}\left[\sum_{k=1}^{\infty} | \alpha_{k,l}|^2\right]^{1/2} \\ &\qquad \leqq [K_1(x_1, x_1)]^{1/2}[K_2(x_2, x_2)]^{1/2}\cdot \left[\sum_{k=1}^{\infty}\sum_{l=1}^{\infty} | \alpha_{k,l}|^2\right]^{1/2}\end{aligned}$$

since the space F_2 possesses the reproducing kernel K_2.

The class of functions g' of the form (6) clearly forms a complete Hilbert space, isomorphic with the space of double sequences $\{\alpha_{k,l}\}$ satisfying (7). The inequality (8) gives us further, for a function g' of type (6),

$$|g'(y_1, y_2)| < K_1(y_1, y_1)^{1/2}K_2(y_2, y_2)^{1/2}\|g'\|',$$

which means that the space of functions of type (6) possesses a reproducing kernel.

It remains to be proved that this space is the completion of the class of all functions f' of type (1) with the norm $\| \quad \|'$. As already the finite sums of type (6) are everywhere dense in the space of all functions of type (6) it is sufficient to prove that every function of type (1) is of type (6). For this, it is enough to prove that every function of type (1) may be approximated as closely as we wish (in respect to the norm $\| \quad \|'$) by finite sums of type (6). Let us consider a representation (1) of the function f'. We can approximate every $f_i^{(k)}$ by a finite linear combination $h_i^{(k)}$ of functions $g_i^{(l)}$ so that $\|h_i^{(k)}\|_i \leqq \|f_i^{(k)}\|_i$, $\|f_i^{(k)} - h_i^{(k)}\|_i \leqq \epsilon$. Before we proceed further, we shall prove for every function f' and any of its representations (1) the inequality

$$\|f'\|' \leqq \sum_{k=1}^{n} \|f_1^{(k)}\|_1 \cdot \|f_2^{(k)}\|_2.$$

In fact,

$$\begin{aligned}\|f'\|'^2 = (f', f')' &= \sum_{k=1}^{n}\sum_{l=1}^{n} (f_1^{(k)}, f_1^{(l)})_1 (f_2^{(k)}, f_2^{(l)})_2 \\ &\leqq \sum_{k=1}^{n}\sum_{l=1}^{n} \|f_1^{(k)}\|_1 \|f_1^{(l)}\|_1 \cdot \|f_2^{(k)}\|_2 \|f_2^{(l)}\|_2 \\ &= \left[\sum_{k=1}^{n} \|f_1^{(k)}\|_1 \|f_2^{(k)}\|_2\right]^2.\end{aligned}$$

Continuing with the proof of the approximation we consider the functions

$$h'(x_1, x_2) = \sum_{k=1}^{n} h_1^{(k)}(x_1) f_2^{(k)}(x_2),$$

$$g'(x_1, x_2) = \sum_{k=1}^{n} h_1^{(k)}(x_1) h_2^{(k)}(x_2).$$

It is clear that h' is of type (1) and that g' is at the same time of type (1) and (6), which can be seen by developing the functions $h_i^{(k)}$ as linear combinations of $g_i^{(l)}$. Denoting by M the maximum of all $\|f_i^{(k)}\|_i$, we obtain

$$\|f' - g'\|' \leqq \|f' - h'\|' + \|h' - g'\|',$$

$$\begin{aligned}\|f' - h'\|' &= \sum_{k=1}^{n} \|(f_1^{(k)}(x_1) - h_1^{(k)}(x_1)) f_2^{(k)}(x_2)\|' \\ &\leqq \sum_{k=1}^{n} \|f_1^{(k)} - h_1^{(k)}\|_1 \cdot \|f_2^{(k)}\|_2 \leqq \sum_{k=1}^{n} M\epsilon = nM\epsilon,\end{aligned}$$

$$\|h' - g'\|' = \left\| \sum_{k=1}^{n} h_1^{(k)}(x_1)((f_2^{(k)}(x_2) - h_2^{(k)}(x_2)) \right\|'$$

$$\leqq \sum_{k=1}^{n} \|h_1^{(k)}\|_1 \cdot \|f_2^{(k)} - h_2^{(k)}\|_2 \leqq \sum_{k=1}^{n} M\epsilon = nM\epsilon.$$

Finally, we obtain

$$\|f' - g'\|' \leqq 2nM\epsilon,$$

which proves our assertion.

The class of all functions of type (6) with the norm given by (7) forms the direct product $F' = F_1 \otimes F_2$. As it is obtained by functional completion of the class of functions of type (1), this class being independent of the choice of orthogonal systems $\{g_1^{(k)}\}$ and $\{g_2^{(k)}\}$, the class F' is also independent of the choice of these systems (see the uniqueness of functional completion in §4).

We shall now prove the following theorem:

THEOREM I. *The direct product $F' = F_1 \otimes F_2$ possesses the reproducing kernel $K'(x_1, x_2, y_1, y_2) = K_1(x_1, y_1)K_2(x_2, y_2)$.*

The proof is immediate. Firstly, as a function of x_1, x_2, K' is of the form (1) and so belongs to F'. Secondly, for any function g' of the form (6) we have

$$g'(y_1, y_2) = \sum \sum \alpha_{k,l}(g_1^{(k)}(x_1), K_1(x_1, y_1))_1(g_2^{(l)}(x_2), K_2(x_2, y_2))_2$$
$$= (g'(x_1, x_2), K'(x_1, x_2, y_1, y_2))'$$

which completes the proof.

From Theorem I we see immediately that the kernel $K(x, y) = K_1(x, y)K_2(x, y)$ is a p. matrix as the restriction of the kernel $K'(x_1, x_2, y_1, y_2)$ to the subset $E_1 \subset E'$ consisting of the "diagonal" elements $\{x, x\}$ of E'. Further, from the theorem of §5 we obtain the class of functions and the norm corresponding to the kernel K. Thus we have the following theorem.

THEOREM II. *The kernel $K(x, y) = K_1(x, y)K_2(x, y)$ is the reproducing kernel of the class F of the restrictions of all functions of the direct product $F' = F_1 \otimes F_2$ to the diagonal set E_1 formed by all the elements $\{x, x\} \in E'$. For any such restriction f, $\|f\| = \min \|g'\|'$ for all $g' \in F'$, the restriction of which to the diagonal set E_1 is f.*

REMARK. Let $\{g_1^{(k)}\}$ be a complete orthonormal system in F_1. Then every function $f \in F$ is representable as a series

$$f(x) = \sum_1^{\infty} f_2^{(k)}(x)g_1^{(k)}(x), \; f_2^{(k)} \in F_2, \; \sum_1^{\infty} \|f_2^{(k)}\|_2^2 < \infty.$$

Among all such representations of $f(x)$ there exists one (and only one) which gives its minimum to the sum $\sum \|f_2^{(k)}\|_2^2$. This minimum is equal to $\|f\|^2$.

We can apply Theorem II to a class F and its conjugate $\overline{F}$. The product of the corresponding kernels is $|K(x, y)|^2 = K(x, y)K(y, x)$ and the corresponding class may be obtained from the remark above.

9. **Limits of reproducing kernels.** We shall consider two cases: (A) essentially, the case of a decreasing sequence of classes $F_1 \supset F_2 \supset \cdots$ with a decreasing sequence of kernels $K_1 \gg K_2 \gg K_3 \gg \cdots$; (B) essentially, the case of an increasing sequence of classes and kernels.

A. *The case of a decreasing sequence.* Let $\{E_n\}$ be an increasing sequence of sets, E their sum

(1) $$E = E_1 + E_2 + \cdots, \quad E_1 \subset E_2 \subset \cdots.$$

Let F_n, $n = 1, 2, \cdots$, be a class of functions defined in E_n. For a function $f_n \in F_n$ we shall denote by f_{nm}, $m \leqq n$, the *restriction* of f_n to the set $E_m \subset E_n$ ($f_{nn} = f_n$). We shall suppose then that the classes F_n form a *decreasing sequence* in the sense

(2) $$\textit{for every } f_n \in F_n \textit{ and every } m \leqq n, \qquad f_{nm} \in F_m.$$

Suppose further that the norms $\| \ \|_n$ defined in F_n form an *increasing sequence* in the sense

(3) $$\textit{for every } f_n \in F_n \textit{ and every } m \leqq n, \qquad \|f_{nm}\|_m \leqq \|f_n\|_n.$$

Finally, we suppose that every F_n possesses a reproducing kernel $K_n(x, y)$.

The case of all sets E_n equal, $E_1 = E_2 = \cdots = E$, is not excluded. Clearly, in this case $f_{nm} = f_n$, $F_n \subset F_m$, and, following Theorem II of §7, it is enough to suppose the existence of $K_1(x, y)$ in order to deduce the existence of all K_n and to obtain the property $K_n \ll K_m$ for $m < n$.

In the general case we have to introduce the restrictions K_{nm} of K_n to the set E_m ($m \leqq n$). By the theorem of §5, K_{nm} is the r.k. of the class F_{nm} of all restrictions f_{nm} for $f_n \in F_n$. The norm in F_{nm} is given by

$$\|f_{nm}\|_{nm} = \min \|f_n'\|_n \qquad \textit{for all } f_n' \textit{ with } f_{nm}' = f_{nm}.$$

From (3) we get

$$\|f_{nm}\|_{nm} \geqq \|f_{nm}\|_m$$

and consequently, by Theorem II of §7,

(4) $$K_{nm} \ll K_m, \qquad m < n.$$

We shall now prove the following theorem.

THEOREM I. *Under the above assumptions on the classes F_n, the kernels K_n converge to a kernel $K_0(x, y)$ defined for all x, y in E. K_0 is the r.k. of the class F_0 of all functions f_0 defined in E such that 1° their restrictions f_{0n} in E_n*

belong to F_n, $n=1, 2, \cdots$, 2° $\lim_{n=\infty} \|f_{0n}\|_n < \infty$. *The norm of* $f_0 \in F_0$ *is given by* $\|f_0\|_0 = \lim_{n=\infty} \|f_{0n}\|_n$.

REMARK. Condition 1° implies, following (3), that $\lim_{n=\infty}\|f_{0n}\|_n$ exists, but it may be infinite.

Proof. The convergence of K_n to K is to be understood in this way: any two points x, y in E belong to all E_n starting from some E_{n_0} on. Consequently $K_n(x, y)$ are defined for $n > n_0$ and we have to prove $\lim_{n=\infty} K_n(x, y) = K(x, y)$.

For fixed $y \in E_k$ and $k \leqq m \leqq n$ we have, following our assumptions,

$$
(5) \quad \begin{aligned}
\|K_{mk}(x, y) - K_{nk}(x, y)\|_k^2 &\leqq \|K_m(x, y) - K_{nm}(x, y)\|_m^2 \\
&= K_m(y, y) - \overline{K_{nm}(y, y)} - K_{nm}(y, y) + \|K_{nm}(x, y)\|_m^2 \\
&\leqq K_m(y, y) - 2K_n(y, y) + \|K_n(x, y)\|_n^2 \\
&= K_m(y, y) - K_n(y, y).
\end{aligned}
$$

From (4) it follows that $K_m - K_{nm}$ is a p.d. matrix. Therefore $K_m(y, y) - K_{nm}(y, y) = K_m(y, y) - K_n(y, y) \geqq 0$, and the sequence $\{K_m(y, y)\}_{m \geqq k}$ is a decreasing sequence of non-negative numbers. Consequently it is a convergent sequence. (5) shows then that the functions $K_{mk}(x, y) \in F_k$, for fixed k and $m \to \infty$, converge strongly in F_k to some function $\phi_k(x) \in F_k$. This involves $\lim_{m=\infty} K_{mk}(x, y) = \lim_{m=\infty} K_m(x, y) = \phi_k(x)$ for every $x \in E_k$.

Since for every x, y in E we can choose a k so that x and y belong to E_k, it is clear that $K_m(x, y)$ converge and that the limit $K_0(x, y)$ does not depend upon the choice of k. The function $\phi_k(x)$ is clearly the restriction $K_{0k}(x, y)$ of K_0 to E_k. Consequently $K_{mk}(x, y)$, for fixed y, converges strongly in F_k to $K_{0k}(x, y)$ which belongs to F_k. We then obtain from (5), by taking $n \to \infty$,

$$
(6) \quad \|K_{mk}(x, y) - K_{0k}(x, y)\|_k^2 \leqq K_m(y, y) - K_0(y, y);
$$

$$
\begin{aligned}
\|K_{0k}(x, y)\|_k &\leqq \|K_{0k}(x, y) - K_{mk}(x, y)\|_k + \|K_{mk}(x, y)\|_k \\
&\leqq (K_m(y, y) - K_0(y, y))^{1/2} + \|K_m(x, y)\|_m \\
&\leqq (K_m(y, y) - K_0(y, y))^{1/2} + (K_m(y, y))^{1/2}
\end{aligned}
$$

and for $m \to \infty$, $\|K_{0k}(x, y)\|_k^2 \leqq K_0(y, y)$.

Therefore for each $y \in E$, $K_0(x, y)$, as function of x, belongs to the class F_0 of our theorem.

Let us now prove that the class F_0 is a Hilbert space.

F_0 is linear, since $\|\alpha f_{0n} + \beta g_{0n}\|_n \leqq |\alpha| \, \|f_{0n}\|_n + |\beta| \, \|g_{0n}\|_n$.

$\|f_0\|_0^2$ is a quadratic form, since

$$
\begin{aligned}
\|\alpha f_0 + \beta g_0\|_0^2 &= \lim_{n=\infty} \|\alpha f_{0n} + \beta g_{0n}\|_n^2 \\
&= \lim_{n=\infty} [\alpha\bar{\alpha}\|f_{0n}\|_n^2 + \alpha\bar{\beta}(f_{0n}, g_{0n})_n + \bar{\alpha}\beta(g_{0n}, f_{0n})_n + \beta\bar{\beta}\|g_{0n}\|_n^2].
\end{aligned}
$$

Since the quadratic form in square brackets converges for all values of the complex variables α, β, it converges to a form of the same kind. This proves that $\|f_0\|_0^2$ is a quadratic form and also that

(7) $$(f_0, g_0)_0 = \lim_{n=\infty} (f_{0n}, g_{0n})_n.$$

It remains to be proved that F_0 is complete.

Take a Cauchy sequence $\{f_0^{(n)}\} \subset F_0$. The inequality $\|f_{0k}^{(m)} - f_{0k}^{(n)}\|_k \leqq \|f_{0l}^{(m)} - f_{0l}^{(n)}\|_l$, for $l > k$, gives, when $l \to \infty$, $\|f_{0k}^{(m)} - f_{0k}^{(n)}\|_k \leqq \|f_0^{(m)} - f_0^{(n)}\|_0$. Hence, $\{f_{0k}^{(m)}\}_{m=1,2,\cdots}$ is a Cauchy sequence in F_k, $\lim_{m=\infty} f_{0k}^{(m)} = \psi_k \in F_k$. It is clear that ψ_k is the restriction to E_k of a function ψ_0 defined in E. We have

$$\|f_{0k}^{(m)} - \psi_{0k}\|_k = \lim_{n=\infty} \|f_{0k}^{(m)} - f_{0k}^{(n)}\|_k \leqq \lim_{n=\infty} \|f_0^{(m)} - f_0^{(n)}\|_0.$$

This allows us to prove that $\psi_0 \in F_0$, since it gives a bound for $\|\psi_{0k}\|_k$ independent of k, namely

$$\|\psi_{0k}\|_k \leqq \|f_{0k}^{(m)}\|_k + \lim_{n=\infty} \|f_0^{(m)} - f_0^{(n)}\|_0 \leqq \|f_0^{(m)}\|_0 + \lim_{n=\infty} \|f_0^{(m)} - f_0^{(n)}\|_0.$$

On the other hand, it shows that

$$\|f_0^{(m)} - \psi_0\|_0 = \lim_{k=\infty} \|f_{0k}^{(m)} - \psi_{0k}\|_k \leqq \lim_{n=\infty} \|f_0^{(m)} - f_0^{(n)}\|_0$$

and consequently $\lim_{m=\infty} \|f_0^{(m)} - \psi_0\|_0 = 0$. This achieves the proof of completeness.

We have still to prove the reproducing property of K_0. To this effect take any $f_0 \in F_0$, $y \in E$. For sufficiently large n we have

$$f_0(y) = f_{0n}(y) = (f_{0n}(x), K_n(x, y))_n = (f_{0n}(x), K_{0n}(x, y))_n$$
$$+ (f_{0n}(x), K_n(x, y) - K_{0n}(x, y))_n.$$

For $n \to \infty$, the first scalar product in the last member converges, by formula (7), to $(f_0, K_0(x, y))_0$. The second scalar product converges to 0; in fact, by formula (6) (with $k = m = n$), it is in absolute value smaller than

$$\|f_{0n}\|_n K_n(x, y) - K_{0n}(x, y)\|_n \leqq \|f_0\|_0 (K_n(y, y) - K_0(y, y))^{1/2}.$$

This achieves the proof of our theorem.

B. *The case of an increasing sequence.* Let $\{E_n\}$ be a decreasing sequence of sets, E their intersection

(8) $$E = E_1 \cdot E_2 \cdot \cdots, \quad E_1 \supset E_2 \supset \cdots.$$

Let F_n be a class of functions defined in E_n. As before, we define the restriction f_{nm}, for $f_n \in F_n$, but now m has to be greater than n. We suppose then that F_n form an *increasing sequence*

(9) *for every* $f_n \in F_n$ *and every* $m \geqq n$, $\quad f_{nm} \in F_m$.

We suppose further that the norms $\| \ \|_n$ form a *decreasing sequence*

(10) *for every* $f_n \in F_n$ *and every* $m \geqq n$, $\quad \|f_{nm}\|_m \leqq \|f_n\|_n$.

Finally, we suppose that every F_n possesses a r.k. $K_n(x, y)$.

Now, even for all E_n equal, we cannot deduce the existence of all kernels K_n from the existence of one of them.

As in the case A, we get for the restrictions K_{nm} of K_n the formula

(11) $$K_{nm} \ll K_m, \qquad \textit{for } m > n.$$

For $y \in E$, $\{K_m(y, y)\}$ is an *increasing* sequence of positive numbers. Its limit may be infinite. We define, consequently,

(12) $$E_0 = \textit{set of } y \in E, \textit{ such that } K_0(y, y) = \lim_{m=\infty} K_m(y, y) < \infty.$$

For an illustration of this point consider Bergman's kernels K_n for a decreasing sequence of domains E_n. If the intersection E of the domains E_n is composed of a closed circle with an exterior segment attached to it, the set E_0 will be composed of all interior points of the circle.

We suppose that E_0 is not empty and define the limit-class of the classes F_n in the following way: let F_0 be the class of all restrictions f_{n0} of functions $f_n \in F_n$ ($n=1, 2, \cdots$) to the set E_0. From (10), we know that the sequence $\{\|f_{nk}\|_k\}_{k=n,n+1,\cdots}$ is decreasing and we can define

(13) $$\|f_{n0}\|_0 = \lim_{k=\infty} \|f_{nk}\|_k.$$

As in case A we prove that $\|f_{n0}\|_0^2$ is a positive quadratic form. This form is positive definite since from $\|f_{n0}\|_0 = 0$ it follows that for any $y \in E_0$, $|f_{n0}(y)| = |f_{nk}(y)| = |(f_{nk}(x), K_k(x, y))_k| \leqq \|f_{nk}\|_k (K_k(y, y))^{1/2} \to \|f_{n0}\|_0 (K_0(y, y))^{1/2} = 0$, that is, $f_{n0} = 0$. Consequently $\| \ \|_0$ is a norm in F_0.

In general F_0 will not be complete. In order that F_0 admit of a functional completion with a reproducing kernel, there are two conditions to be fulfilled which are given in the theorem of §4. The first one is that for every y there exists a constant M_y so that

(14) $$|f_{n0}(y)| \leqq M_y \|f_{n0}\|_0 \qquad \textit{for all } f_{n0} \in F_0.$$

Let us remark that the functions f_{n0} may be considered as defined in the whole set E (taking the restriction of f_n to E). Then, for every $y \in E$, the condition (14) is equivalent to $K_0(y, y) = \lim K_n(y, y) < \infty$. In fact, from the latter condition it follows in the same manner as above that $|f_{n0}(y)| \leqq \|f_{n0}\|_0 (K_0(y, y))^{1/2}$, that is, (14) with $M_y = (K_0(y, y))^{1/2}$. From (14), by taking $f_n(x) = K_n(x, y)$ we get $K_n(y, y) \leqq M_y \|K_{n0}(x, y)\|_0 \leqq M_y \|K_n(x, y)\|_n = M_y (K_n(y, y))^{1/2}$, that is, $K_n(y, y) \leqq M_y^2$.

Consequently, condition (14), that is, condition 1° of §4, is assured by our restriction to the set E_0. But the condition 2° of §4 is in general not assured.

We may illustrate this by the counter-example of §4. Take all the E_n equal to the set E of this counter-example. As class F_n we take the n-dimensional subspace of the class F introduced there, consisting of all polynomials of degree $\leqq n$. Each F_n is a complete space and its r.k. K_n converges to the Bergman kernel of the circle, restricted to E. Consequently $E_0=E$, $F_0=F$ and $\| \quad \|_0$ is clearly the norm introduced there.

To overcome this difficulty we can proceed as indicated in §4. We complete F_0 by ideal elements; in the completed Hilbert space $\overline{F}_0$ we choose an additional set E' of ideal elements such that the functions of F_0, extended to E_0+E', with the same norm as in F_0, form a space admitting a functional completion leading to a class $\tilde{F}_0$ with a reproducing kernel $\tilde{K}_0$.

Following the theorem of §5, we can return now to our set E_0 by restricting the functions of $\tilde{F}_0$ to E_0. If we take in the restricted class the norm defined in the theorem of §5, we shall get as r.k. the restriction of $\tilde{K}_0$ to E_0. The restricted class F_0^* and its norm $\| \quad \|_0^*$ can then be described, in terms of the space F_0 and its norm $\| \quad \|_0$, in the following way.

$f_0^* \in F_0^*$ if there is a Cauchy sequence $\{f_0^{(n)}\} \subset F_0$ such that

$$(15) \qquad f_0^*(x) = \lim_{n=\infty} f_0^{(n)}(x) \qquad \textit{for every } x \in E_0,$$

$$(16) \qquad \|f_0^*(x)\|_0^* = \min \lim_{n=\infty} \|f_0^{(n)}\|_0,$$

the minimum being taken for all Cauchy sequences $\{f_0^{(n)}\} \subset F_0$ satisfying (15). There exists at least one Cauchy sequence for which the minimum is attained. Such sequences will be called *determining* f_0^*.

The scalar product corresponding to $\| \quad \|_0^*$ is defined by

$$(17) \qquad (f_0^*, g_0^*)_0^* = \lim_{n=\infty} (f_0^{(n)}, g_0^{(n)})_0$$

for any two Cauchy sequences $\{f_0^{(n)}\}$ and $\{g_0^{(n)}\}$ determining f_0^* and g_0^*.

It is important to note that formula (17) is still valid when *only one* of the sequences $\{f_0^{(n)}\}$, $\{g_0^{(n)}\}$ is determining, the other satisfying only (15).

All these facts about the space F_0^* and its norm and scalar product are easily obtained when we form, as in §5, the subspace $\tilde{F}_0^0 \subset \tilde{F}_0$ of all $\tilde{f}_0 \in \tilde{F}_0$ vanishing in E_0. The complementary subspace $\tilde{F}_0' = \tilde{F}_0 \ominus \tilde{F}_0^0$ is then in an isomorphic correspondence with F_0^*, $\tilde{f}_0' \to f_0^*$, where f_0^* is the restriction of $\tilde{f}_0'$ to E_0. Further,

$$\|f_0^*\|_0^* = \|\tilde{f}_0'\|_{\tilde{0}}, \qquad (f_0^*, g_0^*)_0^* = (\tilde{f}_0', \tilde{g}_0')_{\tilde{0}},$$

where $\| \quad \|_0^\sim$ and $(\, , \,)_0^\sim$ are the norm and scalar product in $\tilde{F}_0$.

A Cauchy sequence $\{f_0^n\} \subset F_0$ converges in $\tilde{F}_0$ to a function $\bar{f}_0$. In the set E_0 it converges everywhere to a function $f_0^* \in F_0^*$ which is the restriction of $\bar{f}_0$ to E_0. f_0^* is also the restriction of $\bar{f}_0' =$ projection of $\bar{f}_0$ on $\tilde{F}_0'$. Since $\bar{f}_0 - \bar{f}_0' \in \tilde{F}_0^0$, we get

$$\lim_{n\to\infty} \|f_0^{(n)}\|_0^2 = \|\bar{f}_0\|_0^{\sim 2} = \|\bar{f}_0'\|_0^{\sim 2} + \|\bar{f}_0 - \bar{f}_0'\|_0^{\sim 2} \geqq \|f_0^*\|_0^{*2}.$$

The equality here is attained if $\{f_0^{(n)}\}$ converges in $\tilde{F}_0$ to $\bar{f}_0'$. Such a Cauchy sequence we have called as determining f_0^*. If now two Cauchy sequences $\{f_0^{(n)}\}$ and $\{g_0^{(n)}\}$ converge everywhere in E_0 to f_0^* and g_0^*, they converge in $\tilde{F}_0$ to vectors $\bar{f}_0$ and $\bar{g}_0$ whose restrictions to E_0 are f_0^* and g_0^*. If $\bar{f}_0'$ and $\bar{g}_0'$ are projections of $\bar{f}_0$ and $\bar{g}_0$ on $\tilde{F}_0'$, then $(f_0^*, g_0^*)_0^* = (\bar{f}_0', \bar{g}_0')_0^\sim$, but $\lim (f_0^{(n)}, g_0^{(n)})_0 = (\bar{f}_0, \bar{g}_0)_0^\sim$. If one of the sequences, say $\{f_0^{(n)}\}$, determines its limit in E_0, then $\bar{f}_0 = \bar{f}_0'$, and $(f_0^*, g_0^*)_0^* = (\bar{f}_0', \bar{g}_0')_0^\sim = (\bar{f}_0', \bar{g}_0)_0^\sim = \lim(f_0^{(n)}, g_0^{(n)})_0$.

The space F_0^* being completely defined we prove the following theorem.

THEOREM II. *The restrictions $K_{n0}(x, y)$ for every fixed $y \in E_0$ form a Cauchy sequence in F_0. They converge to a function $K_0^*(x, y) \in F_0^*$ which is the reproducing kernel of F_0^*.*

Proof. By an argument similar to the one used in (5) we obtain for $n \leqq m \leqq k$

$$\|K_{mk}(x, y) - K_{nk}(x, y)\|_k^2 \leqq K_m(y, y) - K_n(y, y). \tag{18}$$

Taking $k \to \infty$, we have

$$\|K_{m0}(x, y) - K_{n0}(x, y)\|_0^2 \leqq K_m(y, y) - K_n(y, y). \tag{19}$$

This proves, together with (12), that $\{K_{n0}(x, y)\}$ is a Cauchy sequence in F_0. By property (14) this sequence converges for every $x \in E_0$ to a function $K_0^*(x, y)$ which, by definition (15), belongs to F_0^*.

It remains to prove the reproducing property of K_0^*. To this effect take any $f_0^* \in F_0^*$ and a Cauchy sequence $\{f_0^{(n)}\} \subset F_0$ determining f_0^*. Each $f_0^{(n)}$ is a restriction of some $f_{k_n} \in F_{k_n}$, $f_0^{(n)} = f_{k_n0}$. By (13) there exists an increasing sequence $m_1 < m_2 < \cdots$ such that

$$m_n > k_n, \qquad \|f_{k_n m_n}\|_{m_n}^2 - \|f_{k_n 0}\|_0^2 \leqq \frac{1}{n^2}. \tag{20}$$

It is clear that $\{K_{m_n 0}(x, y)\}$ is also a Cauchy sequence converging to $K_0^*(x, y)$. Consequently, from (17) it follows that

$$(f_0^*(x), K_0^*(x, y))_0^* = \lim_{n=\infty} (f_{k_n 0}(x), K_{m_n 0}(x, y))_0. \tag{21}$$

We may now write

$$(f_{k_n0}(x), K_{m_n0}(x, y))_0 = (f_{k_nm_n}(x), K_{m_n}(x, y))_{m_n}$$
(22)
$$- [(f_{k_nm_n}(x), K_{m_n}(x, y))_{m_n} - (f_{k_n0}(x), K_{m_n0}(x, y))_0].$$

The square bracket is of the form $[(g, h)_{m_n} - (g_0, h_0)_0]$ for g, h of F_{m_n} (g_0, h_0 are restrictions of g, h to E_0). This is a bilinear form in g, h and the corresponding quadratic form $(g, g)_{m_n} - (g_0, g_0)_0 = \|g\|^2_{m_n} - \|g_0\|^2_0$ is positive (following (10) and (13)). Consequently the Cauchy-Schwarz inequality is valid for this form and in the case of the square bracket of (22) it gives in connection with (20)

$$| [\cdots] | \leqq [\|f_{k_nm_n}\|^2_{m_n} - \|f_{k_n0}\|^2_0]^{1/2} [\|K_{m_n}(x, y)\|^2_{m_n} - \|K_{m_n0}(x, y)\|^2_0]^{1/2}$$
$$\leqq \frac{1}{n} \|K_{m_n}(x, y)\|_{m_n} = \frac{1}{n} K_{m_n}(y, y)^{1/2}.$$

For $n \to \infty$ this converges to 0, since $K_{m_n}(y, y) \nearrow K_0(y, y) < \infty$. Therefore, (21) and (22) yield

$$(f_0^*(x), K_0^*(x, y))_0^* = \lim_{n=\infty} (f_{k_nm_n}(x), K_{m_n}(x, y))_{m_n} = \lim_{n=\infty} f_{k_nm_n}(y)$$
$$= \lim f_{k_n0}(y) = \lim f_0^{(n)}(y) = f_0^*(y),$$

which is the reproducing property of K_0^*.

REMARK. A particularly simple case is the one where the class F_0 with the norm $\| \ \|_0$ happens to be a subspace of a class F possessing a reproducing kernel. Then, condition 2° of §4 is clearly satisfied; F_0^* is the functional completion of F_0, the norm $\| \ \|_0^*$ is an extension of the norm $\| \ \|_0$, and F_0^* is simply the closure in F of F_0.

A trivial case of this kind is one where the F_n form an increasing sequence of subspaces of a class F with a r.k. Hence $E_1 = E_2 = \cdots = E = E_0$ and F_0^* is the closure of the sum $\sum F_n$.

10. **Construction of a r.k. by resolution of identity.** Let us give a brief résumé of the essential properties of resolutions of identity in a Hilbert space $\mathfrak{H}$ (for a complete study of resolutions of identity, especially in connection with the theory of operators, see M. H. Stone [1]). For simplicity's sake we shall suppose here that the space $\mathfrak{H}$ is *separable*.

We call a resolution of identity a class $\{P_\lambda\}$ of projections in $\mathfrak{H}$, depending on a real parameter λ, $-\infty < \lambda < +\infty$, and having the following properties:

1. P_λ is a projection on a closed subspace $\mathfrak{H}_\lambda \subset \mathfrak{H}$, increasing with λ: $\mathfrak{H}_{\lambda'} \subset \mathfrak{H}_\lambda$ for $\lambda' < \lambda$.

2. $P_\lambda \to 0$ for $\lambda \to -\infty$; $P_\lambda \to I$ (identity operator) for $\lambda \to +\infty$.

For any open interval $\Delta: \lambda' < \lambda < \lambda''$, we define

(1) $$\Delta\mathfrak{H} = \mathfrak{H}_{\lambda''} \ominus \mathfrak{H}_{\lambda'}, \qquad \Delta P = P_{\lambda''} - P_{\lambda'}.$$

ΔP is the projection on $\Delta\mathfrak{H}$.

For any decomposition of the real axis into intervals $\Delta_k=(\lambda_k, \lambda_{k+1})$, $-\infty\leftarrow \cdots <\lambda_{-2}<\lambda_{-1}<\lambda_0<\lambda_1<\lambda_2< \cdots \rightarrow+\infty$, we have, obviously,

$$I = \sum_k \Delta_k P, \tag{2}$$

the series converging in the sense of strong limit for operators.

A real number λ_0 belongs to the *spectrum* of $\{P_\lambda\}$ if $\Delta P\neq 0$ for every interval Δ containing λ_0. The numbers belonging to the spectrum form a closed set.

For any real θ and any decreasing sequence of intervals Δ_n containing θ and converging to θ there exist the limits

$$\delta_\theta P = \lim \Delta_n P, \qquad \delta_\theta\mathfrak{H} = \lim \Delta_n\mathfrak{H}, \tag{3}$$

the second limit being the intersection of the decreasing sequence of subspaces $\Delta_n\mathfrak{H}$. These limits do not depend on the choice of Δ_n and $\delta_\theta P$ is the projection on $\delta_\theta\mathfrak{H}$. Only for an enumerable set of θ, say $\{\theta_k\}$, is $\delta_\theta P\neq 0$.

If we have $\delta_{\theta_n}P=I$, which means $\delta_{\theta_1}\mathfrak{H}+\delta_{\theta_2}\mathfrak{H}+ \cdots =\mathfrak{H}$, we say that the spectrum of $\{P_\lambda\}$ is *discrete*.

If, for all θ, $\delta_\theta P=0$, we say that the spectrum of $\{P\}$ is *continuous* (often called *purely continuous*).

In our applications we shall meet, essentially, only discrete spectra or continuous spectra.

It has been proved (theorem of Hellinger-Hahn) that for any spectrum there exist finite or infinite systems $\{f_n(\lambda)\}$ of elements $f_n(\lambda)\in\mathfrak{H}$, depending on λ, such that if we denote by Δf_n the difference $f_n(\lambda'')-f_n(\lambda')$, we have

(a) *for* $m\neq n$, $(\Delta_1 f_m, \Delta_2 f_n)=0$ *for any intervals* Δ_1, Δ_2.

(b) $(\Delta_1 f_n, \Delta_2 f_n)=0$ *for any non-overlapping intervals* Δ_1, Δ_2.

(c) *For every interval* Δ, *the elements* $\Delta_1 f_n$, $n=1, 2, \cdots$ *for all* $\Delta_1\subset\Delta$, *belong to the subspace* $\Delta\mathfrak{H}$ *and form a complete system in* $\Delta\mathfrak{H}$.

The minimal number of elements in such a system $\{f_n(\lambda)\}$ is called the *multiplicity* of the spectrum. The spectrum is called *simple* if the multiplicity $=1$, that is, if there exists such a system with only one element $f_1(\lambda)$.

In the case of a discrete spectrum the multiplicity is the maximal dimension of the subspaces $\delta_\theta\mathfrak{H}$.

If a system of elements $\{f_n(\lambda)\}$ satisfies (a) and (b) and instead of (c) satisfies the weaker condition

(c′) *The elements* Δf_n *for* $n=1, 2, \cdots$ *and for all intervals* Δ *form a complete system in* $\mathfrak{H}$,

then the system $\{f_n(\lambda)\}$ determines a corresponding resolution of identity $\{P_\lambda\}$ (for which it satisfies (c)) in the following manner:

$\mathfrak{H}_\lambda$ *is the subspace generated by all the* Δf_n, $n=1, 2, \cdots$, $\Delta=(\lambda', \lambda'')$ *with* $\lambda''<\lambda$.

This resolution of identity is continuous to the left, that is, $\mathfrak{H}_\lambda = \lim \mathfrak{H}_{\lambda'}$ for $\lambda' \nearrow \lambda$. This condition (or the right side continuity) is usually accepted, for reasons of convenience, as an additional condition on resolutions of identity.

For any system $\{f_n(\lambda)\}$ satisfying (a), (b), and (c), it is seen that $\|f_n(\lambda)\|^2$ is a non-decreasing function $\mu_n(\lambda)$, $\Delta\mu_n = \mu_n(\lambda'') - \mu_n(\lambda') = \|\Delta f_n\|^2$. We consider the measure μ_n, introduced on the real axis by $\mu_n(\lambda)$, which leads to the Lebesgue-Stieltjes integral $\int \Phi(\lambda) d\mu_n(\lambda)$. It has been proved that for every $u \in \mathfrak{H}$ there exists the limit

$$(4) \qquad \phi_n(\lambda) = \lim_{\lambda' \nearrow \lambda,\, \lambda'' \searrow \lambda} \frac{(u, \Delta f_n)}{\|\Delta f_n\|^2} = \frac{d(u, f_n(\lambda))}{d\mu_n(\lambda)},$$

for all λ with exception of a set of μ_n-measure 0.

We have further

$$(5) \qquad \|u\|^2 = \sum_n \int_{-\infty}^{\infty} |\phi_n(\lambda)|^2 d\mu_n.$$

$$(6) \qquad (u, v) = \sum_n \int_{-\infty}^{\infty} \phi_n(\lambda)\overline{\psi_n(\lambda)} d\mu_n, \quad \text{where} \quad \psi_n(\lambda) = \frac{d(v, f_n(\lambda))}{d\mu_n(\lambda)}.$$

Let us now apply the above considerations to the construction of a r.k. We suppose that our Hilbert space $\mathfrak{H}$ is a class of functions defined in E with a r.k. $K(x, y)$.

For a given resolution of identity $\{P_\lambda\}$ every subspace $\Delta\mathfrak{H}$ will have a r.k. which we shall denote by $\Delta K(x, y)$. The kernel ΔK determines the projection ΔP by the equation

$$(7) \qquad \Delta P f = f_1(y) = (f(x), \Delta K(x, y)), \qquad \textit{for any } f \in \mathfrak{H}.$$

The kernel K corresponds to the identity I and following (2) we have

$$(8) \qquad K(x, y) = \sum_k \Delta_k K(x, y),$$

for any decomposition $\{\Delta_k\}$ of the real axis. The series in (8) converges absolutely. In fact, following (2), the series $K(y, z) = IK(y, z) = \sum \Delta_k PK(y, z) = \sum (K(x, z), \Delta_k K(x, y))_x = \sum \Delta_k K(y, z)$ as function of y is strongly convergent. It converges then in the ordinary sense for every y, in particular for $y = z$. Thus, $K(z, z) = \sum \Delta_k K(z, z) < \infty$, $\Delta_k K(z, z) \geqq 0$. Consequently

$$\sum |\Delta_k K(x, y)| \leqq \sum (\Delta_k K(x, x))^{1/2} (\Delta_k K(y, y))^{1/2} \leqq [\sum \Delta_k K(x, x) \cdot \sum \Delta_k K(y, y)]^{1/2}.$$

If the resolution of identity $\{P_\lambda\}$ has a discrete spectrum $\{\theta_k\}$ and if the r.k. of $\delta_{\theta_k}\mathfrak{H}$ is denoted by $\delta_{\theta_k}K$, then we have again an absolutely convergent representation

$$K(x, y) = \sum_k \delta_{\theta_k} K(x, y). \tag{9}$$

An especially important case which is most often applied is one where the spectrum is simple. Then the subspaces $\delta_{\theta_k}\mathfrak{H}$ are one-dimensional and each is generated by a function $g_k(x)$ which we can suppose normalized, $\|g_k\| = 1$. The functions $g_k(x)$ form a complete orthonormal system in $\mathfrak{H}$. The kernels $\delta_{\theta_k}K(x, y)$ are given by $g_k(x)\overline{g_k(y)}$, and (9) takes the form of the well known development of the kernel in an orthonormal complete system

$$K(x, y) = \sum_k g_k(x)\overline{g_k(y)}, \tag{10}$$

which for a long time was taken as a basis of the definition of a r.k.

Suppose now that the resolution of identity $\{P_\lambda\}$ is given by a system of functions $f_n(\lambda) \equiv f_n(x, \lambda)$ satisfying (a), (b), and (c'). Following (4) we define for $u = K(x, y)$ (considered as function of x), the functions

$$\Phi_n(y, \lambda) = \lim_{\Delta\to\lambda} \frac{(K(x, y), \Delta f_n)}{\Delta\mu_n} = \lim \frac{\overline{f_n(y, \lambda'') - f_n(y, \lambda')}}{\mu_n(\lambda'') - \mu_n(\lambda')} = \frac{\overline{df_n(y, \lambda)}}{d\mu_n(\lambda)} \tag{11}$$

From (6) and (5) we then obtain

$$K(y, z) = (K(x, z), K(x, y)) = \sum_n \int_{-\infty}^{\infty} \Phi_n(y, \lambda)\overline{\Phi_n(z, \lambda)}d\mu_n, \tag{12}$$

$$K(y, y) = \sum_n \int_{-\infty}^{\infty} |\Phi_n(y, \lambda)|^2 d\mu_n. \tag{13}$$

The series and the integrals in (12) are absolutely convergent because of (13).

The function $\Phi_n(y, \lambda)$ is in general defined for each y only almost everywhere in λ in the sense of the measure μ_n. Nevertheless, in most applications it turns out to be a continuous function of λ. In spite of this, $\Phi_n(y, \lambda_0)$, as function of y for a fixed λ_0, will not in general belong to $\mathfrak{H}$.

11. Operators in spaces with reproducing kernels(9). In a class F forming a Hilbert space with a r.k. K, the bounded operators admit of an interesting representation.

The notation $L_x K(x, y)$ indicates that the operator is applied to $K(x, y)$ as function of x and that the resulting function is considered as function of x (but it will depend also on y which will act in the transformation as a parameter). It is then clear what is meant by $L_z\overline{K(x, z)}$, $L_x L_z' \overline{K(x, z)}$, and so on. The notation $Lf(x)$ is clear and we may also write $Lf(x_0)$ if x_0 is a particular value of x. Consider the adjoint operator L^* (that is, the operator for which $(Lf, g) = (f, L^*g)$). Take the transform

(9) The developments of this section are closely related with the work and ideas of E. H. Moore.

$$(1)\qquad \Lambda(x, y) = L_x^* K(x, y).$$

Λ is a function of the two points x, y. As function of x it belongs to F. Take then for any $f \in F$ the scalar product $(f(x), \Lambda(x, y))_x = (f(x), L_x^* K(x, y))_x = (Lf(x), K(x, y))_x = Lf(y)$,

$$(2)\qquad Lf(y) = (f(x), \Lambda(x, y)).$$

In this way, to each bounded operator there corresponds a kernel $\Lambda(x, y)$ which for every y, as function of x, belongs to F. The operator is represented in terms of the kernel by formula (2).

Let us now find the kernel $\Lambda^*(x, y)$ corresponding to the adjoint operator L^*. We have $(L^*)^* = L$ and thus

$$(L_x K(x, z), K(x, y)) = (K(x, z), L_x^* K(x, y)),$$

$$(\Lambda^*(x, z), K(x, y)) = \overline{(\Lambda(x, y), K(x, z))},$$

$$(3)\qquad \Lambda^*(y, z) = \overline{\Lambda(z, y)}.$$

It is clear that to $L_1 + L_2$ or αL correspond $\Lambda_1 + \Lambda_2$ and $\bar{\alpha}\Lambda$ respectively. We shall now find the kernel Λ corresponding to the composition $L = L_1 L_2$. Since $(L_1 L_2)^* = L_2^* L_1^*$, we have

$$\begin{aligned}\Lambda(y, z) &= (L_1 L_2)_y^* K(y, z) = L_{2y}^* L_{1y}^* K(y, z)\\ &= L_{2y}^* \Lambda_1(y, z) = (\Lambda_1(x, z), \Lambda_2^*(x, y)) = (\Lambda_1(x, z), \overline{\Lambda_2(y, x)})\end{aligned}$$

$$(4)\qquad \Lambda(y, z) = (\Lambda_1(x, z), \overline{\Lambda_2(y, x)}) \qquad \textit{for } L = L_1 L_2.$$

Let us note the following properties resulting immediately from (1)–(4):

$$(5)\qquad \begin{aligned}((f(x), \Lambda(x, y))_x, g(y))_y &= (f(x), (g(y), \overline{\Lambda(x, y)})_y)_x\\ &= (f(x), \overline{(\Lambda(x, y), g(y))_y})_x.\end{aligned}$$

(6) *The symmetry of L is equivalent to the hermitian symmetry of Λ:*

$$\Lambda(x, y) = \overline{\Lambda(y, x)}.$$

We prove now the following property:

(7) *The operator L is positive if and only if Λ is a p. matrix.*

In fact, L positive means that for every $f \in F$, $(Lf, f) \geqq 0$. For $f = \sum_1^n \zeta_k K(x, y_k)$ we then get

$$\begin{aligned}&\sum_i \sum_j \zeta_i \bar{\zeta}_j (L_x K(x, y_i), K(x, y_j))\\ &\qquad = \sum \sum \zeta_i \bar{\zeta}_j (\Lambda^*(x, y_i), K(x, y_j)) = \sum \sum \Lambda^*(y_j, y_i) \zeta_i \bar{\zeta}_j\\ &\qquad = \sum \sum \overline{\Lambda(y_i, y_j)} \zeta_i \bar{\zeta}_j > 0.\end{aligned}$$

This proves that $\bar{\Lambda}$ and thus Λ is also a p. matrix. It also proves the well known fact that a positive operator is always symmetric.

If now Λ is a p. matrix, we see that $(Lf, f) \geqq 0$ will be satisfied for all f of the form $\sum_1^n \zeta_k K(x, y_k)$. As these functions form a dense set in F, every function $f \in F$ may be approximated by them and we get $(Lf, f) \geqq 0$ by a passage to the limit.

In generalizing the notation of §7 we shall write $\Lambda_1 \ll \Lambda_2$ or $\Lambda_2 \ll \Lambda_1$ *for any two kernels* if $\Lambda_2 - \Lambda_1$ is a p. matrix.

THEOREM I. *For an arbitrary kernel* $\Lambda(x, y)$, *hermitian symmetric* (*that is,* $\Lambda(x, y) = \overline{\Lambda(y, x)}$), *the necessary and sufficient condition that it correspond to a bounded symmetric operator with lower bound* $\geqq m > -\infty$ *and upper bound* $\leqq M < +\infty$ *is that* $mK \ll \Lambda \ll MK$.

Proof. *Necessity.* If L is the corresponding symmetric operator with bounds not less than m and not greater than M, we have

$$m(f, f) \leqq (Lf, f) \leqq M(f, f) \qquad \text{for every } f \in F.$$

It follows that $((L - mI)f, f) \geqq 0$ and $((MI - L)f, f) \geqq 0$, that is, the operators $L - mI$ and $MI - L$ are positive. Therefore, from (7) we obtain that $\Lambda - mK$ and $MK - \Lambda$ are p. matrices.

Sufficiency. The condition of the theorem is clearly equivalent to

$$0 \ll \frac{1}{M - m}(\Lambda - mK) \ll K.$$

This means that the kernel $K_1 = (1/(M - m)(\Lambda - mK)$ is a p. matrix and is $\ll K$. Therefore it is a reproducing kernel of a class F_1 with the norm $\| \ \|_1$ and following Theorem I, §7, $F_1 \subset F$ and $\|f_1\|_1 \geqq \|f_1\|$ for $f_1 \in F_1$. Then, as in Theorem III, §7, the operator

$$L_1 f(y) = (f(x), K_1(x, y))$$

is a positive operator in F with a bound not greater than 1, that is,

$$0 \leqq (L_1 f, f) \leqq (f, f).$$

This operator corresponds to $K_1(x, y)$ by its definition. Consequently, to $\Lambda = (M - m)K_1 + mK$ there corresponds the operator $L = (M - m)L_1 + mI$ and the last inequalities give

$$m(f, f) \leqq (mIf, f) + ((M - m)L_1 f, f) \leqq M(f, f),$$

$$m(f, f) \leqq (Lf, f) \leqq M(f, f).$$

An arbitrary kernel Λ is representable in a unique way in the form

(8) $$\Lambda = \Lambda_1 + i\Lambda_2, \quad \Lambda_1 \textit{ and } \Lambda_2 \textit{ hermitian symmetric.}$$

Namely, we have

$$(9)\qquad \begin{aligned} \Lambda_1(x, y) &= \frac{1}{2}(\Lambda(x, y) + \overline{\Lambda(y, x)}), \\ \Lambda_2(x, y) &= \frac{1}{2i}(\Lambda(x, y) - \overline{\Lambda(y, x)}). \end{aligned}$$

The necessary and sufficient condition in order that Λ correspond to a bounded operator is, clearly, that Λ_1 and Λ_2 correspond to such operators. To the last kernels we can apply Theorem I.

We now consider convergence of operators. The three simplest notions of limit for bounded operators are the following: the *weak limit*, w. $\lim_{n=\infty} L_n = L$, if $L_n u$ converges weakly to Lu for every $u \in F$; the *strong limit*, str. $\lim L_n = L$, if $L_n u$ converges strongly to Lu; the *uniform limit*, un. $\lim L_n = L$, if $\|L_n - L\| \to 0$, where $\|\ \ \|$ for operators denotes their bound.

It is clear that weak convergence follows from strong convergence and that strong convergence follows from the uniform one.

It is known that w. $\lim L_n = L$ involves the boundedness of all $\|L_n\|$ and the inequality $\|L\| \leqq \lim \inf. \|L_n\|$.

THEOREM II. *If $L =$ w. $\lim L_n$, then for the corresponding kernels we have $\Lambda(x, y) = \lim \Lambda_n(x, y)$ for every x, y in E. If $L =$ un. $\lim L_n$, then Λ_n converges uniformly to Λ in every set of couples (x, y) for which $K(x, x)$ and $K(y, y)$ are uniformly bounded.*

The first part follows immediately, by the definition of weak convergence from

$$\begin{aligned} \Lambda(x, y) &= (\Lambda(z, y), K(z, x))_z = (L_z^* K(z, y), K(z, x))_z \\ &= (K(z, y), L_z K(z, y))_z = \lim (K(z, y), L_{nz} K(z, x))_z = \lim \Lambda_n(x, y). \end{aligned}$$

The second part follows easily from

$$\begin{aligned} |\Lambda(x, y) - \Lambda_n(x, y)| &= |(K(z, y), (L - L_n)_z K(z, x))| \\ &\leqq \|K(z, y)\|_z \|(L - L_n)_z K(z, x)\|_z \\ &\leqq \|K(z, y)\|_z \|L - L_n\| \, \|K(z, x)\|_z \\ &= \|L - L_n\| (K(x, x)\, K(y, y))^{1/2}. \end{aligned}$$

Consider now two orthonormal complete systems in F, $\{g'_m\}$ and $\overline{\{g''_n\}}$ (in particular we may have $g'_m = g''_m$). The double system $\{g'_m(x)\overline{g''_n(y)}\}$ is a complete orthonormal system in the direct product $F \otimes \overline{F}$.

If $\Lambda(x, y)$ belongs to the direct product we know that it is representable by an absolutely convergent double series

$$(10)\qquad \Lambda(x, y) = \sum_{m,n} \alpha_{mn} g'_m(x) \overline{g''_n(y)}$$

where the coefficients α_{mn} satisfy $\sum_{m,n} |\alpha_{mn}|^2 < \infty$ and are given by

$$(11) \qquad \alpha_{mn} = (g_n''(y), (g_m'(x), \Lambda(x, y))_x)_y = (g_n''(y), Lg_m'(y)).$$

THEOREM III. *For every bounded operator L, the series in* (10) *with coefficients given by* (11) *is convergent for every x and y in the sense*

$$(12) \qquad \Lambda(x, y) = \lim_{p,q=\infty} \sum_{m=1}^{p} \sum_{n=1}^{q} \alpha_{mn} g_m'(x) \overline{g_n''(y)}.$$

The $\Lambda(x, y)$ *belonging to the direct product* $F \otimes \overline{F}$ *correspond to operators with finite norm.*

Let P_p' and P_q'' be the projections on the subspaces generated by $g_1', g_2', \cdots, g_p'$ and $g_1'', g_2'', \cdots, g_q''$. It is clear that str. $\lim_{p=\infty} P_p' = I$, str. $\lim_{q=\infty} P_q'' = I$. Consequently, for any u, v in F

$$(13) \qquad \lim_{p,q=\infty} (P_q''v, LP_p'u) = (v, Lu).$$

If we take now $v = K(z, y)$ and $u = K(z, x)$ as functions of z, we get $P_q''v = P_{qz}''K(z, y) = \sum_1^q g_n''(z)\overline{g_n''(y)}$, $P_p'u = \sum_1^p g_m'(z) g_m'(x)$ and (11) and (13) then lead directly to (12).

The norm of an operator L is given by $\mathfrak{N}(L) = \sum_{m=1}^{\infty} \|Lg_m\|^2$ for any orthonormal complete system $\{g_n\}$. It is independent of the choice of this system and may be finite or infinite. From (11) it is clear that

$$Lg_m'(y) = \sum_{n=1}^{\infty} \overline{\alpha_{mn}} g_n''(y)$$

by development in the system $\{g_n''\}$. Consequently, $\|Lg_m'(y)\|^2 = \sum_{n=1}^{\infty} |\alpha_{mn}|^2$ and $\mathfrak{N}(L) = \sum_{m=1}^{\infty} \sum_{n=1}^{\infty} |\alpha_{mn}|^2$ which proves the second part of our theorem.

12. **The reproducing kernel of a sum of two closed subspaces.** Let F be a class with a r.k. K. We know that the r.k.'s of closed subspaces of F correspond to the projections on these subspaces.

The problem of expressing the r.k. of the sum $F_1 \oplus F_2$ of two closed subspaces in terms of the r.k.'s K_1 and K_2 of these subspaces is therefore reduced to the problem of expressing the projection P on $F_1 \oplus F_2$ in terms of the projections P_1 and P_2 on F_1 and F_2.

In order to obtain this we shall at first prove the identity

$$(1) \qquad \begin{aligned} &[(P - P_1)(P - P_2)]^m \\ &= P - \sum_{k=1}^{m} P_1(P_2P_1)^{k-1} + P_2(P_1P_2)^{k-1} - (P_2P_1)^k - (P_1P_2)^k] - (P_2P_1)^m. \end{aligned}$$

We shall use the known properties of projections, namely: $P_1 = P_1P = PP_1 = P_1^2$, $P_2 = P_2P = PP_2 = P_2^2$,

$$P_1(P-P_1)=(P-P_1)P_1=0, \qquad P_2(P-P_2)=(P-P_2)P_2=0.$$

Then, denoting the expression $(P-P_1)(P-P_2)$ by Q we have

$$Q^kP_1=Q^{k-1}(P-P_1)(P-P_2)P_1=Q^{k-1}(P-P_1)(P_1-P_2P_1)$$
$$=-Q^{k-1}P_2P_1+Q^{k-1}P_1P_2P_1.$$

If $k>1$, the first term is $-Q^{k-2}(P-P_1)(P-P_2)P_2P_1=0$, and we have

$$Q^kP_1=Q^{k-1}P_1P_2P_1, \qquad k>1.$$

For $k=1$ we obtain

$$QP_1=-P_2P_1+P_1P_2P_1.$$

Finally, we obtain by induction

(2) $$Q^kP_1=-(P_2P_1)^k+P_1(P_2P_1)^k.$$

Further, we have

(3) $$Q=(P-P_1)(P-P_2)=P-P_1-P_2+P_1P_2.$$

This gives

$$Q^n=Q^{n-1}(P-P_1-P_2+P_1P_2)=Q^{n-1}P-Q^{n-1}P_1-Q^{n-1}P_2+Q^{n-1}P_1P_2.$$

Since $Q^{n-1}P=Q^{n-1}$, $Q^{n-1}P_2=0$, we get from (2)

$$\begin{aligned} Q^n &= Q^{n-1}-Q^{n-1}P_1+Q^{n-1}P_1P_2 \\ &= Q^{n-1}-\left[-(P_2P_1)^{n-1}+P_1(P_2P_1)^{n-1}\right] \\ &\quad +\left[-(P_2P_1)^{n-1}P_2+P_1(P_2P_1)^{n-1}P_2\right] \\ &= Q^{n-1}+(P_2P_1)^{n-1}-P_1(P_2P_1)^{n-1}-P_2(P_1P_2)^{n-1}+(P_1P_2)^n. \end{aligned}$$

This expression is valid for $n\geqq 2$. Adding these equations side by side for $n=m, m-1, \cdots, 2$, and using the formula for Q given in (3), we obtain the required formula for Q^m. This formula may be written in the form

(4) $$\begin{aligned} P &= [(P-P_1)(P-P_2)]^m+(P_2P_1)^m \\ &\quad +\sum_{k=1}^{m}\left[P_1(P_2P_1)^{k-1}+P_2(P_1P_2)^{k-1}-(P_2P_1)^k-(P_1P_2)^k\right]. \end{aligned}$$

We shall prove now that for $m\to\infty$, $(P_2P_1)^m$ converges strongly to the projection P_0 on the intersection F_0 of F_1 and F_2. In order to do so we shall consider the operator $L=P_2P_1$ in the space F_2.

In this space L is a positive operator (and therefore symmetric) with bound not greater than 1. In fact, for $u\in F_2$

$$(P_2P_1u, P_2P_1u)=\|P_2P_1u\|^2\leqq\|P_1u\|^2=(P_1u, P_1u)=(P_1u, u)=(P_1u, P_2u)$$
$$=(P_2P_1u, u)=\|P_1u\|^2\leqq\|u\|^2,$$

$$(5) \qquad 0 \leqq (Lu, Lu) \leqq (Lu, u) \leqq (u, u).$$

Consider now for any $f \in F$ the sequence $\{L^k f\}$. For $k \geqq 1$, $L^k f = P_2 P_1 L^{k-1} f \in F_2$. Putting $u = L^k f$ in (5), we get

$$0 \leqq (L^{k+1}f, L^{k+1}f) \leqq (L^{k+1}f, L^k f) \leqq (L^k f, L^k f),$$

$$0 \leqq (L^{2k+1}f, Lf) \leqq (L^{2k}f, Lf) \leqq (L^{2k-1}f, Lf).$$

Consequently the sequence $\{L^n f, Lf)\}$ is a decreasing sequence of positive numbers and therefore it is convergent. This gives

$$\begin{aligned} \lim_{m,n=\infty} \|L^m f - L^n f\|^2 &= \lim \, [(L^m f, L^m f) - (L^m f, L^n f) \\ &\qquad - (L^n f, L^m f) + (L^n f, L^n f)] \\ &= \lim \, [(L^{2m-1}f, Lf) - 2(L^{m+n-1}f, Lf) + (L^{2n-1}f, Lf)] = 0 \end{aligned}$$

and thus $L^m f$ converges strongly. This means that L^m converges strongly to some bounded operator P_0. We have further $L^{m+1}f = LL^m f = L^m Lf$, which, for $m \to \infty$, gives

$$(6) \qquad LP_0 f = P_0 f = P_0 Lf.$$

Therefore $L^m P_0 f = P_0 f$ and $P_0^2 f = \lim L^m P_0 f = P_0 f$. In the subspace F_2, P_0 as a limit of symmetric operators is symmetric. Together with $P_0^2 f = P_0 f$ it shows that in F_2 the operator P_0 is a projection. It is the projection on the subspace of all $P_0 f$. From (6) we get $\|P_2 P_1 P_0 f\| \leqq \|P_1 P_0 f\| \leqq \|P_0 f\| = \|P_2 P_1 P_0 f\|$. Consequently $\|P_2 P_1 P_0 f\| = \|P_1 P_0 f\| = \|P_0 f\|$, $P_2 P_1 P_0 f = P_1 P_0 f = P_0 f$ and $P_0 f \in F_0 = F_1 \cdot F_2$. Inversely if $u \in F_1 \cdot F_2$, then $Lu = P_2 P_1 u = u$ and $P_0 u = \lim L^n u = u$.

Thus, in F_2, P_0 is the projection on F_0. Then for any $f \in F$, we have by (6) $P_0 f = P_0 Lf = P_0 P_2 P_1 f =$ projection of f on F_0.

In our formula (4), besides the series $\sum$ and the term $(P_2 P_1)^m$ we have still the expression $(P - P_1)(P - P_2)^m$. $P - P_1$ is the projection on $F' \ominus F_1$, and $P \ominus P_2$ is the projection on $F' \ominus F_2$, if we denote $F_1 \oplus F_2$ by F'. Consequently for $m \to \infty$, the last expression converges strongly to the projection on the intersection of $(F' \ominus F_1)$ and $(F' \ominus F_2)$. But this intersection is reduced to the element zero because were there in it any element $u \neq 0$, it would belong to F' and would be orthogonal to F_1 as well as to F_2. Thus, u would be orthogonal to $F_1 \oplus F_2 = F'$ which is impossible.

In this way we finally obtain the desired formula for the projection P:

$$(7) \qquad P = P_0 + \sum_{k=1}^{\infty} [P_1(P_2 P_1)^{k-1} + P_2(P_1 P_2)^{k-1} - (P_2 P_1)^k - (P_1 P_2)^k].$$

The subspace $F_1 \oplus F_2$ is defined as the closure of the subspace $F_1 \dotplus F_2$ composed of all sums $f_1 + f_2$, $f_1 \in F_1$, $f_2 \in F_2$. In general $F_1 \dotplus F_2$ is not a closed subspace.

Formula (7) is especially advantageous when $F_1 \dot{+} F_2$ is closed and thus equal to $F' = F_1 \oplus F_2$.

Let us analyze this case more in detail. It will be convenient to make the non-essential assumption that

$$(8) \qquad F_0 = F_1 \cdot F_2 = (0), \quad \text{that is,} \quad P_0 = 0.$$

The angle between two elements (vectors) $f_1 \neq 0$, $f_2 \neq 0$ is given by $\cos \alpha = \mathrm{Re}\,(f_1, f_2)/\|f_1\|\|f_2\|$. The minimal angle ϕ, $0 \leqq \phi \leqq \pi/2$, between F_1 and F_2 is given by(10)

$$(9) \qquad \cos \phi = \text{l.u.b. Re} \frac{(f_1, f_2)}{\|f_1\|\|f_2\|} \qquad \textit{for } 0 \neq f_1 \in F_1,\ 0 \neq f_2 \in F_2.$$

It is easily seen that, for $f_1 \in F_1$, $f_2 \in F_2$,

$$(10) \qquad |(f_1, f_2)| \leqq \|f_1\|\|f_2\| \cos \phi,$$

$$(11) \qquad \|P_1 f_2\| \leqq \|f_2\| \cos \phi, \qquad \|P_2 f_1\| \leqq \|f_1\| \cos \phi,$$

$$(12) \qquad \|f_1 + f_2\| \geqq \|f_1\| \sin \phi, \qquad \|f_1 + f_2\| \geqq \|f_2\| \sin \phi.$$

In (12), $\sin \phi$ is the greatest constant $c \geqq 0$ for which an inequality of type $\|f_1 + f_2\| \geqq c\|f_1\|$ is true. By a theorem of H. Kober [1] such inequality with $c > 0$ is necessary and sufficient in order that $F_1 \dot{+} F_2$ be closed.

Consequently, we shall know that $F' = F_1 \dot{+} F_2$ if we prove an inequality

$$(13) \qquad \|f_1 + f_2\| \geqq c\|f_1\|,$$

with any $c > 0$. Such a constant is necessarily less than or equal to 1 and it gives always an evaluation of the minimal angle ϕ:

$$(14) \qquad \sin \phi \geqq c > 0.$$

The angle ϕ being positive, the inequalities (11) show that the bounds of the operators $(P_2P_1)^n$ in F_2 or $(P_1P_2)^n$ in F_1 are not greater than $\cos^{2n}\phi$. Formula (7) may now be written in the form

$$(15) \qquad \begin{aligned} P = {} & (P_1 - P_1P_2 + P_1P_2P_1 - P_1P_2P_1P_2 + \cdots) \\ & + (P_2 - P_2P_1 + P_2P_1P_2 - P_2P_1P_2P_1 + \cdots) \end{aligned}$$

and the two series are uniformly convergent to the operators Q_1 and Q_2 which give the decomposition of $f \in F_1 + F_2$ in $f = Q_1 f + Q_2 f$, $Q_1 f \in F_1$, $Q_2 f \in F_2$.

It should be remarked that the decomposition of the series in (7) into the two series (15) is not possible when $\phi = 0$, as the operators Q_1 and Q_2 are then unbounded.

When the series in (15) are used for computation it is very easy to get

(10) The notion of a minimal angle between two subspaces seems to have been first introduced by K. Friedrichs [1].

estimates for the remainder. Usually we shall want to compute, for f and g in F, the value of $(f, Pg) = (f, Q_1g) + (f, Q_2g)$. It is clear that when we stop in the series of Q_1 at the nth term $P_1^{(n)} = (-1)^{n-1}P_1P_2P_1 \cdots$, then the remainder $R_1^{(n)}$ of this series will be given by

$$(16) \qquad R_1^{(n)} = -P_1^{(n)}Q_2 \quad \textit{for odd } n, \qquad R_1^{(n)} = -P_1^{(n)}Q_1 \quad \textit{for even } n.$$

We have similar developments for the second series defining Q_2. Consequently, the error in (f, Q_1g) (for example when n is odd) is given by $(f, R_1^{(n)}g) = -(P_1^{(n)*}f, Q_2g)$

$$(17) \qquad |(f, R_1^{(n)}g)| \leqq \|P_1^{(n)*}f\| \, \|Q_2g\|.$$

By (12) we have $\|Q_2g\| \leqq (1/\sin\phi)\|g\|$. As $P_1^{(n)}$ is already computed, $P_1^{(n)*}$ is known also and we can compute $\|P_1^{(n)*}f\|$. This will give quite a precise evaluation of the error. Without knowing $P_1^{(n)}$ we can evaluate $\|P_1^{(n)*}f\| \leqq \|f\| \cos^{n-1}\phi$.

Even in case $\phi = 0$ we could still evaluate the error in (f, Pg) if Q_1g and Q_2g exist and if we can evaluate their norms.

Still another evaluation of error (preferable as an a priori evaluation), in the case $\phi > 0$, is obtained directly from (4):

$$P - \sum_{k=1}^{m} [\quad] = [(P - P_1)(P - P_2)]^m + (P_2P_1)^m.$$

It can be proved that the minimal angle of $F' \ominus F_1$ and $F' \ominus F_2$ is the same as between F_1 and F_2. Consequently

$$\|[(P - P_1)(P - P_2)]^m\| \leqq \cos^{2m-1}\phi, \qquad \|(P_2P_1)^m\| \leqq \cos^{2m-1}\phi,$$

$$\left\| P - \sum_{k=1}^{m} [\quad] \right\| < 2\cos^{2m-1}\phi,$$

where $\| \quad \|$ signifies bounds of operators.

In case of a sum of more than two subspaces $F' = F_1 \oplus F_2 \oplus F_3 \oplus \cdots$ we can still express the projection on F' in terms of projections on $F_1, F_2, \cdots$, but the formula will be much more complicated than in the case of two subspaces and for this reason may not be as valuable.

Let us consider now the translation of our formulas in terms of the reproducing kernels K_1, K_2, and K' of the classes F_1, F_2, and $F' = F_1 \oplus F_2$. We shall suppose that the classes F_1 and F_2 have no function $\neq 0$ in common, that is, $F_1 \cdot F_2 = (0)$.

To the projections P_1, P_2, P there correspond (in the sense of §11) the kernels K_1, K_2 and K'. To each term in the series (7) or (15) there corresponds a kernel given by the following table of correspondence

$$P \leftrightarrow K'(x, y), \qquad P_1 \leftrightarrow K_1(x, y), \qquad P_2 \leftrightarrow K_2(x, y),$$

$$P_1P_2 \leftrightarrow \Lambda_1(x, y) = (K_1(z, y), K_2(z, x))_z,$$
$$P_2P_1 \leftrightarrow \overline{\Lambda_1(y, x)} = (K_2(z, y), K_1(z, x))_z,$$
$$(P_1P_2)^n \leftrightarrow \Lambda_n(x, y), \qquad (P_2P_1)^n \leftrightarrow \overline{\Lambda_n(y, x)},$$

where

$$\Lambda_n(x, y) = (\Lambda_1(z, y), \overline{\Lambda_{n-1}(x, z)})_z = (\Lambda_{n_1}(z, y), \overline{\Lambda_{n_2}(x, z)})_z, \qquad n_1 + n_2 = n,$$
$$P_1(P_2P_1)^n \leftrightarrow \Lambda_n'(x, y) = \overline{\Lambda_n'(y, x)} = (K_1(z, y), \Lambda_n(z, x))_z,$$
$$P_2(P_1P_2)^n \leftrightarrow \Lambda_n''(x, y) = \overline{\Lambda_n''(y, x)} = (K_2(z, y), \overline{\Lambda_n(x, z)})_z.$$

Formula (15) can now be written in the form

$$K'(x, y) = \sum_{n=1}^{\infty} (\Lambda_{n-1}'(x, y) + \Lambda_{n-1}''(x, y) - \Lambda_n(x, y) - \overline{\Lambda_n(y, x)}) \tag{18}$$
$$= \sum_{n=1}^{\infty} (\Lambda_{n-1}'(x, y) - \Lambda_n(x, y)) + \sum_{n=1}^{\infty} (\Lambda_{n-1}''(x, y) - \overline{\Lambda_n(y, x)}).$$

If we use these series to compute $K'(x, y)$ for given points x and y, we shall represent it in the form

$$K'(x, y) = (K'(z, y), P_zK'(z, x)) = (K(z, y), P_zK(z, x))$$

and apply our evaluation of error to this form.

13. **Final remarks in the general theory.** In the present section we shall collect a number of shorter remarks about the nature of classes of functions with reproducing kernels and of their norms, and concerning some relations between the classes and their r.k.'s.

(A) *Classes of functions for which a r.k. exists. (R.K.)-classes.* Consider a set E and a linear class F of functions defined (and finite) everywhere in E. The problem which arises is to find under what circumstances we can define a norm in F giving to F the structure of a Hilbert space with a r.k. For abbreviation we shall call such classes of functions *(R.K.)-classes.*

THEOREM I. *In order that the class F (not necessarily linear) be contained in a (R.K.)-class it is necessary that there exist an increasing sequence of sets $E_1 \subset E_2 \subset \cdots$, $E = E_1 + E_2 + \cdots$, and for each $f \neq 0$ of F a positive number $N(f)$, so that the functions $f(x)/N(f)$ be uniformly bounded in each E_n.*

In fact if F_1 is the (R.K.)-class containing F, $\|\ \|_1$ and K_1 its norm and kernel, we define as E_n the set of all $y \in E$ with $K_1(y, y) \leq n$ and as $N(f)$ the norm $\|f\|_1$. Then, for each $y \in E_n$ and $f \in F \subset F_1$ we have $|f(y)| = |(f(x), K_1(x, y))_1| \leq \|f\| \|_1 K_1(x, y)\|_1 = N(f)((K_1(y, y))^{1/2}$ and $|f(y)|/N(f) \leq n^{1/2}$.

The necessary condition of Theorem I is not always satisfied even for an enumerable sequence of functions. As an example, consider in the interval $E = (0, 1)$ the sequence of functions $f_n(x) = 1/|x - r_n|$ for $x \neq r_n$ and $f_n(r_n) = 0$.

Here $\{r_n\}$ is a sequence of numbers everywhere dense in E. By an easy topological argument we prove that the sequence of functions f_n does not satisfy the condition of Theorem I.

THEOREM II. *For an enumerable sequence $\{f_n(x)\}$ the condition of Theorem I is also sufficient in order that this sequence be contained in a (R.K.)-class.*

In fact, consider an upper bound $M_m < \infty$ for $|f_n(x)|/N(f)$, $n=1, 2, \cdots$, $x \in E_m$. We write

$$K(x, y) = \sum_{n=1}^{\infty} \frac{1}{2^n M_n^2 N^2(f_n)} f_n(x)\overline{f_n(y)}.$$

Clearly, the series is absolutely convergent and represents a p. matrix. Each term of it

$$K_n(x, y) = \frac{1}{2^n M_n^2 N^2(f_n)} f_n(x)\overline{f_n(y)}$$

is also a p. matrix and $K_n \ll K$. Theorem I of §7 gives then for the corresponding classes $F_n \subset F$. Obviously F_n is the one-dimensional class generated by f_n. Therefore $f_n \in F$ and $\{f_n\} \subset F$.

The condition of Theorem I is certainly not sufficient in general. This may be shown by a simple set-theoretical argument. Let us consider namely the class F_b of all bounded functions on E. We can then take $E_n = E$, $N(f) = \text{l.u.b. } |f(x)|$. If $\aleph$ is the power of E then the power of a (R.K.)-class F is $\aleph^{\aleph_0}$ (as the functions $K(x, y) = h_y(x)$ form a complete system in F). On the other hand, the power of the class F_b is $= \mathfrak{c}^{\aleph}$ ($\mathfrak{c}$ is the power of continuum) and for $\aleph > \aleph_0$, $\mathfrak{c}^{\aleph} > \aleph^{\aleph_0}$.

(B) *Convergence in classes with reproducing kernels.* Consider a class F with a r.k. K. We know that if f_n converges strongly to f in F, then it converges uniformly in every subset of E where $K(x, x)$ is uniformly bounded. Therefore the sequence $\{f_n\}$ satisfies the condition

(1) *$f_n(x) \to f(x)$ for every $x \in E$, the convergence being uniform in every set of an increasing sequence of sets $E_1 \subset E_2 \subset \cdots$ with $E = E_1 + E_2 + \cdots$.*

Consider now the class Φ of all functions defined in E. In Φ we can introduce a notion of limit as follows:

(2) $$f(x) = \Phi\text{-lim } f_n(x), \quad \textit{if condition } (1) \textit{ is satisfied.}$$

It is clear that in general the sequence of sets E_n will depend on the sequence $\{f_n\}$. We can now formulate the following theorem.

THEOREM III. *In every class F with a r.k., the strong convergence of $f_n(x)$ to $f(x)$ involves Φ-lim $f_n(x) = f(x)$.*

It should be noted that the weak convergence in F does not involve in general the Φ-convergence. But there are important cases where even weak convergence involves Φ-convergence. Such cases were considered in §2, (5).

(C) *Relations between (R.K.)-classes and corresponding norms and reproducing kernels.* To a p. matrix there corresponds a uniquely determined class and norm, but to a (R.K)-class there correspond infinitely many norms giving to it the structure of a Hilbert space with a r.k. Consequently to a (R.K.)-class there correspond also infinitely many p. matrices which are r.k.'s of the class for convenient norms.

If the norm $\|\ \ \|$ corresponds to a (R.K.)-class F, the norm $\|\ \ \|_1 = c\|\ \ \|$, $c>0$, obviously also corresponds to F and the corresponding r.k.'s K and K_1 satisfy

$$K_1(x, y) = \frac{1}{c^2} K(x, y).$$

In fact, the scalar product $(\ ,\ )_1$ is clearly $=c^2(\ ,\ )$ and thus $f(y) = (f(x), K(x, y)) = c^2(f(x), (1/c^2)K(x, y)) = (f(x), (1/c^2)K(x, y))_1$.

We shall now have to apply an important theorem of S. Banach [1] in the theory of linear transformations.

Let T be a linear transformation of a linear subspace F' of a complete space F on a linear subspace F_1' of a complete space F_1. The subspaces F' and F_1' are not necessarily closed. The transformation T is called *closed* if from $\{f_n\} \subset F'$, $f_n \to f \in F$, and $Tf_n \to f_1 \in F_1$ follows $f \in F'$, $f_1 \in F_1'$, and $Tf = f_1$.

BANACH'S THEOREM. *If T is a closed linear transformation of F' on F_1', $F' \subset F$, $F_1' \subset F_1$, F and F_1 complete normed vector spaces and if F' is a closed subspace of F, then T is continuous and consequently bounded (that is, there exists a $M>0$ with $\|Tf\|_1 \leq M\|f\|$). The image F_1' is either $=F_1$ or of first category in F_1.*

Before we apply this theorem we shall prove the following lemma:

LEMMA. *Let F_1 and F_2 be classes with r.k.'s and let F_0 be their intersection $F_1 \cdot F_2$. The correspondence transforming $f \in F_0$ considered as belonging to F_1 into f considered as belonging to F_2 is a closed linear transformation.*

In fact, suppose that $\{f_n\} \subset F_0$ and that f_n converges strongly to f' in F_1 and to f'' in F_2. Following Theorem III

$$f'(x) = \Phi\text{-}\lim f_n(x) = f''(x).$$

Therefore, $f' = f'' \in F_0$, which proves the lemma.

THEOREM IV. *Let F and $F_1 \subset F$ be (R.K.)-classes and $\|\ \ \|$, $\|\ \ \|_1$ some norms corresponding to F and F_1. Then there exists a constant $M>0$ such that $\|f\| \leq M\|f\|_1$ for $f \in F_1$.*

In fact the identical transformation of F_1 considered as subspace of F_1

on F_1 as subspace of F is closed (following the lemma), F_1 is a closed subspace of F_1, and thus our theorem follows immediately from Banach's theorem.

COROLLARY IV$_1$. *Let* $\| \ \|$ *and* $\| \ \|_1$ *be two norms corresponding to the same (R.K.)-class F. There exist two positive constants m and M such that* $m\|f\| \leqq \|f\|_1 \leqq M\|f\|$, *for* $f \in F$.

Theorem IV, together with Theorems I and II from §7 and with the remark that to norm $M\| \ \|$ corresponds the kernel $(1/M^2)K$, gives immediately the corollaries:

COROLLARY IV$_2$. *Let K and* K_1 *be two p. matrices, F and* F_1 *the corresponding classes. In order that* $F_1 \subset F$ *it is necessary and sufficient that there exists a positive constant M such that* $K_1 \ll MK$.

COROLLARY IV$_3$. *Under the hypotheses of corollary* IV$_2$, *in order that* $F_1 = F$ *it is necessary and sufficient that there exist two positive constants m and M such that* $mK \ll K_1 \ll MK$.

The second part of Banach's theorem together with our lemma leads to the following remark which belongs to the subject matter of section (A).

REMARK. If $\{F_n\}$ is a strictly increasing sequence of (R.K.)-classes, then their sum $F = \sum F_n$ is not a (R.K.)-class. In fact, were there a norm $\| \ \|$ in F giving it the structure of a Hilbert space with r.k., the subspaces $F_n \subset F$ would be of first category in F and therefore F would be of first category in itself which is impossible.

(D) *Connection with existence domains in a Hilbert space.* We shall now use the notion introduced recently by J. Dixmier [1] of domains of existence in a Hilbert space. A linear subset of a Hilbert space is called a domain of existence, d.e., if there exists a closed linear transformation defined in this subspace and transforming it into a subspace of another Hilbert space (which, in particular, may be identical to the first Hilbert space). The d.e.'s D in a given Hilbert space $\mathfrak{H}$ with norm $\| \ \|$ may be characterized by the following property: there exists a norm $\| \ \|_1$ defined in D giving it the structure of a Hilbert space and satisfying

$$\|h\|_1 \geqq \|h\| \qquad \text{for every } h \in D. \tag{3}$$

In fact, if D is a d.e., then we consider the linear closed transformation T of D into a subspace of some Hilbert space $\mathfrak{H}'$, with the norm $\| \ \|'$. It is then clear that the norm $\| \ \|_1$ defined by

$$\|h\|_1^2 = \|h\|^2 + \|Th\|'^2$$

gives D the character of a complete Hilbert space which satisfies condition (3).

On the other hand, suppose that a norm $\| \ \|_1$ is defined in D satisfying (3) and giving D the character of a complete Hilbert space. Then the correspondence transforming any element of D, considered as a subspace of $\mathfrak{H}$, into the same element considered in the Hilbert space D (with norm $\| \ \|_1$) is obviously a closed transformation and D is therefore a d.e.

Using Theorem II from §7 and Theorem IV of section (C) we prove now immediately the following theorem.

THEOREM V. *If a class of functions F forms a Hilbert space with a reproducing kernel, then for any linear subclass $F_1 \subset F$, the necessary and sufficient condition in order that F_1 be a (R.K.)-class is that F_1 be a d.e. in F.*

If we have two classes of functions F_1, F_2, with reproducing kernels, we can combine these two classes in different ways in order to form new classes. Let us consider in particular the following classes of functions: $F_0 = F_1 \cdot F_2$ and $F = F_1 \dotplus F_2$.

THEOREM VI. *If F_1 and F_2 are (R.K.)-classes, then the same is true of the classes $F_1 \cdot F_2$ and $F_1 \dotplus F_2$.*

Proof. The linearity of the classes is obvious. We take firstly the intersection F_0. With any norms $\| \ \|_1$ and $\| \ \|_2$ corresponding to F_1 and F_2 we define the norm in F_0 by the equation

$$\|f\|^2 = \|f\|_1^2 + \|f\|_2^2.$$

This norm clearly defines a quadratic metric in F_0 satisfying all the required properties. For instance, the completeness of F_0 results immediately from the lemma of section (C).

As $\|f\| \geq \|f\|_1$ for $f \in F_0$, Theorem II, §7 gives then the existence of a r.k. for F_0.

In the case of the sum, $F = F_1 \dotplus F_2$, we may apply the theorem of §6 which states that K_1 and K_2 being the r.k.'s with the norms $\| \ \|_1$ and $\| \ \|_2$ of F_1 and F_2, $K_1 + K_2$ is the reproducing kernel of our class F.

Besides the operations of $\cdot$ and $\dotplus$, we can also introduce the direct product $F_1 \otimes F_2$ as defined in §8 as another operation leading to a (R.K.)-class when F_1 and F_2 are (R.K.)-classes. The class $F_1 \otimes F_2$ however is defined not in E but in the product set $E \times E$. If we take its restriction to the diagonal set of all pairs $\{x, x\}$, we get a class of functions defined in E. It can be proved that this class does not depend on the choice of norms in F_1 and F_2 as long as the norms give to F_1 and F_2 the structure of a Hilbert space with a r.k.

PART II. EXAMPLES

1. Introductory remarks. In this part we shall give examples showing how our general theory may be applied in particular cases and to what kind of results it leads. With a few exceptions we will not go into the details of calculation and will not give in explicit form the formulas and relations obtainable by our general methods.

We shall treat essentially two kinds of kernels: the Bergman's kernels $K(z, z_1)$ and the harmonic kernels $H(z, z_1)$.

(1) *Bergman's kernels.* These kernels correspond to a domain D in the space of n complex variables $z=(z^{(1)}, z^{(2)}, \cdots, z^{(n)})$. We consider the class $\mathfrak{A}\equiv\mathfrak{A}_D$ of all analytic regular functions in D with a finite norm given by

$$\|f\|^2 = \iint_D \cdots \iint |f(z^{(1)}, z^{(2)}, \cdots, z^{(n)})|^2 dx^{(1)}dy^{(1)}dx^{(2)}dy^{(2)} \cdots dx^{(n)}dy^{(n)},$$

where $z^{(k)}=x^{(k)}+iy^{(k)}$.

The class $\mathfrak{A}$ possesses a reproducing kernel $K\equiv K_D$—the Bergman's kernel corresponding to D.

In our examples we shall consider essentially the case of plane domains D. If D is multiply-connected we shall consider also the reduced Bergman's kernel $K'(z, z_1)$, which is the reproducing kernel of the subspace $\mathfrak{A}'$ of $\mathfrak{A}$ consisting of all functions of $\mathfrak{A}$ with a uniform integral $\int^z f dz$. If D is of finite connection n, the complementary subspace $\mathfrak{A}\ominus\mathfrak{A}'$ is $(n-1)$-dimensional and is generated by n functions $w_n'(z)$ (between which there is the linear relation $\sum w_k'=0$) defined in the following way: if B_k, $k=1, 2, \cdots, n$, are the boundary components of the boundary B of D, w_k' is the derivative (which is uniform) of the multiform analytic function w_k whose real part is the harmonic measure u_k of D corresponding to B_k, that is, the harmonic function regular in D, equal to 1 on B_k, and vanishing on all the other components B_i.

The functions w_k' belong always to $\mathfrak{A}$ and are orthogonal to $\mathfrak{A}'$. We have the relation

$$K(z, z_1) = K'(z, z_1) + \sum_{i,j} c_{ij} w_i'(z)\overline{w_j'(z_1)},$$

where $\sum$ is the r.k. of $\mathfrak{A}\ominus\mathfrak{A}'$. Consequently the matrix $\{c_{ij}\}$ is definite positive (see §3) and it is the conjugate inverse of the Gramm's matrix $\{(w_i', w_j')\}$.

Bergman's kernels possess an important property of invariance: in case of domains in the space of n variables $z^{(1)}, \cdots, z^{(u)}$, if T represents D *pseudo-conformally* on D', then

$$K_{D'}(z', z_1')\partial T(z)\overline{\partial T(z_1)} = K_D(z, z_1).$$

Here, $z'=T(z)$, $z_1'=T(z_1)$, and $\partial T(z)$ is the Jacobi determinant of T. In the case of domains in the plane, if $t(z)$ represents D conformally on D', this formula takes the form

$$K_{D'}(z', z_1')\cdot t'(z)\overline{t'(z_1)} = K_D(z, z_1).$$

The importance of the Bergman kernels lies in the possibility they offer of generalizing different theorems on analytic functions of one complex variable to functions of several complex variables (such as Schwarz's lemma, distortion theorems, representative domains in pseudo-conformal mappings).

In the case of one variable almost all the important conformal mappings are expressible in terms of these kernels. For instance if D is a simply connected domain, the mapping function $\zeta = f(z, z_0)$ which represents D on a circle $|\zeta| < R$ in such a way that the point $z_0 \in D$ goes into $\zeta = 0$ and $f'(z_0, z_0) = 1$ is given by

$$f(z, z_0) = \frac{1}{K(z_0, z_0)} \int_{z_0}^{z} K(t, z_0)dt.$$

(2) *Harmonic kernels*. Consider in a plane domain D (we could consider also a domain in n-dimensional space) the class $\mathfrak{B} \equiv \mathfrak{B}_D$ of all regular harmonic functions (in general complex-valued) with a finite norm given by

$$\|h\|^2 = \iint_D |h|^2 dxdy, \qquad z = x + iy.$$

This class possesses a reproducing kernel which will be denoted by $H(z, z_1)$.

It should be remarked that another harmonic kernel is often considered, namely the one which corresponds to the Dirichlet metric

$$\|h\|^2 = \iint_D [|h'_x|^2 + |h'_y|^2]dxdy.$$

This kernel is easily expressible by Bergman's kernel and consequently does not present any additional difficulties to the ones encountered in the study and computation of Bergman's kernels.

The situation is different for the kernel H. Even for very simple domains (for instance for a rectangle) there is no known explicit expression of H even in the form of an infinite development. (We disregard here the developments in terms of a complete non-orthogonal system which are always possible to establish for a r.k., but whose coefficients are quotients of determinants of growing orders.)

One reason for the greater difficulty of the investigation of the kernels $H(z, z_1)$ as compared to Bergman's kernels lies in the fact that H has no such invariancy property vis-à-vis conformal transformations as have Bergman's kernels.

The interest of the kernel H lies in its connection with the biharmonic problem which governs the question of equilibrium of elastic plates.

The kernel H gives a simple expression for a function $u(z)$ such that $u = \partial u/\partial n = 0$ on the boundary B of D and $\Delta\Delta u = \phi$ in D, for a given function ϕ.

Supposing that we know a function ψ such that $\Delta\psi=\phi$ (we can take as ψ the logarithmic potential of ϕ: $\psi(z)=(1/\pi)\iint_D \log|z-z'|\phi(z')dx'dy'$) we get for u the expression (where g is the ordinary Green's function)

$$u(z) = -\iint_D g(z, z')dx'dy'\left[\psi(z') - \iint_D H(z', z'')\psi(z'')dx''dy''\right].$$

The Green's function $g_{II}(z, z_1)$ of the biharmonic problem, satisfying $\Delta\Delta g_{II}=0$ for $z\neq z_1$, $g_{II}=\partial g_{II}/\partial n=0$ on the boundary, is given by

$$g_{II}(z, z_1) = \iint_D g(z, z')g(z', z_1)dx'dy' - \iint_D g(z, z')dx'dy' \iint_D H(z', z'')g(z'', z_1)dx''dy''.$$

These formulas were essentially noticed already by S. Zaremba [2].

2. **Comparison domains.** Consider two domains in the plane, D and D', $D\subset D'$. The kernels $K_{D'}$ or $H_{D'}$, restricted to the domain D, are reproducing kernels of classes $\mathfrak{A}^0$ or $\mathfrak{B}^0$ formed by the restrictions of functions from $\mathfrak{A}_{D'}$ or $\mathfrak{B}_{D'}$. As any analytic or harmonic function vanishing in D vanishes everywhere, any function f_0 of $\mathfrak{A}^0$ or $\mathfrak{B}^0$ is a restriction of only one function f from $\mathfrak{A}_{D'}$ or $\mathfrak{B}_{D'}$ and, following §5, Part I, the norm $\|f_0\|^0=\|f\|'$. It is then clear that every $f_0\in\mathfrak{A}^0$ belongs to $\mathfrak{A}_D$ and that $\|f_0\|^0\geqq\|f_0\|$.

We can apply Theorem II of §7, I, which gives

(1) $\overset{0}{K}_{D'} \ll K_D$, *$\overset{0}{K}_{D'}$ being the restriction of $K_{D'}$ to D.*

In the same way we get

$$\overset{0}{H}_{D'} \ll H_D. \tag{2}$$

If the kernel K_D is known, we get immediately the well known estimates for the kernel $K_{D'}$:

$$K_{D'}(z, z) \leqq K_D(z, z), \quad |K_{D'}(z, z_1)| \leqq (K_D(z, z)K_D(z_1, z_1))^{1/2} \tag{3}$$

for points z and z_1 belonging to D.

But the relation (1) (or (2)) allows much better estimates. Suppose that the kernel $K_{D'}$ is known. For two points z and z_1 in D take domains D'' and D_1'' such that $z\in D''\subset D$, $z_1\in D_1''\subset D$ and that the kernels $K_{D''}$ and $K_{D_1''}$ be known (for instance circles). Then, from (1), we get $K_D(z, z)\leqq K_{D_1''}(z, z)$, $K_D(z_1, z_1)\leqq K_{D_1''}(z_1, z_1)$,

$$\begin{aligned} &|K_D(z, z_1) - K_{D'}(z, z_1)|^2 \\ &\qquad\leqq [K_D(z, z) - K_{D'}(z, z)][K_D(z_1, z_1) - K_{D'}(z_1, z_1)] \\ &\qquad\leqq [K_{D''}(z, z) - K_{D'}(z, z)][K_{D_1''}(z_1, z_1) - K_{D'}(z_1, z_1)]. \end{aligned} \tag{4}$$

If we consider a boundary-point t where the boundary has a finite curvature and if we fix z_1 and move z towards t, the estimate (3) will grow like $1/|z-t|$. The estimate (4) by a convenient choice of the comparison domains D', D'', and D_1'' will give a bound for $|K_D(z, z_1)|$ growing only like $1/|z-t|^{1/2}$.

To show the interest of this improvement, take D simply-connected and consider the conformal mapping of D on a circle $|\zeta| < R$ given by

$$\zeta = \frac{1}{K(z_0, z_0)} \int_{z_0}^{z} K(z, z_0)dz.$$

Our problem will be to compute the point τ on the circumference $|\zeta| = R$ corresponding to t on the boundary B. As the kernel K is not known we approximate it by a development in orthogonal functions. This development may converge fairly quickly inside the domain but in general it will not converge on the boundary and will converge less and less well the nearer we come to the boundary.

To calculate τ we have to integrate from z_0 to the point t on the boundary. We cannot integrate term by term the development of K as it does not converge on the boundary. What we do then is to find, with the help of the estimate (4), a point z_1 near t for which the integral $\int_{z_1}^{t}|K(z, z_0)|dz$ is sufficiently small. We can integrate the development of K term by term from z_0 to z_1 and obtain as good an approximation of τ as we wish.

It is clear that with the estimate (3) we would not be able to do this.

3. **The difference of kernels.** As we saw in §2, in $D \subset D_1$, the kernels $K \equiv K_D$ and $K_1 \equiv K_{D_1}$ satisfy the relation $K_1 \ll K$ (the kernel K_1 being restricted to D). To illustrate the developments of §7, I, let us investigate the class of functions F_2 corresponding to the p. matrix K_2 given by

(1) $$K_2(z, z_1) = K(z, z_1) - K_1(z, z_1) \qquad z \text{ and } z_1 \text{ in } D.$$

Following the notation in the proof of Theorem II, §7, I (where $F = \mathfrak{A}$, $F_1 = \mathfrak{A}_1$), we introduce the operator L in $\mathfrak{A}$ by

(2) $$Lf = f_1(z_1) = (f, K_1(z, z_1)) = \iint_D f(z)\overline{K_1(z, z_1)}dxdy.$$

If we consider the Hilbert space $\mathfrak{H}_1$ of all functions $u(z)$ in square integrable in D_1 with the norm

$$\|u\|_1^2 = \iint_D |u(z)|^2 dxdy,$$

the general property of r.k.'s as projections shows that $f_1(z)$, as function in D_1, is the projection on $\mathfrak{A}_1$ of the function $\bar{f}(z) = f(z)$ in D and $= 0$ in $D_1 - D$. Consequently,

$$\|f_1(z)\| \leqq \|f_1(z)\|_1 \leqq \|\bar{f}(z)\|_1 = \|f(z)\|.$$

The second inequality may become equality for $f(z)\neq 0$ only in the case when D_1-D is of two-dimensional measure 0 (for instance when D differs from D_1 only by some slits). We will exclude this case and consequently

$$\|f_1(z)\| = \|Lf\| < \|f\| \qquad \text{for } f \neq 0. \tag{3}$$

We introduce then the operator L' by

$$L'^2 = I - L.$$

The subspace F_0 is here reduced to 0 as $0\neq f=Lf$ is impossible in view of (3). Therefore $F'=F\equiv\mathfrak{A}$ and the only possibilities for the class F_2 are: 1°, $F_2=\mathfrak{A}$ or 2°, F_2 is a dense subspace of $\mathfrak{A}$.

The first case represents itself always when D is completely interior to D_1 ($\overline{D}\subset D_1$). In fact, the operator L is then completely continuous. To prove this we take a sequence $\{g^{(n)}\}\subset\mathfrak{A}$ converging weakly to $g\in\mathfrak{A}$. The functions $\mathfrak{g}^{(n)}$ converge then weakly in $\mathfrak{H}_1$ to $\mathfrak{g}$ and their projections $g_1^{(n)}$ on $\mathfrak{A}_1$ converge weakly to g_1. But the weak convergence in $\mathfrak{A}_1$ involves *uniform* convergence of $g_1^{(n)}(z)$ towards $g_1(z)$ in any *closed* subset of D_1, in particular in $\overline{D}$ (see section (5), §2, I). When we restrict the functions $g_1^{(n)}(z)$ and $g_1(z)$ to D they become the transforms $Lg^{(n)}$ and Lg. Therefore, the uniform convergence of $g_1^{(n)}$ to g_1 in $\overline{D}$ involves the strong convergence of $Lg^{(n)}$ to Lg in the space $\mathfrak{A}$.

Following our remarks after Theorem III, §7, I, the class $F_2=F'=\mathfrak{A}$. To get the norm $\|f\|_2$ corresponding to the kernel K_2, we have to find the solution $g(z)$ of the equation

$$g - Lg = f$$

which exists and is unique for every $f\in\mathfrak{A}$. Then

$$\|f\|_2^2 = \|g\|^2 - \|g_1\|_1^2$$

where

$$g_1(z_1) = Lg = \iint_D g(z)K_1(z, z_1)dxdy.$$

Let us note that the operator L which in general has a bound not greater than 1 has here a bound less than 1.

To illustrate the second case we have to take a domain having common boundary points with the boundary of D_1. It seems probable that for every such domain D we shall be in the second case (at least if one of the boundary components of D arrives at the boundary of D_1).

To prove that we are in the second case, we have to show that the operator L has a bound $=1$. Then L cannot be completely continuous (as the bound is not attained). The class F_1 is a proper subclass of $\mathfrak{A}$. The class F_2' of all functions

$$f_2(z) = g(z) - g_1(z), \qquad \text{where } g_1 = Lg,\ g \in \mathfrak{A},$$

is a proper subspace of F_2, dense in F_2. For such a function f_2, the norm in F_2 is given by

$$\|f_2\|_2^2 = \|g\|^2 - \|g_1\|_1^2.$$

The class F_2 in the metric of $\mathfrak{A}$ is a dense subspace of $\mathfrak{A}$. There are functions $f\in\mathfrak{A}$ which do not belong to F_2 and for which, a fortiori, the equation $f=g-Lg$ has no solution $g\in\mathfrak{A}$. (There may be such a solution, analytic in D but not in square integrable in D.)

For two explicitly given domains $D\subset D_1$, it may not be easy to prove that we are in the second case by using the property that the bound of L is 1. We can transform this property into another one, more easily proved.

To this effect we shall consider for any function $f_1\in\mathfrak{A}_1$ the quotient

$$Q(f_1) = \|f_1\|^2/\|f_1\|_1^2. \tag{4}$$

LEMMA. *In order that the bound of L be 1 it is necessary and sufficient that there exist functions f_1 with $Q(f_1)$ as near 1 as we wish.*

To prove this lemma we remark firstly that the l.u.b. $Q(f_1)$ for $f_1\in\mathfrak{A}_1$ is the same as the l.u.b. $Q(Lf)$ for $f\in\mathfrak{A}$. In fact the Lf form a dense subspace in the space $\mathfrak{A}_1$ (otherwise there would be a $g_1\in\mathfrak{A}_1$, $g_1\neq 0$, with $(g_1, Lf)_1=0$ for every $f\in\mathfrak{A}$. Therefore $(g_1, f)=(g_1, Lf)_1=0$ and $g_1=0$ in D which involves $g_1=0$ in D_1). Consequently, there is for every $f_1\in\mathfrak{A}_1$ a function $f\in\mathfrak{A}$ with $\|f_1-Lf\|_1$ as small as we wish. As $\|f_1-Lf\|\leqq\|f_1-Lf\|_1$ it is clear that $Q(Lf)$ will approach $Q(f_1)$ as nearly as we wish.

Now, $Q(Lf)$ can be represented as

$$Q(Lf) = \frac{(Lf, Lf)}{(Lf, Lf)_1} = \frac{(Lf, Lf)}{(Lf, f)}\,. \tag{5}$$

Our lemma amounts to the equivalence of the two properties, for any α, $0<\alpha\leqq 1$:

$$(Lf, Lf) \leqq \alpha^2(f, f) \qquad \textit{for all } f \in \mathfrak{A}, \tag{6'}$$

$$(Lf, Lf) \leqq \alpha(Lf, f) \qquad \textit{for all } f \in \mathfrak{A}. \tag{6''}$$

From (6″) follows (6′):

$$(Lf, Lf) \leqq \alpha(Lf, f) \leqq \alpha(Lf, Lf)^{1/2}(f, f)^{1/2}$$

$$(Lf, Lf)^{1/2} \leqq \alpha(f, f)^{1/2}, \qquad (Lf, Lf) \leqq \alpha^2(f, f).$$

From (6′) follows

$$(Lf, f) \leqq \alpha(f, f) \qquad \textit{for all } f \in \mathfrak{A}. \tag{6'''}$$

In fact, we have

$$(Lf, f) \leqq (Lf, Lf)^{1/2}(f, f)^{1/2} \leqq \alpha(f, f)^{1/2}(f, f)^{1/2} = \alpha(f, f).$$

From (6''') follows (6''): we use the fact that L is positive which gives $(Lf, g) \leqq (Lf, f)^{1/2}(Lg, g)^{1/2}$. Then

$$(Lf, Lf) \leqq (Lf, f)^{1/2}(LLf, Lf)^{1/2} \leqq (Lf, f)^{1/2}\alpha^{1/2}(Lf, Lf)^{1/2},$$

$$(Lf, Lf)^{1/2} \leqq \alpha^{1/2}(Lf, f)^{1/2}, \qquad (Lf, Lf) \leqq \alpha(Lf, f).$$

This proves our lemma.

We shall apply the lemma to two domains $D \subset D_1$ having a common boundary point which belongs also to the boundary of the exterior of D. We suppose further that at this boundary point the boundaries of D and D_1 have a common tangent. We can take the common boundary point as the origin O and the inner normal of the boundaries at this point as the positive axis.

It is then easily verified that the functions

$$f_n(z) = \frac{1}{(nz + 1)^2}, \qquad n = 1, 2, \cdots,$$

for sufficiently great values of n, belong to $\mathfrak{A}_1$ and that they satisfy the asymptotic equation

$$\|f_n\| \sim \|f_n\|_1 \qquad \text{for } n \to \infty.$$

This shows that the l.u.b. $Q(f) = 1$ and that we are in the second case.

Let us consider now another kind of example which we excluded till now. Namely, we shall suppose that the domain D differs from D_1 only by a number of slits of finite length. It is then immediately seen that for a function $f_1 \in \mathfrak{A}_1$, considered as belonging to $\mathfrak{A}_1^0 \subset \mathfrak{A}$ ($\mathfrak{A}_1^0$ = class $\mathfrak{A}_1$ restricted to D), we have

$$\|f_1\| = \|f_1\|_1.$$

Consequently $\mathfrak{A}_1^0$ is a closed linear subspace of $\mathfrak{A}$ and the kernel $K(z, z_1) - K_1(z, z_1)$ corresponds to the subspace $\mathfrak{A} \ominus \mathfrak{A}_1^0$.

4. The square of a kernel introduced by Szegö. We shall now give an application of Theorem II, §8, I. Consider a domain in the plane with a sufficiently smooth boundary (for simplicity's sake we may suppose that the boundary curves are analytic). For this domain we shall consider a kernel, first introduced by Szegö [1], which we shall denote by $k(z, z_1)$. This kernel corresponds to the class S of all analytic functions which possess in square integrable boundary values. As the functions are not necessarily continuous on the boundary we have to specify the meaning of the boundary values of the functions. We shall suppose that for such a function $f(z)$, its integral $F(z)$ is a continuous function in the closed domain (but $F(z)$ may be a multiform function). Then we suppose that for any two points t_1, t_2 of the same boundary curve the difference $F(t_2) - F(t_1)$ may be represented by the integral

$$\int_{t_1}^{t_2} f(t)dt,$$

where $f(t)$ is defined almost everywhere on the boundary curves and is in square integrable there. The integral is taken over an arc of the boundary curve going from t_1 to t_2.

The function $f(t)$ will be considered as the *boundary value* of $f(z)$ in the boundary-point t. $f(t)$ is completely determined by $f(z)$ with exception of a boundary set of linear measure zero. The existence of boundary values of $f(z)$ in our sense is equivalent to the absolute continuity of $F(z)$ on the boundary. With the boundary values taken in this sense we define a norm in class S by the equation

$$\|f\|^2 = \sum \int_{C_\nu} |f(t)|^2 ds,$$

where C_ν are the boundary curves and ds is the element of length on the curve. For the functions of class S it is easily proved that the Cauchy theorem is still valid in the form

$$f(z) = \sum \int_{C_\nu} \frac{f(t)}{t - z} dt.$$

From this it is immediately seen that the class possesses a reproducing kernel which is the kernel $k(z, z_1)$ introduced by Szegö. Quite recently, in his thesis, P. Garabedian [1] proved that

$$k^2(z, z_1) = \frac{1}{4\pi} K(z, z_1) + \sum_{i,j} \alpha_{i,j} w_i'(z) w_j'(z_1) . \tag{1}$$

where K is the Bergman kernel for the domain and w_i' are the functions introduced in §1.

In cases of simply-connected domains Garabedian's formula takes a very simple form, namely:

$$k^2(z, z_1) = \frac{1}{4\pi} K(z, z_1).$$

In this case Theorem II, §8, I, gives a property of analytic functions which seems to have been unnoticed even for this simple case. Every function $f(z)$ in square integrable in the domain D is representable in infinitely many ways by a series

$$f(z) = \sum \phi_k(z)\psi_k(z) \tag{2}$$

where the functions $\phi_k(z)$ and $\psi_k(z)$ are in square integrable on the boundary. In addition, we have the formula

$$(3) \qquad 4\pi \int\int_D |f(z)|^2 dxdy = \min \sum_k \sum_l \int_C \phi_k(t)\overline{\phi_l(t)}ds \int_C \psi_k(t)\overline{\psi_l(t)}ds,$$

the minimum being extended to all representations of $f(z)$ in the form (2). In particular, we have the inequality

$$4\pi \int\int_D |\phi(z)\psi(z)|^2 dxdy \leqq \int_C |\phi(t)|^2 ds \int_C |\psi(t)|^2 ds.$$

In the case of a multiply-connected domain the problem is a little more complicated because of the presence of the functions w_k'. It can be proved then that if we replace the Bergman's kernel K by the reduced kernel K' (see §1) we have again a formula similar to (1):

$$(4) \qquad k^2(z, z_1) = \frac{1}{4\pi} K'(z, z_1) + \sum_{i,j} \beta_{i,j} w_i'(z)\overline{w_j'(z_1)}$$

where, now, the $\beta_{i,j}$ are the coefficients of a positive quadratic hermitian form, which means that $\sum_{i,j}$ represents a positive matrix. Consequently, the functions of the class corresponding to the kernel k^2 are sums of functions in square integrable in D with a uniform integral (which form the class belonging to K') and of a linear combination of the w_k' (which form the class corresponding to $\sum_{i,j}$). The functions w_k' are in square integrable in D, and thus every function belonging to the class of k^2 is in square integrable in D. Conversely, it can be proved that every function in square integrable over D belongs to the class of k^2. By Theorem II, §8, I, we know that the functions belonging to k^2 are of the form (2), but we will not be able to obtain a formula like (3) for the case of a simply-connected domain. However, a more complicated formula generalizing (3) does exist. In the present case of a multiply-connected domain, we have still the property that the product $\phi(z)\psi(z)$ is in square integrable in D if ϕ and ψ are in square integrable on the boundary, but the inequality between the integrals will not be as simple as in the case of a simply-connected domain.

5. **The kernel $H(z, z_1)$ for an ellipse.** We shall construct the kernel $H(z, z_1)$ for the ellipse D

$$(1) \qquad D: \quad \frac{x^2}{a^2} + \frac{y^2}{b^2} = 1, \qquad a > b,$$

by use of an orthonormal complete system(11).

(11) The system and the corresponding expression for the kernel were communicated to us by A. Erdélyi. It should be noted that the system was already introduced by S. Zaremba [1] who also noticed that it is orthogonal and complete in the Dirichlet metric.

We write

$$a = h \cosh \epsilon, \qquad b = h \sinh \epsilon, \tag{2}$$

$$z = x + iy = h \cosh \zeta, \qquad \zeta = \xi + i\eta. \tag{3}$$

(3) gives us a conformal transformation of the rectangle R, $0<\xi<\epsilon$, $-\pi<\eta<\pi$ on the ellipse (1) with the rectilinear slit $-a<x<h$. Consider in the rectangle R the analytic function $p_n(z) = \sinh n\zeta/\sinh \zeta$ (positive integer n). It is immediately seen that in the variable z the function is a polynomial of degree $n-1$. Consequently, all the polynomials in the variable z can be expressed as finite linear combinations of these polynomials for $n = 1, 2, 3, \cdots$. Since the harmonic polynomials form a complete system in our class $\mathfrak{B}$ of harmonic functions, the real and imaginary parts of $p_n(z)$ also form a complete system. On the other hand, it is easy to verify that these real and imaginary parts form an orthogonal system. This verification is made in an easy way by performing the integration in the rectangle R instead of the domain D. Using the formula

$$\iint_D f(x, y)dxdy = h^2 \int_0^\epsilon \int_{-\pi}^{+\pi} f(x, y) \, | \sinh \zeta |^2 d\xi d\eta$$

one verifies easily that the sequence $\phi_n(z)$ defined by

$$
\begin{aligned}
\phi_{2n-2} &= \frac{2}{h}\left(\frac{n}{\pi}\right)^{1/2} (\sinh 2n\epsilon + n \sinh 2\epsilon)^{-1/2} \operatorname{Re} p_n(z), \qquad n = 1, 2, \cdots, \\
\phi_{2n-3} &= \frac{2}{h}\left(\frac{n}{\pi}\right)^{1/2} (\sinh 2n\epsilon - n \sinh 2\epsilon)^{-1/2} \operatorname{Im} p_n(z), \qquad n = 2, 3, \cdots,
\end{aligned}
\tag{4}
$$

is orthonormal. We can then write the kernel of our class in the form

$$H(z, z_1) = \frac{4}{h^2\pi} \sum_{n=1}^{\infty} n \left\{ \frac{\operatorname{Re} p_n(z) \operatorname{Re} p_n(z_1)}{\sinh 2n\epsilon + n \sinh 2\epsilon} + \frac{\operatorname{Im} p_n(z) \operatorname{Im} p_n(z_1)}{\sinh 2n\epsilon - n \sinh 2\epsilon} \right\}. \tag{5}$$

6. Construction of $H(z, z_1)$ for a strip. Our next example will use the theorem of the limit of kernels for decreasing sequences of classes (see Theorem I, §9, I). It will be at the same time an example of a representation of a kernel by use of a resolution of identity in a Hilbert space (see §10, I). Consider the kernel of §1 and suppose that in the ellipse

$$\frac{x^2}{a^2} + \frac{y^2}{b^2} = 1$$

the axis $b<a$ remains fixed, while $a \to \infty$. The ellipse in the limit will become the horizontal strip $|y| < b$. It is clear that our theorem on the limit of kernels applies in this case and we get the kernel $H(z, z_1)$ for the strip as a limit of the kernels corresponding to ellipses. Before performing this passage to the limit, we shall at first make a few preliminary remarks.

As in the preceding paragraph, we consider the quantities h and ϵ given by $b=h \sinh \epsilon$, $a=h \cosh \epsilon$, b being fixed and $a\to\infty$. We see immediately that $h\to\infty$ and $\epsilon\to 0$ in such a way that $h\epsilon\to b$.

In the conformal mapping, $z=h \cosh \zeta$ we introduce a new variable ζ' by the equation

$$\epsilon\zeta' = \zeta - \frac{\pi}{2} i.$$

Then $z=h \sin i\epsilon\zeta'$, and the ellipse with the slit along the real axis going from $+h$ to $-a$ is transformed in the rectangle

$$R_\epsilon, \quad 0 < \xi' < 1, \quad -\frac{3\pi}{2\epsilon} < \eta' < \frac{\pi}{2\epsilon}\cdot$$

Consider a point z in the horizontal strip $|y| < b$. For sufficiently large values of h the ellipse will contain z. Suppose that $\operatorname{Im} z>0$, then the corresponding ζ' will lie in the upper half of the rectangle R_ϵ and for $h\to\infty$, $\zeta'\to z/bi$. The point $\zeta''=\bar{\zeta}'-\pi i/\epsilon$ will then correspond to the conjugate point $\bar{z}$, so that $(\zeta''+\pi i/\epsilon)\to\bar{z}i/b$. If we now return to the formula for the kernel from §5 and replace the function $p_n(z)$ by the expression $\sinh n\zeta/\sinh \zeta$, then replace ζ by $\epsilon\zeta'+\pi i/2$ and separate the series which expresses the kernel H (see (5), §5) into parts corresponding to even and odd indices, we can write the kernel in the form of a sum of four series:

$$\frac{4}{\pi h}\sum_{n \text{ odd}} \frac{n}{h} \frac{\operatorname{Re}\dfrac{\cos in\epsilon\zeta'}{\cos i\epsilon\zeta'}\operatorname{Re}\dfrac{\cos in\epsilon\zeta_1'}{\cos i\epsilon\zeta_1'}}{\sinh 2n\epsilon + n \sinh 2\epsilon}$$

$$+\frac{4}{\pi h}\sum_{n \text{ odd}} \frac{n}{h} \frac{\operatorname{Im}\dfrac{\cos in\epsilon\zeta'}{\cos i\epsilon\zeta'}\operatorname{Im}\dfrac{\cos in\epsilon\zeta_1'}{\cos i\epsilon\zeta_1'}}{\sinh 2n\epsilon - n \sinh 2\epsilon}$$

$$+\frac{4}{\pi h}\sum_{n \text{ even}} \frac{n}{h} \frac{\operatorname{Re}\dfrac{\sin in\epsilon\zeta'}{\cos i\epsilon\zeta'}\operatorname{Re}\dfrac{\sin in\epsilon\zeta_1'}{\cos i\epsilon\zeta_1'}}{\sinh 2n\epsilon + n \sinh 2\epsilon}$$

$$+\frac{4}{\pi h}\sum_{n \text{ even}} \frac{n}{h} \frac{\operatorname{Im}\dfrac{\sin in\epsilon\zeta'}{\cos i\epsilon\zeta'}\operatorname{Im}\dfrac{\sin in\epsilon\zeta_1'}{\cos i\epsilon\zeta_1'}}{\sinh 2n\epsilon - n \sinh 2\epsilon}\cdot$$

For $h\to\infty$, if we denote n/h by t and if we then notice the asymptotic formulas $h\epsilon\sim b$, $n\epsilon\sim bt$, $n \sin 2\epsilon\sim 2bt$, it is immediately seen that the series represent approximating Riemannian sums for the integrals

$$\frac{2}{\pi}\int_0^\infty t\frac{\text{Re}\cos ibt\zeta'\,\text{Re}\cos ibt\zeta_1'}{\sinh 2bt+2bt}dt+\frac{2}{\pi}\int_0^\infty t\frac{\text{Im}\cos ibt\zeta'\,\text{Im}\cos ibt\zeta_1'}{\sinh 2bt-2bt}dt,$$

$$+\frac{2}{\pi}\int_0^\infty t\frac{\text{Re}\sin ibt\zeta'\,\text{Re}\sin ibt\zeta_1'}{\sinh 2bt+2bt}dt+\frac{2}{\pi}\int_0^\infty t\frac{\text{Im}\sin ibt\zeta'\,\text{Im}\sin ibt\zeta_1'}{\sinh 2bt-2bt}dt,$$

when we subdivide the infinite interval $(0, +\infty)$ into equal intervals of length $2/h$ and in each interval take the value of the integrated function in the center (for the first two series) or in the right end (for the two last series).

The convergence of these sums towards the integrals for $h\to\infty$ is easily verified and in this way we obtain for the kernel of the horizontal strip the expression given by the above four integrals where we replace ζ' and ζ_1' by the values z/bi and z_1/bi:

$$H(z, z_1) = \frac{2}{\pi}\int_0^\infty \frac{\text{Re}\cos zt\,\text{Re}\cos z_1t+\text{Re}\sin zt\,\text{Re}\sin z_1t}{\sinh 2bt+2bt}tdt$$

$$+\frac{2}{\pi}\int_0^\infty \frac{\text{Im}\cos zt\,\text{Im}\cos z_1t+\text{Im}\sin zt\,\text{Im}\sin z_1t}{\sinh 2bt-2bt}tdt.$$

This integral representation of the kernel corresponds to a resolution of identity in the Hilbert space $\mathfrak{B}$ corresponding to the strip. This resolution of identity has a quadruple spectrum (see §10, I) defined by the following four functions $f_k(z, \lambda): f_k(z, \lambda)=0$ for $\lambda\leqq 0$, for $\lambda>0$

$$f_1(z,\lambda) = \int_0^\lambda \text{Re}\cos ztdt = \text{Re}\frac{\sin z\lambda}{z},\qquad f_2(z,\lambda) = \text{Re}\frac{1-\cos z\lambda}{z},$$

$$f_3(z,\lambda) = \text{Im}\frac{\sin z\lambda}{z},\qquad f_4(z,\lambda) = \text{Im}\frac{1-\cos z\lambda}{z}.$$

It is easily verified that these functions satisfy the conditions (a) and (b) from §10, I. The condition (c′) results from the fact that the functions f_k determine the r.k. $H(z, z_1)$ of the class $\mathfrak{B}$ by the formula above.

7. Limits of increasing sequences of kernels. In the preceding section we had an example of a limit of decreasing sequence of classes and kernels. We shall give here an example of an increasing sequence of kernels.

Consider the Bergman's kernels K_n for a decreasing sequence of simply-connected domains D_n such that $\overline{D}_{n+1}\subset D_n$ and $E=\lim D_n=D_1\cdot D_2\cdot\cdots$ consists of the two closed circles $\overline{C}_1$: $|z-2|\leqq 1$, and $\overline{C}_2$: $|z+2|\leqq 1$, with the segment of the real axis I: $-1\leqq x\leqq 1$, $y=0$.

Following the general theory of §9, I, we have to consider the set E_0 where $K_0(z, z)=\lim K_n(z, z)<\infty$.

Every point of the open circle C_1 belongs to E_0. In fact, if $K_{(1)}$ is the Bergman's kernel for C_1, by the method of comparison domains (see §2) we get

$K_n \ll K_{(1)}$ in C_1 and consequently $K(z, z) = \lim K_n(z, z) \leqq K_{(1)}(z, z)$ for z in C_1. The same is true for $z \in C_2$. We are going to prove that

$$(1) \qquad E_0 = C_1 + C_2.$$

We have to show that for $z_0 \in E - E_0$, $\lim K_n(z_0, z_0) = \infty$. We use again a domain of comparison. We take a closed line L defining an exterior domain S containing E and such that by a convenient translation we can approach L as near as we wish to z_0 without touching E. The domain S will contain all D_n from some n onwards. For these n, $K_n \gg K_S$, therefore $\lim K_n(z_0, z_0) \geqq K_S(z_0, z_0)$. The translation of the line L (and the domain S) will have on $K_S(z_0, z_0)$ the effect as if the domain S were fixed and the point z_0 were moving towards the boundary L. As for Bergman's kernels of domains of finite connection we have the theorem that $K(z, z)$ goes uniformly to $+\infty$ when z approaches the boundary ([12]), we arrive at the result, $\lim K_n(z_0, z_0) = +\infty$.

For $z_0 \in E - E_0$, with exception of $z_0 = \pm 1$, we can take as L a sufficiently small circumference. For $z_0 = \pm 1$, a circumference cannot do. We take then for L the boundary of a square, for instance, for $z_0 = +1$, we take the square: $-\epsilon < x < 1 - \epsilon$, $\epsilon < y < 1 + \epsilon$.

Using the notation of §9, I, (B), we see immediately that the class F_0 with the norm $\| \;\; \|_0$ is here a subspace of the class F of all functions $f(z)$ defined in the two open circles C_1 and C_2, analytic and regular in each (but $f(z)$ in C_1 is not necessarily an analytic continuation of $f(z)$ in C_2), with a finite norm

$$\|f\|^2 = \iint_{C_1} |f(z)|^2 dxdy + \int_{C_2} |f(z)|^2 dxdy.$$

Consequently, the condition 2° of §4, I is satisfied and we obtain a functional completion F_0^* of F_0. Then, it is easily proved, by using general approximation theorems of analytic functions, that the space F_0^* coincides with F.

The r.k. of F is immediately seen to be given by:

$$K(z, z_1) = 0 \textit{ if } z \textit{ and } z_1 \textit{ belong to different circles,}$$

$$K(z, z_1) = K_{(i)}(z, z_1) \textit{ if } z \textit{ and } z_1 \textit{ belong to } C_i.$$

Similar results can be obtained if the domains D_n, $\overline{D}_{n+1} \subset D_n$ have an arbitrary intersection $E = D_1 \cdot D_2 \cdot \cdots$. The set E_0 is then the set of all interior points of E.

8. **Construction of reproducing kernels by the projection-formula of §12,**

([12]) This theorem is proved by using conformal mapping and the invariancy property of Bergman's kernels. For the kernels H for which the invariancy property is not true, a similar theorem has been proved only for domains with sufficiently smooth boundaries.

I. The formula of §12, giving the r.k. of a sum of two closed subspaces, may serve in many cases for the construction of kernels. We shall indicate here a few cases when it can be applied.

(1) *The expression for $H(z, z_1)$ in terms of $K(z, z_1)$.* Consider for a domain D in the plane the classes $\mathfrak{A}$ and $\mathfrak{B}$. In spite of the similarity of the metrics in the two classes, the relationship between their kernels K and H does not seem a simple one. We shall get such a relation by using the formula (18), §12, I.

To this effect we remark firstly that the class $\mathfrak{B}$ is a class of complex valued functions, that is, every function h of the class is representable in the form $h = h_1 + ih_2$, h_1 and h_2 harmonic and real-valued. Consequently, the class $\mathfrak{A}$ is a linear closed subspace of the class $\mathfrak{B}$. Also the class $\overline{\mathfrak{A}}$ of all *antianalytic* functions (that is, conjugate $\overline{f(z)}$ of analytic functions f) with a similar norm is a closed linear subspace of $\mathfrak{H}$. On the other hand, it is clear that every function $h \in \mathfrak{H}$ is representable as a sum

$$h(z) = f_1(z) + \overline{f_2(z)}$$

with two analytic functions f_1 and f_2. These functions are uniquely determined with the exception of additive constants. In general, in this decomposition, the functions f_1 and $\overline{f_2}$ may be of infinite norm, that is, they may not belong to our classes $\mathfrak{A}$ and $\overline{\mathfrak{A}}$. But when the boundary of D is not too irregular it may be proved that a dense subclass of $\mathfrak{B}$ is decomposable in this form with f_1 and $\overline{f_2}$ belonging to $\mathfrak{A}$ and $\overline{\mathfrak{A}}$ which means that $\mathfrak{B} = \mathfrak{A} \oplus \overline{\mathfrak{A}}$. In order to avoid the indetermination of the above decomposition of $h \in \mathfrak{B}$ into $f_1 + \overline{f_2}$ because of the additive constant, we shall admit to the class $\overline{\mathfrak{A}}$ only functions $\bar{f}$ satisfying the condition

$$\iint_D \bar{f}\, dx\, dy = 0.$$

With $\overline{\mathfrak{A}}$ so fixed, the condition $\mathfrak{A} \cdot \overline{\mathfrak{A}} = (0)$ is satisfied. Consequently, we can apply our formula to calculate the kernel H.

We obtain for H a development of the form

$$H(z, z_1) = \sum_{n=1}^{\infty} \left(\Lambda'_{n-1}(z, z_1) + \Lambda''_{n-1}(z, z_1) - \Lambda_n(z, z_1) - \overline{\Lambda_n(z_1, z)}\right) \tag{1}$$

where the Λ's are expressible in terms of the r.k.'s of $\mathfrak{A}$ and $\overline{\mathfrak{A}}$ following the formulas from §12. Since the kernel for $\mathfrak{A}$ is $K(z, z_1)$, the kernel for $\overline{\mathfrak{A}}$ is immediately seen to be $\overline{K(z, z_1)} + \text{const}$. The constant is easy to determine and we get for the kernel $\overline{K}$ of $\overline{\mathfrak{A}}$ the expression $\overline{K}(z, z_1) = K(z_1, z) - 1/\sigma$ where σ is the area of D. All the terms of the development of H are then expressible by repeated integration in terms of the kernel K alone.

This development can serve to compute H when K is known. It will be especially useful when the domain is such that the subspaces $\mathfrak{A}$ and $\overline{\mathfrak{A}}$ of $\mathfrak{B}$

form a positive minimal angle. This is equivalent to the fact that there exists a constant $m>0$ such that

$$m\|f\| \leqq \|f+\overline{f_1}\| \qquad \textit{for any } f \in \mathfrak{A}, \overline{f_1} \in \overline{\mathfrak{A}}.$$

It is easy to see that the last condition is equivalent to the following property of harmonic, real-valued functions $h \in \mathfrak{B}$:

$$\|\tilde{h}\| \leqq c\|h\|, \qquad \textit{for some constant } c > 0,$$

where $\tilde{h}$ is the conjugate harmonic function of h (that is, $h+i\tilde{h}$ is analytic). This property can be proved for domains with a fairly smooth boundary (continuous tangent), with at most a finite number of angular points with positive angles (see K. Friedrichs, [1]). Consequently we can apply (1) to compute the kernel H for a rectangle, which computation would be especially useful in view of applications to rectangular elastic plates.

(2) We shall describe now a manner of applying our formula to calculate the kernels K.

Let C be a closed set in the plane (not necessarily bounded) of two-dimensional measure 0. Usually C will be composed of a finite number of arcs or closed curves. C decomposes the plane in a certain number of domains $D_1, D_2, \cdots$ which together form an open set D. In D we consider the class of functions $\mathfrak{A}_D$ formed by all functions which in each of the domains D_n are analytic and regular. We call such functions locally analytic and the set C their singular set(13).

We shall suppose further that the functions of $\mathfrak{A}_D$ have a finite norm given by

$$\|f\|^2 = \iint_D |f(z)|^2 dxdy. \tag{2}$$

The class $\mathfrak{A}_D$ possesses a r.k. K_D which is simply expressible by the Bergman's kernels K_{D_n} in the following way:

$$\begin{aligned} K_D(z, z_1) &= K_{D_n}(z, z_1) \ \textit{if } z \textit{ and } z_1 \textit{ belong to } D_n, \\ K_D(z, z_1) &= 0 \ \textit{if } z \textit{ and } z_1 \textit{ belong to two different } D_n\textit{'s}. \end{aligned} \tag{3}$$

There is no loss in generality if we suppose that C is bounded (we can use conformal mappings to reduce every case to this one). We shall denote by D_0 the domain D_n containing ∞.

Consider a decomposition of C in two closed sets

$$C = C^{(1)} + C^{(2)}. \tag{4}$$

To every $C^{(i)}$ corresponds the complementary open set $D^{(i)}$ and we have

$$D = D^{(1)} \cdot D^{(2)}. \tag{5}$$

(13) The author introduced and investigated this kind of function in his thesis in Paris [2].

Every function $f(z)$ locally analytic in D is decomposable into two functions $f(z) = f_1(z) + f_2(z)$ each of which is locally analytic in the corresponding domain $D^{(i)}$ (see N. Aronszajn [2]). Two such decompositions differ by a function with the singular set $C^{(1)} \cdot C^{(2)}$.

The classes $\mathfrak{A}_{D^{(i)}}$ are clearly linear closed subspaces of $\mathfrak{A}_D$ (when we restrict their functions to the set D).

The intersection of $\mathfrak{A}_{D^{(1)}}$ and $\mathfrak{A}_{D^{(2)}}$ is equal to the class $\mathfrak{A}_{D^*}$ where D^* is the complementary set of $C^* = C^{(1)} \cdot C^{(2)}$.

The class $\mathfrak{A}_{D^*}$ is reduced to 0 when the intersection C^* is a finite or an enumerable set.

The equality

$$\mathfrak{A}_D = \mathfrak{A}_{D^{(1)}} \oplus \mathfrak{A}_{D^{(2)}} \tag{6}$$

is not always true.

For simplicity's sake we suppose now that C is not equal to 0 and is composed of a finite number of rectifiable arcs or closed curves, not reduced to isolated points, and that the same is true of $C^{(1)}$ and $C^{(2)}$.

It can then be proved that *the necessary and sufficient condition in order that* (6) *be true is that* $C^* = C^{(1)} \cdot C^{(2)} \neq 0$.

Further, if $C^* = 0$, then $\mathfrak{A}_D \ominus [\mathfrak{A}_{D^{(1)}} \oplus \mathfrak{A}_{D^{(2)}}]$ is a one-dimensional space generated by the function $w'(z)$ defined as follows: let $h(z) \equiv h_{C^{(1)}, C^{(2)}}$ be the locally harmonic function in D, taking the value 1 on $C^{(1)}$ and 0 on $C^{(2)}$. Denote then by $w(z)$ the function locally analytic in D (but not necessarily single-valued in the multiply-connected components D_n of D), whose real part is $h(z)$. The derivative $w'(z) = h'_x - ih'_y$ is single-valued in D and belongs to $\mathfrak{A}_D$. $h(z)$ is called the harmonic measure in D corresponding to $C^{(1)}$.

From what we said, it results that we shall be able to calculate the kernel K_D by the formulas of §12 in the following cases:

(1) When $C^* \neq 0$ and $\mathfrak{A}_{D^*} \neq 0$ we can apply the general formula (7), §12, I, translating operators into kernels, in particular P, P_0, P_1, and P_2 into K_D, K_{D^*}, $K_{D^{(1)}}$, and $K_{D^{(2)}}$. To calculate K_D we have then to know K_{D^*}, $K_{D^{(1)}}$, and $K_{D^{(2)}}$.

(2) When $C^* \neq 0$ and $\mathfrak{A}_{D^*} = (0)$ we can apply formula (18) §12, I.

This case contains interesting particular cases; for instance in the case of a simply-connected domain bounded by a polygonal line C. We can decompose C into one side $C^{(1)}$ and a polygonal line $C^{(2)}$ having one side less than C. This gives an inductive process to calculate the kernel K_D of the open set complementary to C. By this process we obtain, at the same time, the two Bergman's kernels for the interior and exterior of C. This is especially interesting as these kernels will give the conformal mapping functions of the domain on a circle.

Another case in point is one when to a slit $C^{(1)}$ in the plane we add a rectilinear segment $C^{(2)}$, having only one point in common with $C^{(1)}$. This case may be of interest for the variational methods, as they were used by Loewner [1]; for instance, in the problem of coefficients of schlicht functions.

(3) When $C^* = 0$, then $\mathfrak{A}_{D^*} = 0$. We can apply formula (18), §12, I, to compute the kernel of $\mathfrak{A}_{D^{(1)}} \oplus \mathfrak{A}_{D^{(2)}}$ if $K_{D^{(1)}}$ and $K_{D^{(2)}}$ are known. Then, if we know the function $w'(z)$, we add the function $(1/\|w'\|^2)w'(z)\overline{w'(z_1)}$ to the obtained kernel and arrive at the kernel K_D. The classes $\mathfrak{A}_{D^{(1)}}$ and $\mathfrak{A}_{D^{(2)}}$ have, in this case, a positive minimal angle for which a lower estimate can be obtained (using the constant m from the inequality $m\|f_1\| \leqq \|f_1 + f_2\|$ as in §12, I) by the use of methods developed by the author in [3].

If we consider the reduced classes $\mathfrak{A}'_D$ of functions of $\mathfrak{A}_D$ with single-valued integrals in every component domain D_n of D, we shall obtain a r. k. K'_D expressible by a formula similar to (3) in terms of the reduced Bergman's kernels K'_{D_n}. For the reduced classes we have always $\mathfrak{A}'_D = \mathfrak{A}'_{D^{(1)}} \oplus \mathfrak{A}'_{D^{(2)}}$ and in the case $C^* = 0$, we shall calculate K'_D directly by formula (18).

Bibliography

N. Aronszajn

1. *Sur les invariants des transformations dans le domaine de n variables complexes*, C. R. Acad. Sci. Paris vol. 197 (1933) p. 1579, and vol. 198 (1934) p. 143.
2. *Sur les décompositions des fonctions analytiques uniformes et sur leurs applications*, Acta Math. vol. 65 (1935) p. 1. (Thesis, Paris.)
3. *Approximation des fonctions harmoniques et quelques problèmes de transformation conforme*, Bull. Soc. Math. France vol. 67 (1939).
4. *La théorie générale des noyaux reproduisants et ses applications*, Première Partie, Proc. Cambridge Philos. Soc. vol. 39 (1944) p. 133.
5. *Les noyaux pseudo-reproduisants. Noyaux pseudo-reproduisants et complétion des classes hilbertiennes. Complétion fonctionnelle de certaines classes hilbertiennes.—Propriétés de certaines classes hilbertiennes completées*, C. R. Acad. Sci. Paris vol. 226 (1948) p. 456, p. 537, p. 617, p. 700.
6. *Reproducing and pseudo-reproducing kernels and their application to the partial differential equations of physics*, Studies in partial differential equations of physics, Harvard University, Graduate School of Engineering, Cambridge, Massachusetts, 1948.

S. Banach

1. *Theorie des opérations linéaires*, Monografje Matematyczne, Warsaw, 1932.

S. Bergman

1. *Ueber die Entwicklung der harmonischen Funktionen der Ebene und des Raumes nach Orthogonalfunktionen*, Math. Ann. vol. 86 (1922) pp. 238–271. (Thesis, Berlin, 1921.)
2. *Ueber die Bestimmung der Verzweigungspunkte eines hyperelliptischen Integrals aus seinen Periodizitätsmoduln mit Anwendungen auf die Theorie des Transformators*, Math. Zeit. vol. 19 (1923) pp. 8–25.
3. *Ueber die Bestimmung der elastischen Spannungen und Verschiebungen in einem konvexen Körper*, Math. Ann. vol. 98 (1927) pp. 248–263.
4. *Ueber unendliche Hermitesche Formen, die zu einem Bereich gehören, nebst Anwendungen auf Fragen der Abbildung durch Funktionen von zwei komplexen Veränderlichen*, Berichte Berliner Mathematische Gesellschaft vol. 26 (1927) pp. 178–184, and Math. Zeit. vol. 29 (1929) pp. 640–677.
5. *Ueber die Existenz von Repräsentantenbereichen*, Math. Ann. vol. 102 (1929) pp. 430–446.

6. *Ueber den Wertevorrat einer Funktion von zwei komplexen Veränderlichen*, Math. Zeit. vol. 36 (1932) pp. 171–183.
7. *Ueber die Kernfunktion eines Bereiches und ihr Verhalten am Rande*, J. Reine Angew. Math. vol. 169 (1933) pp. 1–42, and vol. 172 (1934) pp. 89–128.
8. *Sur quelques propriétés des transformations par un couple des fonctions de deux variables complexes*, Atti della Accademia Nazionale dei Lincei, Rendiconti (6) vol. 19 (1934) pp. 474–478.
9. *Zur Theorie von pseudokonformen Abbildungen*, Rec. Math. (Mat. Sbornik) N.S. vol. 1 (1936) pp. 79–96.
10. *Zur Theorie der Funktionen, die eine lineare partielle Differentialgleichung befriedigen*, Rec. Math. (Mat. Sbornik) N.S. vol. 2 (1937) pp. 1169–1198.
11. *Partial differential equations, advanced topics*, Brown University, Providence, 1941.
12. *A remark on the mapping of multiply-connected domains*, Amer. J. Math. vol. 68 (1946) pp. 20–28.
13. *Funktions satisfying certain partial differential equations of elliptic type and their representation*, Duke Math. J. vol. 14 (1947) pp. 349–366.
14. *Sur les fonctions orthogonales de plusieurs variables complexes avec les applications à la théorie des fonctions analytiques*, Mémorial des Sciences Mathématiques, vol. 106, Paris, 1947.

S. Bergman and M. Schiffer
1. *A representation of Green's and Neumann's functions in the theory of partial differential equations of second order*, Duke Math. J. vol. 14 (1947) pp. 609–638.
2. *On Green's and Neumann's functions in the theory of partial differential equations*, Bull. Amer. Math. Soc. vol. 53 (1947) pp. 1141–1151.
3. *Kernel functions in the theory of partial differential equations of elliptic type*, Duke Math. J. vol. 15 (1948) pp. 535–566.

S. Bochner
1. *Ueber orthogonale Systeme analytischer Funktionen*, Math. Zeit. vol. 14 (1922) pp. 180–207.
2. *Vorlesungen ueber Fouriersche Integrale*, Leipzig, 1932.
3. *Hilbert distances and positive definite functions*, Ann. of Math. vol. 42 (1941) pp. 647–656.

J. Dixmier
1. *Sur une classe nouvelle de variétés linéaires et d'opérateurs linéaires de l'espace d'Hilbert. Propriétés géometriques des domaines d'existence des opérateurs linéaires fermés de l'espace de Hilbert. Definition des opérateurs linéaires de l'espace d'Hilbert par leurs domaines d'existence et des valeurs*, C. R. Acad. Sci. Paris vol. 223 (1946) p. 971, and vol. 224 (1947) p. 180, p. 225.

K. Friedrichs
1. *On certain inequalities and characteristic value problems for analytic functions and for functions of two variables*, Trans. Amer. Math. Soc. vol. 41 (1937) pp. 321–364.

B. Fuchs
1. *Ueber einige Eigenschaften der pseudokonformen Abbildungen*, Rec. Math. (Mat. Sbornik) N.S. vol. 1 (1936) pp. 569–574.
2. *Ueber geodätische Mannigfaltigkeiten einer bei pseudokonformen Abbildungen invarianten Riemannschen Geometrie*, Rec. Math. (Mat. Sbornik) N.S. vol. 2 (1937) pp. 567–594.

P. Garabedian
1. *Schwarz' lemma and the Szegö kernel function*, Trans. Amer. Math. Soc. vol. 67 (1949) pp. 1–35.

P. Garabedian and M. Schiffer
1. *Identities in the theory of conformal mapping*, Trans. Amer. Math. Soc. vol. 65 (1948) pp. 187–238.

I. GELFAND and D. RAIKOFF

1. *Irreducible unitary representation of arbitrary locally bicompact groups*, Rec. Math. (Mat. Sbornik) vol. 13 (1943) p. 316.

R. GODEMENT

1. *Sur les fonctions de type positif. Sur les propriétés ergodiques des fonctions de type positif. Sur les partitions finies des fonctions de type positif. Sur certains opérateurs définis dans l'espace d'une fonction de type positif. Sur quelques propriétés des fonctions de type positif définies sur un groupe quelconque*, C. R. Acad. Sci. Paris vol. 221 (1945) p. 69, p. 134, and vol. 222 (1946) p. 36, p. 213, p. 529.
2. *Les fonctions de type positif et la théorie des groupes*, Trans. Amer. Math. Soc. vol. 63 (1948) pp. 1–84. (Thèse, Paris.)

J. KOBER

1. *A theorem on Banach spaces*, Compositio Math. vol. 7 (1939) pp. 135–140.

C. LOEWNER

1. *Untersuchungen über schlichte konforme Abbildungen des Einheitskreises* I, Math. Ann. vol. 89 (1923) pp. 103–121.

J. MERCER

1. *Functions of positive and negative type and their connection with the theory of integral equations*, Philos. Trans. Roy. Soc. London Ser. A vol. 209 (1909) pp. 415–446.
2. *Sturm-Louville series of normal functions in the theory of integral equations*, Philos. Trans. Roy. Soc. London Ser. A vol. 211 (1911) pp. 111–198.

E. H. MOORE

1. *On properly positive Hermitian matrices*, Bull. Amer. Math. Soc. vol. 23 (1916) p. 59.
2. *General analysis*, Memoirs of the American Philosophical Society, Part I, 1935, Part II, 1939.

Z. NEHARI

1. *The kernel function and canonical conformal maps*, Duke Math. J. (1949).
2. *On analytic functions possessing a positive real part*, Duke Math. J. (1949).

J. v. NEUMANN and F. J. MURRAY

1. *On rings of operators*, Ann. of Math. vol. 37 (1936) pp. 116–229.

J. v. NEUMANN and I. J. SCHOENBERG

1. *Fourier integrals and metric geometry*, Trans. Amer. Math. Soc. vol. 50 (1941) pp. 226–251.

M. SCHIFFER

1. *The kernel function of an orthonormal system*, Duke Math. J. vol. 13 (1945) pp. 529–540.
2. *An application of orthonormal functions in the theory of conformal mappings*, Amer. J. Math. vol. 70 (1948) pp. 147–156.

I. J. SCHÖNBERG

1. *Metric spaces and positive definite functions*, Trans. Amer. Math. Soc. vol. 44 (1938) pp. 522–536.
2. *Positive definite functions on spheres*, Duke Math. J. vol. 9 (1942) pp. 96–108.

M. H. STONE

1. *Linear transformations in Hilbert space*, Amer. Math. Soc. Colloquium Publications, vol. 15.

G. SZEGÖ

1. *Ueber orthogonale Polynome, die zu einer gegebenen Kurve der komplexen Ebene gehoeren*, Math. Zeit. vol. 9 (1921) pp. 218–270.

A. WEIL

1. *L'intégration dans les groupes topologiques et ses applications*, Actualités Scientifiques et Industrielles, vol. 869, Paris, 1940.

H. WELKE

1. *Ueber die analytischen Abbildungen von Kreiskörpern und Hartogsschen Bereichen*, Math. Ann. vol. 103 (1930) pp. 437–449. (Thesis, Münster, 1930.)

K. Zarankiewicz

1. *Ueber ein numerisches Verfahren zur konformen Abbildung zweifach zusammenhängender Gebiete*, Zeitschrift für angewandte Mathematik und Mechanik vol. 14 (1934) pp. 97–104.

S. Zaremba

1. *L'équation biharmonique et une classe remarquable de fonctions fondamentales harmoniques*, Bulletin International de l'Académie des Sciences de Cracovie (1907) pp. 147–196.
2. *Sur le calcul numérique des fonctions demandées dans le problème de Dirichlet et le problème hydrodynamique*, Bulletin International de l'Académie des Sciences de Cracovie (1908) pp. 125–195.

Oklahoma Agricultural and Mechanical College,
Stillwater, Okla.

2

Reprinted from *Rocky Mt. J. Math.* **6**:409–434 (1976)

POSITIVE DEFINITE FUNCTIONS AND GENERALIZATIONS, AN HISTORICAL SURVEY

JAMES STEWART

1. Introduction. A complex-valued function f of a real variable is said to be *positive definite* (abbreviated as p.d.) if the inequality

$$\sum_{i,j=1}^{n} f(x_i - x_j)\xi_i\bar{\xi}_j \geqq 0 \tag{1.1}$$

holds for every choice of $x_1, \cdots, x_n \in R$(the real numbers) and $\xi_1, \cdots, \xi_n \in C$(the complex numbers). In other words,the matrix

$$[f(x_i - x_j)]_{i,j=1}^{n} \tag{1.2}$$

is positive definite (strictly speaking we should say positive semi-definite or non-negative definite) for all n, no matter how the x_i's are chosen. A synonym for positive definite function is *function of positive type*.

For example, the function $f(x) = \cos x$ is p.d. because

$$\begin{aligned}\sum_{i,j=1}^{n} \cos(x_i - x_j)\xi_i\bar{\xi}_j &= \sum_{i,j=1}^{n} (\cos x_i \cos x_j + \sin x_i \sin x_j)\xi_i\bar{\xi}_j \\ &= \left| \sum_{i=1}^{n} \xi_i \cos x_i \right|^2 + \left| \sum_{i=1}^{n} \xi_i \sin x_i \right|^2 \geqq 0.\end{aligned}$$

Likewise it is easily verified directly that $e^{i\lambda x}$ is p.d. for real λ, but it is not so straightforward to see that such functions as $e^{-|x|}$, e^{-x^2}, and $(1 + x^2)^{-1}$ are p.d. These and other examples are discussed in § 3.

Positive definite functions and their various analogues and generalizations have arisen in diverse parts of mathematics since the beginning of this century. They occur naturally in Fourier analysis, probability theory, operator theory, complex function-theory, moment problems, integral equations, boundary-value problems for partial differential equations, embedding problems, information theory, and other areas. Their history constitutes a good illustration of the words of Hobson [51, p. 290]: "Not only are special results, obtained independently of one another, frequently seen to be really included in

Received by the editors on March 6, 1975.

some generalization, but branches of the subject which have developed quite independently of one another are sometimes found to have connections which enable them to be synthesized in one single body of doctrine. The essential nature of mathematical thought manifests itself in the discernment of fundamental identity in the mathematical aspects of what are superficially very different domains." Fourier [36] put it more succinctly: "[Mathematics] compares the most diverse phenomena and discovers the secret analogies which unite them."

To cite a specific instance, Mathias and the other early workers with p.d. functions of a real variable were chiefly concerned with Fourier transforms and apparently did not realize that more than a decade previously Mercer and others had considered the more general concept of positive definite kernels $K(x, y)$ (satisfying (1.1) with $f(x_i - x_j)$ replaced by $K(x_i, x_j)$) in research on integral equations. I have likewise found that present-day mathematicians working with some of the manifestations of p.d. functions are unaware of other closely related ideas. Thus one of the purposes of this survey is to correlate some of the more important generalizations of p.d. functions with the hope of making them better known. For example, probabilists are acquainted on the one hand with p.d. functions (in the guise of characteristic functions) and on the other hand with the Kolmogorov or Lévy-Khintchine formula for the logarithm of the characteristic function of an infinitely divisible random variable, but probably very few of them realize that the latter functions are examples of a significant generalization of p.d. functions, namely functions with a finite number of negative squares, and that Krein's integral representation for such functions may be of use to them.

It is not possible to discuss all of the analogues and generalizations of p.d. functions in this article; there are simply too many of them. Those to which we devote an entire section are p.d. sequences (which arose first), p.d. functions on groups, integrally p.d. functions, p.d. distributions, p.d. kernels, functions with a finite number of negative squares, and Schoenberg's functions which are p.d. in metric spaces. Some other generalizations are mentioned in the final section.

2. **Positive definite sequences.** The concept of a p.d. sequence was inspired by a problem of Carathéodory in complex function-theory. Contained in his paper [16], which appeared in 1907, was the following problem: What are necessary and sufficient conditions on the coefficients of the power series representation

$$w = f(z) = 1 + \sum_{k=1}^{\infty} (a_k + ib_k) z^k$$

of an analytic function f in order that it map the unit disk $|z| < 1$ into the right half-plane $\mathrm{Re}(w) > 0$? Carathéodory's answer was that, for each $n = 1, 2, 3, \cdots$, the point $(a_1, b_1, a_2, b_2, \cdots, a_n, b_n)$ in R^{2n} should lie in the smallest convex set containing the curve with parametric representation

$$(2\cos\theta, -2\sin\theta, 2\cos 2\theta, -2\sin 2\theta, \cdots, 2\cos n\theta,$$
$$-2\sin n\theta), 0 \leqq \theta \leqq 2\pi.$$

In 1911 Toeplitz [115] noticed that Carathéodory's conditions could be reformulated algebraically in terms of the non-negativity of certain Hermitian forms:

$$\sum_{i,j=1}^{n} c_{i-j}\xi_i\bar{\xi}_j \geqq 0, \text{ for } n = 1, 2, \cdots, \tag{2.1}$$

where $c_0 = 2, c_k = a_k - ib_k, c_{-k} = a_k + ib_k$. Any sequence $\{c_n\}$ which satisfies (2.1) is called *positive definite.*

Within a year several of the ablest mathematicians of the day published papers offering alternative proofs of the Carathéodory-Toeplitz results and pointing out connections with other areas of mathematics. In particular F. Riesz [90] saw the application to systems of integral equations, Herglotz [46] established the connection with the trigonometric moment problem, Carathéodory himself [17] considered the series expansion of positive harmonic functions, and further related papers were written by Fischer [35], Schur [102], and Frobenius [37]. Of these, the paper of Herglotz has turned out to have the most far-reaching consequences. The trigonometric moment problem can be stated as follows: Given a sequence $\{c_n\}_{-\infty}^{\infty}$ of complex numbers, what are necessary and sufficient conditions for the existence of a bounded non-decreasing function σ on $[-\pi, \pi]$ such that $\{c_n\}$ is the sequence of Fourier-Stieltjes coefficients of σ, i.e., $c_n = \int_{-\pi}^{\pi} e^{in\theta}\, d\sigma(\theta)$ for every integer n? Herglotz solved this problem by proving that a necessary and sufficient condition is the non-negativity of the Toeplitz forms (2.1), i.e., the positive-definiteness of $\{c_n\}$. We shall see that this theorem of Herglotz has many important analogues and generalizations.

For proofs, further details, and applications of p.d. sequences to problems in analysis and probability theory, we refer the reader to the books of Akhiezer and Krein [2], Akhiezer [1], and Grenander and Szegö [44]. Fan [33] has established various properties of p.d. sequences (without using Herglotz's Theorem) by representing them in terms of stationary sequences of vectors in a Hilbert space.

3. Positive definite functions of a real variable. Mathias, in 1923, [69], was the first person to define and study the properties of p.d. functions of a real variable. Motivated by the results of Carathéodory and Toeplitz he defined a complex-valued function f on R to be positive definite if

$$(3.1) \qquad f(-x) = \overline{f(x)}, \text{ for } x \in R,$$

and the Hermitian form

$$(3.2) \qquad \sum_{i,j=1}^{n} f(x_i - x_j)\xi_i\bar{\xi}_j \geqq 0,$$

for every choice of $x_1, \cdots, x_n \in R$ and $\xi_1, \cdots, \xi_n \in C$.

Condition (3.1) is superfluous, as F. Riesz, [91], pointed out. To see this, set $n = 2, x_1 = 0, x_2 = x, \xi_1 = 1$, and $\xi_2 = \xi$ in (3.2). Then

$$(3.3) \qquad (1 + |\xi|^2)f(0) + \xi f(x) + \bar{\xi} f(-x) \geqq 0$$

for every $\xi \in C$, and so $\xi f(x) + \bar{\xi} f(-x)$ is real for every $\xi \in C$. Setting $\xi = 1$, we have that $f(x) + f(-x)$ is real; setting $\xi = i$, we see that $i(f(x) - f(-x))$ is real. Thus (3.1) holds.

Mathias [69] observed the following elementary properties of positive definite functions:

I. If f is p.d., then so is $\bar{f}$.
II. If $f_1, \cdots, f_n$ are p.d. and $c_i \geqq 0$ then $f(x) = \sum_{i=1}^{n} c_i f_i(x)$ is p.d.
III. If each f_n is p.d., then so is $f(x) = \lim_{n\to\infty} f_n(x)$.
IV. Any p.d. function f is bounded, and in fact, $|f(x)| \leqq f(0)$.
V. The product of p.d. functions is p.d.

The first three properties are immediate consequences of the definition. The fourth follows from (3.3) by choosing ξ so that $\xi f(x) = -|f(x)|$. The fifth is a consequence of Schur's theorem [101] that the product of p.d. matrices is p.d.

The main theorem of Mathias, slightly rephrased, is that if f is p.d., then its Fourier transform $\hat{f}(t) = \int_{-\infty}^{\infty} f(x)e^{-itx}\,dx$ is non-negative (provided that it exists). Conversely if f satisfies the Fourier inversion formula and $\hat{f}(t) \geqq 0$, then f is p.d. His proof makes use of the analogous result of Carathéodory and Toeplitz [115] for Fourier series.

We can make use of part of this theorem of Mathias to give some examples of p.d. functions. The functions $f_1(x) = e^{-|x|}, f_2(x) = e^{-x^2}$, $f_3(x) = (1 + x^2)^{-1}$, and $f_4(x) = 1 - |x|$ for $|x| \leqq 1, f_4(x) = 0$ for $|x| > 1$, are all p.d. because their Fourier transforms are positive and integrable. In fact it can be shown (see, e.g., Schoenberg [97]) that $f_\alpha(x)$

$= \exp[-|x|^{\alpha}]$ is p.d. if and only if $0 \leqq \alpha \leqq 2$. (Further examples are provided by Pólya's criterion [84]: Any real, even, continuous function f which is convex on $(0, \infty)$, i.e., $f(\frac{1}{2}(x_1 + x_2)) \leqq (\frac{1}{2})[f(x_1) + f(x_2)]$, and satisfies $\lim_{x \to \infty} f(x) = 0$, is p.d.)

Mathias did not prove the analogue of the theorem of Herglotz. That had to wait until 1932 when Bochner [11] proved the celebrated theorem which bears his name: *If f is a continuous p.d. function on R, then there exists a bounded non-decreasing function V on R such that f is the Fourier-Stieltjes transform of V, i.e.,*

$$f(x) = \int_{-\infty}^{\infty} e^{i\alpha x}\, dV(\alpha) \tag{3.4}$$

holds for all x.

The converse of this theorem is easy to prove, for if f has this form, then

$$\begin{aligned}\sum_{i,j=1}^{n} f(x_i - x_j)\xi_i\bar{\xi}_j &= \sum_{i,j=1}^{n} \left\{ \int_{-\infty}^{\infty} \exp[i(x_i - x_j)\alpha]\, dV(\alpha) \right\} \xi_i\bar{\xi}_j \\ &= \int_{-\infty}^{\infty} \left| \sum_{j=1}^{n} \xi_j \exp(ix_j\alpha) \right|^2 dV(\alpha) \geqq 0.\end{aligned}$$

Bochner's Theorem itself is not so easy to establish, but in view of its great importance many, quite different, proofs have been given. (In particular, it can be deduced from the theorem of Herglotz. See, e.g., [54, p. 137].) The generalization to functions of several real variables was also given by Bochner [12].

In 1933, F. Riesz [91] extended Bochner's Theorem by proving that if we merely assume the measurability of a p.d. function f, then, for almost all x, $f(x)$ is equal to the Fourier-Stieltjes transform of a bounded non-decreasing function. An interesting refinement of this theorem of Riesz was given in 1956 by Crum [24] who showed that if f is a measurable p.d. function, then $f = p + r$, where p is a Fourier-Stieltjes integral, and the remainder function r is equal to zero almost everywhere (as Riesz had shown) and is also positive definite itself.

In order to indicate some of the many applications of Bochner's Theorem, we first mention that Bochner himself showed in 1933 [12] how it can be used to deduce and generalize much of the harmonic analysis of Wiener [120, 121].

About the same time both Bochner [13] and F. Riesz [91] showed how Bochner's Theorem can be applied to prove another important theorem of the same period, namely Stone's Theorem of 1930 [**110**, **111**] on one-parameter groups of unitary operators. Let U_t, $-\infty < t < \infty$, be a group of unitary operators on a Hilbert space H, i.e., $U_0 = I$, the identity operator, and $U_{s+t} = U_s U_t$, and assume that $(U_t x, y)$ is a continuous function of t for every $x, y \in H$. Stone's Theorem asserts that there is a unique self-adjoint operator A on H with canonical resolution of the identity E such that $U_t = \int_{-\infty}^{\infty} e^{it\lambda}\, dE_\lambda$ $(\equiv e^{itA})$ in the sense that $(U_t x, y) = \int_{-\infty}^{\infty} e^{it\lambda}\, d(E_\lambda x, y)$, for $x, y \in H$. The proofs of Bochner and Riesz use the fact that for any $x \in H$ the function $f(t) = (U_t x, x)$ is p.d. Indeed

$$\sum_{i,j=1}^{n} (U_{t_i - t_j} x, x)\xi_i \bar{\xi}_j = \left(\sum_{i=1}^{n} \xi_i U_{t_i} x,\ \sum_{j=1}^{n} \xi_j U_{t_j} x \right) \geqq 0.$$

Stone's Theorem, in turn, has applications to quantum mechanics and ergodic theory. (See Riesz and Sz.-Nagy [**92**, §§138–139] for Bochner's proof of Stone's Theorem and applications.)

It is perhaps true to say that the area of mathematics in which the largest number of people are familiar with p.d. functions and Bochner's Theorem is that of probability theory. The Fourier-Stieltjes transform of the distribution function of a random variable is called a *characteristic function*, and so, by virtue of Bochner's Theorem, f is a characteristic function if and only if f is continuous, p.d., and $f(0) = 1$. Although characteristic functions can be traced back as far as Laplace and Cauchy, it was Paul Lévy [**63**, **64**] who first exploited systematically the fact that characteristic functions are in general easier to work with than distribution functions. This is especially true in connection with sums of independent random variables (which correspond to products of characteristic functions) and convergence of sequences of random variables (in part because for sequences $\{f_n\}$ of p.d. functions, $f_n \to f$ a.e. if and only if $f_n \to f$ uniformly on every finite interval; see §4). Thus it is not surprising that the Central Limit Problem (the problem of convergence of laws of sequences of sums of random variables) was solved with the aid of p.d. functions.

Another occurrence of p.d. functions in probability theory is in the theory of stationary stochastic processes. Khintchine [**55**] used Bochner's Theorem to show that $R(t)$ is the correlation (covariance) function of a continuous stationary stochastic process if and only if it is of the form $R(t) = \int_{-\infty}^{\infty} \cos t\alpha\, dV(\alpha)$ where V is bounded and non-decreasing. See Fan [**34**] for this and other connections between p.d. functions and probability theory. Further connections can be found in Yaglom [**122**] and Schreiber, Sun and Barucha-Reid [**100**].

4. **Positive definite functions on groups.** With the advent of harmonic analysis on groups, and especially the Banach algebra approach to the subject in the 1940's, the central role of positive definite functions in Fourier analysis became apparent. The classical treatises on Fourier series and integrals of Zygmund [125] and Titchmarsh [114] had certainly managed to thrive without ever mentioning such functions. However in their present-day counterparts which deal with Fourier series and integrals, e.g., Edwards [27] and Katznelson [54], p.d. functions do have a role to play. Furthermore, virtually every book which treats harmonic analysis on groups gives prominence to p.d. functions. In some of these books, e.g., Loomis [67] and Rudin [93], p.d. functions play a very fundamental role indeed. The inversion formulas for the Fourier transform are based on the analogue of Bochner's Theorem, and then the duality theorem and Plancherel's Theorem are deduced. Consequently, in such treatments everything depends on p.d. functions.

The definition given in § 1 generalizes easily. A complex-valued function f defined on an arbitrary group G is *positive definite* if the inequality

$$\sum_{i,j=1}^{n} f(x_j^{-1}x_i)\xi_i\bar{\xi}_j \geqq 0 \tag{4.1}$$

holds for every choice of $x_i \in G$ and $\xi_i \in C$. Let P denote the set of all continuous p.d. functions on G. For the case where G is a locally compact abelian group (LCAG), Bochner's Theorem was generalized in 1940 (almost simultaneously) by Weil [118], Povzner [85], and Raikov [86] as follows. *A character* $\mathfrak{x}$ of G is a homomorphism of G into the circle group, i.e., it satisfies $\mathfrak{x}(xy) = \mathfrak{x}(x)\mathfrak{x}(y)$ and $|\mathfrak{x}(x)| = 1$ for $x, y \in G$. The dual group $\hat{G}$ is the group of all continuous characters of G under pointwise multiplication with the topology of uniform convergence on compact subsets of G. The Weil-Povzner-Raikov Theorem says that if $f \in P$, then there is a positive bounded measure μ on $\hat{G}$ such that f is the Fourier-Stieltjes transform of μ, i.e.,

$$f(x) = \int_{\hat{G}} \mathfrak{x}(x)\, d\mu(\mathfrak{x}). \tag{4.2}$$

Let $B(G)$ be the algebra of all finite linear combinations of functions in P. Then (4.2) in conjunction with the Jordan decomposition theorem shows that $B(G)$ is precisely the set of all Fourier-Stieltjes transforms of bounded measures on $\hat{G}$.

Since every continuous character on R is of the form $\mathfrak{x}(x) = e^{i\alpha x}$ for some $\alpha \in R$, we can consider $\hat{R} = R$, and (4.2) reduces to Bochner's

Theorem for $G = R$. Similarly if $G = Z$, the group of integers, then every continuous character is of the form $\mathfrak{X}(n) = e^{in\theta}$ for some $\theta \in [-\pi, \pi]$, and (4.2) becomes Herglotz's Theorem.

Just as for $G = R$, many proofs of (4.2) have since appeared. In addition to the proofs in current texts, most of which, like Raikov's, use Banach algebra techniques, we cite in particular the proof of Cartan and Godement [19] which uses the Krein-Milman Theorem, the proof of Bingham and Parthasarathy [9] which uses probabilistic methods, and the proof of Bucy and Maltese [15] which uses the Choquet Representation Theorem. See also Phelps [82] and Choquet [20].

The proofs of Weil and Raikov referred to above also generalized Riesz's extension of Bochner's Theorem, i.e., any measurable p.d. function on a LCAG can be written as $f = p + r$ where p is the Fourier-Stieltjes transform of a positive measure and $r = 0$ locally almost everywhere. We have seen that for $G = R$, the residual function r is actually p.d. itself (Crum's Theorem, 1956). For an arbitrary LCAG this result is in fact a consequence of an earlier theorem (1950) of Segal and von Neumann [104, Thm. 2] on unitary representations, though they did not explicitly point this out. In 1960, Devinatz [26] made it explicit and gave a direct proof. A far-reaching generalization of this fact was given by de Leeuw and Glicksberg [62]. They showed that an arbitrary p.d. function (not necessarily measurable) on an arbitrary topological group G can be expressed as $f = p + r$, where $p \in P$ and r is a p.d. function which averages uniformly to 0 at e in the sense that $0 \in \bigcap_{V \in \mathcal{V}} \mathcal{C}(R_V r)$, where $\mathcal{V}$ is the set of all neighborhoods of the identity e of G, $\mathcal{C}$ denotes the closed convex hull in the set of all bounded functions on G, and the partial orbit $R_V r = \{R_g r : g \in V\}$, where $R_g r(h) = r(hg)$.

The importance of p.d. functions does not diminish when we turn our attention to non-abelian groups. The importance stems in part from the intimate connection among p.d. functions, unitary representations of the group, and positive functionals on the group algebra.

Let U be a unitary representation of a locally compact group G, i.e., each $U(x)$ is a unitary operator on a Hilbert space H, $U(e) = I$, and $U(xy) = U(x)U(y)$, for $x, y \in G$. Then, as in § 3, the function

$$f(x) = (U(x)\xi, \xi) \tag{4.3}$$

is p.d. for any $\xi \in H$. Conversely if f is p.d. Gelfand and Raikov [38] showed how to construct a unitary representation U which satisfies (4.3). Let H_0 be the set of functions ϕ on G such that $\phi(x) = 0$ except for finitely many x. If $(\phi, \psi) = \sum_{s,t} f(t^{-1}s)\phi(s)\overline{\psi(t)}$ then $(\phi, \phi) \geqq 0$.

Let H be the completion of H_0/N, where $N = \{\phi : (\phi, \phi) = 0\}$, and define U via $(U(x)\phi)(y) = \phi(xy^{-1})$. Then U is a unitary representation of G, and if ξ corresponds to the function which is equal to 1 at e and is zero elsewhere, we have $(U(x)\xi, \xi) = \sum_{s,t} f(t^{-1}s)\xi(xs)\overline{\xi(t)} = f(x)$. Both halves of this correspondence between unitary representations and p.d. functions are useful. For example, Ambrose [4] and Godement [41] used the first half, together with the Weil-Povzner-Raikov Theorem, to generalize Stone's Theorem to the situation where U is a continuous unitary representation of a LCAG. Conversely, the other half can be used to deduce (4.2) from the theorem of Ambrose and Godement. (See Nakano [79] for $G = R$ and Nakamura and Umegaki [78] in general.)

In order to describe another application, we introduce an ordering on P as follows: $\phi_1 > \phi_2 \Leftrightarrow \phi_1 - \phi_2 \in P$. A function $\phi_1 \in P$ is said to be *elementary* if the only functions $\phi_2 \in P$ with $\phi_1 > \phi_2$ are of the form $\phi_2 = \lambda\phi_1$, where $\lambda \in C$. It was proved by Gelfand and Raikov in 1943 [38], and independently by Godement [42], that ϕ is elementary if and only if the corresponding unitary representation U is irreducible (i.e., if S is a closed subspace of H with $U(x)S \subset S$ for all $x \in G$, then $S = H$ or $S = \{0\}$). Gelfand and Raikov expoited this correspondence to prove their famous theorem that every locally compact group admits "sufficiently many" irreducible unitary representations. Indeed it was Gelfand and Raikov who pointed out the full significance of the central role that p.d. functions play in analysis on locally compact groups.

Let $P_0 = \{\phi \in P : \phi(e) = 1\}$, the set of normalized functions in P. (If G is abelian, then $\hat{G}$ is the set of all elementary functions in P_0.) Then P_0 is a compact convex subset of $L^\infty(G)$ in the weak topology, and, by identifying the extreme points of P_0 as precisely the elementary functions in P_0, Gelfand and Raikov [38] used the Krein-Milman Theorem to show that any $f \in P$ is the weak limit in L^∞ of functions of the form $\sum \lambda_i\phi_i$, where $\lambda_i \geqq 0$, $\sum \lambda_i \leqq f(e)$, and the ϕ_i are elementary functions in P_0.

There have been various extensions of Bochner's Theorem to non-abelian locally compact groups. For the details we refer the reader to Godement [42, p. 52] and, for a more explicit version in the case where G is compact, to Krein [60, § 7]. (See also Hewitt and Ross [49, p. 334] for Krein's version.)

Convergence in P holds some surprises. Raikov [88] (and independently Yoshizawa [123]) proved that the mere assumption of pointwise convergence of a sequence of functions $f_n \in P$ to a function $f \in P$ implies that $f_n \to f$ uniformly on compact subsets of a locally compact group. For an historical discussion of the relationship between such theorems and the various Cramér-Lévy convergence theorems of the 1920's and 1930's see McKennon [**70**, p. 62].

Several authors have recently shown interest in p.d. functions and analogues of Bochner's Theorem on groups which are abelian but not locally compact. For instance, the underlying additive group of an infinite-dimensional Banach space is such a group. We cite in particular the papers of Minlos [**74**] who deals with nuclear spaces (see also Gelfand and Vilenkin [**39**, p. 350]), Sazonov [**96**] and Gross [**45**] who deal with Hilbert spaces, and Waldenfels [**117**] who also deals with vector spaces. Shah [**106**] obtains a Bochner-type theorem for abelian groups on which the only restriction is the existence of a "quasi-invariant" measure.

5. **The extension problem.** In 1940, M. Krein [**57**] posed the following problem. Suppose that f is continuous and p.d. on the finite interval $[-A, A]$, i.e., the inequality (1.1) holds whenever $\xi_i \in C$ and $0 \leqq x_1 < x_2 < \cdots < x_n \leqq A$. Can f be extended to a continuous p.d. function on R, i.e., does there exist $g \in P$ such that $g(x) = f(x)$ for $-A \leqq x \leqq A$? Krein answered this question in the affirmative, and so by Bochner's Theorem, any such f is a Fourier-Stieltjes transform. In the same year, Raikov [**87**] proved this fact directly.

In the same paper, Krein also showed that the extension need not be unique, and, by employing methods reminiscent of those used in the classical moment problems, he gave several criteria for uniqueness of the extension. An example is the following result. Let B_A be the set of entire functions g which are bounded on the real axis and satisfy

$$\limsup_{r\to\infty} \frac{\log M(r)}{r} \leqq A, \text{ where } M(r) = \max\{|g(z)| : |z| \leqq r\}.$$

If f is p.d. on $[-A, A]$, it has a representation $f(x) = \int e^{itx}\, dF(t)$, for $|x| \leqq A$, by the theorems of Krein and Bochner. The functional $\Phi_f(g) = \int_{-\infty}^{\infty} g(t)\, dF(t)$ on B_A turns out to be independent of F. Krein proved that the p.d. extension of f is unique if there exists $g \in B_A$, $g \geqq 0$, $g \not\equiv 0$, satisfying $\Phi_f(g) = 0$. He also gave the example $f(x) = 1 - |x|$, $|x| \leqq A$, which is p.d. on $[-A, A]$ if and only if $0 < A \leqq 2$; the extension is unique for $A = 2$ but not unique for $0 < A < 2$.

Many authors have since given other criteria for uniqueness and tried to classify all possible extensions in the case of non-uniqueness. For example, both Akutowicz [**3**] and Devinatz [**25**] gave criteria in terms of the self-adjointness of certain operators on Hilbert spaces.

Part of the interest in such questions stems from probability theory. Several authors in the 1930's gave examples of distinct characteristic functions which coincide on a finite interval, and this, of course, implies the non-uniqueness of the extension problem for p.d. functions. Esseen [31] and Lévy [65] treated the extension problem from the point of view of probability theory. See also Loève ([66, p. 212].

Chover [21] has dealt with an application of the extension problem to information theory. He showed that under certain conditions there is a unique extension of a p.d. function f which maximizes the entropy of the extension measure, i.e., among all the measures μ for which $f(x) = \int e^{ixt}\, d\mu(t)$, $|x| \leqq A$, there is one which carries the minimum amount of additional information.

The extension problem can be formulated in a more general context. If S is any subset of a group G, we say that f is positive definite on $S^{-1}S = \{y^{-1}x : x, y \in S\}$ if the inequality (4.1) holds whenever $\xi_i \in C$ and $x_i \in S$.

Rudin [94] has shown that the analogue of Krein's Theorem (on the existence of an extension if S is an interval of the real line) is false in higher-dimensional Euclidean spaces. Specifically, if S is an n-dimensional cube in R^n, where $n \geqq 2$, then there is a function which is p.d. on $S^{-1}S$ but has no p.d. extension to all of R^n. However, Rudin [95] has also shown that such extensions *do* exist in R^n if S is a ball instead of a cube and if the functions are radial. As far as the other classical groups are concerned, see Devinatz [25, § 7] for the non-existence of extensions when G is the circle group and Rudin [94] for the existence of extensions when $G = Z$ and non-existence when $G = Z^n$ and $n \geqq 2$.

For more general groups the interest lies in the case where S is a subgroup of G (and so, $S^{-1}S = S$). Actually it is not hard to see that any function which is p.d. on a subgroup G_0 can be extended so as to be p.d. on G simply by defining it to be zero outside of G_0. However the real interest lies in continuous p.d. extensions. Hewitt and Ross [49, p. 364] have shown that if G_0 is closed and G is either compact or LCA, then the problem of extending continuous p.d. functions from G_0 to G always has a solution. McMullen [71] has proved the same result for the case where G is locally compact and G_0 is compact. McMullen's monograph may be consulted for further results in this direction.

6. Integrally positive definite functions. The early workers in the field realized that for many purposes it is convenient to replace the inequality (1.1) by its integral analogue

$$\int_{-\infty}^{\infty} \int_{-\infty}^{\infty} f(x-y)\phi(x)\overline{\phi(y)}\, dx dy \geqq 0, \tag{6.1}$$

where the function ϕ ranges over L^1 or C_c (the continuous functions with compact support). Indeed if f is continuous, then (1.1) is equivalent to (6.1). (See Bochner [11] and also § 8.)

In view of this situation, Cooper [23] proposed the following definition in 1960. Given a set F of complex-valued functions on R, he called a function f *positive definite for* F if the integral in (6.1) exists as a Lebesgue integral and is non-negative for every $\phi \in F$. Let us denote by $P(F)$ the class of all functions which are p.d. for F. Clearly $F_1 \subset F_2$ implies that $P(F_1) \supset P(F_2)$. It turns out that $P(L^1)$ is identical, up to sets of measure zero, with the class of ordinary continuous p.d. functions. However, $P(C_c)$ is a much more extensive class of functions; ordinary p.d. functions are necessarily bounded, whereas the functions in $P(C_c)$ may be unbounded.

Cooper showed that $P(C_c) = P(L_c^p)$ for every $p \geqq 2$, where L_c^p is the set of functions in L^p with compact support. Furthermore, if $1 \leqq p \leqq 2$ and $q = p/2(p-1)$, then any function in $P(L_c^2)$ which is locally in L^q is in $P(L_c^p)$. (The converse is false; see Stewart [108].) These results indicate that for all practical purposes the gamut of p.d. functions is spanned by the classes $P(L_c^p)$, where $1 \leqq p \leqq 2$; as p increases from 1 to 2, $P(L_c^p)$ increases from the smallest class of p.d. functions to the largest such class. (However the examples at the end of this section and in § 9 show that $P(F)$ can be larger when F is *substantially* smaller than the usual function classes C_c, C_c^∞, L_c^p, etc.)

Cooper's principal result is a generalization of Bochner's Theorem for these integrally p.d. functions: if $f \in P(C_c)$, then there is a non-decreasing function V, not necessarily bounded, such that the equation $f(x) = \int e^{i\alpha x}\, dV(\alpha)$ holds in the sense of Cesàro summability almost everywhere. Furthermore the function V must satisfy $V(\alpha) = o(\alpha)$ as $\alpha \to \pm\infty$. For example, if we take $V(\alpha) = \sqrt{\alpha}$ for $\alpha \geqq 0$, and $V(\alpha) = 0$ for $\alpha < 0$, then

$$f(x) = \int_{-\infty}^{\infty} e^{i\alpha x}\, dV(\alpha) = \begin{cases} \dfrac{1+i}{2} \sqrt{\dfrac{\pi}{2x}}, & \text{for } x > 0 \\[2ex] \dfrac{1-i}{2} \sqrt{\dfrac{\pi}{-2x}}, & \text{for } x < 0 \end{cases}$$

is p.d. for C_c, but it is not an ordinary p.d. function, since it is unbounded.

Functions of this wider class $P(C_c)$ had been studied earlier under certain restrictions. Weil [118] had considered those functions in $P(C_c)$ which belong to L^p for some $p \geqq 1$, whereas Cooper [22] had investigated the functions in $P(C_c)$ which satisfy the condition $\int_0^h |f(x)|\, dx = 0(|h|^{\alpha})$ as $h \to 0$, where $0 \leqq \alpha \leqq 1$.

The notion of integrally p.d. functions makes sense on any locally compact group G. If integration with respect to left-invariant Haar measure on G is denoted by dx, then $P(F)$ consists of those f for which $\int\int f(y^{-1}x)\phi(x)\overline{\phi(y)}\, dxdy$ exists and is non-negative for every $\phi \in F$. In recent papers by Hewitt and Ross [48], Edwards [28], and Rickert [89], constructions have been given on non-discrete locally compact groups for functions in $P(C_c)$ which are not in L^{∞} and, therefore, not almost everywhere equal to the ordinary continuous p.d. functions. A Bochner-type theorem, which generalizes Cooper's Theorem on the one hand and the Weil-Povzner-Raikov Theorem on the other hand, was proved by Stewart [107, Thm. 4.2]. Any $f \in P(C_c)$ on a LCAG is the Fourier-Stieltjes transform (in a suitable summability sense) of a positive measure μ, possibly unbounded, on $\hat{G}$. Furthermore, μ must satisfy $\mu(\hat{x} + K) \to 0$ as $\hat{x} \to \infty$, where K is any compact subset of G.

It would be interesting to find integral representation theorems for functions in $P(F)$ which would reflect any kind of symmetry that the class F might possess. For instance, if $G = R$ and E denotes the even functions in C_c, then any even continuous function $f \in P(E)$ is of the form

$$f(x) = \int_0^{\infty} \cos \lambda x \, d\mu_1(\lambda) + \int_0^{\infty} \cosh \lambda x \, d\mu_2(\lambda), \tag{6.2}$$

where μ_1 and μ_2 are positive measures, μ_1 is finite, and μ_2 is such that the second integral converges. (See Gelfand and Vilenkin [39, p. 197] where the result is attributed to Krein.) More generally, if $G = R^n$ and F is symmetric with respect to a group of rotations (or more general linear transformations), is there an analogue of the formula (6.2) for functions in $P(F)$ which conveys the symmetry of F? Partial answers have been given by Nussbaum [81] (for the orthogonal group) and Tang [113].

7. **Distributions.** Schwartz [103] has extended the theory of p.d. functions to distributions. Let C_c^{∞} be the space of infinitely differentiable functions with compact support on R, and give C_c^{∞} the topology usual for the theory of distributions, i.e., $\phi_n \to 0$ in $C_c^{\infty} \Leftrightarrow$ the supports of all the ϕ_n's lie in a common compact set, and ϕ_n and all its derivatives converge uniformly to 0. It T is a distribution, i.e., a continuous linear functional on C_c^{∞}, then T is called positive definite if $T(\phi * \phi^*) \geqq 0$ for all $\phi \in C_c$, where $\phi^*(x) = \overline{\phi(-x)}$ and $\phi * \psi$ denotes convolution: $\phi * \psi(x) = \int_{-\infty}^{\infty} \phi(x-y)\psi(y)\, dy$. In order to see why this definition can be considered as a generalization of p.d. functions, let us

consider the distribution T_f which is associated with any locally integrable function f:

$$T_f(\phi) = \int_{-\infty}^{\infty} f(x)\phi(x)\,dx.$$

If f is p.d., or more generally is in $P(C_c^{\infty})$, then the associated distribution T_f is p.d. according to Schwartz's definition, because

$$\begin{aligned} T_f(\phi * \phi^*) &= \int_{-\infty}^{\infty} f(x)\phi * \phi^*(x)\,dx \\ &= \int_{-\infty}^{\infty} f(x)\,dx \int_{-\infty}^{\infty} \phi(x+y)\overline{\phi(y)}\,dy \\ &= \int_{-\infty}^{\infty}\int_{-\infty}^{\infty} f(x-y)\phi(x)\overline{\phi(y)}\,dxdy \geqq 0. \end{aligned}$$

The analogue of Bochner's Theorem is Schwartz's representation for a p.d. distribution as the Fourier transform of a positive tempered measure μ, i.e., $T(\phi) = \int_{-\infty}^{\infty}\hat{\phi}(x)\,d\mu(x)$ where $\hat{\phi}$ is the Fourier transform of ϕ, and, for some $p \geqq 0$, $\int_{-\infty}^{\infty} d\mu(x)/(1+|x|^2)^p < \infty$. This Bochner-Schwartz Theorem has been extended to distributions on LCA groups by Maurin and Wawrzyńczyk [**116**].

Positive definite distributions have found applications in the theory of generalized random processes. See Gelfand and Vilenkin [**39**, Ch. 3].

8. **Kernels.** We can generalize the notion of a p.d. function by replacing $f(x_i - x_j)$ in (1.1) by $K(x_i, x_j)$. If $K(x, y)$ is any complex-valued function on R^2, we call K a *positive definite kernel* if

$$\sum_{i,j=1}^{n} K(x_i, x_j)\xi_i\bar{\xi}_j \geqq 0 \tag{8.1}$$

holds whenever $x_i \in R$ and $\xi_i \in C$. Although this concept is, of course, more general than that of p.d. functions, it appeared earlier. In fact p.d. kernels, as defined by (8.1), seem to have arisen first in 1909 in a paper by Mercer [**72**] on integral equations, and, although several other authors made use of this concept in the following two decades, none of them explicitly considered kernels of the form $K(x, y) = f(x - y)$, i.e., p.d. functions. Indeed Mathias and Bochner seem not to have been aware of the study of p.d. kernels.

Mercer's work arose from Hilbert's paper of 1904 [**50**] on Fredholm integral equations of the second kind:

(8.2) $$f(s) = \phi(s) - \lambda \int_a^b K(s, t)\phi(t)\, dt.$$

In particular, Hilbert had shown that

(8.3) $$\int_a^b \int_a^b K(s, t)x(s)x(t)\, dsdt = \sum \frac{1}{\lambda_n} \left[\int_a^b \psi_n(s)x(s)\, ds \right]^2,$$

where K is a continuous real symmetric $[K(s, t) = K(t, s]$ kernel, x is continuous, $\{\psi_n\}$ is a complete system of orthonormal eigenfunctions, and the λ_n's are the corresponding eigenvalues of (8.2), i.e., $\psi_n(s) = \lambda_n \int_a^b K(s, t)\psi_n(t)\, dt$. Hilbert defined a "definite" kernel as one for which the double integral $J(x) = \int_a^b \int_a^b K(s, t)x(s)x(t)\, dsdt$ satisfies $J(x) > 0$ except for $x(s) \equiv 0$.

The original object of Mercer's paper [72] was to characterize the kernels which are definite in the sense of Hilbert, but Mercer soon found that the class of such functions was too restrictive to characterize in terms of determinants. He therefore defined a continuous real symmetric kernel $K(s, t)$ to be of *positive type* if $J(x) \geqq 0$ for all real continuous functions x on $[a, b]$, and he proved that (8.1) is a necessary and sufficient condition for a kernel to be of positive type. (In view of (8.3) a necessary and sufficient condition for a kernel to be p.d. is that all its eigenvalues be positive.) Mercer then proved that for any continuous p.d. kernel the expansion

$$K(s, t) = \sum \frac{\psi_n(s)\psi_n(t)}{\lambda_n}$$

holds absolutely and uniformly.

At about the same time, W. H. Young [124], motivated by a different question in the theory of integral equations, showed that for continuous kernels (8.1) is equivalent to $J(x) \geqq 0$ for all x in $L^1[a, b]$. In a later paper, Mercer [73] considered unbounded kernels of positive type, thereby anticipating the integrally p.d. functions considered in § 6.

E. H. Moore [75, 76] initiated the study of a very general kind of p.d. kernel. If E is an abstract set, he called functions $K(x, y)$ defined on $E \times E$ "positive Hermitian matrices" if they satisfy (8.1) for all $x_i \in E$. Moore was interested in a generalization of integral equations and showed that to each such K there is a Hilbert space H of functions such that, for each $f \in H$, $f(y) = (f(x), K(x, y))$. This property is called the *reproducing property* of the kernel and turns out to have importance in the solution of boundary-value problems for elliptic partial differential equations. For an account of reproducing kernels, with further indications of their applications, see Aronszajn [6], especially the historical introduction.

Another line of development in which p.d. kernels played a large role was the theory of harmonics on homogeneous spaces as begun by E. Cartan [18] in 1929 and continued by Weyl [119] and Ito [53]. The most comprehensive theory of p.d. kernels on homogeneous spaces is that of Krein [60] which includes as special cases not only the work of the above three authors but also the work of Gelfand and Raikov (described in § 4) on p.d. functions and irreducible unitary representations of locally compact groups. Krein was able to represent a broad class of kernels on homogeneous spaces in the form $K(x, y) = \int_T Z(x, y; t)\, d\sigma(t)$, where σ is bounded and positive on a certain space T, and each $Z(x, y; t)$ is a "zonal" kernel (analogous to the elementary p.d. functions). As special cases of this very comprehensive representation we can mention the Weil-Povzner-Raikov Theorem and Schoenberg's results described in § 10, specifically equations (10.1) and (10.2). See Hewitt [47; pp. 145–149] for an exposition of this work of Krein.

In probability theory p.d. kernels arise as covariance kernels of stochastic processes. (See Loève [66, p. 466].)

9. **Functions with a finite number of negative squares.** In this section we discuss a generalization of p.d. functions which is due to M. Krein. Although certain special cases had been dealt with earlier, it was in 1959 [61] that Krein formulated the concept of functions with k negative squares. These are the complex-valued functions which are Hermitian, in the sense that $\overline{f((-x)} = f(x)$, and such that the form (1.1) has at most k negative squares (when reduced to diagonal form) for every choice of n and $x_1, \cdots, x_n \in R$, and at least one of these forms has exactly k negative squares. In other words the matrix (1.2) has at most k negative eigenvalues no matter how the x_i's are chosen, and has exactly k negative eigenvalues for some choice of the x_i's. For example, the function $f(x) = \cosh x$ has one negative square because

$$\sum \cosh(x_i - x_j)\xi_i\bar{\xi}_j = |\sum \xi_i \cosh x_i|^2 - |\sum \xi_i \sinh x_i|^2.$$

Krein proved that if f is a continuous function with k negative squares, then there is a positive measure μ and a polynomial Q of degree k such that

$$f(x) = h(x) + \int_{-\infty}^{\infty} \frac{e^{i\lambda x} - S(x, \lambda)}{|Q(\lambda)|^2}\, d\mu(\lambda), \tag{9.1}$$

where h is a solution of the differential equation

$$\bar{Q}\left(-i\frac{d}{dx}\right)Q\left(-i\frac{d}{dx}\right)h(x) = 0, \; (\bar{Q}(\lambda) = Q(\bar{\lambda})),$$

and $S(x, \lambda)$ is a regularizing correction (compensating for the real zeros of Q). Notice that in the case $k = 0$ the definition reduces to that of a p.d. function, and Krein's integral representation (9.1) becomes Bochner's Theorem. The proof makes use of Pontryagin's Theorem on invariant subspaces associated with self-adjoint operators in spaces with indefinite scalar product.

Krein had earlier given an integral representation for continuous functions with one negative square in connection with the problem of the continuation of screw lines in infinite-dimensional Lobachevski space. See [59] or [52, Theorem 6.2] for a simplified version of (9.1) in the case $k = 1$. The generalization of (9.1) to functions of several variables appears in Gorbachuk [43].

Iohvidov and Krein [52, Theorem 5.2] proved an integral representation analogous to (9.1) for sequences with k negative squares which reduces to the theorem of Herglotz when $k = 0$.

Functions with a finite number of negative squares have applications to probability theory, in particular to infinitely divisible distribution laws and, therefore, to stochastic processes with stationary increments. A random variable is said to be infinitely divisible if for every positive integer n it can be expressed as the sum of n independent and identically distributed random variables. We shall show that the logarithm $f(x)$ of the characteristic function $g(x)$ of such a random variable has at most one negative square. For any n, $g(x) = [g_n(x)]^n$, where g_n is a characteristic function, and hence is p.d., i.e., $\exp[f(x)/n]$ is p.d. But the product of p.d. functions is p.d., and so $\exp[(m/n)f(x)]$ is p.d. Since the limit of p.d. functions is p.d., it follows that $\exp[tf(x)]$ is p.d. whenever $t \geqq 0$. Thus for any $x_1, \cdots, x_n \in R$ and $t \geqq 0$, we have

$$\begin{aligned} 0 &\leqq \sum_{i,j=1} \exp[tf(x_i - x_j)]\,\xi_i\bar{\xi}_j \\ &= |\sum \xi_i|^2 + t\sum f(x_i - x_j)\xi_i\bar{\xi}_j \\ &\quad + \frac{t^2}{2}\sum \exp[\theta_{ij}tf(x_i - x_j)]\,f^2(x_i - x_j)\xi_i\bar{\xi}_j, \end{aligned}$$

where $0 < \theta_{ij} < 1$. By letting $t \to 0$, we see from this that $\sum f(x_i - x_j)\,\xi_i\bar{\xi}_j \geqq 0$ whenever $\sum \xi_i = 0$, and so f has at most one negative square. If we apply Krein's formula (9.1) with $k = 1$, we obtain Kolmogorov's formula

$$\text{(9.2)} \qquad \log g(x) = i\gamma x + \int_{-\infty}^{\infty} \frac{e^{i\lambda x} - 1 - i\lambda x}{\lambda^2}\, d\mu(\lambda)$$

for the logarithm of the characteristic function of an infinitely divisible distribution with finite variance. Formula (9.2) was first proved by Kolmogorov [56] in 1932, but, of course, not by the above method.

One can attempt to generalize the concept of functions with a finite number of negative squares in the same manner as described in §§ 4, 6, 7, and 8 for p.d. functions. There is no problem in formulating the definition and elementary properties of such functions on general groups (see [109]), but no one seems to have found an analogue of Krein's integral representation for groups. However the results of §§ 6, 7, 8 do have analogues. Shah Tao-Shing [105] has characterized distributions with k negative squares by a formula which generalizes both Krein's formula (9.1) and the Bochner-Schwartz Theorem. (Such distributions are useful in generalized random processes. See [39, chapters 2, 3].) Gorbachuk [83, 43] has generalized Krein's work by considering kernels $K(x, y)$ with k negative squares. Stewart [109] has enlarged Krein's class of functions with k negative squares and generalized (9.1) in a sense similar to that in which Cooper's results (§ 6) on integrally p.d. functions extend those of Bochner.

In this connection we mention that although functions with k negative squares are clearly a *generalization* of p.d. functions, in another sense they can be regarded as a *special case* of p.d. functions. The explanation of this paradox lies in the fact that if f has k negative squares, then it satisfies the inequality

$$\iint f(x-y)\bar{Q}\left(i\frac{d}{dx}\right)\phi(x)\overline{\bar{Q}\left(i\frac{d}{dx}\right)\phi(y)}\, dxdy \geqq 0$$

for every $\phi \in C_c^{\infty}$, the infinitely differentiable functions with compact support, where Q is the polynomial in (9.1), and so, in the notation of § 6, $f \in P(F)$, where $F = \bar{Q}(id/dx)C_c^{\infty}$.

10. Metric spaces. In 1938 Schoenberg published a certain generalization of real p.d. functions which arose from the problem of isometrically embedding metric spaces in Hilbert space [97, 98, 99] and the related problem of determining the screw lines and screw functions of Hilbert space [80].

Schoenberg calls a set S a *quasi-metric space* if it has a distance function d with the following two properties: (i) $d(P, P') = d(P', P) \geqq 0$, (ii) $d(P, P) = 0$, where $P, P' \in$ S. A real, continuous, even function f, defined in the range of values of $\pm\, d(P, P')$, is called *positive definite in* S if for any $P_i \in$ S and $\xi_i \in R$, we have $\sum_{i,j=1}^{n} f(d(P_i, P_j))\xi_i\xi_j \geqq 0$.

(Notice that, for $S = R$, this agrees with the definition of the ordinary continuous real p.d. functions.) For example, the equation

$$\exp\left(-\sum_1^m x_j^2\right) = \int \cdots \int \exp\left(-\tfrac{1}{4}\sum_1^m u_j^2\right) \exp\left(i \sum_1^m x_j u_j\right) du_1 \cdots du_m$$

shows that the function $f(t) = e^{-t^2}$ is p.d. in R^m, and, hence, in a real Hilbert space. One connection between these p.d. functions and the embedding problem is Schoenberg's result [97] that a separable quasi-metric space is isometrically embeddable in a real Hilbert space if and only if the function $f(t) = e^{-\lambda t^2}$ is p.d. in S for every $\lambda > 0$.

Schoenberg has proved a number of interesting integral representation theorems for such functions. For example, in [98] any function p.d. in R^m was shown to be of the form

(10.1) $$f(t) = \int_0^\infty \Omega_m(tu)\, d\alpha(u),$$

where α is bounded and non-decreasing, and Ω_m is essentially a Bessel function:

$$\Omega_m(t) = \Gamma\left(\frac{m}{2}\right) \left(\frac{2}{t}\right)^{(1/2)(m-2)} J_{(1/2)(m-2)}(t).$$

In particular $\Omega_1(t) = \cos t$, $\Omega_2(t) = J_0(t)$, $\Omega_3(t) = (\sin t)/t$, and so (10.1) reduces to Bochner's Theorem for real functions when $m = 1$. By letting $m \to \infty$ in (10.1), the representation

(10.2) $$f(t) = \int_0^\infty e^{-t^2u^2}\, d\alpha(u)$$

is deduced for functions p.d. in Hilbert space.

Let S_m denote the unit sphere in R^{m+1} and $d(P, P')$ denote spherical distance. Schoenberg's representation for functions p.d. in S_2 [99] is $f(t) = \sum_{n=0}^{\infty} a_n P_n(\cos t)$, where $a_n \geqq 0$, $\sum a_n < \infty$, and P_n is a Legendre polynomial. A similar formula, in terms of ultraspherical polynomials, holds for functions p.d. in S_m and was extended by Bochner [14] from spheres to compact spaces with transitive groups of transformations using the generalized spherical harmonics of Cartan and Weyl. Again, letting $m \to \infty$ gives the general form of functions p.d. in the unit sphere in Hilbert space: $f(t) = \sum_{n=0}^{\infty} a_n \cos^n t$, where

$a_n \geqq 0$, $\sum a_n < \infty$. For recent extensions of Schoenberg's work, see Bingham [10] and the references therein.

Finally we note Einhorn's discovery [29, 30] of the extreme paucity of functions p.d. in $C[0, 1]$, the space of continuous functions on $[0, 1]$ with the supremum metric. The only such functions are the positive constants.

11. **Other generalizations.** A complete list of the current generalizations of p.d. functions would be very long and exhibit great variety. In this final section we give brief mention to some of the more interesting ideas in this list which have not already been covered.

(i) Krein [58] and Berezanski [8] have constructed a theory which generalizes Bochner's in that $e^{i\alpha x}$ is replaced by eigenfunctions of differential (and more general) operators. Their integral representation theorem includes as special cases not only Bochner's Theorem but also Bernstein's Theorem on the representation of completely monotone functions as Laplace-Stieltjes integrals.

(ii) Positive functionals on Banach algebras with involution (see, e.g., Loomis [67, p. 96]) can be considered as a generalization of p.d. functions in view of the 1-1 correspondence established by Gelfand and Raikov [38] between P and the positive functionals on the group algebra $L^1(G)$. Any functional of the form $L(\phi) = \int f(x)\phi(x)\,dx$, where $f \in P$, and dx denotes integration with respect to Haar measure, is a positive functional (cf. § 7), and, conversely, any positive functional on $L^1(G)$ is of this form. Furthermore there is a version of the Weil-Povzner-Raikov Theorem which is valid for positive functionals on certain algebras. See [67, Thm. 26I] and Lumer [68].

(iii) There is a natural generalization from p.d. functions to p.d. measures. The inequality (6.1) can be rewritten as $\int \phi * \phi^*(x) f(x)\,dx \geqq 0$, for all $\phi \in C_c$ (cf. § 7). In view of this, a measure is said to be positive definite if $\int \phi * \phi^*(x)\,d\mu(x) \geqq 0$ for all $\phi \in C_c$. Such measures have been studied by Godement [42] and Argabright and Gil de Lamadrid [5].

(iv) So far all the generalizations that we have discussed have been numerical-valued functions, but several authors have considered functions with more general range. Falb and Haussmann [32] have given a representation like Bochner's for p.d. functions with values in a Banach space. Of particular interest are p.d. functions on a group whose values are operators in a Hilbert space, and such functions have been studied by Naimark [77] and Sz.-Nagy [112]. There are two possible ways of defining such functions. For the connection between the definitions and further references see the exposition in § 2 of

Berberian [7]. In the appendix to [92], Sz.-Nagy considers p.d. operator-valued functions on semigroups with involution and gives applications to contraction operators and extensions of operators.

References

1. N. Akhiezer, *The Classical Moment Problem and some Related Questions in Analysis*, Hafner, New York, 1965.

2. N. Akhiezer and M. Krein, *Some Questions in the Theory of Moments*, Amer. Math. Soc. Translations of Math. Monographs, Vol. 2, Providence, R.I., 1962.

3. E. J. Akutowicz, *On extrapolating a positive definite function from a finite interval*, Math. Scand. 7 (1959), 157–169.

4. W. Ambrose, *Spectral resolution of groups of unitary operators*, Duke Math J. **11**(1944), 589–595.

5. L. Argabright and J. Gil de Lamadrid, *Fourier analysis of unbounded measures on locally compact abelian groups*, Memoirs Amer. Math. Soc. **145**, Providence, R.I., 1974.

6. N. Aronszajn, *Theory of reproducing kernels*, Tran. Amer. Math. Soc. **68** (1950), 337–404.

7. S. K. Berberian, *Naimark's moment theorem*, Michigan Math. J. **13**(1966), 171–184.

8. Yu. M. Berezanski, *A generalization of a multidimensional theorem of Bochner*, Soviet Math. Dokl. **2**(1961), 143–147.

9. M. S. Bingham and K. R. Parthasarathy, *A probabilistic proof of Bochner's Theorem on positive definite functions*, J. London Math. Soc. **43**(1968), 626–632.

10. N. H. Bingham, *Positive definite functions on spheres*, Proc. Camb. Phil. Soc. **73**(1973), 145–156.

11. S. Bochner, *Vorlesungen über Fouriersche Integrale*, Akademische Verlagsgesellschaft, Leipzig, 1932.

12. ——, *Monotone Funktionen*, Stieltjessche Integrale und harmonische Analyse, Math. Ann. **108**(1933), 378–410. [References 11 and 12 were translated by M. Tenenbaum and H. Pollard as: *Lectures on Fourier Integrals* (with the author's supplement on Monotonic Functions, Stieltjes Integrals and Harmonic Analysis), Annals of Mathematics Studies No. 42, Princeton Univ. Press, 1959.]

13. ——, *Spektralzerlegung linearer Scharen unitärer Operatoren*, Sitzungsberichte Preuss. Akad. Wiss. phys.-math. Kl., (1933), 371–376.

14. ——, *Hilbert distances and positive definite functions*, Ann. of Math. **42**(1941), 647–656.

15. R. S. Bucy and G. Maltese, *Extreme positive definite functions and Choquet's representation theorem*, J. Math. Anal. Appl. **12**(1965), 371–377.

16. C. Carathéodory, *Über den Variabilitätsbereich der Koeffizienten von Potenzreihen*, die gegebene Werte nicht annehmen, Math. Ann. **64**(1907), 95–115.

17. ——, *Über den Variabilitätsbereich der Fourierschen Konstanten von positiven harmonischen Funktionen*, Rend. Circ. Mat. Palermo 32(1911), 193–217.

18. E. Cartan, *Sur la détermination d'un système orthogonal complet dans un espace de Riemann symétrique clos*, Rend. Circ. Mat. Palermo 53(1929), 217–252.

19. H. Cartan and R. Godement, *Théorie de la dualité et analyse harmonique dans les groupes abéliens localement compacts*, Ann. Sci. Ecole Norm. Sup. (3), **64**(1947), 79–99.

20. G. Choquet, *Deux exemples classiques de représentation intégrale,* Enseignement Math. (2), **15**(1969), 63–75.

21. J. Chover, *On normalized entropy and the extensions of a positive-definite function,* J. Math. Mech. **10**(1961), 927–945.

22. J. L. B. Cooper, *Fourier-Stieltjes integrals,* Proc. London Math. Soc. (2), **51**(1950), 265–284.

23. ——, *Positive definite functions of a real variable,* Proc. London Math. Soc. (3), **10**(1960), 53–66.

24. M. Crum, *On positive-definite functions,* Proc. London Math. Soc. (3), **6**(1956), 548–560.

25. A. Devinatz, *On the extensions of positive definite functions,* Acta Math. **102**(1959), 109–134.

26. ——, *On measurable positive definite operator functions,* J. London Math. Soc. 35(1960), 417–424.

27. R. E. Edwards, *Fourier Series: a Modern Introduction,* 2 vols., Holt, Rinehart and Winston, New York, 1967.

28. ——, *Unbounded integrally postitive definite functions,* Studia Math. 33(1969), 185–191.

29. S. J. Einhorn, *Functions positive definite in the space C,* Amer. Math. Monthly 75(1968), 393.

30. ——, *Functions positive definite in* $C[0, 1]$, Proc. Amer. Math. Soc. 22(1969), 702–703.

31. C. G. Esseen, *Fourier analysis of distribution functions, Acta Math.* 77 (1945), 1–125.

32. P. L. Falb and U. Haussmann, *Bochner's Theorem in infinite dimensions,* Pacific J. Math. 43(1972), 601–618.

33. K. Fan, *On positive definite sequences,* Ann. of Math. 47(1946), 593–607.

34. ——, *Les Fonctions Définies-Positives et les Fonctions Complètement Monotones,* Gauthier-Villars, Paris, 1950.

35. E. Fisher, *Über das Carathéodory'sche Problem,* Potenzreihen mit positivem reelen Teil betreffend, Rend. Circ. Mat. Palermo 32(1911), 240–256.

36. J. Fourier, *Théorie Analytique de la Chaleur,* Discours Préliminaire, Didot, Paris, 1822, p. xxiii.

37. G. Frobenius, *Ableitung eines Satzes von Carathéodory aus einer Formel von Kronecker,* Sitzungsberichte Preuss. Akad. Wiss., (1912), 4–15.

38. I. M. Gelfand and D. A. Raikov, *Irreducible unitary representations of locally bicompact groups* (in Russian), Mat. Sbornik N.S., 13(1943), 301–316. English translation: Amer. Math. Soc. Translations Ser. 2, 36(1964), 1–15.

39. I. M. Gelfand and N. Vilenkin, *Generalized Functions,* Vol. 4, Applications of Harmonic Analysis, Academic Press, New York, 1964.

40. B. V. Gnedenko, *Sur une propriété caractéristique des lois infiniment divisibles,* Bull. Univ. d'Etat Moscou, Ser. Internat. Sec. A, 1, no. 5, (1937), 9–15.

41. R. Godement, *Sur une généralisation d'un théorème de Stone,* C. R. Acad. Sci. Paris, 218(1944), 901–903.

42. ——, *Les fonctions de type positif et la théorie des groupes,* Trans. Amer. Math. Soc. 83(1948), 1–84.

43. V. Gorbachuk, *On integral representations of Hermitian-indefinite forms (the case of several variables)* (in Russian), Ukrain, Mat. Ž. 16(1964), 232–236.

44. U. Grenander and G. Szegö, *Toeplitz Forms and their Applications,* Univ. of Calif. Press, Berkeley, 1958.

45. L. Gross, *Harmonic Analysis on Hilbert Space,* Memoirs Amer. Math. Soc. 46, Providence, R.I., 1963.

46. G. Herglotz, *Über Potenzreihen mit positivem, reelen Teil im Einheitskreis,* Leipziger Berichte, math.-phys. Kl. 63(1911), 501–511.

47. E. Hewitt, *A survey of abstract harmonic analysis, Some Aspects of Analysis and Probability,* pp. 101–168. Surveys in Applied Mathematics, Vol. 4, Wiley, New York, 1958.

48. E. Hewitt and K. A. Ross, *Integrally positive-definite functions on groups,* Studia Math. 31(1968), 145–151.

49. ——, *Abstract Harmonic Analysis* Vol. II, Springer-Verlag, New York, 1970.

50. D. Hilbert, *Grundzüge einer allgemeinen Theorie der linearen Integralgleichungen* I, Gött. Nachrichten, math.-phys. Kl. (1904), 49–91.

51. E. W. Hobson, *Presidential address to Brit. Ass. for the Advancement of Science,* Sec. A, Nature 84(1910), 284–291.

52. I. Iohvidov and M. Krein, *Spectral theory of operators in spaces with an indefinite metric* II (in Russian), Trudy Moskov. Mat. Obšč. 8(1959), 413–496. English translation: Amer. Math. Soc. Translations Ser. 2, 34(1963), 283–373.

53. S. Ito, *Positive definite functions on homogeneous spaces,* Proc. Japan Acad. 26(1950), 17–28.

54. Y. Katznelson, *An Introduction to Harmonic Analysis,* Wiley, New York, 1968.

55. A. Khintchine, *Korrelationstheorie der stationaren stochastischen Prozesse,* Math. Ann. 109(1934), 604–615.

56. A. Kolmogorov, *Sulla forma generale di un processo stocastico omogeneo,* Atti Acad. Naz. Lincei Rend. Cl. Sci. Fis. Mat. Nat. (6), 15(1932), 805–808, 866–869.

57. M. Krein, *Sur le problème du prolongement des fonctions hermitiennes positives et continues,* Dokl. Akad. Nauk SSSR 26(1940), 17–22.

58. ——, *On a general method for decomposing positive-definite kernels into elementary products* (in Russian), Dokl. Akad. Nauk SSSR 53(1946), 3–6.

59. ——, *Screw lines in infinite-dimensional Lobachevski space and the Lorentz transformation* (in Russian), Ups. Mat. Nauk 3(1948), 158–160.

60. ——, *Hermitian-positive kernels on homogeneous spaces* I *and* II (in Russian), Ukrain. Mat. Ž. 1(1949), 64–98 and 2(1950), 10–59. English translation: Amer. Math. Soc. Translations Ser. 2, 34(1963), 69–164.

61. ——, *Integral representation of a continuous Hermitian-indefinite function with a finite number of negative squares* (in Russian), Dokl. Akad. Nauk SSSR 125(1959), 31–34.

62. K. de Leeuw and I. Glicksberg, *The decomposition of certain group representations,* J. Analyse Math. 15(1965), 135–192.

63. P. Lévy, *Calcul des Probabilités,* Gauthier-Villars, Paris, 1925.

64. ——, *Théorie de l'Addition des Variables Aléatoires,* Gauthier-Villars, Paris, 1937.

65. ——, *Quelques problèmes non résolus de la théorie des fonctions caractéristiques,* Annali Mat. Pura e Appl. (4), 53(1961), 315–332.

66. M. Loève, *Probability Theory,* 2nd ed., Van Nostrand, Princeton, N.J., 1960.

67. L. Loomis, *An Introduction to Abstract Harmonic Analysis,* Van Nostrand, Princeton, N.J., 1953.

68. G. Lumer, *Bochner's Theorem, states, and the Fourier transform of measures*, Studia Math. 46(1973), 135–140.

69. M. Mathias, *Über positive Fourier-Integrale*, Math. Zeit, 16(1923), 103–125.

70. K. McKennon, *Multipliers, Positive Functionals, Positive-Definite Functions, and Fourier-Stieltjes Transforms*, Memoirs Amer. Math. Soc. **111**, Providence, R.I., 1971.

71. J. McMullen, *Extensions of Positive-Definite Functions*, Memoirs Amer. Math. Soc. 117, Providence, R.I., 1972.

72. J. Mercer, *Functions of positive and negative type and their connection with the theory of integral equations*, Philos. Trans. Roy. Soc. London, Ser. A, 209(1909), 415–446.

73. ——, *Linear transformations and functions of positive type*, Proc. Roy. Soc. London, Ser. A, 99(1921), 19–38.

74. R. A. Minlos, *Generalized random processes and their extension to a measure* (in Russian), Trudy Moskov. Mat. Obšč. 8(1959), 497–518. English translation: Selected Translations in Math. Stat. and Prob. 3(1963), 291–313.

75. E. H. Moore, *On properly positive Hermitian matrices*, Bull. Amer. Math. Soc. 23(1916), 59, 66–67.

76. ——, *General Analysis, Part* I, Memoirs Amer. Philos. Soc. 1, Philadelphia, 1935.

77. M. A. Naimark, *Positive definite operator functions on a commutative group* (in Russian, English summary), Izv. Akad. Nauk SSSR, Ser. Mat. 7(1943), 237–244.

78. M. Nakamura and H. Umegaki, *A remark on theorems of Stone and Bochner*, Proc. Japan Acad. 27(1951), 506–507.

79. H. Nakano, *Reduction of Bochner's Theorem to Stone's Theorem*, Ann. of Math. (2), 49(1948), 279–280.

80. J. Von Neumann and I. J. Schoenberg, *Fourier integrals and metric geometry*, Trans Amer. Math. Soc. 50(1941), 226–251.

81. A. E. Nussbaum, *Integral representation of functions and distributions* positive definite relative to the orthogonal group, Trans. Amer. Math. Soc. **175** (1973), 355–387.

82. R. Phelps, *Lectures on Choquet's Theorem*, Van Nostrand, Princeton, N.J., 1966.

83. V. Plyuschcheva (Gorbachuk), *On integral representations of continuous Hermitian-indefinite forms* (in Russian), Dokl. Akad. Nauk SSSR, 145(1962), 534–537.

84. G. Pólya, *Remarks on characteristic functions*, Proc. First Berkeley Conf. on Math. Stat. and Prob., pp. 115–123, Univ. of Calif. Press, Berkeley, 1949.

85. A. Povzner, *Über positive Funktionen auf einer Abelschen Gruppe*, Dokl. Akad. Nauk SSSR 28(1940), 294–295.

86. D. A. Raikov, *Positive definite functions on commutative groups with an invariant measure*, Dokl. Adad. Nauk SSSR 28(1940), 296–300.

87. ——, *Sur les fonctions positivement définies*, Dokl. Akad. Nauk SSSR 26(1940), 860–865.

88. ——, *On various types of convergence of positive definite functions* (in Russian), Dokl. Akad. Nauk SSSR 58(1947), 1279–1282.

89. N. W. Rickert, *A note on positive definite functions*, Studia Math. 36 (1970), 223–225.

90. F. Riesz, *Sur certains systèmes singuliers d'équations intégrales*, Ann. Sci. Ecole Norm. Sup. 28(1911), 33–62.
91. ——, *Über Sätze von Stone und Bochner*, Acta Sci. Math. Szeged 6(1933), 184–198.
92. F. Riesz and B. Sz.-Nagy, *Functional Analysis*, Ungar, New York, 1955.
93. W. Rudin, *Fourier Analysis on Groups*, Interscience New York, 1962.
94. ——, *The extension problem for positive definite functions*, Ill. J. Math. 7(1963), 532–539.
95. ——, *An extension theorem for positive-definite functions*, Duke Math. J. 37(1970), 49–53.
96. V. Sazonov, *A note on characteristic functionals*, Theory Prob. and Appl. 3(1958), 188–192.
97. I. J. Schoenberg, *Metric spaces and positive definite functions*, Trans. Amer. Math. Soc. 44(1938), 522–536.
98. ——, *Metric spaces and completely monotone functions*, Ann. of Math. 39(1938), 811–841.
99. ——, *Positive definite functions on spheres*, Duke Math. J. 9(1942), 96–108.
100. B. M. Schreiber, T. C. Sun and A. T. Barucha-Reid, *Algebraic models for probability measures associated with stochastic processes*, Trans. Amer. Math. Soc. 158(1971), 93–105.
101. I. Schur, *Bemerkungen zur Theorie der beschränkten Bilinearformen mit unendlich vielen Veränderlichen*, J. Reine Angew. Math. 140(1911), 1–28.
102. ——, *Über einen Satz von C. Carathéodory*, Sitzungsberichte Preuss. Akad. Wiss. (1912), 4–15.
103. L. Schwartz, *Théorie des Distributions*, Vol. 2, Actualités Sci. Indust., no. 1122, Hermann, Paris, 1951.
104. I. Segal and J. von Neumann, *A theorem on the representation of semi-simple Lie groups*, Ann. of Math. (2), 52(1950), 509–517.
105. T.-S. Shah, *On conditionally positive-definite generalized functions*, Scientia Sinica 11(1962), 1147–1168.
106. ——, *Positive definite functions on commutative topological groups*, Scientia Sinica 14(1965), 653–665.
107. J. Stewart, *Unbounded positive definite functions*, Canad. J. Math. 21 (1969), 1309–1318.
108. ——, *Counterexample to a conjecture on positive definite functions*, Canad. J. Math. 24(1972), 926–929.
109. ——, *Functions with a finite number of negative squares*, Canad. Math. Bull. 15(1972), 399–410.
110. M. Stone, *Linear transformations in Hilbert space* III, Proc. Nat. Acad. Sci. U.S.A. 16(1930), 172–175.
111. ——, *On one-parameter unitary groups in Hilbert space*, Ann. of Math. (2), 33(1932), 643–648.
112. B. Sz.-Nagy, *Trnasformations de l'espace de Hilbert, fonctions de type positif sur un groupe*, Acta Sci. Math. Szeged 15 (1954), 104–114.
113. L. M. Tang, *Symmetrically positive definite functions*, M.Sc. Thesis, McMaster University, Hamilton, Ontario, 1974.
114. E. C. Titchmarsh, *Introduction to the Theory of Fourier Integrals*, Oxford University Press, 1937.
115. O. Toeplitz, *Über die Fourier'sche Entwickelung positiver Funktionen*, Rend. Circ. Mat. Palermo 32(1911), 191–192.

116. A. Wawrzyńczyk, *On tempered distributions and Bochner-Schwartz Theorem on arbitrary locally compact abelian groups*, Colloq. Math. 19(1968), 305–318.

117. W. von Waldenfels, *Positiv definite Funktionen auf einem unendlich dimensionalen Vektorraum*, Studia Math., 30(1968), 153–162.

118. A. Weil, *L'Intégration dans les Groupes Topologiques et ses Applications*, Actualités Sci. Indust., no. 869, Hermann, Paris, 1940.

119. H. Weyl, *Harmonics on homogeneous manifolds*, Ann. of Math. (2), 35 (1934), 486–499.

120. N. Wiener, *Generalized harmonic analysis*, Acta Math. 55(1930), 117–258.

121. ——, *The Fourier Integral and Certain of its Applications*, Cambridge Univ. Press, 1933.

122. A. M. Yaglom, *Positive-definite functions and homogeneous random fields on groups and homogeneous spaces*, Societ Math. Dokl. 1(1960), 1402–1405.

123. H. Yoshizawa, *On some types of convergence of positive definite functions*, Osaka Math. J. 1(1949), 90–94.

124. W. H. Young, *A note on a class of symmetric functions and on a theorem required in the theory of integral equations*, Philos. Trans. Roy. Soc. London, Ser. A, 209(1909), 415–446.

125. A. Zygmund, *Trigonometrical Series*, Monografje Matematyczne, Warsaw, 1935.

Department of Mathematics, McMaster University, Hamilton, Ontario.

Part II

GENERAL STATISTICAL APPLICATIONS

Editor's Comments on Papers 3 Through 7

In the late 1940s, Loève [1, Sect. 12] established the first link between reproducing kernels and stochastic processes. He recognized that the covariance function of a second-order stochastic process is a reproducing kernel and vice versa. He also stated the basic congruence relation between the Hilbert space of random variables spanned by a stochastic process and the RKHS determined by the covariance of the process. Loève's results were based on Aronszajn's 1943 paper [2] and can be read in English in *Probability Theory II* [3, p. 156].

Matters remained at this stage until Parzen, in the late 1950s, applied Loève's results to problems in statistical signal processing—in particular to estimation (including regression) and detection problems. Parzen clearly demonstrated that a RKHS provides an elegant, general, unified framework for least-squares estimation of random variables, minimum variance unbiased estimation of regression coefficients, and detection of known signals in Gaussian noise. In fact, solutions to all these problems can be expressed in terms of an appropriate RKHS inner product.

The original source for most of Parzen's results is a Stanford University technical report [4]. In the period 1961–1963 Parzen also published several papers that cover more or less the same ground.

Paper 3 is included here as representative of this group; it and the others [4–7] have been collected in *Time Series Analysis Papers* [8].

Meanwhile, in Czechoslovakia, Hájek was proceeding along much the same lines, unaware of the work of Parzen, Loève, and even Aronszajn. In a remarkable article presented here as Paper 4, Hájek establishes the basic congruence relation and shows that the estimation and detection problems mentioned previously can be solved by inverting the basic congruence map. Detailed procedures are given for carrying out this inversion for stationary processes with rational spectral densities. The reader will recognize in Hájek's Section 2 the reproducing property and other aspects of RKHS theory, even though Aronszajn's terminology is not used.

In Paper 5, Parzen surveys a wide range of RKHS applications in statistical signal processing and stochastic process theory. Of particular interest are Sections 3 and 7 in which the structural equivalences among problems in estimation, control, and approximation are discussed. These equivalences have been developed further since 1970 and will be treated in greater detail later in this work, along with many of the other applications Parzen reviews.

Since solutions to the statistical problems discussed in Papers 3, 4, and 5 can be expressed in terms of RKHS inner products, general methods are needed for the evaluation of the inner product in the RKHS determined by a given covariance function (reproducing kernel). Two papers that address this question conclude Part II. Paper 6 by Weiner describes a gradient procedure for iteratively evaluating RKHS inner products. This method, which was first suggested in less general terms by Parzen [5], is particularly useful when the relevant covariance function is known only numerically. Paper 7, by Kailath et al., contains formulas for RKHS inner products when the reproducing kernel is the covariance of a random process with a white noise component in one of its derivatives. This important class of processes includes those generated by passing white noise through a finite-dimensional linear system. The authors show that the form of the inner product is determined by the innovations representation of the process.

REFERENCES

1. M. Loève, Second Order Random Functions (in French), Appendix to *Stochastic Processes and Brownian Motion,* P. Lévy, Gauthier-Villars, Paris, 1948, 365p.
2. N. Aronszajn, The Theory of Reproducing Kernels and Their Applications (in French), *Cambridge Philos. Soc. Proc.* **39:**133–153 (1943).
3. M. Loève, *Probability Theory II,* 4th ed., Springer-Verlag, Berlin, 1978, 389p.

4. E. Parzen, *Statistical Inference on Time Series by Hilbert Space Methods I,* Technical Report 23, Statistics Department, Stanford University, Stanford, Calif., 1959, 130p.
5. E. Parzen, Regression Analysis of Continuous Parameter Time Series, in *Proceedings of the Fourth Berkeley Symposium on Mathematical Statistics and Probability,* vol. 1, J. Neyman, ed., University of California Press, Berkeley and Los Angeles, 1961, pp. 469–489.
6. E. Parzen, A New Approach to the Synthesis of Optimal Smoothing and Prediction Systems, in *Mathematical Optimization Techniques,* R. Bellman, ed., University of California Press, Berkeley and Los Angeles, 1963, pp. 75–108.
7. E. Parzen, Extraction and Detection Problems and Reproducing Kernel Hilbert Spaces, *J. Soc. Ind. Appl. Math. Control* **1**:35–62 (1962).
8. E. Parzen, *Time Series Analysis Papers,* Holden-Day, San Francisco, 1967, 565p.

3

Reprinted from *Ann. Math. Stat.* **32**:951-989 (1961)

AN APPROACH TO TIME SERIES ANALYSIS[1]

BY EMANUEL PARZEN

Stanford University

Summary. It may fairly be said that modern time series analysis is a subject which embraces three fields which while closely related have tended to develop somewhat independently. These fields are (i) statistical communication and control theory, (ii) the probabilistic (and Hilbert space) theory of stochastic processes possessing finite second moments, and (iii) the statistical theory of regression analysis, correlation analysis, and spectral (or harmonic) analysis of time series. In this paper it is my aim to show the close relation between these fields and to summarize some recent developments.

The topics discussed are (i) stationary time series and their statistical analysis, (ii) prediction theory and the Hilbert space spanned by a time series, and (iii) regression analysis of time series with known covariance function. In particular, I describe a new approach to prediction and regression problems using reproducing kernel Hilbert spaces.

1. Introduction. A set of observations arranged chronologically is called a time series. Time series are observed in connection with quite diverse phenomena, and by a wide variety of researchers, such as (1) the economist observing yearly wheat prices, (2) the geneticist observing daily egg production of a certain breed of hen, (3) the meteorologist studying daily rainfall in a given city, (4) the physicist studying the ambient noise level at a given point in the ocean, (5) the aerodynamicist studying atmospheric turbulence gust velocities, (6) the electronic engineer studying the internal noise of a radio receiver, and so on.

Time series analysis constitutes one of the most important tools of the economist. Consider the prices or quantities of commodities traded on an exchange. The record of prices or quantities over time may be represented as a fluctuating function (or wiggly record). The analysis of such economic time series is a problem of great interest to economists desiring to explain the dynamics of economic systems and to speculators desiring to forecast prices.

Techniques of time series analysis have long been used in science and engineering (for example, to smooth data and to search for "periodicities" [6]). The theory and practice of time series analysis is assuming new importance in the space age since a wide variety of problems involving communication and/or control (involving such diverse problems as the automatic tracking of moving

---

Received September 27, 1960.

[1] An address presented on August 25, 1960 at the Stanford meetings of the Institute of Mathematical Statistics by invitation of the IMS Committee on Special Invited Papers. This paper was prepared with the partial support of the Office of Naval Research (Nonr-225-21). Reproduction in whole or part is permitted for any purpose of the United States Government.

objects, the reception of radio signals in the presence of natural and artificial disturbances, the reproduction of sound and images, the design of guidance systems, the design of control systems for industrial processes, and the analysis of any kind of record representing observation over time) may be regarded as problems in time series analysis.

To represent a time series, one proceeds as follows. The set of time points at which measurements are made is called T. The observation made at time t is denoted by $X(t)$. The set of observations $\{X(t), t \varepsilon T\}$ is called a time series.

In regard to the index set T, there are cases of particular importance. One may be observing (i) a discrete parameter time series $X(t)$, in which case one assumes T is a finite set of points written $T = \{1, 2, \cdots, N\}$, (ii) a continuous parameter time series, in which case T is a finite interval written $T = \{t: 0 \leqq t \leqq L\}$, (iii) a multiple (discrete or continuous parameter) time series $\{(X_1(t), \cdots, X_k(t)), t \varepsilon T'\}$ which may be written as a time series $\{(X(t), t \varepsilon T\}$ with index set $T = \{(j, t): j = 1, \cdots, k \text{ and } t \varepsilon T'\}$, or (iv) a space field $X(x, y, z, t)$ defined on space-time which is a function of three coordinates of position and one coordinate of time.

The basic idea of the statistical theory of analysis of a time series $\{X(t), t \varepsilon T\}$ is to regard the time series as being an observation made on a family of random variables $\{X(t), t \varepsilon T\}$; that is, for each t in T, $X(t)$ is an observed value of a random variable. A family of random variables $\{X(t), t \varepsilon T\}$ is called a *stochastic process*. An observed time series $\{X(t), t \varepsilon T\}$ is thus regarded as an observation (or, in a different terminology, a realization) of a stochastic process $\{X(t), t \varepsilon T\}$.

It has been pointed out by various writers (see, for example, Neyman [34]) that there are two broad categories of statistical problems: problems of stochastic model building for natural phenomena and problems of statistical decision making. These two categories of problems are well illustrated in the analysis of economic time series; some study time series in order to understand the mechanism of the economic system while others study time series with the simple aim of being able to forecast, for example, stock market prices. In general, it may be said that the aims of time series analysis are

(1) to understand the mechanism generating the time series,

(2) to predict the behavior of the time series in the future. To attack either of these problems, one adopts a model for the time series.

A model often adopted for the analysis of an observed time series $\{X(t), t \varepsilon T\}$ is to regard $X(t)$ as the sum of two functions:

$$X(t) = m(t) + Y(t), \qquad t \varepsilon T. \tag{1.1}$$

We call $m(t)$ the mean value function and $Y(t)$ the fluctuation function.

The stochastic process $Y(t)$ is assumed to possess finite second moments, and to have zero means and *covariance kernel*

$$K(s, t) = E[Y(s)Y(t)]. \tag{1.2}$$

In addition it is often assumed that $Y(t)$ is a *normal process* in the sense that for every finite subset $\{t_1, \cdots, t_n\}$ of T, the random variables $Y(t_1), \cdots, Y(t_n)$ are jointly normally distributed.

The mean value function

$$m(t) = E[X(t)] \tag{1.3}$$

is assumed to belong to a known class M of functions. For example, M may be the set of all linear combinations of q known functions $w_1(t), \cdots, w_q(t)$; then, for t in T,

$$m(t) = \beta_1 w_1(t) + \cdots + \beta_q w_q(t), \tag{1.4}$$

for some coefficients $\beta_1, \cdots, \beta_q$ to be estimated. Other possible assumptions often made concerning the mean value function $m(t)$ are as follows: (i) $m(t)$ represents a *systematic oscillation*,

$$m(t) = \sum_{j=1}^{q} A_j \cos(\omega_j t + \varphi_j) \tag{1.5}$$

in which the amplitudes A_j, the angular frequencies ω_j, and the phases φ_j are constants, some of which are given and the rest are unknown and are to be estimated; (ii) $m(t)$ represents a polynomial trend,

$$m(t) = \sum_{j=0}^{q-1} \beta_j t^j, \tag{1.6}$$

an assumption often adopted if $m(t)$ represents the trajectory [given by $m(t) = x_0 + vt + \frac{1}{2}at^2$, say] of a moving object, or (iii) $m(t)$ is the sum of a systematic oscillation and a polynomial trend, an assumption traditionally adopted in treating economic time series.

Early workers in time series analysis sought to explain the dependence between successive observations of a time series $X(t)$ by assuming that $X(t)$ [sometimes written X_t] was generated by a scheme of the following kind:

$$X_t = m(t) + Y_t \tag{1.7}$$

where $m(t)$ represents a systematic oscillation of the form of (1.5) and the fluctuations $Y_1, \cdots, Y_n$ are assumed to be independent, normal random variables with mean 0 and common unknown variance σ^2.

The model given by (1.7) is called the scheme of *hidden periodicities* and was first introduced by Schuster ([47], [48]). The method used to estimate the frequencies ω_j (or, equivalently, the periods $2\pi/\omega_j$) is called *periodogram analysis*. The problem of tests of significance in periodogram analysis ([11], [18]) played an important role in the early history of time series analysis.

Approaches to time series analysis which seem to be more fruitful than periodogram analysis (see Kendall [22]) were pioneered by Yule and Slutsky in the 1920's.

Yule's researches [60] led to the notion of the autoregressive scheme, in which a time series X_t is assumed to be generated as a linear function of its past values, plus a random shock; in symbols, for some integer m (called the order of the autoregressive scheme) and constants $a_1, \cdots, a_m$

$$X_t = a_1 X_{t-1} + \cdots + a_m X_{t-m} + \eta_t$$

in which the sequence $\{\eta_t\}$ consists of independent identically distributed random variables. In particular, Yule showed that an autoregressive scheme of order 2 provided a better model for sunspots than did the scheme of hidden periodicities.

Slutsky's researches [51] led to the notion of a *moving average scheme*, in which a time series X_t is assumed to be generated as a finite moving average of a sequence of independent and identically distributed random variables $\{\eta_t\}$; in symbols, for some integer m and constants $a_0, \cdots, a_m$

$$X_t = a_0 \eta_t + a_1 \eta_{t-1} + \cdots + a_m \eta_{t-m} .$$

Slutsky showed that moving averages exhibit properties of disturbed periodicity and consequently can be used as a model for oscillatory time series. In particular, Slutsky proved the Sinusoidal Limit Theorem which showed that a sine wave could be approximated by a moving average scheme.

In the 1930's and 1940's, the probabilistic theory of stationary time series was developed, first as a result of the development of ergodic theory and then as a result of prediction theory. That the autoregressive and moving average schemes may be interpreted as special cases of the theory of stationary processes was pointed out by Wold [58] in 1938 (see [57], p. 169). Thus the link was established between the statistical theory of analysis of time series and the probabilistic theory of the structure of time series. In the last twenty years, an extensive literature has developed exploring this link.

It may fairly be said that modern time series analysis is a subject which embraces three fields which while closely related have tended to develop somewhat independently. These fields are (i) statistical communication and control theory ([26], [32]) (ii) the probabilistic (and Hilbert space) theory of stochastic processes possessing finite second moments ([9], Chaps. 9–12; [28], Chap. 10), and (iii) the statistical theory of regression analysis, correlation analysis, and spectral (or harmonic) analysis of time series ([13], [17], [23], [58]). In this paper it is my aim to show the close relation between these fields and to summarize some recent developments with which I have been closely associated. The contents of the paper are as follows.

(I) *Stationary time series and their statistical analysis.* While it is a fiction to regard an observed time series as having zero means, it is mathematically convenient to consider the analysis of time series under this assumption. Consequently, one may consider the analysis of an observed time series $\{X(t), t \in T\}$ with vanishing mean value function and *unknown* covariance function.

It has long been traditional among physical scientists to regard time series as arising from a superposition of sinusoidal waves of various amplitudes, frequencies, and phases. In the theory of time series analysis and statistical communications theory, a central role is played by the notion of the spectrum of a time series. For the time series analyst, the spectrum represents a basic tool for determining the mechanism generating an observed time series. For the communication theorist, the spectrum provides the major concept in terms of which to analyze the effect of passing stochastic processes (representing either signal or noise) through linear (and, to some extent, non-linear) devices. *Spectral* (or *harmonic*) *analysis* is concerned with the theory of the decomposition of a time series into sinusoidal components. For many time functions $\{X(t), -\infty < t < \infty\}$, such a decomposition is provided by the Fourier transform

$$S(\omega) = \int_{-\infty}^{\infty} e^{-it\omega} X(t)\, dt.$$

Unfortunately, no meaning can be attached to this integral for many stochastic processes $\{X(t), -\infty < t < \infty\}$ since their *sample functions are nonperiodic undamped functions* and therefore do not belong to the class of functions dealt within the usual theories of Fourier series and Fourier integrals. Nevertheless, it is possible to define a notion of *harmonic analysis of stochastic processes* (that is, a method of assigning to each frequency ω a measure of its contribution to the "content" of the process) as was first shown by N. Wiener [56] and A. Khintchine [24]. Among the stochastic processes which possess a harmonic analysis, *stationary* processes are most important since a time series may be represented as a superposition of sinusoidal waveforms with "independent amplitudes" if and only if it is stationary (see Section 4 for a more precise form of this assertion). In Section 2, some basic results concerning stationary processes are summarized.

Much of the recent statistical literature on time series analysis has been concerned with questions of statistical inference on stationary time series and especially with

(i) deriving the exact and asymptotic distributions of various estimates of the covariance functions $R(v)$ and the normalized covariance (or correlation) function $\rho(v) = R(v)/R(0)$ of a stationary time series,

(ii) fitting stationary time series by mechanisms (such as autoregressive schemes or moving average schemes) which are completely specified except for a finite number of parameters, and with estimating the parameters of such schemes,

(iii) estimating (and forming confidence sets) for the spectral density function and spectral distribution function of a stationary time series.

For many purposes, it is preferable to estimate the spectrum of a stationary time series rather than its correlation function, since many aspects of a stationary time series are best understood in terms of its spectrum. The spectrum enables one to (i) investigate the physical mechanism generating a time series, (ii) determine the behavior of a dynamic linear system in response to random excitations, and (iii) possibly simulate a time series. Other uses of the spectrum are as operational means (i) of transmitting or detecting signals, (ii) of classifying records of phenomena such as brain waves, (iii) of studying radio propagation phenomena, and (iv) of determining characteristics of control systems. The theory of statistical spectral analysis is too extensive to be reviewed here. For surveys of this theory, see Bartlett [3], Hannan [17], Blackman and Tukey [4], Jenkins [20], Parzen [41], Rosenblatt [45], and Tukey [53].

A comprehensive survey of the results available on topics (i) and (ii) has been given recently by Hannan [17]. Other comprehensive reviews are given by Bartlett [3], Moran [33], and Wold [57], Chap. 11; in these reviews one may find references to the work of M. S. Bartlett, H. E. Daniels, J. Durbin, E. J. Hannan, G. M. Jenkins, M. H. Quenouille, A. M. Walker, G. S. Watson, and P. Whittle. In Section 8, I have attempted to give an introduction to some of the large sample results available on topics (i) and (ii).

(II) *Prediction theory and the Hilbert space spanned by a time series.* A basic problem in time series analysis is that of *minimum mean square error linear prediction.* Let Z be an unobserved random variable with finite second moment. Let $\{X(t), t \varepsilon T\}$ be an observed time series. One seeks that random variable, linear in the observations, whose mean square distance from Z is smallest. In other words, if one desires to predict the value of Z, on the basis of having observed the values of the time series $\{X(t), t \varepsilon T\}$, one method might be to take that linear functional in the observations, denoted by $E^*[Z \mid X(t), t \varepsilon T]$, whose mean square error as a predictor is least. (The symbol E^* is used to denote a predictor because in the case of jointly normally distributed random variables, the best linear predictor $E^*[Z \mid X(t), t \varepsilon T]$ coincides with the conditional expectation $E[Z \mid X(t), t \varepsilon T]$; for an elementary discussion of this fact, see Parzen ([38], p. 387). Indeed, it should be noted that in any event the conditional expectation $E[Z \mid X(t), t \varepsilon T]$ can be defined as the minimum mean square error *non-linear* predictor.)

The prediction problem has provided a framework in terms of which many problems of statistical communication theory have come to be formulated. The pioneering work on prediction theory was done by Wiener [56a] and Kolmogorov [25] who were concerned with a stationary time series which had been observed over a semi-infinite interval of time. They sought predictors which had minimum mean square over all possible linear predictors. Wiener showed how the solution of the prediction problem could be reduced to the solution of the so-called Wiener-Hopf integral equation, and gave a method (spectral factorization) for the solution of the integral equation. Simplified methods of solution of this equation in the practically important special case of rational spectral density functions were given by Zadeh and Ragazzini [61] and Bode and Shannon [5]. Zadeh and Ragazzini [62] also treated the problem of regression analysis of time series with stationary fluctuation function, by reducing the problem to one involving the solution of a Wiener-Hopf equation. There then developed an extensive literature, seeking to treat prediction and smoothing problems involving a finite time of observation and non-stationary time series. The methods employed were either to reduce the problem to the solution of a suitable integral equation (generalization of the Wiener-Hopf equation) or to employ expansions (in a series of suitable eigen functions) of the time series involved.

As a result of these developments, prediction theory has turned out to provide a theory of the structure of time series and to provide mathematical tools for the solutions of other problems besides the prediction problem, especially regression problems. In Section 3, it is shown how the prediction problem leads naturally to the introduction of the important notion of the *Hilbert space spanned by a time series* which plays a central role in modern time series analysis. In Sections 4 and 6, I describe an approach to prediction and regression problems (in terms of reproducing kernel Hilbert spaces) which may be called coordinate free, and which by the introduction of suitable coordinate systems contains previous approaches as special cases.

The approach I take seems to me to be a rigorous version of an approach that is being developed in the Soviet Union by V. S. Pugachev [44]. Pugachev has in recent years advanced a point of view, which he calls the method of canonic representations of random functions, for which in one of his articles [43] he makes the following claim. "The results of this article, together with the results of [previous] papers, permits us to state that the method of canonic representations of random functions is the foundation of the modern statistical theory of optimum systems." It is my feeling that reproducing kernel Hilbert spaces provide a more powerful and more elegant means of achieving in a unified manner the results which Pugachev has sought to unify by the method of canonic representations.

(III) *Regression analysis of time series with known covariance function:* Let the observed time series be of the form of (1.1) with unknown mean value function $m(t)$ and known covariance function $K(s, t)$. Various methods of forming estimates of $m(t)$ are available. The most important methods are classical least squares estimation and minimum variance linear unbiased estimation. In the case of normally distributed observations, one has in addition the methods of maximum likelihood estimation and minimum variance unbiased estimation. In Sections 6 and 7, it is shown how Hilbert space techniques may be used to form explicit expressions for these estimates in terms of certain so-called reproducing kernel inner products.

There are, of course, large numbers of important problem areas of time series analysis which have not been mentioned in the foregoing such as (i) the problem of the distribution of zero-crossings and extrema of a time series (see references, see Longuet-Higgins [30] and Slepian [50]), (ii) the problem of the asymptotic efficiency of various classes of estimates of regression coefficients (see Grenander and Rosenblatt [13], Hebbe [19], and Striebel [52]), (iii) the use of filters to eliminate or extract trend or other components of a time series, and (iv) the distribution of various functionals of a time series, such as quadratic forms.

Further, the statistical analysis of multiple time series is not discussed. The relations that exist between different time series is on the whole a problem of greater interest than the relations that exist within a single time series. The results which exist under categories (I), (II), and (III) for univariate time series can be formally extended to multiple time series. However, many new problems arise which have not been thoroughly investigated.

A word should be said about the references given at the end of this paper. I have given a representative list rather than a complete list. Fortunately, a complete list of references will soon be available. The International Statistical Institute is compiling a bibliography on *Time Series and Stochastic Processes* which is to list and classify books and papers published, in the years 1900–1959, on both theory and applications. A bibliography (Parzen [42a]) of American publications has been compiled at Stanford for inclusion in the I.S.I. bibliography; a limited number of copies of this bibliography are available, and may be obtained by writing to the author. A bibliography is also given by Deming [8].

2. Stationary time series. A discrete parameter time series

$$\{X(t), t = 0, \pm 1, \cdots\}$$

or a continuous parameter time series $\{X(t), -\infty < t < \infty\}$ is said to be (weakly or wide-sense) stationary if the product moment

$$E[X(s)X(s + t)] = R(t)$$

is a function only of t. One calls $R(\cdot)$ the covariance function of the stationary time series.

It was shown by Khintchine [24] in 1933 that in the continuous parameter case there exists a non-decreasing bounded function $F(\omega)$, defined for $-\infty < \omega < \infty$, such that

$$R(t) = \int_{-\infty}^{\infty} e^{it\omega}\, dF(\omega), \qquad -\infty < t < \infty, \tag{2.1}$$

if it is assumed that $R(\cdot)$ is continuous at $t = 0$. Wold [58] in 1938 showed that in the discrete parameter case there exists a non-decreasing bounded function $F(\omega)$, defined for $-\pi \leqq \omega \leqq \pi$, such that

$$R(t) = \int_{-\pi}^{\pi} e^{it\omega}\, dF(\omega), \qquad t = 0, \pm 1, \cdots \tag{2.2}$$

The function $F(\omega)$ is called the *spectral distribution function* of the time series. Like a probability distribution function, $F(\omega)$ can be uniquely written as the sum,

$$F(\omega) = F_d(\omega) + F_{sc}(\omega) + F_{ac}(\omega),$$

of three distribution functions with the following properties. The function $F_{ac}(\omega)$ is absolutely continuous and is the integral of a non-negative function $f(\omega)$ called the *spectral density function* of the time series. The function $F_d(\omega)$ is a purely discontinuous (or discrete or step) function:

$$F_d(\omega) = \sum_{\omega_j \leqq \omega} \Delta F(\omega_j)$$

where $\{\omega_j\}$ are the discontinuity points of $F(\omega)$, and $\Delta F(\omega) = F(\omega + 0) - F(\omega - 0)$. Finally, $F_{sc}(\omega)$ is a singular continuous function.

It is usually assumed that physically observed time series have a spectral distribution function of the following form:

$$F(\omega) = \sum_{\substack{\omega' \text{ such} \\ \text{that } \Delta F(\omega') > 0 \\ \text{and } \omega' \leqq \omega}} \Delta F(\omega') + \int_{-\infty}^{\omega} f(\omega')\, d\omega'$$

where (i) the spectral density function $f(\omega)$ has the property that it is an integrable non-negative function which is continuous except at a finite number of points where it has finite left-hand and right-hand limits, and (ii) the set of frequencies at which the spectral jump function (or spectral mass function) $\Delta F(\omega)$ is positive contains at most countably infinite many points distributed on the real line in such a way that in any finite interval there are only a finite number of points of positive spectral mass. If these conditions are satisfied, we say that the time series has a *mixed spectrum*. If the spectral density function vanishes for all ω, we say that the time series has a *discrete* spectrum. If the spectral jump function $\Delta F(\omega)$ vanishes for all ω, we say that the time series has a *continuous* spectrum.

In terms of the spectral distribution function one can characterize various representations (or models) for a stationary time series $X(t)$. For example, it may be shown that a discrete parameter time series with a mixed spectrum whose spectral density function satisfies the condition

$$\int_{-\pi}^{\pi} \log f(\omega)\, d\omega > -\infty$$

may be written

$$X(t) = \sum_{\nu} A_\nu e^{it\omega_\nu} + \sum_{\nu=0}^{\infty} c_\nu \eta(t-\nu) \tag{2.3}$$

for suitable sequences of frequencies $\{\omega_\nu\}$, constants $\{c_\nu\}$, and uncorrelated random variables $\{A_\nu\}$ and $\{\eta_\nu\}$. In view of (2.3) one sees that the scheme of hidden periodicities and the scheme of moving averages may be viewed as a special kind of stationary process. Similarly, it may be shown that an autoregressive scheme (where the η_t are uncorrelated rather than independent) corresponds to a stationary time series whose distribution function is absolutely continuous and whose spectral density function is of the form

$$f(\omega) = \left[2\pi \left| \sum_{k=0}^{m} b_k\, e^{ik\omega} \right|^2 \right]^{-1}$$

for suitable constants $b_0, \cdots, b_m$. To prove these assertions, one uses the Hilbert space representation theory described in Section 4.

3. The problem of minimum mean square error linear prediction. In order to show existence and uniqueness, and to obtain conditions characterizing, the best linear predictor, we need to introduce the notion of a Hilbert space. (For a discussion of Hilbert space theory see any suitable text, such as Halmos [16].)

DEFINITION 3A. By an abstract Hilbert space is meant a set H whose members $u, v, \cdots$ are usually called vectors or points which possesses the following properties.

(I) H is a linear space [that is, for any vectors u and v in H, and real number a, there exist vectors, denoted by $u + v$ and au respectively, which satisfy the usual algebraic properties of addition and multiplication; also there exists a zero vector 0 with the usual properties under addition].

(II) H is an inner product space [that is, to every pair of points u and v in H there corresponds a real number, written (u, v) and called the inner product of u and v, possessing the following properties: for all points u, v, and w in H, and every real number a,

(i) $(au, v) = a(u, v)$

(ii) $(u + v, w) = (u, w) + (v, w)$

(iii) $(v, u) = (u, v)$

(iv) $(u, u) > 0$ if and only if $u \neq 0$].

(III) H is a complete metric space under the norm $\|u\| = (u, u)^{\frac{1}{2}}$ [that is, if $\{u_n\}$ is a sequence of points such that $\|u_m - u_n\| \to 0$ as $m, n \to \infty$ then there is a vector u in H such that $\|u_n - u\|^2 \to 0$ as $n \to \infty$].

In order to define the notion of the Hilbert space spanned by a time series, we first define the notion of the Hilbert space spanned by a family of vectors.

DEFINITION 3B. Let T be an index set, and let $\{u(t), t \varepsilon T\}$ be a family of members of a Hilbert space H. The linear manifold spanned by the family $\{u(t), t \varepsilon T\}$, denoted $L(u(t), t \varepsilon T)$, is defined to be the set, consisting of all vectors u in H which may be represented in the form $u = \sum_{i=1}^{n} c_i u(t_i)$ for some integer n, some constants $c_1, \cdots, c_n$, and some points $t_1, \cdots, t_n$ in T. The Hilbert space spanned by the family $\{u(t), t \varepsilon T\}$, denoted $V(u(t), t \varepsilon T)$ [or $L_2(u(t), t \varepsilon T)$ if H is the space of square integrable functions on some measure space], is defined to be the set of vectors which either belong to the linear manifold $L(u(t), t \varepsilon T)$ or may be represented as a limit of vectors in $L(u(t), t \varepsilon T)$. If $V(u(t), t \varepsilon T)$ coincides with H, we say that $\{u(t), t \varepsilon T\}$ *spans* H.

DEFINITION 3C. The Hilbert space spanned by a time series $\{X(t), t \varepsilon T\}$, denoted by $L_2(X(t), t \varepsilon T)$, is defined to consist of all random variables U which are either finite linear combinations of the random variables $\{X(t), t \varepsilon T\}$ or are limits of such finite linear combinations in the norm corresponding to the inner product defined on the space L_2 of square integrable random variables by

$$(U, V) = E[UV]. \tag{3.1}$$

In words, $L_2(X(t), t \varepsilon T)$ consists of all linear functionals in the time series.

We next state without proof the projection theorem for an abstract Hilbert space.

PROJECTION THEOREM. *Let H be an abstract Hilbert space, let M be a Hilbert subspace of H, let v be a vector in H, and let v^* be a vector in M. A necessary and* *sufficient condition that v^* is the unique vector in M satisfying*

$$\|v^* - v\| = \min_{u \text{ in } M} \|u - v\| \tag{3.2}$$

is that

$$(v^*, u) = (v, u) \qquad \textit{for every } u \textit{ in } M. \tag{3.3}$$

The vector v^* satisfying (3.2) *is called the projection of v onto M, and will here be written $E^*[v \mid M]$.*

In the case that M is the Hilbert space spanned by a family of vectors $\{x(t), t \varepsilon T\}$ in H, we write $E^*[v \mid x(t), t \varepsilon T]$ to denote the projection of v onto M. In this case, a necessary and sufficient condition that v^* satisfy (3.3) is that

$$(v^*, x(s)) = (v, x(s)) \qquad \text{for every } s \text{ in } T. \tag{3.4}$$

We are now in a position to solve the problem of obtaining an explicit expression for the minimum mean square error linear prediction $E^*[Z \mid X(t), t \varepsilon T]$. From (3.4), with H equal to the Hilbert space L_2 of all square integrable random variables, and $v = Z$, it follows that the optimum linear predictor is the unique random variable in $L_2(X(t), t \varepsilon T)$ satisfying, for all s in T,

$$E[E^*[Z \mid X(t), t \varepsilon T]X(s)] = E[ZX(s)]. \tag{3.5}$$

Equation (3.5) may look more familiar if we consider the special case of an interval $T = \{t: a \leqq t \leqq b\}$. If one writes heuristically

$$\int_a^b X(t)w(t)\,dt \tag{3.6}$$

to represent a random variable in $L_2(X(t), t \varepsilon T)$, then (3.5) states that the weighting function $w^*(t)$ of the best linear predictor

$$E^*[Z \mid X(t), t \varepsilon T] = \int_a^b w^*(t)X(t)\,dt, \tag{3.7}$$

must satisfy the generalized Wiener-Hopf equation

$$\int_a^b w^*(t)K(s, t)\,dt = \rho_Z(s), \qquad a \leqq s \leqq b \tag{3.8}$$

where we define

$$K(s, t) = E[X(s)X(t)] \tag{3.9}$$

$$\rho_Z(t) = E[ZX(t)]. \tag{3.10}$$

There is an extensive literature concerning the solution of the integral equation in (3.8); see [39] for references. In my opinion, however, this literature is concerned with an unnecessarily hard problem, as well as one in which the very formulation of the problem makes it difficult to be rigorous. The integral equation in (3.8) possesses a solution only if one interprets $w^*(t)$ as a generalized function which includes terms which are Dirac delta functions and derivatives of delta functions.

It seems to me that a simple reinterpretation of (3.8) avoids all these difficulties. Let us not regard (3.8) as an integral equation for the weighting function $w^*(t)$. Rather, let us compare (3.7) and (3.8). These equations say that *if one can find a representation for the function $\rho_Z(s)$ in terms of linear operations on the functions $\{K(s, t), t \varepsilon T\}$, then the minimum mean square error linear predictor $E^*[Z \mid X(t), t \varepsilon T]$ can be written in terms of the corresponding linear operations on the time series $\{X(t), t \varepsilon T\}$*. It should be emphasized that the most important linear operations are integration and differentiation. Consequently, the problem of finding the best linear predictor is not one of solving an integral equation, but is one of hunting for a linear representation of $\rho_Z(t)$ in terms of the covariance kernel $K(s, t)$. A general method of finding such representations will be discussed in Sections 4 and 5.

We illustrate the ideas involved by considering a simple example.

EXAMPLE 3A. Consider a stationary time series $X(t)$, with covariance kernel

$$K(s, t) = Ce^{-\beta|t-s|}, \tag{3.11}$$

which one has observed over a finite interval of time, $a \leqq t \leqq b$. Suppose that one desires to predict $X(b + c)$, for $c > 0$. Now, for $a \leqq t \leqq b$,

$$\rho(t) = E[X(t)X(b + c)] = Ce^{-\beta(b+c-t)} = e^{-\beta c}K(b, t). \tag{3.12}$$

In view of (3.12), by the interpretation of (3.7) and (3.8) just stated, it follows that

$$E^*[X(b + c) \mid X(t), a \leqq t \leqq b] = e^{-\beta c}X(b). \tag{3.13}$$

4. Hilbert space representations of time series. In the decade of the 1940's, probabilists began to employ Hilbert space methods to clarify the structure of time series (see [21] and [27]). Among the fundamental theorems proved in this period were the spectral representation theorem for stationary time series, and the Karhunen-Loève representation for random functions of second order on a finite interval. Various workers (especially Grenander [12]) have made use of these representation theorems in treating problems of statistical inference on time series. A representation theorem which does not seem to have found any application is one due to Loève ([27], p. 338) which shows that there is a very intimate connection between time series (random functions of second order) and reproducing kernel Hilbert spaces. It turns out, in my opinion, that reproducing kernel Hilbert spaces are the natural setting in which to solve problems of statistical inference on time series. In this section we define the notion of a Hilbert space representation of a time series and show how this notion may be used to explicitly solve the prediction problem.

The definition we give of the notion of a Hilbert space representation of a time series is based on the following theorem (for proof, see Parzen [37] or [40]).

BASIC CONGRUENCE THEOREM. *Let H_1 and H_2 be two abstract Hilbert spaces.*

Denote the inner product between two vectors u_1 and u_2 in H_1 by $(u_1, u_2)_1$. Similarly, denote the inner product between two vectors v_1 and v_2 in H_2 by $(v_1, v_2)_2$. Let T be an index set. Let $\{u(t), t \varepsilon T\}$ be a family of vectors which span H_1. Similarly, let $\{v(t), t \varepsilon T\}$ be a family of vectors which span H_2. Suppose that, for every s and t in T,

$$(4.1) \qquad (u(s), u(t))_1 = (v(s), v(t))_2 .$$

Then there exists a congruence (a one-one inner product preserving linear mapping) ψ from H_1 onto H_2 which has the property that

$$(4.2) \qquad \psi(u(t)) = v(t), \qquad t \text{ in } T.$$

DEFINITION 4A. A family of vectors $\{f(t), t \varepsilon T\}$ in a Hilbert space H is said to be a representation of a time series $\{X(t), t \varepsilon T\}$ if, for every s and t in T,

$$(4.3) \qquad (f(s), f(t))_H = K(s, t) = E[X(s)X(t)].$$

Then there is a congruence (a one-one inner product preserving linear mapping) ψ from $V(f(t), t \varepsilon T)$ onto $L_2(X(t), t \varepsilon T)$ satisfying

$$(4.4) \qquad \psi(f(t)) = X(t)$$

and every random variable U in $L_2(X(t), t \varepsilon T)$ may be written

$$(4.5) \qquad U = \psi(g)$$

for some unique vector g in $V(f(t), t \varepsilon T)$.

We next show that the representation of a time series as a stochastic integral is best viewed as a Hilbert space representation.

DEFINITION 4B. We call $(Q, \mathbf{B}, \mu)$ a measure space if Q is a set, $\mathbf{B}$ is a σ-field of subsets of Q, and μ is a measure on the measurable space $(Q, \mathbf{B})$. We denote by $L_2(Q, \mathbf{B}, \mu)$ the Hilbert space of all B-measurable real valued functions defined on Q satisfying

$$(4.6) \qquad (f, f)_\mu = \int_Q f^2 \, d\mu < \infty.$$

DEFINITION 4C. Let $(Q, \mathbf{B}, \mu)$ be a measure space, and, for every B in $\mathbf{B}$, let $Z(B)$ be a random variable. The family of random variables $\{Z(B), B \varepsilon \mathbf{B}\}$ is called an *orthogonal random set function* with covariance kernel μ if, for any two sets B_1 and B_2 in $\mathbf{B}$,

$$(4.7) \qquad E[Z(B_1)Z(B_2)] = \mu(B_1B_2),$$

where, as usual, B_1B_2 denotes the intersection of B_1 and B_2.

The Hilbert space $L_2(Z(B), B \varepsilon \mathbf{B})$ of random variables spanned by an orthogonal random set function may be defined, as was the Hilbert space spanned by a time series, to be the smallest Hilbert subspace of the Hilbert space of all square integrable random variables containing all random variables U of the form $U = \sum_{i=1}^{n} c_i Z(B_i)$ for some integer n, subfamily $\{B_1, \cdots, B_n\} \subset B$, and real constants $c_1, \cdots, c_n$. On the other hand, $L_2(Q, \mathbf{B}, \mu)$ may be described as the Hilbert space spanned under the norm (4.6) by the family of indicator functions $(I_B, B \in \mathbf{B})$, where the indicator function I_B of B is defined by $I_B(q) = 1$ or 0 according as $q \in B$ or $q \notin B$. Now for any B_1 and B_2 in $\mathbf{B}$,

$$(4.8) \qquad (I_{B_1}, I_{B_2})_\mu = \mu(B_1 B_2) = E[Z(B_1)Z(B_2)].$$

Therefore, by the Basic Congruence Theorem, there is a congruence ψ from $L_2(Q, \mathbf{B}, \mu)$ onto $L_2(Z(B), B \in \mathbf{B})$ such that for any $B \in \mathbf{B}$,

$$(4.9) \qquad \psi(I_B) = Z(B).$$

This fact justifies the following definition of the stochastic integral.

DEFINITION 4D. Let $(Q, \mathbf{B}, \mu)$ be a measure space and let $\{Z(B), B \in \mathbf{B}\}$ be an orthogonal random set function with covariance kernel μ. For any function f in $L_2(Q, \mathbf{B}, \mu)$ one defines the stochastic integral of f with respect to $\{Z(B), B \in \mathbf{B}\}$, denoted $\int_Q f\, dZ$, by

$$(4.10) \qquad \int_Q f\, dZ = \psi(f),$$

where ψ is the congruence from $L_2(Q, \mathbf{B}, \mu)$ onto $L_2(Z(B), B \in \mathbf{B})$ determined by (4.9).

We are now in a position to state our version of Karhunen's theorem (see [13], p. 29).

THEOREM 4A. *Let $\{X(t), t \in T\}$ be a time series with covariance kernel K. Let $\{f(t), t \in T\}$ be a family of functions in a space $L_2(Q, \mathbf{B}, \mu)$, such that for all s, t inT*

$$(4.11) \qquad K(s, t) = \int_Q f(s)f(t)\, d\mu.$$

Then $\{f(t), t \in T\}$ is a representation for $\{X(t), t \in T\}$.

If, further, $\{f(t), t \in T\}$ spans $L_2(Q, \mathbf{B}, \mu)$, then there is an orthogonal random set function $\{Z(B), B \in \mathbf{B}\}$ with covariance kernel μ such that

$$(4.12) \qquad X(t) = \int_Q f(t)\, dZ, \qquad t \in T,$$

and every random variable U in $L_2(X(t), t \in T)$ may be represented

$$(4.13) \qquad U = \int_Q g\, dZ$$

for some unique function g in $L_2(Q, \mathbf{B}, \mu)$.

PROOF. Let ψ be the congruence from $L_2(f(t), t \in T)$ onto $L_2(X(t), t \in T)$ satisfying (4.4). If $\{f(t), t \in T\}$ spans $L_2(Q, \mathbf{B}, \mu)$, define, for $B \in \mathbf{B}$, $Z(B) = \psi(I_B)$. It is immediate that $\{Z(B), B \in \mathbf{B}\}$ is an orthogonal random set function with covariance kernel μ. By the definition of the stochastic integral, (4.12) is merely another way of writing the fact that $X(t) = \psi(f(t))$.

Theorem 4A, together with (2.2) and (2.1), yields the following fundamental result.

SPECTRAL REPRESENTATION THEOREM FOR STATIONARY TIME SERIES. *A discrete parameter time series* $\{X(t),\ t = 0, \pm 1, \cdots\}$ *is weakly stationary if and only if for some Lebesgue-Stieltjes measures* μ *on the interval* $Q = \{\lambda: -\pi \leqq \lambda \leqq \pi\}$ *the complex exponentials* $\{e^{i\lambda t},\ t = 0 \pm 1, \cdots\}$ *form a representation for the time series in* $L_2(Q, \mathbf{B}, \mu)$ *where* $\mathbf{B}$ *is the* σ*-field of Borel subsets of* Q. *Then there exists an orthogonal random set function* $\{Z(B),\ B \,\varepsilon\, \mathbf{B}\}$ *such that*

$$X(t) = \int_{-\pi}^{\pi} e^{it\lambda} Z(d\lambda), t = 0, \pm 1, \cdots . \tag{4.14}$$

A similar theorem holds for continuous parameter time series with

$$Q = \{\lambda: -\infty < \lambda < \infty\}.$$

The representation of a time series as an integral with respect to an orthogonal random set function is not a natural representation, since one may choose such representations of a time series in a multitude of ways. Indeed, if $(Q, \mathbf{B}, \mu)$ is a measure space such that $L_2(X(t), t \,\varepsilon\, T)$ and $Q_2(Q, \mathbf{B}, \mu)$ have the same dimension, there are many families $\{f(t), t \,\varepsilon\, T\}$ of functions in $L_2(Q, \mathbf{B}, \mu)$ which are a representation for $\{X(t),\ t \,\varepsilon\, T\}$. What one desires is a family $\{f(t),\ t \,\varepsilon\, T\}$ of familiar functions [such as the family of complex exponentials $e^{i\lambda t}$, which are a representation in a suitable space $L_2(Q, \mathbf{B}, \mu)$ for a stationary time series]. I believe there is a natural representation in terms of which to solve problems of statistical inference on time series, namely the representation of a time series with covariance kernel K by the functions $\{K(\cdot, t), t \,\varepsilon\, T\}$ in the reproducing kernel Hilbert space $H(K)$.

DEFINITION 4E. A Hilbert space H is said to be a reproducing kernel Hilbert space, with reproducing kernel K, if the members of H are functions on some set T, and if there is a kernel K on $T \otimes T$ having the following two properties; for every t in T (where $K(\cdot, t)$ is the function defined on T, with value at s in T equal to $K(s, t)$):

$$K(\cdot, t) \,\varepsilon\, H \tag{4.15}$$

$$(g, K(\cdot, t))_H = g(t) \tag{4.16}$$

for every g in H.

Intuitively, a reproducing kernel Hilbert space is a Hilbert space which contains a function playing the role of the Dirac delta function $\delta(t)$. It should be recalled that, for square integrable functions $f(\cdot)$,

$$\int_{-\infty}^{\infty} f(s)\delta(s - t)\, ds = f(t).$$

Consequently, the kernel $K(s, t) = \delta(s - t)$ satisfies (4.16). However it does not satisfy (4.15), and therefore is not truly a reproducing kernel.

THEOREM 4B (Moore-Aronsjazn-Loève [1], [27]). *The covariance kernel* K *of a time series generates a unique Hilbert space, which we denote by* $H(K)$, *of which* K *is the reproducing kernel.*

Since $K(s, t) = (K(\cdot, s), K(\cdot, t))_{H(K)} = E[X(s)X(t)]$ we immediately obtain the following important theorem.

THEOREM 4C. *Let $\{X(t), t \varepsilon T\}$ be a time series with covariance kernel K. Then the family $\{K(\cdot, t), t \varepsilon T\}$ of functions in $H(K)$ is a representation for $\{X(t), t \varepsilon T\}$. Given a function g in $H(K)$, we denote by $(X, g)_K$ or $(g, X)_K$ the random variable U in $L_2(X(t), t \varepsilon T)$ which corresponds to g under the congruence which maps $K(\cdot, t)$ into $X(t)$. We then have the following formal relations: for every t in T, and g, h in $H(K)$,*

$$(4.17) \qquad \begin{aligned} (X, K(\cdot, t))_K &= X(t) \\ E[(X, h)_K(X, g)_K] &= (h, g)_K \end{aligned}$$

where we hereafter write $(h, g)_K$ for $(h, g)_{H(K)}$.

The next theorem shows the relationship between the reproducing kernel Hilbert space representation of a time series, and the representation of a time series by an orthogonal decomposition of the form of (4.12).

THEOREM 4D. *Let K be a covariance kernel. If there exist a measure space $(Q, \mathbf{B}, \mu)$, and a family of functions $\{f(t), t \varepsilon T\}$ in $L_2(Q, \mathbf{B}, \mu)$ such that (4.11) holds, then the reproducing kernel Hilbert space $H(K)$ corresponding to the covariance kernel K may be described as follows: $H(K)$ consists of all functions g, defined on T, which may be represented in the form*

$$(4.18) \qquad g(t) = \int_Q g^* f(t) \, d\mu$$

for some (necessarily unique) function g^* in the Hilbert subspace $L_2(f(t), t \varepsilon T)$ of $L_2(Q, \mathbf{B}, \mu)$ spanned by the family of functions $\{f(t), t \varepsilon T\}$, with norm given by

$$(4.19) \qquad \|g\|^2 = \int_Q |g^*|^2 \, d\mu.$$

If $\{f(t), t \varepsilon T\}$ spans $L_2(Q, \mathbf{B}, \mu)$, so that $X(t)$ has an orthogonal decomposition (4.12), then we may write

$$(4.20) \qquad (X, g)_K = \int_Q g^* \, dZ.$$

PROOF. Verify that the set H of functions of the form of (4.18), with norm given by (4.19), is a Hilbert space satisfying (4.15) and (4.16).

THEOREM 4E. (General solution of the prediction problem.) *Let $\{X(t), t \varepsilon T\}$, be a time series with covariance kernel $K(s, t)$, and let $H(K)$ be the corresponding reproducing kernel Hilbert space. Between $L_2(X(t), t \varepsilon T)$ and $H(K)$ there exists a one-one inner product preserving linear mapping under which $X(t)$ and $K(\cdot, t)$ are mapped into one another. Denote by $(h, X)_K$ the random variable in $L_2(X(t), t \varepsilon T)$ which corresponds under the mapping to the function $h(\cdot)$ in $H(K)$. Then the general solution to the prediction problem may be written as follows. If Z is a random variable with finite second moment, and if*

$$(4.21) \qquad \rho_Z(t) = E[ZX(t)],$$

then

$$(4.22)\qquad E^*[Z \mid X(t), t \varepsilon T] = (\rho_Z, X)_K$$

with mean square error of prediction given by

$$(4.23)\qquad E[\mid Z - E^*[Z \mid X(t), t \varepsilon T] \mid^2] = E \mid Z \mid^2 - (\rho_Z, \rho_Z)_K .$$

Theorem 4E represents a coordinate free solution of the prediction problem. The usual methods of explicitly writing optimum predictors, using either eigenfunction expansions, Green's functions (impulse response function), or (power) spectral density functions, are merely methods of writing down the reproducing kernel inner product corresponding to the covariance kernel $K(s, t)$ of the observed time series.

The validity of Theorem 4E follows immediately from the definition of the concepts involved. However, it may be instructive to give a proof of the theorem, using the following properties of the mapping $(h, X)_K$. For any functions g and h in $H(K)$ and random variables Z with finite second moment it holds that

$$(4.24)\qquad E[(h, X)_K(g, X)_K] = (h, g)_K$$

$$(4.25)\qquad E[Z(h, X)_K] = (\rho_Z, h)_K ,$$

in which $\rho_Z(t) - E[ZX(t)]$. Now a random variable in $L_2(X(t), t \varepsilon T)$ may be written $(h, X)_K$ for some h in $H(K)$. Consequently the mean square error between any linear functional $(h, X)_K$ and Z may be written

$$(4.26)\qquad \begin{aligned} E[\mid (h, X)_K - Z \mid^2] &= E[(h, X)_K^2] + E[Z^2] - 2E[Z(h, X)_K] \\ &= E[Z^2] + (h, h)_K - 2(\rho_Z, h)_K \\ &= E[Z^2] - (\rho_Z, \rho_Z)_K + (h - \rho_Z, h - \rho_Z)_K . \end{aligned}$$

From (4.26) it is immediate that $(\rho_Z, X)_K$ is the minimum mean square error linear predictor of Z, with mean square prediction error equal to $E[Z^2] - (\rho_Z, \rho_Z)_K$. The proof of Theorem 4E is complete.

5. Examples of reproducing kernel Hilbert space representations. In this section we give the reproducing kernel Hilbert space representation of a time series $\{X(t), t \varepsilon T\}$ under a variety of standard assumptions.

EXAMPLE 5A. Suppose $T = \{1, 2, \cdots, N\}$ for some positive integer N, and that the covariance kernel K is given by a symmetric positive definite matrix $\{K_{ij}\}$ with inverse $\{K^{ij}\}$. The corresponding reproducing kernel space $H(K)$ consists of all N-dimensional vectors $f = (f_1, \cdots, f_N)$ with inner product

$$(5.1)\qquad (f, g)_K = \sum_{s,t=1}^{N} f_s K^{st} g_t .$$

To prove (5.1) one need only verify that the reproducing property holds: for $u = 1, \cdots, N$,

$$(f, K_{\cdot u})_K = \sum_{s,t=1}^{N} f_s K^{st} K_{tu} = \sum_{s=1}^{N} f_s \delta(s, u) = f_u .$$

The inner product may also be written as a ratio of determinants:

$$(5.2) \qquad (f, g)_K = -\begin{vmatrix} K_{11} & \cdots & K_{1N} & f_1 \\ \vdots & \cdots & \vdots & \vdots \\ K_{N1} & & K_{NN} & f_N \\ g_1 & & g_N & 0 \end{vmatrix} \div \begin{vmatrix} K_{11} & \cdots & K_{1N} \\ \vdots & & \\ K_{N1} & \cdots & K_{NN} \end{vmatrix}.$$

To prove (5.2) one again need only verify the reproducing property. In the case in which the covariance matrix K is singular, one may define the corresponding reproducing kernel inner product in terms of the *pseudo-inverse* of the matrix K (see Greville [15] for a discussion of the notion of pseudo-inverse).

Although (5.1) provides a formula for the reproducing kernel inner product in terms of the inverse of the covariance matrix, it is to be emphasized that one need not necessarily invert the covariance matrix in order to find the reproducing kernel inner product. The point of introducing the reproducing kernel inner product is that the inversion of the covariance matrix is usually an intractable problem, and one should look instead to evaluate that for which one would use the inverse K^{-1}; namely, the evaluation of inner products in the reproducing kernel space. Various iterative methods of evaluating these inner products can be given (see [39] or [40]). This observation is undoubtedly not as important in the case of discrete parameter time series as it is in the case of multiple time series and continuous parameter time series.

EXAMPLE 5B. *Autoregressive schemes* (*discrete parameter*). A discrete parameter weakly stationary time series $X(t)$ is said to satisfy an autoregressive scheme of order m if $X(t)$ is the solution of the stochastic difference equation

$$(5.3) \qquad L_t X(t) = \sum_{k=0}^{m} a_k X(t-k) = \eta(t)$$

where $a_0, \cdots, a_m$ are given constants, and $\{\eta(t)\}$ is an orthonormal sequence of random variables. We now show that given observations $\{X(t), t = 1, 2, \cdots, N\}$ the reproducing kernel Hilbert space $H(K)$ corresponding to the covariance kernel K of the observations consists of all N-vectors $f = ((f(1), \cdots, f(N))$ with inner product given by

$$(5.4) \qquad (f, g)_K = \sum_{t=m+1}^{N} \{L_t f(t)\}\{L_t g(t)\} + \sum_{j,k=1}^{m} d_{jk} f(j) g(k)$$

where the matrix $D = \{d_{jk}\}$ has an inverse $D^{-1} = \{d^{jk}\}$ with general term

$$(5.5) \qquad d^{jk} = K(j-k) = E[X(j)X(k)].$$

In the case that $N \geqq 2m$, an explicit expression for d_{jk} is given by

$$(5.6) \qquad d_{jk} = \sum_{u=1}^{\min(j,k)} \{a_{j-u}a_{k-u} - a_{u+m-j}a_{u+m-k}\}.$$

In particular for a first order autoregressive scheme and $N \geqq 2$

$$(5.7)\qquad (f, g)_K = (a_0^2 - a_1^2)f(1)g(1) + \sum_{t=2}^{N} \{a_0f(t) + a_1f(t-1)\}\{a_0g(t) + a_1g(t-1)\}.$$

For a second order autoregressive scheme and $N \geqq 4$

$$(5.8)\qquad \begin{aligned}(f, g)_K = {} & (a_0^2 - a_2^2)\{f(1)g(1) + f(2)g(2)\} \\ & + (a_0a_1 - a_1a_2)\{f(1)g(2) + f(2)g(1)\} \\ & + \sum_{t=3}^{N} \{a_0f(t) + a_1f(t-1) + a_2f(t-2)\} \\ & \cdot\{a_0g(t) + a_1g(t-1) + a_2g(t-2)\}.\end{aligned}$$

One can give a purely algebraic proof of (5.4). However a simpler proof can be given if one uses certain facts from probability theory. Let us suppose that $X(1), \cdots, X(N)$ are jointly normally distributed random variables with covariance matrix $K_N = \{K_{s,t}\}$ with inverse matrix $K_N^{-1} = \{K^{st}\}$. Then the joint probability density function of $X(1), \cdots, X(N)$ may be written

$$(5.9)\qquad f_{X(1),\cdots,X(N)}(x_1, \cdots x_n) = \{(2\pi)^N \mid K_N \mid\}^{-\frac{1}{2}} \exp\{-\tfrac{1}{2}(x, x)_{K_N}\}$$

where $|K_N|$ is the determinant of K_N, and the inner product $(x, x)_{K_N}$ is defined by the right hand side of (5.1). On the other hand, if $X(1), \cdots, X(N)$ satisfy the difference equation $L_tX(t) = \eta(t)$, where $\eta(1), \cdots, \eta(N)$ are independent normal random variables with means 0 and variance 1, then

$$(5.10)\qquad \begin{aligned} & f_{X(1),\cdots,X(m),\eta(m+1),\cdots,\eta(N)}(x_1, \cdots, x_m, y_{m+1}, \cdots, y_N) \\ & \quad = \{(2\pi)^N |K_m|\}^{-\frac{1}{2}} \exp[-\tfrac{1}{2}\{(x, x)_{K_m} + \sum_{j=m+1}^{N} y_j^2\}].\end{aligned}$$

Transforming from

$$(X(1), \cdots, X(m), \eta(m+1), \cdots, \eta(N)) \text{ to } (X(1), \cdots, X(N))$$

by the linear transformation $L_tX(t) = \eta(t), t = m+1, \cdots, N$, it follows from (5.10) that

$$(5.11)\qquad \begin{aligned} & f_{X(1),\cdots,X(N)}(x_1, \cdots, x_N) = \{(2\pi)^N|K_m|\}^{-\frac{1}{2}}a_0^{N-m} \\ & \quad \cdot \exp\left[-\frac{1}{2}\left\{(x, x)_{K_m} + \sum_{j=m+1}^{N} |L_t x_t|^2\right\}\right].\end{aligned}$$

Comparing (5.9) and (5.11) it follows that for any N-vector x

$$(5.12)\qquad (x, x)_{K_N} = (x, x)_{K_m} + \sum_{j=m+1}^{N} |L_tx_t|^2$$

which is equivalent to (5.4).

To prove (5.6), define the function $e_j(t)$ by $e_j(t) = 1$ or 0 according as $t = j$ or $t \neq j$. Since the time series $X(t)$ is stationary,

$$(e_j, e_k)_K = (e_{N-j+1}, e_{N-k+1})_K. \tag{5.13}$$

For $1 \leqq j, k \leqq m$, defining $a_j = 0$ for $j < 0$ or $j > m$,

$$\begin{aligned}(e_{N-j+1}, e_{N-k+1})_K &= \sum_{t=m+1}^{N} L_t(e_{N-j+1})L_t(e_{N-k+1}) \\ &= \sum_{t=m+1}^{N} a_{j+t-N-1}a_{k+t-N-1} \\ &= \sum_{u=1}^{N-m} a_{j-u}a_{k-u}\end{aligned} \tag{5.14}$$

while

$$\begin{aligned}(e_j, e_k)_K &= d_{jk} + \sum_{t=m+1}^{N} L_t(e_j)L_t(e_k) \\ &= d_{jk} + \sum_{t=m+1}^{N} a_{t-j}a_{t-k} \\ &= d_{jk} + \sum_{u=1}^{N-m} a_{u+m-j}a_{u+m-k}.\end{aligned} \tag{5.15}$$

From (5.13), (5.14), and (5.15), we obtain (5.6).

From (5.4) and (5.6) one may obtain the inverse matrix of the covariance matrix of an autoregressive scheme (see Siddiqui [49] and references cited there).

EXAMPLE 5C: *Autoregressive schemes (continuous parameter)*. We next consider the reproducing kernel Hilbert space corresponding to the covariance kernel of an autoregressive scheme $X(t)$ observed over a finite interval $a \leqq t \leqq b$.

A continuous parameter stationary time series $X(t)$ is said to be an autoregressive scheme of order m if its covariance function $R(u) = E[X(t)X(t+u)]$ may be written (see Doob [D1], p. 542)

$$R(s-t) = \int_{-\infty}^{\infty} \frac{e^{i(s-t)\omega}}{2\pi \left| \sum_{k=0}^{m} a_k (i\omega)^{m-k} \right|^2} \, d\omega \tag{5.16}$$

where the polynomial $\sum_{k=0}^{m} a_k z^{m-k}$ has no zeros in the right half of the complex z-plane. It may be shown that given observations of such a time series over a finite interval $a \leqq t \leqq b$, the corresponding reproducing kernel Hilbert space contains all functions $h(t)$ on $a \leqq t \leqq b$ which are continuously differentiable of order m. The reproducing kernel inner product is given by

$$(h, g)_K = \int_a^b (L_t h)(L_t g)\, dt + \sum_{j,k=0}^{m-1} d_{j,k} h^{(j)}(a) g^{(k)}(a) \tag{5.17}$$

where

$$L_t h = \sum_{k=0}^{m} a_k h^{(m-k)}(t) \tag{5.18}$$

$$\{d_{j,k}\}^{-1} = \left\{ \frac{\partial^{j+k}}{\partial t^j \partial u^k} R(t-u) \bigg|_{t=a, u=a} \right\}. \tag{5.19}$$

The first and second autoregressive schemes are of particular importance.

A stationary time series $X(t)$ is said to satisfy a first order autoregressive scheme if it is the solution of a first order linear differential equation whose input is white noise $\eta'(t)$ (the symbolic derivative of a process $\eta(t)$ with independent stationary increments):

$$(dX/dt) + \beta X = \eta'(t). \tag{5.20}$$

It should be remarked that from a mathematical point of view (5.20) should be written

$$dX(t) + \beta X(t) dt = d\eta(t). \tag{5.21}$$

Even then, by saying that $X(t)$ satisfies (5.20) or (5.21) we mean that

$$X(t) - \int_{-\infty}^{t} H(t-s)\, d\eta(s) \tag{5.22}$$

where $H(t-s) = e^{-\beta(t-s)}$ is the one-sided Green's function of the differential operator $L_t f = f'(t) + \beta f(t)$.

The covariance function of the stationary time series $X(t)$ is

$$R(t-u) = (1/2\beta) e^{-\beta|u-t|}. \tag{5.23}$$

The corresponding reproducing kernel Hilbert space $H(K)$ contains all differentiable functions. The inner product is given by

$$(h, g)_k = \int_a^b (h' + \beta h)(g' + \beta g)\, dt + 2\beta h(a) g(a). \tag{5.24}$$

More generally, corresponding to the covariance function

$$K(s, t) = Ce^{-\beta|s-t|} \tag{5.25}$$

the reproducing kernel inner product is

$$\begin{aligned} (h, g)_K &= \frac{1}{2\beta C} \left\{ \int_a^b (h' + \beta h)(g' + \beta g)\, dt + 2\beta h(a) g(a) \right\} \\ &= \frac{1}{2\beta C} \int_a^b (h'g' + \beta^2 hg)\, dt + \frac{1}{2C} \{h(a)g(a) + h(b)g(b)\}. \end{aligned} \tag{5.26}$$

The random variable $(h, X)_K$ in $L_2(X(t), a \leqq t \leqq b)$ corresponding to $h(\cdot)$ in

$H(K)$ may be written

$$(5.27)\quad (h, X)_K = \frac{1}{2\beta C}\left\{\beta^2 \int_a^b h(t)X(t)\,dt + \int_a^b h'(t)\,dX(t)\right\} + \frac{1}{2C}\{h(a)X(a) + h(b)X(b)\}.$$

Note that $X'(t)$ does not exist in any rigorous sense; consequently we write $dX(t)$ where $X'(t)\,dt$ seems to be called for. It can be shown that (5.27) makes sense. In the case that $h(\cdot)$ is twice differentiable, one may integrate by parts and write

$$(5.28)\quad \int_a^b h'(t)\,dX(t) = h'(b)X(b) - h'(a)X(a) - \int_a^b X(t)h''(t)\,dt.$$

A stationary time series $X(t)$ is said to satisfy a second order autoregressive scheme if it is the solution of a second order linear differential equation whose input is white noise $\eta'(t)$:

$$(5.29)\quad (d^2X/dt^2) + 2\alpha(dX/dt) + \gamma^2 X = \eta'(t).$$

If $\omega^2 = \gamma^2 - \alpha^2 > 0$, the covariance function of the time series is

$$(5.30)\quad R(t - u) = \frac{e^{-\alpha|u-t|}}{4\alpha\gamma^2}\left\{\cos\omega(u - t) + \frac{\alpha}{\omega}\sin\omega|u - t|\right\}.$$

The corresponding reproducing kernel Hilbert space contains all twice differentiable functions on the interval $a \leqq t \leqq b$ with inner product

$$(5.31)\quad (h, g)_K = \int_a^b (h'' + 2\alpha h' + \gamma^2 h)(g'' + 2\alpha g' + \gamma^2 g)\,dt + 4\alpha\gamma^2 h(a)g(a) + 4\alpha h'(a)g'(a).$$

To write an expression for $(h, X)_K$, one uses the same considerations as in (5.27).

Other examples of reproducing kernel Hilbert spaces are given in [39] and [40].

6. Regression analysis of time series with known covariance function. The theory of regression analysis (and of the general linear hypothesis) plays a central role in statistical theory. In this section we show how to solve certain standard problems of regression analysis in cases in which the observations possess properties of dependence or continuity. For a discussion of the history and literature of regression analysis the reader is referred to Wold [58].

The classical problem of regression analysis may be posed as follows. Given (i) observations $X(t)$, $t = 1, \cdots, N$, with known covariance kernel

$$(6.1)\quad K(s, t) = \text{Cov}\,[X(s), X(t)]$$

and mean value function $m(t) = E[X(t)]$ of the form

$$(6.2)\quad m(t) = \beta_1 w_1(t) + \cdots + \beta_q w_q(t)$$

where $w_1(\cdot), \cdots, w_q(\cdot)$ are known functions, and $\beta_1, \cdots, \beta_q$ are unknown real numbers, and (ii) a linear function

$$\psi(\beta) = \psi_1\beta_1 + \cdots + \psi_q\beta_q \tag{6.3}$$

of the parameters, where $\psi_1, \cdots, \psi_q$ are known constants. Estimate $\psi(\cdot)$ by an estimate which (i) is *linear* in the observations in the sense that it is of the form $\sum_{t=1}^{N} c_t X(t)$ for some real numbers $c_1, \cdots, c_N$, (ii) is an *unbiased* estimate of $\psi(\cdot)$ in the sense, that for all $\beta = (\beta_1, \cdots, \beta_q)$,

$$E_\beta\left[\sum_{t=1}^{N} c_t X(t)\right] = \sum_{t=1}^{N} c_t m(t) = \sum_{j=1}^{q} \beta_j \sum_{t=1}^{N} c_t w_j(t) = \psi(\beta), \tag{6.4}$$

and (iii) has variance

$$\operatorname{Var}\left[\sum_{t=1}^{N} c_t X(t)\right] = \sum_{s,t=1}^{N} c_s K(s, t) c_t \tag{6.5}$$

equal to the minimum variance of any unbiased linear estimate.

The problem of finding the minimum variance unbiased linear estimate of a linear parametric function $\psi(\beta)$ can be posed as a problem involving the minimization of a *quadratic form* subject to linear restraints. Define $K = \{K(s, t)\}$,

$$W = \begin{bmatrix} w_1(1) & \cdots & w_q(1) \\ \vdots & \cdots & \vdots \\ w_1(N) & \cdots & w_q(N) \end{bmatrix}, \quad \psi = \begin{bmatrix} \psi_1 \\ \vdots \\ \psi_q \end{bmatrix}, \quad c = \begin{bmatrix} c_1 \\ \vdots \\ c_N \end{bmatrix} \tag{6.6}$$

and let c' denote the transpose of a (column) vector c. The unbiasedness condition (6.4) can be stated in matrix form as

$$c'W = \psi'. \tag{6.7}$$

The problem of finding the minimum variance unbiased linear estimate can now be posed as follows: find the vector c which minimizes the quadratic form $c'Kc$, subject to the constraints $c'W = \psi'$ (compare Bush and Olkin [7]).

THEOREM 6A. *Let K be a positive definite $n \times n$ symmetric matrix, W be an $n \times q$ matrix, and ψ a q-vector. Assume that*

$$V = W'K^{-1}W \tag{6.8}$$

is non-singular. The n-vector c^* which minimizes the quadratic form $c'Kc$ among all n-vectors c satisfying $W'c = \psi$ is given by

$$c^* = K^{-1}WV^{-1}\psi \tag{6.9}$$

and the minimum value of the quadratic form is given by

$$c^{*\prime}Kc^* = \psi'V^{-1}\psi. \tag{6.10}$$

PROOF. One easily verifies that the vector c^* defined by (6.9) satisfies the restraint $W'c = \psi$, and that (6.10) holds. To complete the proof we show that for any n-vector c such that $c'W = \psi'$ it holds that

$$c'Kc \geqq \psi'V^{-1}\psi. \tag{6.11}$$

Now for any q-vector z, letting $y = Wz$,

$$c'Kc \geqq \frac{(c'y)^2}{y'K^{-1}y} = \frac{(c'Wz)^2}{z'Vz} = \frac{(\psi'z)^2}{z'Vz}. \tag{6.12}$$

Taking the supremum of the right side of (6.12) over all q-vectors z, one obtains (6.11), since

$$\sup_z [(\psi'z)^2/z'Vz] = \psi'V^{-1}\psi. \tag{6.13}$$

From Theorem 6A, one immediately obtains Theorem 6B.

THEOREM 6B. *The minimum variance linear unbiased estimate of a parametric function* $\psi(\beta)$ *is*

$$\psi^* = c^{*\prime}X = \psi'V^{-1}(W'K^{-1}X). \tag{6.14}$$

The variance of ψ^* *is given by*

$$\operatorname{Var}[\psi^*] = c^{*\prime}Kc^* = \psi'V^{-1}\psi. \tag{6.15}$$

In particular, the vector $\beta^{*\prime} = (\beta_1^*, \cdots, \beta_q^*)$ *of minimum variance unbiased linear estimates of* $\beta_1, \cdots, \beta_q$ *may be written*

$$\beta^* = V^{-1}(W'K^{-1}X) \tag{6.16}$$

with covariance matrix

$$\{\operatorname{Cov}[\beta_i^*, \beta_j^*]\} = V^{-1}. \tag{6.17}$$

The foregoing treatment of the problem of regression analysis with known covariance function depended very much on the assumptions that there were only a finite number of observations, and that the matrices K and V were non-singular. We now show how to relax these assumptions by using the reproducing kernel Hilbert space representation of a time series. The results we now state include as special cases the results which were first obtained by Grenander ([12], [14]).

Let $\{X(t), t \varepsilon T\}$ be a time series whose covariance kernel $K(s, t) = \operatorname{Cov}[X(s), X(t)]$ is known and whose mean value function $m(t) = E[X(t)]$ is only assumed to belong to a known class M. Let $H(K)$ be the reproducing kernel Hilbert space corresponding to K. Assume that M is a subset of $H(K)$. It may be shown that between $L_2(X(t), t \varepsilon T)$ and $H(K)$ there exists a one-one linear mapping with the following properties: if $(h, X)_K$ denotes the random variable in $L_2(X(t), t \varepsilon T)$ which corresponds under the mapping to the function h in $H(K)$, then for every t in T, and h and g in $H(K)$,

$$(K(\cdot, t), X)_K = X(t), \tag{6.18}$$

$$E_m[(h, X)_K] = (h, m)_K, \qquad \text{for all } m \text{ in } M, \tag{6.19}$$

$$\operatorname{Cov}[(h, X)_K, (g, X)_K] = (h, g)_K. \tag{6.20}$$

The subscript m on an expectation operator is written to indicate that the expectation is computed under the assumption that $m(\cdot)$ is the true mean value function.

If T is finite, and K is non-singular, then $(h, X)_K = h'K^{-1}X$. For other examples of $(h, X)_K$, see Section 5.

We are interested in estimating various functionals $\psi(m)$ of the true mean value function $m(\cdot)$ by estimates which (i) are linear in the observations $\{X(t), t \varepsilon T\}$ in the sense that they belong to $L_2(X(t), t \varepsilon T)$, (ii) are *unbiased* and (iii) have *minimum variance* among all linear unbiased estimates. A functional $\psi(m)$ is said to be linearly estimable if it possesses an unbiased linear estimate $(g, X)_K$. Since

$$E_m[(g, X)_K] = (g, m)_K = \psi(m), \qquad \text{for all } m \text{ in } M \tag{6.21}$$

it follows that $\psi(m)$ is linearly estimable if and only if there exists a function g in $H(K)$ satisfying (6.21). Now the variance of a linear estimate is given by

$$\text{Var}\,[(g, X)_K] = (g, g)_K . \tag{6.22}$$

Consequently finding the minimum variance unbiased linear estimate $\psi^* = (g^*, X)_K$ of $\psi(m)$ is equivalent to finding that function g^* in $H(K)$ which has minimum norm among all functions g satisfying the restraint (6.21). To find the vector g^* with minimum norm it suffices to find any vector g satisfying (6.21). Then the projection

$$g^* = E^*[g \mid \bar{M}], \tag{6.23}$$

of g onto the smallest Hilbert subspace $\bar{M}$ containing M, satisfies (6.21) and has m nimum norm among all vectors satisfying (6.21).

THEOREM 6C. *The uniformly minimum variance unbiased linear estimate ψ^* of a linearly estimable function $\psi(m)$ is given by*

$$\psi^* = (E^*[g \mid \bar{M}], X)_K \tag{6.24}$$

with variance

$$\text{Var}\,[\psi^*] = \|E^*[g \mid \bar{M}]\|_K^2 , \tag{6.25}$$

where g is any function satisfying (6.21), *$\bar{M}$ is the smallest Hilbert subspace of $H(K)$ containing M, and $E^*[g \mid \bar{M}]$ denotes the projection onto $\bar{M}$ of g. In particular, the uniformly minimum variance unbiased linear estimate $m^*(t)$ of the value $m(t)$ at a particular point t of the mean value function $m(\cdot)$ is given by*

$$m^*(t) = (E^*[K(\cdot, t) \mid \bar{M}], X)_K \tag{6.26}$$

since

$$m(t) = (K(\cdot, t), m)_K . \tag{6.27}$$

In the special case that M consists of all functions $m(t)$ of the form of (6.2), and the matrix

$$V = \begin{bmatrix} (w_1, w_1)_K & \cdots & (w_1, w_q)_K \\ \vdots & & \\ (w_q, w_1)_K & \cdots & (w_q, w_q)_K \end{bmatrix} \tag{6.28}$$

is non-singular, then

$$\beta^* = V^{-1}\begin{bmatrix}(w_1, X)_K \\ \vdots \\ (w_q, X)_K\end{bmatrix}. \tag{6.29}$$

One may write an explicit formula for the minimum variance unbiased linear estimate ψ^* of a linear parametric function $\psi(\beta) = \psi_1\beta_1 + \cdots + \psi_q\beta_q$ as follows, where $V_{ij} = (w_i, w_j)_K$;

$$\psi^* = -\begin{vmatrix} V_{11} & \cdots & V_{1q} & (X, w_1)_K \\ \vdots & \cdots & \vdots & \vdots \\ V_{q1} & \cdots & V_{qq} & (X, w_q)_K \\ \psi_1 & \cdots & \psi_q & 0 \end{vmatrix} \div \begin{vmatrix} V_{11} & \cdots & V_{1q} \\ \vdots & \cdots & \vdots \\ V_{q1} & \cdots & V_{qq}\end{vmatrix}. \tag{6.30}$$

It should be noted that the proof of Theorem 6C is exactly the same in spirit as the proof of the Gauss-Markov theorem given in Scheffé ([46], p. 14). The point of Theorem 6C is that it enables one to develop a theory of regression analysis and analysis of variance for cases in which one has an infinite number of observations. In particular, we state the analogues of certain basic results on simultaneous confidence intervals (Scheffé [46], p. 68) and hypothesis testing (Scheffé [46], p. 31).

Hypothesis testing and simultaneous confidence bands for mean value functions. If the time series $X(t)$ is assumed to be normal, or if all linear functionals $(h, X)_K$ may be assumed to be approximately normally distributed, then one may state a confidence band for the entire mean value function $m(\cdot)$ as follows. Given a confidence level α, let $C_q(\alpha)$ denote the α percentile of the χ^2 distribution with q degrees of freedom; in symbols, $P[\chi_q^2 \geqq C_q(\alpha)] = \alpha$.

We now show that if the smallest space $\bar{M}$ containing all mean value functions has finite dimension q, then

$$m^*(t) - [C_q(\alpha)]^{\frac{1}{2}}\sigma[m^*(t)] \leqq m(t) \leqq m^*(t) + [C_q(\alpha)]^{\frac{1}{2}}\sigma[m^*(t)] \quad \text{for all } t \text{ in } -\infty < t < \infty \tag{6.31}$$

is a simultaneous confidence band for all values of the mean value function with a level of significance not less that α; that is, if $m(\cdot)$ is the true mean value function then (6.31) holds with a probability greater than or equal to α.

To prove (6.31) we prove more generally the following theorem.

THEOREM 6C. (Simultaneous confidence interval of significance level α for all estimable functions (m, g).) *If $\bar{M}$ has dimension q then for all m in $\bar{M}$*

$$P_m\left[\sup_{g \varepsilon H(K)} \frac{|(X, E^*[g \mid \bar{M}])_K - (m, g)_K|^2}{\operatorname{Var}[(X, E^*[g \mid \bar{M}])_K]} \leqq C_q(\alpha)\right] = \alpha. \tag{6.32}$$

PROOF. Let $w_1, \cdots, w_q$ be orthonormal functions which span $\bar{M}$. Then we may write $m = \beta_1 w_1 + \cdots + \beta_q w_q$ where $\beta_j = (m, w_j)$ is a function of m. Further, $(m, g)_K = \alpha_1\beta_1 + \cdots + \alpha_q\beta_q$, $(X, E^*[g \mid \bar{M}])_K = \alpha_1\beta_1^* + \cdots + \alpha_q\beta_q^*$,

$\operatorname{Var}[(X, E^*[g \mid \bar{M}])_K] = \sum_{j=1}^q \alpha_j^2$, where $\alpha_j = (w_j, g)_K$ and $\beta_j^* = (X, w_j)_K$. Next the random variable appearing in (6.32) is equal to

$$\sup_{-\infty<\alpha_1,\cdots,\alpha_q<\infty} \frac{\left|\sum_{j=1}^q \alpha_j(\beta_j^* - \beta_j)\right|^2}{\sum_{j=1}^q \alpha_j^2} = \sum_{j=1}^q (\beta_j^* - \beta_j)^2 \tag{6.33}$$

which is distributed as χ_q^2 (compare Scheffé [46], p. 416).

Similarly one may prove the following theorem.

THEOREM 6D. *Given a q-dimensional subspace M of $H(K)$, and a q'-dimensional subspace M' of M, to test the composite null hypothesis $H_0 : m(\cdot) \varepsilon M'$, against the composite alternative hypothesis, $H_1 : m(\cdot) \varepsilon M$, one may use the statistic*

$$\Delta = \|m_M^*(t) - m_{M'}^*(t)\|_k^2 \tag{6.34}$$

where $m_M^*(t)$ $[m_{M'}^*(t)]$ denotes the minimum variance unbiased linear estimate of $m(t)$ under the hypothesis $H_1[H_0]$. Under H_0, Δ is distributed as χ^2 with $q - q'$ degrees of freedom.

In the special case that M consists of all functions $m(t)$ of the form (6.2), and M' consists of all functions in M for which $\beta_j = 0$ for $j = q' + 1, \cdots, q$, then the statistic Δ may be written

$$\Delta = \sum_{j=q'+1}^q \delta_j \tag{6.35}$$

where, defining $V_{ij} = (w_i, w_j)_K$,

$$\begin{aligned}\delta_j &= \frac{|(w_j - E^*[w_j \mid w_1, \cdots, w_{j-1}], X)_K|^2}{\|w_j - E^*[w_j \mid w_1, \cdots, w_{j-1}]\|_K^2} \\ &= \begin{vmatrix} V_{11} & \cdots & V_{1,j-1} & (w_1, X)_K \\ \vdots & & & \\ V_{j1} & \cdots & V_{j,j-1} & (w_j, X)_K \end{vmatrix}^2 \div \begin{vmatrix} V_{11} & \cdots & V_{1j} \\ \vdots & \cdots & \vdots \\ V_{j1} & \cdots & V_{jj} \end{vmatrix} \begin{vmatrix} V_{11} & \cdots & V_{1,j-1} \\ \\ V_{j-1,1} & \cdots & V_{j-1,j-1} \end{vmatrix}\end{aligned} \tag{6.36}$$

The reader may find it illuminating to write out (6.36) in the case that $q = 2$ and $q' = 1$.

Regression analysis when the covariance function of the observations is only known up to a constant factor. Suppose that the covariance function of the time series $\{X(t), t \varepsilon T\}$ is of the form

$$\operatorname{Cov}[X(s), X(t)] = \sigma^2 K(s, t)$$

where the kernel $K(s, t)$ is known and σ^2 is an unknown positive constant, and that the mean value function $m(t) = E[X(t)]$ is known to belong to a set M which is a subspace (of dimension q) of $H(K)$, the reproducing kernel Hilbert space corresponding to K. Theorem 6C continues to hold, except that (6.25) should be replaced by

$$\operatorname{Var}_\sigma[\psi^*] = \sigma^* \|E^*[g \mid M]\|_K^2 . \tag{6.25$'$}$$

The variance of the estimate ψ^* depends on the unknown parameter σ^2. Therefore one needs to estimate σ^2 in order to know $\text{Var}_\sigma[\psi^*]$. To discuss the estimation of σ^2, we need to distinguish between the case in which the index set T is finite and the case in which T is infinite.

If T is finite, the time series $\{X(t), t \varepsilon T\}$, regarded as a function of t, may be shown to belong to $H(K)$. Further if n is the dimension of $H(K)$, then for all possible mean value functions $m(t)$ and values of σ^2

$$\begin{aligned} E[\|X(t) - m(t)\|_K^2] &= n\sigma^2 \\ E[\|m^*(t) - m(t)\|_K^2] &= q\sigma^2 \\ E[\|X(t) - m^*(t)\|_K^2] &= (n-q)\sigma^2. \end{aligned} \tag{6.37}$$

Therefore

$$\sigma^{*2} = (n-q)^{-1} \|X(t) - m^*(t)\|_K^2 \tag{6.38}$$

is an unbiased estimate of σ^2 (which in the case of normally distributed observations is independent of $m^*(t)$).

If T is infinite, it is possible to estimate σ^2 exactly by forming a sequence of estimates of the form of (6.38) based on a monotone sequence of finite subsets $\{T_n\}$ of T whose limit is dense in T.

7. The probability density functional of a normal process. The prediction and regression problems considered in the foregoing have all involved linear estimates chosen according to a criterion expressed in terms of mean square error. Nevertheless the mathematical tools developed continue to play an important role if one desires to employ other criteria of statistical inference. All modern theories of statistical inference take as their starting point the idea of the probability density function of the observations. Thus in order to apply any principle of statistical inference to problems of time series analysis, it is first necessary to develop the notion of the probability density function (or functional) of a stochastic process. In this section we state a result showing how one may write a formula for the probability density function of a stochastic process which is normal.

Given a normal time series $\{X(t), t \varepsilon T\}$ with known covariance function

$$K(s, t) = \text{Cov}\,[X(s), X(t)] \tag{7.1}$$

and mean value function $m(t) = E[X(t)]$, let P_m be the probability measure induced on the space of sample functions of the time series. Next, let m_1 and m_2 be two functions, and let P_1 and P_2 be the probability measures induced by normal time series with the same covariance kernel K, and mean value functions equal to m_1 and m_2 respectively. By the Lebesgue decomposition theorem it follows that there is a set N of P_1-measure 0 and a non-negative P_1-integrable function, denoted by dP_2/dP_1, such that for every measurable set B of sample functions

$$P_2(B) = \int_B (dP_2/dP_1)\, dP_1 + P_2(BN). \tag{7.2}$$

If $P_2(N) = 0$, then P_2 is absolutely continuous with respect to P_1, and dP_2/dP_1 is called the probability density function of P_2 with respect to P_1. Two measures which are absolutely continuous with respect to one another are called *equivalent*. Two measures P_1 and P_2 are said to be *orthogonal* if there is a set N such that $P_1(N) = 0$ and $P_2(N) = 1$.

It has been proved, independently by various authors under various hypotheses (for references, see [40], Section 4), that two normal probability measures are either equivalent or orthogonal. From the point of view of obtaining an explicit formula for the probability density function, the following formulation of this theorem is useful.

THEOREM 7A (Parzen [37], [40]). *Let P_m be the probability measure induced on the space of sample functions of a time series $\{X(t), t \in T\}$ with covariance kernel K and mean value function m. Assume that either* (i) *T is countable or* (ii) *T is a separable metric space, K is continuous, and the stochastic process $\{X(t), t \in T\}$ is separable. Let P_0 be the probability measure corresponding to the normal process with covariance kernel K and mean value function $m(t) = 0$. Then P_m and P_0 are equivalent or orthogonal, depending on whether m does or does not belong to the reproducing kernel Hilbert space $H(K)$. If m belongs to $H(K)$, then the probability density functional of P_m with respect to P_0 is given by*

$$f(X, m) = dP_m/dP_0 = \exp\{(X, m)_K - (\tfrac{1}{2})(m, m)_K\}. \tag{7.3}$$

Using the concrete formula for the probability density functional of a normal process provided by (7.3), there is no difficulty in applying the concepts of classical statistical methodology to problems of inference on normal time series. In particular the following theorem may be proved.

THEOREM 7B. *Let $\{X(t), t \in T\}$ be a normal time series, satisfying the assumptions of Theorem* 7A *with known covariance kernel $K(s, t) = \text{Cov}[X(s), X(t)]$, whose mean value function is only assumed to belong to a known class M. If M is a finite dimensional subspace of the reproducing kernel space $H(K)$, then the maximum likelihood estimate $m^*(\cdot)$, defined as that estimate in the space M of admissible mean value functions such that*

$$f(X, m^*) = \max_{m \in M} f(X, m), \tag{7.4}$$

exists and is given at each t in T by the right hand side of (6.26).

If M is an infinite dimensional space, then a maximum likelihood estimate does not exist. This is not too surprising, since M is not compact in this case. However, an estimate does exist which is the uniformly minimum variance unbiased linear estimate of the value $m(t)$ at a particular time t of the mean value function; this estimate is given by (6.26).

The theory of reproducing kernel Hilbert spaces turns out to provide a natural tool for treating problems of minimum variance unbiased estimation (see Parzen [37]). Further work along these lines in the case of normal time series is being done by Ylvisaker ([59]).

8. Correlation analysis of regression free stationary time series. In this section, we state some results for discrete parameter time series (a more comprehensive survey is given by Hannan [17]). Many of the results stated may be extended to continuous parameter time series.

We consider a discrete parameter time series $\{X(t), t = 1, 2, \cdots\}$, with zero means, which is weakly stationary of order 4 in the sense that its covariance function

$$R(v) = E[X(t)X(t+v)] \tag{8.1}$$

and its fourth cumulant function

$$\begin{aligned} Q(v_1, v_2, v_3) &= E[X(t)X(t+v_1)X(t+v_2)X(t+v_3)] \\ &\quad - R(v_1)R(v_2-v_3) - R(v_2)R(v_1-v_3) - R(v_3)R(v_1-v_2) \end{aligned} \tag{8.2}$$

are independent of t.

Example: *Linear Processes.* A discrete parameter time series $X(t)$ is said to be a linear process, if it may be represented

$$X(t) = \sum_{\alpha=-\infty}^{\infty} w(t-\alpha)\eta(\alpha) \tag{8.3}$$

where $\sum_{\alpha=-\infty}^{\infty} |w(\alpha)| < \infty$, and $\{\eta(\alpha), \alpha = 0, \pm 1, \cdots\}$ is a sequence of independent identically distributed random variables with zero means, finite fourth cumulant λ_4, and second cumulant λ_2. A linear process $X(t)$ is weakly stationary up to order 4, with covariance function, spectral density function, and fourth cumulant function satisfying

$$\begin{aligned} &R(v) = \lambda_2 \sum_{\alpha=-\infty}^{\infty} w(\alpha)w(\alpha+v), \qquad f(\lambda) = \frac{\lambda_2}{2\pi}\left| \sum_{\alpha=-\infty}^{\infty} w(\alpha)e^{-i\lambda\alpha} \right|^2, \\ &Q(v_1, v_2, v_3) = \lambda_4 \sum_{\alpha=-\infty}^{\infty} w(\alpha)w(\alpha+v_1)w(\alpha+v_2)w(\alpha+v_3), \\ &\sum_{u=-\infty}^{\infty} Q(v_1, u, u+v_2) = \alpha R(v_1)R(v_2), \qquad \alpha = \frac{\lambda_4}{(\lambda_2)^2}. \end{aligned} \tag{8.4}$$

Correlation analysis is concerned with estimating the covariance function $R(v)$, and the normalized covariance (or correlation) function

$$\rho(v) = R(v)/R(0) \tag{8.5}$$

of a stationary time series.

Given observations $\{X(t), t = 1, 2, \cdots, N\}$, one can form the sample covariance function, for $|v| \leqq N-1$,

$$R_N(v) = \frac{1}{N} \sum_{t=1}^{N-|v|} X(t)X(t+|v|) \tag{8.6}$$

which has mean

$$E[R_N(v)] = [1 - (|v|/N)]R(v). \tag{8.7}$$

As an estimate of $R(v)$, $R_N(v)$ is biased (although asymptotically unbiased). Consequently if we are interested in estimating $R(v)$ it may be preferable to take as our estimate

$$R_N^u(v) = [N/(N - |v|)]R_N(v).$$

Many authors have advocated the use of the unbiased estimate $R_N^u(v)$ in preference to the biased estimate $R_N(v)$. However, it appears to me that $R_N(v)$ is preferable to $R_N^u(v)$ for two reasons: (i) $R_N(v)$ is a positive definite function of v, which is not the case of $R_N^u(v)$; (ii) the mean square error of $R_N(v)$ as an estimate of $R(v)$ is in general less than that of $R_N^u(v)$. That (i) holds is immediate. That (ii) holds is shown in Parzen [42]. It will be seen that for theoretical purposes it is certainly more useful to consider $R_N(v)$ rather than $R_N^u(v)$.

Using the large of large numbers proved in Parzen ([38], pp. 419–420), one may prove the following theorems on consistency of the sample covariance function.

THEOREM 8A. *The sample covariance function of a weakly stationary time series is consistent in quadratic mean, in the sense that, for* $v = 0, 1, \cdots$,

$$\lim_{N\to\infty} E\,|\,R_N(v) - R(v)\,|^2 = 0 \tag{8.8}$$

if the time series is weakly stationary of order 4, *and satisfies (for* $v = 0, 1, \cdots$)

$$\lim_{N\to\infty} \frac{1}{N} \sum_{s=0}^{N-1} R^2(s) = 0 \tag{8.9}$$

$$\lim_{N\to\infty} \frac{1}{N} \sum_{s=0}^{N-v-1} Q(v, s, v+s) = 0. \tag{8.10}$$

THEOREM 8B. *The sample covariance function of a weakly stationary time series is strongly consistent, in the sense that, for each* $v = 0, 1, \cdots$,

$$P[\lim_{N\to\infty} R_N(v) = R(v)] = 1 \tag{8.11}$$

if the time series is weakly stationary of order 4 *and satisfies for positive constants* C *and* q

$$\frac{1}{N} \sum_{s=0}^{N-1} R^2(s) \leq CN^{-q} \qquad \textit{for all } N \tag{8.12}$$

$$\frac{1}{N} \left| \sum_{s=0}^{N-v-1} Q(v, s, v+s) \right| \leq CN^{-q} \qquad \textit{for all } N. \tag{8.13}$$

In particular, (8.12) and (8.13) hold if it is assumed that

$$\sum_{v=-\infty}^{\infty} |R(v)| < \infty \tag{8.14}$$

$$\sum_{v_1, v_2, v_3, =-\infty}^{\infty} |Q(v_1, v_2, v_3)| < \infty. \tag{8.15}$$

We next obtain expressions for the asymptotic covariance of the sample covariance function (for proofs of the following theorem see Bartlett [2], [3] or Parzen [36]).

THEOREM 8C. *Let $X(t)$ be a time series weakly stationary of order 4, with absolutely summable covariance and fourth cumulant functions (that is, (8.14) and (8.15) hold). Then the sample covariance function $R_N(v)$ has asymptotic covariance, for any non-negative integers v_1 and v_2,*

$$\lim_{N\to\infty} N \operatorname{Cov}[R_N(v_1), R_N(v_2)] = D(v_1, v_2) \tag{8.16}$$

where we define

$$\begin{aligned} D(v_1, v_2) = \sum_{u=-\infty}^{\infty} \{R(u)R(u + v_2 - v_1) \\ + R(u)R(u + v_2 + v_1) + Q(v_1, u, u + v_2)\}. \end{aligned} \tag{8.17}$$

For a linear process with spectral density function $f(\cdot)$

$$\begin{aligned} D(v_1, v_2) = 4\pi \int_{-\pi}^{\pi} \cos \lambda v_1 \cos \lambda v_2 f^2(\lambda) d\lambda \\ + \alpha \int_{-\pi}^{\pi}\int_{-\pi}^{\pi} \cos \lambda v_1 \cos \lambda v_2 f(\lambda_1) f(\lambda_2)\, d\lambda_1\, d\lambda_2 . \end{aligned} \tag{8.18}$$

In particular, the variance of $R_N(v)$ is approximately given by

$$\operatorname{Var}[R_N(v)] = \frac{4\pi}{N} \int_{-\pi}^{\pi} \cos^2 \lambda v\, f^2(\lambda)\, d\lambda + \frac{\alpha}{N} R^2(v) \geqq \frac{2+\alpha}{N} R^2(v). \tag{8.19}$$

The mean square error of $R_N(v)$ as an estimate of $R(v)$ is given by

$$E\,|\,R_N(v) - R(v)\,|^2 = \operatorname{Var}[R_N(v)] + \left(\frac{v}{N}\right)^2 R^2(v). \tag{8.20}$$

It was empirically observed by M. G. Kendall that the sample covariance function (traditionally called the *observed correlogram*) fails to damp down to 0 for increasing values of v, although the true covariance function $R(v)$ does damp down to 0 as v tends to ∞. This fact is borne out theoretically by (8.19) and (8.20), which show that the coefficient of mean square error $E\,|\,R_N(v) - R(v)\,|^2 / R^2(v)$ is of the order of $1/N$ for all lags v of the sample covariance function.

One may state in a variety of ways conditions under which *the sample covariance function $R_N(v)$ is asymptotically normal* in the sense that for every choice of lags $v_1, \cdots, v_k$ and real numbers $u_1, \cdots, u_k$,

$$\begin{aligned} E[\exp i\{u_1 N^{\frac{1}{2}}(R_N(v_1) - E[R_N(v_1)]) + \cdots + u_k N^{\frac{1}{2}}(R_N(v_k) - E[R_N(v_k)])\}] \\ \to \exp\left[-\frac{1}{2}\left\{\sum_{i,j=1}^{k} u_i D(v_i, v_j) u_i\right\}\right] \end{aligned} \tag{8.21}$$

as $N \to \infty$ (see Walker [54], Lomnicki and Zaremba [29], Parzen [35]). In particular, (8.21) holds if $X(t)$ is a linear process.

As an estimate of the correlation function $\rho(v)$ we take the sample correlation function

(8.22) $$\rho_N(v) = R_N(v)/R_N(0).$$

We do not discuss here the question of the best definition of the correlation function for short series. For a discussion of this problem, and references to the literature, see Weinstein [55].

By standard large sample statistical theory one readily obtains, from (8.21) and (8.16), the following theorem.

THEOREM 8D. *If $X(t)$ is a linear process, then $\rho_N(v)$ is asymptotically normal with asymptotic covariances satisfying, as $N \to \infty$,*

(8.23) $$NE[\rho_n(v_1) - \rho(v_1), \rho_N(v_2) - \rho(v_2)] \to d(v_1, v_2)$$

where we define

(8.24) $$d(v_1, v_2) = 4\pi \int_{-\pi}^{\pi} d\lambda\, \bar{f}^2(\lambda)\{\cos \lambda v_1 - \rho(v_1)\}\{\cos \lambda v_2 - \rho(v_2)\},$$

(8.25) $$\bar{f}(\lambda) = \frac{f(\lambda)}{R(0)}.$$

REMARK. It should be noted that while the variance of the sample covariance function $R_N(v)$ of a linear process depends on α, the variance of the sample correlation function $\rho_N(v)$ does not.

PROOF. Using only the first few terms of the Taylor series expansion one obtains that

(8.26) $$\frac{x}{y} - \frac{x_0}{y_0} = \frac{y}{y_0^3}\{(y - y_0)x_0 - (x - x_0)y_0\} + 0(|x - x_0|^2 + |y - y_0|^2)$$

Consequently, if X_n, Y_n, and Z_n are sequences of random variables, and x_0, y_0, and z_0 are constants, such that

$$n^{\frac{1}{2}}(X_n - x_0), \qquad n^{\frac{1}{2}}(Y_n - y_0), \qquad n^{\frac{1}{2}}(Z_n - z_0)$$

are jointly asymptotically normal it follows that

$$n^{\frac{1}{2}}\left(\frac{X_n}{Y_n} - \frac{x_0}{y_0}\right), \qquad n^{\frac{1}{2}}\left(\frac{Z_n}{Y_n} - \frac{z_0}{y_0}\right)$$

are jointly asymptotically normal with asymptotic covariance satisfying

(8.27) $$\begin{aligned} y_0^4\, n \operatorname{Cov}\left[\frac{X_n}{Y_n} - \frac{x_0}{y_0}, \frac{Z_n}{Y_n} - \frac{z_0}{y_0}\right] &\to x_0 z_0 E[(Y_n - y_0)^2] \\ + y_0^2 E[(X_n - x_0)(Z_n - z_0)] &- x_0 y_0 E[(Y_n - y_0)(Z_n - z_0)] \\ &- z_0 y_0 E[(Y_n - y_0)(X_n - x_0)]. \end{aligned}$$

Applying these results to the present case it follows that the sample correlations

$\rho_N(v)$ are asymptotically normal with asymptotic covariances satisfying

$$\begin{aligned} &R^4(0)N \operatorname{Cov}[\rho_N(v_1) - \rho(v_1), \rho_N(v_2) - \rho(v_2)] \\ (8.28) \quad &= R(v_1)R(v_2)D(0,0) + R^2(0)D(v_1, v_2) - R(0)R(v_1)D(0, v_2) \\ &\qquad - R(0)R(v_2)D(0, v_1). \end{aligned}$$

From (8.28) and (8.18), one obtains (8.24).

We are now in a position to obtain confidence intervals for, or test hypotheses about, a correlation coefficient $\rho(v)$. From Theorem 8D it follows that the sample correlation coefficient $\rho_N(v)$ may be regarded as being normally distributed with mean $\rho(v)$ and variance equal to $d(v)/N$ where we define

$$(8.29) \qquad d(v) = 4\pi \int_{-\pi}^{\pi} d\lambda\, \bar{f}^2(\lambda)\{\cos \lambda v - \rho(v)\}^2.$$

Now $d(v) \leqq 16\pi \int_{-\pi}^{\pi} d\lambda \bar{f}^2(\lambda)$; further, for large values of v, approximately $d(v) = 2\pi \int_{-\pi}^{\pi} d\lambda \bar{f}^2(\lambda)$. One thus sees that in order to obtain bounds for $d(v)$ one must have a knowledge of the quantity

$$(8.30) \qquad d = 2\pi \int_{-\pi}^{\pi} d\lambda \frac{f^2(\lambda)}{R^2(0)} = \frac{1}{R^2(0)} \sum_{v=-\infty}^{\infty} R^2(v).$$

In the study of both correlation analysis and spectral analysis of stationary time series it will be found that the quantity d arises frequently as information which one requires about the time series under consideration in order to carry out various statistical procedures. A satisfactory estimate of d from observations $\{X(t), t = 1, 2, \cdots, N\}$ is provided by

$$(8.31) \qquad d_N = \frac{1}{2R_N^2(0)} \sum_{v=-(N-1)}^{N-1} R_N^2(v).$$

If one does not desire to compute the sample covariance function for all $v = 0, 1, \cdots, N$ then one may take, for any θ in $0 < \theta \leqq 1$,

$$(8.32) \qquad d_{N,[\theta N]} = \frac{1}{(1 + 2\theta - \theta^2)R_N^2(0)} \sum_{v=-[\theta N]}^{[\theta N]} R_N^2(v)$$

as an estimate of d. The properties of the estimates d_N and $d_{N,[\theta N]}$ have been extensively investigated by Lomnicki and Zaremba [29]; among other things they show that d_N is a consistent estimate of d which in the case of a linear process has an asymptotic variance not dependent on the residuals $\{\eta(\alpha)\}$.

An alternate approach to the problem of investigating the mechanism generating a time series is to attempt to fit the time series by a finite parameter scheme (such as an autoregressive scheme or a moving average scheme). Here we consider only the problem of fitting an autoregressive scheme which has the most developed theory (for recent work on fitting moving average schemes, see Durbin [10]).

THEOREM 8E. *In order that a stationary time series $X(t)$ with covariance func-*

tion $R(v)$, satisfy the autoregressive scheme of order m,

$$X(t) = a_1X(t-1) + \cdots + a_mX(t-m) + \eta(t) \tag{8.33}$$

where $\eta(t)$ are a sequence of orthogonal random variables (with common variance σ^2) representing the innovation at time t so that

$$E[X(s)\eta(t)] = 0 \qquad \text{for } s < t \tag{8.34}$$

it is necessary and sufficient that the covariance function $R(v)$ satisfy the difference equation

$$R(u) = a_1R(u-1) + \cdots + a_mR(u-m) \qquad \text{for } u > 0 \tag{8.35}$$

while for $u = 0$

$$R(0) = a_1R(1) + \cdots + a_mR(m) + \sigma^2. \tag{8.36}$$

REMARK. Equations (8.35) are called the Yule-Walker equations, after G. Udny Yule and Sir Gilbert Walker who first obtained relations of this kind (see Wold [58], especially pp. 104–5 and pp. 140–146).

PROOF. Verify that (8.33) and (8.35) are each equivalent to the assertion that the minimum means square error linear predictor of $X(t)$, given its infinite past, depends only on the finite past $X(t-1), \cdots, X(t-m)$; in symbols, for all t

$$\begin{aligned} &E^*[X(t) \mid X(t-1), \cdots, X(t-m), \cdots] \\ &\qquad = a_1X(t-1) + \cdots + a_mX(t-m). \end{aligned} \tag{8.36}$$

We may use the fact that the covariance of a stationary autoregressive scheme satisfies the difference equation (8.35) to obtain expressions for the constants $a_1, \cdots, a_m$ in terms of correlations; (8.35) with $u = 1, \cdots, m$ yields m equations which may be written in matrix form

$$\begin{bmatrix} \rho(0) & \rho(1) & \cdots & \rho(m-1) \\ \rho(1) & \rho(0) & \cdots & \rho(m-2) \\ \vdots & & & \\ \rho(m-1) & \rho(m-2) & \cdots & \rho(0) \end{bmatrix} \begin{bmatrix} a_1 \\ a_2 \\ \vdots \\ a_m \end{bmatrix} = \begin{bmatrix} \rho(1) \\ \rho(2) \\ \vdots \\ \rho(m) \end{bmatrix}. \tag{8.37}$$

Consistent asymptotically normal estimates $a_1^{(N)}, \cdots, a_m^{(N)}$, of $a_1, \cdots, a_m$ respectively, may be obtained from observations $\{X(t), t = 1, 2, \cdots, N\}$ by forming consistent asymptotically normal estimates $\rho_N(v)$ of $\rho(v)$ and defining $a_1^{(N)}$ to be the solutions of

$$\begin{bmatrix} \rho_N(0) & \rho_N(1) & \cdots & \rho_N(m-1) \\ \rho_N(1) & \rho_N(0) & \cdots & \rho_N(m-2) \\ \vdots & \vdots & \cdots & \vdots \\ \rho_N(m-1) & \rho_N(m-2) & \cdots & \rho_N(0) \end{bmatrix} \begin{bmatrix} a_1^{(N)} \\ a_2^{(N)} \\ \vdots \\ a_m^{(N)} \end{bmatrix} = \begin{bmatrix} \rho_N(1) \\ \rho_N(2) \\ \vdots \\ \rho_N(m) \end{bmatrix}.$$

It may be shown, using standard techniques of large sample statistical theory, that if the estmates $R_N(v)$ satisfy (8.21), then the estimates $a_i^{(N)}$ satisfy

$$E[\exp i\{u_1(N^{\frac{1}{2}})(a_1^{(N)} - a_1) + \cdots + u_m N^{\frac{1}{2}}(a_m^{(N)} - a_m)\}] \rightarrow \exp\left[-\frac{1}{2}\sum_{i,j=1}^{m} u_i \sigma^2 C_{ij} u_j\right] \tag{8.39}$$

where $\{C_{ij}\}$ is the inverse matrix of the m by m matrix whose (i, j)th entry is $\rho(i - j)$. As an *estimate* $\{C_{ij}^{(N)}\}$ of $\{C_{ij}\}$ one may take the inverse matrix of $\{\rho_N(i - j)\}$, and as an estimate of σ^2 one may take

$$\sigma_N^2 = \frac{1}{N - m}\sum_{t=m+1}^{N} \{X(t) - a_1^{(N)}X(t-1) - \cdots - a_m^{(N)}X(t-m)\}^2. \tag{8.40}$$

In words, (8.39) says that the usual theorems of regression analysis apply asymptotically to the problem of estimating the autoregressive coefficients, even though the regression functions $X(t - 1), \cdots, X(t - m)$ represent lagged values of the observed time series $X(t)$. This fact was first shown by Mann and Wald [31] whose paper is a fundamental contribution to the theory of time series analysis.

To prove (8.39), we write (8.37) and (8.38) in alternate form as follows. Define $a_0 = a_0^{(N)} = 1$. Then, for $i = 1, 2, \cdots, m$

$$\sum_{j=0}^{m} a_j \rho(i - j) = 0 \tag{8.37'}$$

$$\sum_{j=0}^{m} a_j^{(N)} \rho_N(i - j) = 0. \tag{8.38'}$$

Therefore for $i = 1, \cdots, m$

$$\sum_{j=0}^{m} \rho_N(i - j)\{a_j^{(N)} - a_j\} = \sum_{j=0}^{m} a_j\{\rho(i - j) - \rho_N(i - j)\}. \tag{8.41}$$

From (8.41) one may deduce (8.39).

EXAMPLE. Let us write out the foregoing formulas for the case of an autoregressive scheme of order 2. Then (8.38) may be written

$$\begin{aligned} a_1^{(N)} \quad\quad\;\; + a_2^{(N)}\rho_N(1) &= \rho_N(1) \\ a_1^{(N)}\rho_N(1) + a_2^{(N)} \quad\quad\;\; &= \rho_N(2). \end{aligned} \tag{8.42}$$

The estimates $a_1^{(N)}$ and $a_2^{(N)}$ are explicitly given by

$$\begin{aligned} a_1^{(N)} &= \frac{\rho_N(1)\{1 - \rho_N(2)\}}{1 - \rho_N^2(1)} \\ a_2^{(N)} &= \frac{\rho_N(2) - \rho_N^2(1)}{1 - \rho_N^2(1)}. \end{aligned} \tag{8.43}$$

The estimated covariance matrix of $\{\text{Cov}\,[a_i^{(N)}, a_j^{(N)}]\}$ is given by

$$\{\text{Cov}\,[a_i^{(N)}, a_j^{(N)}]\}_{\text{est}} = \frac{1}{N}\sigma_N^2 \begin{bmatrix} 1 & \rho_N(1) \\ \rho_N(1) & 1 \end{bmatrix}^{-1}. \tag{8.44}$$

To test the null hypothesis that the time series obeys an autoregressive scheme or order 1 against the alternative hypothesis that it obeys an autoregressive scheme of order 2 one uses the statistic

$$\delta = \frac{|a_2^{(N)}|^2}{\text{Var}\,[a_2^{(N)}]} = \frac{N\{\rho_N(2) - \rho_N^2(1)\}^2}{\sigma_N^2\{1 - \rho_N^2(1)\}} \tag{8.45}$$

which under the null hypothesis is distributed as χ^2 with 1 degree of freedom. One may similarly give a test of the null hypothesis that the time series obeys an autoregressive scheme of order q' against the alternative hypothesis that it obeys an autoregressive scheme of order q (greater than q').

For an excellent review of both the small and large sample theory of goodness of fit tests for autoregressive schemes, we refer the reader to the monograph by E. J. Hannan [17]. For references to recent work on explosive stochastic difference equations, see Rao [44a].

REFERENCES

[1] N. Aronszajn, "Theory of reproducing kernels," *Trans. Amer. Math. Soc.*, Vol. 68 (1950), pp. 337–404.

[2] M. S. Bartlett, "On the theoretical specification and sampling properties of autocorrelated time series," *J. Roy. Stat. Soc., Suppl.*, Vol. 8 (1946), pp. 27–41.

[3] M. S. Bartlett, *An Introduction to Stochastic Processes*, Cambridge, 1955.

[4] R. B. Blackman and J. W. Tukey, *The Measurement of Power Spectra from the Point of View of Communications Engineering*, New York, Dover Publications, 1959.

[5] H. W. Bode and C. E. Shannon, "A simplified derivation of linear least squares smoothing and prediction theory," *Proc. I.R.E.*, Vol. 38 (1950), pp. 417–425.

[6] C. E. P. Brooks and N. Carruthers, *Handbook of Statistical Methods in Meteorology*, London, H. M. Stationary Office, 1953.

[7] K. A. Bush and I. Olkin, "Extrema of quadratic forms with applications to statistics," *Biometrika*, Vol. 46 (1959), pp. 483–486.

[8] Lola S. Deming, "Selected bibliography of statistical literature, 1930–1957: II. Time series," *J. Res. Nat. Bur. Standards*, Vol. 64B (1960), pp. 69–76.

[9] J. L. Doob, *Stochastic Processes*, New York, John Wiley and Sons, 1953.

[10] J. Durbin, "Efficient estimation of parameters in moving average models," *Biometrika*, Vol. 46 (1959), pp. 306–316.

[11] R. A. Fisher, "Tests of significance in harmonic analysis," *Proc. Roy. Soc.*, Ser. A. Vol. 125 (1929), p. 54.

[12] U. Grenander, "Stochastic processes and statistical inference," *Arkiv fur Mathematik*, Vol. 1 (1950), pp. 195–277.

[13] U. Grenander and M. Rosenblatt, *Statistical Analysis of Stationary Time Series*, New York, John Wiley and Sons, 1957.

[14] U. Grenander and G. Szego, *Toeplitz Forms and their Applications*, Berkeley, University of California Press, 1958.

[15] T. N. E. Greville, "Some applications of the pseudo inverse of a matrix," *SIAM Review*, Vol. 2 (1960), pp. 15–22.

[16] P. Halmos, *Introduction to Hilbert Space*, Chelsea, New York, 1951.

[17] E. J. Hannan, *Time Series Analysis*, Methuen, London, 1960.

[18] H. O. Hartley, "Tests of significance in harmonic analysis," *Biometrika*, Vol. 36 (1949), pp. 194–201.
[19] M. P. Heble, "Some problems in time series analysis," *J. Math. and Mechanics*, Vol. 9 (1960), pp. 951–987.
[20] G. M. Jenkins, "General considerations in estimation of spectra," *Technometrics*, Vol. 3 (1961), pp. 133–166.
[21] K. Karhunen, "Über linear methoden in der Wahrscheinlichkeitsrechnung," *Ann. Acad. Sci. Fennicae*, I. Math.-Physica, No. 37 (1947).
[22] M. G. Kendall, "On the analysis of oscillatory time series," *J. Roy. Stat. Soc.*, Vol. 106 (1945), p. 91.
[23] M. G. Kendall, *Advanced Theory of Statistics*, Vol. II, London, Charles Griffin, 1946.
[24] A. Khintchine, "Korrelationstheorie der stationaren stochastischen Prozesse," *Math. Ann.*, Vol. 109 (1933), pp. 604–615.
[25] A. Kolmogorov, "Interpolation and extrapolation," *Bull. Acad. Sci. U.R.S.S.*, Ser. Math., Vol. 5 (1941), pp. 3–14.
[26] J. H. Lanning and R. H. Battin, *Random Processes in Automatic Control*, New York, McGraw Hill Book Co., 1956.
[27] M. Loève, "Fonctions aléatoires du second ordre," Suplement to P. Lévy, *Processus Stochastiques et Mouvement Brownien*, Paris, Gauthier-Villars, 1948.
[28] M. Loève, *Probability Theory*, New York, D. Van Nostrand, 1955.
[29] Z. A. Lomnicki and S. K. Zaremba, "On some moments and distributions occurring in the theory of linear stochastic processes," *Monatshefte für Mathematik*, Part I, Vol. 61 (1957), pp. 318–358; Part II, Vol. 63 (1959), pp. 128–168.
[30] M. S. Longuet-Higgins, "On the intervals between successive zeros of a random function," *Proc. Roy. Soc.*, Ser. A, Vol. 246 (1958), pp. 99–118.
[31] H. B. Mann and A. Wald, "On the statistical treatment of linear stochastic difference equations," *Econometrica*, Vol. 11 (1943), pp. 173–220.
[32] D. Middleton, *Statistical Communication Theory*, New York, McGraw Hill Book Co., 1960.
[33] P. A. P. Moran, "Random processes in economic theory and analysis," *Sankhyā*, Vol. 21 (1959), pp. 99–126.
[34] J. Neyman, "Current problems of mathematical statistics," *Proceedings of the International Congress of Mathematicians*, Amsterdam, 1954, Volume 1.
[35] E. Parzen, "A central limit theorem for multilinear stochastic processes," *Ann. Math. Stat.*, Vol. 28 (1957), pp. 252–255.
[36] E. Parzen, "On consistent estimates of the spectrum of a stationary time series," *Ann. Math. Stat.*, Vol. 28 (1957), pp. 329–348.
[37] E. Parzen, "Statistical inference on time series by Hilbert space methods, I," Department of Statistics, Stanford University, Technical Report No. 23, January 2, 1959.
[38] E. Parzen, *Modern Probability Theory and its Applications*, John Wiley and Sons, 1960.
[39] E. Parzen, "A new approach to the synthesis of optimal smoothing and prediction systems," Department of Statistics, Stanford University, Technical Report No. 34, July 15, 1960. (To be published in the *Proceedings of a Symposium on Optimization*, edited by R. Bellman.)
[40] E. Parzen, "Regression analysis of continuous parameter time series, II," Department of Statistics, Stanford University, Technical Report No. 35, July 28, 1960. (To be published in the *Proceedings of the Fourth Berkeley Symposium on Mathematical Statistics and Probability Theory*, edited by J. Neyman.)
[41] E. Parzen, "Mathematical considerations in estimation of spectra," *Technometrics*, Vol. 3 (1961), pp. 167–190.
[42] E. Parzen, "Estimating the covariance function of a stationary time series," to be published.

[42a] E. Parzen and G. R. Hext, "A Bibliography of American Publications on Stochastic Processes and Time Series Analysis Published in the years 1900–1959 (Preliminary Report)," Stanford University Technical Report, 1961.

[43] V. S. Pugachev, "A method of solving the basic integral equation of statistical theory of optimum systems in finite form" (in Russian), *J. Applied Math. and Mechanics*, Vol. 23 (1959), pp. 3–14; English translation, pp. 1–16.

[44] V. S. Pugachev, *Theory of Random Functions and its Application to Automatic Control Problems* (in Russian), 2nd ed., State Publishing House of Theoretical Technical Literature (Gostekhizdat), Moscow, 1960.

[44a] M. M. Rao, "Consistency and limit distributions of estimators of parameters in explosive stochastic difference equations," *Ann. Math. Stat.*, Vol. 32 (1961), pp 195–218.

[45] M. Rosenblatt, "Statistical analysis of stochastic processes with stationary residuals," *Probability and Statistics*, edited by U. Grenander, John Wiley and Sons, New York, 1959, pp. 246–273.

[46] H. Scheffé, *Analysis of Variance*, John Wiley and Sons, New York, 1959.

[47] A. Schuster, "On the investigation of hidden periodicities with application to a supposed 26-day period of meteorological phenomena," *Terr. Mag.*, Vol. 3 (1898), p. 13.

[48] A. Schuster, "On the periodicities of sunspots," *Phil. Trans.*, Ser. A., Vol. 206 (1906), p. 69.

[49] M. M. Siddiqui, "On the inversion of the sample covariance matrix in a stationary autoregressive process," *Ann. Math. Stat.*, Vol. 29 (1958), pp. 585–588.

[50] D. Slepian, "First passage problem for a particular Gaussian process," *Ann. Math. Stat.*, Vol. 32 (1961), pp. 610–612.

[51] E. Slutsky, "The summation of random causes as the source of cyclic processes," *Econometrika*, Vol. 5 (1937), p. 105 (translation of a 1927 paper).

[52] C. T. Striebel, "Efficient estimation of regression parameters for certain second order stationary processes," Ph.D. dissertation, University of California, 1960.

[53] J. W. Tukey, "An introduction to the measurement of spectra," *Probability and Statistics*, edited by U. Grenander, John Wiley and Sons, New York, 1959, pp. 300–330.

[54] A. M. Walker, "The asymptotic distribution of serial correlation coefficients for autoregressive processes with dependent residuals," *Proc. Camb. Phil. Soc.*, Vol. 50 (1954), pp. 60–64.

[55] Abbott S. Weinstein, "Alternative definitions of the serial correlation coefficient in short autoregressive sequences," *J. Amer. Stat. Ass.*, Vol. 53 (1958), pp. 881–892.

[56] N. Wiener, "Generalized harmonic analysis," *Acta. Math.*, Vol. 55 (1930), p. 117.

[56a] N. Wiener, *The Extrapolation, Interpolation, and Smoothing of Stationary Time Series*, John Wiley and Sons, New York, 1949.

[57] H. Wold, *Demand Analysis*, John Wiley and Sons, New York, 1953.

[58] H. Wold, *A Study in the Analysis of Stationary Time Series*, Uppsala, 2nd ed., 1954 (with an appendix by Peter Whittle).

[59] D. Ylvisaker, "On a class of covariance kernels admitting a power series expansion," to be published.

[60] G. U. Yule, "On a method of investigating periodicities in disturbed series with special reference to Wolfer's sunspot numbers," *Phil. Trans.*, Ser. A, Vol. 226 (1927), p. 267.

[61] L. A. Zadeh and J. R. Ragazzini, "An extension of Wiener's theory of prediction," *J. Applied Physics*, Vol. 21 (1950), pp. 645–655.

[62] L. A. Zadeh and J. R. Ragazzini, "Optimum filters for the detection of signals in noise," *Proc. I. R. E.*, Vol. 40 (1952), pp. 1223–1231.

4

Reprinted from *Czech. Math. J.* **12**:404–444 (1962)

ON LINEAR STATISTICAL PROBLEMS IN STOCHASTIC PROCESSES

JAROSLAV HÁJEK, Praha

(Received August 2, 1960)

A unified theoretical basis is developed for the solution of such problems as prediction and filtration (including the unbiased ones), estimation of regression parameters, establishing probability densities for Gaussian processes etc. The results are applied in deriving explicit solutions for stationary processes. The paper is a continuation of [17], [18], [19] and [20].

1. Introduction and summary. Linear statistical problems have been treated in very many papers. Most of them are referred to in extensive monographies [28] and [29]. No doubt, the topic has attracted so much attention because of its great practical and theoretical interest. The aim of the present paper is to contribute to developing a unified theory and to provide explicit results for some particular classes of stationary processes.

In Section 2 we introduce closed linear manifolds generated by random variables x_t and covariances R_{ts}, respectively, and study their interplay. We also discuss the abstract feature of a linear problem and enumerate some possible applications. In Section 4 we analyse conditions under which the solution of a linear problem may be interpreted in terms of individual trajectories (i.e. not only as a limit in the mean). Section 5 contains explicit solutions for a finite segment of a stationary process with a rational spectral density. With respect to previous papers [7], [8] and [9] on this topic, our results exhaust all possibilities, are more explicit, and we indicate when they may be interpreted in terms of individual trajectories. The last two sections are devoted to Gaussian processes. We define strong equivalency of normal (i.e. Gaussian) distributions of a stochastic process and study the determinant and quadratic form defining the probability density of a normal distribution with respect to another one, stongly equivalent to it. In Section 7 we present explicit probability densities for stationary processes with rational spectral densities.

2. Basic concepts and preliminary considerations. Let $\{x_t, t \in T\}$ be an arbitrary stochastic process with finite second moments. Suppose that the mean value $\mathsf{E}x_t$ vanishes, $t \in T$, and that the covariances of x_t and x_s equal $\mathsf{E}x_t\bar{x}_s = R(x_t, x_s) = R_{ts}$, $t \in T$, $s \in T$. Let $\mathscr{X}$ be the closed linear manifold of random variables consisting of finite linear combinations $\sum c_\nu x_{t_\nu}$, $t_\nu \in T$, and of their limits in the mean. Obviously, the covariance of any two random variables $x, y \in \mathscr{X}$, say $R(x, y)$, is uniquely determined by R_{ts}, $t \in T$, $s \in T$, and the mean value of any random variable $x \in \mathscr{X}$ equals 0. The variance of x will be denoted either by $R(x, x)$ or by $R^2(x)$. The closed linear manifold $\mathscr{X}$ is a Hilbert space with the norm $R(x)$ and inner product $R(x, y)$. As usually, random variables x, y such that $R(x - y) = 0$ are considered as identical.

Let U be the following mapping of $\mathscr{X}$ in the space of complex-valued functions, φ_t, $t \in T$:

$$(Uv)_t = R(x_t, v) \quad (v \in \mathscr{X};\ t \in T)\,, \tag{2.1}$$

where $R(x_t, v) = \varphi_t$ is considered as a function of $t \in T$. Let Φ be the set of functions φ such that $\varphi = Uv$ for some $v \in \mathscr{X}$, i.e. such that $\varphi_t = R(x_t, v)$. Obviously $Uv_1 = Uv_2$ implies $v_1 = v_2$, so that there is an inverse operator U^{-1}, $U^{-1}\varphi = v$, $\varphi \in \Phi$. If we introduce in Φ the norm $Q(\varphi) = R(U^{-1}\varphi)$ and the inner product

$$Q(\psi, \varphi) = R(U^{-1}\varphi, U^{-1}\psi) \quad (\varphi, \psi \in \Phi)\,, \tag{2.2}$$

the mapping U will be unitary (i.e. isometric and one-to-one) and Φ also will be a Hilbert space.

If $R^s = R_{ts}$ is considered as a function of t only, and s is fixed, we have $U^{-1}R_s = x_s$, and

$$Q(R^s, R^t) = R_{ts} \quad (t, s \in T)\,. \tag{2.3}$$

Obviously

$$\varphi_t = Q(\varphi, R^t) \quad (t \in T)\,, \tag{2.4}$$

so that $Q(\varphi^n) \to 0$ implies $\varphi_t^n \to 0$ in every point $t \in T$. Clearly, Φ consists of finite linear combinations $\sum c_\nu R^{t_\nu}$ and of their limits in the Q-norm.

Definition 2.1. We shall say that $\mathscr{X}$ and Φ, or, more explicitely, $(\mathscr{X}, R)$ and (Φ^x, Q^x), are closed linear manifolds generated by random variables x_t and covariances R_{ts}, respectively.

To any bounded linear operator A defined on $\mathscr{X}$ there corresponds a bounded linear operator $\bar{A}$ on Φ defined by the following relation:

$$(\bar{A}\varphi)_t = R(Ax_t, U^{-1}\varphi) \quad (\varphi \in \Phi)\,, \tag{2.5}$$

where U^{-1} is the inverse of the unitary mapping (2.1) of $\mathscr{X}$ on Φ. We may also write $\bar{A}\varphi = UA^*U^{-1}\varphi$, where A^* is the adjoint of A. Also conversly, to any linear operator $\bar{A}$ defined on Φ, there corresponds an operator A on $\mathscr{X}$ defined by the relation $Ax = U^{-1}\bar{A}^*Ux$. Obviously, $\overline{AB} = \bar{B}\bar{A}$. In what follows we shall omitt the bar so that any bounded linear operator A will be considered as defined on both spaces Φ and $\mathscr{X}$. We have to bear in mind, of course, that AB in $\mathscr{X}$ must be interpreted as BA in Φ, and vice versa.

Let $\mathscr{X}_0$ be the set of all finite linear combinations $\sum c_\nu x_{t_\nu}$. Let $\mathscr{X}^+$ be the closure of $\mathscr{T}_0$ with respect to a covariance R^+, and A be an operator in $\mathscr{X}^+$. Let R be another covariance. If

$$R(x, y) = R^+(Ax, Ay), \quad (x, y \in \mathscr{X}_0) \tag{2.6}$$

then

$$R(x) \leqq kR^+(x), \tag{2.7}$$

where $k = \|A\|$ and $\|A\|$ is the norm of A. Conversly, (2.7) implies existence of a bounded, positive and symmetric linear operator A in $\mathscr{X}^+$ such that (2.6) holds. Actually, if y is fixed, then $R(y, x)$ represents a linear functional in $\mathscr{X}^+$, and, consequently $R(y, x) = R^+(z, x)$, $x \in \mathscr{X}^+$. On putting $z = By$ and $A = B^{1/2}$, we get the needed result. Obviously

$$\|B\| = \sup_{x \in \mathscr{X}_0} \frac{R(x, x)}{R^+(x, x)}. \tag{2.8}$$

Definition 2.2. If (2.7) is satisfied, we say that the R-norm is dominated by the R^+-norm.

Lemma 2.2. *Let A be a bounded linear operator in the closed liner manifold $\mathscr{Z}$ generated by random variables z_t, $t \in T$, and let $\mathscr{X}$ be the closed linear manifold generated by random variables $x_t = Az_t$, $t \in T$. Let (Φ^z, Q^z) and (Φ^x, Q^x) be closed linear manifolds generated by covariances $R^+_{ts} = R(z_t, z_s)$ and $R_{ts} = R(x_t, x_s)$, respectively.*

Then $\Phi^x \subset \Phi^z$. Moreover, Φ^x consists of functions expressible in the form $\varphi = A\chi$, $\chi \in \Phi^z$, and

$$Q^x(\psi, \varphi) = Q^z(A_x^{-1}\psi, A_x^{-1}\varphi), \tag{2.9}$$

where the function, $A_x^{-1}\varphi$, $\varphi \in \Phi^x$, is uniquely determined by the conditions

$$A(A_x^{-1}\varphi)_t = \varphi_t \quad \text{and} \quad (A_x^{-1}\varphi)_t = R(z_t, v), \quad v \in \mathscr{X}. \tag{2.10}$$

Proof. If $\varphi \in \Phi^x$, then $\varphi_t = R(Az_t, v) = AR(z_t, v) = (A\chi)_t$; and conversly $AR(z_t, v) = R(x_t, v) \in \Phi^x$. If we suppose that $v \in \mathscr{X}$, then v is determined by $R(x_t, v)$ uniquely, and $(A_x^{-1}\varphi)_t = R(z_t, v)$ is a unique solution of (2.10). The proof is accomplished.

Remark 2.1. From Lemma 2.2 it follows that $\Phi^x = \Phi^z$ if R and R^+ dominate each other.

The values φ_t of every $\varphi \in \Phi$ may be considered as values of a linear functional $f(x) = R(x, U^{-1}\varphi)$ in the points $x = x_t$, $t \in T$. On applying the well-known extension theorem ([32], § 35), we get the following result.

Lemma 2.3. *A function φ_t belongs to Φ if and only if there exist a finite constant k such that for any linear combination,*

$$|\sum c_\nu \varphi_{t_\nu}| \leqq kR(\sum c_\nu x_{t_\nu}). \tag{2.11}$$

Example 2.1. Let the family $\{x_t, t \in T\}$ be finite, $\{x_t, t \in T\} = \{x_1, \ldots, x_n\}$, and let the matrix (R_{ij}), $R_{ij} = R(x_i, x_j)$, $1 \leqq i, j \leqq n$, be regular. Then Φ consists of all functions $\varphi = \varphi_i$, $1 \leqq i \leqq n$, and

$$Q(\psi, \varphi) = \sum\sum \psi_i \bar{\varphi}_j Q_{ij}, \tag{2.12}$$

where (Q_{ij}) is the inverse of $(R_{ij}, (Q_{ij}) = (R_{ij})^{-1}$. Moreover

$$U^{-1}\varphi = \sum\sum x_i \bar{\varphi}_j Q_{ij}. \tag{2.13}$$

Now, consider the $x_i's$ as functions of an elementary event $\omega \in \Omega$. Then, for a fixed ω, the sample sequence (trajectory) $x_1(\omega), \ldots, x_n(\omega)$, represents a function of i, $1 \leqq i \leqq n$. If we denote the latter function by x^ω, then (2.13) may be rewritten in a form dual to (2.1)

$$(U^{-1}\varphi)_\omega = Q(x^\omega, \varphi). \tag{2.14}$$

Proof. In view of (2.3) and (2.4), the formulas (2.12) and (2.13) are clearly true for $\varphi = R^t$ and $\psi = R^s$, $1 \leqq t, s \leqq n$, and the general case may be obtained on putting $\varphi = \sum_{\nu=1}^{n} c_\nu R^\nu$, $\psi = \sum_{\nu=1}^{n} d_\nu R^\nu$.

The determination of Φ and $Q(\psi, \varphi)$, and the solution of the equation $Uv = \varphi$ constitute what will be called a linear problem. If T is a finite set, then, as we have seen in Example 2.1, the linear problem is equivalent to inverting the covariance matrix.

Remark 2.2. If $\mathsf{E}x_t$ does not vanish, then $R(x, y) = \mathsf{E}xy - \mathsf{E}x\mathsf{E}y$, and, generally, $R(x) = 0$ is compactible with $\mathsf{E}x \neq 0$. Consequently, $R(x - y) = 0$ does not imply that $x = y$ with probability 1. This difficulty does not arise, if $\varphi_t = \mathsf{E}x_t$, $t \in T$, belongs to Φ, because then $\mathsf{E}x \neq 0$ only if $R(x) > 0$, in view of

$$\mathsf{E}x = R(x, U^{-1}\varphi). \tag{2.15}$$

So, if necessary, the above condition $\mathsf{E}x_t \equiv 0$ may be replaced by a more general condition $\mathsf{E}x_t \in \Phi$. If, and only if, $\mathsf{E}x_t \in \Phi$, $\mathsf{E}x$ represents a linear functional on $(\mathscr{X}, R)$.

Now, let us show in what kinds of applications the linear problems appear.

Application 2.1. Let y be a random variable not belonging to $\mathscr{X}$. The projection of y on $\mathscr{X}$, say Proj y, is given by

$$\text{Proj } y = U^{-1}R(x_t, y). \tag{2.16}$$

In particular circumstances, the projection is called prediction, interpolation, filtration, or a regression estimate.

Application 2.2. Let $\varphi_1, \ldots, \varphi_m$ be known linearly independent functions belonging to Φ, and let the mean value of the process depend on an unknown vector $\alpha = (\alpha_1, \ldots, \alpha_m)$, where $\alpha_1, \ldots, \alpha_m$ are arbitrary real or complex numbers, so that

$$\mathsf{E}_\alpha x_t = \sum_{j=1}^{m} \alpha_j \varphi_{jt}. \tag{2.17}$$

Let us choose linear estimates $\hat{\alpha}_j$ of the α_js, which are best according to an arbitrary criterion with the following property: $\mathsf{E}_\alpha\hat{\Theta} = \mathsf{E}_\alpha\hat{\Theta}'$ for any α and $R(\hat{\Theta}) < R(\hat{\Theta}')$ imply that $\hat{\Theta}$ is a better estimate then $\hat{\Theta}'$ (recall that $R^2(.)$ denotes the variance). Then the vector $\hat{\alpha} = (\hat{\alpha}_1, \ldots, \hat{\alpha}_m)$, where the $\hat{\alpha}_j$s are the best linear estimates of the α_js, has the form

$$\hat{\alpha} = Cv\,, \tag{2.18}$$

where $v = (U^{-1}\varphi_1, \ldots, U^{-1}\varphi_m)$ and C is a matrix which depends on other properties of the mentioned criterion. If we postulate that the estimate should be unbiassed and of minimum variance, then $C = B^{-1}$, where $B = (Q(\varphi_j, \varphi_k))$. If we postulate that the mean value of $\mathsf{E}_\alpha|\hat{\alpha}_j - \alpha_j|^2$ with respect to an apriori distribution of $\alpha_1, \ldots, \alpha_m$ should be minimum, then $C = (B + D^{-1})^{-1}$, where $D = (\mathsf{E}_{\text{ap}}\alpha_j\bar{\alpha}_k)$ and $\mathsf{E}_{\text{ap}}(.)$ denotes the apriori mean value. In the former case B^{-1} also represents the covariance matrix of the vector $(\hat{\alpha}_1, \ldots, \hat{\alpha}_m)$, and in the latter case $(B + D^{-1})^{-1}$ also represents the matrix with elements $\mathsf{E}_{\text{ap}}\mathsf{E}_\alpha(\hat{\alpha}_j - \alpha_j)\overline{(\hat{\alpha}_k - \alpha_k)}$. The main idea of the proof is as follows: From (2.15) it follows that $\mathsf{E}_\alpha x = \sum_{j=1}^m \alpha_j R(x, U^{-1}\varphi_j)$, so that projection of any estimate $\hat{\Theta}$ on the subspace $\mathscr{X}_m$ spanned by $U^{-1}\varphi, \ldots, U^{-1}\varphi_m$ has the same mean value as $\hat{\Theta}$ for any α, and, if $\hat{\Theta}$ does not belong to $\mathscr{X}_m$, has a smaller variance. See [20] and [28].

Application 2.3. Suppose that (2.17) still holds and that y is a random variable not belonging to $\mathscr{X}$ such that $\mathsf{E}_\alpha y = \sum_{j=1}^m \alpha_j c_j$ where c_j are known constants. Let us choose an estimate $\hat{y} \in \mathscr{X}$ of y according to a criterion with the following property: $\mathsf{E}_\alpha\hat{y} = \mathsf{E}_\alpha\hat{y}'$ for any α and $R(\hat{y} - y) < R(\hat{y}' - y)$ implies that $\hat{y}$ is a better estimate then $\hat{y}'$. (Obviously, this situation is a generalisation of one considered in Application 2.2.) Then the best linear estimate of y equals

$$\hat{y} = y_0 + \sum_{j=1}^m \hat{\alpha}_j(c_j - Q(\varphi_j, \varphi_0))\,, \tag{2.19}$$

where the $\hat{\alpha}_j$s have been defined in Application 2.2, and $y_0 = U^{-1}\varphi_0$, where $\varphi_{0t} = R(x_t, y)$, $t \in T$. If we postulate that $\mathsf{E}_\alpha\hat{y} = \mathsf{E}_\alpha y$ for any α and that $R(\hat{y} - y)$ should be minimum, then the best unbiassed linear estimate (2.19) has the following property:

$$R^2(\hat{y} - y) = R^2(y_0 - y) + R^2\Big(\sum_{j=1}^m \hat{\alpha}_j(c_j - Q(\varphi_j, \varphi_0))\Big) \tag{2.20}$$

(see [20] and [28]).

Application 2.4. Let P and P^+ be two normal distributions of a real stochastic process $\{x_t, t \in T\}$ defined by a common covariance $R_{ts} = R^+_{ts}$ and mean values φ_t and $\varphi^+_t \equiv 0$, respectively. If $\varphi \in \Phi$, then P is absolutely continuous with respect to P^+ and

$$\frac{\mathrm{d}\mathsf{P}}{\mathrm{d}\mathsf{P}^+} = \exp\{U^{-1}\varphi - \tfrac{1}{2}Q(\varphi, \varphi)\}\,. \tag{2.21}$$

If φ does not belong to Φ, then P and P^+ are perpendicular (mutually singular). The proof follows from the fact that $U^{-1}\varphi$ is a sufficient statistic for the pair (P, P^+), as is shown in [19]. The variance of $U^{-1}\varphi$ equals $Q(\varphi, \varphi)$ with respect to both P and P^+. In view of (2.16), the mean value of $U^{-1}\varphi$ equals $R(U^{-1}\varphi, U^{-1}\varphi) = Q(\varphi, \varphi)$, if P holds true, and, obviously, equals 0, if P^+ holds true. Now,

$$-\tfrac{1}{2}[U^{-1}\varphi - Q(\varphi, \varphi)]^2/Q(\varphi, \varphi) + \tfrac{1}{2}[U^{-1}\varphi]^2/Q(\varphi, \varphi) = U^{-1}\varphi - \tfrac{1}{2}Q(\varphi, \varphi),$$

which proves (2.21). The case of different covariances will be treated in Sections 6 and 7.

Application 2.5. A somewhat different application is given by the following Lemma: A difference of two covariances $R^+_{ts} - R_{ts}$ represents again a covariance, if and only if R^+ dominates R and the operator B determined by $R(x, y) = R^+(Bx, y)$ has the norm smaller or equal 1.

Proof: If $\|B\| \leqq 1$, then $I - B$ is a positive operator, so that $R^+_{ts} - R_{ts} = R((I - B)x_t, x_s)$ is a covariance. Conversly, if $R^+_{ts} - R_{ts}$ is a covariance, then $R^+(x, x) \geqq R(x, x)$ i.e. R^+ dominates R, and the norm of the operator B defined by $R(x, y) = R(Bx, y)$ is smaller or equal 1, in view of (2.8). Especielly, $R_{ts} - \varphi_t\bar{\varphi}_s$ is a covariance, if and only if $\varphi \in \Phi$ and $Q(\varphi) \leqq 1$. This corollary generalizes a result by A. V. Balakrishnan [21].

3. Stochastic integrals. Let $\mathscr{G}$ be a Borel field of measurable subsets Λ of T, and $\mu = \mu(\Lambda)$ be a σ-finite measure on $\mathscr{G}$. Let $Y_\Lambda = Y(\Lambda)$, be an additive random set function defined on subsets of finite measure and such that $\mathsf{E}Y_\Lambda = 0$ and

$$R(Y_\Lambda, Y_{\Lambda'}) = \mu(\Lambda \cap \Lambda') \quad (\Lambda, \Lambda' \in \mathscr{G}). \tag{3.1}$$

The closed linear manifold $\mathscr{Y}$ generated by random variables Y_Λ consits of random variables expresible in the form

$$v = \int_T h_t\, Y(\mathrm{d}t), \tag{3.2}$$

where h_t is a quadratically integrable function, $h \in \mathscr{L}^2(\mu)$. The stochastic integral (3.2) is defined as a limit in the mean (see J. L. Doob [33]).

The closed linear manifold generated by covariances (3.1) consists of all σ-additive set functions $v(\Lambda)$ such that $v(\lambda) = \int_\Lambda f\, \mathrm{d}\mu$ and

$$\int_T |f|^2\, \mathrm{d}\mu < \infty. \tag{3.3}$$

We put $f = \mathrm{d}v/\mathrm{d}\mu$ and denote $\mu(\mathrm{d}t)$ briefly by $\mathrm{d}\mu$. Moreover, we have

$$Q(v_1, v_2) = \int_T \left(\frac{\mathrm{d}v_1}{\mathrm{d}\mu}\right)\overline{\left(\frac{\mathrm{d}v_2}{\mathrm{d}\mu}\right)}\, \mathrm{d}\mu, \tag{3.4}$$

and the equation $v(\Lambda) = R(Y_\Lambda, v)$ is solved by (3.2), where $h = \mathrm{d}v/\mathrm{d}\mu$.

It often is preferable to introduce the formal derivative $y_t = (\mathrm{d}Y/\mathrm{d}\mu)_t$ (so called white noise) and to write $\int h_t y_t \mu(\mathrm{d}t)$ and $R(y_t, v)$ instead of $\int h_t\, Y(\mathrm{d}t)$ and $[(\mathrm{d}/\mathrm{d}\mu)$

$R(Y_A, v)]$, respectively. Unless the point t has a positive measure, y_t itself has no meaning, but the integrals $\int hy \, d\mu$, $h \in \mathscr{L}^2(\mu)$, and covariances $R(y_t, v)$, $t \in T$, are well-defined in the above sense. We confine ourselves to this remark without entering into the theory of random distributions [12].

$\mathscr{L}^2(\mu)$, with the usual inner product $(h, g) = \int h\bar{g} \, d\mu$, may be considered as the closed linear manifold generated by the covariance function of the white noise y_t (formally, $R(y_t, y_s) = 0$ if $t \neq s$, and $R(y_t, y_t) = 1/\mu(dt)$). The relations

$$R(y_t, v) = h_t \quad \text{and} \quad v = \int \bar{h}_t y_t \, \mu(dt) \tag{3.5}$$

define a unitary transformation $Uv = h \, (U^{-1}h = v)$ of $\mathscr{Y}$ on $\mathscr{L}^2(\mu)$ (of $\mathscr{L}^2(\mu)$ on $\mathscr{Y}$).

Now consider an operator K in $\mathscr{L}^2(\mu)$, generated by a kernel $K(t, s)$, which is measurable on $\mathscr{G} \times \mathscr{G}$ and quadratically integrable w. r. t. $\mu \times \mu$. We have

$$(Kh)_t = \int K(t, s) \, h_s \, \mu(ds) . \tag{3.6}$$

This operator may be carried over to $\mathscr{Y}$ according to the formula $Kv = U^{-1}K^*Uv$, where K^* is generated by $K^*(t, s) = \overline{K(s, t)}$. So, if $v = \int \bar{h} y \, d\mu$, then

$$Kv = \int (K^*h) \, y \, d\mu . \tag{3.7}$$

The domain of definition of the operator K in $\mathscr{Y}$ may be extended to include the white noise y_t by puting

$$x_t = Ky_t = \int K(t, s) \, y_s \, \mu(ds) . \tag{3.8}$$

Then we also may write,

$$Kv = \int \bar{h}(Ky) \, d\mu = \int \bar{h}_t x_t \, d\mu , \tag{3.9}$$

where the integral is defined in the weak sense (see [3]), i.e. as a random variable ξ belonging to the closed linear manifold $\mathscr{X}$ spanned by random variables $x_t = Ky_t$, $t \in T$, and such that

$$R(y_\tau, \xi) = \int R(y_\tau, x_t) \, h_t \, d\mu .$$

Actually, we have $R(y_\tau, x_t) = \overline{K(t, \tau)} = K^*(\tau, t)$, so that

$$\int R(y_\tau, x_t) \, h_t \, d\mu = (K^*h)_\tau = R(y_\tau, Kv) ,$$

where the last equality follows from (3.7).

The equality

$$\int \overline{(K^*h)} \, y \, d\mu = \int \bar{h}(Ky) \, d\mu , \tag{3.10}$$

obtained from (3.7) and (3.9), will now be generalized for arbitrary bounded operators. Let $X(\Lambda) = AY(\Lambda)$, $\Lambda \in \mathcal{G}$, where $Y(\Lambda)$ is an additive random set function satisfying (3.1) and A is a bounded operator defined in the closed linear manifold $\mathcal{Y}$ generated by $\{Y(\Lambda), \Lambda \in \mathcal{G}\}$. Put, by definition,

$$\int g(t)\, X(\mathrm{d}t) = A \int g(t)\, Y(\mathrm{d}t) \tag{3.11}$$

for any $g \in \mathcal{L}_2(\mu)$. Then

$$\int g(t)\, AY(\mathrm{d}t) = \int (A^*g)_t\, Y(\mathrm{d}t)\,. \tag{3.12}$$

In fact, (3.12) is a direct consequence of the identity $A = U^{-1}A^*U$, where U and U^{-1} are unitary operators defined by (2.5).

Application 3.1. Let $\{x_t, t \in T\}$ be the process expressed by (3.8). Then 1° the closed linear manifold Φ^x generated by covariances $R(x_t, x_s) = R(Ky_t, Ky_s)$, $t \in T$, $s \in T$, consists of functions expressible in the form $\varphi_t = (Kh)_t$, $h \in \mathcal{L}^2(\mu)$, 2° the inner product in Φ^x, say Q^x, is given by

$$Q^x(\psi, \varphi) = \int (K_x^{-1}\psi)_t\, \overline{(K_x^{-1}\varphi)_t}\, \mathrm{d}\mu\,, \tag{3.13}$$

and 3° the equation $\varphi_t = R(x_t, v)$ is solved by

$$v^\varphi = \int (K_x^{-1}\varphi)_t\, Y(\mathrm{d}t)\,. \tag{3.14}$$

The formula (3.14) is based on Lemma 2.2, where $K_x^{-1}\varphi$ is defined as a function such that $KK_x^{-1}\varphi = \varphi$ and $(K_x^{-1}\varphi)_t = R(y_t, x)$, where x belongs to the closed linear manifold $\mathcal{X}$ generated by random variables x_t, $t \in T$. If $\mathcal{X} = \mathcal{Y}$ then $K_x^{-1} = K^{-1}$ is an ordinary inverse of K. Actually, then $0 \equiv \varphi_t = R(x_t, v)$ implies $v \equiv 0$, and hence $(Kh)_t \equiv 0$ if and only if $h_t \equiv 0$.

Application 3.2. The operator K^* is generated by the kernel $K^*(t, s) = \overline{K(s, t)}$, and the operator KK^* by the kernel

$$R(x_t, x_s) = \int K(t, \tau)\, \overline{K(s, \tau)}\, \mu(\mathrm{d}\tau)\,. \tag{3.15}$$

The equality of both sides in (3.15) follows from (3.8). If the function φ is expressible in the form $\varphi = KK^*h$, $h \in \mathcal{L}^2(\mu)$, i.e.

$$\varphi_t = \int R(x_t, x_s)\, h_s\, \mu(\mathrm{d}s)\,, \tag{3.16}$$

then the equation $\varphi_t = R(x_t, v)$ is solved by

$$v = \int \bar{h}_t x_t\, \mathrm{d}\mu = \int \overline{((KK^*)^{-1}\varphi)_t}\, x_t\, \mathrm{d}\mu\,. \tag{3.17}$$

Actually, $v \in \mathcal{X}$ and $\varphi_t = R(x_t, v)$ follows directly from (3.16) and the week definition of the integral $\int \bar{h}_t x_t\, \mathrm{d}\mu$.

Example 3.1. Let $\{x_t, t \leqq t_0\}$ be a semi-infinite segment of a regular stationary process. As is well-known (J. L. Doob [30], Chap. XII, § 5), there exist a uniquely determined quadratically integrable function $c(\tau)$ vanishing for $\tau < 0$, and a process with uncorrelated increments Y_t, $E\,|\mathrm{d}Y_t|^2 = \mathrm{d}t$, such that $x_t = \int_{-\infty}^{t} c(t - s)\,\mathrm{d}Y_s$, and the closed linear manifolds $\mathscr{X}$ and $\mathscr{Y}$ spanned by

$$\{x_t, t \leqq t_0\} \quad \text{and} \quad \left\{y_t = \frac{\mathrm{d}Y_t}{\mathrm{d}t}, \quad t \leqq t_0\right\}$$

respectively, coincide, $\mathscr{X} = \mathscr{Y}$. So we may apply the preceeding results with $K(t, s) = c(t - s)$, $\mu(\mathrm{d}t) = \mathrm{d}t$ and $T = (-\infty, t_0]$. In the present case equation (3.16) is called the Wiener-Hopf equation. The functions given by (3.16), however, do not exhaust the whole set Φ^x of functions φ, for which the linerar problem $\varphi_t = R(x_t, v)$ may be solved.

For a moment, let us consider the whole process x_t, $-\infty < t < \infty$, and denote the usual Fourier-Plancherel transform by F. We know that $F^* = F^{-1}$, which implies, in view of (3.12), that $x_t = \int [Fc(t - .)]_\lambda\, \mathrm{d}(FY)_\lambda$, where, as is well-known, $[Fc(t - .)]_\lambda = e^{it\lambda} a(\lambda)$. On puting $\mathrm{d}Z_\lambda = a(\lambda)\,\mathrm{d}(FY)_\lambda$, we get the well-known spectral representation $x_t = \int e^{it\lambda}\,\mathrm{d}Z_\lambda$, where $\mathsf{E}\,|\mathrm{d}Z_\lambda|^2 = |a(\lambda)|^2\,\mathrm{d}\lambda$, $|a(\lambda)|^2$ being the spectral density.

Example 3.2. Let Y_t be a process with uncorrelated increments such that $E\,|\mathrm{d}Y_t|^2 = \mathrm{d}t$, $-1 \leqq t \leqq t_0$. We are interested in $x_t = Y_t - Y_{t-1}$, $0 \leqq t \leqq t_0$, which is a stationary process with correlation function $R(x_t, x_{t-\tau}) = \max(0, 1 - |\tau|)$. If we add the random variables $x_t = Y_t - Y_{t-1}$, $-1 \leqq t \leqq 0$, we obtain $\mathscr{X} = \mathscr{Y}$. The kernel $K(t, s)$ is apparent from relations

$$(3.18) \qquad \begin{aligned} x_t &= \int_{-1}^{t} y_s\,\mathrm{d}s\,, \quad \text{if} \quad -1 \leqq t \leqq 0\,, \\ &= \int_{t-1}^{t} y_s\,\mathrm{d}s\,, \quad \text{if} \quad 0 \leqq t \leqq t_0\,. \end{aligned}$$

Let $K_0^{-1}\varphi$ be a function such that $(KK_0^{-1}\varphi)_t = \varphi_t$, $0 \leqq t \leqq t_0$, and that $(K_0^{-1}\varphi)_t = R(y_t, v)$, where v belongs to the closed linear manifold $\mathscr{X}_0$ generated by random variables $\{x_t, 0 \leqq t \leqq t_0\}$. We may find, after some computations, that

$$(3.19) \qquad (K_0^{-1}\varphi)_t = \varphi'_t + \varphi'_{t-1} + \varphi'_{t-2} + \dots,$$

where the sum stops when the index falls into the interval $[-1, 0)$ and φ'_t has been extended so that

$$(3.20) \qquad \begin{aligned} \varphi'_t &= \frac{1}{N+1}(c - N\varphi'_{t+1} - (N-1)\,\varphi'_{t+2} - \dots - \varphi'_{t+N}) \quad \text{for} \quad t_0 - N - 1 < t < 0, \\ &= \frac{1}{N+2}(c - (N+1)\,\varphi'_{t+1} - N\varphi'_{t+2} - \dots - \varphi'_{t+N+1}) \\ &\qquad \text{for} \quad -1 < t < t_0 - N - 1\,, \end{aligned}$$

where N is the greatest integer not exceeding t_0 and

$$c = \frac{(N + 1)(\varphi_0 + \varphi_{t_0}) + \ldots + (\varphi_N + \varphi_{t_0 - N})}{2N - t_0 + 2}. \tag{3.21}$$

For $\varphi_t \equiv 1$ we could obtain the best linear estimate [15] of a constant mean value.

For $\varphi_t = R(x_{t_0+\tau}, x_t) = \max(0, 1 - |t_0 + \tau - t|)$ and $t_0 = N$,

we would obtain the best predictor [16].

The same method could be applied to processes expressed generally by $x_t = \sum_{k=0}^{m} b_{m-k} Y_{t-k}$. In all cases Φ^x consists of absolutely continuous functions with a quadratically integrable derivative.

4. Solutions expressed in terms of individual trajectories. In what follows we shall assume that the random variables $x_t = x_t(\omega)$ are defined on a probability space $(\Omega, \mathcal{F}, \mathsf{P})$. The covariances and finite-dimensional distributions of random variables $x_t(\omega)$ are not influenced by any adjustment (modification) of every $x_t(\omega)$ on a ω-subset having probability 0. The properties of individual trajectories $x^\omega = x_t(\omega)$, $t \in T$, however, may be changed very substantially.

Theorem 4.1. *The process $x_t = x_t(\omega)$ defined by (3.8) may be adjusted on ω-subsets having probability zero, so that the solution of the equation $\varphi_t = R(x_t, v)$, φ_t expressible as $\varphi_t = \int R(x_t, x_s)\, h_s\, d\mu$, $h \in \mathcal{L}^2(\mu)$, is given by*

$$v(\omega) = \int_T \bar{h}_t x_t(\omega)\, d\mu, \tag{4.1}$$

where for almost every ω the integral on the right side is to be understood in the usual Lebesgue-Stieltjes sense.

Proof. The compactness of KK^* in $\mathcal{L}^2(\mu)$ follows from (3.6) by standard arguments. Let $\chi_n(t)$ and $\kappa_n (n \geqq 1)$ be the eigen-elements and corresponding eigen-values of KK^*, respectively. The equality of some κ_n's is not excluded. We first show that $\sum \kappa_n < \infty$. As the system $\{\chi_n(t), n \geqq 1\}$ is complete and orthonormal, we have

$$\sum_1^\infty \kappa_n = \sum_1^\infty \int |K^* \chi_n|^2\, d\mu = \int \sum_1^\infty \left| \int K^*(t, s) \chi_n(s)\, \mu(ds) \right|^2 \mu(dt) = \iint |K(s, t)|^2\, \mu(ds)\, \mu(dt), \tag{4.2}$$

where the last expression is finite according to our assumptions.

Introduce the unitary transforms U and U^{-1} defined by $(Uv)_t = R(y_t, v)$. Obviously, the random variables $v_n = U^{-1}\chi_n$ form a complete orthonormal system in $\mathcal{Y}$. Moreover,

$$R(Kv_m, Kv_n) = R(v_m, K^* K v_n) = Q(KK^* \chi_n, \chi_m) = \kappa_n Q(\chi_n, \chi_m),$$

so that $\{Kv_n, n \geqq 1\}$ is a orthogonal system with $R(Kv_n, Kv_n) = \kappa_n, n \geqq 1$. In view of (3.9), $Kv_n \in \mathscr{X}$, where $\mathscr{X}$ is the closed linear manifold spanned by random variables $x_t = Ky_t, t \in T$. If $R(y, Kv_n) = 0, n \geqq 1$, then $R(K^*y, v_n) = 0, n \geqq 1$, so that $K^*y \equiv 0$ and $(KUy)_t = (UK^*y)_t \equiv 0$. Consequently,

$$R(x_t, y) = \int K(t, s)\,(Uy)_s \,\mathrm{d}\mu = (KUy)_t = 0\,,$$

$t \in T$, which shows that $y \perp Kv_n, n \geqq 1$, only if $y \perp \mathscr{X}$. We conclude that $\{Kv_n, n \geqq 1\}$, where Kv_n with $\kappa_n = 0$ may be omitted, forms complete system in $\mathscr{X}$.

Consequently, for every $t \in T$, we have

$$x_t = \sum_1^\infty Kv_n \frac{R(x_t, Kv_n)}{\kappa_n} = \sum_1^\infty Kv_n \frac{R(y_t, K^*Kv_n)}{\kappa_n} = \sum_1^\infty Kv_n \,.\, \chi_n(t)\,, \tag{4.3}$$

where the sum converges in the mean. Further, the sequence $\{\sum_1^N \chi_n(t)\, Kv_n(\omega), N \geqq 1\}$ converges in the $(\mu \times \mathrm{P})$ – mean on $T \times \Omega$ because

$$\int_T \int_\Omega \Big|\sum_{N+1}^{N+p} \chi_n(t)\, Kv_n(\omega)\Big|^2 \,\mathrm{dP}\,\mathrm{d}\mu = \sum_{N+1}^{N+p} \kappa_n \tag{4.4}$$

and $\sum_1^\infty \kappa_n < \infty$. Consequently, we may draw a subsequence such that

$$\lim_{k\to\infty} \sum_1^{N_k} \chi_n(t)\, Kv_n(\omega) = \bar{x}_t(\omega) \tag{4.5}$$

exist for almost all (t, ω) with respect to $\mu \times \mathrm{P}$. Let T_0 be a subset of T such that $\mu(T - T_0) = 0$ and that for $t \in T_0$ the limit (4.5) exists with probability one. On puting

$$\begin{aligned} \tilde{x}_t(\omega) &= \bar{x}_t(\omega)\,, \quad \text{for} \quad t \in T_0\,, \\ &= x_t(\omega)\,, \quad \text{for} \quad t \in T - T_0\,, \end{aligned} \tag{4.6}$$

and recalling that the limits (4.3) and (4.5) coincide with probability 1 for $t \in T_0$, we conclude that $\{\tilde{x}_t(\omega), t \in T\}$ is an equivalent modification of the process $\{x_t(\omega), t \in T\}$.

Now $\int(\sum_1^\infty |Kv_n|^2)\,\mathrm{dP} = \sum_1^\infty \kappa_n < \infty$, so that

$$\sum_1^\infty |Kv_n(\omega)|^2 < \infty \tag{4.7}$$

with probability one. Without any loss of generality, we may assume that (4.7) holds true for any $\omega \in \Omega$. Then $\sum_1^\infty \chi_n(t)\, Kv_n(\omega)$ for every fixed ω represents an orthonormal expansion of $\tilde{x}_t(\omega)$ in $\mathscr{L}^2(\mu)$. So, if $v(\omega)$ is given by (4.1), where $x_t(\omega) \equiv \tilde{x}_t(\omega)$, we get

$$\begin{aligned} v(\omega) &= \sum_1^\infty Kv_n(\omega) \int_T \hbar_t \chi_n(t)\,\mathrm{d}\mu = \sum_1^\infty \kappa_n^{-1} Kv_n(\omega)\,. \\ &\cdot \int_T \hbar_t (KK^*\chi_n)_t \,\mathrm{d}\mu = \sum_1^\infty \kappa_n^{-1} Kv_n(\omega) \int_T \bar{\varphi}_t \chi_n(t)\,\mathrm{d}\mu \end{aligned} \tag{4.8}$$

and, in view of (4.3),

$$R(x_t, v) = \sum_1^\infty \chi_n(t) \int_T \bar{\varphi}_t \chi_n(t)\, d\mu = \varphi_t\,, \tag{4.9}$$

which concludes the proof.

Remark 4.1. Let x_{t_1}, $t_1 \in T_1$, be arbitrary random variables from the closed linear manifold $\mathscr{X}$ spanned by random variables x_t, $t \in T$. Clearly, the extended process $\{x_t, t \in T \cup T_1\}$ spans the same closed linear manifold as the process $\{x_t, t \in T\}$. However, the class of random variables v expressible in the form

$$v = \int_{T \cup T_1} \bar{h}_t x_t\, d\mu$$

never is smaller that the class of random variables $v = \int_T \bar{h}_t x_t\, d\mu$, and usually will be larger. By this device, we may enlarge considerably the class of functions φ expressibles as $\varphi_t = \int R(x_t, x_s)\, \mu(ds)$, permitting the solution to be written in the form (4.1) For example, if the process is originaly defined on a segment $0 \leqq t \leqq T$, the added random variables may be derivatives in the endpoints (see the next section). In the extreme case we could include in the set $\{x_t, t \in T \cup T_1\}$ every random variable from $\mathscr{X}$, i.e. put $\{x_t, t \in T \cup T_1\} = \mathscr{X}$.

Another way to enlarge the class of functions φ for which the solution is expressible in terms of individual trajectories consists in replacing the white noise y_t by an ordinary random process z_t, $t \in T$. We shall suppose that z is the closed linear manofold spanned by the z_ts and K is a compact linear operator in $\mathscr{Z}$. Let Φ^z, Φ^x and Φ° be closed linear manifolds generated by covariances $R(z_t, z_s)$, $R(Kz_t, Kz_s)$ and $R(K^*Kz_t, K^*Kz_s)$, respectively. From Lemma 2.2. it follows that

$$\Phi^\circ \subset \Phi^x \subset \Phi^z\,. \tag{4.10}$$

Let Q^z and Q^x be the inner product in Φ^z and Φ^x, respectively.

Lemma 4.1. *If $\varphi \in \Phi^\circ$ and $\psi \in \Phi^x$, then*

$$Q^x(\psi, \varphi) = Q^z(\psi, (KK^*)^{-1}\varphi)\,, \tag{4.11}$$

where $(KK^*)^{-1}\varphi$ is any function such that $KK^*(KK^*)^{-1}\varphi = \varphi$.

Proof. Clearly, $Q^z(\psi, (KK^*)^{-1}\varphi) = Q(K^{-1}\psi, K^*(KK^*)^{-1}\varphi)$, where $K^{-1}\psi$ is any solution if the equation $K\chi = \psi$. In particular, we may take the solution $K_x^{-1}\psi$ considered in Lemma 2.2. Now if $(KK^*)^{-1}\varphi = R(z_t, v)$, then $K^*(KK^*)^{-1}\varphi = R(z_t, Kv)$, where Kv belongs to the closed linear manifold $\mathscr{X}$ spanned by the random variables $x_t = Kz_t$, $t \in T$. Consequently, $K^*(KK^*)^{-1}\varphi = K_x^{-1}\varphi$ where K_x^{-1} is defined by (2.10). So, we have $Q^z(\psi, (KK^*)^{-1}\varphi) = Q^z(K_x^{-1}\psi, K_x^{-1}\varphi)$, which gives (4.11), in accordance with (2.9). The proof is finished.

Note that the right side of (4.11) is an extention of Q^x, originally defined on $\Phi^x \times \Phi^x$, on $\Phi^z \times \Phi^\circ$. Also observe that in Lemma 2.2. we need a particular choice of K^{-1}, $K^{-1} = K_x^{-1}$, whereas $(KK^*)^{-1}$ may be chosen in any way, if non-unique.

Now we shall extend the formula (4.1) to the case when the white noise y_t is replaced by an ordinary process z_t, $t \in T$.

Theorem 4.2. *Suppose that the above-considered compact operator K in $\mathscr{X}$ is such that the eigen-values κ_n of K^*K form a convergent series, $\sum_1^\infty \kappa_n < \infty$. Then the process $x_t = Kz_t$, $t \in T$, may be adjusted so that almost every trajectory $x^\omega = x_t(\omega)$ belongs to Φ^z, and the solution of the equation $\varphi_t = R(x_t, v)$, $\varphi \in \Phi^0$, equals*

$$v(\omega) = Q^z(x^\omega, (KK^*)^{-1}\varphi). \tag{4.12}$$

Proof. We may proceed quite similarly as in proving Theorem 4.1, with the only exception that the limit

$$\sum_1^\infty Kv_n(\omega)\,\chi_n(t) = \sum_1^\infty Kv_n(\omega)\,R(z_t, v_n) = R(z_t, \sum_1^\infty v_n Kv_n(\omega)) = \bar{x}_t(\omega) \tag{4.13}$$

exist for every t and each ω satisfying (4.7). Other arguments need no changes.

The formula (4.12) presumes that almost every trajectory $x^\omega = x_t(\omega)$ belongs to Φ^z. The following theorem shows that in the Gaussian case x^ω cannot belong to Φ^z almost surely, unless the operator K has the property assumed in Theorem 4.2, so that this theorem cannot be strengthened.

Theorem 4.3. *Let $x_t(\omega)$ and $z_t(\omega)$ be Gaussian processes related by a bounded linear operator K, $x_t = kz_t$, $t \in T$, and Φ^z be the closed linear manifold generated by covariances $R(z_t, z_s)$. Then, for an equivalent modification of the process $x_t(\omega)$,*

$$P(x^\omega \in \Phi^z) = 1, \tag{4.14}$$

if and only if the operator K^*K is compact and the series of its eigen-values κ_n is convergent, $\sum_1^\infty \kappa_n < \infty$.

Proof. Sufficiency follows from Theorem 4.2. Necessity. If (4.14) holds true, then there exist a function $w(\omega, \omega')$ representing a random variable from $\mathscr{X}$ for ω fixed, and such that

$$x_t(\omega) = \int z_t(\omega')\,w(\omega, \omega')\,P(d\omega') = Kz_t(\omega). \tag{4.15}$$

If $\{z_n\}$ converges weakly to 0, then

$$\lim_{n\to\infty} x_n(\omega) = \lim_{n\to\infty} \int z_n(\omega')\,w(\omega, \omega')\,P(d\omega') = 0$$

for almost every ω, and, concequently, $x_n \to 0$ in probability. If the x_n's are Gaussian, then

$$P(|x_n| \geqq \varepsilon) \geqq 2 \int_{\varepsilon/R(x_n)}^{\infty} (2\pi)^{-1/2} e^{-\frac{1}{2}\tau^2}\, d\tau,$$

so that convergence in probability implies convergence in the mean, $R(x_n) \to 0$. It means that operator K transforms weakly convergent sequences in strongly convergent ones, and consequently is compact (see [32], § 85). Obviously, the operator K^*K also is compact. Let κ_n be the non-zero eigen-values of K^*K, and let $v_n(\omega)$ be the corresponding eigen-elements. We have

$$\sum_1^\infty \kappa_n = \sum_1^\infty R(Kv_n, Kv_n) = \mathsf{E}\{\sum_1^\infty |Kv_n(\omega)|^2\} \tag{4.16}$$

and, for almost all ω,

$$\sum_1^\infty |Kv_n(\omega)|^2 = \sum_1^\infty |R(v_n, w^\omega)|^2 = R(w^\omega, w^\omega) < \infty, \tag{4.17}$$

where $w^\omega = w(\omega, .)$. The random variables Kv_n are uncorrelated, because $R(Kv_n, Kv_m) = R(K^*Kv_n, v_m) = \kappa_n R(v_n, v_m)$. If we suppose that they are Gaussian, then they are independent. It remains to show that finiteness of (4.17) for almost all ω implies finiteness of (4.16), or, equivalently, that infiniteness of (4.16) would imply infiniteness of (4.17) with positive probability. If $\sum_1^\infty \kappa_n = \infty$, we may form an infinite sequence of partial sums

$$\xi_k = \sum_{n_k+1}^{n_{k+1}} Kv_n$$

such that each partial sum consist either (i) of a unique elements Kv_n such that $\kappa_n \geqq 1$, or (ii) of several elements such that $\kappa_n \leqq 1$ and $\sum_{n_k+1}^{n_k} \kappa_n \geqq 4$. The ξ_k's have mean values $\mathsf{E}\xi_k = \sum_{n_k+1}^{n_{k+1}} \kappa_n$ and variances $R^2(\xi_k) = 2\sum_{n_k+1}^{n_{k+1}} \kappa_n^2$ so that

$$\mathsf{P}(\xi_k > \tfrac{1}{4}) = 2\int_{1/2}^{\infty} (2\pi)^{-1/2} e^{-\frac{1}{2}\tau^2}\, d\tau > \tfrac{1}{3}$$

in the case (i) and

$$\mathsf{P}(\xi_k > \tfrac{1}{4}) \geqq 1 - \frac{2\sum\kappa_n^2}{|\sum\kappa_n - \frac{1}{4}|} \geqq 1 - \frac{2\sum\kappa_n}{|\sum\kappa_n - \frac{1}{4}|^2} \geqq \tfrac{1}{3} \quad (n_k < n \leqq n_{k+1})$$

in the case (ii). So $\sum_1^\infty \mathsf{P}(\xi_k > \frac{1}{4}) = \infty$ where the ξ_k's are independent. On applying to the wellknown lemma (Borel-Cantelli), we get that with probability 1 the event $\xi_k > \frac{1}{4}$ takes place for an infinite number of indices k. It means that $\sum \xi_k = \sum Kv_n = \infty$ with probability one. So (4.17) cannot hold unless $\sum \kappa_n < \infty$. The proof is finished.

5. Application to stationary processes with rational spectral density. Let us consider a finite segment $\{x_t, 0 \leqq t \leqq T\}$ of a second-order stationary process with spectral density of form

$$f(\lambda) = \frac{1}{2\pi}\left|\sum_{k=0}^{n} a_{n-k}(i\lambda)^k\right|^{-2} \quad (-\infty < \lambda < \infty). \tag{5.1}$$

where the constants a_{n-k} are real and chosen so that all roots of $\sum a_{n-k}\lambda^k = 0$ have negative real parts. Put

$$X_t = \int_0^t x_s \, \mathrm{d}s \quad (0 \leqq t \leqq T), \tag{5.2}$$

and

$$Y_t = \sum_{k=0}^{n} a_{n-k} X_t^{(k)} \quad (0 \leqq t \leqq T), \tag{5.3}$$

where $X_t^{(k)} = (\mathrm{d}^k/\mathrm{d}t^k)\, X_t$. Of course, $X_t^{(k)} = x_t^{(k-1)}$ for $k > 0$.

Lemma 5.1. *The process Y_t, defined by (5.3), has uncorrelated increments and $E|\mathrm{d}Y_t|^2 = \mathrm{d}t$. $Y_t - Y_0$ is uncorrelated with random variables $x_0, x_0', \dots, x_0^{(n-1)}$, and the closed linear manifold $\mathscr{Y}$ generated by $\{Y_t - Y_0, 0 \leqq t \leqq T, x_0, \dots, x_0^{(n-1)}\}$ equals the closed linear manifold $\mathscr{X}$ generated by $\{x_t, 0 \leqq t \leqq T\}$, $\mathscr{Y} = \mathscr{X}$.*

Proof. On making use of the well-known unitary mapping W, defined by $Wx_t = e^{it\lambda}$, and remembering that $R(x, y) = \int Wx \, Wy \, f(\lambda) \, \mathrm{d}\lambda$ we get from (5.2) and (5.3) that

$$W(Y_t - Y_0) = \frac{e^{it\lambda} - 1}{i\lambda} \sum_{k=0}^{n} a_{n-k}(i\lambda)^k \tag{5.4}$$

and $Wx_0^{(j)} = (i\lambda)^j$, $0 \leqq j \leqq n - 1$. Consequently,

$$R(Y_t - Y_0, Y_s - Y_0) = \frac{1}{2\pi} \int_{-\infty}^{\infty} \frac{(e^{it\lambda} - 1)(e^{-is\lambda} - 1)}{\lambda^2} \, \mathrm{d}\lambda = \min(t, s), \tag{5.5}$$

which proves that Y_t has uncorrelated increments and $\mathrm{E}|\mathrm{d}Y_t|^2 = \mathrm{d}t$. Further, we have

$$R(Y_t - Y_0, x_0^{(j)}) = \frac{1}{2\pi} \int_{-\infty}^{\infty} \frac{(e^{it\lambda} - 1)(-i\lambda)^{j-1}}{\sum a_{n-k}(-i\lambda)^k} \, \mathrm{d}\lambda = 0, \tag{5.6}$$

because, according to our assumption, the function under the integral sign is analytic in the upper hyperplane, including the real line, and is of order $|\lambda|^{j-1-n}$. Finally, if $v \in \mathscr{X}$ and $v \perp y_t$, $0 \leqq t \leqq T$, and $v \perp x_0^{(j)}$, $0 \leqq j \leqq n - 1$, then the function $\varphi_t = R(x_t, v)$ satisfies the relations $(L\varphi)_t = 0$, $0 \leqq t \leqq T$, $\varphi_0^{(j)} = 0$, $0 \leqq j \leqq n - 1$, i.e. $\varphi_t = 0$, $0 \leqq t \leqq T$ and $v \equiv 0$. The proof is concluded.

Let $\mathscr{L}_n^2$ denote the set of complex-value functions of $t \in [0, T]$ possessing quadratically integrable derivatives up to the order n. Introduce in $\mathscr{L}_n^2$ the linear operators

$$(L\varphi)_t = \sum_{k=0}^{n} a_{n-k} \varphi_t^{(k)}, \tag{5.7}$$

$$(L^*\varphi)_t = \sum_{k=0}^{n} (-1)^k \, a_{n-k} \varphi_t^{(k)}, \tag{5.8}$$

where a_{n-k} are taken from (5.1). Introduce in $\mathscr{L}_{2n}^2$ the operator

$$(L^*L\varphi)_t = \sum_{k=0}^{n} \sum_{j=0}^{n} (-1)^k a_{n-k} a_{n-j} \varphi_t^{(j+k)} = (-1)^n a_0^2 \varphi_t^{(2n)} + \sum_{k=1}^{n} (-1)^k A_{n-k} \varphi_t^{(2k)} \tag{5.9}$$

Theorem 5.1. *Let $\{x_t, t \in T\}$ be a finite segment of stationary process with spectral density (5.1). Let (Φ^x, Q^x) be closed linear manifold generated by covariances $R(x_t, x_s)$, $0 \leqq t, s \leqq T$.*

Then, in above notation, $\Phi^x = \mathscr{L}_n^2$ and

(5.10)
$$Q^x(\psi, \varphi) = \int_0^T (L\psi)_t (L\bar{\varphi}_t)\,\mathrm{d}t + \sum_{\substack{0 \leqq j,\ k \leqq n-1 \\ j+k \text{ even}}} \sum \psi_0^{(j)} \bar{\varphi}_0^{(k)} 2 \sum_{i=\max(0, j+k+1-n)}^{\min(j,k)} (-1)^{j-i} a_{n-i} a_{n+i-j-k-1},$$

or, equivalently,

(5.11)
$$Q^x(\psi, \varphi) = a_0^2 \int_0^T \psi_t^{(n)} \bar{\varphi}_t^{(n)}\,\mathrm{d}t + \sum_{k=1}^{n} A_{n-k} \int_0^T \psi_t^{(k)} \bar{\varphi}_t^{(k)}\,\mathrm{d}t + \\ + \sum_{0 \leqq j,\ k \leqq n-1} \sum [\psi_T^{(j)} \bar{\varphi}_T^{(k)} + \psi_0^{(j)} \bar{\varphi}_0^{(k)}] \sum_{i=\max(0, j+k+1-n)}^{\min(j,k)} (-1)^{j-i} a_{n-i} a_{n+i-j-k-1}.$$

If $\varphi \in \mathscr{L}_{2n}^2$, we may write

(5.12)
$$Q^x(\psi, \varphi) = \int_0^T \psi_t (L^* L\bar{\varphi})_t\,\mathrm{d}t + \sum_{j=0}^{n-1} \psi_T^{(j)} \sum_{k=0}^{n-1-j} (-1)^k (L\bar{\varphi})_T^{(k)} a_{n-j-k-1} + \\ + \sum_{j=0}^{n-1} \psi_0^{(j)} \sum_{k=0}^{n-1-j} (-1)^j (L^* \bar{\varphi})_0^{(k)} a_{n-k-j-1}.$$

The solution $v^\varphi = U^{-1}\varphi$ of the equation $\varphi_t = R(x_t, v)$, $\varphi \in \Phi^x$, is given by (5.10) or (5.11) on substituting $\mathrm{d}X_t^{(k)}$ for $\psi_t^{(k)}\,\mathrm{d}t$, $0 \leqq k \leqq n$, and defining the integrals in the sense of Section 3.1. Random variables $x_t(\omega)$ may be adjusted on zero-probabality ω-subsets so that the trajectories $x^\omega = x_t(\omega)$ belong to $\mathscr{L}_{n-1}^2$ with probality 1 and for $\varphi \in \mathscr{L}_{n+1}^2$, $v^\varphi(\omega)$ is given by (5.10) or (5.11) on substituting $x_t^{(k)}(\omega)$ for $\psi_t^{(k)}$, $0 \leqq k \leqq n-1$, and

$$x_T^{(n-1)}(\omega)\,\bar{\varphi}_T^{(j)} - x_0^{(n-1)}(\omega)\,\bar{\varphi}_0^{(j)} - \int_0^T x_t^{(n-1)}(\omega)\,\bar{\varphi}_t^{(j+1)}\,\mathrm{d}t \quad \text{for} \quad \int_0^T x_t^{(n)} \varphi_t^{(j)}\,\mathrm{d}t.$$

If $\varphi \in \mathscr{L}_{2n}^2$, $v^\varphi(\omega)$ is given by

(5.13)
$$v^\varphi(\omega) = \int_0^T x_t(\omega)\,(L^* L\bar{\varphi})_t\,\mathrm{d}t + \sum_{j=0}^{n-1} x_T^{(j)}(\omega) \sum_{k=0}^{n-1-j} (-1)^k (L\bar{\varphi})_T^{(k)} a_{n-j-k-1} + \\ + \sum_{j=0}^{n-1} x_0^{(j)}(\omega) \sum_{k=0}^{n-1-j} (-1)^j (L\bar{\varphi})_0^{(k)} a_{n-j-k-1}.$$

Proof. From Lemma 5.1 it follows that every element $v \in \mathscr{X}$ is of form

(5.14)
$$v = v_1 + v_2 = \sum_{j=0}^{n-1} c_j x_0^{(j)} + \int_0^T g(t)\,\mathrm{d}Y_t$$

where $\int |g(t)|^2 \, dt < \infty$, and the random variable $v_1 = \sum c_j x_0^{(j)}$ is uncorrelated with dY_t, $0 \leqq t \leqq T$, and $v_2 = \int g_t \, dY_t$ is uncorrelated with $x_0^{(j)}$, $0 \leqq j \leqq n-1$. If $\varphi_t = R(x_t, v)$, then

$$\varphi_0^{(j)} = R(x_0^{(j)}, v) = R(x_0^{(j)}, v_1) \quad (0 \leqq j \leqq n-1), \tag{5.15}$$

and, in view of (5.3),

$$(L\varphi)_t = \frac{d}{dt} R(Y_t, v) = R(y_t, v_2) \quad (0 \leqq t \leqq T), \tag{5.15'}$$

where $y_t = dY_t/dt$ is the white noise and $L\varphi \in \mathscr{L}^2$. Consequently, if $\varphi \in \Phi^x$, then $\varphi \in \mathscr{L}_n^2$. Now, let $\varphi \in \mathscr{L}_n^2$, and chose the constants $c_0, \ldots, c_{n-1}$ and the function g_t such that (5.15) and (5.15′) hold true for v given by (5.14). A direct determination of $c_0, \ldots, c_{n-1}$ would insolve cumbersome inversion of the matrix $(R(x_0^{(j)}, x_0^{(k)}))_{j,k-1}^n$. For this reason, we first establish g_t and, subsequently, we find $c_0, \ldots, c_{n-1}$ by an indirect method. In accordance with (3.5), (5.15′) is solved by

$$v_2 = \int_0^T (L\bar{\varphi})_t \, dY_t = \int_0^T (L\bar{\varphi})_t \, d(LX)_t \,. \tag{5.16}$$

The last form is based on (5.3). On inserting (5.16) in (5.14), we obtain

$$v = \sum_{j=0}^{n-1} c_j x_0^{(j)} + \int_0^T (L\bar{\varphi})_t \, d(LX)_t \,. \tag{5.17}$$

Now consider a unitary and selfadjoint operator V defined in $\mathscr{X}$ by $Vx_t = x_{T-t}$. On carrying over V to Φ^x, we obtain $(V\varphi)_t = R(Vx_t, v^\varphi) = R(x_{T-t}, v^\varphi) = \varphi_{T-t}$. If $\varphi_{T-t} = \varphi_t$, $0 \leqq t \leqq T$, we have $V\varphi = \varphi$ and $v = V^* v = Vv$ for $v = U_x^{-1}\varphi$ (i.e. for v such that $\varphi_t = R(x_t, v)$). Consequently, if $\varphi_{T-t} = \varphi_t$, the formula (5.17) must be invariant with respect to the substitution of x_{T-t} for x_t, i.e. of $(L^*X)_{T-t}$ for $(LX)_t$ and $(-1)^j x_T^{(j)}$ for $x_0^{(j)}$. This means that, for $\varphi_{T-t} = \varphi_t$, we have

(5.18)
$$v = \int_0^T (L\bar{\varphi})_t \, d(L^*X)_{T-t} + \sum_{j=0}^{n-1} c_j (-1)^j x_T^{(j)} = \int_0^T (L^*\varphi)_t \, d(L^*X)_t + \sum_{j=0}^{n-1} c_j (-1)^j x_T^{(j)},$$

where the last expression is obtained from the preceeding one on substituting $(L\bar{\varphi})_t = (L^*\bar{\varphi})_{T-t}$, which is justified because $\varphi_{T-t} = \varphi_t$. If is easy to show that

$$\int_0^T (L\bar{\varphi})_t \, d(LX_t) = \tag{5.19}$$

$$= \int_0^T (L^*\bar{\varphi})_t \, d(L^*X)_t + \sum_{\substack{j+k \text{ odd} \\ k>j}}^{n-1} \sum^{n-1} a_{n-k} a_{n-j} \sum_{v=0}^{k-j-1} (-1)^v [\bar{\varphi}_T^{(j+v)} x_T^{(k-j-v)} - \bar{\varphi}_0^{(j+v)} x_0^{(k-j-v)}] =$$

$$= \int_0^T (L^*\bar{\varphi})_t \, d(L^*X)_t + \sum_{j+k \text{ even}}^{n-1} \sum^{n-1} (x_t^{(j)} \bar{\varphi}_t^{(k)} - x_0^{(j)} \bar{\varphi}_0^{(k)}) \sum_{i=\max(0,j+k+1-n)}^{\min(j,k)} (-1)^{j-i} a_{n-i} a_{n+i-j-k-1} \,.$$

On comparing (5.17), (5.18) and (5.19), we can easily see that

$$(5.20)\quad v_1 = \sum_{j=0}^{n-1} c_j x_0^{(j)} = \sum_{j+k\ \text{even}}^{n-1} \sum^{n-1} x_0^{(j)} \overline{\varphi_0^{(k)}} 2 \sum_{i=\max(0,j+k+1-n)}^{\min(j,k)} (-1)^{j-i} a_{n-i} a_{n+i-j-k-1}\,,$$

which inserted in (5.17) gives the needed result. Since the condition. $\varphi_{T-t} = \varphi_t$ imposes no restrictions on $\varphi_0, \ldots, \varphi_0^{(n-1)}$, (5.20) represents a general solution of (5.15).

As we know, $f(x) = R(x, v^{\varphi})$ is linear functional attaining the value φ_t at the point $x = x_t$. Consequently, $Q(\psi, \varphi) = R(v^{\varphi}, v^{\psi})$ is obtained from (5.17) simply by replacing x_t by ψ_t. Bearing in mind (5.20), we thus obtain (5.10). Formulas (5.11) and (5.12) are obtained from (5.10) simply by integration per partes, the details of which we shall not reproduce here. In deriving (5.12) we may apply Green's formula and the principle of symetry, assuming that $\varphi_{T-t} = \varphi_t$.

In order to show that (5.13) may be considered as a special case of (4.1), let us extend the parameter set by adding parameter-values $01, \ldots, 0n$ and $T1, \ldots, Tn$, and putting $x_{0i} = x_0^{(i-1)}$ and $x_{Ti} = x_T^{(i-1)}$, $1 \leqq i \leqq n$. The white noise y_t will be extended so that y_{0i} are any orthonormal linear combinations of $x_{0i} = x_0^{(i-1)}$, and $y_{T1}, \ldots, y_{Tn}$ are any orthonormal random variables uncorrelated with x_t, $0 \leqq t \leqq T$ (or with y_t, $0 \leqq t \leqq T$, and $y_{01}, \ldots, y_{0n}$). Defining the linear operator K by $x_t = Ky_t$, $0 \leqq t \leqq T$, $t = 01, \ldots, 0n, T1, \ldots, Tn$, we can see that for $0 \leqq t \leqq T$, L coincides with K_x^{-1} defined by (2.10). In our case $K_x^{-1}\varphi$ simply is such a solution of $K(.) = \varphi$ that vanishes for $t = T1, \ldots, Tn$. After this extention of the set of parameter-values, (5.13) is equivalent to (3.17), where μ is defined as Lebesque measure for $0 \leqq t \leqq T$, and $\mu(0i) = \mu(Ti) = 1$, $1 \leqq i \leqq n$. Moreover, (5.13) is equivalent (4.1). The only purpose of adding points $T1, \ldots, Tn$ was to enlarge the range of the operator KK^* so that it contains all functions from $\mathscr{L}_{2n}^2$ (see Remark 4.1).

Finally, if $n > 1$ then the derivative $x_t' = z_t$ exist. Considering the operator K defined by

$$x_t = Kz_t = x_0 + \int_0^t z_s \, ds\,,$$

and applying Theorem 4.2, we readily see that $x_t(\omega)$ may be adjusted so that $x^{\omega} \in \mathscr{L}_{n-1}^2$ with probability 1, and that, for $\varphi \in \mathscr{L}_{n+1}^2$, the solution of $\varphi_t = R(x_t, v)$ has the form mentioned in the theorem. The proof is finished.

Remark 5.1. From (5.20) it follows that the inverse of $(R(x_0^{(j)}, x_0^{(k)}))_{j,k=0}^{n-1}$ consists of elements

$$(5.21)\quad D_{jk} = 2 \sum_{i=\max(0,j+k+1-n)}^{\min(j,k)} (-1)^{j-i} a_{n-i} a_{n+i-j-k-1} \quad \text{for } j+k \text{ even}\,,$$

$$= 0 \qquad \text{for } j+k \text{ odd.}$$

Now, we shall consider a general rational spectral density

$$(5.22)\qquad g(\lambda) = \frac{1}{2\pi}\left|\frac{\sum_{k=0}^{m} b_{m-k}(i\lambda)^k}{\sum_{k=0}^{n} a_{n-k}(i\lambda)^k}\right|^2 \qquad (-\infty < \lambda < \infty;\ m < n),$$

where the a's and b's are real and all roots of equations $\sum a_{n-k}\lambda^k = 0$ and $\sum b_{m-k}\lambda^k = 0$ have negative real parts. As is well-known, if x_t is a process with spectral density (5.1), then the process

$$(5.23)\qquad z_t = \sum_{k=0}^{m} b_{m-k} x_t^{(k)} \quad (m < n)$$

has the spectral density the (5.22).

Let us add $2m$ parameter-values $01, \ldots, 0m, T1, \ldots, Tm$ and define $z_{0i} = x_{0i} = x_0^{(i-1)}$, $z_{Ti} = x_{Ti} = x_T^{(i-1)}$, $1 \leqq i \leqq m$. Then the operator H defined by $x_t = Hz_t$, $0 \leqq t \leqq T$, $t = 01, \ldots, 0m, T1, \ldots, Tm$ is bounded and may be expressed by

$$(5.24)\qquad x_t = \sum_{i=1}^{m} x_0^{(i-1)} h_i(t) + \int_0^T H(t, s)\, z_s \,\mathrm{d}s \quad (0 \leqq t \leqq T),$$

where $h_j(t)$ are m linearly independent solutions of the equation $\sum b_{m-k}h(t) = 0$, and $H(t, s)$ may be established by well-known methods [31] of the theory of linear differential equations. The differential operator involved in (5.23) will be denoted shortly by M,

$$(5.25)\qquad M = \sum_{k=0}^{m} b_{m-k} \frac{\mathrm{d}^k}{\mathrm{d}t^k}.$$

Further, we put

$$M^* = \sum_{k=1}^{m} b_{m-k}(-1)^k \frac{\mathrm{d}^k}{\mathrm{d}t^k} \quad \text{and} \quad MM^* = \sum_{k=1}^{m}\sum_{j=1}^{m} b_{m-k}b_{m-j}(-1)^k \frac{\mathrm{d}^{k+j}}{\mathrm{d}t^{k+j}}.$$

Theorem 5.2. *Let $\{z_t, 0 \leqq t \leqq T\}$ be a finite segment of stationary process with spectral density (5.22). Let (Φ^z, Q^z) the closed linear manifold generated by covariances $R(z_t, z_s)$.*

Then $\Phi^z = \mathscr{L}^2_{n-m}$ and the solution $v^\chi = U_z^{-1}\chi$ of the equation $\chi_t = R(z_t, v)$ is the same as the solution of the equation $\psi_t = R(x_t, v)$, where x_t is given by (5.24) and ψ_t is uniquely determined by

$$(5.26)\qquad (M\psi)_t = \chi_t \quad \textit{and} \quad Q^x(\psi, h_j) = 0 \quad (1 \leqq j \leqq m),$$

where M is given by (5.25), $h_1, \ldots, h_m$ are m independent solutions of $\sum b_{m-k}h(t) = 0$ and Q^x is given by (5.10) or (5.11). The inner product Q^z in Φ^z is given by $Q^z(v, \chi) = Q^x(\xi, \psi)$ where Q^x is given by (5.10), ψ is determined by (5.26), and ξ also is determined by (5.26) on substituting v for χ.

Random variables $z_t(\omega)$ may be adjusted on zero-probability ω-subsets so that the trajectories $z^\omega = R_t(\omega)$ belong to $\mathscr{L}^2_{n-m-1}$ with probability 1, and that the in-

volved integral and differential operators may be applied to individual trajectories for $\chi \in \mathscr{L}^2_{n-m+1}$. If $\chi \in \mathscr{L}^2_{2n-2m}$, then the solution is given by

$$(5.27)\quad v^{\chi}(\omega) = \int_0^T z_t(\omega)\,(L^*L\varphi)_t\,dt + \sum_{j=0}^{n-m-1} z_T^{(j)}(\omega) \sum_{k=0}^{n-1-j} (-1)^z (L\varphi)_T^{(k)} a_{n-j-k-1} + $$
$$+ \sum_{j=0}^{n-m-1} z_0^{(j)}(\omega) \sum_{k=0}^{n-1-j} (-1)^j (L^*\varphi)_0^{(k)} a_{n-j-k-1}\,,$$

where φ_t is uniquely determined by

$$(5.28)\qquad (MM^*\varphi)_t = \chi_t\,,$$

and

$$(5.29)\qquad \sum_{k=0}^{n-1-j} (-1)^k (L\varphi)_T^{(k)} a_{n-j-k-1} = 0 \quad (n-m \leqq j \leqq n-1)\,,$$

$$(5.30)\qquad \sum_{k=0}^{n-1-j} (-1)^j (L^*\varphi)_T^{(k)} a_{n-j-k-1} = 0 \quad (n-m \leqq j \leqq n-1)\,.$$

Proof. Variances generated by spectral density (5.22) and variances generated by the spectral density (5.1), where $n \equiv n - m$, dominate each other, so that the closed linear manifolds generated by respective covariances coincide (cf. Remark 2.1). So, in view of Theorem 5.1, $\Phi^z = \mathscr{L}^2_{n-m}$.

Now, if $R(x_t, v) = \varphi_t$, and φ_t satisfies (5.26), then obviously $R(z_t, v) = \chi_t$. Further, it is easy to verify on the basis of (5.24) that $Q^x(\varphi, h_j) = 0$, $1 \leqq j \leqq m$, guarantees that the solution v belongs to the closed linear manifold $\mathscr{Z}$ generated by random variables z_t, $0 \leqq t \leqq T$.

Further, if $\chi \in \mathscr{L}^2_{2n-2m}$, and the conditions (5.29) and (5.30) are satisfied, then the operator M may be applied to the right side of (5.13) term by term under the integral sign which shows that $Mv^{\varphi} = v^{\chi}$, where v^{χ} is given by (5.27). Consequently $R(z_t, v^{\chi}) = MM^*R(x_t, v\varphi) = (MM^*\varphi)_t = \chi_t$, $0 \leqq t \leqq T$, in accordance with (5.27).

The assertions concerning individual trajectories are implied by Theorems 4.1 and 4.2 as in the preceeding theorem.

Remark 5.2. If $h_1, \ldots, h_m$ are the solutions of $Mh = 0$, then the solutions of $MM^*h = 0$ are $h_1, \ldots, h_1, h_1^*, \ldots, h_m^*$ where $h^*(t) = h(T-t)$. Let φ_0 be particular solution of $MM^*\varphi = \chi$. Then

$$\varphi = \varphi_0 + \sum_1^m (c_j h_j + d_j h_j^*), \quad \text{where} \quad c_1, \ldots, c_m, d_1, \ldots, d_m$$

are determined by (5.29) and (5.30). If $\chi_t = \chi_{T-t}$, then $c_j = d_j$, and if $\chi_t = -\chi_{T-t}$, then $c_j = -d_j$. In both the cases the number of unknown constants is m, and not $2m$, and they are determined by (5.29) or (5.30). Denoting $\chi_t^* = \chi_{T-t}$, we have

$$(5.31)\qquad \varphi = \varphi^\circ + \tfrac{1}{2}\sum_{j-1}^m \{c_j(\chi + \chi^*)\,[h_j + h_j^*] + c_j(\chi - \chi^*)\,[h_j - h_j^*]\}\,,$$

where $c_j(v)$ denote the constants corresponding to a function v.

Example 5.1. If $M\varphi = \varphi' + \beta\varphi$, $\beta > 0$, then the solution of (5.28) is given by

$$\varphi_t = \frac{1}{2\beta}\int_0^T e^{-|t-s|\beta}\chi_s\,ds + \tfrac{1}{2}(c_1 + c_2)\,e^{-t\beta} + \tfrac{1}{2}(c_1 - c_2)\,e^{-(T-t)\beta}, \tag{5.31'}$$

where

$$c_1 = \frac{\sum_{1\leq j\leq n/2} \beta^{2j-2}\sum_{k=0}^{n-2j} a_{n-2j-k}(\chi_0^{(k)} + \chi_T^{(k)}) - L(-\beta)\int_0^T (e^{-t\beta} + e^{-(T-t)\beta})\,\chi_t\,dt}{L(\beta) + L(-\beta)\,e^{-T\beta}}, \tag{5.32}$$

and

(5.33)

$$c_2 = \frac{\sum_{1\leq j\leq n/2} \beta^{2j-2}\sum_{k=0}^{n-2j} a_{n-2j-k}(\chi_0^{(k)} - (-1)^k\chi_T^{(k)}) - L(-\beta)\int_0^T (e^{-t\beta} - e^{-(T-t)\beta})\,\chi_t\,dt}{L(\beta) - L(-\beta)\,e^{-T\beta}},$$

where

$$L(\beta) = \sum_{k=0}^{n} a_{n-k}\beta^k .$$

Moreover,

$$(L^*L\varphi)_t = \varphi_t L^*L(\beta) + \frac{L^*L\left(\dfrac{d}{dt}\right) - L^*L(\beta)}{\beta^2 - \left(\dfrac{d}{dt}\right)^2}\,\chi_t, \tag{5.34}$$

where $L^*L(\beta) = L(-\beta)\,L(\beta)$ and d/dt denotes the differential operator. We omit the details.

Integral-valued parameter t. The spectral density of an n-th order Markovian process with integral-valued t is given by

$$f(\lambda) = \frac{1}{2\pi}\left|\sum_{k=0}^{n} a_{n-k}e^{i\lambda k}\right|^{-2} \quad (-\pi \leqq \lambda \leqq \pi), \tag{5.35}$$

where a_{n-k} are real and such that all roots of $\sum_{k=0}^{n} a_{n-k}\lambda^k = 0$ are greater that 1 in absolute value.

It may be easily shown that the random variables

$$y_t = \sum_{k=0}^{n} a_{n-k}x_{t-k} \quad (n \leqq t \leqq N) \tag{5.36}$$

are uncorrelated mutually as well as with $x_0, \ldots, x_{n-1}$, and have unit variance, $R(y_t) = 1$. If we consider a finite segment, of the process, $\{x_t, 0 \leqq t \leqq N\}$, $N \geqq 2n$, then Φ^x consist of all complex-valued functions φ_t, $0 \leqq t \leqq N$, and it may by shown by the previous method that

$$Q^x(\psi, \varphi) = \sum_{t=0}^{N}\sum_{s=0}^{N} \psi_t\overline{\varphi}_s Q^x_{ts}, \quad \text{where for } |t - s| > n \;\; Q^x_{ts} = 0 .$$

and for $|t-s| \leqq n$

$$(5.37)\quad Q^x_{ts} = \sum_{i=0}^{\min[N-t,N-s,n-|t-s|]} a_{n-i}a_{n-i-|t-s|}\,, \quad \text{if } \max(t,s) > N-n\,,$$
$$= \sum_{i=0}^{n-|t-s|} a_{n-i}a_{n-i-|t-s|}\,, \quad \text{if } n \leqq t,\ s \leqq N-n\,,$$
$$= \sum_{i=0}^{\min[t,s,n-|t-s|]} a_{n-i}a_{n-i-|t-s|}\,, \quad \text{if } \min(t,s) < n\,.$$

The solution of $\varphi_t = R(x_t, v)$ is given by

$$(5.38)\qquad v^\varphi = \sum_{t=0}^{N}\sum_{s=0}^{N} x_t\overline{\varphi}_s Q^x_{ts}\,,$$

which may be put in a form analogous to (5.13)

$$(5.39)\quad v^\varphi = \sum_{t=n}^{N-n} x_t(L^*L\varphi)_t + \sum_{t=N-n+1}^{N} x_t \sum_{k=0}^{N-t}(L\varphi)_{t+k}a_{n-k} + \sum_{t=0}^{n-1} x_t \sum_{k=0}^{t}(L^*\varphi)_{t-k}a_{n-k}\,,$$

where

$$(L\varphi)_t = \sum_{k=0}^{n} a_{n-k}\varphi_{t-k} \quad \text{and} \quad (L^*\varphi)_t = \sum_{k=0}^{n} a_{n-k}\varphi_{t+k}\,.$$

We also find that the inverted covariance matrix $(R(x_j, x_k))^{n-1}_{j,k=0}$ consists of the following elements

(5.40)

$$D_{jk} = \sum_{i=0}^{\min[j,k,n-|j-k|]} a_{n-i}a_{n-i-|j-k|} - \sum_{i=n-\max(j,k)}^{n-|j-k|} a_{n-i}a_{n-i+|j-k|} \quad (0 \leqq j,\ k \leqq n-1)\,.$$

If we have a process $\{z_t, 0 \leqq t \leqq N\}$ possessing a general rational spectral density

$$(5.41)\qquad g(\lambda) = \frac{1}{2\pi}\left|\frac{\sum_{k=0}^{m} b_{m-k}e^{i\lambda k}}{\sum_{k=0}^{m} a_{n-k}e^{i\lambda k}}\right|^2 \qquad (-\pi \leqq \lambda \leqq \pi)\,,$$

where both the a's and b's are real and such that the roots of $\sum_{k=0}^{m} a_{n-k}\lambda^k = 0$ and $\sum_{k=0}^{m} b_{m-k}\lambda^k = 0$ lie outside of the unit circle, we may proceed as we did in the continuous-parameter case.

First, we may consider, the process $\{x_t, -m \leqq t \leqq N\}$ having the spectral density (5.35) related to z_t by the difference equation

$$(5.42)\qquad z_t = \sum_{k=0}^{m} b_{m-k}x_{t-k} = (Mx)_t \quad (0 \leqq t \leqq N)\,.$$

Then the equation $\chi_t = R(z_t, v)$ is solved by

$$v^\varphi = \sum_{t=-m}^{N}\sum_{s=-m}^{N} x_t\overline{\psi}_t Q^{ts}_x$$

where ψ is uniquely determined by $(M\psi)_t = \sum_{k=0}^{m} b_{m-k}\,\psi_{t-k} = \chi_t, 0 \leqq t \leqq N$, and $Q^x(\psi, h_j) = 0$ for m arbitrary linearly independent solutions of $(Mh)_t = 0, 0 \leqq t \leqq N$. Similarly $Q^z(v, \chi) = Q^x(\xi, \psi)$, where ξ is an arbitrary solution of the equation $M\xi = v$.

Second, we may consider the process $\{x_t, -m \leqq t \leqq N + m\}$ and the adjoint operator

$$(M^*\varphi)_t = \sum_{k=0}^{m} b_{m-k}\varphi_{t+k}\,.$$

Then we obtain the following analogues of equation (5.27):

$$v = \sum_{t=0}^{N} z_t(L^*L\varphi)_t \quad (m \geqq n)\,, \tag{5.43}$$

and

$$v = \sum_{t=n-m}^{N-n+m} z_t(L^*L\varphi)_t + \sum_{t=N-n+m}^{N} z_t \sum_{k=0}^{N-t+m} (L\varphi)_{t+k}a_{n-k} + \sum_{t=0}^{n-m-1} z_t \sum_{k=0}^{t+m} (L^*\varphi)_{t-k}a_{n-k}\,. \tag{5.44}$$

$$(m < n)\,,$$

where φ_t is uniquely determined by $(MM^*\varphi)_t = \chi_t$, $0 \leqq t \leqq N$, and

$$\sum_{k=0}^{r} (L^*\varphi)_{r-m-k}a_{n-k} = 0 \quad (r = 0, \ldots, m-1)\,, \tag{5.45}$$

and

$$\sum_{k=0}^{r} (L\varphi)_{N+m-r+k}a_{n-k} = 0 \quad (r = 0, \ldots, m-1)\,. \tag{5.46}$$

Cf. (5.29) and (5.30). Remark (5.2) is also valid for the present case. If $n = 0$, we get results for the moving-average scheme.

6. Strong equivalency of normal distributions. Let us first consider two normal distributions P and P^+ of a random sequence $\{v_n, n \geqq 1\}$ defined by vanishing mean values $\mathsf{E}v_n = \mathsf{E}^+v_n = 0$, $n \geqq 1$, and covariances

$$\begin{aligned} R(v_n, v_m) &= 0\,, \quad R^+(v_n, v_m) = 0\,, \quad \text{if} \quad n \neq m\,, \\ R(v_n, v_n) &= 1\,, \quad R^+(v_n, v_n) = \frac{1}{\lambda_n} \qquad (n \geqq 1)\,. \end{aligned} \tag{6.1}$$

The J-divergence of P a P^+ restricted to the vector $\{v_1, \ldots, v_n\}$, say J_n, equals (see [18])

$$J_n = \tfrac{1}{2}\sum_{i=1}^{n} \frac{(1-\lambda_i)^2}{\lambda_i}\,, \tag{6.2}$$

and, consequently, $J_\infty = \lim_{n\to\infty} J_n < \infty$, if and only if

$$\sum_{n=1}^{\infty} (1-\lambda_n)^2 < \infty\,. \tag{6.3}$$

So, according to [18], P and P^+ defined on the Borel field generated by $\{v_n, n \geqq 1\}$

are equivalent, $P \sim P^+$, if and only if (6.3) is true. If $P \sim P^+$, then it follows from theory of martingales ([30], Th. 4.3, Ch. VIII) that

$$\frac{dP}{dP^+} = \exp\{-\tfrac{1}{2}\sum_1^\infty [v_n^2(1 - \lambda_n) + \log \lambda_n]\}, \tag{6.4}$$

where the sum converges with probability 1. In general, however, we cannot write

$$\frac{dP}{dP^+} = (\prod_1^\infty \lambda_n)^{-1/2} \exp\{-\tfrac{1}{2}\sum_1^\infty v_n^2(1 - \lambda_n)\}, \tag{6.5}$$

unless the product $\prod_1^\infty \lambda_n$ is absolutely convergent. The well-known necessary and sufficient condition for absolute convergence of $\prod_1^\infty \lambda_n$ is $\lambda_n \neq 0$ and

$$\sum_1^\infty |1 - \lambda_n| < \infty, \tag{6.6}$$

which is stronger then (6.3). If (6.6) is satisfied, then almost surely

$$\sum_1^\infty v_n^2(\omega)\,|1 - \lambda_n| < \infty, \tag{6.7}$$

because $E\{\sum_1^\infty v_n^2(\omega)\,|1 - \lambda_n|\} \leqq \sum_1^\infty |1 - \lambda_n| < \infty$.

Thus (6.6) is a necessary and sufficient condition for dP/dP^+ being of the form (6.5), where the product $\prod_1^\infty \lambda_n$ is absolutely convergent and the quadratic form $\sum_1^\infty v_n^2(1 - \lambda_n)$ is absolutely convergent with probability 1. This leads us to the notion of strongly equivalent normal distributions introduced in the following.

Definition 6.1. Two covariances R and R^+ will be called strongly equivalent, if they dominate each other (Definition 2.2), the operator B defined by $R(x, y) = R^+(Bx, y)$ has purely point spectrum, and the non-unity eigen-values λ_n of B satisfy the condition (6.6). Two normal distribution P and P^+ defined by strongly equivalent covariances R a R^+ and by vanishing mean values will be called strongly equivalent.

Lemma 6.1. *The Probability density of a normal distribution* P *with respect to another normal distribution* P^+, *strongly equivalent to* P, *is given by* (6.5) *where* v_n *and* λ_n *denote the eigen-elements and eigen-values of the operator* B *mentioned in Definition* 6.1. *The* v_n'*s are normed by* $R(v_n, v_n) = 1$, $n \geqq 1$.

Proof. Clear.

Unfortunately, the right side of (6.5) scarcely may be considered as an ultimate expression for dP/dP^+, because the eigen-values and eigen-elements are difficult to establish even if dP/dP^+ may be found explicitely (see Section 7). This leads us to derive a theory of the product $\prod_1^\infty \lambda_n$ and of the quadratic form $\sum_1^\infty v_n^2(1 - \lambda_n)$.

We begin with the following:

Definition 6.2. Let H be a compact symmetric operator, and κ_n be the non-zero eigen-values of H. If $\sum_1^\infty |\kappa_n| < \infty$, we say that H has a finite trace, and put $\operatorname{tr} H = \sum_1^\infty \kappa_n$.

Definition 6.3. Let $B - I$ be a compact symmetric operator and λ_n be the non-unity eigen-values of B. If the product $\prod_1^\infty \lambda_n$ is absolutely convergent, we say that B has a determinant, and put $\det B = \prod_1^\infty \lambda_n$.

Obviously, B has a determinant if and only if the eigen-values of B are different from 0 and $B - I$ has a finite trace. If $\{v_n\}$ is the orthonormal system of eigen-elements of B, corresponding to the non-zero eigen-values, and $\{x_n\}$ is another orthonormal system in $(\mathscr{X}, R^+)$, then

$$\sum_1^\infty |(Hx_n, x_n)| = \sum_{n=1}^\infty \Big|\sum_{j=1}^\infty (Hx_n, v_j)(x_n, v_j)\Big| = \sum_{n=1}^\infty \Big|\sum_{j=1}^\infty \kappa_j |(x_n, v_j)|^2\Big| \leqq$$
$$\leqq \sum_{j=1}^\infty |\kappa_j| \sum_{n=1}^\infty |(x_n, v_j)|^2,$$

i.e.

$$\sum_1^\infty |(Hx_n, x_n)| \leqq \sum_1^\infty |\kappa_j|, \tag{6.8}$$

where $(.,.) = R^+(.,.)$ and $H = B - I$.

Denote by $\log B$ the operator which has the same eigen-elements as the operator B but eigen-values $\log \lambda_n$ instead of λ_n. Denoting the eigen-elements of B by v_n, we have

$$(Bx, x) = \sum_1^\infty \lambda_n |(x, v_n)|^2 \quad \text{and} \quad (\log Bx, x) = \sum_1^\infty \log \lambda_n |(x, v_n)|^2 .$$

On making use of Jensen inequality, we get for $(x, x) = 1$

$$\log(Bx, x) \geqq (\log Bx, x) \quad [(x, x) = 1] . \tag{6.9}$$

Let $\mathscr{X}_B$ be the subspace of $\mathscr{X}$ spanned by eigen-elements of B corresponding to non-unity eigen-values. $\mathscr{X}_B$ is separable even if $\mathscr{X}$ is not separable. For any orthonormal system $\{x_n\}$ complete in $\mathscr{X}_B$ we have

$$\log \det B = \operatorname{tr} \log B = \sum_1^\infty (\log Bx_n, x_n) \leqq \sum_1^\infty \log(Bx_n, x_n),$$

i.e.

$$\det B \leqq \prod_{n=1}^\infty (Bx_n, x_n), \tag{6.10}$$

where the absolute convergence on the right side is guaranted by (6.8).

Let $\mathscr{X}_n$ be a n-dimensional subspace of $\mathscr{X}$, and let I_n^+ denote the projection on $\mathscr{X}_n$. The contraction of B on $\mathscr{X}_n$ say B_n, will be defined as follows.

$$B_n = I + I_n^+(B - I)\, I_n^+ \,. \tag{6.11}$$

Obviously $(B_n x, x) = (Bx, x)$, if $x \in \mathscr{X}_n$, and $(B_n x, x) = (x, x)$, if $x \perp \mathscr{X}_n$. The operator B_n is well-defined on $\mathscr{X}$ as well as on $\mathscr{X}_n$ and in both cases the determinant Det B_n is the same. If $\mathscr{X}_n$ is spanned by linearly independent elements $x_1, \ldots, x_n$, then

$$\det B_n = \frac{|(Bx_i, x_j)|}{|(x_i, x_j)|} \quad (i, j = 1, \ldots, n)\,, \tag{6.12}$$

where $|a_{ij}|$ is the ordinary determinant of a matrice (a_{ij}). Relation (6.12) will be clear, if we take for $x_1, \ldots, x_n$ the eigen-elements of B_n.

Theorem 6.1. *Let $\mathscr{X}_1 \subset \mathscr{X}_2 \subset \ldots$ be a sequence of finite-dimensional subspaces of $\mathscr{X}$ such that the smallest subspace containing $\bigcup_1^\infty \mathscr{X}_n$ contains all eigen-elements of B corresponding to non-unity eigen-values. Let B_n be defined by* (6.11). *Then*

$$\det B = \lim_{n \to \infty} \det B_n \,. \tag{6.13}$$

Proof. Let v_j and $\lambda_j \neq 1$ be the eigen-elements and eigen-values of B, respectively. For any $\varepsilon > 0$ we may chose l so that $\sum_{l+1}^{\infty} |\lambda_j - 1| < \varepsilon$ and then N so that in $\mathscr{X}_N$ there exists an orthonormal system $x_1, \ldots, x_l$ such that

$$\sum_1^l |((B - I)\, x_j, x_j)| \geqq \sum_1^l |((B - I)\, v_j, v_j)| - \varepsilon = \sum_1^l |\lambda_j - 1| - \varepsilon \,. \tag{6.14}$$

In view of (6.8) we have for any orthonormal system $\{z_n\}$, such that $z_1 = x_1, \ldots, z_l = x_l$,

$$\sum_{l+1}^{\infty} |((B - I)\, z_n, z_n)| \leqq \sum_1^{\infty} |\lambda_n - 1| - \sum_1^l |((B - I)\, z_n, z_n)| < 2\varepsilon \,. \tag{6.15}$$

Now let $n > N$, and $c^{-1} = \min_{1 \leqq j \leqq l} (Bx_j, x_j)$. Let $\{v_{nj}\}$ and $\{v_j\}$ be the eigen-elements of B_n and B, respectively. From (6.10), (6.14) and (6.15) it follows that

$$\det B_n = \prod_{j=1}^{n} (Bv_{nj}, v_{nj}) \geqq (1 - 2\varepsilon) \det B \tag{6.16}$$

and, in view of $\sum_{l+1}^{\infty} |\lambda_j - 1| < \varepsilon$,

$$\begin{aligned} \det B &\geqq (1 - \varepsilon) \prod_1^l (Bv_j, v_j) \geqq (1 - \varepsilon)(1 - c\varepsilon) \prod_1^l (Bx_j, x_j) \geqq \\ &\geqq (1 - \varepsilon)(1 - 2\varepsilon)(1 - c\varepsilon) \prod_1^n (Bx_j, x_j) \geqq (1 - 3\varepsilon - c\varepsilon) \det B_n \,, \end{aligned} \tag{6.17}$$

where $x_1, \ldots, x_n$ is an orthonormal system from $\mathscr{X}_n$ such that $x_1, \ldots, x_l$ satisfy (6.14).

Now we may let $\varepsilon \to 0$ and $c^{-1} \geqq c_0^{-1} > 0$, where c_0^{-1} is independent of ε. The proof is completed.

Now we shall study the ratio $\det B/\det B_n$, where B_n is given by (6.11), or more generally, $\det B/\det B_c$, where $B_c = I + I_c^+(B - I) I_c^+$, I_c^+ denoting the projection on a subspace $\mathscr{X}_c$ (not necessarily finite-dimensional) of $\mathscr{X}$. More precisely, I_c^+ denotes the projection on $\mathscr{X}_c$ with respect to the covariance R^+. The projection with respect to the covariance $R(x, y) = R^+(Bx, y)$, say I_c, will be generally different from I_c^+. Now, let us introduce the following "conditional" covariances

$$R^+(x, y \mid \mathscr{X}_c) = R^+(x - I_c^+ x, y - I_c^+ y), \tag{6.18}$$

$$R(x, y \mid \mathscr{X}_c) = R(x - I_c x,\ y - I_c y). \tag{6.19}$$

If $R^+(. \mid \mathscr{X}_c)$ dominates $R(. \mid \mathscr{X}_c)$, denote by B^c the operator defined by

$$R(x, y \mid \mathscr{X}_c) = R^+(B^c x, y \mid \mathscr{X}_c). \tag{6.20}$$

Theorem 6.2. *If* $\det B$ *exists, then the determinants* $\det B_c$ *and* $\det B^c$ *also exist, and*

$$\det B = \det B_c \det B^c. \tag{6.21}$$

Proof. First suppose that we have $n + m$ random variables $x_1, \ldots, x_n, x_{n+1}, \ldots, \ldots, x_{n+m}$ which are independent with unit variances, if P^+ is true, and have an arbitrary non-singular normal distribution with covariance matrix $R = (R_{ij})$, if P is true. Let $R_n = (R_{ij})_{i,j=1}^n$ and $R^n = (R_{ij}^n)_{i,j=n+1}^{n+m}$, where R_{ij}^n is the conditional (partial) covariance of x_i and x_j $(i, j > n)$ for given $x_1, \ldots, x_n$. In this case (6.21) is equivalent to

$$|R| = |R_n|\,|R^n|, \tag{6.22}$$

where $|\cdot|$ denotes the determinant of the corresponding matrix. Equation (6.22) may be proved as follows: We introduce random variables $z_i = x_i$ $(1 \leqq i \leqq n)$ and $z_i = x_i - I_n x_i$ $(n + 1 \leqq i \leqq n + m)$, where I_n is the projection on the subspace spanned by $x_1, \ldots, x_n$. In matrix notation $z = Ax$, $A = \{a_{ij}\}$, where $|A| = 1$, because $a_{ij} = 0\,(i < j)$ and $a_{ii} = 1\,(1 \leqq i \leqq n+m)$. The covariance matrix of random variables z_i equals

$$ARA^* = \begin{pmatrix} R_n & O \\ O & R^n \end{pmatrix}, \tag{6.23}$$

from which it follows that $|R| = |A|\,|R|\,|A^*| = |ARA^*| = |R_n|\,|R^n|$.

The general case will be obtained by Theorem 6.1. We take random variables $y_1, \ldots, y_m$ from $\mathscr{X} \ominus \mathscr{X}_c$ and n random variables $x_1, \ldots, x_n$ from $\mathscr{X}_c$ so that the closed linear manifold spanned by $x_1, \ldots, x_n$ contains the projections $I_c y_i$ and $I_c^+ y_i$ of y_i on $\mathscr{X}_c$. Then we let $n \to \infty$, and subsequently $m \to \infty$ so that the closed linear manifold spanned by $\{y_1, y_2, \ldots, x_1, x_2, \ldots\}$ contains all eigen-elements of B with non-unity eigen-values. The proof is finished.

Now we shall study the quadratic form $\sum_1^\infty v_n^2(1 - \lambda_n)$ appearing in (6.5).

Theorem 6.3. *Let* P *and* P^+ *be two normal distributions of a stochastic process* $\{x_t, t \in T\}$. *Let* $\int x_t \, dP = \int x_t \, dP^+ = 0$, $t \in T$. *Assume that* $x_t = Kz_t$, $t \in T$, *where* z_t *is a stochastic process or a white noise, and* K *is a compact operator such that* K^*K *has a finite trace. Let* (Φ, Q), (Φ^+, Q^+), (Φ^z, Q^z) *be closed linear manifolds generated by covariances* $R(x_t, x_s)$, $R^+(x_t, x_s)$ *and* $R(z_t, z_s)$, *respectively. Let* B *be the linear operator defined by* $R(x, y) = R^+(Bx, y)$. *Assume that* $\Phi^+ = \Phi$ *and for some constant* C

$$|Q((I - B)\varphi, \psi)| \leqq C Q^z(\varphi) Q^z(\psi) \quad (\varphi, \psi \in \Phi^x). \tag{6.24}$$

Then P *and* P^+ *are strongly equivalent. Moreover, there exist a unique extention* $\widehat{Q - Q^+}$ *of* $Q - Q^+$ *from* $\Phi \times \Phi$ *on* $\Phi^z \times \Phi^z$, *continuous in the* Q^z*-norm, and we have*

$$\frac{dP}{dP^+} = (\det B)^{-1/2} e^{-\frac{1}{2}\widehat{Q - Q^+}(x^\omega, x^\omega)}. \tag{6.25}$$

where $x^\omega = x_t(\omega)$ *is the trajectory of the process modified according to Theorem* 4.2 *or* 4.1.

Proof. $I - B$ has a finite trace because K^*K has a finite trace and because (6.24) holds, where $Q^z(\varphi) = Q(K\varphi)$. Since $\Phi^+ = \Phi$, the covariances R a R^+ dominate each the other and B has all eigen-values different from 0. Hence B has a determinant, and P a P^+ are strongly equivalent.

In Φ the operator B is defined by $Q^+(\psi, \varphi) = Q(B\psi, \varphi)$, in view of Lemma 2.2. Let $\chi_n(t)$ be the eigen-elements of B corresponding to non-unity eigenvalues, $\lambda_n \neq 1$. We first show that $\chi_n = KK^*h_n$, where $h_n \in \Phi^z$. In view of (6.24), we have

$$|Q(\psi, \chi_n)| = \frac{|Q(\psi, (I - B)\chi_n)|}{|1 - \lambda_n|} \leqq C \frac{Q^z(\chi_n)}{|1 - \lambda_n|} Q^z(\psi), \tag{6.26}$$

which shows that $Q(\psi, \chi_n)$ is a linear functional on (Φ^z, Q^z). So $Q(\psi, \chi_n) = Q^z(\psi, h_n)$ for some $h_n \in \Phi^z$, which implies, in accordance with (4.11), that $\chi_n = KK^*h_n$ $(Q^x \equiv Q)$.

Now from (6.24) it follows that $(Q - Q^+)(\varphi, \varphi) = Q^z(A\varphi, \varphi)$, where A is a bounded operator in Φ^z. Consequently, if $\varphi_n \to \varphi$, where $\varphi_n \in \Phi$ and $\varphi \in \Phi^z$, there exist a limit

$$\widehat{Q - Q^+}(\varphi, \varphi) = \lim_{n \to \infty} (Q - Q^+)(\varphi_n, \varphi_n) = Q^z(A\varphi, \varphi), \tag{6.27}$$

which represents a unique continuous extention of $Q - Q^+$ on $\Phi^z \times \Phi^z$. Because

$$(Q - Q^+)(\chi_n, \chi_n) = Q((1 - \lambda_n)\chi_n, \chi_n) = Q^z((1 - \lambda_n)\chi_n, h_n),$$

we conclude that $A\chi_n = (1 - \lambda_n) h_n$. So, on developing $Q - Q^+(\varphi, \varphi)$ into a series, we get

$$\widehat{Q - Q^+}(\varphi, \varphi) = \sum_1^\infty |Q(\varphi, \chi_n)|^2 (1 - \lambda_n) = \tag{6.28}$$

$$= \sum_1^\infty \left| \widehat{Q - Q^+}\left(\varphi, \frac{\chi_n}{1 - \lambda_n}\right)\right|^2 (1 - \lambda_n) = \sum_1^\infty |Q^z(\varphi, h_n)|^2 (1 - \lambda_n).$$

Random variables v_n satisfying $\chi_n(t) = R(x_t, v_n)$ are eigen-elements of B in $\mathscr{X}$. Because $\chi_n = KK^*h_n$, from Theorem 4.2 or 4.1 we have

$$v_n(\omega) = Q^z(x^\omega, h_n). \tag{6.29}$$

On comparing (6.25) and (6.29), we get $\widehat{Q - Q^+}(x^\omega, x^\omega) = \sum_1^\infty v_n^2(\omega)(1 - \lambda_n)$. Returning to the equation (6.5) and noting that $\det B = \prod_1^\infty \lambda_n$, we see that the proof is completed.

7. Probability densities for stationary Gaussian processes. We begin with the case of an integral-valued parameter $t = 0, 1, \ldots, N$, which is easier. The probability densities may be taken with respect to Lebesgue measure, because the number of random variables is finite.

Theorem 7.1. (i) *The probability density of a finite part $\{x_t, 0 \leqq t \leqq N\}$ of a Gaussian stationary process with vanishing mean values and covariances generated by the spectral density* (5.35) *is given by*

$$p(x_0, \ldots, x_N) = |Q_{ts}|^{-1/2} \exp\left(-\tfrac{1}{2}\sum_{t=0}^{N}\sum_{t=0}^{N} Q_{ts}x_tx_s\right) = \tag{7.1}$$

$$= |D_{jk}|^{1/2} a_0^{N-n+1} \exp\left[-\tfrac{1}{2}\sum_{j=0}^{n-1}\sum_{k=0}^{n-1} D_{jk}x_jx_k - \tfrac{1}{2}\sum_{t=n}^{N}\left(\sum_{s=0}^{n} a_k x_{t-k}\right)^2\right],$$

where Q_{ts}, $0 \leqq t, s \leqq N$, and D_{jk}, $0 \leqq j, k \leqq n - 1$ are given by (5.37) *and* (5.40), *respectively, and $|.|$ denotes the determinant.*

(ii) *The probability density of a finite part $\{z_t, 0 \leqq t \leqq N\}$ of a stationary Gaussian process with vanishing mean values and covariances generated by the spectral density* (5.41) *is given by*

$$p(z_0, \ldots, z_N) = |D_{jk}|^{1/2} \, |R(x_i, x_h | z_0, \ldots, z_N)|^{1/2} \, a_0^{N+m+1-n} b_0^{-N-1} \, . \tag{7.2}$$

$$. \exp\left[-\tfrac{1}{2}\sum_{j=0}^{n-1}\sum_{k=0}^{n-1} D_{jk}\check{x}_{j-m}\check{x}_{k-m} - \tfrac{1}{2}\sum_{t=n-m}^{N}\left(\sum_{k=0}^{n} a_k\check{x}_{t-k}\right)^2\right],$$

where $x_{-m}, \ldots, x_N$ is a process considered in (i) *and such that $z_t = \sum b_k x_{t-k}$, and $R(x_i, x_h \mid z_0, \ldots, z_N)$, $-m \leqq i, h \leqq -1$, are conditional* (*partial*) *covariances of x_i, x_h, when $z_0, \ldots, z_N$ are fixed. Moreover, $\check{x}_{-m}, \ldots, \check{x}_N$ is the solution of equations $z_t = \sum b_k \psi_{t-k}$ satisfying the conditions*

$$\sum_{-m}^{N}\sum_{-m}^{N} Q_{ts}\psi_t h_s = 0$$

for some m linearly independet solutions of $\sum b_k h_{t-k} = 0$.

Proof. (i) As Q_{ts} given by (5.37) represent elements of inverted covariance matrix $R = (R_{ts})$, the first expression for $p(x_0, \ldots, x_N)$ in (7.1) is clear. If we put $R_n = (R(x_i, x_j))_{i,j=0}^{n-1}$, then, in accordance with Theorem 6.2,

$$|Q_{ts}| = |R_n|^{-1}\left[\prod_{t=n}^{N} R^2(x_t - I_{t-1}x_t)\right]^{-1}, \tag{7.3}$$

where $I_{t-1}x_t$ is the projection of x_t on the subspace spanned by the random variables $x_0, \ldots, x_{t-1}$. However, we have $|R_n|^{-1} = |D_{jk}|$ and $R^2(x_t - I_{t-1}x_t) = a_0^{-2}$, as $y_t = a_0(x_t - I_{t-1}x_t)$, where y_t is given by (5.36). Thus we obtain the determinant of the second expression for $p(x_0, \ldots, x_N)$ in (7.1). The quadratic form of the second expression is a mere transcription of $\sum\sum Q_{ts}x_t x_s$, and is based on the fact that the y_t's given by (5.36) are independent of each other and of $x_0, \ldots, x_{n-1}$.

(ii) First we derive the determinant. Put $z_t = x_t$, if $-m \leqq t \leqq -1$, and $z_t = \sum b_k x_{t-k}$, if $0 \leqq t \leqq N$. This transformation may be denoted in the matrix form as $z = Cx$, where the matrix $C = \{c_{ij}\}$ is such that $c_{ij} = 0$, $i < j$, and $c_{ii} = 1$, if $-m \leqq i \leqq 1$, and $c_{ii} = b_0$, if $0 \leqq i \leqq N$. Consequently $|C| = b_0^{N+1}$ and

$$|R(z_t, z_s)| = b_0^{2N+2}|R(x_t, x_s)|\,, \quad -m \leqq t,\ s \leqq N\,.$$

Now we know from (i) that

$$|R(x_t, x_s)| = |D_{jk}|^{-1}\, a_0^{-2(N+m+1-n)}$$

which gives

$$|R(z_t, z_s)| = |D_{jk}|^{-1}\, a_0^{-2(N+m+1-n)}\, b_0^{2(N+1)} \quad (-m \leqq t,\ s \leqq N)\,. \tag{7.4}$$

Finally, according to Theorem 6.2,

$$|R(z_t, z_s)| = |R(z_{t'}, z_{s'})|\;|R(x_i, x_h \mid z_0, \ldots, z_N)|\,. \tag{7.5}$$
$$.\,(-m \leqq t,\ s \leqq N;\ 0 \leqq t',\ s' \leqq N;\ -m \leqq i,\ h \leqq -1)\,,$$

where $x_i = z_i$, if $-m \leqq i \leqq -1$. On combining (7.4) and (7.5), we get for $|R(z_{t'}, z_{s'})|$, $0 \leqq t', s' \leqq N$, the expression appearing in (7.2). The quadratic form in (7.2) follows from the form of $Q^z(v, \chi)$ described below equation (5.42).

Now we proceed to the case of a continuous t, and first consider the n-th order Markovian processes. The probability density cannot be taken with respect to Lebesgue measure, because the system of random variables is infinite. The dominating distribution $\mathsf{P}^+ = \mathsf{P}^+_{n,a}$ of $\{x_t, 0 \leqq t \leqq T\}$ used in the next theorem will be defined by the following conditions:

(7.6) the vector $(x_0, x_0', \ldots, x_0^{(n-1)})$ is distributed according to n-dimensional Lebesgue measure; $x_t^{(n-1)}$ is a Gaussian process with independent increments such that

$$\mathsf{E}\,|\mathrm{d}x_t^{(n-1)}|^2 = a^{-2}\,\mathrm{d}t\,;$$

$x_t^{(n-1)} - x_0^{(n-1)}$ is independent of $(x_0, \ldots, x_0^{(n-1)})$; the mean values vanish.

Theorem 13.1. *Let $\{x_t, 0 \leqq t \leqq T\}$ be a finite segment of a stationary Gaussian process with vanishing mean values and with covariances generated by spectral density (5.1). Then the distribution of $\{x_t, 0 \leqq t \leqq T\}$, say P, is strongly equivalent to $\mathsf{P}^+ = \mathsf{P}^+_{n,a_0}$ defined by (7.6), and*

$$\text{(7.7)} \qquad \frac{\mathrm{dP}}{\mathrm{dP}^+} = |D_{jk}|^{1/2} \exp\left\{\frac{1}{2}\frac{a_1}{a_0} T - \tfrac{1}{2}\sum_{k=0}^{n-1} A_{n-k} \int_0^T |x_t^{(k)}(\omega)|^2 \,\mathrm{d}t - \right.$$

$$\left. - \tfrac{1}{4}\sum_{j+k \text{ even}}^{n-1} \sum^{n-1} [x_T^{(j)} x_T^{(k)} + x_0^{(j)} x_0^{(k)}]\, D_{jk}\right\},$$

where $D_{j,k}$, $0 \leqq j, k \leqq n - 1$ are given by (5.21) and A_{n-k} by (5.9).

Proof. Let (Φ^+, Q^+) be the closed linear manifold generated by covariances $R^+(x_t, x_s)$ corresponding to the P^+_{n,a_0} – distribution. It is easy to see that $\Phi^+ = \mathscr{L}^2_n$ and

$$\text{(7.8)} \qquad Q^+(\psi, \varphi) = Q^+_{n,a_0}(\psi, \varphi) = a_0^2 \int_0^T \psi_t^{(n)} \bar{\varphi}_t^{(n)} \,\mathrm{d}t \quad (\psi, \varphi \in \mathscr{L}^2_n).$$

Consequently, for Q given by (5.11), where we have made use of (5.21), we get

$$\text{(7.9)} \qquad (Q - Q^+)(\psi, \psi) = \sum_{k=0}^{n-1} A_{n-k} \int_0^T |\psi_t^{(k)}|^2 \,\mathrm{d}t +$$

$$+ \tfrac{1}{2}\sum_{j=0}^{n-1} \sum_{k=0}^{n-1} [\psi_T^{(j)} \bar{\psi}_T^{(k)} + \psi_0^{(j)} \bar{\psi}_0^{(k)}]\, D_{jk}\,.$$

Now let α_1 be a root of $L(\lambda) = \sum a_{n-k}\lambda^k = 0$, and let $L_1(\lambda) = L(\lambda)/(\lambda - \alpha_1)$. Then $z_t = x' + \alpha_1 x_t$ is a $(n - 1)$-th order Markovian process, if $n > 1$, and a white noise, if $n = 1$. We have

$$x_t = x_0 e^{\alpha_1 t} + e^{\alpha_1 t} \int_0^t e^{-\alpha_1 t} z_s \,\mathrm{d}s = K z_t\,,$$

where K^*K has a finite spure. Obviously Q^z satisfies (6.24). Consequently, according to Theorem 6.3, P and $\mathrm{P}^+ = \mathrm{P}^+_{n,a_0}$ are strongly equivalent. Since $\Phi^z = \mathscr{L}^2_{n-1}$, $Q - Q^+$ may be extended to $\mathscr{L}^2_{n-1}$. The right side of (7.9) is, however, adjusted so that it directly represents the extention of $Q - Q^+$ to $\mathscr{L}^2_{n-1}$. Now, in view of (6.25), we only have to substitute x_t for ψ_t in (7.9), which yields he quadratic form of (7.7).

It is now necessary to find the determinant. The determinant corresponding to the vector $(x_0, \ldots, x_0^{(n-1)})$ equals $|D_{jk}|$, D_{jk} given by (5.21). This determinant is to be multiplied by the determinant of the operator B such that

$$\text{(7.10)} \quad R(x_t^{(n-1)}, x_s^{(n-1)} \mid x_0, \ldots, x_0^{(n-1)}) = R^+(Bx_t^{(n-1)}, x_s^{(n-1)} \mid x_0, \ldots, x_0^{(n-1)})\,.$$

Let B_N be the restriction of B ot the subspace spanned by random variables

$$\text{(7.11)} \qquad u_i = x_{iT/N}^{(n-1)}\,, \quad (i = 1, \ldots, N)\,.$$

According to Theorem 6.2, we have

$$\text{(7.12)} \qquad \det B_N = \prod_{i=1}^{N} \frac{R(u_i - u_i^0, u_i - u_i^0)}{R^+(u_i - u_i^+, u_i - u_i^+)}\,,$$

where u_i^0 and u_i^+ are the projections of u_i on the subspace spanned by $(x_0, \ldots, x_0^{(n-1)},$

$u_1, \ldots, u_{i-1})$ with respect to the R-covariance and R^+-covariance, respectively. According to the assumptions (7.6), $u_i^+ = u_{i-1}$ and

$$R^+(u_i - u_i^+, u_i - u_i^+) = \frac{T}{N} a_0^{-2}. \tag{7.13}$$

If R-covariance holds true, the situation is more complicated. First we find the projection of u_i, say $\bar{u}_i$, on the subspace spanned by $\{x_t, 0 \leqq t \leqq [(i-1)/N]\, T\}$. Because the process $\{x_{T-t}, 0 \leqq t \leqq T\}$ has the same distribution as $\{x_t, 0 \leqq t \leqq T\}$, we may write (5.10) in the following equivalent form:

in the following equivalent form:

$$Q(\psi, \varphi) = \int_0^T (L^*\bar{\varphi})_t\, (L^*\psi)_t\, dt + \sum_{j=0}^{n-1} \sum_{k=0}^{n-1} \psi_T^{(j)} \bar{\varphi}_T^{(k)} D_{jk}, \tag{7.14}$$

where D_{jk} is given by (5.21). When looking for $\bar{u}_i$ we have to put $T \equiv [(i-1)/N]\,.\,T$ and $\varphi_t = R(x_t, u_i)$. However,

$$(L^*\varphi)_t = L^*R(x_t, u_i) = R(L^*x_t, u_i) = 0, \quad 0 \leq t \leq [(i-1)/N]\,.\,T, \tag{7.15}$$

because L^*x_t is a white noise independent of x_s, $s > t$, similarly as Lx_t was a white noise independent of x_s, $s < t$. This means that for $\varphi_t = R(x_t, u_i)$

$$Q(\psi, \varphi) = \sum_{j=0}^{n-1} \sum_{k=0}^{n-1} \psi_{(i-1)T/N}^{(j)} \bar{\varphi}_{(i-1)T/N}^{(k)} D_{jk}.$$

If we replace ψ_t by x_t and substitute

$$\varphi_t^{(k)} = \frac{\partial^k}{\partial t^k} R(x_t, x_{iT/N}^{(n-1)}) = (-1)^k R_{t-iT/N}^{(n-1+k)},$$

where $R_{t-s} = R_{ts}$, we get

$$\begin{aligned} \bar{u}_i &= \sum_{j=0}^{n-1} x_{(i-1)T/N}^{(j)} \sum_{k=0}^{n-1} (-1)^k R_{T/N}^{(n-1+k)} D_{jk} = \\ &= \sum_{j=0}^{n-1} x_{(i-1)T/N}^{(j)} \sum_{k=0}^{n-1} (-1)^k \left[R_0^{(n-1+k)} + \frac{T}{N} R_{0+}^{(n+k)} + O(N^{-2}) \right] D_{jk}. \end{aligned} \tag{7.16}$$

Now

$$\begin{aligned} \sum_{k=0}^{n-1} (-1)^k R_0^{(n-1+k)} D_{jk} &= 1, \quad \text{if} \quad j = n-1, \\ &= 0, \quad \text{if} \quad h < n-1, \end{aligned} \tag{7.17}$$

because $((-1)^k R_0^{(j+k)})$ is the covariance matrix of the vector $(x_t, \ldots, x_t^{(n-1)})$, and (D_{jk}) is its inverse. Moreover, in view of $L^*{}_t R_{s-t}^{(k)} = 0$, $t < s$, we have

$$R_{0+}^{(n+k)} = -\frac{1}{a_0} \sum_{h=0}^{n-1} a_{n-k} R_{0+}^{(h+k)} \quad (0 \leqq k \leqq n) \tag{7.18}$$

which, in connection with (7.17) gives

$$\sum_{k=0}^{n-1}(-1)^k R_{0+}^{(n+k)} D_{jk} = -\frac{a_{n-j}}{a_0}, \quad 0 \leqq j \leqq n-1. \tag{7.19}$$

If we insert (7.17) and (7.19) into (7.16), we get

$$\bar{u}_i = u_{i-1} - \frac{T}{N}\sum_{j=0}^{n-1}\frac{a_{n-j}}{a_0}x_{(i-1)T/N}^{(j)} + 0(N^{-2}). \tag{7.20}$$

From (7.20) it follows that

$$\begin{aligned} & R(u_i - \bar{u}_i, u_i - \bar{u}_i) = \\ &= 2(1-(-1)^n R_{T/N}^{(2n-2)}) + \frac{2T}{N}\sum_{j=0}^{n-1}\frac{a_{n-j}}{a_0}.(-1)^j(R_{T/N}^{(n-1+j)} - R_0^{(n-1+j)}) + \\ &+ \frac{T^2}{N^2}\sum_{j=0}^{n-1}\sum_{k=0}^{n-1}\frac{a_{n-j}a_{n-k}}{a_0^2}(-1)^j R_{0+}^{(j+k)} + 0(N^{-3}) = \\ &= 2(-1)^n\frac{T}{N}R_{0+}^{(2n-1)} + (-1)^n\frac{T^2}{N^2}R_{0+}^{(2n)} + \frac{2T^2}{N^2}\sum_{j=0}^{n-1}\frac{a_{n-j}}{a_0}(-1)^j R_{0+}^{(n+j)} + \\ &+ \frac{T^2}{N^2}\sum_{j=0}^{n-1}\sum_{k=0}^{n-1}\frac{a_{n-j}a_{n-k}}{a_0^2}(-1)^j R_{0+}^{(j+k)} + 0(N^{-3}) = \\ &= 2(-1)^n\frac{T}{N}R_{0+}^{(2n-1)} + \frac{T^2}{N^2}\sum_{j=0}^{n-1}\frac{a_{n-j}}{a_0}(-1)^j R_{0+}^{(n+j)} + 0(N^{-3}) = \\ &= 2(-1)^n\frac{T}{N}R_{0+}^{(2n-1)} - 2(-1)^n\frac{T^2}{N^2}\frac{a_1}{a_0}R_{0+}^{(2n-1)} + 0(N^{-3}) = \\ &= \frac{T}{N}a_0^{-2}\left(1 - \frac{T}{N}\frac{a_1}{a_0}\right) + 0(N^{-3}), \end{aligned} \tag{7.21}$$

where we have made use of (7.18), and of the adjoint relation

$$(-1)^n R_{0-}^{(n+k)} = -\sum_{k=0}^{n-1}(-1)^h\frac{a_{n-h}}{a_0}R_{0-}^{(h+k)},$$

together with $R_{0\pm}^{(2j+1)} = 0$, $j < n-1$, $R_{0+}^{(2j)} = R_{0-}^{(2j)} = R_0^{(2j)}$, $0 \leqq j \leqq n-1$, and

$$R_{0+}^{(2n-1)} = -R_{0-}^{(2n-1)} = (-1)^n \tfrac{1}{2}a_0^{-2}.$$

If we chose the projection u_i^0 of u_i on the subspace spanned by $(x_0, \ldots, x_0^{(n-1)}, u_1, \ldots \ldots, u_{i-1})$, we cannot make use of random variables $x_{(i-1)T/N}^{(j)}$, $0 \leqq j \leqq n-2$, appearing in (7.20). However, we may approximate them by sums

$$\hat{x}_{(i-1)T/N}^{(j)} = \sum_{j=0}^{n-1} x_0^{(j)} c_j + \sum_{k=1}^{i-1} u_k d_k,$$

so that

$$R^2(x_{(i-1)T/N}^{(j)} - \hat{x}_{(i-1)T/N}^{(j)}) = 0(N^{-1}).$$

For example we may put

$$\hat{x}^{(n-2)}_{(i-1)T/N} = x_0^{(n-2)} + \sum_{k=1}^{i-1} u_k \frac{T}{N}$$

etc. Consequently, we have

$$R(u_i - u_i^0, u_i - u_i^0) = \frac{T}{N} a_0^{-2} \left(1 - \frac{T}{N}\frac{a_1}{a_0}\right) + 0(N^{-3}). \tag{7.22}$$

So, in accordance with Theorem 6.1 and with (7.13) and (7.22),

$$\det B = \lim_{\substack{n\to\infty \\ N=2^n}} \det B_N = \lim_{\substack{n\to\infty \\ N=2^n}} \prod_{i=1}^{N} \frac{\frac{T}{N} a_0^{-2}\left(1 - \frac{T}{N}\frac{a_1}{a_0}\right) + 0(N^{-3})}{\frac{T}{N} a_0^{-2}} = e^{-a_1/a_0 T}, \tag{7.23}$$

which concludes the proof.

In the case of a general rational spectral density the leading term of $Q^z(\chi, \chi)$ is

$$a_0^2 b_0^{-2} \int_0^T |\chi_t^{(n-m)}|^2 \, dt = Q^+(\chi, \chi),$$

$$Q^z(\chi, \chi) = a_0^2 b_0^{-2} \int_0^T |\chi_t^{(n-m)}|^2 \, dt + \ldots \tag{7.24}$$

and $Q^z(\chi, \chi) - Q^+(\chi, \chi)$ is dominated in the sense of (6.29) by $\bar{Q}(\chi, \chi)$ corresponding to any rational spectral density with $\bar{n} - \bar{m} = n - m - 1$. Without entering into details let us present the following

Theorem 7.3. *Let $\{z_t, 0 \leq t \leq T\}$ be a finite segment of a stationary Gaussian process with vanishing mean values and with covariances generated by spectral density (5.22). Then the distribution of $\{z_t, 0 \leqq t \leqq T\}$, say P, is strongly equivalent to $\mathsf{P}^+ = \mathsf{P}^+_{n-m, a_0/b_0}$ and*

$$\frac{d\mathsf{P}}{d\mathsf{P}^+} = |D_{jk}|^{1/2} b_0^{m-n} |R(x_0^{(i)}, x_0^{(h)} | z_t, 0 \leqq t \leqq 1)|^{1/2} . \tag{7.25}$$

$$. \exp\left\{\frac{1}{2}\left(\frac{a_1}{a_0} - \frac{b_1}{b_0}\right) T - \tfrac{1}{2}\widehat{Q^z - Q^+}(x^{\omega}, x^{\omega})\right\}.$$

where $Q^+ = Q^+_{n-m, a_0/b_0}$ is given by (7.8), and the D_{jk}'s, $0 \leqq j, k \leqq n-1$, are given by (5.21). Further, x_t is a process considered in Theorem 7.2 such that $z_t = \sum_{k=0}^{m} b_{m-k} x_t^{(k)}$, and $R(x_0^{(i)}, x_0^{(h)} \mid z_t, 0 \leqq t \leqq T)$, $0 \leqq i, h \leqq m-1$, are conditional (partial) covariances of $x_0^{(i)}$ and $x_0^{(h)}$, when $\{z_t, 0 \leqq t \leqq T\}$ is fixed.

Proof. The assertions concerning the quadratic form follow from Theorems 5.2

and 6.3 and from the above discussion. As for the determinant, let us go back to formula (7.7). If the dominating distribution P^+ were modified so that

$$\sum_{k=0}^{m} b_{m-k} x_t^{(m-m-1+k)} = Y_t$$

has independent increments and $\mathsf{E}^+|\mathrm{d}Y_t|^2 = a_0^2 b_0^{-2}$, then the only change in the determinant would consist in replacing

$$\exp\left\{\frac{1}{2}\frac{a_1}{a_0}T\right\} \quad \text{by} \quad \exp\left\{\frac{1}{2}\left(\frac{a_1}{a_0} - \frac{b_1}{b_0}\right)T\right\}.$$

This could be proved by arguments similar to those used in the proof of Theorem 7.2. Now, if we put $z_t = \sum_{k=0}^{m} b_{m-k} x_t^{(k)}$, we can easily see that P^+ – distribution of z_t is the one used in (7.25). Further the transformation $(z_0^{(-m)}, \ldots, z^{(n-m-1)}) = A(x_0, \ldots, x_0^{(n-1)})$, defined by

$$(7.26) \qquad \begin{aligned} z_0^{(j)} &= \sum_{k=0}^{m} b_{m-k} x_0^{(j+k)}, &&\text{if } 0 \leqq j \leqq n-m-1, \\ &= x_0^{(j+m)} &&\text{if } -m \leqq j \leqq -1 \end{aligned}$$

has the determinant $|A| = b_0^{n-m}$. Now replacing x_0 by z_0 amounts to multiplying the determinant by b_0^{n-m}, and excluding random variables $x_0^{(j+m)} = z_0^{(j)}$, $-m \leqq j \leqq -1$, amounts to dividing the determinant by the factor $|R(x_0^{(i)}, x_0^{(h)} \mid z_t, 0 \leqq t \leqq T)|$, which follows from Theorem 6.2. The proof is finished.

Remark 7.1. Obviously, the following limit exists:

$$(7.27) \qquad \lim_{T\to\infty} |R(x_0^{(i)}, x_0^{(h)} \mid z_t, 0 \leqq t \leqq T)| = |R(x_0^{(i)}, x_0^{(h)} \mid z_t, 0 \leqq t < \infty)|.$$

Remark 7.2. Consider a stationary Gaussian process $\{x_t, 0 \leqq t \leqq T\}$ with correlation function $R(\tau) = \max(0, 1 - |\tau|)$. From Example 3.2 it follows, after some computations, that, for $0 < T < 1$,

$$(7.28) \qquad Q(\varphi, \varphi) = \frac{1}{2}\frac{(\varphi_0 + \varphi_T)^2}{2 - T} + \frac{1}{2}\int_0^T |\varphi_t'|^2 \,\mathrm{d}t.$$

Consequently, the respective distribution, say P, is strongly equivalent to $\mathsf{P}^+ = \mathsf{P}^+_{1,2}$ defined by (7.6), and

$$(7.29) \qquad \frac{\mathrm{d}\mathsf{P}}{\mathrm{d}\mathsf{P}^+} = \text{const} \exp\left[-\frac{1}{4}\frac{(x_0 + x_T)^2}{2 - T}\right].$$

If, however, $1 < T < 2$, we get

$$(7.30) \qquad \begin{aligned} Q(\varphi, \varphi) = {} & \frac{1}{6}\frac{(2\varphi_0 + 2\varphi_T + \varphi_1 + \varphi_{T-1})^2}{4 - T} + \frac{1}{2}\int_{T-1}^{1} |\varphi_t'|^2 \,\mathrm{d}t + \\ & + \frac{2}{3}\int_1^T (|\varphi_t'|^2 + |\varphi_{t-1}'|^2 + \varphi_t'\varphi_{t-1}')^2 \,\mathrm{d}t, \end{aligned}$$

which shows that the distribution is not equivalent to any distribution we have met with.

If we introduce two parameters by puting $R(\tau) = d^2.\ \max(0, 1 - |\tau|/a)$, then for $0 < T < a$ and $P^+ = P^+_{1,2d^2/a}$,

$$\frac{dP}{dP^+} = \left(\frac{2a - T}{2a}\right)^{1/2} . \exp\left[-\frac{1}{4}\frac{(x_0 + x_T)^2}{d^2(2 - T/a)}\right]. \tag{7.31}$$

The determinant was established as follows:

If we put $\varphi_t = R(s + \Delta - t) = d^2(1 - (s + \Delta - t)/a)$, $0 \leqq t \leqq s$, then

$$Q(\varphi, \varphi) = d^2(1 - 2\Delta a^{-1} + 2\Delta^2 a^{-1}(2a - s)^{-1})$$

gives the variance of the projection of $x_{s+\Delta}$ on the subspace spanned by $\{x_t, 0 \leqq t \leqq s\}$, so that the residual variance equals $\Delta 2d^2 a^{-1}(1 - \Delta(2a - s)^{-1})$. Then we may proceed with a development similar to that used in evaluating of (7.23). Note that $x_0 + x_T$ is a sufficient statistic for estimating a. As is well-known

$$2d^2a^{-1} = \underset{n\to\infty}{\text{l. i. m.}} \sum_{i=1}^{n} (x_{(i-1)T/n} - X_{iT/n})^2 .$$

Remark 7.3. From (7.7) it follows that the vector

$$\left\{\int_0^T |x_t|^2\, dt, \ldots, \int_0^T |x_t^{(n-1)}|^2\, dt, x_0, \ldots, x_0^{(n-1)}, x_T, \ldots, x_T^{(n-1)}\right\}$$

represents a sufficient statistics for all n-th order Markovian processes with fixed a_0. In the case of a general rational spectral density with $m > 0$ apparently no sufficient statistic exists, which would not be equivalent to the whole process $\{z_t, 0 \leqq t \leqq 1\}$. See Example 7.3.

Remark 7.4. We have proved, by the way, that distributions P_1 and P_2 corresponding to two rational spectral densities are strongly equivalent, if $n_1 - m_1 = n_2 - m_2$ and $a_{01}b_{02} = a_{02}b_{01}$, and perpendicular in other cases. This result (with equivalence instead of strong equivalence) has been announced by V. F. Pisarenko [24].

Example 7.1. If $n = 1$, we get

$$R(\tau) = (2a_0a_1)^{-1}\, e^{-(a_1/a_0)|\tau|},$$

and

$$\frac{dP}{dP^+} = (2a_0a_1)^{1/2} \exp\left\{\frac{1}{2}\frac{a_1}{a_0}T - \tfrac{1}{2}a_1^2 \int_0^T |x_t|^2\, dt\right\}, \tag{7.32}$$

where $x_t = x_t(\omega)$. See also [25].

Example 7.2. If

$$f(\lambda) = \frac{1}{2\pi}\frac{a_0^{-2}}{(\lambda^2 + \alpha_1^2)(\lambda^2 + a_2^2)} = \frac{1}{2\pi}\frac{a_0^{-2}}{|(i\lambda)^2 + (i\lambda)(\alpha_1 + \alpha_2) + \alpha_1\alpha_2|^2}, \tag{7.33}$$

$$(\alpha_2 > \alpha_1 > 0)$$

then $a_1/a_0 = \alpha_1 + \alpha_2$, $a_2/a_0 = \alpha_1\alpha_2$. Consequently,

$$R(\tau) = \frac{1}{2a_0^2}\left(\frac{1}{\alpha_1}e^{-\alpha_1|\tau|} - \frac{1}{\alpha_2}e^{-\alpha_2|\tau|}\right), \tag{7.34}$$

$D_{12} = 0$, $D_{00} = 2a_2a_1 = 2(\alpha_1 + \alpha_2)\,\alpha_1\alpha_2 a_0^2$, $D_{11} = 2a_0a_1 = 2(\alpha_1 + \alpha_2)\,a_0^2$, and

$$\frac{\mathrm{dP}}{\mathrm{dP}^+} = 2(\alpha_1 + \alpha_2)\,(\alpha_1\alpha_2)^{1/2}\,a_0^2 \exp\left[\tfrac{1}{2}(\alpha_1 + \alpha_2)\,T)\right]. \tag{7.35}$$

$$.\exp\left\{\tfrac{1}{2}(\alpha_1^2 + \alpha_2^2)\,a_0^2\int_0^T |x_t'|^2\,\mathrm{d}t - \tfrac{1}{2}\alpha_1^2\alpha_2^2a_0^2\int_0^T |x_t|^2\,\mathrm{d}t + \right.$$

$$\left. + \tfrac{1}{2}(x_0'^2 + x_T'^2)\,(\alpha_1 + \alpha_2)\,a_0^2 - \tfrac{1}{2}(x_0^2 + x_T^2)\,(\alpha_1 + \alpha_2)\,\alpha_1\alpha_2a_0^2\right\}.$$

Example 7.3. Consider a general spectral density with $n = 2$ and $m = 1$,

$$g(\lambda) = \frac{b_0^2(\lambda^2 + \beta^2)}{|a_0(i\lambda)^2 + a_1(i\lambda) + a_2|^2} \tag{7.36}$$

and put $L(\beta) = a_0\beta^2 + a_1\beta + a_2$, $L^*(\beta) = a_0\beta^2 - a_1\beta + a_2$ and $LL^*(\beta) = L(\beta)L^*(\varepsilon) = a_0^2\beta^4 + (2a_0a_2 - a_1^2)\,\beta^2 + a_2^2$, etc. In order to find $Q^z(\chi, \chi)$, let us first suppose that $\chi \in \mathscr{L}_2^2$, and use the form of $Q^z(\chi, \chi)$ resulting from the right side of (5.27) after substituting χ_t for $z_t(\omega)$. On obtaining φ_t from formula (5.31), we get

$$\varphi_t = \frac{1}{2\beta}\int_0^T e^{-|t-s|\beta}\,\chi_s\,\mathrm{d}s + c_1e^{-t\beta} + c_2e^{(\tau-t)\beta}, \tag{7.37}$$

where

(7.38)

$$c_1 = \frac{a_0L(\beta)\,(\chi_0 - L^*(\beta)\,\frac{1}{2\beta}\int_0^T e^{-t\beta}\chi_t\,\mathrm{d}t) - L^*(\beta)\,e^{-T\beta}(\chi_T - L^*(\beta)\,\frac{1}{2\beta}\int_0^T e^{-(T-t)\beta}\chi_t\,\mathrm{d}t)}{LL(\beta) - L^*L^*(\beta)\,e^{-2T\beta}}$$

and

(7.39)

$$c_2 = \frac{a_0L(\beta)\,(\chi_T - L^*(\beta)\frac{1}{2\beta}\int_0^T e^{-(T-t)\beta}\chi_t\,\mathrm{d}t) - L^*(\beta)\,e^{-T\beta}(\chi_0 - L^*(\beta)\frac{1}{2\beta}\int_0^T e^{-t\beta}\,\chi_t\,\mathrm{d}t)}{LL(\beta) - L^*L^*(\beta)\,e^{-2T\beta}}.$$

Now in view of (5.29) and (5.30) $(L\varphi)_T = (L^*\varphi)_0 = 0$, so that

$$Q^z(\chi, \chi) = \int_0^T \chi_t(LL^*\varphi)_t\,\mathrm{d}t + \chi_Ta_0[a_0\varphi_T''' + a_1\varphi_T'' + a_2\varphi_T'] +$$

$$+ \chi_0a_0[a_0\varphi_0''' - a_1\varphi_0'' + a_2\varphi_0'].$$

On using the relations $\varphi_t'' = -\chi_t + \beta\varphi_t$ and (7.37), we get, after some transformations, that

(7.40) $$Q^z(\chi, \chi) = \left(\frac{a_0}{b_0}\right)^2 \int_0^T |\chi_t'|^2 \, dt + (a_1^2 - 2a_0a_2 - \beta^2 a_0^2)\, b_0^{-2} \int_0^T |\chi_t|^2 \, dt +$$

$$+ \left(\frac{a_0}{b_0}\right)^2 LL^*(\beta) \int_0^T \int_0^T e^{-|t-s|\beta} \chi_t \chi_s \, dt\, ds +$$

$$+ (|\chi_0|^2 + |\chi_T|^2)\, b_0^{-2} \left(a_0 a_1 - \beta a_0^2 \frac{LL(\beta) + L^*L^*(\beta)\, e^{-2T\beta}}{LL(\beta) - L^*L^*(\beta)\, e^{-2T\beta}}\right) -$$

$$- \tfrac{1}{2}\beta \left(\frac{a_0}{b_0}\right)^2 \left(\left|\frac{1}{\beta}\int_0^T e^{-t\beta}\chi_t \, dt\right|^2 + \left|\frac{1}{\beta}\int_0^T e^{-(T-t)\beta}\chi_t \, dt\right|^2 \right) \frac{LLL^*L^*(\beta)}{LL(\beta) - L^*L^*(\beta)\, e^{-2T\beta}} +$$

$$+ 2\left(\frac{a_0}{b_0}\right)^2 \left(\chi_T \int_0^T e^{-(T-t)\beta}\chi_t \, dt + \chi_0 \int_0^T e^{-t\beta}\chi_t \, dt\right) \frac{LLL^*(\beta)}{LL(\beta) - L^*L^*(\beta)\, e^{-2T\beta}} +$$

$$+ 4\beta \left(\frac{a_0}{b_0}\right)^2 \left(\chi_0\chi_T + L^*L^*(\beta) \frac{1}{2\beta}\int_0^T e^{-t\beta}\chi_t \, dt \, \frac{1}{2\beta}\int_0^T e^{-(T-t)\beta}\chi_t \, dt\right).$$

$$\frac{LL^*(\beta)\, e^{-T\beta}}{LL(\beta) - L^*L^*(\beta)\, e^{-2T\beta}} - 2\left(\frac{a_0}{b_0}\right)^2 \chi_0 \int_0^T e^{-(T-t)\beta}\chi_t \, dt +$$

$$+ \chi_T \int_0^T e^{-t\beta}\chi_t \, dt \Bigg) \frac{LL^*L^*(\beta)\, e^{-T\beta}}{LL(\beta) - L^*L^*(\beta)\, e^{-2T\beta}}.$$

If $T \to \infty$ the terms involving $e^{-T\beta}$ become negligible. The quadratic form (7.40), obviously, is well-defined for any $\chi \in \mathscr{L}_1^2$. On putting

(7.41) $$\widehat{Q^z - Q^+}(\chi, \chi) = Q^z(\chi, \chi) - \left(\frac{a_0}{b_0}\right)^2 \int_0^T |\chi_t'|^2 \, dt,$$

$\widehat{Q^z - Q^+}$ will be well-defined for any $\chi \in \mathscr{L}_0^2$. The probability density will equal

(7.42) $$\frac{dP}{dP^+} = 2b_0^{-1} a_1 (a_0 a_2)^{1/2} R(x_0 \quad z_t,\ 0 \leqq t \leqq T) \,.\, \exp\left[\frac{1}{2}\left(\frac{a_1}{a_0} - \beta\right) T\right].$$

$$.\, \exp\{-\tfrac{1}{2}\widehat{Q^z - Q^+}(x^\omega, x^\omega)\},$$

where $\widehat{Q^z - Q^+}$ is given by (7.41) and (7.40). The conditional variance of x_0 equals

(7.43) $$R^2(x_0 \mid z_t,\ 0 \leqq t \leqq T) = R^2(x_0) - Q^z(v, v),$$

where $R^2(x_0)$ is the absolute variance and $v_t = R(x_0, z_t) = b_0 R(x_0, x_t' + \beta x_t) \equiv$ $\equiv b_0[R'(t) + \beta R(t)]$. In the special case considered in Example 7.2, we have

(7.44) $$R^2(x_0) = \frac{1}{2}\frac{|\alpha_1 - \alpha_2|}{a_0^2 \alpha_1 \alpha_2}, \quad v_t = \frac{1}{2a_0^2}\left(e^{-\alpha_1 t}\frac{\beta - \alpha_1}{\alpha_1} - e^{-\alpha_2 t}\frac{\beta - \alpha_2}{\alpha_2}\right).$$

References

[1] *А. Н. Колмогоров:* Стационарные последовательности в гильбертовском пространстве. Бюлл. МГУ, *2*, № 6, 1941, 1—40.

[2] *N. Wiener:* Extrapolation, interpolation and smoothing of stationary time series. Cambridge-New York, 1949.

[3] *Kari Karhunen:* Über die lineare Methoden in der Wahrscheinlichkeitsrechnung. Ann. Acad. Sci. Fennicae, Ser. A, I. Math. Phys., *37*, 1947, 3—79.

[4] *U. Grenander:* Stochastic processes and statistical inference. Arkiv för Matematik, *1*, 1949, 195—277.

[5] *U. Grenander:* On empirical spectral analysis of stochastic processes. Arkiv för Matematik, *1*, 1952, 503—531.

[6] *U. Grenander* nad *M. Rosenblatt:* Statistical analysis of stationary time series. Stockholm, 1956.

[7] *L. A. Zadeh* and *R. Ragazzini:* Extension of Wiener's theory of prediction. Journ. Appl. Phys. *21*, 1950, 645—655.

[8] *C. L. Dolph* and *M. A. Woodbury:* On the relations between Green's functions and covariance of certain stochastic processes. TAMS, *72*, 1952, 519—550.

[9] *А. М. Яглом:* Экстраполирование, интерполирование и фильтрация стационарных случайных процессов с рациональной плотностю. Труды Моск. Мат. Общ., *4*, 1955, 333—374.

[10] *А. М. Яглом:* Явные формулы для экстраполяции, фильтрации и вычисления количества информации в теории гауссовскик стохастических процессов. Trans. Sec. Prague Conf. Inf. Th. etc., Praha 1960, 251—262.

[11] *И. М. Гельфанд* и *А. М. Яглом:* О вычислении количества информации о случайной функции, содержащейся в другой такой функции. Успехи мат. наук, *XII*, 1957, 3—52.

[12] *И. М. Гельфанд:* Обобщеные случайные процессы. ДАН 100, № *5* (1955), 853—856.

[13] *А. В. Скороход:* О дифференцируемости мер соответствующих случайным процессам. I. Процессы с независимими Приращениями. Теория вероятностей, *II*, 1957, 417—442. II. Марковские процессы, Теория вероятностей, *V*, 1960, 45—53.

[14] *А. В. Скороход:* Про одну задачу статистики гауссiвских процесiв. Доповiдi Академиiї наук УРСР, *9*, 1960, 1167—69.

[15] *Я. Гаек* (J. Hájek): Линейная оценка средней стационарного случайного процесса с выпуклой корреляционной функцией. Czech. Math. Journal, *6*, 1956, 94—117.

[16] *J. Hájek:* Predicting a stationary process when the correlation function is convex. Czech. Math. Journ., *8*, 1958, 150—154.

[17] *J. Hájek:* A property of *J*-divergences of marginal probability distributions. Czech. Math. Journ., *8*, 1958, 460—463.

[18] *Я. Гаек* (J. Hájek): Об одном свойстве нормальных распределений произвольного случайного процесса. Czech. Math. Journal, *8*, 1958, 610—618.

[19] *J. Hájek:* On a simple linear model in Gaussian processes. Trans. Second. Prague Conf. Inf. Th. etc., 1960, 185—197.

[20] *J. Hájek:* On linear estimation theory for an infinite number of observations. Теория вероятностей, *VI*, 1961, 182—193.

[21] *A. V. Balakrishnan:* On a charakterization of covariances. Ann. Math. Stat., *30*, 1959, 670—675.

[22] *G. Kallianpur:* A problem in optimum filtering with finite date. Ann. Math. Stat., *30*, 1959, 659—669.

[23] *Ли-Хен-Вон:* К теории оптимальной фильтрации сигнала при наличии внутренных шумов. Теория вероятностей, *IV*, 1959, 458—464.

[24] *В. Ф. Писаренко:* Об абсолютной непрерывности мер, соответствующих гауссовским процессам с рациональной спектральной плотностью. Теория вероятностей, *IV*, 1959, 481—481.

[25] *Charlotte T. Striebel:* Densities for stochastic processes. Ann. Math. Stat., *30*, 549—567, 1959.

[26] *М. С. Пинскер:* Энтропия, скорость создания энтропии и энтропийная устойчивость гауссовских случайных величин и процессов. ДАН СССР, *133*, № 3, 1960, 531—533.

[27] *В. С. Пугачев:* Эффективный метод нахождения Бейесова решения. Trans. Sec. Prague Conf. Inf. Th. etc., Praha, Nakl. ČSAV, 1959, 531—540.

[28] *В. С. Пугачев:* Теория случайных функций и ее применения к задачам автоматического управления, второе изд. Москва, 1960.

[29] *D. Middleton,* Introduction to statistical communication theory. London, 1960.

[30] *J. L. Doob:* Stochastic processes. New York-Toronto 1953, (Russian translation). Moskva, 1956.

[31] *E. A. Coddington* and *N. Levinson:* Theory of ordinary diferential equations. New York-Toronto-London 1955, Moskva (Russian translation) 1958.

[32] *F. Riesz* et *B. Sz.-Nagy:* Lecons d'analyse fonctionnelle. Budapest 1953, Moskva (Russian translation) 1954.

Резюме

О ЛИНЕЙНЫХ СТАТИСТИЧЕСКИХ ПРОБЛЕМАХ В СТОХАСТИЧЕСКИХ ПРОЦЕССАХ

ЯРОСЛАВ ГАЕК (Jaroslav Hájek), Прага

В работе развита объединенная теория линейных статистических проблем. в том числе предсказывания, фильтрации, оценок коэффициентов регрессии и определения плотности вероятности одной гауссовской меры по другой.

Пусть $\{x_t, t \in T\}$ — какой-нибудь стохастический процесс такой, что $E|x_t|^2 < \infty$ и $Ex_t = 0$, $t \in T$. Пусть $\mathscr{X}$ — пространство Гильберта, элементами которого являются линейные комбинации случайных величин x_t и их пределы по норме $\|x\| = [E|x|^2]^{1/2}$. Пространству $\mathscr{X}$ можно поставить в соответствие пространство Φ, элементами которого являются комплексные функции $\varphi_t(t \in T)$ такие, что для определенного элемента $v \in \mathfrak{X}$

$$\varphi_t = Ex_t\bar{v} \quad (t \in T)\,. \tag{2.1}$$

Притом норма $[Q(\varphi, \varphi)]^{1/2}$ в пространстве Φ дана соотношением $Q(\varphi, \varphi) = E|v|^2$, где v — элемент $\mathscr{X}$, соответствующий функции φ в смысле уравнения (2.1). Связанные с процессом $\{x_t\}$ линейные проблемы состоят, вообще, в отыскании Φ и $Q(\varphi, \varphi)$ и в решении уравнения (2.1) относительно v.

В § 2 определяются основные свойства пространств $\mathscr{X}$ и Φ а также здесь формулируются и решаются в общем виде основные типы линейных проблем. § 4 посвящается паре процессов $\{x_t\}$ и $\{y_t\}$, связанных соотношением $x_t = Ky_t$.

где K – линейный оператор. Доказывается, что для гауссовских процессов $\{x_t\}$ можно осуществить в функциях вида $\psi_t = Ey_t\bar{v}$, $v \in y$, тогда и только тогда, если оператор K^*K – ядерный. В § 5 дается в явном виде решение линейных проблем для стационарных процессов с рациональной спектральной плотностью на конечном интервале.

В § 6 две гауссовские меры P и P^+ названы сильно эквивалентными, если линейный оператор, переводящий одну ковариантную функцию в другую, является ядерным (для эквивалентности достаточно, чтобы оператор был типа, Гильберта-Шмидта). Здесь выводятся две теоремы о поведении определителя этого оператора и достаточное условие для того, чтобы $\mathrm{d}P/\mathrm{d}P^+$ было функцией квадратичной формы, определенной на выборочных функциях. В § 7 эти общие теоремы применяются к стационарным процессам с рациональной спектральной плотностью или с функцией корреляции $R(\tau) = \max(0, 1 - |\tau|)$. Например, выводится, что для гауссовской меры P, индуцированной спектральной плотностью

$$f(\lambda) = \frac{1}{2\pi} \left|\sum_{k=0}^{n} a_{n-k}(i\lambda)^k\right|^{-2} = \frac{1}{2\pi} \left|\sum_{k=0}^{n} A_{n-k}\lambda^{2k}\right|^{-1}$$

имеет место равенство

$$\frac{\mathrm{d}P}{\mathrm{d}P^+} = |D_{jk}|^{1/2} \exp\left\{\frac{1}{2}\frac{a_1}{a_0} T - \frac{1}{2}\sum_{k=0}^{n-1} A_{n-k} \int_0^T |x_t^{(k)}(\omega)|^2 \,\mathrm{d}t - \right.$$
$$\left. - \frac{1}{4} \sum_{j+k}^{n-1} \sum_{\text{четные}}^{n-1} \left[x_T^{(j)} x_T^{(k)} + x_0^{(j)} x_0^{(k)}\right] D_{jk}\right.,$$

где D_{jk} даны уравнением (5.21), $|D_{jk}|$ означает определитель, а распределение $P^+ = P^+_{n,a_0}$ определяется на стр. 433.

Editor's Note

Following are transliterations of authors' names from the reference list: 1, A. N. Kolmogorov; 9, 10, and 11, A. M. Jaglom; 11 and 12, I. M. Gel'fand; 13 and 14, A. V. Skorokhod; 23, Li-Hen-Won; 24, V. F. Pisarenko; 26, M. S. Pinsker; 27 and 28, V. S. Pugachev.

5

Reprinted by permission from pages 1-37 of *12th Biennial Seminar Canadian Mathematical Congress Proc.*, R. Pyke, ed., Canadian Mathematical Congress, Montreal, 1970, 279p.

Statistical inference on time series by RKHS methods

Emanuel Parzen[1]

State University of New York at Buffalo

Dedicated to the beloved memory of Alfred Rényi

The theory of time series is studied by probabilists (under such names as Gaussian processes and generalized processes), by statisticians (who are mainly concerned with modelling discrete parameter time series by finite parameter schemes), and by communication and control engineers (who are mainly concerned with the extraction and detection of signals in noise). The aim of this review is to outline the unifying role of reproducing kernel Hilbert spaces (RKHS) in the theory of time series. There are 13 sections (which are divided into an introduction and 4 chapters). The section headings are:

0. The role of RKHS in time series analysis.
1. Time series, means and covariances.
2. RKHS representations of time series.
3. Interpolation with minimum norm.
4. Filtering and parameter estimation.
5. A general model of linear statistical inference on time series.
6. RKHS solution of the zero mean linear statistical inference problem.
7. Equivalence between time series parameter estimation, control theory, and approximation theory.
8. Integral representation theorem.
9. Hilbert space indexed time series.
10. Normal Markov processes.
11. Equivalence and orthogonality of normal processes.
12. Covariance kernels of normal probability density functionals.

[1] Prepared with the support of an Office of Naval Research contract at Stanford University and a Fellowship at the IBM Systems Research Institute, New York, N.Y. 10017.

0. The role of RKHS in time series analysis. Among the aspects of time series analysis for which RKHS seem to provide (simple, general, and elegant) formulations are:

1. Estimation of parameters of linear models.
2. Regression analysis and design of experiments.
3. Relations of time series analysis to approximation theory.
4. Probability density functionals of Gaussian processes.
5. Minimum variance unbiased estimation.
6. Representation of stochastic processes.
7. Properties of the probability measures on linear spaces induced by Gaussian processes.
8. Limit theorems for stochastic processes.

It would take a book to explain the theory and applications of RKHS in time series analysis. My aim in these lectures can only be to provide an introduction, which in turn is inevitably biased by my own interest in questions of *stostatistics* (statistical inference on stochastic processes).

It seems to me fruitful to regard the theory of time series as composed of three broad areas, which we call *statistical theory*, *probability theory: structural* and *probability theory: distributional.*

In the **statistical theory** of time series, one imagines that an observed time series $Y(\cdot) = \{Y(t), t \in T\}$ has the representation

$$Y(t) = Z(t;\theta) + N(t)$$

where, for each t, $Z(t;\theta)$ is a random variable which is a function of a parameter θ, which could itself be a stochastic process. One seeks to estimate, test hypotheses, or make a decision about the parameter θ from the data $Y(\cdot)$. Often the first step in statistical inference of a parameter θ is to *define and obtain the probability density functional* (called by statisticians the *likelihood* function) of a sample path of $Y(\cdot)$ given a fixed parameter value θ. When θ is a random element, this is the same as the *conditional probability density functional* of $Y(\cdot)$ given θ. These notions can be made precise, using the notion of a Radon–Nikodým derivative, and one is then able to treat the problem of statistical inference on time series within the framework of the theory of statistical inference originally developed for independent observations.

In the **probability theory: structural** of time series, one seeks to regard a time series $\{Y(t), t \in T\}$ as a function chosen, in accordance with a probability measure $P[\cdot]$, from a space Ω_T whose members are functions on an index set T. For each t, $Y(t)$ is a random variable (function) with domain Ω_T whose value [denoted $Y(t, \omega)$] at a point ω in Ω_T is defined by $Y(t, \omega) = \omega(t)$. One seeks to (i) determine the spaces Ω_T on whose "measurable" subsets the time series $Y(\cdot)$ determines a probability measure, and (ii) characterize and *represent the "measurable" functions* $f(\omega)$ having prescribed properties, such as f is linear in the sense that

$$f(\omega_1 + \omega_2) = f(\omega_1) + f(\omega_2),$$

or f is square integrable.

The **probability theory: distributional** of time series is concerned with evaluating (exactly or approximately) the distribution of random variables $f(\omega)$ which arise as functionals of a time series.

In all these areas of time series analysis (especially for normal processes) RKHS provide elegant and general formulations of results and proofs. One reason for the central role of RKHS is that a basic tool in all these areas is the theory of equivalence and singularity of normal measures (probability measures corresponding to normal processes), and this theory seems to me to be most simply expressed in terms of RKHS.

This review is divided into four chapters. Chapter 1 (§§ 1, 2, 3) defines a time series and introduces their RKHS representation. Chapter 2 (§§ 4, 5, 6, 7) is an introduction to the statistical theory of time series analysis, emphasizing recent work on the relations of this theory to optimization. Chapter 3 (§§ 8, 9, 10) surveys the main examples of RKHS. Chapter 4 (§§ 11, 12) very briefly states the main results of the theory of equivalence and singularity of normal measures.

As far as the mathematical theory of RKHS is concerned, it should be noted that credit for the notion of a reproducing kernel is due to Stefan Bergman who introduced the notion in his 1920 Berlin thesis [see Bergman (1950) for references to the extensive work of Bergman]. The Hilbert space theory of RKHS is due to Aronszajn [see Aronszajn (1950)], although related ideas appear in the work of Kreĭn (1949). A more abstract theory of reproducing kernels has been developed by Laurent Schwartz (1964). Aside from the writing of the above authors, the main mathematical sources of information on RKHS are the research papers of Devinatz [see Devinatz (1959) for references to his work] and a text in German by Meschkowski (1962).

Chapter 1. Time Series and RKHS

1. Time series, means and covariances. In order to define what we mean by a time series, we must first define "stochastic process." There are three main approaches one can take to defining the notion of a stochastic process.

(1) One can define a stochastic process to be an *indexed collection of random variables* which we write $\{Y(t), t \in T\}$ or $\{Y(t, \cdot), t \in T\}$. For each t

(in the index set T) $Y(t)$ or $Y(t, \cdot)$ is a random variable (real valued measurable function) on a probability space $(\Omega, \mathscr{A}, P)$. The value of $Y(t)$ or $Y(t, \cdot)$ at a point ω in Ω is denoted $Y(t, \omega)$.

(2) One can define a stochastic process to be a *probability measure on a function space* whose members are functions (possibly generalized functions) on an index set T. More generally, one can define a stochastic process to be a probability space $(V, \mathscr{B}_V, P_V)$ where V is a topological linear space (especially a Banach space or a Hilbert space), $\mathscr{B}_V$ is the sigma-field of topological Borel sets (the smallest sigma-field of subsets of V containing all open sets), and P_V is a probability measure with domain $\mathscr{B}_V$.

(3) One can define a stochastic process to be a *Banach or Hilbert space valued random element* Y; in symbols, $Y:(\Omega, \mathscr{A}, P) \to (V, \mathscr{B}_V)$ and $Y^{-1}(\mathscr{B}_V) \subset \mathscr{A}$, where $(\Omega, \mathscr{A}, P)$ is an abstract probability space, V is a Banach or Hilbert space, and $\mathscr{B}_V$ is the sigma-field of topological Borel sets in V.

It is my view that, in developing the theory of time series, it is most convenient to begin with the first point of view towards stochastic processes, since it can be shown that a stochastic process defined from the second and third points of view can be transformed into an equivalent process defined as an indexed collection of random variables.

From the point of view of "data analysis" a stochastic process is usually regarded as a model of a set of observations of a measured quantity over a period of time; it is natural in this case to adopt the view that a stochastic process is a family of random variables. In problems of "signal extraction" or "parameter estimation" one seeks an estimator of a vector θ which represents an unknown signal or parameter; regarding θ as a *random* vector is often convenient (and leads to so-called Bayesian estimators of θ). When θ is an infinite dimensional random vector it is a stochastic process.

We use the phrase "time series" to describe a stochastic process which is being discussed as a *family* of random variables which, further, are assumed to have finite second moments and are being studied from the *point of view of the behavior of their mean m and covariance K*. The definition of mean and covariance of a stochastic process is given below [it depends on the point of view one has adopted in defining the process; compare (1) and (4.29)].

Formal definition. A *time series* is a family $Y(\cdot) = \{Y(t), t \in T\}$ of random variables with finite second moments. The *mean value function m* and *covariance kernel K* of a time series are functions on T and $T \otimes T$ respectively defined by (for all s, t in T)

$$(1) \qquad m(t) = E[Y(t)], \qquad K(s, t) = \operatorname{Cov}[Y(s), Y(t)].$$

A time series is called *normal* if for every integer n, n points $t_1, \ldots, t_n$

and real numbers $u_1, \ldots, u_n$, the random variable $\sum_{j=1}^{n} u_j Y(t_j)$ is normal, or equivalently

$$(2)\quad E\left[\exp i \sum_{j=1}^{n} u_j Y(t_j)\right] = \exp\left\{i \sum_{j=1}^{n} u_j m(t_j) - \frac{1}{2} \sum_{j,k=1}^{n} u_j u_k K(t_j, t_k)\right\}.$$

We will see that it is often convenient to think of a time series Y as indexed by a Hilbert space H^0, with inner product $(x, y)_{H^0}$, in the sense that the inner product $(Y, h)_{H^0}$ between Y and a fixed vector h in H^0 is not a true inner product but symbolizes a random variable indexed by h. There are many Hilbert spaces for which such "random inner products" can be defined. The basic technique for defining such "random inner products" is to define a correspondence between elements in H^0 and random variables in the Hilbert space denoted

$$(3)\qquad H_Y = H(Y(t), t \in T) = L_2(Y(t), t \in T),$$

called the *Hilbert space spanned by the time series* $Y(\cdot)$, and defined as the smallest Hilbert space of random variables containing $Y(t)$ for each $t \in T$. A random variable U is called a *simple* linear functional of $Y(\cdot)$ if it is of the form

$$(4)\qquad U = \sum_{j=1}^{n} c_j Y(t_j)$$

for some integers n, reals $c_1, \ldots, c_n$, and indices $t_1, \ldots, t_n$. It can be shown that every random variable in H_Y is either a simple linear functional of $Y(\cdot)$ or is a limit in quadratic mean of simple linear functionals of $Y(\cdot)$. Intuitively we think of H_Y as the space of all linear functionals of $Y(\cdot)$.

2. RKHS representations of time series. The existence of a reproducing kernel Hilbert space (RKHS) representation of a time series was noted by Loève (1948): their systematic use for statistical inference was discussed by me in 1958 at the International Congress of Mathematicians and was first published in 1959 (reprinted in Parzen (1967), p. 253). Here we summarize the basic ideas and notation with emphasis on obtaining a representation of a random variable which is a linear functional of a stochastic process $\{\theta(s), s \in S\}$.

A kernel Q, with domain $S \otimes S$, is called *nonnegative definite* if for every integer n, points $s_1, \ldots, s_n$ in S, and real constants $c_1, \ldots, c_n$

$$(1)\qquad \sum_{i,j=1}^{n} c_i Q(s_i, s_j) c_j \geqq 0.$$

A kernel Q is called *symmetric* if $Q(s_1, s_2) = Q(s_2, s_1)$.

If Q is the covariance kernel of a time series, it is symmetric and nonnegative definite.

According to a basic theorem of E. H. Moore and N. Aronszajn [see Aronszajn (1950)] every symmetric nonnegative definite kernel Q possesses a unique reproducing kernel Hilbert space (RKHS) with reproducing kernel Q, denoted $H(Q)$ or RKHS(Q), defined as follows: $H(Q)$ is a Hilbert space of functions f on S with the properties, for all $s \in S$ and $f \in H(Q)$,

(I) $Q(\cdot, s) \in H(Q)$,

(II) $f(s) = (f, Q(\cdot, s))_{H(Q)}$.

In (I), $Q(\cdot, s)$ is the function on S with value at $s' \in S$ equal to $Q(s', s)$.

Let $\{\theta(s), s \in S\}$ be a normal time series with mean $\bar{\theta}$ and covariance Q. Then the process $\tilde{\theta}(s) = \theta(s) - \bar{\theta}(s)$ has zero mean and covariance kernel

$$Q(s_1, s_2) = E[\tilde{\theta}(s_1)\tilde{\theta}(s_2)].$$

It can be shown that (i) every function f in $H(Q)$ can be represented

$$f(s) = E[\tilde{\theta}(s)U] \quad \text{for some unique } U \text{ in } L_2(\tilde{\theta}(s), s \in S) \tag{2}$$

where U has zero mean and variance satisfying

$$\|f\|^2_{H(Q)} = E[U^2]; \tag{3}$$

(ii) under the assumption that $\bar{\theta} \in H(Q)$, we can set up a one-one correspondence between $H(Q)$ and $L_2(\theta(s), s \in S)$ such that

$$\theta(s) \leftrightarrow Q(\cdot, s),$$

$$\sum c_i \theta(s_i) \leftrightarrow \sum c_i Q(\cdot, s_i),$$

$$U = \text{l.i.m.} \sum c_i^{(n)} \theta(s_i^{(n)}) \leftrightarrow f = \text{l.i.n.} \sum c_i^{(n)} Q(\cdot, s_i^{(n)})$$

where l.i.m. denotes "limit in mean square" and l.i.n. denotes "limit in norm."

The random variable $U \in L_2(\theta(s), s \in S)$ corresponding to $f \in H(Q)$ we denote by $(f, \theta)_{\sim H(Q)}$ or $f^{\sim}$; we call the first notation a "*congruence inner product*" because (when $H(Q)$ is infinite dimensional) it is not an inner product between two functions in $H(Q)$ but rather is the value at f of a *congruence* mapping on $H(Q)$ to $L_2(\theta(s), s \in S)$ which maps $Q(\cdot, s)$ into $\theta(s)$.

One can show that under the assumptions:

(i) θ is a time series with mean $\bar{\theta}$ and covariance Q,

(ii) $\bar{\theta} \in H(Q)$,

$$\begin{aligned} E[(f, \theta)_{\sim H(Q)}] &= (f, \bar{\theta})_{H(Q)}, \\ \text{Cov}[(f_1, \theta)_{\sim H(Q)}, (f_2, \theta)_{\sim H(Q)}] &= (f_1, f_2)_{H(Q)}, \end{aligned} \tag{4}$$

for every $f, f_1, f_2 \in H(Q)$.

In terms of the congruence inner product any random variable $Z(t)$ which is a linear function of $\{\theta(s), s \in S\}$ can be expressed

(5) $$Z(t) = (a(t, \cdot), \theta)_{\sim H(Q)} \quad \text{for some } a(t, \cdot) \in H(Q).$$

Note that $a(t, \cdot)$ is a function with domain S; when (5) holds we sometimes write

(6) $$Z(t) = (a(t, s), \theta(s))_{\sim H(Q)}.$$

When (5) holds, we call $a(t, \cdot)$ the *representor* of $Z(t)$; intuitively it is obtained from $\{Q(\cdot, s), s \in S\}$ by the same linear operations by which $Z(t)$ is obtained from $\{\theta(s), s \in S\}$. As an example

(7) $$Z(t) = \int_S A(t, s)\theta(s)\, ds \quad \text{if and only if } a(t, \cdot) = \int_S A(t, s)Q(\cdot, s)\, ds.$$

3. Interpolation with minimum norm. One aim of this review is to provide an introduction to recent work on the relations between time series, control theory, and approximation theory, stimulated by the work of Kimeldorf and Wahba (1968), (1969). An important link in this study is the use of RKHS to provide a convenient notation for describing the solution to the following problem of abstract Hilbert space theory which at first glance has no relation to time series.

Problem. Find the vector in a Hilbert space H^0 which has minimum norm among all vectors satisfying a prescribed set of linear constraints.

We call this problem *the problem of interpolation with minimum norm.* Let H^0 be a real Hilbert space, T an index set, and $\{\Psi(t), t \in T\}$ a family of vectors in H^0. Let Y or $Y(\cdot)$ denote a real valued function on T, and

$$C = \{U \in H^0 : Y(t) = (\Psi(t), U)_{H^0} \text{ for all } t \text{ in } T\}.$$

Find the element in C, denoted $\hat{U}$, satisfying

$$\|\hat{U}\|^2_{H^0} = \min_{U \in C} \|U\|^2_{H^0}.$$

To solve this problem, we denote by H_Ψ, or $H(\Psi(t), t \in T)$, the Hilbert subspace of H^0 spanned by the family $\{\Psi(t), t \in T\}$; it consists of all finite linear combinations (for some n, reals $c_1, \ldots, c_n$ and indices $t_1, \ldots, t_n$)

$$v = \sum_{i=1}^{n} c_i \Psi(t_i)$$

and limits in norm of such finite linear combinations. One can verify that the vector in C with minimum norm is the unique vector $\hat{U}$ in H_Ψ which is in C. By the solution to the problem of interpolation with minimum norm we mean an explicit formula or algorithm for $\hat{U}$.

When the index set T is finite, say $T = \{1, 2, \ldots, n\}$, an expression for the unique vector $\hat{U}$ in the intersection of C and H_Ψ is obtained as follows: $\hat{U}$ is of the form

$$\hat{U} = \sum_{s=1}^{n} \lambda_s \Psi(s) \tag{1}$$

and satisfies for all t

$$Y(t) = (\hat{U}, \Psi(t))_{H^0} = \sum_{s=1}^{n} \lambda_s K(s, t) \tag{2}$$

defining

$$K(s, t) = (\Psi(s), \Psi(t))_{H^0}. \tag{3}$$

We call K the *covariance kernel* of the family of vectors $(\Psi(t), t \in T)$; it will play a central role in our discussion.

Define matrices

$$\Psi = \begin{bmatrix} \Psi(1) \\ \cdots \\ \Psi(n) \end{bmatrix}, \qquad Y = \begin{bmatrix} Y(1) \\ \cdots \\ Y(n) \end{bmatrix}, \qquad K = \begin{bmatrix} K(1,1) & \cdots & K(1,n) \\ \cdots & \cdots & \cdots \\ K(n,1) & \cdots & K(n,n) \end{bmatrix}.$$

Assume K to have an inverse K^{-1}. From (1) and (2) we obtain

$$K\lambda = Y, \qquad \lambda = K^{-1}Y, \qquad \hat{U} = \Psi'\lambda = \Psi' K^{-1} Y. \tag{4}$$

Further one can verify that

$$\| U \|_{H^0}^2 = \lambda' K \lambda = Y' K^{-1} Y. \tag{5}$$

To extend these results to an arbitrary index set, we introduce the RKHS $H(K)$ corresponding to the covariance kernel K. Define

$$H(K) = \{ f \text{ on } T : f(s) = (\Psi(s), U)_{H^0}, \text{ all } s \in T, \text{ for some unique } U \text{ in } H_\Psi, \ \| f \|_{H(K)}^2 = \| U \|_{H^0}^2 \}.$$

One can verify that $H(K)$ is a Hilbert space of functions on T, with the properties: first, for all $t \in T$

$$K(\cdot, t) \in H(K) \tag{I}$$

since $K(s, t) = (\Psi(s), \Psi(t))_{H^0}$, second, for all $t \in T$ and $f \in H(K)$

$$f(t) = (f, K(\cdot, t))_{H(K)} \tag{II}$$

since by definition the last inner product equals $(U, \Psi(t))_H$ which equals $f(t)$.

Since

$$K(s, t) = (\Psi(s), \Psi(t))_{H^0} = (K(\cdot, s), K(\cdot, t))_{H(K)} \tag{6}$$

there is a congruence between H_Ψ and $H(K)$ such that $\Psi(t)$ and $K(\cdot, t)$ are transformed into each other; in symbols

$$\Psi(t) \leftrightarrow K(\cdot, t). \tag{7}$$

The element in H_Ψ corresponding to a function f in $H(K)$ under this congruence will be denoted

$$f^\sim \quad \text{or} \quad (f, \Psi)_{\sim H(K)}. \tag{8}$$

We call $(f, \Psi)_{\sim H(K)}$ a *congruence* inner product; it is not a true inner product, but represents the vector in H_Ψ which is the same "linear combination" of $\{\Psi(t), t \in T\}$ that f is of $\{K(\cdot, t), t \in T\}$.

In terms of the congruence inner product notation, the element $\hat{U}$ in H_Ψ satisfying

$$(\hat{U}, \Psi(t))_{H^0} = Y(t), \qquad t \in T, \tag{9}$$

is given by

$$\hat{U} = Y^\sim = (Y, \Psi)_{\sim H(K)} \tag{10}$$

with norm squared

$$\|\hat{U}\|^2_{H^0} = \|Y\|^2_{H(K)} = (Y, Y)_{H(K)}. \tag{11}$$

We consider (10) to be a useful explicit solution to the minimum norm interpolation problem. The congruence inner product may seem to be merely a notation for the solution rather than an explicit solution; in our view it is an explicit solution since many techniques for evaluating congruence inner products are available.

Let us summarize the main properties of congruence inner products: for any f, g in $H(K)$ and t in T,

$$\begin{aligned}((f, \Psi)_{\sim H(K)}, \Psi(t))_{H^0} &= (f, (\Psi(\cdot), \Psi(t))_{H^0})_{H(K)} \\ &= (f, K(\cdot, t))_{H(K)} = f(t),\end{aligned} \tag{12}$$

$$((f, \Psi)_{\sim H(K)}, (g, \Psi)_{\sim H(K)})_{H^0} = (f, g)_{H(K)}. \tag{13}$$

When the Hilbert space H^0 is a function space, for great clarity we write $\Psi(t, \cdot)$ for the elements in H^0; then we are seeking $\hat{U}(\cdot)$ of minimum H^0-norm in the set of U satisfying $(\Psi(t, \cdot), U(\cdot))_{H^0} = Y(t)$. The solution will then be written

$$\hat{U}(\cdot) = (Y(t), \Psi(t, \cdot))_{\sim H(\{(\Psi(s, \cdot), \Psi(t, \cdot))_H\})}. \tag{14}$$

Chapter 2. Parameter Estimation and Optimization

4. Filtering and parameter estimation. The modern era of time series analysis has basically been concerned with developing methods for solving the *filtering* problem: given a time series $\{Y(t), t \in T\}$ which is of the form [called signal $X(\cdot)$ plus noise $N(\cdot)$]

(1) $$Y(t) = X(t) + N(t), \qquad 0 \leqq t \leqq T,$$

compute, for each $t \in T$,

(2) $$\hat{X}(t) \equiv E^p[X(t) | Y(s), s \in T],$$

the "wide sense" conditional mean of $X(t)$ given the entire set of $Y(\cdot)$ values.

The wide sense conditional mean is best defined as the solution to the following optimization problem: Let $H_Y = H(Y(t), t \in T)$ be the Hilbert subspace (of $L_2(\Omega, \mathscr{A}\, P)$, the space of square integrable random variables) spanned by the family $\{Y(t), t \in T\}$. Find $\hat{U}$ in H_Y such that

(3) $$E[|\hat{U} - X(t)|^2] = \min_{U \in H_Y} E[|U - X(t)|^2].$$

We denote $\hat{U}$ by $\hat{X}(t)$ or by $E^p[X(t)| Y(s), s \in T]$ where the superscript p connotes "projection in Hilbert space."

The vector $\hat{X}(t)$ can be shown to be the solution of the normal equations

(4) $$E[\hat{X}(t) Y(s)] = E[X(t)Y(s)], \qquad s \in T.$$

From (4) we write

(5) $$\hat{X}(t) = (E[X(t)Y(s)], Y(s))_{\sim H(\{E[Y(t_1)Y(t_2)]\})}.$$

In words, define the covariance kernel on $T \otimes T$

(6) $$K_Y(t_1, t_2) = E[Y(t_1)Y(t_2)].$$

Let $H(K_Y)$ be the RKHS corresponding to K_Y. Then for each $t \in T$

(7) $$K_{XY}(t, s) = E[X(t)Y(s)] \in H(K_Y)$$

and

(8) $$\hat{X}(\cdot) = (K_{XY}(\cdot, s), Y(s))_{\sim H(K_Y)}.$$

Typically the signal and noise are assumed to be uncorrelated. Denote by R and K the covariance kernel of $N(\cdot)$ and $X(\cdot)$ respectively;

(9) $$R(t_1, t_2) = E[N(t_1)N(t_2)],$$

(10) $$K(t_1, t_2) = E[X(t_1)X(t_2)].$$

Then

(11) $$K_Y = R + K, \qquad K_{XY} = K$$

and

(12) $$\hat{X}(\cdot) = (K(\cdot, s), Y(s))_{\sim H(R+K)}.$$

It may make (12) look more familiar if we relate it to the Wiener Hopf equation. Suppose $T = \{t : a \leqq t \leqq b\}$ and we write very unrigorously

(13) $$\hat{X}(t) = \int_a^b W(t, u) Y(u)\, du.$$

The normal equations (4) become

(14) $$\int_a^b W(t, u)\{K(u, s) + R(u, s)\}\, du = K(t, s), \qquad a \leqq s \leqq b,$$

which is to be solved for the weighting function $W(t, u)$. To write (14) in operator form define an operator $R + K$ on $L_2(a, b)$ as follows: $(R + K)f$ is the function whose value at s is

(15) $$\{(R + K)f\}(s) = \int_a^b f(u)\{K(u, s) + R(u, s)\}\, du.$$

In this notation (14) can be written

(16) $$\{(R + K)W(t, \cdot)\}(s) = K(t, s)$$

with formal solution

(17) $$W(t, u) = \{(R + K)^{-1}K(t, \cdot)\}(u).$$

From (13) and (17) we have (still arguing nonrigorously)

(18) $$\hat{X}(t) = \int_a^b Y(u)\{(R + K)^{-1}K(t, \cdot)\}(u)\, du.$$

Equation (12) is a rigorous restatement of the heuristic equation (18).

There are two important extensions of the basic model (1) of signal plus noise:

(1) *Parameterization of the signal.* In addition to extracting the signal $X(\cdot)$ we desire to estimate some underlying "parameter process" $U(\cdot)$ of which $X(\cdot)$ is a linear functional in a sense to be made precise.

(2) *Transformation of the signal.* The observed series $Y(\cdot)$ is the sum of a process $Z(\cdot)$ and noise $N(\cdot)$,

$$Y(t) = Z(t) + N(t), \tag{19}$$

where $Z(t)$ is a linear transformation of the signal process $X(\cdot)$. We call $Z(\cdot)$ a *transformed signal process.*

Three important examples of parameterization of a signal are

(i) the *regression model*;

$$X(t) = \sum_{j=1}^{q} \beta_j \phi_j(t), \tag{20}$$

in which $X(\cdot)$ is an unknown linear combination of q known functions $\phi_1(\cdot), \ldots, \phi_q(\cdot)$,

(ii) the *inner product model*

$$X(t) = (\Psi(t), U)_{H^0}, \qquad t \in T, \tag{21}$$

in which U is a random element taking values in a Hilbert space H^0, and $\{\Psi(t), t \in T\}$ is a known family of vectors in H^0;

(iii) the *stochastic inner product model*

$$X(t) = (\Psi(t), U)_{\sim H^0}, \qquad t \in T, \tag{22}$$

in which $\{\Psi(t), t \in T\}$ is a known family of elements in a Hilbert space H^0, and where we use the notation $(h, U)_{\sim H^0}$ to denote the value at h (an element in H^0) of a Hilbert space indexed stochastic process $\{U(h), h \in H^0\}$ with finite second moments which is *linear* in the sense that (for all h, h_1, h_2 in H^0)

$$U(h_1 + h_2) = U(h_1) + U(h_2) \quad \text{with probability one}, \tag{23}$$

and is *continuous* in the sense that the mean value function

$$m_U(h) = E[U(h)] = E[(h, U)_{\sim H^0}] \tag{24}$$

and the covariance kernel

$$Q_U(h_1, h_2) = \mathrm{Cov}[U(h_1), U(h_2)] = \mathrm{Cov}[(h_1, U)_{\sim H^0}, (h_2, U)_{\sim H^0}] \tag{25}$$

are continuous functions of their arguments. The implications of these assumptions will be explored at the end of this section.

Three important examples of signal transformation are:

(i) *instantaneous transformation,* when $X(t)$ is a random vector of dimension q, $Z(t)$ is a random vector of dimension r, and

$$Z(t) = H(t)X(t) \tag{26}$$

where $H(t)$ is a known r by q matrix;

(ii) *integral transformation*, when the signal $\{X(\tau), \tau \in T_X\}$ is a process with index set T_X for which one can define the notion of stochastic integral, and

$$Z(t) = \int_{T_X} A(t, \tau)X(\tau)\, d\tau \tag{27}$$

where A is a known kernel with domain $T \otimes T_X$;

(iii) *RKHS transformation*, where $Z(t)$ is assumed only to be a random variable in $H_X = H[X(\tau), \tau \in T_X]$, the Hilbert space of random variables spanned by the signal process $\{X(\tau), \tau \in T_X\}$; then corresponding to $Z(t)$ there is a function $a(t, \cdot)$ in $H(K_X)$, the RKHS corresponding to the covariance kernel K_X of $X(\cdot)$, in terms of which we can write

$$Z(t) = (a(t, \cdot), X)_{\sim H(K_X)}. \tag{28}$$

The RKHS transformation is the most general; other examples of signal transformation can always be represented as an RKHS transformation.

It should be noted that parameterization is in a rough sense the *inverse* of transformation; if $Z(\cdot)$ is considered the signal process, then $X(\cdot)$ is a parameterization of $Z(\cdot)$. We still would employ a parameterization of $X(\cdot)$, since for computational tractability it is usually necessary to parametrize a process $X(\cdot)$ by a white noise process $U(\cdot)$. We conclude this section by discussing the notion of white noise appropriate for our purposes.

Let H^0 be a Hilbert space, and let $\{U(h), h \in H^0\}$ be an H^0-indexed time series which is linear (in the sense that (23) holds) and continuous (in the sense that $m_U(h)$ and $Q_U(h_1, h_2)$ are continuous). Then there exists an element $\mathbf{m}$ in H^0 and a bounded linear operator $Q: H^0 \to H^0$ satisfying

$$m_U(h) = (\mathbf{m}, h)_{H^0}, \qquad Q_U(h_1, h_2) = (Qh_1, h_2)_{H^0}. \tag{29}$$

We call $\mathbf{m}$ the mean vector and Q the covariance operator of the time series $U(\cdot)$.

We call $U(\cdot)$ *white noise* with intensity λ if it has zero mean, $\mathbf{m} = 0$, and covariance operator a multiple of the identity operator I:

$$Q = \lambda I. \tag{30}$$

When $\lambda = 1$, we call $U(\cdot)$ white noise.

Corresponding to a zero mean time series $\{Y(t), t \in T\}$ with covariance kernel Q with RKHS $H(Q)$, there is an $H(Q)$-indexed white noise

$$\{(h, Y)_{\sim H(Q)}, h \in H(Q)\}.$$

5. A general model of linear statistical inference on time series. We seek to extract a signal process $X(\cdot)$ from an observed process $Y(\cdot)$ which is the sum of $Z(\cdot)$ and $N(\cdot)$, where $Z(\cdot)$ is a transformed signal process.

We assume a signal process $\{X(\tau), \tau \in T_X\}$ with the representation

$$X(\tau) = \sum_{j=1}^{q} \beta_j \phi_j(\tau) + (\Psi(\tau, \cdot), U)_{\sim H^0}, \qquad \tau \in T_X, \tag{1}$$

where H^0 is a Hilbert space of functions, $\phi_1, \ldots, \phi_q$ are known functions (satisfying conditions to be specified), $\{\Psi(\tau, \cdot), \tau \in T_X\}$ is a family of functions in H^0, and U is H^0-valued white noise. We call U the parameter process.

For the transformed signal process $\{Z(t), t \in T\}$ we assume the representation

$$Z(t) = \sum_{j=1}^{q} \tau_j A\phi_j(t) + (A\Psi(t, \cdot), U)_{\sim H^0} \tag{2}$$

where $A\phi_1, \ldots, A\phi_q$ are known functions (intuitively $A\phi_j$ is the result of operating on ϕ_j with an operator A, but since we have not specified a space to which ϕ_j belong we cannot specify the sense in which A is an operator in its own right), and $\{A\Psi(t, \cdot), t \in T\}$ is a family of functions in H^0.

The two kinds of parameters in the representation we have assumed for $X(\cdot)$ represent respectively the contribution of initial conditions and ongoing input. The parameters $\beta_1, \ldots, \beta_q$ represent the initial conditions, and the expression in $\beta_1, \ldots, \beta_q$ represent the contribution of initial conditions were there no additional input. The expression in U represents the contribution of the continuous input were the initial conditions zero. For an example of a process with the representation (1) see equation (8.17).

For the observed process $Y(\cdot)$ we assume

$$Y(t) = Z(t) + N(t)$$

where the noise process $N(\cdot)$ is independent of the parameter process $U(\cdot)$, has zero mean, and covariance kernel R:

$$E[N(t)] = 0, \qquad E[N(s)N(t)] = R(s, t). \tag{3}$$

The mean value function and covariance kernel of $Y(\cdot)$, $X(\cdot)$, and $Z(\cdot)$ depend on the attitude we adopt towards the vector parameter

$$\beta = \begin{bmatrix} \beta_1 \\ \cdots \\ \beta_q \end{bmatrix}$$

which can be *non-Bayes* or *Bayes*.

Non-Bayes attitude: β is a vector of unknown constants to be estimated.

Bayes attitude: β is a random vector which is independent of U and $N(\cdot)$, has mean $\bar{\beta}$ and covariance Q:

$$E[\beta] = \bar{\beta}, \qquad E[(\beta - \bar{\beta})(\beta - \bar{\beta})'] = Q. \tag{4}$$

In the non-Bayes case, the mean value function and covariance kernel of $X(\cdot)$, $Z(\cdot)$ and $Y(\cdot)$ are as follows (for fixed values of the parameter β):

$$\begin{aligned}
E_\beta[X(\tau)] &= \sum_{j=1}^{q} \beta_j \phi_j(\tau),\\
\mathrm{Cov}[X(\tau_1), X(\tau_2)] &= (\Psi(\tau_1, \cdot), \Psi(\tau_2, \cdot))_{H^0},\\
E_\beta[Z(t)] &= \sum_{j=1}^{q} \beta_j A\phi_j(t),\\
\mathrm{Cov}[Z(t_1), Z(t_2)] &= (A\Psi(t_1, \cdot), A\Psi(t_2, \cdot))_{H^0},\\
E_\beta[Y(t)] &= \sum_{j=1}^{q} \beta_j A\phi_j(t),\\
\mathrm{Cov}[Y(t_1), Y(t_2)] &= (A\Psi(t_1, \cdot), A\psi(t_2, \cdot))_{H^0} + R(t_1, t_2).
\end{aligned} \tag{5}$$

The subscript β on the expectation operator indicates that the probability measure depends on the value of the parameter β.

In the Bayes case the mean value function and covariance kernel of $X(\cdot)$, $Z(\cdot)$, and $Y(\cdot)$ are as follows:

$$\begin{aligned}
E[X(\tau)] &= \sum_{j=1}^{q} \bar{\beta}_j \phi_j(\tau),\\
\mathrm{Cov}[X(\tau_1), X(\tau_2)] &= \sum_{i,j=1}^{q} \phi_i(\tau_1) Q_{ij} \phi_j(\tau_2) + (\Psi(\tau_1, \cdot), \Psi(\tau_2, \cdot))_{H^0},\\
E[Z(t)] &= \sum_{j=1}^{q} \bar{\beta}_j A\phi_j(t),\\
\mathrm{Cov}[Z(t_1), Z(t_2)] &= \sum_{i,j=1}^{q} A\phi(t_1) Q_{ij} A\phi_j(t_2) + (A\Psi(t_1, \cdot), A\Psi(t_2, \cdot))_{H^0}\\
E[Y(t)] &= \sum_{j=1}^{q} \bar{\beta}_j A\phi_j(t),\\
\mathrm{Cov}[Y(t_1), Y(t_2)] &= \sum_{i,j=1}^{q} A\phi_j(t_1) Q_{ij} A\phi_j(t_2) + (A\Psi(t_1, \cdot), A\Psi(t_2,))_{H^0}\\
&\quad + R(t_1, t_2).
\end{aligned} \tag{6}$$

For ease of exposition let us assume that all stochastic processes are jointly normal so that we need not distinguish between wide sense conditional expectation and strict sense conditional expectations.

When β is Bayes, the statistical inference problem is to find the conditional distribution of β and $U(\cdot) = \{(h, U)_{\sim H^0}, h \in H^0\}$, given the observed series $\{Y(t), t \in T\}$. We denote by β^+ and Q^+ the conditional mean and conditional covariance of β;

$$\text{(7)} \quad \beta^+ = E[\beta | Y(t), t \in T], \qquad Q^+ = E[(\beta - \beta^+)(\beta - \beta^+)' | Y(t), t \in T].$$

We denote by m_U^+ and Q_U^+ the conditional mean value function and conditional covariance kernel of $U(\cdot)$;

$$\text{(8)} \quad \begin{aligned} m_U^+(h) &= E[(h, U)_{\sim H^0} | Y(t), t \in T], \\ Q_U^+(h_1, h_2) &= \operatorname{Cov}[(h_1, U)_{\sim H^0}, (h_2, U)_{\sim H^0} | Y(t), t \in T]. \end{aligned}$$

It should be noted that the unconditional mean value function and covariance kernel of $U(\cdot)$ is

$$\text{(9)} \quad \begin{aligned} m_U(h) &= E[(h, U)_{\sim H^0}] = 0, \\ Q_U(h_1, h_2) &= \operatorname{Cov}[(h_1, U)_{\sim H^0}, (h_2, U)_{\sim H^0}] = (h_1, h_2)_{H^0}. \end{aligned}$$

When β is non-Bayes, we choose as our estimation criterion minimum variance unbiased linear estimation of β and minimum variance unbiased linear prediction of $U(\cdot)$. We briefly define these concepts.

An estimator β^* of the parameter vector β is called unbiased if it is a function only of the observations $Y(\cdot)$ and

$$\text{(10)} \qquad E_\beta[\beta^*] = \beta \quad \text{for all } \beta \text{ in } E_q,$$

where E_q is q-dimensional Euclidean space. β^* is called a linear estimator if each component is a random variable belonging to H_Y. The minimum variance unbiased linear estimator, denoted $\hat{\beta}$, is defined to have minimum covariance matrix among all unbiased linear estimators; in symbols, for all β^* satisfying (10)

$$\text{(11)} \qquad E[(\hat{\beta} - \beta)(\hat{\beta} - \beta)'] \ll E[(\beta^* - \beta)(\beta^* - \beta)'].$$

A predictor θ^* of the random variable $\theta \equiv X(\tau)$ is called an unbiased linear predictor if it belongs to H_Y and

$$E_\beta[\theta^*] = E_\beta[\theta] \quad \text{for all } \beta \text{ in } E_q.$$

The *minimum variance unbiased linear predictor* is the unbiased linear predictor, denoted $\hat{\theta}$, satisfying

$$E[|\hat{\theta} - \theta|^2] = \min_{\theta^*} E[|\theta^* - \theta|^2]$$

where the minimum is taken over all unbiased linear predictors θ^* of θ.

Associated with the minimum variance unbiased estimator $\hat{\beta}$ of β is its covariance matrix,

$$\hat{Q} = E_{\beta}[(\hat{\beta} - \beta)(\hat{\beta} - \beta)'].$$

Associated with the minimum variance unbiased predictors $\{\hat{X}(\tau), \tau \in T_X\}$ of $\{X(\tau), \tau \in T_X\}$ is the covariance kernel

$$\hat{Q}_X(\tau_1, \tau_2) = E_{\beta}[\{\hat{X}(\tau_1) - X(\tau_1)\}\{\hat{X}(\tau_2) - X(\tau_2)\}].$$

Note that $Q_X(\tau_1, \tau_2) = \mathrm{Cov}_{\beta}[X(\tau_1), X(\tau_2)]$.

The non-Bayesian solution to our statistical inference problem consists in giving formulas for $\hat{\beta}$, $\hat{Q}$, $\hat{X}$, $\hat{Q}_X$.

In discussing the solution to the foregoing problems, one can proceed in three stages:

(1) obtain predicators of a signal process X from an observation process Y under the assumption that each has known mean, covariance, and cross-covariance.

(2) under the assumption that means are known linear functions of an unknown vector parameter β, obtain estimators of β,

(3) using the estimators of β [given by step (2)] as if they were the known true values, obtain predictors of X [using step (1)], after establishing the validity of this procedure.

In order not to overload the present paper, we will only discuss step (1); the discussion of steps (2) and (3) will be presented elsewhere.

6. RKHS solution of the zero mean linear statistical inference problem. In studying the solution to the linear statistical inference problem it is useful to begin by omitting β from the problem, by setting $\beta = 0$. We thus consider an observed process

$$Y(t) = (\Phi(t, \cdot), U)_{\sim H^0} + N(t) \tag{1}$$

where U is a H^0-valued white noise and $\{\Phi(t, \cdot), t \in T\}$ is a family of functions in H^0. Note that the function previously denoted $A\Psi(t, \cdot)$ is now denoted $\Phi(t, \cdot)$.

Fix $h \in H^0$; the best (minimum mean square error linear) predictor of $U(h) = (h, U)_{\sim H^0}$, given $\{Y(t), t \in T\}$, is the random variable $m_U^+(h)$ defined as follows: $m_U^+(h)$ is the unique random variable in $H_Y = H(Y(t), t \in T)$ satisfying

$$E[m_U^+(h) Y(t)] = E[(h, U)_{\sim H^0} Y(t)], \qquad t \in T. \tag{2}$$

But

$$E[(h, U)_{\sim H^0} Y(t)] = (\Phi(t, \cdot), h)_{H^0}. \tag{3}$$

Define a mapping Φ from H^0 to functions of t by

(4) $$\Phi h(t) = (\Phi(t, \cdot), h)_{H^0}.$$

Note that (3) implies $\Phi h \in H(R + K)$ for all $h \in H^0$. We will find other Hilbert spaces to which Φh belongs for all $h \in H^0$.

The covariance kernel of the observed process $Y(\cdot)$ is $K(t_1, t_2) + R(t_1, t_2)$; defining

(5) $$K(t_1, t_2) = (\Phi(t_1, \cdot), \Phi(t_2, \cdot))_{H^0};$$

note that K is the covariance kernel of the transformed signal process $Z(t) = (\Phi(t, \cdot), U)_{\sim H^0}$.

The notation is now at hand for expressing the *conditional mean* and *covariance* of $U(\cdot)$ given $Y(\cdot)$:

(6) $$m_U^+(h) = (\Phi h, Y)_{\sim H(R+K)},$$

(7) $$Q_U^+(h_1, h_2) = Q_U(h_1, h_2) - (\Phi h_1, \Phi h_2)_{H(R+K)}.$$

To apply the general results (6) and (7) note that for a signal process $X(\cdot)$ defined by

(8) $$X(t) = (\Psi(t, \cdot), U)_{\sim H^0}$$

the best predictor is

(9) $$\hat{X}(t) = (\Phi\Psi(t, \cdot), Y)_{\sim H(R+K)}$$

with conditional covariance kernel

$$\mathrm{Cov}[X(\tau_1), X(\tau_2) | Y(t), t \in T]$$

(10) $$= (\Psi(\tau_1, \cdot), \Psi(\tau_2, \cdot))_{H^0} - (\Phi\Psi(\tau_1, \cdot), \Phi\Psi(\tau_2, \cdot))_{H(R+K)}.$$

The general results (6) and (7) are expressed in terms of the RKHS corresponding to the covariance kernel $R + K$. The inner product in this space is not usually known to us; for computation it is convenient to express m_U^+ and Q_U^+ in terms of $H(R)$. For this purpose one needs to make a basic assumption on the covariance kernel K:

(11) $$K \in H(R) \otimes H(R).$$

In words, (11) states that K belongs to the RKHS which is the direct product of $H(R)$ and $H(R)$, and is denoted $H(R) \otimes H(R)$.

Direct product RKHS play a basic role in the application of RKHS methods to time series analysis; I believe their first application was by myself [(1963), p. 164, reprinted in (1967), p. 486] to provide a simple formulation of conditions for the equivalence of two normal processes with unequal covariance kernels [see § 11]. For the purposes of the present exposition we use the following definition of the direct product $H(R_1) \otimes H(R_2)$ of the RKHS corresponding to two covariance kernels R_1 and R_2, since it is the most convenient for proving theorems.

Let $H(R_i)$ consist of functions on an index set T_i,

$$\{\phi_\alpha\} \quad \text{be any CONS in } H(R_1),$$

$$\{\psi_\beta\} \quad \text{be any CONS in } H(R_2),$$

where CONS denotes complete orthonormal set of functions; then

$$H(R_1) \otimes H(R_2) = \left\{ h \quad \text{on} \quad T_1 \otimes T_2 : h(t_1, t_2) = \sum_{\alpha,\beta} h_{\alpha\beta}\phi_\alpha(t_1)\psi_\beta(t_1) \right.$$

$$\left. \text{for some double sequence } \{h_{\alpha\beta}\} \text{ such that } \sum_{\alpha,\beta} h_{\alpha\beta}^2 < \infty \right\}.$$

The norm of a function h in $H(R_1) \times H(R_2)$ is given by

$$\|h\|^2_{H(R_1)\otimes H(R_2)} = \sum_{\alpha,\beta} h^2_{\alpha\beta}.$$

A kernel K in $H(R) \otimes H(R)$ generates an operator $\mathbf{K}: H(R) \to H(R)$ by, for every $f \in H(R)$,

$$\mathbf{K}f(t) = (K(\cdot, t), f)_{H(R)}. \tag{12}$$

The operator $\mathbf{K}$ is Hilbert Schmidt (for a definition of the notion of Hilbert Schmidt operator see any text on functional analysis or integral equations; for example, Riesz-Nagy, ((1955), p. 242)).

It is of interest to relate the operator $\mathbf{K}$ to the operator Φ defined by (4) when

$$K(s, t) = (\Phi(s, \cdot), \Phi(t, \cdot))_{H^0}. \tag{13}$$

Therefore (12) can be written (formally)

$$\begin{aligned} \mathbf{K}f(t) &= (K(s, t), f(s))_{H(R)} = ((\Phi(s, u), \Phi(t, u))_{H^0}, f(s))_{H(R)} \\ &= ((\Phi(s, u), f(s))_{H(R)}, \Phi(t, u))_{H^0} = \Phi\Phi^* f(t) \end{aligned} \tag{14}$$

defining $\Phi^*: H(R) \to H^0$ by

$$\Phi^* f(u) = (\Phi(s, u), f(s))_{H(R)}. \tag{15}$$

The notation Φ^* is meant to indicate that Φ^* is the *adjoint* of Φ; see (20).

To make the foregoing assumptions rigorous we make the basic assumption that the functions $\{\Phi(t, \cdot), t \in T\}$ are selected as functions of two variables t and u so that

$$\Phi(t, u) \in H(R) \otimes H^0 \tag{16}$$

or equivalently that

(17) $$\Phi(t, u) = \sum_{\alpha,\beta} \Phi_{\alpha\beta}\phi_\alpha(t)\psi_\beta(u)$$

for some CONS $\{\phi_\alpha\}$ in $H(R)$, CONS $\{\psi_\beta\}$ in H^0, and double sequence $\{\Phi_{\alpha\beta}\}$ satisfying

(18) $$\sum_{\alpha,\beta} \Phi_{\alpha\beta}^2 < \infty;$$

then (4) defines a Hilbert Schmidt operator

(19) $$\Phi : H^0 \to H(R)$$

with adjoint $\Phi^* : H(R) \to H^0$ given by (15), in the sense that for every $h \in H^0$ and $f \in H(R)$

(20) $$(\Phi h, f)_{H(R)} = (h, \Phi^* f)_{H^0}.$$

We briefly outline the proof of these assertions. Write

(21) $$h(u) = \sum_\beta h_\beta \psi_\beta(u), \qquad f(t) = \sum_\alpha f_\alpha \phi_\alpha(t).$$

Then

(22) $$\Phi h(t) = \sum_\beta h_\beta \sum_\alpha \Phi_{\alpha\beta}\phi_\alpha(t) = \sum_\alpha \phi_\alpha(t) \sum_\beta \Phi_{\alpha\beta} h_\beta$$

and

(23) $$(\Phi h, f) = \sum_{\alpha,\beta} f_\alpha \Phi_{\alpha\beta} h_\beta;$$

similarly

(24) $$\Phi^* f(u) = \sum_\alpha f_\alpha \sum_\beta \Phi_{\alpha\beta}\psi_\beta(u) = \sum_\beta \psi_\beta(u) \sum_\alpha f_\alpha \Phi_{\alpha\beta}$$

and

(25) $$(h, \Phi^* f)_{H^0} = \sum_{\alpha,\beta} f_\alpha \Phi_{\alpha\beta} h_\beta.$$

We next prove that (16) implies that

(26) $$K \in H(R) \otimes H(R)$$

and

(27) $$\mathbf{K} = \Phi\Phi^*.$$

To prove (26) and (27) we use (13) and (17) to express $K(s, t)$ in a double series in the CONS $\{\phi_\alpha\}$:

$$K(s, t) = \sum_\beta \sum_{\alpha_1} \Phi_{\alpha_1\beta}\phi_{\alpha_1}(s) \sum_{\alpha_2} \Phi_{\alpha_2\beta}\phi_{\alpha_2}(t) = \sum_{\alpha_1,\alpha_2} K_{\alpha_1\alpha_2}\phi_{\alpha_1}(s)\phi_{\alpha_2}(t) \tag{28}$$

defining

$$K_{\alpha_1\alpha_2} = \sum_\beta \Phi_{\alpha_1\beta}\Phi_{\alpha_2\beta}. \tag{29}$$

The coefficients $\{K_{\alpha_1\alpha_2}\}$ satisfy

$$\sum_{\alpha_1,\alpha_2} |K_{\alpha_1\alpha_2}|^2 < \infty, \qquad \sum_\alpha |K_{\alpha\alpha}| < \infty \tag{30}$$

since

$$|K_{\alpha_1\alpha_2}|^2 \leqq \sum_{\beta_1} \Phi^2_{\alpha_1\beta_1} \sum_{\beta_2} \Phi^2_{\alpha_2\beta_2}, \qquad |K_{\alpha\alpha}| \leqq \sum_\beta \Phi^2_{\alpha\beta}. \tag{31}$$

From (28) and (30) we infer (26). To prove (27), note that

$$\mathbf{K}f(t) = \sum_{\alpha_1} f_{\alpha_1} \sum_{\alpha_2} K_{\alpha_1\alpha_2}\phi_{\alpha_2}(t) \tag{32}$$

while

$$\Phi\Phi^* f(t) = \sum_{\alpha_2} \phi_{\alpha_2}(t) \sum_\beta \phi_{\alpha_2\beta} \sum_{\alpha_1} f_{\alpha_1}\Phi_{\alpha_1\beta}. \tag{33}$$

So far we have merely defined the operator $\mathbf{K}$ and related it to Φ. Next let us state the *basic relation between* $H(R + K)$ *and* $H(R)$ that follows from (26): $H(R + K)$ consists of the same functions as $H(R)$ and for every f and g in $H(R)$

$$(f, g)_{H(R+K)} = ((I + \mathbf{K})^{-1}f, g)_{H(R)}. \tag{34}$$

We outline a proof of (34); let $f_1 = (I + K)^{-1}f$ so that $(I + K)f_1 = f$. One sees that (34) is equivalent to

$$((I + K)f_1, g)_{H(R+K)} = (f_1, g)_{H(R)} \tag{35}$$

for all f_1 and g in $H(R)$. To prove (35) it suffices to prove it for $g = R + K(\cdot, t)$; with this choice of g, both sides of (35) equal $f_1(t) + \mathbf{K}f_1(t) = f(t)$. To summarize the argument, we have shown that the inner product defined on the functions belonging to $H(R + K)$ by the right-hand side of (34) is the correct inner product since it has the reproducing property

$$(f, R + K(\cdot, t))_{H(R+K)} = f(t). \tag{36}$$

The facts are at hand to conclude our discussion of the zero mean linear statistical inference problem. Assuming (16), the conditional mean and covariance of $U(\cdot)$ given $Y(\cdot)$ is given by

$$m_U^+(h) = ((I + \Phi\Phi^*)^{-1}\Phi h, Y)_{\sim H(R)}, \tag{37}$$

$$Q_U^+(h_1, h_2) = Q_U(h_1, h_2) - ((I + \Phi\Phi^*)^{-1}\Phi h_1, \Phi h_2)_{H(R)}. \tag{38}$$

A comment is necessary on the meaning of $(f, Y)_{\sim H(R)}$ for any $f \in H(R)$. For f a simple function in the sense that (for some integer n, reals c_i, and indices t_i)

$$f = \sum_{i=1}^{n} c_i R(\cdot, t_i) \tag{39}$$

we define

$$(f, Y)_{\sim H(R)} = \sum_{i=1}^{n} c_i Y(t_i). \tag{40}$$

For f a limit in $H(R)$-norm of simple functions f_n, we define

$$(f, Y)_{\sim H(R)} = \lim_n (f_n, Y)_{\sim H(R)} \tag{41}$$

where this limit is in the Hilbert space of random variables with inner product corresponding to the covariance kernel $R + K$. For f in (39) and random variable in (40), one can verify that

$$E|(f, Y)_{\sim H(R)}|^2 = \sum_{i,j} c_i c_j \{R(t_i, t_j) + K(t_i, t_j)\} = ((I + \mathbf{K})f, f)_{H(R)}. \tag{42}$$

One can show that $\{(f, Y)_{\sim H(R)}, f \in H(R)\}$ is a Hilbert space indexed collection of random variables with mean zero and covariance operator $I + \mathbf{K}$.

Finally, we turn to an important interpretation of (37) which will enable us to show the equivalence of the solution of various approximation and control problems. In this study a central role will be played by the operator identity

$$(I + \Phi\Phi^*)^{-1}\Phi = \Phi(I + \Phi^*\Phi)^{-1}; \tag{43}$$

to verify this identity multiply on the left by $I + \Phi\Phi^*$ and multiply on the right by $I + \Phi^*\Phi$.

Using (43) rewrite the formula (37) for m_U^+:

$$m_U^+(h) = (\Phi(I + \Phi^*\Phi)^{-1}h, Y)_{\sim H(R)}. \tag{44}$$

To take the next step we must act as if the inner product in (44) is a real inner product in $H(R)$, rather than a congruence inner product. Then we can express it as an inner product in H^0:

(45) $$m_U^+(h) = (h, (I + \Phi^*\Phi)^{-1}\Phi^* Y)_{H^0}.$$

Our formula for Q_U^+ can be rigorously rewritten in terms of inner products in H^0:

(46) $$\begin{aligned} Q_U^+(h_1, h_2) &= (h_1, h_2)_{H^0} - (\Phi^*\Phi(I + \Phi^*\Phi)^{-1}h_1, h_2)_{H^0} \\ &= ((I + \Phi^*\Phi)^{-1}h_1, h_2)_{H^0}. \end{aligned}$$

The parameter process $U(\cdot)$ regarded as a random element with values in H^0 thus has conditional mean

(47) $$\hat{U} = (I + \Phi^*\Phi)^{-1}\Phi^* Y$$

and a conditional covariance operator equal to $(I + \Phi^*\Phi)^{-1}$. In the next section we show that $\hat{U}$, which at this stage has no rigorous meaning since the sample path of the process Y never belongs to $H(R)$, is the solution to a suitable optimization problem.

7. Equivalence between time series parameter estimation, control theory, and approximation theory. There is a remarkable equivalence between the solutions to the problems of *regression* analysis, *approximation* by spline functions, *control* of linear plants with quadratic cost criterion, Kalman *filtering*, solution of *integral equations*, and *time series* parameter estimation.

The first observation that leads to this equivalence is to note that all the foregoing problems (in their simplest versions) are of the form: "solve" for U in the linear system

(1) $$Y = \Phi U$$

where Y is the observation or measurement (and is therefore considered a known function) and Φ is a known *linear operator*. To make this problem more precise we must specify the domain and range spaces of the operator Φ. We find it clearer to leave open the range space of Φ and to specify it as a collection $\{\Phi(t), t \in T\}$ of elements in a Hilbert space H^0. Then we seek to find that U in H^0 satisfying

(2) $$Y(t) = (\Phi(t), U)_{H^0}, \qquad t \in T.$$

We continue to write the linear system (2) intuitively as $Y = \Phi U$.

There are various senses in which one can solve the linear system (2) for U; our basic classification is in terms of

(3)
interpolation vs. smoothing,
non-Bayes unknown vs. Bayes unknown.

These names are a blend of ideas from the theory of approximation (of a function given its values at a set of points) and the theory of statistical inference. It is beyond the scope of this paper to define these terms; we wish merely to indicate the solutions to $Y = \Phi U$ to which they lead:

	Interpolation	*Smoothing*
Non-Bayes U	$U = \Phi^{-1}Y$	$U = (\Phi^*\Phi)^{-1}\Phi^* Y.$
Bayes U	$U = \Phi^*(\Phi\Phi^*)^{-1}Y$	$U = (\lambda I + \Phi^*\Phi)^{-1}\Phi^* Y = \Phi^*(\lambda I + \Phi\Phi^*)^{-1}Y$

The problem we desire to consider in detail is the optimization problem: given a Hilbert space H^0, and index set T, a family of elements $\{\Phi(t), t \in T\}$ in H^0, and RKHS $H(R)$ consisting of functions on T, with reproducing kernel R, and a function $Y \in H(R)$, find U in H^0 to minimize (for a fixed positive constant λ)

$$J(U) = \|Y - \Phi U\|^2_{H(R)} + \lambda \|U\|^2_{H^0}. \tag{4}$$

In order to guarantee that

$$\Phi U(t) = (\Phi(t), U)_{H^0} \tag{5}$$

is a function in $H(R)$ we assume that Φ belongs to $H(R) \otimes H^0$, regarded as a function of two variables. Then $\Phi: H^0 \to H(R)$ is a Hilbert Schmidt operator with adjoint $\Phi^*: H(R) \to H^0$. To minimize $J(U)$ one writes it

$$\begin{aligned} J(U) &= \|Y\|^2_{H(R)} - 2(Y, \Phi U)_{H(R)} + (\Phi U, \Phi U)_{H(R)} + \lambda(U, U)_{H^0} \\ &= \|Y\|^2_{H(R)} - 2(\Phi^* Y, U)_{H^0} + ((\lambda I + \Phi^*\Phi)U, U)_{H^0} \end{aligned} \tag{6}$$

which is minimized by

$$\hat{U} = (\lambda I + \Phi^*\Phi)^{-1}\Phi^* Y \tag{7}$$

with

$$\begin{aligned} J(\hat{U}) &= \|Y\|^2_{H(R)} - ((\lambda I + \Phi^*\Phi)\hat{U}, \hat{U})_{H^0} \\ &= \|Y\|^2_{H(R)} - (\Phi Y, \Phi(\lambda I + \Phi^*\Phi)^{-1}\Phi^* Y)_{H(R)} \\ &= \|Y\|^2_{H(R)} - (Y, (\lambda I + \Phi\Phi^*)^{-1}\Phi\Phi^* Y)_{H(R)} \\ &= \lambda((\lambda I + \Phi\Phi^*)^{-1}Y, Y)_{H(R)}. \end{aligned} \tag{8}$$

Comparing equation (47) of the previous section with equation (7) of this section one sees that the two answers are equivalent. It is widely accepted among engineers concerned with filtering theory that the solution to the time series parameter estimation can be found by solving a suitable optimization problem of the form (4); see for example Bensoussan (1969). However, no rigorous proof of this equivalence has been given in general. I believe that a rigorous proof can be given using the ideas we have discussed in this paper.

CHAPTER 3. EXAMPLES OF RKHS

8. Integral representation theorem. The basic approach to finding the RKHS associated with a covariance kernel K on $T \otimes T$ is to find an integral representation for K of the form

(1) $$K(s, t) = \int_\Lambda g(s, \lambda) g(t, \lambda) \mu(d\lambda)$$

where μ is a measure and $\{g(t, \cdot), t \in T\}$ is a family of functions in $L_2(\mu) = \{\text{measurable } f \text{ with domain } \Lambda\colon \int_\Lambda f^2(\lambda)\, d\lambda < \infty\}$. When (1) holds, $H(K)$ consists of all functions f on T of the form

(2) $$f(t) = \int_\Lambda F(\lambda) g(t, \lambda) \mu(d\lambda)$$

for some unique F in $H_g = H(g(t, \cdot), t \in T)$, the Hilbert subspace of $L_2(\mu)$ spanned by $\{g(t, \cdot), t \in T\}$. The RKHS norm of f is given by

(3) $$\|f\|^2_{H(K)} = \|F\|^2_{L_2(\mu)}.$$

To prove (2) we merely note that

(4) $$K(s, t) = (K(\cdot, s), K(\cdot, t))_{H(K)} = (g(s, \cdot), g(t, \cdot))_{L_2(\mu)}$$

implies that there is a congruence (one-to-one inner product preserving mapping) between $H(K)$ and H_g such that $K(\cdot, t)$ and $g(t, \cdot)$ transform into each other. If $f \in H(K)$ and $F \in L_2(\mu)$ are transforms of each other,

(5) $$f(t) = (f, K(\cdot, t))_{H(K)} = (F, g(t, \cdot))_{L_2(\mu)}$$

which proves (2).

To illustrate the use of the integral representation approach, let us consider the Wiener process $\{W(t), 0 \leqq t < T\}$ which has covariance kernel

(6) $$K_W(s, t) = \min(s, t).$$

This kernel has integral representation

(7) $$K(s, t) = \int_0^T (s - u)^0_+ (t - u)^0_+ \, du$$

where for $k \geqq 0$

$$x_+^k = 0 \quad \text{if } x \leqq 0, \qquad (8)$$
$$= x^k \quad \text{if } x > 0$$

so that $(t - u)_+^0 = 1$ or 0 according as $u < t$ or $u \geqq t$.

From (7) it follows that $H(K)$ consists of functions f of the form

$$f(t) = \int_0^T F(u)(t - u)_+^0 \, du = \int_0^t F(u)\, du, \qquad 0 \leqq t \leqq T, \tag{9}$$

which are indefinite integrals of functions F which are square integrable on the interval $0 \leqq t \leqq T$; the norm of f is

$$\| f \|_{H(K)}^2 = \int_0^T |F(u)|^2 \, du. \tag{10}$$

Formally $F(t) = f'(t)$.

Next consider a process $\{X(t), 0 \leqq t \leqq 1\}$ with covariance kernel

$$K(s, t) = \int_0^1 (q!)^{-2}(s - \lambda)_+^q (t - \lambda)_+^q \, d\lambda \tag{11}$$

where q is a fixed integer. It follows that $H(K)$ consists of all functions f of the form

$$f(t) = \int_0^1 F(\lambda) \frac{1}{q!}(t - \lambda)_+^q \, d\lambda \tag{12}$$

for some square integrable function F. Formally

$$f^{(q+1)}(t) = F(t) \tag{13}$$

and $H(K)$ consists of functions f whose $(q + 1)$st derivative is square integrable. The process $X(\cdot)$ can be represented in terms of the Wiener process $W(\cdot)$ by

$$X(t) = \int_0^1 \frac{1}{(q - 1)!}(t - \lambda)_+^{q-1} W(\lambda)\, d\lambda; \tag{14}$$

in terms of white noise $\dot{W}(\lambda)$, the formal derivative of $W(\lambda)$. One can represent $X(t)$ by

$$X(t) = \int_0^1 \frac{1}{q!}(t - \lambda)_+^q \dot{W}(\lambda)\, du. \tag{15}$$

The process $X(\cdot)$ defined by (15) satisfies the formal stochastic differential equation

(16) $$X^{(q)}(t) = \dot{W}(t);$$

in words, the qth derivative of $X(\cdot)$ is white noise. The most general solution of (16) is

(17) $$X(t) = \beta_1 + \beta_2 + \ldots + \beta_q t^{q-1} + \int_0^1 \frac{1}{q!}(t-\lambda)_+^q \dot{W}(\lambda)\,d\lambda.$$

This is an example of a process $X(\cdot)$ with representation (5.1).

9. Hilbert space indexed time series. Let H^0 be a separable Hilbert space, and $\{U(h), h \in H^0\}$ an H^0-indexed series with zero mean and covariance operator Q:

(1) $$Q_U(h_1, h_2) = \mathrm{Cov}[U(h_1), U(h_2)] = (Qh_1, h_2)_{H^0}.$$

Let $\Phi = \Phi^* = Q^{1/2}$ so that $Q = \Phi\Phi^*$. Then

(2) $$Q_U(h_1, h_2) = (\Phi h_1, \Phi h_2)_{H^0}.$$

Define

(3) $$H_\Phi = \overline{\Phi H^0} \equiv \text{closure of } \{g : g = \Phi h \text{ for some } h \in H^0\}.$$

The RKHS corresponding to Q is

(4) $$H(Q) = \{f \text{ on } H^0 : f(h) = (\Phi h, F)_{H^0} \text{ for some } f \in H_\Phi, \|f\|^2_{H(Q)} = \|F\|^2_{H^0}\}.$$

In words, the RKHS of an H^0-indexed series is a family of linear functionals, so that $H(Q)$ is set theoretically a subset of H^{0*}, the topological dual space of H^0 (which is congruent to H^0). There is a subset of H^0, denoted $H_\sim(Q)$, with which $H(Q)$ is congruent:

(5) $$H_\sim(Q) = \{g \in H^0 : g = \Phi^* F \text{ for some } F \in H_\Phi, \quad \|g\|^2_{H_\sim(Q)} = \|F\|^2_{H^0}\}.$$

We call $H_\sim(Q)$ a *congruent* RKHS, with reproducing kernel Q. Note that a RKHS is always a space of functions, while a congruent RKHS is a subset of a Hilbert space H^0 which is congruent to a RKHS of functions on H^0.

One can characterize $H_\sim(Q)$ more concretely in the case that Φ is a Hilbert Schmidt operator; then one can find an orthonormal set $\{\phi_j\}$ in H^0 and a sequence of nonnegative constants λ_j such that ϕ_j and λ_j are respectively eigenfunctions and eigenvalues of Q:

(6) $$Q\phi_j = \lambda_j \phi_j, \qquad \Phi\phi_j = \lambda_j^{1/2}\phi_j.$$

Note that for every $h \in H^0$

(7) $$h = \sum (\phi_j, h)_{H^0}\phi_j, \qquad \Phi h = \sum_j \lambda_j^{1/2}(\phi_j, h)_{H^0}\phi_j.$$

Assume further that the range of Φ is dense in H^0 so that $H_\Phi = H^0$ (this assumption is made only for ease of exposition and can be dispensed with). Then $H_\sim(Q)$ consists of

(8) $$\{g \in H^0 : g = \sum \lambda_j^{1/2}(\phi_j, F)_{H^0}\phi_j \text{ for some } f \in H^0, \quad \|g\|^2_{H_\sim(Q)} = \sum |(\phi_j, F)_{H^0}|^2\}.$$

But one can write g in terms of its own Fourier coefficients:

(9) $$g = \sum g_j\phi_j, \qquad g_j = (\phi_j, g)_{H^0}.$$

One concludes that

(10) $$H_\sim(Q) = \left\{ g \in H^0 : g = \sum g_j\phi_j \text{ for some sequence } \{g_j\} \text{ such that } \|g\|^2_{H_\sim(Q)} = \sum \frac{1}{\lambda_j} g_j^2 < \infty \right\}.$$

It should be noted that every Hilbert space H^0 is the congruent RKHS of H^0-indexed white noise; when $Q = I$ we obtain from (5)

(11) $$H_\sim(I) = \{g \in H^0 : g = F \text{ for some } f \in H^0, \ \|g\|^2_{H_\sim(I)} = \|F\|^2_{H^0}\}.$$

L_2-spaces (Hilbert spaces of square integrable functions on some measure space) differ fundamentally from RKHS in the fact that in L_2-spaces convergence in norm of a sequence of functions does *not* imply pointwise convergence of the sequence. The celebrated Dirac delta function $\delta(\cdot)$ is the "reproducing kernel" of the L_2-space of square integrable functions on the real line; symbolically one can consider white noise to have covariance kernel $\delta(t - s)$.

Finally, it should be noted that a Hilbert space whose members are functions on a set T, which is a congruent RKHS in the sense of (10), can be the true RKHS of a time series $\{Y(t), t \in T\}$ with index set T. For example, corresponding to $\{Y(t), 0 \leqq t \leqq 1\}$ with zero means and continuous kernel $Q(s, t)$, we can define an operator Q on $H^0 \equiv L_2[0, 1]$ by

(12) $$Qf(t) = \int_0^1 Q(s, t) f(s)\, ds, \qquad 0 \leqq t \leqq 1.$$

Let $\{\phi_j\}$ and $\{\lambda_j\}$ be the eigenfunctions and positive eigenvalues of the integral equation

(13) $$Q\phi_j(t) = \int_0^1 Q(s, t)\phi_j(s)\,ds = \lambda_j\phi_j(t), \qquad 0 \leqq t \leqq 1.$$

Assume further that Q is positive definite in the sense that for any function f in H^0

(14) $$\int_0^1\int_0^1 Q(s, t)f(s)f(t)\,ds\,dt = 0 \quad \text{implies } f(t) = 0 \text{ a.e.}$$

Then the RKHS corresponding to the kernel $Q(s, t)$ is

(15) $$H(Q) = \left\{g \in H^0 : \|g\|^2_{H(Q)} = \sum \frac{1}{\lambda_j}\left|\int_0^1 \phi_j(t)g(t)\,dt\right|^2 < \infty\right\}$$

which is the right-hand side of (10).

10. Normal Markov processes. A normal process $\{Y(t), 0 \leqq t \leqq T\}$ with zero means and continuous covariance kernel $K(s, t)$ can be shown to be Markov if and only if

(1) $$K(s, t) = g(s)G(\min(s, t))g(t)$$

where

(i) g is nonvanishing, continuous, and $g(0) = 1$, and

(ii) G is continuous and monotone increasing.

Important examples of normal Markov processes are:

Wiener: $K(s, t) = \sigma^2 \min(s, t)$, $g(t) = 1$, $G(t) = \sigma^2 t$;

Stationary Markov: $K(s, t) = \sigma_0^2 e^{-\beta|t-s|}$, $g(t) = e^{-\beta t}$, $G(t) = \sigma_0^2 e^{2\beta t}$;

Pinned Wiener ($T = 1$): $K(s, t) = \min(s, t) - st$, $g(t) = 1 - t$, $G(t) = t/(1 - t)$.

A covariance kernel of the form (1) has the following (Stieltjes) integral representation:

(2) $$K(s, t) = \int_0^T g(s)(s - u)_+^0 g(t)(t - u)_+^0\,dG(u) + g(s)g(t)G(0).$$

Let us assume $G(0) > 0$ since this is slightly more complicated than the case $G(0) = 0$.

From the integral representation theorem, it follows that $H(K)$ consists of all functions f of the form

(3) $$f(t) = \int_0^T g(t)(t - u)_+^0 F(u)\,dG(u) + g(t)F_0G(0)$$

for some constant F_0 and function $F(u)$ in $\{h: \int_0^T |h(u)|^2 \, dG(u) < \infty\}$; the RKHS norm of $f(\cdot)$ is

$$\|f\|^2_{H(K)} = \int_0^T |F(u)|^2 \, dG(u) + F_0^2 G(0). \tag{4}$$

To be more concrete, let us assume that *g and G* are *differentiable*; then (3) yields

$$G'(t)F(t) = \left\{\frac{f(t)}{g(t)}\right\}', \qquad F_0 = \frac{f(0)}{g(0)G(0)} = \frac{f(0)}{G(0)} \tag{5}$$

since $g(0) = 1$. One can now show that $H(K)$ consists of all L_2-differentiable functions $f(\cdot)$ with domain $0 \leqq t \leqq T$, such that

$$\|f\|^2_{H(K)} = \int_0^T \left|\left\{\frac{f(t)}{g(t)}\right\}'\right|^2 \frac{1}{G'(t)} \, dt + |f(0)|^2 \frac{1}{G(0)} < \infty. \tag{6}$$

Formula (6) has been given by several authors; one important reference is Sacks and Ylvisaker (1966), p. 86.

In my view a more physically meaningful formula for $\|f\|^2_{H(K)}$ is found by introducing

$$\rho(t) = g'(t)/g(t), \qquad \sigma^2(t) = g^2(t)G'(t) \tag{7}$$

in terms of which we can write the $H(K)$ inner product

$$\begin{aligned}(f_1, f_2)_{H(K)} = &\int_0^T \{f_1'(t) - \rho(t)f_1(t)\}\sigma^{-2}(t)\{f_2'(t) - \rho(t)f_2(t)\} \, dt \\ &+ f_1(0)G^{-1}(0)f_2(0).\end{aligned} \tag{8}$$

When one considers problems of statistical inference for normal Markov processes, it appears that $\rho(t)$ and $\sigma^2(t)$ are the natural parameters to estimate; for this reason we regard them as the natural parameters in terms of which to express the RKHS inner product.

CHAPTER 4.
PROBABILITY DENSITY FUNCTIONALS OF NORMAL PROCESSES

In this review many areas of time series analysis to which RKHS have been applied can only be indicated by reference to the literature. RKHS have been used to:

1. Characterize the support of the probability measure on a Banach space induced by a normal time series [see Kallianpur (1970a)];

2. Represent an arbitrary nonlinear functional of a normal process (Cameron-Martin expansion and multiple Wiener integral expansion) [see Neveu (1968), Kallianpur (1970b), Hida-Ikeda (1967)];

3. Represent a normal process in terms of white noise processes representing "innovations" in the process [see Hida (1960)];

4. Obtain zero-one laws for normal processes [see Kallianpur (1970c)];

5. Determine conditions for convergence in distribution of stochastic processes [see Brown (1969)];

6. Obtain statistical designs and quadrature formulas [see Sacks and Ylvisaker (1966), (1968), (1969) and Wahba (1969a)];

7. Establish equivalence of time series prediction and spline functions [see Kimeldorf and Wahba (1968, 1969)];

8. Provide algorithms for numerical solution of integral and differential equations [see Wahba (1969b)];

9. Provide coordinate free approach to statistical communication theory [see Kailath (1969)].

It was mentioned in the introduction (§ 0) that a major reason for the usefulness of RKHS is that a basic tool in many of the foregoing problem areas is the theory of equivalence and singularity of normal measures. The aim of this chapter is to indicate how the main results of this theory can be simply expressed in terms of RKHS.

11. Equivalence and orthogonality of normal processes. Given two probability measures P_1 and P_0 on a measurable space $(\Omega, \mathscr{A})$ we say that

(i) P_1 is absolutely continuous with respect to P_0, denoted $P_1 \ll P_0$, if

$$\text{for all } A \in \mathscr{A}, \qquad P_0[A] = 0 \text{ implies } P_1[A] = 0. \tag{1}$$

or equivalently if there exists a measurable function p such that for all $A \in \mathscr{A}$

$$P_1[A] = \int_A p\, dP_0; \tag{2}$$

(ii) P_1 and P_0 are equivalent, denoted $P_1 \equiv P_0$, if $P_1 \ll P_0$ and $P_0 \ll P_1$;

(iii) P_1 and P_0 are orthogonal, denoted $P_1 \perp P_0$, if there exists a set A in $\mathscr{A}$ such that

$$P_1[A] = 1 \quad \text{and} \quad P_0[A] = 0. \tag{3}$$

When (1) holds, we call p the Radon-Nikodým derivative of P_1 with respect to P_0, denoted $p = dP_1/dP_0$.

To apply these notions to a stochastic process $Y(\cdot)$ we first define its *probability distribution.*

Given a stochastic process with index set T, such as $\{Y(t), t \in T\}$, by the *probability distribution* induced by the process we mean the probability space $(\Omega_T, \mathscr{B}_T, P_Y)$, where

Ω_T is the linear space of all real valued functions on T,

$\mathscr{B}_T$ is the sigma field of cylinder sets of Ω_T, defined as the smallest sigma field containing all sets of the form $\{\omega \in \Omega_T : \omega(t) \leqq x\}$ for any $t \in T$ and real number x,

P_Y is a probability measure with domain $\mathscr{B}_T$ such that

$$P_Y[\{\omega \in \Omega_T : \omega(t_1) \leqq x_1, \ldots, \omega(t_n) \leqq x_n\}] = P[Y(t_1) \leqq x_1, \ldots, Y(t_n) \leqq X_n]$$

for any integers n, points $t_1, \ldots, t_n$ in T and real numbers $x_1, \ldots, x_n$.

When considering the relations between normal probability measures, we denote by $P_{m,R}$ the probability measure on $(\Omega_T, \mathscr{B}_T)$ which is the probability distribution of a normal time series $Y(\cdot) = \{Y(t), t \in T\}$ with mean m and covariance R. The following notions are needed:

(i) $H(R)$ or RKHS(R), the reproducing kernel Hilbert space corresponding to R;

(ii) $H(R) \otimes H(R)$, the direct product RKHS;

(iii) equivalence class of R defined to be the set of all covariance kernels R, which are equivalent to R.

We define two covariance kernels R_1 and R_2 to be equivalent, denoted $R_1 \equiv R_2$, if there exist constants C_1 and C_2 such that

$$C_1 \sum_{i,j=1}^{n} c_i c_j R_1(t_i, t_j) \leqq \sum_{i,j=1}^{n} c_i c_j R_2(t_i, t_j)$$

$$\leqq C_2 \sum_{i,j=1}^{n} c_i c_j R_1(t_i, t_j)$$

for every integer n, indices $t_1, \ldots, t_n$ in T, and reals $c_1, \ldots, c_n$.

The basic facts about probability density functionals of normal processes are expressed in the following theorems for any mean value functions m, m_1, m_2 and covariance kernels R, R_1, R_2.

DICHOTOMY THEOREM. *Either* $P_{m_1,R_1} \equiv P_{m_2,R_2}$ *or* $P_{m_1,R_1} \perp P_{m_2,R_2}$; *see Feldman* (1958) *and Hajek* (1958).

Equivalence of normal processes with unequal means, equal covariances. $P_{m,R} \equiv P_{0,R}$ if and only if $m \in H(R)$; then

$$p = \frac{dP_{m,R}}{dP_{0,R}} = \exp\{(m, Y)_{\sim H(R)} - \tfrac{1}{2}\|m\|^2_{H(R)}\}. \tag{4}$$

Equivalence of normal processes with equal means, unequal covariances. $P_{0,R_1} \equiv P_{0,R_2}$ if and only if

(5) $$R_1 \equiv R_2 \quad \text{and} \quad R_2 - R_1 \in H(R_1) \otimes H(R_1);$$

the probability density functional is of the form

$$p = dP_{0,R_2}/dP_{0,R_1} = Ce^J$$

where C is a constant and J is a random variable which is a *quadratic form* in the sense that it belongs to the Hilbert space spanned by the family of random variables $\{Y(s)Y(t) - R_1(s,t), (s,t) \in T \otimes T\}$. One can express p in a variety of ways; see Shepp (1965), Varberg (1966), Golosov (1966), Neveu (1968), Rozanov (1968), Kailath (1969).

It seems difficult to claim credit for a new formula for the probability density functional, but let me note without proof what I believe to be a new version of the formula for p in the case when both normal measures correspond to Markov processes on $0 \leqq t \leqq T$, so that the representation (10.1) holds:

(6) $$\begin{aligned} R_1(s,t) &= g_1(s)G_1(\min(s,t))g_1(t), \\ R_2(s,t) &= g_2(s)G_2(\min(s,t))g_2(t) \end{aligned}$$

where, for $i = 1, 2$, $g_i(\cdot)$ is nonvanishing, continuous, and $g_i(0) = 1$, and $G_i(\cdot)$ is continuous and monotone increasing. Assume $g_i(\cdot)$ and $G_i(\cdot)$ have continuous derivatives, and define the "natural parameters" [used in (10.8)]

(7) $$\rho_i(t) = g_i'(t)/g_i(t), \qquad \sigma_i^2(t) = g_i^2(t)G_i'(t).$$

A necessary and sufficient condition for equivalence of the normal measure is

(8) $$\sigma_1^2(t) = \sigma_2^2(t) = \sigma^2(t) \quad \text{for all } t$$

and

(9) $$G_1(0), G_2(0) \quad \text{both positive or both zero.}$$

Then, in the case that $G_1(0)$ and $G_2(0)$ are both positive

(10) $$\begin{aligned} p = \frac{G_2(0)}{G_1(0)} &\exp\left[-\tfrac{1}{2}Y^2(0)\left\{\frac{1}{G_2(0)} - \frac{1}{G_1(0)}\right\}\right] \\ &\cdot\exp\left[\int_0^T \sigma^{-2}(\tau)[\rho_2(\tau) - \rho_1(\tau)]Y(\tau)\,dY(\tau)\right. \\ &\qquad \left. - \frac{1}{2}\int_0^T \sigma^{-2}(\tau)[\rho_2^2(\tau) - \rho_1^2(\tau)]Y^2(\tau)\,d\tau\right] \end{aligned}$$

where the first integral in the exponent is an Itô stochastic integral and the second integral is an ordinary stochastic integral.

A good summary of the diverse RKHS conditions available for equivalence and singularity of normal processes is given by Golosov-Tempelman (1969).

12. Covariance kernels of probability density functionals. A field in which many different reproducing kernel Hilbert spaces appear simultaneously in elegant interplay is the theory of unbiased minimum variance estimation of parameters of normal processes (see Parzen (1967), p. 336). This theory is now ripe for practical application because of the availability of the following basic formulas for the covariance kernels of probability density functionals.

THEOREM: UNEQUAL MEANS [PARZEN (1959)]. *Let m_1 and m_2 belong to $H(R)$. Then*

$$E_{0,R}\left[\frac{dP_{m_1,R}}{dP_{0,R}}\frac{dP_{m_2,R}}{dP_{0,R}}\right] = \exp(m_1, m_2)_{H(R)} \tag{1}$$

where $E_{0,R}$ denotes expectation with respect to $P_{0,R}$.

THEOREM: UNEQUAL COVARIANCES [DUTTWEILER (1970)]. *Let R, R_1 and R_2 be equivalent covariances. For $i = 1, 2$, $R_i \equiv R_0$ and $dP_{0,R_i}/dP_{0,R}$ is square integrable with respect to $P_{0,R}$ if and only if*

$$\|R_i - R\|_{H(R)\otimes H(R)} \leqq 1. \tag{2}$$

When (2) *holds for $i = 1, 2$, a formula for*

$$E_{0,R}\left[\frac{dP_{0,R_1}}{dP_{0,R}}\frac{dP_{0,R_2}}{dP_{0,R}}\right], \tag{3}$$

is given by Duttweiler in his thesis.

REFERENCES

1. N. Aronszajn, *Theory of reproducing kernels*, Trans. Amer. Math. Soc. **68** (1950), 337–404. MR **14**, 479.

2. A. Bensoussan, *Sur l'identification et le filtrage de systèmes gouvernés par des équations aux derivées partielles*, Institut de Recherche d'Informatique et d'Automatique, Cahier no. 1, Rocquencourt, France, 1969.

3. S. Bergman, *The kernel function and conformal mapping*, Math. Surveys, no. 5, Amer. Math. Soc., Providence, R. I., 1950. MR **12**, 402.

4. M. Brown, *Convergence in distribution of stochastic integrals*, Ann. Math. Statist. **41** (1970) (to appear).

5. J. Capon, *Radon-Nikodým derivatives of stationary Gaussian measures*, Ann. Math. Statist. **35** (1964), 517–531. MR **28** #4584.

6. A. Devinatz, *On the extensions of positive definite functions*, Acta Math. **102** (1959), 109–134. MR **22** #875.

7. D. Duttweiler, Electrical Engineering Department Ph.D. Thesis, Stanford University, Stanford, Calif., 1970.

8. J. Feldman, *Equivalence and perpendicularity of Gaussian processes*, Pacific J. Math. **8** (1958), 699–708. MR **21** #1546.

9. ———, *A clarification concerning certain equivalence classes of Gaussian processes on an interval*, Ann. Math. Statist. **39** (1968), 1078–1079. MR **37** #965.

10. Ju. I. Golosov, *On Gaussian measures equivalent to Gauss-Markov measures*, Dokl. Akad. Nauk SSSR **166** (1966), 263–266 = Soviet Math. Dokl. **7** (1966), 48–52. MR **33** #3365.

11. Ju. I. Golosov and A. A. Tempelman, *On equivalence of measures corresponding to Gaussian vector-valued functions*, Dokl. Akad. Nauk SSSR **184** (1969), 1271–1274 = Soviet Math. Dokl. **10** (1969), 228–232.

12. J. Hájek, *On a property of normal distributions of any stochastic process*, Czechoslovak Math. J. **8** **(83)** (1958), 610–617; English transl., Selected Transl. Math. Stat. and Prob., vol. 1, Amer. Math. Soc., Providence, R. I., 1961, pp. 245–252. MR **21** #3045; MR **22** #7167.

13. ———, *On linear statistical problems in stochastic processes*, Czechoslovak Math. J. **(12)** **87** (1962), 404–444. MR **27** #2070.

14. C. J. Henrich, *Equivalence and Radon-Nikodým derivatives of Gaussian measures* (to appear).

15. T. Hida, *Canonical representations of Gaussian processes and their applications*, Mem. Coll. Sci. Univ. Kyoto. Ser. A. Math. **33** (1960/61), 109–155. MR **22** #10012.

16. T. Hida and N. Ikeda, *Analysis on Hilbert space with reproducing kernel arising from multiple Wiener integral*, Proc. Fifth Berkeley Sympos. Math. Statist. and Prob. (Berkeley, Calif., 1965/66), vol. 2, part 1, Univ. of California Press, Berkeley, Calif., 1967, pp. 117–143. MR **36** #2214.

17. T. Kailath, *On measures equivalent to Wiener measure*, Ann. Math. Statist. **38** (1967), 261–263. MR **34** #3675.

18. ———, *Likelihood ratios for Gaussian processes*, IEEE Trans. Information Theory (to appear).

19. G. Kallianpur, *Abstract Wiener processes and their reproducing kernel Hilbert spaces*, Z. Wahrscheinlichkeitstheorie und Verw. Gebiete (to appear).

20. ———, *The role of reproducing kernel Hilbert spaces in the study of Gaussian processes*, Advances in Probability **2** (1970) (to appear).

21. ———, *Zero-one laws for Gaussian processes*, Trans. Amer. Math. Soc. **149** (1970), 199–211.

22. G. Kallianpur and H. Oodaira, *The equivalence and singularity of Gaussian measures*, Proc. Sympos. Times Series Analysis (Brown University, Providence, R.I., 1962), Wiley, New York, 1963, pp. 279–291. MR **26** #7013.

23. G. S. Kimeldorf and G. Wahba, *A correspondence between Bayesian estimation on stochastic processes and smoothing by splines*, Math. Res. Center Technical Summary Report No. 947, University of Wisconsin, Madison, Wis., 1968.

24. G. S. Kimeldorf and G. Wahba, *Spline functions and stochastic processes*, Math. Res. Center Technical Summary Report No. 969, University of Wisconsin, Madison, Wis., 1969.

25. M. G. Kreĭn, *Hermitian positive kernels on homogeneous spaces*. I, II, Ukrain. Mat. Ž. **1** (1949), no. 4, 64–98; Ukrain. Mat. Ž. **2** (1950), no. 1, 10–59; English transl., Amer. Math. Soc. Transl. (2) **34** (1963), 69–164. MR **12**, 719; MR **14**, 480.

26. R. Le Page, *Estimation of parameters in signals of known forms and an isometry related to unbiased estimation*, Ph.D. Thesis, University of Minnesota, Minneapolis, Minn., 1967.

27. M. Loève, "Fonctions aléatoires du second ordre," Appendix to P. Lévy, *Processes stochastiques et mouvement Brownien*, Gauthier-Villars, Paris, 1948. MR **10**, 551.

28. H. Meschkowski, *Hilbertsche Räume mit Kernfunktion*, Die Grundlehren der math. Wissenschaften, Band 113, Springer-Verlag, Berlin and New York, 1962. MR **25** #4326.

29. J. Neveu, *Processus aléatoires Gaussiens*, Séminaire de Math. Supérieures, no. 34, Presses Univ. Montréal, 1968.

30. H. Oodaira, *The equivalence of Gaussian stochastic processes*, Statistics Department Technical Report and Ph.D. Thesis, Michigan State University, East Lansing, Mich., 1963.

31. E. Parzen, *Statistical inference on time series by Hilbert space methods*, Statistics Department Technical Report No. 23, Stanford University, Stanford, Calif., 1959.

32. ———, *An approach to time series analysis*, Ann. Math. Statist. **32** (1961), 951–989. MR **26** #874.

33. ———, *Probability density functionals and reproducing kernel Hilbert spaces*, Proc. Sympos. Time Series Analysis (Brown University, Providence, R. I., 1962), Wiley, New York, 1963, pp. 155–169. MR **26** #7119.

34. ———, *Time series analysis papers*, Holden-Day, San Francisco, Calif., 1967. MR **36** #6091.

35. F. Riesz and B. Sz.-Nagy, *Functional analysis*, 2nd ed., Akad. Kiadó, Budapest, 1953; English transl., Ungar, New York, 1955. MR **15**, 132; MR **17**, 175.

36. Ju. A. Rozanov, *On the density of Gaussian distributions and Wiener-Hopf integral equations*, Teor. Verojatnost. i Primenen. **11** (1966), 170–179 = Theor. Probability Appl. **11** (1966), 152–160. MR **33** #3343.

37. ———, *Infinite dimensional Gaussian distributions*, Trudy Mat. Inst. Steklov. **108** (1968) = Proc. Steklov Inst. Math. **108** (1968) (to appear).

38. J. Sacks and N. D. Ylvisaker, *Designs for regression problems with correlated errors*, Ann. Math. Statist. **37** (1966), 66–89. MR **33** #826.

39. ———, *Designs for regression problems with correlated errors; many parameters*, Ann. Math. Statist. **39** (1968), 49–69. MR **36** #3484.

40. ———, *Designs for regression problems with correlated errors*. III (to appear).

41. L. Schwartz, *Sous-espaces hilbertiens d'espaces vectoriels topologiques et noyaux associés (noyaux reproduisants)*, J. Analyse Math. **13** (1964), 115–256. MR **31** #3835.

42. I. E. Segal, *Distributions in Hilbert space and canonical systems of operators*, Trans. Amer. Math. Soc. **88** (1958), 12–41. MR **21** #1545.

43. L. A. Shepp, *Radon–Nikodým derivatives of Gaussian measures*, Ann. Math. Statist. **37** (1966), 321–354. MR **32** #8408.

44. D. E. Varberg, *On Gaussian measures equivalent to Wiener measure*. II, Math. Scand. **18** (1966), 143–160. MR **35** #3729.

45. G. Wahba, *A note on the regression design of Sacks and Ylvisaker*, Statistics Department Technical Report No. 29, Stanford University, Stanford, Calif., 1969.

46. ———, *On the numerical solution of Fredholm integral equations of the first kind*, Statistics Department Technical Report No. 217, University of Wisconsin, Madison, Wis., 1969.

47. A. M. Yaglom (Jaglom), *On the equivalence and perpendicularity of two Gaussian probability measures in functions space*, Proc. Sympos. Time Series Analysis (Brown University, Providence, R. I., 1962), Wiley, New York, 1963, pp. 327–346. MR **26** #5635b.

48. N. D. Ylvisaker, *On linear estimation for regression problems on time series*, Ann. Math. Statist. **33** (1962), 1077–1084. MR **27** #2068.

49. ———, *Lower bounds for minimum covariance matrices in time series regression problems*, Ann. Math. Statist. **35** (1964), 362–368. MR **28** #4636.

6

THE GRADIENT ITERATION IN TIME SERIES ANALYSIS*

HOWARD J. WEINER†

1. Introduction. Let $\{X_t, t \in T\}$ be a real time series with finite second moments and let $K(s, t) = E[X_s X_t]$. Denote the reproducing kernel Hilbert space corresponding to K by $H(K; T)$ and the corresponding inner product by $(f, g)_K$ for $f, g \in H(K; T)$. Parzen [1], [2], [3] has given an extensive treatment of these concepts and has expressed various time series results in terms of the inner products $(h, h)_K$ and corresponding random variable $(X, h)_K$. It is the purpose of this note to prove that the gradient (steepest descent) method of iteratively evaluating reproducing kernel Hilbert space inner products $(h, h)_K$ and corresponding random variables $(X, h)_K$ is convergent to the respective true values of these quantities.

Balakrishnan [4] has considered a norm minimization problem in control theory using the gradient iteration with a boundedness condition on the norms of the iterates.

We will require the following concepts and notation. Let G be a Hilbert space of functions on T, where T is henceforth taken to be an interval on the real line, and it is assumed that G has a computationally convenient norm. Define the operator A on G as follows: for $g \in G, Ag(t) = (g, K(\cdot, t))_G$. Then A is linear, self-adjoint, nonnegative-definite and completely continuous from G into G. The set $\{\lambda_\nu\}$, $\nu = 1, 2, \cdots$, of eigenvalues of A arranged in decreasing order, converges to zero. Let $\{\varphi_\nu\}$, $\nu = 1, 2, \cdots$, be the corresponding eigenfunctions. $H(K; T)$ is the Hilbert subspace of G spanned by the $\{\varphi_\nu\}$ such that $\sum_{\nu=1}^{\infty} h_\nu^2/\lambda_\nu < \infty$, where we denote $h_\nu = (h, \varphi_\nu)_G$. An important relation is that $(u, Av)_K = (u, v)_G$.

2. Motivation of the gradient iteration. We wish to generate a sequence of functions $\{H_n\}_{n=1}^{\infty}$ by a simple iterative scheme such that

$$\lim_{n\to\infty} E \mid (X, h)_K - (H_n, X)_G \mid^2 = 0, \tag{1}$$

$$(h, h)_K = \lim_{n\to\infty} (H_n, AH_n)_G, \tag{2}$$

without a knowledge of eigenvalues and eigenfunctions.

For example, let $(f, g)_G = \int_0^T f(s)g(s)w(s)\, ds$, with $w > 0$ a known weight function. Then in the quadratic mean,

* Received by the editors February 7, 1964, and in revised form February 2, 1965.
† Department of Statistics, Stanford University, Stanford, California. Now at Department of Mathematics, University of California, Davis, California.

$$(X, h)_K \sim \int_0^T H_n(s) X_s \, w(s) \, ds,$$

where $\{X_s, \; s \in [0, T]\}$ are observations. Also

$$(h, h)_K \sim \int_0^T \int_0^T H_n(s) K(s, t) H_n(t) w(s) w(t) \, ds \, dt.$$

As shown by Parzen [1], for (1) and (2) to be satisfied, it suffices that we find $\{H_n\}_{n=1}^{\infty}$ such that $\| r_n \|_K^2 \to 0$, where $r_n \equiv h - AH_n$. Expanding,

$$\begin{aligned} \| r_n \|_K^2 - (h, h)_K &= (AH_n, AH_n)_K - 2(h, AH_n)_K \\ &= (H_n, AH_n)_G - 2(h, H_n)_G. \end{aligned}$$

We wish to find the maximum rate of change of $\| r_n \|_K^2$.

$$\nabla_{H_n} \| r_n \|_K^2 = -2r_n.$$

Hence the direction of maximum change of $\| r_n \|_K^2$ is along r_n.

This motivates the following choice of iteration:

$$H_0 \in H_K \text{ arbitrary}, \qquad H_{n+1} = H_n + \alpha_n r_n,$$

where $\{\alpha_n\}_{n=1}^{\infty}$ are constants whose choice is motivated by the following.

$$\begin{aligned} \Delta \| r_n \|_K^2 = \| r_n \|_K^2 - \| r_{n+1} \|_K^2 &= \| r_n \|_K^2 - \| h - A(H_n + \alpha_n r_n) \|_K^2 \\ &= \| r_n \|_K^2 - \| r_n - \alpha_n A r_n \|_K^2 \\ &= 2\alpha_n (r_n, Ar_n)_K - \alpha_n^2 (Ar_n, Ar_n)_K \\ &= 2\alpha_n (r_n, r_n)_G - \alpha_n^2 (r_n, Ar_n)_G. \end{aligned}$$

To maximize $\Delta \| r_n \|_K^2$ with respect to α_n, set

$$\frac{\partial \Delta \| r_n \|_K^2}{\partial \alpha_n} = 0,$$

obtaining

$$\alpha_n = \frac{(r_n, r_n)_G}{(r_n, Ar_n)_G} \equiv a_n,$$

where we note that if $r_n \neq 0$,

$$(r_n, r_n)_G = \sum_{\nu=1}^{\infty} r_{n\nu}^2 > 0,$$

and there exists ν_0 such that $r_{n\nu_0} \neq 0$, so that

$$(r_n, Ar_n)_G = \sum_{\nu=1}^{\infty} \lambda_\nu r_{n\nu}^2 > 0;$$

hence a_n, as chosen above, is well defined.

We compute

$$\left.\frac{\partial^2 \Delta \| r_n \|_K^2}{\partial \alpha_n^2}\right|_{a_n} = -2(r_n, Ar_n)_G = -2 \| Ar_n \|_K^2 < 0.$$

Hence at each step, for a given value of $\| r_n \|_K^2$, choosing $\alpha_n = a_n$ insures that we obtain the maximum value of $\Delta \| r_n \|_K^2$ among the class of all iterations of the form $H_0 \in H_K$ arbitrary, $H_{n+1} = H_n + \alpha_n r_n$.

Thus we have motivated the gradient or steepest descent iterative procedure (see [5]): $H_0 \in H_K$ arbitrary, $H_{n+1} = H_n + a_n r_n$, $r_n = h - AH_n$, $a_n = (r_n, r_n)_G/(r_n, Ar_n)_G$.

As desired, this procedure is self-contained in that one does not require a knowledge of eigenvalues and eigenfunctions of A.

3. Convergence of the gradient iteration. The basic theorem we want to prove is that if $\{H_n\}_{n=1}^{\infty}$ are generated by the gradient iteration, and $r_n = h - AH_n$, then

$$\| r_n \|_K^2 \downarrow 0.$$

To do this, it is first necessary to prove a lemma which is a strengthened version of a theorem in [1]. We state it in the following way.

LEMMA. *Let* $\{\alpha_n\}_{n=1}^{\infty}$ *satisfy:*

(i) $0 < \alpha_n < 2/\lambda_{\max}$, *where* $\lambda_{\max}$ *is the largest eigenvalue of* A;

(ii) $\sum_{n=1}^{\infty} \alpha_n = \infty$.

Define the iteration $H_0 \in H_K$ *arbitrary,* $H_{n+1} = H_n + \alpha_n r_n$, $r_n = h - AH_n$. *Then*

$$\| r_n \|_K^2 \downarrow 0.$$

Proof.

(a) *Monotonicity.*

$$\begin{aligned}\Delta \| r_n \|_K^2 = \| r_n \|_K^2 - \| r_{n+1} \|_K^2 &= \| r_n \|_K^2 - \| h - A(H_n + \alpha_n r_n) \|_K^2 \\ &= 2\alpha_n (r_n, Ar_n)_K - \alpha_n^2 (Ar_n, Ar_n)_K \\ &= 2\alpha_n (r_n, r_n)_G - \alpha_n^2 (r_n, Ar_n)_G \geqq 0,\end{aligned}$$

because

$$\frac{(r_n, r_n)_G}{(r_n, Ar_n)_G} = \frac{\sum_{\nu=1}^{\infty} r_{n\nu}^2}{\sum_{\nu=1}^{\infty} r_{n\nu}^2 \lambda_\nu} \geqq \frac{\sum_{\nu=1}^{\infty} r_{n\nu}^2}{\lambda_{\max} \sum_{\nu=1}^{\infty} r_{n\nu}^2} = \frac{1}{\lambda_{\max}},$$

so that

$$\frac{2(r_n, r_n)_G}{(r_n, Ar_n)_G} \geqq \frac{2}{\lambda_{\max}} \geqq \alpha_n,$$

the last by hypothesis. The montonicity is proved.

(b) *Convergence.*

$$\begin{aligned} r_{n+1} &= h - AH_{n+1} = h - A(H_n + \alpha_n r_n) \\ &= h - AH_n - \alpha_n A r_n = r_n - \alpha_n A r_n \\ &= (I - \alpha_n A) r_n \\ &= \prod_{K=1}^{n} (I - \alpha_K A) r_0 = \prod_{K=1}^{n} (I - \alpha_K A) g, \end{aligned}$$

where we let $g = r_0 \in H_K$. Since $g(t) = \sum_{\nu=1}^{\infty} g_\nu \varphi_\nu(t)$,

$$Ag(t) = \sum_{\nu=1}^{\infty} g_\nu \lambda_\nu \varphi_\nu(t),$$

$$\prod_{K=1}^{n} (I - \alpha_K A) g = \sum_{\nu=1}^{\infty} g_\nu \prod_{K=1}^{n} [I - \alpha_K \lambda_\nu] \varphi_\nu,$$

$$\| r_{n+1} \|_K^2 = \left\| \prod_{K=1}^{n} (I - \alpha_K A) g \right\|_K^2 = \sum_{\nu=1}^{\infty} \frac{g_\nu^2}{\lambda_\nu} \prod_{K=1}^{n} (1 - \alpha_k \lambda_\nu)^2.$$

By the inequality on α_K,

$$-1 < 1 - \frac{2\lambda_\nu}{\lambda_{\max}} \leqq 1 - \alpha_K \lambda_\nu < 1 \Rightarrow (1 - \alpha_K \lambda_\nu)^2 < 1.$$

Hence for any integer N,

$$\| r_{n+1} \|_K^2 \leqq \sum_{\nu=1}^{N} \frac{g_\nu^2}{\lambda_\nu} \prod_{K=1}^{n} (1 - \alpha_K \lambda_\nu)^2 + \sum_{\nu > N} \frac{g_\nu^2}{\lambda_\nu}.$$

For fixed N, let $n \to \infty$. Since $\sum_{n=1}^{\infty} \alpha_n = \infty$ and $(1 - \alpha_K \lambda_\nu)^2 < 1$, the first term $\to 0$. Let $N \to \infty$. $g \in H_K \Rightarrow \sum_{\nu=1}^{\infty} g_\nu^2 / \lambda_\nu < \infty$, and so the second term is the tail of a convergent series and it goes to zero. The lemma is thus proved.

THEOREM. *Let $\{H_n\}$ be generated by the gradient method. If $r_n = h - AH_n$, then*

$$\| r_n \|_K^2 \downarrow 0.$$

Proof.

(a) *Monotonicity.* This follows from

$$\Delta \| r_n \|_K^2 = \| r_n \|_K^2 - \| r_{n+1} \|_K^2 = \| r_n \|_K^2 - \| h - A(H_n + a_n r_n) \|_K^2$$

$$= \| r_n \|_K^2 - \| r_n - a_n A r_n \|_K^2 = 2a_n(r_n , Ar_n)_K - a_n^2(Ar_n , Ar_n)_K$$

$$= 2 \frac{(r_n , r_n)_G^2}{(r_n , Ar_n)_G} - \frac{(r_n , r_n)_G^2}{(r_n , Ar_n)_G} = \frac{(r_n , r_n)_G^2}{(r_n , Ar_n)_G} > 0.$$

(b) *Convergence.* We prove the convergence by comparing the gradient iteration with an iteration defined by the previous lemma. Choose $H_0 \in H_K$ arbitrary. Then $r_0 = h - AH_0$ is the same for both types of iteration, and $\| r_0 \|_K^2$ is the common initial error.

Choose an α such that $0 < \alpha < 2/\lambda_{\max}$. Compute $a_0 = (r_0 , r_0)_G / (r_0 , Ar_0)_G$. By the maximum property of the gradient iteration,

$$\Delta_G \| r_0 \|_K^2 \geqq \Delta_L \| r_0 \|_K^2,$$

where $\Delta_G \| r_0 \|_K^2 = \| r_0 \|_K^2 - \| r_1 \|_K^2$ computed by the gradient method; similarly, $\Delta_L \| r_0 \|_K^2$ is computed using the iteration of the lemma with the above α.

In the next step, suppose, using $H_2 = H_1 + \alpha r_1$ in the lemma iteration,

$$\Delta_G \| r_1 \|_K^2 + \Delta_G \| r_0 \|_K^2 \geqq \Delta_L \| r_1 \|_K^2 + \Delta_L \| r_0 \|_K^2;$$

then we use the lemma iteration as it stands. If not, since $\Delta_L \| r_1 \|_K^2$ is a quadratic continuous function of α, there exists an α_1, $0 < \alpha_1 < \alpha$, such that

$$\Delta_G \| r_1 \|_K^2 + \Delta_G \| r_0 \|_K^2 = \Delta_L \| r_1 \|_K^2 + \Delta_L \| r_0 \|_K^2;$$

that is, $H_2 = H_1 + \alpha_1 r_1$ in the altered iteration of the lemma, since for $\alpha_1 = 0$, $\Delta_L \| r_1 \|_K^2 = 0$ and

$$\Delta_G \| r_1 \|_K^2 + \Delta_G \| r_0 \|_K^2 > \Delta_L \| r_0 \|_K^2.$$

Now, in the lemma iteration, use $H_3 = H_2 + \alpha r_2$. Again by the maximum property of the gradient iteration,

$$\sum_{i=0}^{2} \Delta_G \| r_i \|_K^2 \geqq \sum_{i=0}^{2} \Delta_L \| r_i \|_K^2.$$

Compute, in the lemma iteration, $H_4 = H_3 + \alpha r_3$. If

$$\sum_{i=0}^{3} \Delta_G \| r_i \|_K^2 \geqq \sum_{i=0}^{3} \Delta_L \| r_i \|_K^2 ,$$

let the lemma iteration be as it is up to this stage. If not, choose an α_3, $0 < \alpha_3 < \alpha$, such that in the lemma iteration, with $H_4 = H_3 + \alpha_3 r_3$,

$$\sum_{i=0}^{3} \Delta_G \| r_i \|_K^2 = \sum_{i=0}^{3} \Delta_L \| r_i \|_K^2 .$$

Continuing in this way, we see that at least all $\alpha_{2n+1} = \alpha$. Further, $0 < \alpha_\nu \leqq \alpha < 2/\lambda_{\max}$ and

$$\sum_{\nu=1}^{\infty} \alpha_\nu \geqq \sum_{\nu=1}^{\infty} \alpha_{2\nu+1} = \sum_{\nu=1}^{\infty} \alpha = \infty .$$

Hence $\sum_{\nu=1}^{\infty} \alpha_\nu = \infty$ and the conditions of the hypothesis of the lemma are fulfilled. Hence in the constructed lemma iteration,

$$\| r_n^{(L)} \|_K^2 \downarrow 0.$$

But since at each stage

$$\sum_{i=0}^{n} \Delta_G \| r_i \|_K^2 \geqq \sum_{i=0}^{n} \Delta_L \| r_i \|_K^2 ,$$

denoting now by $r_n^{(G)}$ the error at the nth stage using the gradient iteration, we thus have

$$\| r_n^{(G)} \|_K^2 \leqq \| r_n^{(L)} \|_K^2 .$$

Hence

$$\| r_n^{(G)} \|_K^2 \downarrow 0,$$

which was to be proved.

REFERENCES

[1] E. Parzen, *Regression analysis of continuous parameter time series*, Proceedings of the Fourth Berkeley Symposium on Probability and Mathematical Statistics, vol. 1, University of California Press, Berkeley, 1961, pp. 469–489.

[2] ———, *Extraction and detection problems and reproducing kernel Hilbert spaces*, J.SIAM Control Ser. A, 1 (1962), pp. 35–62.

[3] ———, *Statistical inference on time series by Hilbert space methods I*, Department of Statistics, Stanford University, Tech. Rpt. 23, 1959.

[4] A. V. Balakrishnan, *An operator theoretic formulation of a class of control problems and a steepest descent method of solution*, J.SIAM Control Ser. A, 1 (1963), pp. 109–127.

[5] M. R. Hestenes and E. Stiefel, *Methods of conjugate gradients in solving linear systems*, J. Res. Nat. Bur. Standards, 49 (1952), pp. 409–436.

7

Some Relations Among RKHS Norms, Fredholm Equations, and Innovations Representations

THOMAS KAILATH, FELLOW, IEEE, ROGER T. GEESEY, MEMBER, IEEE, AND HOWARD L. WEINERT, STUDENT MEMBER, IEEE

***Abstract*—We first show how reproducing kernel Hilbert space (RKHS) norms can be determined for a large class of covariance functions by methods based on the solution of a Riccati differential equation or a Wiener–Hopf integral equation. Efficient numerical algorithms for such equations have been extensively studied, especially in the control literature. The innovations representations enter in that it is they that suggest the form of the RKHS norms. From the RKHS norms, we show how recursive solutions can be obtained for certain Fredholm equations of the first kind that are widely used in certain approaches to detection theory. Our approach specifies a unique solution: moreover, the algorithms used are well suited to the treatment of increasing observation intervals.**

I. Introduction

THERE are by now several problems in signal detection [1]–[3], estimation [3]–[6], statistical design of experiments [7], and spline fitting [8], [9] in which reproducing kernel Hilbert space (RKHS) concepts play a useful role. The defining characteristic of the RKHS is the RKHS norm, and several methods have been used to calculate it for various covariances, see, e.g., [1]–[6]. Certain iterative numerical methods, based on steepest descents, have also been suggested [4]. In this paper, we shall present a different class of numerical methods, based on the solution of a Riccati differential equation or a Wiener–Hopf integral equation. Our results are largely for covariances of lumped processes, i.e., processes generated by passing a white-noise process through finite lumped linear time-variant systems. This class includes almost all processes for which RKHS norms have been explicitly calculated in the literature; nevertheless, we should stress that our goal here is efficient numerical algorithms and not only simple closed-form expressions. The particular first-order Riccati matrix differential equations that arise in our algorithms have been widely studied in connection with space-flight applications; see, e.g., [10].

Our derivation is based on certain results obtained by us, [11]–[14], on innovation representations (IR). The IR of a stochastic process is a representation of it as the output of a causal and causally invertible filter driven by white noise. As explained in [15] and elsewhere, the IR has several applications to deterministic and statistical system problems. Here we shall use it to obtain the RKHS norms, by use of an RKHS theorem that relates such norms to L_2-norms (cf., [2, Theorem 9] or [3, lemma 2.2]). However, as is typical for RKHS, the derivation is somewhat irrelevant, since the final formulas can be checked by testing whether they have the reproducing property. The formulas for the RKHS norms are presented in Section II.

To turn to integral equations, we note that in an earlier paper [11] it was shown how Wiener–Hopf and Riccati equations are closely related to the recursive solution of certain Fredholm equations of the second kind. An easy consequence of our formulas for RKHS norms will be the derivation, following a procedure suggested in [2], of similar results for certain Fredholm equations of the first kind.

We should emphasize our belief that in many problems the use of such equations of the first kind can and should be avoided and that often a proper formulation of the problem will express the solution in terms of an RKHS norm or of the solution of a Fredholm equation of the second kind. Nevertheless, we discuss equations of the first kind because of their prevalence in much of the literature on detection theory, see, e.g., [16] and [17]. We should mention that Wood *et al.* [18] have recently obtained some results that are a special case of ours (cf., the discussion of (44) at the end of Section III), but with somewhat more effort. (The reader may in fact find it useful to glance at that discussion before going on.)

II. RKHS Norms for Some Processes

We shall show how the RKHS norms for certain types of processes can be computed via the solution of certain Riccati differential equations. There is by now an extensive literature on the solution of such equations. We shall actually begin with certain somewhat more general results that will be specialized later to cover lumped processes specified either by their covariance functions or by known lumped models driven by white noise. For notational simplicity, we shall confine our presentation to scalar processes.

The proofs of the various theorems follow directly from [2, Theorem 9] and from the results on innovations representations presented in [11] and [12] for Theorems 1

Manuscript received February 26, 1971; revised December 29, 1971. This work was supported in part by the Applied Mathematics Division of the Air Force Office of Scientific Research under Contract AF 44-620-69-C-0101, in part by the JSEP at Stanford University under Contract N-00014-67-A-0112-0044 in part and by NSF Contract GK-31627.

T. Kailath is with the Department of Electrical Engineering, Stanford University, Stanford, Calif. 94305.
R. T. Geesey is with the HQ Aeronautical Chart and Information Center (ACIC), St. Louis, Mo. 63118.
H. L. Weinert is with the Stanford Electronics Laboratories, Stanford, Calif. 94305.

and 4, [13] for Theorems 2 and 3, and [14] for Theorems 5 and 6. (The existence and uniqueness of the solutions to the various Riccati equations are proved in these references.) However, the final results can be directly checked to see whether they have the reproducing property. This is done in the Appendix, so that we shall not give any proofs in this section.

The forms of the RKHS norms in Theorems 1–3 and 4–6 are suggested by the following heuristic considerations. A white noise v has covariance function

$$R_v(t,s) = \delta(t - s)$$

and

$$\|m\|_{R_v}^2 = \int_0^T m^2(t)\, dt.$$

If a nonwhite noise, say n, can be reduced to a white-noise process by a deterministic (linear or nonlinear) operation, say $n = \mathscr{L}v$ then we would expect that for

$$E[n(t)n(s)] = R(t,s), \qquad \|m\|_R^2 = \|\mathscr{L}^{\#}m\|_{R_v}^2,$$

where $\mathscr{L}^{\#}$ is a pseudoinverse of $\mathscr{L}$. This is in fact true, provided proper attention is paid to the kind of pseudoinverse that is used (cf., [3, Lemma 2.2], [2, Theorem 9 and discussion]). If, however, the operation $\mathscr{L}$ is causal and causally invertible, so that we have the innovations representation of $n(\cdot)$, then $\mathscr{L}$ has a unique inverse, which can be used for $\mathscr{L}^{\#}$. In general, there is no reason for using any special $\mathscr{L}$, but the form used in the innovations representation has several nice properties, interpretations, and applications (cf., [11]–[14], [18], and [19] and the references therein) and innovations representations have been worked out for many covariances. Therefore, we have used them here in order to deduce our RKHS norms, though again we repeat that this is not the only way. However, the forms we give are designed to be convenient for recursive calculation, unlike most previously obtained forms. A very simple example is given in Section II [cf., (33) and (34)].

Some words of explanation may also be in order concerning the separation of our presentation into two sections —Section II-A treating covariances that have delta-function terms and Section II-B without such terms. In the usual RKHS framework, covariances are required to be continuous, so that delta functions as in (1) following cannot, strictly speaking, be permitted. On the other hand, white-noise processes and delta-function covariances are certainly often used in communication theory. In this situation, the demands of rigor can be met either by working with RKHS of functionals (see, e.g., [2]) or by regarding the use of the white noise and delta functions as a convenient shorthand for problems with integrated white-noise, i.e., orthogonal-increment processes with the corresponding integrated covariance

$$t\Lambda s = \min(t,s) = \int_0^t \int_0^s \delta(u - v)\, du\, dv.$$

As an example, when we state that

$$H(\delta(t - s)) = \left\{m(\cdot): \int_0^1 m^2(t)\, dt < \infty\right\},$$

this is to be regarded as a shorthand for the equivalent statement

$$H(t\Lambda s) = \left\{M(\cdot): M(t) = \int_0^t m(s)\, ds, \int_0^1 m^2(s)\, ds < \infty\right\}.$$

For this simple example, the notational gain is slight; however it becomes more significant for the covariances described in Theorems 1–3. We also point out that with the convention just described, Theorems 1–3 are essentially special cases of Theorems 4–6 (with $\alpha = 1$ and $r^2(t) \equiv 1$); the major difference is that [see Remark 1 after (31)] we have taken advantage of the notational compactness to introduce explicitly a feedback matrix $F(\cdot)$. We repeat that our major reason for presenting Theorems 1–3 separately is that many communication problems are described in the language of those theorems.

A. Covariances with Delta Functions

Theorem 1: Suppose a process has a covariance function

$$R(t,s) = \delta(t - s) + K(t,s), \tag{1}$$

where

$$\int_I \int_I K^2(t,s)\, dt\, ds < \infty, \qquad I = [0,T]$$

and $R(t,s)$ is strictly positive-definite on $L_2(I \times I)$.

Let $h(\cdot,\cdot)$ be the unique square-integrable solution of the Wiener–Hopf type of the integral equation

$$h(t,s) + \int_0^t h(t,\tau)K(\tau,s)\, d\tau = K(t,s), \qquad 0 \le s \le t \le T. \tag{2}$$

Then the RKHS norm in $H(R)$ can be written as

$$\|m\|_{H(R)}^2 = \int_I \left[m(t) - \int_0^t h(t,s)m(s)\, ds\right]^2 dt. \tag{3}$$

Theorem 2—Known Lumped Model: Suppose a process $y(\cdot)$ has a known lumped model of the form[1]

$$\dot{\theta}(t) = F(t)\theta(t) + G(t)u(t), \qquad \theta(0) = \theta_0 \tag{4a}$$

$$y(t) = H(t)\theta(t) + v(t), \qquad E\theta_0 = 0, \qquad E\theta_0\theta_0' = \Pi_0 \tag{4b}$$

$$Eu(t)u'(s) = Q(t)\delta(t - s),$$
$$Ev(t)v(s) = \delta(t - s)$$
$$Eu(t)v(s) = C(t)\delta(t - s),$$
$$Eu(t)\theta_0' \equiv 0. \tag{4c}$$

[1] By now state-variable models are sufficiently common in communication theory that, as in control theory, we shall not explicitly indicate that $F(\cdot)$, $G(\cdot)$, $H(\cdot)$, $\theta(\cdot)$, $u(\cdot)$, θ_0, and Π_0 can be matrices of appropriate dimensions. We do remark that primes will be used to denote transposes and that, as stated earlier, for further notational convenience $y(\cdot)$ and $v(\cdot)$ will be scalar.

Then the RKHS norm can be written

$$\|m\|_{H(R)}^2 = \int_0^T [m(t) - H(t)\phi(t)]^2\, dt, \tag{5}$$

where $\phi(\cdot)$ is the solution of the differential equation

$$\dot{\phi}(t) = F(t)\phi(t) + K(t)[m(t) - H(t)\phi(t)],$$
$$\phi(0) = 0 \tag{6a}$$

and

$$K(t) = P(t)H'(t) + G(t)C(t), \tag{6b}$$

where $P(\cdot)$ is the solution of the matrix Riccati differential equation

$$\dot{P} = FP + PF' - KK' + GQG', \qquad P(0) = \Pi_0. \tag{7}$$

Theorem 3—Known Lumped Covariance: Suppose the function $K(\cdot,\cdot)$ in Theorem 1 has the form

$$K(t,s) = \begin{cases} \sum_1^n m_i(t)n_i(s), & t \geq s \quad (8a)\\ \sum_1^n m_i(s)n_i(t), & t \leq s \quad (8b)\end{cases}$$

$$= M(t \vee s)N(t \wedge s), \quad \text{(say)} \tag{8c}$$

where we use the notations

$$t \vee s = \max(t,s), \qquad t \wedge s = \min(t,s). \tag{9}$$

Also assume that $K(t,s)$ is continuous in (t,s).

Let $F(\cdot)$ be an arbitrary continuous matrix function with a fundamental matrix Φ defined by

$$\frac{d\Phi(t,s)}{dt} = F(t)\Phi(t,s), \qquad \Phi(0,0) = I \tag{10}$$

and define

$$M_0(t) = M(t)\Phi(t_0,t), \qquad N_0(t) = \Phi(t,t_0)N(t), \tag{11}$$

where t_0 is an arbitrary time instant. Then the RKHS norm can be written

$$\|m\|_{H(R)}^2 = \int_0^T [m(t) - M_0(t)\phi(t)]^2\, dt, \tag{12}$$

where

$$\dot{\phi} = F\phi + K[m - M_0\phi], \qquad \phi(0) = 0 \tag{13a}$$
$$K = N_0 - \Sigma M_0' \tag{13b}$$

and $\Sigma(\cdot)$ obeys the matrix Riccati equation

$$\dot{\Sigma} = F\Sigma + \Sigma F' + [N_0 - \Sigma M_0'][N_0 - \Sigma M_0']'$$
$$\Sigma(0) = 0. \tag{14}$$

Remark 1: The matrix $F(\cdot)$ can be chosen arbitrarily. For stationary processes, $F(\cdot)$ can be any matrix such that $\det(sI - F)\cdot\det(-sI - F)$ gives the denominator of the Laplace transform of $\delta(t - s) + K(t,s)$. For nonstationary processes, the choice $F(\cdot) \equiv 0$ may be convenient. The choice of $F(\cdot)$ determines some of the internal stability properties of the system (13).

B. Colored Noise Covariances

Suppose a process $y(t)$, $0 \leq t \leq T$, has $(\alpha - 1)$ mean-square derivatives and that the random variables $\{y(0),\cdots, y^{(\alpha-1)}(0)\}$ are linearly independent. Then (cf., Shepp [19, theorem 7]) the covariance function of the process can be uniquely written as

$$R(t,s) = \sum_0^{\alpha-1} C_i(t)C_i(s) + R_0(t,s), \tag{15}$$

where

i) R_0 is a covariance;
ii) $(\partial^{2i}/\partial t^i\, \partial s^i)R_0(t,s)|_{0,0} = 0, \quad i = 0,\cdots,\alpha - 1;$
iii) $C_j^{(i)}(0) = 0, \quad i < j, \quad C_i^{(i)}(0) > 0, \quad i = 0,\cdots,\alpha - 1.$

Suppose also that

$$R_0^{(\alpha,\alpha)} = \frac{\partial^{2\alpha}}{\partial t^\alpha\, \partial s^\alpha} R_0(t,s) = r^2(t)\delta(t - s) + L_0(t,s) \tag{16}$$

$$\int_I\int_I L_0^2(t,s)\, dt\, ds < \infty, \qquad I = [0,T] \tag{17a}$$

and

$$R_0^{(\alpha,\alpha)} \text{ is strictly positive-definite on } L_2[I \times I]. \tag{17b}$$

Theorem 4: If the covariance function R has the preceding properties, then the RKHS norm can be written

$$\|m\|_{H(R)}^2 = Q_m'\Gamma Q_m + \int_0^T \left[m_0^{(\alpha)}(t)r^{-1}(t) - \int_0^t g_0(t,s)m_0^{(\alpha)}(s)r^{-1}(s)\, ds\right]^2 dt, \tag{18}$$

where

$$m_0(t) = m(t) - \sum_{i,j=0}^{\alpha-1} R^{(0,i)}(t,0)\Gamma_{ij}m^{(j)}(0) \tag{19a}$$

$$\Gamma = \text{inverse of the covariance matrix of } \{y(0),\cdots,y^{(\alpha-1)}(0)\} \tag{19b}$$

$$Q_m' = \{m(0),m^{(1)}(0),\cdots,m^{(\alpha-1)}(0)\} \tag{19c}$$

and $g_0(t,s)$ is the unique square-integrable solution of the Wiener–Hopf type of integral equation[2]

$$r(s)r(t)g_0(t,s) + \int_0^t r(t)r^{-1}(\tau)g_0(t,\tau)L_0(\tau,s)\, d\tau = L_0(t,s),$$
$$0 \leq s \leq t \leq T. \tag{20}$$

Theorem 5—Known Lumped Model: Suppose a process $y(\cdot)$ has a known lumped model of the form

$$\dot{\chi}(t) = \psi(t)u(t), \qquad \chi(0) = \chi_0 \tag{21a}$$

$$y(t) = A(t)\chi(t), \qquad Eu(t)u'(s) = I\delta(t - s) \tag{21b}$$

$$E\chi_0 = 0, \qquad E\chi_0\chi_0' = \Pi_0, \qquad Eu(t)\chi_0' \equiv 0, \tag{21c}$$

[2] It should be evident that considerable notational simplicity can be obtained by the normalization $r^2(t) \equiv 1$. This will be done in Section III.

where $\psi(\cdot)$, $A(\cdot)$, and Π_0 are matrices of appropriate dimensions. Suppose also that the system (21) has relative order α in the sense that there exists a finite integer α and a function $r(\cdot)$ such that $A(\cdot)$ is α-differentiable and the product $Q_A(\cdot)\psi(\cdot)$ has the form

$$\psi'(t)Q_A'(t) = [0\cdots \vdots\, r(t)],$$

where

$$Q_A'(t) = \{A'(t),A'^{(1)}(t),\cdots,A'^{(\alpha-1)}(t)\}$$

and $\psi(\cdot)$, $r(\cdot)$, $r^{-1}(\cdot)$, and $A^{(\alpha)}(\cdot)r^{-1}(\cdot)$ are all continuous functions. It can be verified that these restrictions on the model yield processes whose covariance satisfies the conditions of Theorem 4.

Then the RKHS norm of the process has the form

$$\|m\|_{H(R)}^2 = Q_m'[Q_A\Pi_0Q_A']^{-1}Q_m + \int_0^T [m^{(\alpha)}(t)r^{-1}(t) - A^{(\alpha)}(t)r^{-1}(t)\phi(t)]^2\, dt, \tag{22}$$

where

$$\dot{\phi}(t) = K(t)[m^{(\alpha)}(t)r^{-1}(t) - A^{(\alpha)}(t)r^{-1}(t)\phi(t)]$$
$$\phi(0) = \Pi_0Q_A'[Q_A\Pi_0Q_A']^{-1}Q_m \tag{23}$$

and

$$K(t) = P_r(t)A'^{(\alpha)}(t)r^{-1}(t) + \psi(t), \tag{24}$$

where $P_r(\cdot)$ is the unique positive-definite solution of the Riccati equation,

$$\dot{P}_r = -\psi A^{(\alpha)}P_r r^{-1} - P_r A'^{(\alpha)}\psi' r^{-1} - P_r A'^{(\alpha)}A^{(\alpha)}P_r r^{-2} \tag{25a}$$

$$P_r(0) = \Pi_0 - \Pi_0Q_A'[Q_A\Pi_0Q_A']^{-1}Q_A\Pi_0. \tag{25b}$$

Theorem 6—Known Lumped Covariance: Suppose a process has a covariance function

$$R(t,s) = A(t \vee s)B(t \wedge s), \tag{26}$$

where $A(\cdot)$ is an $1 \times n$ matrix and $B(\cdot)$ is a $n \times 1$ matrix. Also suppose that there exists a finite nonnegative integer α, called the relative order of the process, such that

i)
$$A(\cdot) \text{ and } B(\cdot) \text{ have } \alpha \text{ continuous derivatives.} \tag{27a}$$

ii)
$$A^{(i-1)}(t)B^{(i)}(t) - B'^{(i-1)}(t)A'^{(i)}(t) = 0,$$
$$0 \le t \le T, \qquad i = 1, 2,\cdots, \alpha - 1, \qquad \text{if } \alpha > 1. \tag{27b}$$

iii)
$$A^{(\alpha-1)}(t)B^{(\alpha)}(t) - B'^{(\alpha-1)}(t)A'^{(\alpha)}(t) = r^2(t) > 0, \qquad 0 \le t \le T. \tag{27c}$$

iv) Q_AQ_B is nonsingular, where

$$Q_A' = \{A'(0),A'^{(1)}(0),\cdots,A'^{(\alpha-1)}(0)\}$$
$$Q_B = \{B(0),B^{(1)}(0),\cdots,B^{(\alpha-1)}(0)\}.$$

It can be verified that these restrictions yield a covariance function that satisfies the conditions of Theorem 4. It is also easy to see that a stationary process with a rational power-spectral density function will satisfy the preceding conditions, with the relative order α being equal to one-half the difference between the degrees of the denominator and numerator polynomials of the spectral density. Then the RKHS norm can be written

$$\|m\|_{H(R)}^2 = Q_m'[Q_AQ_B]^{-1}Q_m + \int_0^T [m^{(\alpha)}(t) - A^{(\alpha)}(t)\phi(t)]^2r^{-2}(t)\, dt, \tag{28}$$

where

$$\dot{\phi}(t) = K(t)[m^{(\alpha)}(t) - A^{(\alpha)}(t)\phi(t)]r^{-1}(t),$$
$$\phi(0) = Q_B[Q_AQ_B]^{-1}Q_m \tag{29}$$

and

$$K(t) = [B^{(\alpha)}(t) - \Sigma(t)A'^{(\alpha)}(t)]r^{-1}(t), \tag{30}$$

where $\Sigma(\cdot)$ obeys the Riccati equation

$$\dot{\Sigma} = [B^{(\alpha)} - \Sigma A'^{(\alpha)}][B^{(\alpha)} - \Sigma A'^{(\alpha)}]'r^{-2},$$
$$\Sigma(0) = Q_B[Q_AQ_B]^{-1}Q_B'. \tag{31}$$

Remark 1: By introducing a matrix $F(\cdot)$ as in Theorem 3, we can recast the equations for ϕ into a feedback form, viz., $\dot{\phi} = F\phi + \tilde{K}[m^{(\alpha)} - A^{(\alpha)}\phi]$. This may be desirable for stationary processes.

Remark 2: By certain algebraic transformations we can replace the Riccati equations (25) and (31) by equations of a lower order, viz., $n - \alpha$. The procedure for doing this is described in [14].

Example 1: Since no convenient general closed-form solutions are available for the nonlinear Riccati equation, the preceding formulas are clearly not in closed form; however, their numerical evaluation has been extensively studied (see, e.g., [10]). For purposes of illustration, however, we shall choose a covariance for which explicit formulas can be obtained. Thus let

$$R(t,s) = (1 - |t - s|)/2, \qquad 0 \le t, s \le T \le 1. \tag{32}$$

We can apply Theorem 4, with the identifications

$$\alpha = 1, \qquad \Gamma = 2, \qquad C_0(t) = 2^{-1/2}(1 - t)$$
$$R_0(t,s) = t \wedge s - (ts/2), \qquad r^2(t) \equiv 1, \qquad L_0(t,s) = -\tfrac{1}{2}.$$

The equation

$$g_0(t,s) + \int_0^t g_0(t,r)\cdot(-\tfrac{1}{2})\, dr = -\tfrac{1}{2}, \qquad 0 \le s \le t \le T$$

has the unique square-integrable solution

$$g_0(t,s) = (t - 2)^{-1}, \qquad 0 \le s \le t \le T.$$

Therefore, we have, after some simplification,

$$\|m\|^2 = 2m^2(0) + \int_0^T \{\dot{m}(t) - (t - 2)^{-1}[m(t) + m(0)]\}^2\, dt. \tag{33}$$

This is not the usual form for this RKHS norm, see, e.g., [2] and [3]. The usual form is

$$\|m\|^2 = (2 - T)^{-1}[m(0) + m(T)]^2 + \int_0^T \dot{m}^2(t)\, dt, \quad (34)$$

which can be obtained from (33) by integration by parts.

We observe that the form (33) has the property that if T is increased, from T to T_1, say, the new norm can be obtained by adding to the old norm the contribution of the integral term over T to T_1. This convenient updating property is not shared by the form (34). The computational advantage of this updating property can become overwhelmingly significant in several applications, even though (as pointed out by a referee) in our very simple example the nonrecursive form (34) is actually simpler to use. The reason is that the integral term in (34) and in (33) involves only a local operation on the signal $m(\cdot)$; the recursive form has a clear advantage if the integral term involves a general weighting of $m(\cdot)$.

It should be mentioned that (33) can also be calculated by Theorem 6 together with the technique [14] of obtaining a Riccati equation of reduced order. For this we can write (following Geesey [20])

$$\tfrac{1}{2}(1 - |t - s|) = A(t \vee s)B(t \wedge s), \qquad 0 \le t, s \le T,$$

where

$$A(t) = \tfrac{1}{2}\{1 - t, 1\}, \qquad B'(t) = \{1, t\}. \quad (35)$$

The reduced-order Riccati equation for this problem is of the form [20]

$$\dot{P}(t) = +P^2(t), \qquad P(0) = +\tfrac{1}{2},$$

which has the explicit solution

$$P(t) = -(t - 2)^{-1}, \qquad 0 \le t < 2.$$

The reader may find it of interest to calculate by the methods of this paper the RKHS norms of Examples 1–6 in [2, sect. II].

III. Fredholm Equations of the First Kind

We shall consider equations of the kind

$$\int_0^T R(t,s)a(s)\, ds = m(t), \qquad 0 \le t \le T, \quad (36)$$

where $R(t,s)$ is a known covariance function obeying conditions necessary for one of Theorems 4–6 to hold and $m(t)$ is a known function belonging to the RKHS of $R(t,s)$.

One often seeks if at all possible to avoid such equations because, for well-known reasons, their numerical solution is difficult. Thus though there are known conditions under which, for example, unique square-integrable solutions exist, these conditions are usually unnecessarily restrictive (see e.g., the discussion in [2, sect. V-C]). Some generality can be recovered by allowing delta functions and their derivatives in the solution, but this freedom, of course, makes direct numerical solution, e.g., on a digital computer, almost impossible. Moreover, it is now not clear how to resolve the question of uniqueness. We shall present a method of solution that defines a unique solution, a solution that will suffice for the detection problems as posed, e.g., in [16] and [17]. Moreover, the numerical difficulties are perhaps somewhat mitigated.

To present our solution, we begin by noting that in detection theory, (36) is connected with the following detection problem

$$H_1 : x(t) = m(t) + n(t)$$
$$H_0 : x(t) = n(t), \qquad 0 \le t \le T < \infty,$$

where $m(\cdot)$ is a known signal and $n(\cdot)$ is a zero-mean Gaussian random process with covariance $R(t,s)$. The likelihood ratio (LR) for this problem is

$$\text{LR} = \exp\left\{\int_0^T x(t)a(t)\, dt - \tfrac{1}{2}\int_0^T m(t)a(t)\, dt\right\},$$

where $a(\cdot)$ is the solution of (36).

An alternative formula for the LR (see [2]) is

$$\text{LR} = \exp\{\langle x,m\rangle_{H(R)} - \tfrac{1}{2}\|m\|^2_{H(R)}\},$$

where norms and inner products are taken in the RKHS determined by $R(t,s)$. As suggested in [2], by comparing these two expressions for the LR, we can make the correspondence

$$\langle x,m\rangle_{H(R)} = \int_0^T x(t)a(t)\, dt$$

from which we can obtain the solution of the Fredholm equation by letting $x(t) = \delta(t - s)$, where $0 \le s \le T$. In this case,

$$\langle \delta,m\rangle_{H(R)} = a(s).$$

Now let us suppose that $R(\cdot,\cdot)$ is a covariance as described by (15)–(17), for which we have shown that the RKHS inner product is given by[3] [cf., (18) and (19)]

$$\langle x,m\rangle_{H(R)} = Q_m'\Gamma Q_x + \int_0^T \left[m_0^{(\alpha)}(t) - \int_0^t g_0(t,s)m_0^{(\alpha)}(s)\, ds\right] \cdot \left[x_0^{(\alpha)}(t) - \int_0^t g_0(t,u)x_0^{(\alpha)}(u)\, du\right] dt, \quad (37)$$

where $g_0(t,s)$ is the unique square-integrable solution of the Wiener–Hopf-type integral equation

$$g_0(t,s) + \int_0^t g_0(t,\tau)L_0(\tau,s)\, d\tau = L_0(t,s), \qquad 0 \le s \le t \le T < \infty \quad (38)$$

and

$$\Gamma = [R_{ij}]^{-1}, \qquad R_{ij} = \frac{\partial^{i+j}}{\partial t^i\, \partial s^j} R(t,s)|_{t=0=s}.$$

[3] In the remainder of this section, the normalization $r^2(t) \equiv 1$ is assumed.

Since we are dealing with a finite integration interval, we cannot substitute $x(t) = \delta(t - s)$ directly in (37) because boundary terms will be lost if we do so. Instead we first write (37) as

$$\begin{aligned}\langle x,m\rangle_{H(R)} &= Q_m'\Gamma Q_x + \langle (I - g_0)m_0^{(\alpha)}, (I - g_0)x_0^{(\alpha)}\rangle_{L_2[0,T]} \\ &= Q_m'\Gamma Q_x + \langle (I - g_0)D^{(\alpha)}m_0, (I - g_0)D^{(\alpha)}x_0\rangle_{L_2[0,T]} \\ &= Q_m'\Gamma Q_x + \langle x_0, D^{(\alpha)*}(I - g_0^*) \cdot (I - g_0)D^{(\alpha)}m_0\rangle_{L_2[0,T]}, \quad (39)\end{aligned}$$

where the asterisk denotes the adjoint operator. For the integral operator g_0, we have $g_0^*(s,t) = g_0(t,s)$. To find the adjoint of the differential operator $D^{(\alpha)}$ we proceed as follows (see, e.g., [21, pp. 148 and 149]):

$$\langle x_0, D^{(\alpha)*}u\rangle = \langle u, D^{(\alpha)}x_0\rangle, \quad \text{by definition of the adjoint}$$
$$= \int_0^T u(t)x_0^{(\alpha)}(t)\, dt.$$

Integrating by parts α times, we have

$$\langle x_0, D^{(\alpha)*}u\rangle = \sum_{n=1}^{\alpha} (-1)^{n-1}u^{(n-1)}(t)x_0^{(\alpha-n)}(t)\Big|_{t=0}^{t=T} + (-1)^\alpha \int_0^T x_0(t)u^{(\alpha)}(t)\, dt. \quad (40)$$

The boundary terms in (40) comprise what is called the bilinear concomitant, which was first introduced by Lagrange in 1762 [22, p. 124]. Using (40) we can write (39) as

$$\begin{aligned}\langle x,m\rangle_{H(R)} &= Q_m'\Gamma Q_x + \sum_{n=1}^{\alpha} (-1)^{n-1}x_0^{(\alpha-n)}(t) \cdot \left[m_0^{(\alpha+n-1)}(t) - \frac{d^{n-1}}{dt^{n-1}}\int_0^T H(t,v)m_0^{(\alpha)}(v)\, dv\right]\Bigg|_{t=0}^{t=T} \\ &\quad + (-1)^\alpha \int_0^T x_0(t)\left[m_0^{(2\alpha)}(t) - \frac{d^\alpha}{dt^\alpha}\int_0^T H(t,v)m_0^{(\alpha)}(v)\, dv\right] dt, \quad (41)\end{aligned}$$

where $I - H = (I - g_0^*)(I - g_0)$. $H(t,s)$ is the Fredholm resolvent of $L_0(t,s)$ (cf., [11]) and can be directly defined as the unique square-integrable solution of the resolvent equation

$$H(t,s) + \int_0^T H(t,r)L_0(r,s)\, dr = L_0(t,s), \qquad 0 \le t, s \le T. \quad (42)$$

Now if we let $x(t) = \delta(t - s)$ in (41), we shall have (note that the second term vanishes at $t = 0$, since $x_0^{(k)}(0) = 0$, $0 \le k \le \alpha - 1$)

$$\begin{aligned}\langle \delta,m\rangle_{H(R)} &= a(s) = Q_m'\Gamma Q_\delta + \sum_{n=1}^{\alpha} (-1)^{n-1} \cdot [\delta^{(\alpha-n)}(T - s) - A^{(\alpha-n)}(T)Q_B\Gamma Q_\delta] \\ &\quad \cdot \left[m_0^{(\alpha+n-1)}(t) - \frac{d^{n-1}}{dt^{n-1}}\int_0^T H(t,v)m_0^{(\alpha)}(v)\, dv\right]\Bigg|_{t=T} \\ &\quad + (-1)^\alpha \int_0^T [\delta(t - s) - A(t)Q_B\Gamma Q_\delta] \cdot \left[m_0^{(2\alpha)}(t) - \frac{d^\alpha}{dt^\alpha}\int_0^T H(t,v)m_0^{(\alpha)}(v)\, dv\right] dt, \quad (43)\end{aligned}$$

where

$$Q_\delta' = \{\delta(-s), \cdots, \delta^{(\alpha-1)}(-s)\}.$$

Example 2: Let $R(t,s) = (1 - |t - s|)/2$, $0 \le t, s \le T \le 1$. We make the following identifications as in Example 1:

$$\alpha = 1, \qquad \Gamma = 2, \qquad C_0(t) = 2^{-1/2}(1 - t)$$
$$R_0(t,s) = t \wedge s - (ts/2), \qquad L_0(t,s) = -\tfrac{1}{2},$$
$$m_0(t) = m(t) - (1 - t)m(0).$$

For this problem, it is easy to solve the resolvent equation (42) to obtain

$$H(t,s) = -(2 - T)^{-1}, \qquad 0 \le t, s \le T.$$

Then (43) yields the solution of the integral equation (36) as

$$\begin{aligned}a(s) &= -\ddot{m}(s) + \delta(T - s) \cdot [\dot{m}(T) + (2 - T)^{-1}\{m(T) + m(0)\}] \\ &\quad + \delta(s)[-\dot{m}(0) + (2 - T)^{-1}\{m(T) + m(0)\}].\end{aligned}$$

Of course, since the RKHS norm for this example is already known, we can directly integrate by parts in $\langle x,m\rangle_{H(R)}$ as obtained from (33) or (34) to obtain the above formula.

We stress again that the real power of the method is shown in examples where analytical solutions are difficult or impossible. In this regard it should be noted that if

$$R(t,s) = A(t \vee s)B(t \wedge s),$$

where $A(\cdot)$ and $B(\cdot)$ are as in Theorem 6, then $g_0(t,s)$ (and, therefore, $H(t,s)$) can be found by solving an associated Riccati equation (cf., [11]) as follows:

$$g_0(t,s) = A^{(\alpha)}(t)\psi(t,s)K(s), \qquad t \ge s$$
$$\frac{d\psi}{dt}(t,s) = -K(t)A^{(\alpha)}(t)\psi(t,s), \qquad \psi(0,0) = I,$$

where $K(t)$ satisfies (30) and (31).

As stated earlier, for the special separable kernel of Theorem 6, Wood *et al.* [18, eq. (34)–(37)] have obtained results that are equivalent to our (43). However, it is not obvious how their results or derivation can be extended to cover the more general class of covariances that we have treated. [Incidentally, we may note that the covariance in our Examples 1 and 2 is not a lumped covariance when we consider it as zero if $|t - s| > 1$. It is perhaps not so widely known that, as apparently first noted by Geesey [20], for $T < 1$, there is the equivalent lumped representation described in (35).] In fact, as we have noted before in other contexts (see, e.g., [11]–[15]) it seems to us that by tackling the more general problem directly, we are able to obtain results for the special case of lumped covariances more easily than do Wood *et al.* Thus our method is i) start with certain innovations factorization results that lead to ii) certain RKHS norm formulas (whose validity can be checked, if desired, by direct calculation, iii) integrate by parts in these formulas, and iv) suitably plug in a delta function. Our procedure leads directly to the unique solution that is appropriate for the detection theory problem. We cannot seem to give as compact a description of the technique of Wood *et al.*, despite the fact that they too start with certain factorization results (but for the special lumped covariance of Theorem 6).

We should note also that our presentation in this section is basically an elaboration of the relationship, discussed at length in [2], between the RKHS and other approaches to the problem of detecting a signal in Gaussian noise. The innovations representation (IR) of such noise and the RKHS norm for it are clearly two important structural descriptors that must be related and knowledge of which must certainly be helpful in many problems associated with such noise. We have used the relationship here in order to calculate the RKHS norm from the IR, but one can also go in the reverse direction, though not as easily.

Appendix

Proof of Theorems in Section II

The theorems in Section II present certain RKHS norms. The proof that these norms are correct basically only requires verification of the reproducing property,

$$\langle m(\cdot),R(\cdot,t)\rangle_{H(R)} = m(t), \qquad 0 \le t \le T.$$

In our opinion the verification is a relatively straightforward calculation, but is included at the suggestion of a referee. The interesting part is how the proper norms were guessed and this was explained in Sections I and II.

Verification of Theorem 4: We shall show that (18) has the reproducing property; i.e., that

$$\langle m(\cdot),R(\cdot,t)\rangle_{H(R)} = m(t) \tag{44}$$

for all $t \in [0,T]$ and all $m(\cdot) \in H(R)$. It is actually first necessary to check that $R(\cdot,t) \in H(R)$, but the calculations involved in showing (44) will also easily prove this.

For notational simplicity, we shall also assume that $r(t) \equiv 1$ and shall set

$$\langle m(\cdot),R(\cdot,t)\rangle_{H(R)} = f(t).$$

Then we shall show that

$$f^{(\alpha)}(t) = m^{(\alpha)}(t), \qquad 0 \le t \le T$$

and that

$$f^{(i)}(0) = m^{(i)}(0), \qquad 0 \le i \le \alpha - 1,$$

which will prove that $f(t) \equiv m(t)$ and thus establish (44). With an obvious operator notation,

$$f^{(\alpha)}(t) = \frac{d^\alpha}{dt^\alpha} Q_R'\Gamma Q_m + \langle (I - g_0)m_0^{(\alpha)}, (I - g_0)R_0^{(\alpha,\alpha)}\rangle \tag{45}$$

where

$$Q_R' = \{R^{(0,0)}(0,t),R^{(1,0)}(0,t),\cdots,R^{(\alpha-1,0)}(0,t)\}.$$

Under the hypotheses of the theorem, the innovations representation for the process $y_0^{(\alpha)}$ with covariance $R_0^{(\alpha,\alpha)}$ is

$$y_0^{(\alpha)} = (I - g_0)^{-1}\dot{\nu},$$

where $\dot{\nu}$ is white noise with unit intensity (see [11], [12], and [20]). Thus

$$R_0^{(\alpha,\alpha)} = (I - g_0)^{-1}(I - g_0^*)^{-1}.$$

Substituting this into (45), we have

$$\begin{aligned} f^{(\alpha)}(t) &= \frac{d^\alpha}{dt^\alpha} Q_R'\Gamma Q_m + (I - g_0)^{-1}(I - g_0^*)^{-1}(I - g_0^*)(I - g_0)m_0^{(\alpha)} \\ &= \frac{d^\alpha}{dt^\alpha} Q_R'\Gamma Q_m + m_0^{(\alpha)}(t) \qquad (46) \\ &= \frac{d^\alpha}{dt^\alpha} Q_R'\Gamma Q_m + m^{(\alpha)}(t) - \frac{d^\alpha}{dt^\alpha} Q_R'\Gamma Q_m = m^{(\alpha)}(t), \qquad (47) \end{aligned}$$

where we have used (19a).

Furthermore, since $R_0^{(\alpha,K)}(\cdot,0) = 0$ for $0 \le k \le \alpha - 1$,

$$f^{(K)}(0) = \frac{d^K}{dt^K} Q_R'\Gamma Q_m = m^{(k)}(0) \tag{48}$$

using the definitions of Q_R', Γ, and Q_m. Equations (47) and (48) imply that $f(t) \equiv m(t)$ and (44) is proved.

Some Parenthetical Remarks: Sometimes objections are raised to the use of the white noise $\dot{\nu}$. This can be avoided, at the cost of some notation, by working with the integrated process ν. We now illustrate how this can be done. We write the innovations representation, not for $y_0^{(\alpha)}$, but for $y_0^{(\alpha-1)}$ as

$$y_0^{(\alpha-1)}(t) = \int_0^t (I - g_0)^{-1}\, d\nu = \int_0^t (I + k)\, d\nu,$$

where $\nu(\cdot)$ has covariance $t\wedge s$ and k is a Volterra operator with a square-integrable kernel. From this representation we obtain, again using an obvious operator notation,

$$\begin{aligned} R_0^{(\alpha-1,\alpha-1)}(s,t) &= \int_0^t\int_0^s (I + k)(I + k^*) \\ &= t\wedge s + \int_0^t\int_0^s (k + k^* + kk^*), \end{aligned}$$

where k^* is the adjoint of k, i.e., it has kernel

$$k^*(u,v) = k(v,u).$$

[Note that $k^*(u,v) = 0$ for $u > v$.]

We also note that

$$R_0^{(\alpha,\alpha-1)} = \frac{\partial}{\partial s}(t\wedge s) + \int_0^t (k + k^* + kk^*).$$

Now if

$$\langle m(\cdot),R(\cdot,t)\rangle_{H(R)} = f(t)$$

and if we set

$$q = (I - g_0^*)(I - g_0)m_0^{(\alpha)}$$

then

$$f^{(\alpha-1)}(t) - \frac{d^{\alpha-1}}{dt^{\alpha-1}} Q_R' \Gamma Q_m = \langle (I - g_0^*)(I - g_0) m_0^{(\alpha)}(\cdot), R_0^{(\alpha,\alpha-1)}(\cdot,t) \rangle$$

$$= \int_0^t q(u)\,du + \int_0^t (k + k^* + kk^*) q(u)\,du.$$

But now, differentiating both sides with respect to t, we obtain

$$f^{(\alpha)}(t) - \frac{d^\alpha}{dt^\alpha} Q_R' \Gamma Q_m = (I + k)(I + k^*) q(t) = m_0^{(\alpha)}(t),$$

which is just (46), which we had earlier obtained more directly, if less properly.

Theorem 6: It is shown in [20] that the hypotheses of this theorem yield a covariance function that satisfies the conditions of Theorem 4. To verify the reproducing property, it suffices to show that (28) is equivalent to (18). Then the proof of Theorem 4 can be used. Again taking $r(t) \equiv 1$, (26) implies

$$\Gamma = [Q_A Q_B]^{-1}. \tag{49}$$

Now (19a) and (26) imply that

$$m^{(\alpha)}(t) = m_0^{(\alpha)}(t) + A^{(\alpha)}(t) Q_B [Q_A Q_B]^{-1} Q_m = m_0^{(\alpha)}(t) + A^{(\alpha)}(t)\phi(0)$$

using (29).

Also, using (29),

$$\dot{\phi} = -KA^{(\alpha)}\phi + Km^{(\alpha)} = -KA^{(\alpha)}\phi + Km_0^{(\alpha)} + KA^{(\alpha)}\phi(0)$$

$$= -KA^{(\alpha)}(\phi - \phi(0)) + Km_0^{(\alpha)},$$

where K satisfies (30) and (31). Now let $\theta = \phi - \phi(0)$. Then

$$\dot{\theta} = -KA^{(\alpha)}\theta + Km_0^{(\alpha)}, \qquad \theta(0) = 0.$$

Thus

$$\theta(t) = \int_0^t \psi(t,s) K(s) m_0^{(\alpha)}(s)\,ds \triangleq \psi K m_0^{(\alpha)},$$

where

$$\frac{d}{dt}\psi(t,s) = -K(t) A^{(\alpha)}(t) \psi(t,s), \qquad \psi(0,0) = I.$$

But (see [11]) $g_0 = A^{(\alpha)}\psi K$, in operator notation, where ψ and K are defined above. Therefore

$$m^{(\alpha)} - A^{(\alpha)}\phi = m^{(\alpha)} - A^{(\alpha)}\theta - A^{(\alpha)}\phi(0) = m_0^{(\alpha)} - A^{(\alpha)}\theta$$

$$= m_0^{(\alpha)} - A^{(\alpha)}\psi K m_0^{(\alpha)}$$

$$= (I - g_0) m_0^{(\alpha)}. \tag{50}$$

Equations (49) and (50) show that (28) is equivalent to (18).

Theorem 5: As before, we need only show that (22) is equivalent to (28), since the reproducing property is proved for (28). References [12] and [20] show that the covariance of the process satisfying the conditions of Theorem 5 can be expressed in the form (26), where $A(\cdot)$ and $B(\cdot)$ satisfy the conditions of Theorem 6. In other words, by rewriting (23)–(25b) in terms of the covariance parameters in (26), we will obtain (29)–(31). Therefore (22) is the same norm as (28). The relevant equations are [20]

$$\Pi_0 Q_A' = Q_B, \qquad P_r = \Pi - \Sigma$$

$$\Pi A'^{(\alpha)} + \psi = B^{(\alpha)}.$$

Theorems 1, 2, and 3: When stated in their integrated forms, a phrase whose meaning should be clear from the remarks in the proof of Theorem 4, it will be recognized that Theorems 1, 2, and 3 are the same as Theorems 4, 5, and 6 for $\alpha = 1$.

References

[1] E. Parzen, "Probability density functionals and reproducing kernel Hilbert spaces," in *Proc. Symp. Time Series Analysis*, M. Rosenblatt, Ed. New York: Wiley, 1963.

[2] T. Kailath, "RKHS approach to detection and estimation problems—Part I: Deterministic signals in Gaussian noise," *IEEE Trans. Inform. Theory*, vol. IT-17, pp. 503–549, Sept. 1971.

[3] J. Hájek, "On linear statistical problems in stochastic processes," *Czech. Math. J.*, vol. 12, pp. 404–444, Dec. 1962.

[4] E. Parzen, "A new approach to the synthesis of optimal smoothing and prediction systems," in *Mathematical Optimization Techniques*, R. Bellman, Ed. Berkeley, Calif.: Univ. California Press, 1963.

[5] ——, "An approach to time series analysis," *Ann. Math. Statist.*, vol. 32, pp. 951–989, 1961.

[6] D. Ylvisaker, "Lower bounds for minimum covariance matrices in time series regression problems," *Ann. Math. Statist.*, vol. 35, pp. 362–368, 1964.

[7] J. Sacks and D. Ylvisaker, "Designs for regression problems with correlated errors," *Ann. Math. Statist.*, vol. 37, pp. 66–69, 1966.

[8] G. Kimeldorf and G. Wahba, "A correspondence between Bayesian estimation on stochastic processes and smoothing by splines," *Ann. Math. Statist.*, vol. 41, pp. 495–502, 1970.

[9] H. Weinert and T. Kailath, "Recursive spline interpolation and least-squares estimation," submitted to *Ann. Math. Statist.*

[10] C. T. Leondes, Ed., *Theory and Applications of Kalman Filtering.* Paris: NATO Advisory Group for Aerospace R&D, 1970.

[11] T. Kailath, "Fredholm resolvents, Wiener–Hopf equations, and Riccati differential equations," *IEEE Trans. Inform. Theory*, vol. IT-15, pp. 665–672, Nov. 1969.

[12] R. Geesey and T. Kailath, "Applications of the canonical representation of second-order processes to estimation and detection in colored noise," in *Proc. Symp. Computer Processing in Communications*, Polytech. Inst. Brooklyn, Brooklyn, N.Y., Apr. 1969.

[13] T. Kailath and R. Geesey, "An innovations approach to least squares estimation—Part IV: Recursive estimation given the lumped covariance functions," *IEEE Trans. Automat. Contr.*, vol. AC-16, pp. 720–727, Dec. 1971.

[14] ——, "An innovations approach to least squares estimation—Part V: Recursive estimation in colored noise," submitted to *IEEE Trans. Automat. Contr.*

[15] T. Kailath, "The innovations approach to detection and estimation theory," *Proc. IEEE*, vol. 58, pp. 680–695, May 1970.

[16] C. W. Helstrom, *Statistical Theory of Signal Detection*, 2nd ed. London: Pergamon, 1968.

[17] H. L. Van Trees, *Detection, Estimation and Modulation Theory, Part I.* New York: Wiley, 1968.

[18] M. G. Wood, J. B. Moore, and B. D. O. Anderson, "Study of an integral equation arising in detection theory," *IEEE Trans. Inform. Theory*, vol. IT-17, pp. 677-686, Nov. 1971.

[19] L. A. Shepp, "Radon–Nikodym derivatives of Gaussian measures," *Ann. Math. Statist.*, vol. 37, pp. 321–354, 1966.

[20] R. Geesey, "Proper canonical representations of second-order processes," Ph.D. dissertation, Dep. Elec. Eng., Stanford Univ., Stanford, Calif., 1969.

[21] B. Friedman, *Principles and Techniques of Applied Mathematics.* New York: Wiley, 1956.

[22] E. L. Ince, *Ordinary Differential Equations.* New York: Dover, 1956.

Part III

APPLICATIONS TO DETECTION AND ESTIMATION

Editor's Comments on Papers 8 Through 14

One of the problems in which RKHS methods have had great success is that of detecting a known signal in Gaussian noise. Using simple geometric ideas, a coordinate-free solution can be obtained in which the likelihood ratio is expressed in terms of a RKHS inner product. A simple test for nonsingularity of detection also results from this approach. Furthermore, the problems inherent in the use of Karhunen-Loève expansions or first-kind integral equations are avoided. Parzen [1] was the first to apply RKHS to detection and to derive the likelihood ratio and nonsingularity condition for the known signal problem. In Paper 4, Hájek independently obtained the same results using different arguments (see also Kailath [2]).

In Paper 8, Kailath discusses the RKHS approach to this detection problem in great detail, compares it to other approaches, and clearly shows its superiority for computing likelihood ratios, testing for nonsingularity, bounding signal detectability, and determining detection stability. This paper provides a good introduction to RKHS theory with many explicit examples of RKHS, including those associated with multidimensional, multivariable, and generalized random processes.

RKHS methods have proved equally successful in the apparently more difficult problem of detecting a Gaussian signal in Gaussian noise. Again, the earliest work along these lines is that of Parzen and Hájek. In Paper 9, Parzen derives a nonsingularity condition and a likelihood-ratio formula that require the evaluation of an inner product in a direct product RKHS. In Paper 4, Hájek, under a strong nonsingularity condition, shows that the likelihood ratio can be computed in a simpler way, using only the individual RKHS norms. Kailath and Weinert, in Paper 10, show how this important feature can be retained under the ordinary nonsingularity condition. They also present a test that can verify that the likelihood ratio obtained from the formal RKHS expressions is indeed correct. Furthermore, the discussion of general RKHS properties, begun in Paper 8, is continued. In Paper 11, Golosov and Tempel'man give a compendium of equivalent nonsingularity conditions for the Gaussian detection problem with vector-valued observations. Other papers (see [3-17]) using RKHS techniques to study Gaussian detection, or the underlying question of equivalence of Gaussian measures, cover basically the same ground as Papers 4, 9, 10, and 11, but from different points of view and at different levels of generality.

The detection problems discussed so far have all involved Gaussian random processes. The RKHS approach is based on the fact that the statistics of a zero-mean Gaussian process are completely characterized by its covariance function, which is also a reproducing kernel. If the statistics of a non-Gaussian process are completely described by its characteristic functional, then, since the characteristic functional is symmetric and nonnegative-definite and is thus a reproducing kernel [18, 19], much of the above work can be generalized to the non-Gaussian case. This generalization is carried out by Duttweiler and Kailath in Paper 12.

Two papers that use RKHS methods for a variety of parameter estimation or regression problems conclude Part III. In one such problem, a Gaussian random process has an unknown mean that can be expressed as some linear combination of known functions. A minimum variance, unbiased linear estimate of a functional of the mean is desired. In case the process has a known covariance, Parzen and Hájek, in Papers 3, 4, and 5, solve the problem using the basic congruence map (see also Rozanov [17] and Ylvisaker [20, 21]). In

Paper 13, Tempel'man reviews this work then studies the effects of using an approximate covariance. He makes interesting and effective use of some of the lesser known RKHS properties. See Tempel'man [22] for further generalizations. In Paper 14, Duttweiler and Kailath study variance bounds for unbiased estimates of parameters determining the mean or covariance of a Gaussian random process. They also give an explicit formula for estimating the arrival time of a step function in white Gaussian noise. Albuquerque [23] has carried out related work.

One further generalization deserves mention. If, in the regression problem considered by Parzen and Hájek and discussed above, the measurement times are open to choice, the variance of the best estimate can be further reduced. Sacks and Ylvisaker [24] have extensively studied the problem of finding an optimal, or asymptotically optimal, set of measurement times using RKHS methods.Their work has been generalized by Wahba [25, 26] and Hájek and Kimeldorf [27]. The multidimensional case is considered by Ylvisaker [28]. A similar problem of experimental design for system identification is treated by Wahba [29].

REFERENCES

1. E. Parzen, *Statistical Inference on Time Series by Hilbert Space Methods I,* Technical Report 23, Statistics Department, Stanford University, Stanford, Calif., 1959, 130p.
2. T. Kailath, A Projection Method for Signal Detection in Colored Gaussian Noise, *IEEE Trans. Inform. Theory* **IT-13:**441–447 (1967).
3. G. Kallianpur and H. Oodaira, The Equivalence and Singularity of Gaussian Measures, in *Proceedings of the Symposium on Time Series Analysis,* M. Rosenblatt, ed., Wiley, New York, 1963, pp. 279–291.
4. J. Capon, Radon-Nikodym Derivatives of Stationary Gaussian Measures, *Ann. Math. Stat.* **35:**517–531 (1964).
5. J. Capon, Hilbert Space Methods for Detection Theory and Pattern Recognition, *IEEE Trans. Inform. Theory* **IT-11:**247–259 (1965).
6. Ju. I. Golosov, Gaussian Measures Equivalent to Gaussian Markov Measures, *Sov. Math. Dokl.* **7:**48–52 (1966).
7. Ju. I. Golosov, A Method for Evaluating the Radon-Nikodym Derivative of Two Gaussian Measures, *Sov. Math. Dokl.* **7:**1162–1165 (1966).
8. Y. M. Pan, Simple Proofs of Equivalence Conditions for Measures Induced by Gaussian Processes, *Selected Transl. Math. Stat. Probab.* **12:**109–118 (1973).
9. J. Neveu, *Gaussian Random Processes* (in French), University of Montreal Press, Montreal, 1968, 224p.
10. T. I. Mirskaya, A. S. Pabedinskaite, and A. A. Tempel'man, Hilbert Spaces of Certain Reproducing Kernels and the Equivalence of Gaussian Measures, *Selected Transl. Math. Stat. Probab.* **11:**121–131 (1973).
11. W. J. Park, A Multi-Parameter Gaussian Process, *Ann. Math. Stat.* **41:**1582–1595 (1970).

12. W. J. Park, On the Equivalence of Gaussian Processes with Factorable Covariance Functions, *Am. Math. Soc. Proc.* **32:**275–279 (1972).
13. C. J. Henrich, Equivalence and Radon-Nikodym Derivatives of Gaussian Measures, *J. Math. Anal. Appl.* **37:**255–270 (1972).
14. A. F. Gualtierotti and S. Cambanis, Some Remarks on the Equivalence of Gaussian Processes, *J. Math. Anal. Appl.* **49:**226–236 (1975).
15. S. Cambanis, The Relation Between the Detection of Sure and Stochastic Signals in Noise, *Soc. Ind. Appl. Math. J. Appl. Math.* **31:**558–568 (1976).
16. S. D. Chatterji and V. Mandrekar, Equivalence and Singularity of Gaussian Measures and Applications, in *Probabilistic Analysis and Related Topics,* vol. 1, A. T. Barucha-Reid, ed., Academic Press, New York, 1978, pp. 169–197.
17. Ju. A. Rozanov, *Infinite-Dimensional Gaussian Distributions,* American Mathematical Society, Providence, R. I., 1971, 161p.
18. T. Hida and N. Ikeda, Analysis on Hilbert Space with Reproducing Kernel Arising from Multiple Wiener Integral, in *Proceedings of the Fifth Berkeley Symposium on Mathematical Statistics and Probability,* vol. 2, L. M. Le Cam and J. Neyman, eds., University of California Press, Berkeley and Los Angeles, 1967, pp. 117–143.
19. T. Hida, *Stationary Stochastic Processes,* Princeton University Press, Princeton, N. J., 1970, 161p.
20. D. Ylvisaker, On Linear Estimation for Regression Problems on Time Series, *Ann. Math. Stat.* **33:**1077–1084 (1962).
21. D. Ylvisaker, Lower Bounds for Minimum Covariance Matrices in Time Series Regression Problems, *Ann. Math. Stat.* **35:**362–368 (1964).
22. A. A. Tempel'man, On the General Theory of Linear Estimates of the Mean Value, *Sov. Math. Dokl.* **15:**1467–1471 (1974).
23. J. P. A. Albuquerque, The Barankin Bound: A Geometric Interpretation, *IEEE Trans. Inform. Theory* **IT-19:**559–561 (1973).
24. J. Sacks and D. Ylvisaker, Designs for Regression Problems with Correlated Errors, *Ann. Math. Stat.* **37:**66–89 (1966); **39:**49–69 (1968); **41:**2057–2074 (1970).
25. G. Wahba, On the Regression Design Problem of Sacks and Ylvisaker, *Ann. Math. Stat.* **42:**1035–1053 (1971).
26. G. Wahba, Regression Design for Some Equivalence Classes of Kernels, *Ann. Stat.* **2:**925–934 (1974).
27. J. Hájek and G. Kimeldorf, Regression Designs in Autoregressive Stochastic Processes, *Ann. Stat.* **2:**520–527 (1974).
28. D. Ylvisaker, Designs on Random Fields, in *A Survey of Statistical Design and Linear Models,* J. N. Srivastava, ed., North-Holland, Amsterdam, 1975, pp. 593–607.
29. G. Wahba, Parameter Estimation in Linear Dynamic Systems, *IEEE Trans. Autom. Control* **AC-25:**235-238 (1980).

8

Reprinted by permission from *IEEE Trans. Inform. Theory* **IT-17**:530–549 (1971)

RKHS Approach to Detection and Estimation Problems—Part I: Deterministic Signals in Gaussian Noise

THOMAS KAILATH, FELLOW, IEEE

Abstract—First it is shown how the Karhunen–Loève approach to the detection of a deterministic signal can be given a coordinate-free and geometric interpretation in a particular Hilbert space of functions that is uniquely determined by the covariance function of the additive Gaussian noise. This Hilbert space, which is called a reproducing-kernel Hilbert space (RKHS), has many special properties that appear to make it a natural space of functions to associate with a second-order random process. A mapping between the RKHS and the linear Hilbert space of random variables generated by the random process is studied in some detail. This mapping enables one to give a geometric treatment of the detection problem. The relations to the usual integral-equation approach to this problem are also discussed. Some of the special properties of the RKHS are developed and then used to study the singularity and stability of the detection problem and also to suggest simple means of approximating the detectability of the signal. The RKHS for several multidimensional and multivariable processes is presented; by going to the RKHS of functionals rather than functions it is also shown how generalized random processes, including white noise and stationary processes whose spectra grow at infinity, are treated.

I. Introduction

In this and succeeding papers a number of detection and estimation problems, both Gaussian and non-Gaussian, linear and nonlinear, will be studied. A particular kind of Hilbert space called a reproducing-kernel Hilbert space (RKHS) plays a central role in such problems.

In Part I the problem of detecting deterministic signals in colored Gaussian noise is used as a vehicle to introduce and explain the role and the significance of the RKHS. However, even in this simple problem some new results and insights are obtained, including simple yet rigorous derivations with geometric interpretations, simple tests for checking the validity of proposed formulas, and a new method for approximate calculation of the signal detectability. But more important, the understanding of the RKHS gained in this simple context yields important new results in the more difficult problems of Gaussian signal detection (Parts II and III).

In Part II it will be shown that the solution of such problems is determined essentially by the solutions for deterministic signal problems. In Part III it will be shown that, extending the results of [1]–[3], the likelihood ratio (LR) for general Gaussian signals can be written in essentially the same form as the basic deterministic signal formula (15) of the present paper. In later papers the RKHS approach will be extended to non-Gaussian processes and some detection and nonlinear estimation problems will be discussed.

To return to the present paper, Part I, it is repeated that one of its aims is tutorial and preparatory, though, while fulfilling this aim, a comprehensive account of the detection of a known signal from the RKHS point of view is given and the relationship to other approaches is shown.

In Section II the author's first aim is to illustrate that a variety of formulas for the LR that can be obtained by using different types of signal and noise representations can be summarized in a single coordinate-free formula involving RKHS inner products and norms. Explicit formulas for these quantities are given for several covariance functions. A detailed discussion is also given of the relationship between the RKHS method and the more common integral-equation method.

The presence of inner products and norms in the RKHS formula suggests that there should be a direct geometrical interpretation and derivation of the formula for the LR. In Section III we give this derivation, which will also help to explain why the RKHS enters so intimately into this detection problem. In particular, the one-to-one norm-preserving mapping between the RKHS and the Hilbert space of finite-variance linear functionals of a random process is examined in some detail. A good understanding of this mapping is basic to the effective use of the RKHS. An immediate application is made to clarify some questions of rigor associated with the integral-equation method; in so doing it is also shown how the integral-equation technique can be simply and usefully fitted into the RKHS framework.

RKHSs are a special class of Hilbert spaces that have many special properties not shared by all Hilbert spaces, e.g., the space $L_2(0,T)$. These properties are determined by the special form of the inner product for an RKHS. As explained briefly at the beginning of Section IV, the choice of norm is almost irrelevant in finite-dimensional spaces, but can make significant differences in infinite-dimensional spaces. This is why colored-noise problems are trivial when there are only a finite number of observations and why new phenomena appear as an infinite number of observations are reached.

In Section IV some of the more important properties of RKHS are given with emphasis on those especially relevant

Manuscript received October 31, 1968; revised August 11, 1970 and March 2, 1971. This work was supported by the Applied Mathematics Division of the Air Force Office of Scientific Research under Contract AF-49(638)1517 and by the Joint Services Electronics Program at Stanford University under Contract Nonr-225(83). This paper was completed at the Indian Institute of Science, Bangalore, with the assistance of a Fellowship from the John Simon Guggenheim Memorial Foundation.
The author is with the Department of Electrical Engineering, Stanford University, Stanford, Calif. 94305.

to the present detection problem. The proofs of these properties illustrate the simplicity provided by the special form of the RKHS norm. The special properties of RKHS help in the calculation and verification of RKHS norms, and, moreover, they help answer several questions on the properties of the solution to the deterministic signal problem.

In Section V RKHS concepts are applied first for the discussion of some simple tests for nonsingular detection, then for the description of a method for approximate calculation of RKHS norms and signal detectability, and finally for the discussion of the stability of the detection problem. The direct relevance of the RKHS framework to such questions is noteworthy.

In Section VI it is shown how RKHS can be defined and calculated for parameter sets more general than the real line. By this means, the results of Sections I–V can be extended readily to random fields (or multidimensional processes), vector (or multivariable) processes, and generalized random processes. In particular, it is shown how by going to an RKHS of functionals (rather than functions as in the earlier discussions) we can rigorously extend the RKHS approach and results to include the white-noise case. It is also shown how generalized stationary processes are treated, whose power density spectra fall off at infinity not as some even integral power (as for rational spectra) but as a nonintegral power less than 1. The concept of an RKHS of functionals is not only useful for a proper treatment of generalized Gaussian processes, but is also essential for the application of RKHS methods to non-Gaussian processes.

A. Discussion of Previous Work

Here we shall only mention some previous work on the deterministic-signal or known-signal detection; the Gaussian-signal case will be discussed in Part II. A general solution of the known-signal problem was first given by Grenander [4]. Grenander's major tool was an orthogonal series expansion of the random process, usually called the Karhunen–Loève (KL) expansion after Karhunen [5] and Loève [6], although it was also discovered by Kac and Siegert [7] and Kosambi [8]. Grenander's general formulas were in terms of the KL coefficients, though he showed that under additional assumptions (viz., more than those required to ensure nonsingular detection) closed-form expressions could be obtained. A more accessible discussion of some of Grenander's results now appears in several textbooks on detection theory.

The theory of RKHS was first applied to detection and estimation problems by Parzen [9], who gave the basic RKHS formula [(15) of this paper] for the LR, though he derived it via sampling representations of the observed process and not by the geometric method of Section III. This geometric method was given by Hájek [10] and later by Kailath [11]. Hájek [12] has made a deep study of linear statistical problems; however, he was not aware of the explicit connection with RKHS and thus did not fully exploit the many powerful properties of RKHS that are available in the literature, especially in the definitive paper by Aronszajn [13]. RKHSs were first studied in 1910–1920 by Moore [14], in connection with a general theory of integral equations. Krein used them [15], [16] in his fundamental studies on the extension of positive-definite functions. Some related applications were made by Devinatz [17]. They were also encountered and used effectively in the theory of boundary value and related problems by Bergman and Schiffer (cf. [13], which gives additional references). But it was Aronszajn who made a systematic abstract development of the theory in the 1940's, though it was recently discovered that many of the results were independently obtained in the USSR by Povzner [18]. Loève [19] was the first to note that RKHS could be used to provide a representation for second-order random processes.

However, it was the work of Parzen that brought the RKHS to the fore in statistical problems. Since then several other authors have extended Parzen's work and applied it to new problems. In particular, Parzen [20] has recently used the RKHS to clarify some relationships between time series, control theory, and approximation theory. Some references to this work will be provided in Part II. In this and forthcoming papers, the author will pursue more explicitly than Parzen and the others the applications of RKHS to detection and estimation problem. Nevertheless, the reader familiar with Parzen's work will notice his indebtedness to it.

II. Coordinate-Free Solution of Deterministic Signal Problem

Consider the following problem. Given an observation $\{x(t),\, t \in I\}$, determine the LR for the hypotheses

$$h_1: x(t) = m(t) + n(t), \qquad h_0: x(t) = n(t), \qquad t \in I, \quad (1)$$

where $n(\cdot)$ is a sample from a zero-mean second-order Gaussian random process with *continuous* square-integrable covariance function

$$E[n(t)n(s)] = R(t,s), \qquad (t,s) \in I \times I$$

and $m(\cdot)$ is a waveform of known shape and finite energy. The parameter set I can be quite general, as is illustrated in Section VI, but until then it may be taken for concreteness as a subset of the real line. A solution to this problem is well known; it was first given by Grenander [4] and is also discussed in several textbooks. The result is the following. Let

$$x_i = \int_I x(t)\psi_i(t)\, dt, \qquad m_i = \int_I m(t)\psi_i(t)\, dt, \quad (2)$$

where the $\{\psi_i(\cdot)\}$ and $\{\lambda_i\}$ are the eigenfunctions and eigenvalues of $R(\cdot,\cdot)$, viz.,[1]

$$\int_I R(t,s)\psi_i(s)\, ds = \lambda_i\psi_i(t), \qquad t \in I, \qquad i = 1,2,\cdots. \quad (3)$$

If

$$\sum m_i^2/\lambda_i < \infty, \quad (4)$$

[1] Unless otherwise indicated, all summations will run from 1 to infinity.

the detection problem is nonsingular, i.e., incapable of resultion with zero probability of error and

$$\mathrm{LR} = \exp\{\sum x_i m_i/\lambda_i - \tfrac{1}{2}\sum m_i^2/\lambda_i\}. \qquad (5)$$

Note, however, that the solution (2)–(5) is not "coordinate-free." This can be seen most easily as follows. Let $\{\psi_{wi}(\cdot), i = 1,2,\cdots\}$ be solutions of the integral equation

$$\int_I R(t,s)\psi_{wi}(s)w(s)\,ds = \lambda_{wi}\psi_{wi}(t), \qquad t \in I, \qquad (6)$$

where $w(\cdot)$ is a nonnegative weight function. Then if

$$\int_I \int_I R^2(t,s)w(t)w(s)\,dt\,ds < \infty, \qquad (7)$$

it can be shown[2] that another expression for the LR is

$$\mathrm{LR} = \exp\{\sum x_{wi}m_{wi}/\lambda_{wi} - \tfrac{1}{2}\sum m_{wi}^2/\lambda_{wi}\}, \qquad (8)$$

where

$$x_{wi} = \int_I x(t)\psi_{wi}(t)w(t)\,dt,$$

$$m_{wi} = \int_I m(t)\psi_{wi}(t)w(t)\,dt. \qquad (9)$$

These relations show the dependence of the solution upon the arbitrary weight function and hence upon an arbitrary coordinate system $\{\psi_{wi}(\cdot)\}$. In fact, there are several other coordinate-systems that could have been used, e.g., sample values at a dense set in I, or KL coefficients with respect to the eigenfunctions that diagonalize a suitable pair of covariance functions, and so on. It is natural to seek a coordinate-free representation.

A. Integral-Equation Representation

One way of finding such a representation is to notice that if a function $a(\cdot)$ is defined as a solution of the integral equation

$$\int_I R(t,s)a(s)\,ds = m(t), \qquad t \in I, \qquad (10)$$

the expressions (5) and (8) for the LR can be formally written

$$\mathrm{LR} = \exp\left[\int_I a(t)x(t)\,dt - \tfrac{1}{2}\int_I a(t)m(t)\,dt\right]. \qquad (11)$$

The representation (10), (11) has been very fruitful and has yielded most of the known explicit formulas for the LR. By the use of Mercer's formula

$$R(t,s) = \sum \lambda_{wi}\psi_{wi}(t)\psi_{wi}(s), \qquad t,s \in I \times I,$$

it is easy to deduce (10), (11) formally. The difficulty lies in a rigorous proof. In the first place, the integral equation (10) is a so-called Fredholm equation of the first kind, which is in general difficult to solve. Some conditions are known on $m(\cdot)$ under which solutions will exist and be absolutely integrable or square integrable. However, in general the problem is that when $R(t,s)$ depends only upon $t - s$, or when $R(t,s)$ contains singularities like delta functions and their derivatives, the solution $a(\cdot)$ will usually be much "rougher" than $m(\cdot)$. For example,

$$a(t) = \delta(t), \qquad \text{if } R(t,s) = e^{-\alpha|t-s|},\ m(t) = e^{-\alpha t},\ I = [0,T].$$

Because of such possibilities, even numerical solution of (10) is quite difficult. More fundamentally, no general theory for (10) is available, and it is unknown whether a solution will exist even if it is permitted to contain an infinite number of impulsive-function singularities. Similarly, it is not clear how the null space of $R(t,s)$ can be defined, so it is difficult to specify how to obtain a unique solution, or whether uniqueness is important. A related difficulty is that because of the somewhat uncertain state of the theory of (10), it cannot be clearly stated whether a detection problem stated in terms of a given $m(\cdot)$ and $R(\cdot,\cdot)$ will be meaningful or not; that is, $m(\cdot)$ may not be smooth enough for (10) to have a solution (even containing impulsive singularities), but one cannot be sure whether this means that the detection problem is singular, i.e., capable of solution with zero probability of error. Of course, there is also the question whether the LR expressions obtained by using solutions $a(\cdot)$ containing impulsive-function singularities are, in fact, correct. In particular, there is the problem of defining the stochastic integral in (11) when so little can be asserted about the solution $a(\cdot)$. Some of these questions were raised and resolved by Grenander [4] for stationary covariances that have rational Fourier transforms; recently Kadota [22] has carried out a similar justification for a wider class of processes (see also Section III, Example 2). But this problem has not been settled in all generality.

B. RKHS Representation

Some further remarks on the integral-equation method will be made below, but now we wish to turn to another coordinate-free representation that will among other benefits also resolve some of the problems raised above more directly than seems possible by the integral-equation method. This alternative coordinate-free representation is based on the so-called RKHS associated with the covariance $R(\cdot,\cdot)$.

Definition: The RKHS $H(R,I)$ associated with a covariance function $R(\cdot,\cdot)$ on $I \times I$ is a Hilbert space of functions on I whose inner product, say $\langle,\rangle_H$, has for every $m(\cdot) \in H(R,I)$, the "reproducing" property

$$\langle m(\cdot), R(\cdot,t)\rangle_H = m(t), \qquad t \in I.$$

For this relation to be meaningful, it is, of course, also required that

$$R(\cdot,t) \in H(R,I), \qquad t \in I.$$

It is shown later (Section IV) that the RKHS has many special properties, but for the present only the not unex-

[2] The point is that Mercer's theorem, on which the proof of the KL expansion is usually based, is true for all continuous covariances obeying (7) and, in fact, in even more general circumstances (see Zaanen [21]).

pected property that the RKHS $H(R,I)$ is uniquely determined by the covariance function is needed. As a matter of notation, the inner product will also be written, depending upon context as $\langle , \rangle_R$, $\langle , \rangle_I$ or $\langle , \rangle_{H(R,I)}$. It should also be noted here that for Hilbert spaces the inner product can be uniquely determined from the norm $\| \ \|_H^2 \triangleq \langle , \rangle_H$ by the relation

$$4\langle m,n\rangle_H = \|m + n\|_H^2 - \|m - n\|_H^2 \tag{12}$$

so that, for convenience, the RKHS will often be described just in terms of its norm.[3]

To return to the LR problem, first observe that the series that enter in the LR formula (8) can be expressed in RKHS terms. To begin with, it is shown that

$$\sum m_{wi}^2/\lambda_{wi} = \|m\|_H^2. \tag{13}$$

It is well known that the value of the series in (13) determines the "detectability" of the signal $m(\cdot)$ in the noise $n(\cdot)$—for white noise (all λ_{wi} equal) the detectability is proportional to the signal energy (squared length), and (13) shows that for colored noise the detectability depends upon a generalized notion of length. The "geometric" content of (13) suggests that a geometric derivation of the LR formula should be possible; this is done in Section III.

Proof of (13): Let the space of functions $m(\cdot) = \sum m_{wi}\psi_{wi}(\cdot), \sum m_{wi}^2/\lambda_{wi} < \infty$ be made into a Hilbert space $\mathscr{H}$ with inner product

$$Q[a(\cdot),b(\cdot)] = \sum a_{wi}b_{wi}/\lambda_{wi},$$

where a_{wi},b_{wi} are defined in the same way as m_{wi} [see (9)]. But $\mathscr{H}$ is the RKHS associated with $R(t,s)$ because 1)

$$R(\cdot,t) = \sum \lambda_{wi}\psi_{wi}(t)\psi_{wi}(\cdot) \in \mathscr{H}$$

since

$$\sum [\lambda_{wi}\psi_{wi}(t)]^2/\lambda_{wi} = \sum \lambda_{wi}\psi_{wi}^2(t) = R(t,t) < \infty$$

and 2)

$$\begin{aligned} Q[m(\cdot),R(\cdot,t)] &= \sum m_{wi}\cdot \lambda_{wi}\psi_{wi}(t)/\lambda_{wi} \\ &= \sum m_{wi}\psi_{wi}(t) = m(t). \end{aligned}$$

By standard Hilbert-space theory, this last equality may hold only for almost all t. However, it is easy to show that it holds for every t, for by the Schwarz inequality,

$$\begin{aligned} |m(t) - m_N(t)|^2 &= \left| Q\left[R(\cdot,t),\, m(\cdot) - \sum_1^N m_{wi}\psi_{wi}(t)\right]\right|^2 \\ &\le R(t,t)\cdot Q[m(\cdot) - m_N(\cdot),\, m(\cdot) - m_N(\cdot)] \\ &\to 0 \text{ as } N \to \infty \text{ for every } t. \end{aligned}$$

Thus conditions 1) and 2) have been verified for the RKHS, and therefore, by the uniqueness of the RKHS associated with $R(t,s)$, $(t,s) \in I \times I$, the Hilbert space $\mathscr{H}$ must be identical with $H(R,I)$. Therefore,

$$\sum m_{wi}^2/\lambda_{wi} = Q[m,m] = \|m\|_H^2.$$

The Hilbert-space relation (12) between norm and inner product leads one to expect from (13) that the first series, say Λ, in the LR formula (8) can be written

$$\Lambda \triangleq \sum m_{wi}x_{wi}/\lambda_{wi} \overset{?}{=} \langle m,x\rangle_H. \tag{14}$$

However, some care is needed here,[4] because $x(\cdot) = \sum x_{wi}\psi_{wi}(\cdot)$ a.s. does not belong to $H(R,I)$; for if it did, we should have $\sum x_{wi}^2/\lambda_{wi} < \infty$ a.s. Unfortunately, since the $\{x_{wi}\}$ [cf. (9)] are independent zero-mean random variables with variances λ_{wi}, $E[\sum x_{wi}^2/\lambda_{wi}] = \infty$ and hence, by a theorem of Kolmogorov's (Doob [23, p. 108]), $\sum x_{wi}^2/\lambda_{wi} = \infty$ with probability 1. Despite its formal nature, (14) is quite useful because the results it suggests can usually be directly justified in a fairly simple way. The point is that the random variable Λ is always well defined when the detection problem is nonsingular, and (14) should be regarded only as a mnemonic for finding Λ in closed form. Thus, as is illustrated below, often some slight rewriting of (14) to define certain stochastic integrals properly will yield a meaningful random variable that can be directly verified to be appropriate. It will be shown in Section III that for such a verification it is necessary only to check that $E_0[x(t)\Lambda] = m(t)$, $t \in I$, where $E_0(\cdot)$ denotes the expectation under hypothesis h_0.

On the basis of this last remark, it can be seen that the LR can always be rigorously written as

$$\text{LR} = \exp\{\Lambda - \tfrac{1}{2}\|m\|_H^2\}, \tag{15}$$

where Λ is the unique[5] linear square-integrable functional of $x(\cdot)$ that satisfies

$$E_0[x(t)\Lambda] = m(t), \qquad t \in I, \tag{16}$$

where E_0 denotes the expectation under h_0, i.e., with $x(\cdot)$ replaced by $n(\cdot)$. By use of the relation $E_0[\text{LR}] = 1$, or by a more direct method given in Section III [cf. the proof of (42)], it can be shown that

$$E_0|\Lambda|^2 = \|m\|_H^2.$$

Therefore, the LR can also be written as

$$\text{LR} = \exp\{\Lambda - \tfrac{1}{2}E_0|\Lambda|^2\}. \tag{17}$$

By using the formal representation (14), the LR can be written in the more geometrically suggestive way

$$\text{LR} = \exp\{\langle m,x\rangle_H - \tfrac{1}{2}\|m\|_H^2\}. \tag{18}$$

For later reference, it is also noted that the necessary and

[3] For simplicity of notation, it has been assumed in most of this paper that real Hilbert spaces are being dealt with. However, sometimes, especially with stationary processes, complex spaces are used, but the necessary modifications to the formulas, e.g., the use of Hermitian conjugates, are made without comment.

[4] To stress this, Parzen has suggested that inner products as in (14) be written $\langle x,m\rangle_{\sim H}$ and be called congruence inner products; the reason for the name will appear later.

[5] It will not be necessary to distinguish between random variables that differ by a zero-variance random variable; one advantage of the RKHS is that all such random variables map to the zero function so that equivalence classes in the RKHS are not necessary.

sufficient condition for nonsingularity of the detection problem (1) can be expressed [4] as

$$\sum m_{wi}^2/\lambda_{wi} = E_0|\Lambda|^2 = \|m\|_H^2 < \infty. \tag{19}$$

It should also be recorded that the error probability and, more generally, the Bayes risk for the problem (1) are monotonically decreasing functions of $\|m\|_H^2$.

To use the formulas (15)–(19), it is necessary to know the RKHS norms, and, as will be seen later, these norms can be found in several ways. Here some examples in which the norms can be given fairly explicitly are listed. In all these examples, the parameter set I is, for simplicity, a subset of the real line; some more general examples are given in Section VI.

C. Some Examples of RKHS

Example 1—Gauss–Markov Processes: $R(t,s) = f(t)g(s)$, $a \le t \le s \le b$. $H(R,T)$ consists of functions $m(\cdot)$ such that $d/dt\,[m(t)/g(t)]$ is defined almost everywhere and is square integrable and

$$\|m\|_H^2 = \int_a^b \left[\frac{d}{dt}\left\{\frac{m(t)}{g(t)}\right\}\right]^2 \left[\frac{d}{dt}\left\{\frac{f(t)}{g(t)}\right\}\right]^{-1} dt + \frac{m^2(a)}{f(a)g(a)}. \tag{20}$$

There are two important special cases of this example.

Example 2—Wiener Process:

$$R(t,s) = \min(t,s), \qquad a \le (t,s) \le b$$

$$\|m\|_H^2 = \int_a^b \dot{m}^2(t)\,dt + \frac{m^2(a)}{a}, \tag{21}$$

where the dot denotes differentiation.

Example 3—Ornstein–Uhlenbeck Process:

$$R(t,s) = \{\exp - \alpha|t-s|\}/2\alpha, \qquad a \le (t,s) \le b$$

$$\|m\|_H^2 = \int_a^b [\dot{m}(t) + \alpha m(t)]^2\,dt + 2\alpha m^2(a). \tag{22}$$

Example 4—Triangular Covariance: $R(t) = \max(0,1,-|t|)$, $0 \le t \le T \le 1$. $H(R)$ consists of absolutely continuous functions with norm

$$\|m\|_H^2 = \frac{1}{2}\frac{[m(0) + m(T)]^2}{2 - T} + \frac{1}{2}\int_0^T \dot{m}^2(t)\,dt. \tag{23}$$

The norm for $T > 1$ was derived by Hájek [12]. Let $T = N + \delta$, where N is an integer and $0 \le \delta < 1$. Then

$$\|m\|_H^2 = \int_{-1}^{T} \tilde{m}^2(t)\,dt, \tag{24}$$

where

$$\tilde{m}(t) = \dot{m}(t) + \dot{m}(t-1) + \cdots$$

and the sum stops when the argument falls into the interval $(-1,0)$, and where $\dot{m}(\cdot)$ has been extended so that

$$\dot{m}(t) = \frac{1}{N+1}[c - N\dot{m}(t+1) - (N-1)\dot{m}(t+2) - \cdots - \dot{m}(t+N)], \qquad T - N - 1 < t < 0$$

$$= \frac{1}{N+2}[c - (N+1)\dot{m}(t+1) - \cdots - \dot{m}(t+N+1)], \qquad -1 < t < T - N - 1$$

with

$$c = (2N - T + 2)^{-1}\sum_{j=0}^{N}(N+1-j)\cdot[m(j) + m(T-j)].$$

Example 5—Autoregressive Processes:

$$R(t) = \int_{-\infty}^{\infty}\left|\sum_0^N a_k(i\omega)^k\right|^{-2} e^{i\omega t}\frac{d\omega}{2\pi}, \qquad 0 \le t \le T.$$

$H(R)$ consists of functions whose Nth-derivatives are square integrable over $[0,T]$. The norm is

$$\|m\|_H^2 = \int_0^T [L(D)m][L(D)m]\,dt + \underbrace{\sum_{0 \le j,k \le N-1}\sum}_{j+k \text{ even}} m^{(j)}(0)m^{(k)}(0)\,d_{jk}, \tag{25a}$$

where

$$L(D) = \sum_0^N a_k D^k, \qquad D = \frac{d}{dt}, \qquad m^{(j)}(\cdot) = D^j m(\cdot)$$

and

$d_{kj} = d_{jk}$ = the elements of the inverse of the matrix[6]

$$\left[\frac{\partial^{k+j}R(t,s)}{\partial t^k\,\partial s^j}\right]_{t=s=0}$$

$$= 2\sum_{i=1}^{\min(j,N-k+1)} a_{j-i}a_{k+i-1}(-1)^{i+1}, \qquad j \le k,\ j+k \text{ even},$$

$$= 0, \qquad \text{otherwise}. \tag{25b}$$

Example 6—Rational Spectral Densities: The RKHS for stationary noise with a general rational spectral density can also be found, not so explicitly as above, but in terms of the inverse of a certain matrix [12]. If the difference in the degrees of the numerator and denominator polynomials is $2(N - M) > 0$, then $H(R)$ is the space of functions whose $(N - M)$th-derivatives are square integrable.

Example 7—Separable Covariances: The RKHS for a general (i.e., possibly nonstationary) "lumped-circuit" noise (i.e., one with covariance $R(t,s) = \sum_1^N u_i(t)v_i(s)$, $t \le s$, $N < \infty$) be expressed [24] in terms of the solution

[6] The formula (25) also holds, with certain obvious changes, for time-variant autoregressive processes.

of a Riccati differential equation. The advantages of such a result are that the solution can be obtained in a "recursive" form and that a lot of effort, especially in control theory, has gone into studying the Riccati equation.

Example 8—Stationary Processes Over $(-\infty,\infty)$*:*

$$R(t) = \int_{-\infty}^{\infty} \exp(i2\pi\lambda t)\, d\mathcal{S}(\lambda), \qquad -\infty \le t \le \infty. \tag{26a}$$

$H(R,I)$ consists of functions $m(\cdot)$ of the form

$$m(t) = \int_{-\infty}^{\infty} \exp(i2\pi\lambda t) a(\lambda)\, d\mathcal{S}(\lambda),$$

$$\text{with } \|m\|_H^2 = \int_{-\infty}^{\infty} |a(\lambda)|^2\, d\mathcal{S}(\lambda) < \infty. \tag{26b}$$

If the process has a power spectral density function that is uniformly bounded,

$$S(\lambda) = \frac{d\mathcal{S}(\lambda)}{d\lambda} < M < \infty,$$

then $H(R,I)$ can be described as consisting of all square-integrable functions $m(\cdot)$ whose Fourier transforms vanish wherever $S(\lambda)$ does and for which

$$\|m\|^2 = \int_{\Lambda} [|M(\lambda)|^2/S(\lambda)]\, d\lambda < \infty,$$

$$M(\lambda) = \int_{-\infty}^{\infty} m(t) \exp(-i2\pi\lambda t)\, dt, \tag{26c}$$

where

$$\Lambda = \{\lambda: S(\lambda) > 0\}.$$

Example 9—Discrete-Time Processes: Let

$$I = \{1,2,\cdots,N\}$$

and let $\boldsymbol{R}$ be a covariance matrix on $I \times I$. Then $H(\boldsymbol{R})$ consists of N-dimensional vectors with norm

$$\|\boldsymbol{m}\|_H^2 = \boldsymbol{m}'\boldsymbol{R}^{\#}\boldsymbol{m}, \tag{27}$$

where $\boldsymbol{R}^{\#}$ is the so-called Moore–Penrose pseudoinverse of $\boldsymbol{R}$, i.e., $\boldsymbol{R}^{\#}m$ is the solution of $\boldsymbol{Ra} = \boldsymbol{m}$ that has minimum length $\boldsymbol{a}'\boldsymbol{a}$, or, equivalently, that is orthogonal to the null-space of $\boldsymbol{R}$ (see, e.g., [25]). See also the discussion following (28d).

Example 10—General Class of Nonstationary Processes: Suppose that there is a measure space (Ω,β,μ) such that $\{f(t,\cdot),\, t \in I\}$ is in $L_2(\Omega,\beta,\mu)$, i.e., such that

$$\|f\|_{L_2}^2 = \int_{\Omega} |f(t,\lambda)|^2 \mu(d\lambda) < \infty, \qquad t \in I. \tag{28a}$$

Then if

$$R(t,s) = \int_{\Omega} f(t,\lambda) f^*(s,\lambda) \mu(d\lambda), \qquad (t,s) \in I \times I, \tag{28b}$$

the RKHS consists of elements of the form

$$m(t) = \int_{\Omega} f(t,\lambda) a(\lambda) \mu(d\lambda), \tag{28c}$$

where $a(\cdot)$ belongs to $L_2\{f(t,\cdot),\, t \in I\}$, i.e., to the Hilbert subspace of functions that are limits in the norm (28a) of finite linear combinations of the form $\sum c_i f(t_i,\cdot)$. The norm of $m(\cdot)$ is [7]

$$\|m\|_R^2 = \int |a(\lambda)|^2 \mu(d\lambda). \tag{28d}$$

The function $a(\cdot)$ can be described in another useful way. Consider (28c) without any initial restrictions on the solution $a(\cdot)$, except for square integrability with respect to $\mu(d\lambda)$. The solution will clearly be unique if and only if the functions $\{f(t,\cdot),\, t \in I\}$ span $L_2(Q,\beta,\mu)$. If this is not so, what must be done is to choose the solution $a(\cdot)$ that has minimum norm $\int |a(\lambda)|^2 \mu(d\lambda)$. In other words, any portion that is in the nullspace of the $\{f(t,\cdot),\, t \in I\}$ must not be included in the solution $a(\lambda)$. Doing this is the same as requiring $a(\cdot)$ to lie in $L_2\{f(t,\cdot),\, t \in I\}$, which is the requirement originally given.

Several other useful and general examples of RKHS are given, sometimes implicitly in Section III [cf. the discussions encompassed by (37)–(54)] and Section VI.

It is useful to remark that various forms of $\|m\|_H^2$ can be found by integration by parts, assuming $m(\cdot)$ is adequately differentiable. For example, the norm in Example 3 can be rewritten as

$$\|m\|_H^2 = \int_a^b [-\ddot{m}(t) + \alpha^2 m(t)] m(t)\, dt + m(a)[\alpha m(a) - \dot{m}(a)] + m(b)[\alpha m(b) + \dot{m}(b)] \tag{29a}$$

or as

$$\|m\|_H^2 = \int_a^b \dot{m}^2(t)\, dt + \alpha^2 \int_a^b m^2(t)\, dt + \alpha[m^2(a) + m^2(b)]. \tag{29b}$$

D. Computation of LR From (18)

The use of (18) is illustrated by considering a simple, but quite representative example. Let

$$R(t,s) = \frac{1}{2\alpha} e^{-\alpha|t-s|}, \qquad 0 \le (t,s) \le T.$$

The norm $\|m\|_H^2$ is given in Example 3 and by (12) the inner product can be calculated as

$$\langle x,m\rangle_H = \tfrac{1}{4}\{\|x + m\|_H^2 - \|x - m\|_H^2\} = \int_0^T (\dot{m} + \alpha m)(\dot{x} + \alpha x)\, dt + 2\alpha m(0)x(0). \tag{30}$$

However, when $x(\cdot)$ has the given covariance, $\dot{x} + \alpha x$ will contain white Gaussian noise and the above integral will not be well defined. But this integral can be given meaning by writing it as a so-called Wiener stochastic

[7] This result follows from Theorem 9 (Section IV).

integral (cf. Doob [23, ch. 9]), i.e., by setting, with $\phi = \dot{m} + \alpha m$, $\dot{v} = \dot{x} + \alpha x$

$$\int_0^T \phi(t)\dot{v}(t)\,dt \triangleq \int_0^T \phi(t)\,dv(t)$$

$$\triangleq \operatorname{l.i.m.}_{|\Delta t_i| \to 0} \sum \phi(t_i)[v(t_i + \Delta t_i) - v(t_i)]. \quad (31)$$

This integral is well defined for square-integrable $\phi(\cdot)$, a condition that is met in our case if detection is nonsingular, i.e., $\|m\|_H^2 < \infty$. Now it should be directly checked that $\langle x,m \rangle_H$ with the interpretation (31) is in fact equal to Λ. This can be done, as will be shown in Section III, by verifying that

$$E_0[x(s)\langle x,m \rangle_H] = m(s), \qquad 0 \le s \le T.$$

The verification is straightforward and will be omitted.

E. Comparison With Usual Solution

In the formula (30) for Λ, the signal $m(\cdot)$ is required to have only one square-integrable derivative. If $m(\cdot)$ has two square-integrable derivatives, we can integrate by parts [cf. (29a)] to write Λ as

$$\Lambda = \int_0^T x(t)[\alpha^2 m(t) - \ddot{m}(t)]\,dt + x(0)[\alpha m(0) - \dot{m}(0)] + x(T)[\alpha m(T) + \dot{m}(T)]. \quad (32)$$

This is the formula for Λ that is directly given by the usual integral-equation approach that expresses it as

$$\Lambda = \int_0^T x(t)a(t)\,dt,$$

$$\int_0^T R(t,s)a(s)\,ds = m(t), \qquad 0 \le t \le T. \quad (33)$$

The form (32) needs $m(\cdot)$ to be smoother than required for nonsingular detection, but it has the advantage of not requiring a Wiener stochastic integral. However, in many problems the advantage may be only apparent. First, note that with electronic equipment, it is easier to generate $(\dot{m} + \alpha m)$ from $m(\cdot)$ than it is to generate $\ddot{m}(\cdot)$. Second, the fact that $(\dot{x} + \alpha x)$ is white noise is not a source of difficulty because what is needed is the integral functional $\int_0^T [\dot{x}(t) + \alpha x(t)]\phi(t)\,dt$; various methods are known (see Wong and Zakai [26] and Kailath [27]) for approximating even more complicated stochastic integrals where $\phi(\cdot)$ may be random. It may be noted in passing that the same problem has to be faced and is successfully resolved in practice in the implementation of the correlator (or matched filter) receiver for deterministic signals in additive white Gaussian noise. Similar arguments apply in more complicated examples.

Thus it appears that not only do the symmetrical forms for Λ, directly given by use of the RKHS inner product, require fewer assumptions on the signal, but in practice they may be more easily and more directly implementable than nonsymmetrical forms. This last point was suggested to the author by Dr. I. Selin.

F. Integral Equations and RKHS Norms

By comparison of the formulas (11) and (18) for the LR, it would seem to be possible to deduce RKHS norms from the solution of the integral equation and vice versa. For the forward implication, note that

$$\langle x,m \rangle_H = \int_I x(t)a(t)\,dt,$$

where $a(\cdot)$ is the solution of the integral equation

$$\int_I R(t,s)a(s)\,ds = m(t), \qquad t \in I. \quad (10)$$

Conversely, to find $a(\cdot)$ from $\|m\|_H^2$, use the following procedure. Reduce, if possible, any given expression for $\langle q,m \rangle_H$ to a form in which only $q(\cdot)$, but no derivatives and no isolated sample values of $q(\cdot)$ appear. Such a reduction can often be effected by integration by parts and use of the formula

$$q^{(k)}(a) = \int_{-\infty}^{\infty} q(t)\delta^{(k)}(a - t)\,dt, \quad (34)$$

where the superscript denotes differentiation. Now set $q(t) = \delta(t)$ in the reduced form of $\langle q,m \rangle_H$; the result will be a unique solution $a(\cdot)$ to the integral equation, i.e., a solution such that the formal expression $\int_I m(t)a(t)\,dt$ equals the unambiguously defined quantity $\|m\|_H^2$. This is a uniquely defined solution that is appropriate to the detection problem.

For historical reasons, for many covariance functions, the integral-equation solutions were the first to be determined. However, there are some cases in which more explicit RKHS norms were determined first, e.g., the general formulas for the triangular, autoregressive, and separable covariances of Examples 4, 5, and 7. The method used to find the RKHS norms in these examples was basically the so-called innovations representation or whitening-filter method; see Theorem 9 (Section IV). Furthermore, to anticipate a bit, it will be shown in Sections IV and V that the various special properties of RKHS also yield several other ways of computing/approximating RKHS norms. As noted earlier, no comparable general theory apparently exists for Fredholm integral equations of the form (10).

Hopefully, the several detailed comparisons in this section have brought out some of the relationships between the RKHS method and the more familiar integral-equation method. In the rest of this paper, some of the deeper implications of the RKHS method are explored.

III. Direct RKHS Formulation

In the last section the RKHS formula for the LR was obtained somewhat indirectly, via the KL expansion, and it may be asked whether a more direct derivation is possible.

It was also remarked earlier that the presence of inner products and norms in the RKHS formula suggested the possibility of a geometric formulation of the problem. In this section a direct geometric derivation is given by using a projection method, well known for the white-noise case [28], [29, sect. 4.3], to reduce the detection problem (1) to an equivalent one-dimensional problem.

The idea of the method is to project the functions $x(\cdot)$, $m(\cdot)$, and $n(\cdot)$ onto the one-dimensional linear space spanned by the signal in such a way that the components of $n(\cdot)$ and $x(\cdot)$ that are orthogonal to the signal space are "irrelevant" [29] to the detection problem. For white noise, which has a delta-function covariance, such a projection can be obtained by using L_2 inner products

$$\langle f(\cdot),g(\cdot)\rangle_{L_2} = \int_I f(t)g(t)\,dt.$$

Note that formally the Hilbert space $L_2(I)$ has the delta-function as a reproducing kernel; "formally" because the delta-function does not belong to $L_2(I)$, so that $L_2(I)$ is not an RKHS, but it may be regarded as a "pseudo-RKHS."[8]

The white-noise solution suggests that for colored noise with covariance $R(\cdot,\cdot)$, the projection should be done using RKHS inner-products. By this means, the given detection problem,

$$h_1: x(t) = m(t) + n(t), \qquad h_0: x(t) = n(t), \qquad t \in I \quad (35)$$

can be reduced to the one-dimensional problem

$$h_1: x_1 = m_1 + n_1, \qquad h_0: x_1 = n_1, \quad (36)$$

where x_1, m_1, n_1 are the projections

$$x_1 = \frac{\langle x,m\rangle_H}{\|m\|_H^2}, \qquad n_1 = \frac{\langle n,m\rangle_H}{\|m\|_H^2}, \qquad m_1 = \frac{\langle m,m\rangle_H}{\|m\|_H^2} = 1. \quad (37)$$

By linearity, the noise random variable n_1 will be Gaussian with mean zero and variance[9]

$$En_1^2 = \|m\|_H^{-4}E\langle n,m\rangle_H^2$$
$$= \|m\|_H^{-4}\langle m,\langle R,m\rangle_H\rangle_H = \|m\|_H^{-2}. \quad (38)$$

The LR for the one-dimensional problem (36) is

$$\mathrm{LR} = \exp\left\{\frac{x_1m_1}{En_1^2} - \frac{1}{2}\frac{m_1^2}{En_1^2}\right\} = \exp\{\langle x,m\rangle_H - \tfrac{1}{2}\|m\|_H^2\}, \quad (39)$$

which was the formal RKHS expression quoted earlier in (18). As noted there, the mathematically correct counterpart of this formula is (15). Now a rigorous derivation of (15) by the projection method is given in [10]. Only the key step is given here, since the chief aim is to explain why the RKHS enters so intimately into the detection problem.

The key step is to represent the noise $n(\cdot)$ as

$$n(\cdot) = n_1m(\cdot) + n_{\mathrm{rem}}(\cdot), \; E[n_1n_{\mathrm{rem}}(\cdot)] \equiv 0,$$

where n_1 is to be obtained by linear operations on $n(\cdot)$, in order to preserve Gaussianness. The variables x_1 and m_1 are then to be obtained by applying to $x(\cdot)$ and $m(\cdot)$, respectively, the same operations as were used to obtain n_1 from $n(\cdot)$. If all this can be done, the components $n_{\mathrm{rem}}(\cdot)$ and $x_{\mathrm{rem}}(\cdot)$ will be seen to be irrelevant to the decision between h_1 and h_0 and can be ignored. Therefore, the basic problem is the determination of n_1.

Now it is shown in [10], [11] and can also be verified directly that if a random variable u is defined as the solution of the equation

$$E[n(t)u] = m(t), \qquad t \in I, \quad (40)$$

where u is restricted to being a linear functional of the $n(t)$, $t \in I$, then n_1 can be obtained as[10]

$$n_1 = u/E|u|^2. \quad (41)$$

A solution of (40) will be expected to be possible only if the function $m(\cdot)$ is related in some way to the noise-covariance function $R(t,s)$. It is a striking fact that the relationship is the following.

A solution of $E[n(t)u] = m(t)$, $t \in I$, that is a finite-variance linear functional of $\{n(t), t \in I\}$ will exist if and only if $m(\cdot)$ belongs to the RKHS $H(R,I)$ of $\{n(t), t \in I\}$, i.e., if and only if the detection problem is nonsingular. (42)

Proof of (42): Consider the Hilbert space, say $LL_2(n)$, consisting of the random variables $\{n(t), t \in I\}$, all finite linear combinations of these variables, and all their limits in the mean under the norm

$$\|u\|^2 = E|u|^2.$$

The symbol $LL_2(n)$ may be read "linear" $L_2(n)$. In working with non-Gaussian processes and nonlinear operations on Gaussian procpesses, the space $NL_2(n)$ that consists not only of finite linear sums and their limits in the covariance norm, but also of finite products and their limits, is needed. Now the relation (40) immediately defines a mapping in which the random variable u in $LL_2(n)$ is associated with the time function $m(\cdot)$. Call the space of all such time functions $\mathscr{H}$. It is easy to see that if, as is usual, elements in $LL_2(n)$ that differ by a variable of zero variance are taken to be identical, the mapping is one-to-one. Moreover, let the norm in $\mathscr{H}$ be defined in the natural way as

$$\|m\|^2 = Eu^2.$$

With this norm, the space $\mathscr{H}$ will be a Hilbert space because

[8] However, as briefly shown in Section VI, by using generalized functions (Schwartz distributions) a proper RKHS associated with white noise can also be obtained, but it will be an RKHS of functionals rather than functions.

[9] This formal calculation will soon be rigorously verified.

[10] The variance of u will be zero if and only if $m(t) \equiv 0$, $t \in I$; needless to say, this case will be excluded.

$LL_2(n)$ is a Hilbert space. In fact, it is an RKHS because 1) $R(\cdot,t) = En(\cdot)n(t)$ obviously is in $\mathscr{H}$ for every $t \in I$, and 2) if $m(\cdot) = En(\cdot)u$, then $\langle R(\cdot,t),m(\cdot)\rangle = En(t)u = m(t)$. Then by the uniqueness of the RKHS, $\mathscr{H}$ must be the same as $H(R,I)$, which completes the proof of (42).

It can be observed from the proof that

$$Eu^2 = \|m\|_H^2. \quad (43)$$

This fact, combined with (41), will show that

$$n_1 = u/\|m\|_H^2, \qquad u = n_1\|m\|_H^2 \quad (44)$$

and

$$En_1^2 = [\|m\|_H^2]^{-1}, \quad (45)$$

thus verifying the relation obtained formally in (38). It was pointed out earlier that x_1 is obtained by applying to $x(\cdot)$ the same operations applied to get n_1 from $n(\cdot)$. Therefore, if

$$\Lambda \triangleq x_i\|m\|^2, \quad (46)$$

it can be seen that it can be characterized as the unique finite-variance linear functional of $x(\cdot)$ (up to random variables of variance zero) that satisfies [cf. (44)]

$$E_0[x(t)\Lambda] = m(t), \qquad t \in I, \quad (47)$$

which was the relation quoted in Section II.

A. Geometric Interpretation of Singular Detection

The nonsingularity condition (19) also has a simple geometric significance. Note that if $\|m\|_H^2$ is infinite, i.e., $m(\cdot)$ does not belong to $H(R)$, then by (43) u will have infinite variance. This will imply by (45) that n_1 has zero variance, in which case it is obvious that the one-dimensional detection problem (36) will be singular.

It can be seen that the random variable u and the mapping (40) are fundamental to the basic detection problem. They should therefore now be examined in some detail.

B. Mapping Between $H(R,I)$ and $LL_2(n)$

Recall that $LL_2(n)$ is the Hilbert space of all finite-variance linear functionals of $n(\cdot)$, i.e., the Hilbert space containing all finite linear combinations of $\{n(t), t \in I\}$ and all their limits in the norm $\|u\|_{L_2(n)}^2 = E|u|^2$.

Denote by J the mapping that takes a random variable u in $LL_2(n)$ to a function $m(\cdot) \in H(R,I)$ by the relation

$$J: u \to m(\cdot), \qquad En(t)u = m(t), \qquad t \in I. \quad (40)$$

It is easy to see that this mapping is one-to-one (if, as always, random variables are identified whose difference has zero variance) and the proof of (42) shows that it is also isometric (or norm preserving), i.e.,

$$E|u|^2 = \|Ju\|_H^2 = \|m\|_H^2. \quad (48)$$

Therefore, there will exist an inverse map, say J^{-1}, that is also one-to-one and isometric. But given $m(\cdot)$, how can $J^{-1}m(\cdot)$ be evaluated? This is the basic detection problem, and, in fact, also a basic RKHS problem, because its solution will serve to determine the RKHS norm as

$$\|m\|_H^2 = E|J^{-1}m(\cdot)|^2. \quad (49)$$

C. Calculation of $J^{-1}m(\cdot)$

To begin with, it is easy to see that if

$$m(\cdot) = \sum_1^N c_i R(\cdot,t_i), \qquad N < \infty,$$

then

$$J^{-1}m(\cdot) = \sum_1^N c_i n(t_i), \quad (50)$$

and, in general, the trick is to find a representation for $m(\cdot)$ in terms of the $\{R(\cdot,t), t \in I\}$. Such a representation will always be possible at least in principle, because as will be shown in Theorem 1 (Section IV) the $\{R(\cdot,t), t \in I\}$ span $H(R,I)$. Following are some examples.

Example 11—Derivatives of R: If $m(\cdot) = \partial R(\cdot,t)/\partial t$, then $J^{-1}m(\cdot) = n^{(1)}(t)$ and

$$\|m\|^2 = E|n^{(1)}(t)|^2 = R_{11}(t,t), \quad (51)$$

where

$$R_{mn}(t,s) \triangleq \frac{\partial^{m+n}}{\partial x^m\,\partial y^n} R(x,y)\Big|_{\substack{x=t\\y=s}}$$

with $R_{00}(t,s) \triangleq R(t,s)$.

Example 12—Integral-Equation Representations: Suppose that there exist a square-integrable function $a_0(\cdot)$ and constants a_k such that

$$a(\cdot) = a_0(\cdot) + \sum_{k=1}^N a_k\delta^{(k)}(\cdot - t_k) \quad (52)$$

solves the integral equation

$$\int_I R(t,s)a(s)\,ds = m(t), \qquad t \in I. \quad (53)$$

Then

$$J^{-1}m(\cdot) = \int_I n(t)a_0(t) + \sum a_k n^{(k)}(t_k) \quad (54)$$

and

$$\|m\|_H^2 = E|J^{-1}m(\cdot)|^2 = \int_I\int_I R(t,s)a_0(t)a_0(s)\,dt\,ds + \sum_{k,l} a_k a_l R_{k,l}(t_k,t_l). \quad (55)$$

It should be almost needless to say by now that the validity of (54) is established by verifying that

$$E[n(t)\cdot J^{-1}m(\cdot)] = m(t), \qquad t \in I.$$

Kadota [22] has recently presented a generalization of the integral equation (53). He shows that if functions $\{a_{0k}(\cdot)\}$ and constants $\{a_{kl}\}$ can be found such that

$$\sum_{k=0}^N \left[\int_I \frac{\partial^k R(t,s)}{\partial s^k} a_{0k}(s)\,ds + \sum_{l=1}^M a_{kl}\frac{\partial^k R(t_l,s)}{\partial s^k}\right] = m(t), \qquad t \in I, \quad (56)$$

then the LR can be expressed in terms of the $\{a_{0k}(\cdot),a_{kl}\}$ and the derivatives of $x(\cdot)$.

The RKHS mapping shows almost by inspection that for $m(\cdot)$ as in (56),

$$J^{-1}m(\cdot) = \sum_{k=0}^{N} \left[\int_I n^{(k)}(t)a_{0k}(t)\,dt + \sum_{l=1}^{M} a_{kl}n^{(k)}(t_l) \right] \quad (57)$$

from which as noted earlier Λ and $\|m\|_H^2$ can be immediately determined and hence the LR. The RKHS approach also shows that the question of a unique solution to the integral equation (56) is not really relevant. As Parzen [9, p. 448] has stressed, the proper interpretation of (56) is that it is an explicit expression of $m(t)$ as a linear combination of $\{R(\cdot,t)\}$ and thereby immediately suggests what the random variable $v = J^{-1}m(\cdot)$ should be.

D. Indirect Representations

In some problems a direct representation of $m(\cdot)$ in terms of the $\{R(\cdot,t)\}$ may not be obtained, but this is not essential. A representation in terms of certain other functions, say $\{f(t,\lambda),\ \lambda \in \text{some space } \Omega\}$ may be obtained; and if the $J^{-1}f(\cdot,\ )$ are known, $J^{-1}m(\cdot)$ can be found. Following are some examples.

Example 13—Karhunen–Loève Representations: Suppose that

$$m(\cdot) = \sum m_i\psi_i(\cdot), \qquad \sum m_i^2/\lambda_i < \infty, \quad (58)$$

where the $\{m_i\}$ are the Fourier coefficients of $m(\cdot)$ with respect to the eigenfunctions $\{\psi_i(\cdot)\}$ of $R(\cdot,\cdot)$,

$$\int_I R(t,s)\psi_i(s)\,ds = \lambda_i\psi_i(t), \qquad t \in I.$$

It can now be shown fairly easily that

$$\psi_i(\cdot) \in H(R,I) \qquad \langle\psi_i,\psi_j\rangle_H = \delta_{ij}/\lambda_i.$$

Therefore, under the mapping J^{-1}

$$J^{-1}[\lambda_i\psi_i(\cdot)] = J^{-1}\left[\int_I R(\cdot,s)\psi_i(s)\,ds\right]$$
$$= \int_I n(s)\psi_i(s)\,ds = n_i \quad (59)$$

and as expected, because J^{-1} is isometric,

$$E[n_i^2/\lambda_i^2] = 1/\lambda_i = \|\psi_i\|_H^2.$$

Therefore, from (58)

$$J^{-1}m(\cdot) = \sum m_in_i/\lambda_i, \qquad E|J^{-1}m(\cdot)|^2 = \sum m_i^2/\lambda_i.$$

These formulas relate the usual KL approach to the RKHS approach.

E. Representation in Terms of RKHS Norms

In the previous examples, it was shown how certain representations of $m(\cdot)$ in terms of $\{R(\cdot,t),\ t \in I\}$ could be used to determine $J^{-1}m(\cdot)$ and also the RKHS norm. If the RKHS norm is already known, the representation will, of course, be written as

$$m(\cdot) = \langle m(*),R(*,\cdot)\rangle_H,$$

which would suggest that

$$u = J^{-1}m(\cdot) = J^{-1}\langle m(*),R(*,\cdot)\rangle_H$$
$$= \langle m(*),J^{-1}R(*,\cdot)\rangle_H$$
$$= \langle m(*),n(*)\rangle_H. \quad (60)$$

The difficulty with this representation is that, as noted in Section II, $n(\cdot)$ does not belong to $H(R)$. However, it is very suggestive and may be regarded as a mnemonic for guessing u; in all cases known to the author, some slight modifications to make the formal expression $\langle m,n\rangle_H$ mathematically well defined (usually just by writing stochastic integrals properly) will yield the proper result. As usual, this can be checked by seeing whether the u so obtained satisfies the equation $E[n(t)u] = m(t),\ t \in I$. An example of this technique was given in Section II [cf. (30), (31)]; the reader can readily work out several other examples using known RKHS norms. Here the fairly general case of

$$R(t,s) = \int_\Omega f(t,\lambda)f^*(s,\lambda)\mu(d\lambda),$$
$$\text{with } \int_\Omega |f(t,\lambda)|^2\mu(d\lambda) < \infty \quad (61)$$

is treated. Then, as shown in Example 10,

$$H(R,I) = \left\{m(\cdot)\colon m(t) = \int_\Omega f(t,\lambda)a(\lambda)\mu(d\lambda), \int_\Omega |a(\lambda)|^2\mu(d\lambda) < \infty\right\}$$

and (with obvious notation)

$$\langle m_1,m_2\rangle_H = \int_\Omega a_1(\lambda)a_2^*(\lambda)\mu(d\lambda). \quad (62)$$

For reasons explained in the discussion of Example 10, assume that the $\{f(t,\cdot),\ t \in I\}$ span $LL_2(\Omega,\beta,\mu)$. Now to apply the formula (60), write $n(\cdot)$ in the same form as $m(\cdot)$, i.e., to find a function, say $\varepsilon(\cdot)$, such that

$$n(t) = \int_\Omega f(t,\lambda)\varepsilon(\lambda)\mu(d\lambda). \quad (63)$$

Proceed heuristically and examine what properties such an $\varepsilon(\cdot)$ must have. From the requirement that the covariance of $n(\cdot)$ be as given in (61), the following must be obtained:

$$\int f(t,\lambda)f^*(s,\lambda)\mu(d\lambda) = En(t)n(s)$$
$$= \iint f(t,\lambda)f^*(s,\lambda') \cdot E\varepsilon(\lambda)\varepsilon^*(\lambda')\mu(d\lambda)\mu(d\lambda').$$

In order for this to hold, the $\varepsilon(\cdot)$ must satisfy

$$E\varepsilon(\lambda)\varepsilon^*(\lambda') = \delta(\lambda - \lambda'). \quad (64)$$

In other words, $\varepsilon(\cdot)$ must be a white-noise process. But now it should be clear how to make the above discussion more rigorous: the white noise should be integrated! A random orthogonal set function $z(\cdot)$ should be defined on the space (Ω,β,μ) that obeys the relation

$$E[z(B_1)z^*(B_2)] = \mu(B_1 \cap B_2), \qquad B_i \in \beta, \tag{65}$$

and then write $n(\cdot)$ as the stochastic integral

$$n(t) = \int_\Omega f(t,\lambda)z(d\lambda). \tag{66}$$

Then the formula for $J^{-1}m(\cdot)$ will be

$$J^{-1}m(\cdot) = \langle m(\cdot),n(\cdot)\rangle_H = \int_\Omega a(\lambda)z(d\lambda). \tag{67}$$

As was done for the Wiener process in Section II, it can be checked that this expression for $J^{-1}m(\cdot)$ is correct by verifying that $En(t)[J^{-1}m(\cdot)] = m(t)$, for every $t \in I$. Note that (65), (66) could have been obtained directly by appeal to the so-called Karhunen representation theorem (see [30, ch. I]), but the longer route has been taken for tutorial reasons.

The examples in this section have given some useful methods for calculating RKHS norms; other methods can be based on some of the special properties of RKHS. Some of these properties are examined in the next section.

IV. Some Properties of RKHS

In Sections II and III it was shown how the solution to the known-signal detection problem can be expressed in terms of RKHS inner products. In this section some of the special properties of RKHS are explored. These properties arise from the special nature of the RKHS norm, so first a few remarks on this point will be made.

A. Choice of Norm

The most familiar Hilbert-space norm is the L_2 norm, $\|m\|^2 = \int_I |m(t)|^2\, dt$. However, there are many other kinds of Hilbert-space norms, and the properties of the resulting Hilbert spaces can be widely different, even though an L_2 space can be found that is isomorphic to any other separable Hilbert space. The significance of the choice of norm is often overlooked because in finite-dimensional spaces all norms (e.g., $\boldsymbol{m'm}$ or $\boldsymbol{m'Am}$) are essentially equivalent and because it is thought that this continues to hold for infinite-dimensional spaces because of the sampling theorem. However, it does not, and the proper choice of norm can be very helpful for infinite-dimensional spaces. The finite- and infinite-dimensional situations differ, because the closure (or completeness) requirement is trivially met in the finite-dimensional case, but not in the infinite-dimensional case.

B. Some Properties of RKHS

A definitive paper by Aronszajn [13] discusses most of the important properties of RKHS. Some other properties are given in [9] and [12]. In the following a few of these properties are listed, chiefly those that are most relevant to known-signal detection. An elementary acquaintance with the basic definitions of Hilbert-space theory is assumed. The first six theorems are relatively elementary and give various characterizations of $H(R)$. Theorems 7–9 are among the most useful in applications. The theorems can be understood and used without knowing how they are derived. Some illustrations are given in Section V. However, in order to give some idea of the simplifications made possible by the reproducing property of the inner product (see also the remarks after Theorem 9), some of the really elementary proofs are given here.

Theorem 1—Denseness of $\{R(\cdot,t),\, t \in I\}$: $\{R(\cdot,t),\, t \in I\}$ spans $H(R,I)$.

Proof: The only vector orthogonal to $R(\cdot,t), t \in I$, is the zero vector. For if $m(\cdot)$ is such a vector

$$m(t) = \langle m(\cdot),R(\cdot,t)\rangle_H = 0, \qquad \text{all } t \in I.$$

Theorem 2—Pointwise and Norm Convergence: If $\{m_n\}$ is a Cauchy sequence in $H(R,I)$, i.e., if $\|m_n - m_m\|_H \to 0$ as $(n,m) \to \infty$, then the sequence $\{m_n\}$ is also pointwise convergent, i.e., $|m_n(t) - m_m(t)| \to 0$ as $(n,m) \to \infty$ for every $t \in I$. This convergence will be uniform on sets over which $R(t,t)$ is bounded.

Proof: By Schwarz's inequality,

$$\begin{aligned} |m_n(t) - m_m(t)| &= |\langle m_n(\cdot) - m_m(\cdot),R(\cdot,t)\rangle_H| \\ &\le \|m_n - m_m\|_H \|R(\cdot,t)\|_H \\ &= \|m_n - m_m\|_H \sqrt{R(t,t)} \to 0 \text{ as } n,m \to \infty. \end{aligned}$$

Theorem 1 shows that $H(R,I)$ can be regarded as the closure—under an inner product satisfying the reproducing property—of the linear space, say L, of functions of the form $\sum_1^N a_i R(\cdot,t_i)$, the $\{a_i\}$ being real numbers and $\{t_i\} \le I$. Theorem 2 shows that the closure can be obtained by taking pointwise limits of sequences in L rather than limits in the reproducing-kernel norm. This is a very strong property of RKHS, which sets them strikingly apart from general Hilbert spaces. It means, for example, that the elements in $H(R,I)$ are functions defined at every point in I; this is not true in $L_2(I)$, where the elements are not functions, but equivalence classes of functions, each class being formed by functions that are equal "almost everywhere." This property is reflected in Theorem 4 below, where it is shown that the functions in $H(R,I)$ have a certain "smoothness," that is, characteristic of the reproducing kernel $R(t,s)$.

Theorem 3—Alternative Characterizations of RKHS: If $H(R,I)$ is an RKHS with reproducing kernel $R(t,s)$, then for every $m(\cdot) \in H(R,I)$ and every $t \in I$, $|m(t)| \le M_t \cdot \|m\|_H$, where M_t is a finite constant, possibly dependent upon t. Conversely, suppose H is a Hilbert space of functions on I with $|m(t)| \le M_t\|m\|_H$ (with $M_t < \infty$) for every $m(\cdot) \in H$. Then there exists a reproducing kernel $R(t,s)$ such that $m(t) = \langle m(\cdot),R(\cdot,t)\rangle_H$ and H is identical to $H(R,I)$.

(Because of Theorem 3, an RKHS is often called a proper functional Hilbert space.)

Proof: For the first part, write

$$|m(t)| = |\langle m(\cdot),R(\cdot,t)\rangle_H| \le \sqrt{R(t,t)}\cdot\|m\|_H$$

and choose $M_t = R(t,t)$. The proof of the second part requires the Riesz representation theorem (see [13]).

Theorem 4—Smoothness Properties: If $R(t,s)$, $(t,s) \in I \times I$ is continuous on $I \times I$, then the functions in $H(R,I)$ are continuous on I. If the derivative $R_{mm}(t,t)$ exists for $(t,s) \in I \times I$, the functions in $H(R,I)$ are m-times differentiable.

Proof: For $m(\cdot) \in H(R,I)$ and $(t_1,t_2) \in I$,

$$\begin{aligned}|m(t_1) - m(t_2)| &\le |\langle m(\cdot),R(\cdot,t_1) - R(\cdot,t_2)\rangle_H| \\ &\le \|m\|_H\|R(\cdot,t_1) - R(\cdot,t_2)\|_H \\ &= \|m\|_H\{R(t_1,t_1) - 2R(t_1,t_2) + R(t_2,t_2)\}^{1/2} \\ &\to 0 \text{ as } t_1 \to t_2\end{aligned}$$

since $R(\cdot,\cdot)$ is continuous.

Therefore $m(\cdot) \in H(R,I)$ is continuous on I. Next note that

$$\begin{aligned}\dot m(t) &= \lim_{h\to 0}\frac{m(t+h) - m(t)}{h} \\ &= \lim_{h\to 0}\left\langle m(\cdot), \frac{R(\cdot,t+h) - R(\cdot,t)}{h}\right\rangle_H.\end{aligned}$$

But the sequence $\{R(\cdot,t+h) - R(\cdot,t)/h\}$ is readily shown[11] to converge to $R_{01}(t,t)$ if and only if $R_{11}(t,t)$ exists. Therefore, by the continuity of inner products,

$$\dot m(t) = \langle m(\cdot),R_{01}(\cdot,t)\rangle_H.$$

By an argument similar to that for $m(\cdot)$, it can be shown that $\dot m(\cdot)$ is continuous.

For the further smoothness properties, it is possible to proceed by induction, making use of the relation

$$\langle R_{01}(\cdot,t),R_{01}(\cdot,s)\rangle_H = R_{11}(t,s).$$

Theorem 5—Uniqueness of Kernel: If a Hilbert space has two reproducing kernels R and R', they must be identical:

$$\begin{aligned}\|R(\cdot,t) - R'(\cdot,t)\|_H^2 &= \langle R(\cdot,t) - R'(\cdot,t),R(\cdot,t)\rangle_H \\ &\quad -\langle R(\cdot,t) - R'(\cdot,t),R'(\cdot,t)\rangle_H \\ &= R(t,t) - R'(t,t) - R(t,t) + R'(t,t) \\ &= 0.\end{aligned}$$

Theorem 6—Uniqueness of Space: Two Hilbert spaces that have the same reproducing kernels must be identical.

Proof: By definition of the RKHS, for every $(t,s) \in I \times I$,

$$\langle R(\cdot,t),R(\cdot,s)\rangle_{H_1} = R(t,s) = \langle R(\cdot,t),R(\cdot,s)\rangle_{H_2}.$$

From this it can be shown [9, p. 263] that there exists an isometric mapping $\mathscr{L}$ from H_1 onto H_2 such that

$$\mathscr{L}[R(\cdot,s)] = R(\cdot,s).$$

Now if $\tilde m(\cdot)$ in H_2 corresponds under $\mathscr{L}$ to $m(\cdot)$ in H_1, then $m(t) = \tilde m(t)$ for every $t \in I$, because

$$\begin{aligned}m(t) &= \langle m(\cdot),R(\cdot,t)\rangle_{H_1} = \langle \tilde m(\cdot),R(\cdot,t)\rangle_{H_2} \\ &= \tilde m(t), \qquad t \in I.\end{aligned}$$

The next three theorems are among the most useful RKHS theorems.

Theorem 7—Restrictions of the Index Set: Let I' be a subset of I. Then $H(R,I')$ = restrictions to I' of all functions in $H(R,I)$ and if $m \in H(R,I')$, then

$$\|m'\|_{H(R,I')} = \min_m \|m\|_{H(R,I)},$$

where the minimum is taken over all "extensions" of $m'(\cdot)$, i.e., over all functions $m(\cdot) \in H(R,I)$ such that $m'(t) = m(t)$, $t \in I'$.

Proof: A simple proof based on the projection theorem is given in [13]. For variety, a proof is given here based on the mapping (40) between the RKHS and the space $LL_2(n)$. Thus suppose that $m'(\cdot)$ is a function in $H(R_{I'},I')$ and that its image in $LL_2\{n(t), t \in I'\}$ is u, i.e.,

$$E[n(t)u] = m'(t), \qquad t \in I'.$$

However, the obvious extension, say $m_0(\cdot)$, of $m'(\cdot)$ from I to I' is

$$m_0(t) = E[n(t)u], \qquad t \in I.$$

Moreover, $m_0(\cdot)$ clearly belongs to $H(R,I)$; in fact,

$$\|m_0\|_I^2 = Eu^2 = \|m'\|_{I'}.$$

Furthermore, if $m(\cdot)$ is any function in $H(R,I)$, then

$$m(t) = E[n(t)v], \qquad t \in I, \qquad v \in LL_2(n(t), t \in I)$$

and if $m(t) = m'$ for $t \in I'$, it is necessary to have

$$v = u + u^*, \qquad En(t)u^* = 0, \qquad t \in I'.$$

Therefore, u^* is orthogonal to the space $LL_2(n(t), t \in I')$ and, in particular, to the random variable u. Therefore,

$$\|m\|_I^2 = Ev^2 = Eu^2 + Eu^{*2} \ge Eu^2 = \|m'\|_{I'}^2.$$

The theorem is proved. Note that in detection-theoretical terms, the last inequality states the perhaps obvious fact that the signal detectability [cf. (43)] cannot be reduced by going to longer signaling intervals.

Theorem 8—Approximation Theorem: Let the parameter set I be either countable or a separable metric space. Let $\{I_N, N = 1,2,\cdots\}$ be a sequence of subsets of I that is 1) monotone increasing, i.e., $I_1 \subset I_2 \subset \cdots$ and such that 2) the union $\cup_1^\infty I_N$ is either equal to I (if I is countable) or dense in I (if I is a separable metric space). Let $H(R,I)$ be a Hilbert space of functions on I with reproducing kernel $R(\cdot,\cdot)$. Let $R_N(\cdot,\cdot)$ denote the restriction of $R(\cdot,\cdot)$ to I_N and let $H(R_N,I_N)$ denote the corresponding RKHS. Let $m(\cdot)$ be a function on I and let $m_N(\cdot)$ denote its restriction

[11] Apply the criterion that $x = \lim x_n$ iff $\langle x_n,x_m\rangle$ tends to a limit as $n,m \to \infty$. (Cf. Loève [31].)

to I_N. Then 1) if $m(\cdot) \in H(R,I)$,

$$\lim_{N\to\infty} \|m_N\|_{H(R_N,I_N)} = \|m\|_{H(R,I)}$$

and 2)

$$\lim_{N\to\infty} \|m_N\|_{H(R_N,I_N)} < \infty$$

if and only if $m(\cdot) \in H(R,I)$.

The proof requires some properties of wide-sense martingales and can be found in [9, pp. 316–319].

Theorem 9—General Representation Theorem: Suppose a random process $x(t)$, $t \in I$ can be obtained from a white-noise process $v(t)$, $t \in I$ by a linear operation A. Then, if $m \in H(R_x,I)$,

$$\|m\|_{H(R_x)}^2 = \|A^\# m\|_{L_2}^2 \triangleq \int_I |A^\# m|^2 \, dt,$$

where $A^\#$ denotes the Moore–Penrose pseudoinverse of the operator, i.e., $A^\# m$ is the solution u of $Au = m$ that has minimum $L_2(I)$ norm, or equivalently, the solution that is orthogonal to the null space of A.

The proof needs a careful consideration of the domain and range spaces of the operator A; it can be found in [12, Lemma 2.2].

Theorem 9 shows that RKHS norms can be calculated in terms of the known $L_2(I)$ norm; some special cases of this powerful theorem were developed in Sections II and III (see, especially, Example 10).

This is a good time to point out that, by use of these representation theorems, proofs of all RKHS results can be recast in terms of other spaces, e.g., $L_2(I)$, or the sequence space l_2, and so on. However, in so doing the coordinate-free spirit of the RKHS method would often be missing. The few proofs that have been given in this section should show that it is quite easy to work directly with the basic definition without the need for any concrete representation. Certainly, as the example of vector analysis versus Cartesian representations shows, such a claim can be made while still acknowledging the fact that coordinate representations can be very useful for specific calculations.

It should be noted that A is a particular kind of whitening filter for the process $x(\cdot)$. So Theorem 9 provides a rigorous formulation of the whitening-filter approach to the detection problem. It is stressed that not just any whitening filter can be used, a point that has been the source of some confusion in the literature (see [32], [33, pp. 71–74]). This difficulty is also what makes whitening-filter solutions unobvious for processes with "numerator dynamics" (see [12], [34]). Of course, if the original operator A is chosen carefully so that it has a unique inverse, this inverse will coincide with $A^\#$. Such a careful choice of A leads to what has been called the innovations representation or the canonical representation of the process $x(\cdot)$ (see, e.g., [35]), where a specific application is also made to Gaussian detection.

Finally, it is also important to note that there is no overwhelming reason to be confined exclusively to using a white-noise process, or equivalently, the $L_2(I)$ norm in Theorem 9. Any process $z(\cdot)$ with a known RKHS norm can be used instead of $v(\cdot)$, provided, of course, a linear mapping A is known from $z(\cdot)$ to $x(\cdot)$; in fact, such general mappings will be used in Part III.

C. Iterative Evaluation of RKHS Inner Products

In many problems, especially where the covariance function has been empirically determined in graphical or numerical form, it is convenient to have an iterative method of successively computing better and better approximations to the RKHS inner product. Several such schemes can be developed; in fact, in view of Theorem 9, various of the well-known steepest descent and conjugate-gradient methods used in $L_2(T)$ can be easily adapted to RKHS. For example, Parzen [9] has shown that the norm can be computed as

$$\|m\|_R^2 = \lim_{n\to\infty} \int h_n^2(t) \, dt,$$

where

$$h_{n+1}(t) = h_n(t) - \alpha_n \left[\int_I R(t,s) h_n(s) \, ds - m(t) \right]$$

and the $\{\alpha_n\}$ are a suitably chosen sequence of constants. Other algorithms are given by Weiner [36] and in the references therein.

This concludes our first set of RKHS theorems. Several others will be given in later papers. As noted before, some immediate applications of these theorems are to determining the RKHS norms for different covariance functions. In the next section some other applications are considered.

V. Some Further Applications to Deterministic Signal Problem

First the question of how the RKHS leads to simple proofs of certain results [37] on singular detection is discussed and how the RKHS gives a simple way of getting upper bounds on the signal detectability is pointed out [38]. Finally, the "stability" of the detection problem [39], [40] is discussed.

The naturalness of the results and their proofs should be contrasted with the original derivations.

A. Simple Conditions for Nonsingular Detection

According to (19), one condition for nonsingular detection is that $\sum_1^\infty m_{wi}^2/\lambda_{wi} < \infty$. This condition is not generally very useful since it requires computation of the eigenvalues and eigenfunctions of $R(t,s)$. In another paper [37] some more usable conditions for nonsingular detection have been derived in the KL framework. However, these results can also be obtained somewhat more directly by using the properties of the RKHS. As an example, the following is established.

The detection of $m(t)$, $t \in I$, is singular if and only if all

$$m_\infty(t), \quad -\infty < t < \infty, \quad \text{with } m_\infty(t) = m(t), \quad t \in I, \tag{68}$$

are singular over $(-\infty,\infty)$. Or equivalently, the detection of $m(t)$, $t \in I$, will be nonsingular if and only if there exists at least one extension that is nonsingular over $(-\infty,\infty)$.

The "only if" part of the first statement (or equivalently, the "if" part of the second statement) is obviously true (use contradiction!).

For a proof of the "if" part of the first statement, recall that, as shown in Sections II and III, nonsingular detection $\Leftrightarrow m(\cdot) \in H(R,I)$.

It has now been assumed that the covariance function $R(t,s)$, $(t,s) \in I \times I$ can be extended to a covariance function $R_\infty(t,s)$ defined over the whole plane $(-\infty < t,s < \infty)$. In many problems there will be a "natural" extension R_∞. Let $H(R_\infty)$ be the RKHS associated with R_∞. Then (68) is an easy consequence of Theorem 7, according to which if R and R_∞ are related as above, then the elements of $H(R,I)$, say m_I, are restrictions to I of functions m_∞ belonging to $H(R_\infty)$.

The result (68) will not be of much use unless it is easier to show that an extension m_∞ does or does not belong to $H(R_\infty)$ than it is to determine whether or not $m(t)$, $t \in I$, belongs to $H(R,I)$. Fortunately, as is shown in [37], in many cases it is possible to do so. For example, if the noise is stationary with bounded spectral density $S_\infty(f)$, (cf. Example 8), the norm in $H(R_\infty)$ involves only the comparatively easy calculation of Fourier transforms,

$$\|m_\infty\|_{H(R_\infty)}^2 = \int_{-\infty}^{\infty} [|M_\infty(f)|^2/S_\infty(f)]\, df.$$

A simple consequence of this fact is the result that detection in noise with a rational spectral density is nonsingular if and only if the signal has $(N - M)$ mean-square derivatives, where $2(N - M)$ is the difference in the degrees of the numerator and denominator polynomials of the spectral density function. Some other examples are given in [37].

B. Upper Bounds on Detectability

The restrictions theorem also immediately yields another piece of information that was not as obvious in the original KL derivation in [37]. This is that D^2, the signal detectability, can be obtained as

$$D^2 \triangleq \|m_I\|_{H(R,I)}^2 = \min_{m_\infty} \|m_\infty\|_{H(R_\infty)}^2, \tag{69}$$

where the minimum is taken over all extensions of m_I. As first noted in [38], (69) offers the possibility of finding simple upper bounds on $\|m_I\|_{H(R,I)}^2$, which, it may be recalled, defines the "detectability" of the signal $m_I(\cdot)$ in noise with covariance R; the bounds are provided by $\|m_\infty\|_{H(R_\infty)}^2$, where m_∞ is any extension of m_I. If the minimization in (69) can be carried out D^2 can be determined exactly,[12] but even if this cannot be done, a few "educated" trials should give a reasonably good estimate of D^2.

As far as we know, the only previously suggested technique for bounding D^2 was to use the "obvious" extension $\tilde{m}_\infty(t) = m(t)$, $t \in I$, $\tilde{m}_\infty(t) = 0$, $t \notin I$ (see Kelly *et al.* [43]). Then $\int_{-\infty}^{\infty} [|M_\infty(f)|^2/S_\infty(f)]\, df$ was used as an upper bound for D^2. This is a correct upper bound, but, unfortunately, in many cases it may be infinite. The reason is that this $m_\infty(\cdot)$ is often not sufficiently smooth to belong to $H(R_\infty)$. However, the results (68), (69) show that there is always a sufficiently smooth extension, and knowledge of the properties of functions in $H(R_\infty)$ enables one to make educated guesses as to a "good" extension $m_\infty(\cdot)$.

[12] Such methods have actually been used in certain cases (see Ylvisaker [41], Duttweiler [42, ch. 3]).

It should also be pointed out here that in (68), (69) it is not necessary to be restricted to extensions from a finite to an infinite interval. It is possible to extend to any interval I' that includes I. In particular if I is a finite set of points on $[0,T]$, I' can be another finite set on $[0,T]$ that includes every point of I. This version of (68), (69) can be used to give alternative proofs of several results in [43, appendix I], which in turn is closely related to [44]. To mention another application, consider the problem of showing that the detectability of an analytic (e.g., band-limited) signal in noise is the same for observations over any nonzero time interval. This fact seems to be more complicated to prove by KL methods.

C. Stability of Detection

Recently Root [39], [40] studied what may be called the stability of the detection problem (see also [52, p. 331]). The question is the following. It is thought that a detection problem is being solved with hypotheses

$$h_1: x(t) = m(t) + n(t), \qquad h_0: x(t) = n(t), \qquad t \in I, \tag{70}$$

where $m(\cdot)$ is deterministic and $n(\cdot)$ has covariance function $R(t,s)$. But, in fact, the actual detection problem is

$$h_1: x(t) = m_0(t) + n_0(t), \qquad h_0: x(t) = n_0(t), \qquad t \in I, \tag{71}$$

where $m_0(\cdot)$ is also deterministic and $n_0(\cdot)$ has covariance function $R_0(t,s)$. To stay with the deterministic signal problem, assume here that $R(\cdot,\cdot) = R_0(\cdot,\cdot)$. Now ask whether the incorrect assumptions on $m_0(\cdot)$ will cause the detectability D^2 to be much different from the true detectability D_0^2. If the change in the detectability, $D^2 - D_0^2$, tends to zero as $m(\cdot) \to m_0(\cdot)$ (the type of convergence will be specified later), the detection problem will be said to be *stable*. Clearly mathematical models that lead to *unstable* detection problems are not very satisfactory models of physical problems. To analyze the stability of the model, it is necessary to be more precise about the way in which $m(\cdot)$ approaches $m_0(\cdot)$. Let

$$m(\cdot) - m_0(\cdot) = e(\cdot), \qquad \int_I e^2(t)\, dt < \infty. \tag{72}$$

First take $m(\cdot) \to m_0(\cdot)$ in the L_2 sense, i.e., $m(\cdot) \to m_0(\cdot)$ if $\int_I e^2\, dt \to 0$. Root studied the stability of this problem by using the K–L expansion, in terms of which the true signal detectability is (in the notation of Section II)

$$D_0^2 = \sum m_{oi}^2/\lambda_i.$$

On the other hand, from the formula (8) for the LR, it can be seen that the nominal signal detectability will be

$$D^2 = \sum m_{oi} m_i/\lambda_i. \tag{73}$$

It can then be written as

$$D^2 - D_0^2 = \sum m_{oi}(m_i - m_{oi})/\lambda_i \leq \left(\sum m_{oi}^2/\lambda_i^2\right)^{1/2} \left[\int e^2(t)\, dt\right]^{1/2} \quad (74)$$

from which it is clear that if

$$\sum m_{oi}^2/\lambda_i^2 < \infty, \quad (75)$$

then $D^2 \to D_0^2$ as $\int_I e^2\, dt \to 0$ and therefore detection will be "L_2-stable." On the other hand, if (75) is not satisfied, then Root shows by using a theorem of Landau's [45, p. 1] that given any $\delta > 0$, values $\{e_i\}$ can be found such that

$$\sum e_i^2 \leq \delta, \qquad \text{but} \sum m_{oi}e_i/\lambda_i = \infty \quad (76)$$

so that in this case the detection is not L_2-stable. This discussion gives some physical content to the constraint (75), which is often imposed (see [4]) in the KL approach to the detection problem to ensure that the integral equation (10) has a square-integrable solution.

These results are elegant, but irrelevant to the detection problem, which will be nonsingular so long as $\sum m_{0i}^2/\lambda_i < \infty$. The stronger requirement (75) for L_2-stability of detection will often not be met in many simple problems, as the literature on the integral equation (10) will show, and all such problems would be L_2-unstable and apparently physically unsatisfactory. Fortunately, the situation is not so bleak, and it is not necessary to abandon the large class of L_2-unstable problems. The reason is that the constraint $\int e^2\, dt < \infty$ is too weak a requirement on the error $e(\cdot) = m(\cdot) - m_0(\cdot)$. Suppose the noise process is such that almost all its sample functions are continuous and that the true signal $m_0(\cdot)$ is not capable of singular detection in this noise (i.e., $\sum m_{0i}^2/\lambda_i < \infty$). Now if a discontinuous signal $e(\cdot)$ is added to $m_0(\cdot)$, then no matter how small the energy in $e(\cdot)$ is, the signal $m_0(\cdot) + e(\cdot)$ will be perfectly detectable in the given noise. The point is that the error $e(\cdot)$ must be as smooth as the original signal $m_0(\cdot)$, or more precisely it is required that if $m_0(\cdot)$ is not capable of singular detection in the given noise, then neither are $e(\cdot)$ and $m_0(\cdot) + e(\cdot)$. Therefore, it is required that

$$e(\cdot) \in H(R,I), \qquad \text{i.e.,} \sum_1^{\infty} e_i^2/\lambda_i < \infty. \quad (77)$$

Then instead of (74), it is possible to write

$$D^2 - D_0^2 = \sum m_{oi}e_i/\lambda_i \leq (\sum m_{oi}^2/\lambda_i)^{1/2}(\sum e_i^2/\lambda_i)^{1/2}. \quad (78)$$

Now it is clear that if $\sum m_{0i}^2/\lambda_i < \infty$, then $D^2 \to D_0^2$ if $\|e(\cdot)\|_H \to 0$.

It might seem possible to have unstable detection even while keeping $\sum e_i^2/\lambda_i < \infty$. It may be argued (cf. Root [40]) that assuming $\sum m_{0i}^2/\lambda_i^2 = \infty$, then by the theorem of Landau's a sequence $\{e_i, i = 1,\cdots,N, N < \infty\}$ can be found such that $\sum_1^N e_i^2/\lambda_i < \infty$ and $\sum_1^N e_i^2$ is arbitrarily small, but $\sum_1^N m_{0i}e_i/\lambda_i > B$ for any preassigned constant B. Since B can apparently be arbitrarily large, detection will still be unstable. However, B cannot be made arbitrarily large. A reasonable restriction on the nominal signal $m(\cdot)$ is that its detectability $\sum m_i^2/\lambda_i$ be finite, equal to A, say. If this is not true, it is not surprising, as noted earlier, that an arbitrarily large increase in detectability can be obtained. But

$$A > \sum m_i^2/\lambda_i = \sum m_{oi}^2/\lambda_i + \sum_1^N e_i^2/\lambda_i + \sum_1^N 2m_{oi}e_i/\lambda_i.$$

This means that the change in detectability $\sum_1^N m_{0i}e_i/\lambda_i$ must be bounded and cannot be made larger than any preassigned constant B. All that can be said is that the change in detectability can be of the order of $(A - D_0^2)$, but this is not surprising. The important conclusion from this discussion of stability is that it is not sufficient to require that $e(\cdot) \in L_2(I)$; for a proper detection problem it is required that $e(\cdot) \in H(R,I)$.

VI. General Parameter Sets

In the previous sections the parameter set I has been taken tacitly to be an interval of the real line, which means that it has been restricted to one-dimensional processes. However, there is no fundamental reason for such a restriction, and in this section some more general parameter sets are treated explicitly. First consider some examples of multidimensional processes, i.e., processes defined over sets in n-space. Then it is illustrated how vector (or multivariable) processes can be treated as a special case of multidimensional processes. Finally, it is shown how a rigorous RKHS theory can be obtained for generalized random processes (e.g., white noise and its derivatives) by taking the parameter set to be a space of testing functions. This generalization, though simple, is significant because the lack of continuity of the covariance function of white noise does not permit the rigorous application of many of the results presented in previous sections [e.g., the arguments in (1)–(9)].

There are other generalized processes besides white noise that are sometimes used as models for physical processes, e.g., those with power spectral densities that fall off at infinity as $|f|^{-u}$ with $0 \leq u \leq 1$. The RKHS of these processes is also described.

The point of our discussing processes on these more general parameter sets is, of course, that all the results on RKHS and on the detection of a deterministic signal in Gaussian noise will go over to these more general situations. For example, for a signal $m(\cdot)$ in noise with any of the covariance functions discussed below, the necessary and sufficient condition will again be that $m(\cdot)$ belong to the associated RKHS. Thus, in contrast to at least some of the previous literature, no separate discussion of the detection problem will be necessary for the various situations studied below.

A. Multidimensional Processes

Example 14—Cameron–Yeh Process: This process was introduced by Cameron and Yeh [46] in studying certain statistical procedures. It is a kind of two-dimensional Wiener process, defined as a Gaussian process $w(u,v)$,

$(u,v) \in I$, a set in 2-space with mean zero and with covariance function

$$R(u_1,v_1,u_2,v_2) = E\{w(u_1,v_1)w(u_2,v_2)\}$$
$$= (u_1 \wedge u_2)(v_1 \wedge v_2), \qquad (79)$$

where $a \wedge b = \min(a,b)$. For compactness of notation, use $P,Q,\cdots$ to define arbitrary points in the set I and define

$$P \wedge Q = (u_1 \wedge u_2)(v_1 \wedge v_2), \qquad P = (u_1,v_1), \qquad Q = (u_2,v_2).$$

Therefore, (79) can be written more compactly as

$$R(P,Q) = E[W(P)W(Q)] = P \wedge Q. \qquad (80)$$

Now the RKHS associated with the covariance is

$$H(R) = \left\{m(\cdot): m(P) = \int_0^P g(Q)\, dQ, \int_0^P g^2(Q)\, dQ < \infty\right\} \qquad (81)$$

and the RKHS norm is

$$\|m\|_R^2 = \int_I g^2(Q)\, dQ. \qquad (82)$$

This can be checked by verifying the reproducing property

$$\langle m(\cdot),R(\cdot,Q)\rangle_H = \langle m(\cdot,*),(\cdot \wedge u_2)(* \wedge v_2)\rangle_H$$
$$= \int \frac{\partial m(P)}{\partial P} \frac{\partial R(P,Q)}{\partial Q}\, dP$$
$$= \int_I \int_I 1(u_2 - u_1)1(v_2 - v_1) \cdot g(u_1,v_1)\, du_1\, dv_1$$
$$= m(u_2,v_2) = m(Q),$$

where $1(\cdot)$ denotes the Heaviside step function.

Example 15—Lévy's n-Dimensional Brownian Motion: Another important example of a multidimensional process is the n-dimensional Brownian motion in Lévy's sense [47]. This is a zero-mean process with covariance function

$$R(u,v) = \tfrac{1}{2}\{\|u\| + \|v\| - \|u - v\|\}, \qquad (83)$$

where $\|u\|$ = the Euclidean length of u. Such processes have many interesting properties [48]. For example, they are Markovian (in a suitable sense) only when the dimension n is odd. The RKHS consists of n-dimensional functions $m(\cdot)$ such that [49]

$$m(u) = \Omega_{n+1}^{-1/2} \int \cdots \int \|\lambda\|^{-(n+1)/2} \cdot [\exp(i\langle u,\lambda\rangle) - 1] f(\lambda)\, d\lambda, \qquad \int f^2(\lambda)\, d\lambda < \infty, \qquad (84)$$

where Ω_n = the area of the n-dimensional unit sphere. The RKHS norm is

$$\|m\|^2 = \begin{cases} (2\pi)^{-n}\Omega_{n+1} \displaystyle\int \|\Delta^k m(u)\|^2\, du, & n = 4k - 1, \quad (85a) \\ (2\pi)^{-n}\Omega_{n+1} \displaystyle\sum_{i=1}^{n} \int \left\|\frac{\partial}{\partial u_i} \Delta^k m(u)\right\|^2 du, & n = 4k + 1, \quad (85b) \end{cases}$$

where Δ = the n-dimensional Laplacian operator. The properties of this RKHS have been used by Molchan [49] to derive several properties of such processes, especially the Markovian properties noted above. Also note that the parameter set can be more general than Euclidean n-space [50], [51].

B. *Multivariable (or Vector) Processes*

These are processes of the form

$$x'(t) = \{x_1(t),\cdots,x_n(t)\}$$

and have been studied by many authors, using a multivariable generalization of the KL expansion (see [52, sect. 3.7]). However, by a very simple artifice, such processes can be rewritten as scalar multidimensional processes, in which form many properties, e.g., the generalized KL expansion, are obvious. All we have to do is define a new variable

$$\boldsymbol{t} = (t,i), \qquad t \in I,\ i \in \{1,2,\cdots,n\}$$

and replace $x(t)$ by $x(\boldsymbol{t})$. Operations in the $\boldsymbol{t}$-domain have to be suitably defined, but it is usually easy to see what to do. It will be simplest to explain things by some examples.

Example 16—Covariance With a Mercer Expansion: Suppose $0 \le t \le T$ and $i \in \{1,2,\cdots,n\}$; then if

$$R(\boldsymbol{t},\boldsymbol{s}) = Ex(\boldsymbol{t})x(\boldsymbol{s}), \quad \int_I \int_I R^2(\boldsymbol{t},\boldsymbol{s})\, d\boldsymbol{t}\, d\boldsymbol{s} < \infty, \qquad (86)$$

a Mercer's expansion can be obtained for $R(\cdot,\cdot)$ (cf. Zaanen [21] and footnote 2),

$$R(\boldsymbol{t},\boldsymbol{s}) = \sum_i \lambda_i \psi_i(\boldsymbol{t})\psi_i(\boldsymbol{s}), \qquad (87a)$$

where

$$\int_I R(\boldsymbol{t},\boldsymbol{s})\psi_i(\boldsymbol{s})\, d\boldsymbol{s} = \lambda_i \psi_i(\boldsymbol{t}), \qquad (87b)$$

and

$$\int_I \psi_i(\boldsymbol{t})\psi_j(\boldsymbol{t})\, d\boldsymbol{t} = \lambda_i \delta_{ij}. \qquad (87c)$$

Written out in full, (87) reads

$$R(t,k,s,l) = Ex_k(t)x_l(s) = \sum \lambda_i \psi_i(t,k)\psi_i(s,l),$$

where

$$\sum_l \int_0^T R(t,k,s,l)\psi_i(s,l)\, ds = \lambda_i \psi_i(t,k)$$

and

$$\sum_k \int_0^T \psi_i(t,k)\psi_j(t,k)\,dt = \lambda_i\,\delta_{ij}.$$

From the Mercer expansion (87) the KL expansion can be written

$$x(\boldsymbol{t}) = \sum x_i\psi_i(\boldsymbol{t}), \qquad x_i = \int_I x(\boldsymbol{t})\psi_i(\boldsymbol{t})\,d\boldsymbol{t}$$

and the RKHS consists of functions $m(\cdot)$ such that

$$m(\boldsymbol{t}) = \sum m_i\psi_i(\boldsymbol{t}) \tag{88a}$$

with finite RKHS norm

$$\|m\|_R^2 = \sum m_i^2/\lambda_i. \tag{88b}$$

Example 17—Stationary Vector Processes: Suppose that for $-\infty < t, s < \infty$, and $i,j = 1,2,\cdots,n$,

$$R_{ij}(t,s) = Ex_i(t)x_j(s) = \int_{-\infty}^{\infty} \exp\{i2\pi f(t-s)\}S_{ij}(f)\,df. \tag{89}$$

Then $H(R,I)$ consists of functions

$$m(\boldsymbol{t}) \text{ on } I = [\boldsymbol{t} = (t,i): -\infty < t < \infty, i \in \{1,\cdots,n\}]$$

satisfying

$$\int_I m^2(\boldsymbol{t})\,d\boldsymbol{t} = \sum_i \int_{-\infty}^{\infty} m_i^2(t)\,dt < \infty$$

and

$$\|m\|_R^2 = \int_{-\infty}^{\infty} \left[\sum_{i,j} M_i(f)S^{ij}(f)M_j^*(f)\,df < \infty\right], \tag{90}$$

where

$$M_i(f) = \int_{-\infty}^{\infty} m_i(t)\exp\{-i2\pi ft\}\,dt$$

and $[S^{ij}(f)]$ is the inverse of the matrix $[S_{ij}(f)]$.

Multivariable generalizations of all the RKHS described in Sections II and III can also be obtained.

C. RKHS for Generalized Random Processes

Consider a linear space of random variables defined on some probability space (Ω,B,P), e.g., the space $L_2(\Omega,B,P)$ of all zero-mean finite-variance random variables with norm equal to the variance. Then a second-order random process, say $\{x(t,\omega), t \in I\}$, is a mapping that associates with every $t \in I$ a random variable, $x_t(\omega)$ or $x(t,\omega)$ in $L_2(\Omega,B,P)$. All of the discussions in the previous sections have been for such processes and their specializations. White noise is clearly not a second-order process. To get a rigorous theory for white noise, it is necessary to go to the Schwartz–Gelfand–Itô theory of generalized random processes as, for example, discussed in [53, ch. 4]. A brief but excellent discussion is given in [54, appendix II]. A detailed account of this theory cannot be given here, but a rough outline[13] is given that should hopefully suffice to indicate how RKHS can be defined for generalized processes. Once this is done, many of the results of the previous sections can be immediately taken over to white noise and related processes. This will be in contrast to much of the literature where white noise is usually regarded as a singular case (see [56], [57]).

For generalized random processes, the index set I is taken to be a suitable space of "testing" functions $\{\phi(\cdot)\}$, e.g., the space of infinitely differentiable time functions that fall off at $\pm\infty$ faster than any polynomial. This space has a suitable topology, i.e., a notion of convergence, defined on it. A generalized random process on I is a continuous linear mapping that associates with each $\phi \in I$ a random variable $x\phi(\omega)$ or $x(\phi,\omega)$ in $L_2(\Omega,B,P)$. By continuity, it is meant that $x(\phi_i,\omega)$ converges in mean-square to $x(\phi,\omega)$ as ϕ_i tends to ϕ in the topology of I. A generalized random process on a suitable space I can be specified by giving the family of probability-distribution functions

$$F(a_1,\cdots,a_n) = P\{\omega: x_{\phi_1}(\omega) \le a_1,\cdots,x_{\phi_n}(\omega) \le a_n\}$$

for all finite collections $\{\phi_1,\cdots,\phi_n\}$. Certain processes can be described more simply. Thus a generalized process can be specified by giving its *mean-value functional*

$$\tilde{m}(\phi) = Ex(\phi,\omega) = \int x(\phi,\omega)\,dP(\omega), \qquad \phi \in I$$

and its *covariance functional*

$$R(\phi,\psi) = Ex(\phi,\omega)x(\psi,\omega), \qquad \phi,\psi \in I. \tag{91}$$

The space of all continuous linear functionals on I is denoted I': $\tilde{m}(\cdot)$, $R(\cdot,\cdot)$, and $x(\cdot,\omega)$ belong to I'.

A generalized random process is said to be defined on a time interval $[0,T]$ if $\phi_1(t) = \phi_2(t)$ for $0 \le t \le T$ implies $x(\phi_1,\omega) = x(\phi_2,\omega)$ for almost all ω. An important property of generalized processes is that they are infinitely differentiable. The derivative $x^{(1)}(\phi,\omega)$ of $x(\phi,\omega)$ is defined by the relation

$$x^{(1)}(\phi,\omega) = -x(\phi^{(1)},\omega). \tag{92}$$

Example 18—Second-Order Process: A second-order process $\{x(t,\omega), 0 \le t \le T\}$ with locally integrable sample functions can be regarded as a generalized random process by defining

$$x(\phi,\omega) = \int_0^T x(t,\omega)\phi(t)\,dt, \qquad \phi \in I. \tag{93}$$

We have

$$\tilde{m}(\phi) = \int_0^T E[x(t,\omega)]\phi(t)\,dt,$$

$$R(\phi,\psi) = \int_0^T \int_0^T Ex(t)x(s)\phi(t)\psi(s)\,dt\,ds. \tag{94}$$

[13] The discussion is intentionally not very precise, e.g., as to the various spaces of testing functions and their topologies. All omitted facts can be found in [53] and [55].

Example 19—White Noise: A generalized process $x(\phi,\omega)$ is said to be a white noise process on $[0,T]$ if

$$\tilde{m}(\phi) \equiv 0, \qquad R(\phi,\psi) = \int_0^T \phi(t)\psi(t)\,dt. \tag{95}$$

Note that this is consistent with the formal calculation

$$R(\phi,\psi) = \int_0^T \int_0^T E[x(t)x(s)]\phi(t)\psi(s)\,dt\,ds,$$

$$E[x(t)x(s)] = \delta(t-s), \tag{96}$$

which is why white noise can also be formally written as an ordinary process $x(t,\omega)$ (cf. Example 18).

Example 20—Differentiated White Noise: The derivative of a white-noise process is a generalized process with

$$\tilde{m}(\phi) = 0, \qquad R(\phi,\psi) = \int_0^T \phi^{(1)}(t)\psi^{(1)}(t)\,dt. \tag{97}$$

D. RKHS Associated With a Generalized Process

It is possible in the usual way to define the Hilbert space, say $LL_2(x)$, of random variables spanned by the variables $\{x(\phi,\omega),\ \phi \in I\}$. That is, $LL_2(x)$ consists of all finite linear combinations $\sum_1^N c_i x(\phi_i,\omega)$ and limits in the mean of such variables. The norm in LL_2 is, of course, the covariance norm. Now given a random variable, say u, in LL_2, define a functional on I' by

$$E[x(\phi,\omega)u] = m(\phi), \qquad \phi \in I. \tag{98}$$

The set of all such functionals can in turn be made into a Hilbert space H by using the norm

$$\|m\|_H^2 = E|u|^2. \tag{99}$$

It can be verified readily that this norm satisfies

1) $R(\cdot,\phi) \in H$, for all $\phi \in I$,

2) $\langle m(\cdot),R(\cdot,\phi)\rangle_H = m(\phi)$, for all $\phi \in I$ and $m(\cdot) \in H$.

Therefore, the space H is an RKHS of functionals whose reproducing kernel is the functional $R(\cdot,\cdot)$. This space will be denoted $H(R)$ and has, subject to the proper notational modifications, all the properties of RKHS as described in Sections III–V. For example, the analog of Theorem 2 is: Norm convergence of functionals in $H(R)$, i.e., $\|m_n(\phi) - m(\phi)\|_H \to 0$ as $n \to \infty$ implies pointwise convergence, i.e., $|m_n(\phi) - m(\phi)| \to 0$ as $n \to \infty$ for every $\phi \in I$.

We note that this property is not true for RKHS of functions unless the covariance function is continuous, which would exclude white noise. However, by going to the RKHS of functionals, white noise can be included in the theory.

Example 21—RKHS Associated With White Noise: The RKHS of white noise consists of functionals of the form

$$m(\phi) = \int_0^T m(t)\phi(t)\,dt \tag{100a}$$

with

$$\|m\|_H^2 = \int_0^T m^2(t)\,dt < \infty. \tag{100b}$$

Example 22—Detection of Signals in White Noise: Consider the detection problem

$$H_1: x(\phi) = m(\phi) + n(\phi), \qquad H_0: x(\phi) = n(\phi), \tag{101}$$

where $n(\phi)$ is a white Gaussian noise. The arguments of Section III can be carried out again and will yield the same results as for RKHS functions. Thus the detection problem will be nonsingular if

$$m(\phi) \in H(R), \qquad \text{i.e.,} \int_0^T m^2(t)\,dt < \infty, \tag{102}$$

which is of course a well-known fact, albeit often one derived not very rigorously. Similarly the LR can be written [cf. (17)]

$$\text{LR} = \exp(\Lambda - \tfrac{1}{2}\|m\|^2), \tag{103}$$

where Λ is a random variable in LL_2 that satisfies

$$E_0[x(\phi)\Lambda] = m(\phi). \tag{104}$$

To calculate Λ, first use the partly suggestive manipulations

$$m(\phi) = \int_0^T m(t)\phi(t)\,dt, \qquad x(\phi) = \int_0^T x(t)\phi(t)\,dt,$$

$$\Lambda = \langle m(\cdot),x(\cdot)\rangle_H = \int_0^T m(t)x(t)\,dt,$$

where $x(t)$ is the observed generalized process written as an ordinary random process. The resulting formula

$$\text{LR} = \exp\left\{\int_0^T m(t)x(t)\,dt - \tfrac{1}{2}\int_0^T m^2(t)\,dt\right\}$$

is the form in which the LR formula is usually written. To make the formula mathematically meaningful, $\int m(t)x(t)\,dt$ must be written as a Wiener stochastic integral, just as it had to be done in the example in Section II [cf. (30), (31)].

To conclude the discussion of RKHS for generalized random processes, an example will be given of a process with what is sometimes called an "asymptotic" power spectrum. First note that the covariance functional of a *stationary* generalized random process can be written

$$R(\phi,\psi) = \int_{-\infty}^{\infty} \Phi(f)\Psi^*(f)S(f)\,df, \tag{105}$$

where $\Phi(\cdot)$ and $\Psi(\cdot)$ are the Fourier transforms of $\phi(\cdot)$ and $\psi(\cdot)$, and $S(\cdot)$ is an even nonnegative function but, unlike that for an ordinary random process, one that may not be integrable. For example, with white noise, $S(f)$ is a constant. Another example will now be described.

Example 23—[58]: Let

$$S(f) = |f|^{\mu}, \qquad 0 \leq \mu < 1. \tag{106}$$

Then the RKHS consists of functionals of the form

$$m(\phi) = \int_0^T m(t)\phi(t)\,dt, \qquad \int_0^T |m(t)|^{2/(1-\mu)}\,dt < \infty \tag{107}$$

with

$$\|m\|_H^2 = \int_0^T |t^{\mu/2}D^{\mu/2}[m(t)t^{-\mu/2}]|^2\,dt, \tag{108}$$

where $D^{\alpha}[\cdot]$ is the operator of fractional differentiation of order α, i.e.,

$$D^{\alpha}[g] = \frac{1}{\Gamma(1-\alpha)}\frac{d}{dt}\int_0^t (t-s)^{-\alpha}g(s)\,ds.$$

In [58] Molchan and Golosov also give the RKHS for the case $-1 \leq u < 0$ and for processes with covariance functions of the form

$$R(t,s) = |t|^{1+\mu} + |s|^{1+\mu} - |t-s|^{1+\mu}$$

and for certain other processes.

VII. Concluding Remarks

Several examples of RKHS have been given and some of their properties have been noted. In this paper, some applications were made to the problem of detecting a nonrandom signal in Gaussian noise. The various examples of RKHS and the discussion in Section III of the isometric mapping between the space of random variables and the RKHS will be especially useful for further applications.

Acknowledgment

The author would like to acknowledge the friendly influence, stimulation, and encouragement of Prof. E. Parzen.

References

[1] L. A. Shepp, "Radon-Nikodym derivatives of Gaussian measures," *Ann. Math. Statist.*, vol. 37, Apr. 1966, pp. 321–354.

[2] J. Capon, "Hilbert space methods for detection theory and pattern recognition," *IEEE Trans. Inform. Theory*, vol. IT-11, Apr. 1965, pp. 247–259; see also *Ann. Math. Statist.*, vol. 35, June 1964, pp. 517–531.

[3] T. Kailath, "Likelihood ratios for Gaussian processes," *IEEE Trans. Inform. Theory*, vol. IT-16, May 1970, pp. 276–288.

[4] U. Grenander, "Stochastic processes and statistical inference," *Ark. Mat.*, vol. 1, 1950, pp. 195–277.

[5] K. Karhunen, "Über Lineare Methoden in der Wahrscheinlichkeitsrechnung," *Ann. Acad. Sci. Fenn., Ser. A1: Math. Phys.*, vol. 37, 1947, pp. 3–79.

[6] M. Loève, "Sur les fonctions aléatoires stationnaires du second ordre," *Rev. Sci.*, vol. 83, 1945, pp. 297–310; see also *C. R. Acad. Sci.* (Paris), vol. 220, 1945, p. 380, and vol. 222, 1946, p. 489.

[7] M. Kac and A. J. Siegert, "An explicit representation of a stationary Gaussian process," *Ann. Math. Statist.*, vol. 18, 1947, pp. 438–442.

[8] D. D. Kosambi, "Statistics in function space," *J. Indian Math. Soc.*, vol. 7, 1943, pp. 76–88.

[9] E. Parzen, "Statistical inference on time series by Hilbert spaces methods," Appl. Math. Statist. Lab., Stanford Univ., Stanford, Calif., Tech. Rep. 23, 1959. This and other papers by Parzen on RKHS are reprinted in E. Parzen, *Time Series Analysis Papers.* San Francisco: Holden-Day, 1967.

[10] J. Hájek, "On a simple linear model in Gaussian processes," *Trans. 2nd Prague Conf. Information Theory*, 1960, pp. 185–197.

[11] T. Kailath, "A projection method for signal detection in colored Gaussian noise," *IEEE Trans. Inform. Theory*, vol. IT-13, July 1967, pp. 441–447.

[12] J. Hájek, "On linear statistical problems in stochastic processes," *Czech. Math. J.*, vol. 12, Dec. 1962, pp. 404–444.

[13] N. Aronszajn, "Theory of reproducing kernels," *Trans. Amer. Math. Soc.*, vol. 63, May 1950, pp. 337–404.

[14] E. H. Moore, "On properly positive Hermitian matrices," *Bull. Amer. Math. Soc.*, vol. 23, 1916, p. 59; see also *Mem. Amer. Phil. Soc.*, pt. 1, 1935, and pt. 2, 1939.

[15] M. G. Krein, "Sur le problème du prolongement des fonctions hermitiennes positives et continues," *Dokl. Akad. Nauk USSR*, vol. 26, 1940, pp. 17–22.

[16] ——, "Hermitian-positive kernels on homogeneous spaces, I," *Ukrain. Mat. Z.*, vol. 1, 1949, pp. 64–98 (Trans.: *Amer. Math. Soc.*, ser. 2, vol. 34, 1963, pp. 109–168).

[17] a. A. Devinatz, "Integral representations of positive-definite functions, Pt. I," *Trans. Amer. Math. Soc.*, vol. 74, 1953, pp. 56–77. b. ——, "Integral representations of positive-definite functions, Pt. II," *Trans. Amer. Math. Soc.*, vol. 77, 1954, pp. 455–480; also Ph.D. dissertation, Harvard Univ., Boston, Mass., 1950.

[18] A. Ya. Povzner, "A class of Hilbert function spaces," *Dokl. Akad. Nauk USSR*, vol. 68, 1949, pp. 817–820; see also vol. 74, 1950, pp. 13–17.

[19] M. Loève, "Fonctions aléatoires du second ordre," suppl. to P. Lévy, *Processus Stochastiques et Mouvement Brownien.* Paris: Gauthier-Villars, 1948.

[20] E. Parzen, "Statistical inference on time series by RKHS methods," Dep. Statist., Stanford Univ., Stanford, Calif., Tech. Rep. 14, Jan. 1970; see also *Proc. 12th Bienn. Can. Math. Soc. Seminar.* Providence, R.I.: Amer. Math. Soc., to be published.

[21] A. Zaanen, *Linear Analysis.* New York: Interscience, 1953.

[22] T. T. Kadota, "Differentiation of Karhunen-Loève expansion and application to optimum reception of sure signals in noise," *IEEE Trans. Inform. Theory*, vol. IT-13, Apr. 1967, pp. 255–260.

[23] J. L. Doob, *Stochastic Processes.* New York: Wiley, 1953.

[24] T. Kailath, R. Geesey, and H. Weinert, "RKHS norms, Fredholm equations, innovation representations," submitted to *IEEE Trans. Inform. Theory.*

[25] L. A. Zadeh and C. A. Desoer, *Linear Systems—The State-Space Approach.* New York: McGraw-Hill, 1963.

[26] E. Wong and M. Zakai, "On the relation between ordinary and stochastic differential equations and applications to stochastic problems in control theory," in *Proc. 3rd IFAC Congr.*, London, 1966.

[27] T. Kailath, "A general likelihood-ratio formula for random signals in Gaussian noise," *IEEE Trans. Inform. Theory*, vol. IT-15, May 1969, pp. 350–361.

[28] A. V. Balakrishnan, "A contribution to the sphere-packing problem of communication theory," *J. Math. Anal. Appl.*, vol. 3, Dec. 1961, pp. 405–506.

[29] J. M. Wozencraft and I. M. Jacobs, *Principles of Communication Engineering.* New York: Wiley, 1965.

[30] U. Grenander and M. Rosenblatt, *Statistical Analysis of Stationary Time Series.* New York: Wiley, 1959.

[31] M. Loève, *Probability Theory*, 3rd ed. Princeton, N.J.: Van Nostrand, 1963.

[32] I. Selin, "The sequential estimation and detection of signals in normal noise, I," *Inform. Contr.*, vol. 7, 1964, pp. 512–534.

[33] ——, *Detection Theory.* Princeton, N.J.: Princeton Univ. Press, 1965.

[34] R. L. Pickholtz and R. R. Boorstyn, "A recursive approach to signal detection," *IEEE Trans. Inform. Theory*, vol. IT-14, May 1968, pp. 445–450.

[35] R. Geesey and T. Kailath, "Applications of canonical representations to estimation and detection in colored noise," in *Proc. 19th Polytech. Inst. Brooklyn Symp. Computer Processing in Communications.* Brooklyn, N.Y.: Polytechnic Press, 1969.

[36] H. Weiner, "The gradient iteration in time-series analysis," *SIAM J. Appl. Math.*, vol. 13, Dec. 1965, pp. 1096–1101.

[37] T. Kailath, "Some results on singular detection," *Inform. Contr.*, vol. 9, 1966, pp. 130–152.

[38] ——, "Some applications of reproducing kernel Hilbert spaces," in *Proc. 4th Allerton Conf. System Theory*, Allerton, Ill., Oct. 1965.

[39] W. Root, "Stability in signal detection problems," in *Proc. Symp. Applied Mathematics*, vol. 16. Providence, R.I., 1964, pp. 247–263.

[40] ——, "The detection of signals in Gaussian noise," in *Communication Theory*, A. V. Balakrishnan, Ed. New York: McGraw-Hill, 1968, ch. 4.

[41] N. D. Ylvisaker, "Lower bounds for minimum covariance matrices in time series regression problems," *Ann. Math. Statist.*, vol. 33, 1964, pp. 362–368.

[42] D. Duttweiler, "Hilbert space methods for detection and estima-

tion theory," Ph.D. dissertation, Dep. Elec. Eng., Stanford Univ., Stanford, Calif., June 1970.
[43] E. J. Kelly, I. S. Reed, and W. L. Root, "The detection of radar echoes in noise, Part I," *SIAM J. Appl. Math.*, vol. 8, Sept. 1960, pp. 309–341.
[44] J. Chover and J. Feldman, "On positive-definite integral kernels and a related quadratic form," *Trans. Amer. Math. Soc.*, vol. 89, 1958, pp. 92–99.
[45] S. Kacmarz and H. Steinhaus, *Theorie der Orthogonalreihen.* New York: Chelsea, 1951.
[46] J. Yeh, "Wiener measure in a space of functions of two variables," *Trans. Amer. Math. Soc.*, vol. 95, 1960, pp. 433–450; see also vol. 107, 1963, pp. 408–420.
[47] P. Lévy, "Le mouvement Brownien plane," *Amer. J. Math.*, vol. 62, 1940, pp. 487–550.
[48] H. McKean, Jr., "Brownian motion with a several-dimensional time," *Theory Prob. Appl.* (USSR), vol. 8, 1963, pp. 335–354.
[49] G. Molchan, "On some problems concerning Brownian motion in Lévy's sense," *Theory Prob. Appl.* (USSR), vol. 12, 1967, pp. 682–690.
[50] E. A. Morozova and N. N. Chentsov, "P. Lévy's random fields," *Theory Prob. Appl.* (USSR), vol. 1, 1968, pp. 153–156.
[51] A. V. Balakrishnan, "Estimation and detection theory for multiple stochastic processes," *J. Math. Anal. Appl.*, 1960, vol. 386–410.
[52] H. L. Van Trees, *Detection, Estimation and Modulation Theory*, Part 1. New York: Wiley, 1968.
[53] I. M. Gel'fand and N. Ya. Vilenkin, *Generalized Functions*, vol. 4. New York: Academic Press, 1964.
[54] A. M. Yaglom, *Theory of Stationary Random Processes.* Englewood Cliffs, N.J.: Prentice-Hall, 1962.
[55] T. Hida, *Stationary Stochastic Processes.* Princeton, N.J.: Princeton Univ. Press, 1970.
[56] W. Root, "An introduction to the theory of the detection of signals in noise," *Proc. IEEE*, vol. 58, May 1970, pp. 610–623.
[57] T. T. Kadota, "Examples of optimum detection of Gaussian signals and interpretations of white noise," *IEEE Trans. Inform. Theory*, vol. IT-14, Sept. 1968, pp. 725–734.
[58] G. Molchan and Yu. Golosov, "Gaussian stationary processes with asymptotic power spectrum," *Sov. Math. Dokl.*, vol. 10, 1969, pp. 134–137.

9

Reprinted from pages 155–169 of *Proceedings of the Symposium on Time Series Analysis*, M. Rosenblatt, ed., Wiley, New York, 1963, 516p.

Probability Density Functionals and Reproducing Kernel Hilbert Spaces*

Emanuel Parzen, Stanford University

ABSTRACT

The extraction, detection, and prediction of signals in the presence of noise are among the central problems of statistical communication theory. Over the past few years I have sought to develop an approach to those problems that would simultaneously apply to stationary or nonstationary, discrete parameter or continuous parameter, and univariate or multivariate time series and would distinguish between their statistical and analytical aspects. In particular, they would clarify the role played by various widely employed analytical techniques (such as the Wiener-Hopf equation and eigenfunction expansions).

In the development of this approach, two basic concepts are used: the notion of the probability density functional of a time series and the notion of a reproducing kernel Hilbert space. The purpose of this chapter is to sketch the relation between these concepts.

1. THE PROBABILITY DENSITY FUNCTIONAL OF A NORMAL TIME SERIES

Let $[S(t), t \in T]$ and $[N(t), t \in T]$ be time series, called, respectively, the signal process and the noise process. Let Ω be the space of all real-valued functions on T. Let P_N and P_{S+N} be probability measures defined on the measurable subsets B of Ω by

$$P_N[B] = \text{prob}\ \{[N(t),\, t \in T] \in B\} \tag{1}$$

$$P_{S+N}[B] = \text{prob}\ \{[S(t) + N(t),\, t \in T] \in B\}. \tag{2}$$

We are trying to determine, if it exists, a function p on Ω with the property that

$$P_{S+N}[B] = \int_B p\, dP_N. \tag{3}$$

* Prepared under contract Nonr 3440(00) for the Office of Naval Research. Reproduction in whole or in part is permitted for any purpose of the United States Government.

The function p may be called the *probability density functional* of P_{S+N} with respect to P_N in order to emphasize that its argument is a function $[X(t), t \in T]$. It is also denoted $p[X(t), t \in T]$ and called the probability density functional of the signal-plus-noise process

$$X(t) = S(t) + N(t), t \in T, \tag{4}$$

with respect to the noise process $[N(t), t \in T]$. The function p is often written symbolically as a derivative,

$$p = \frac{dP_{S+N}}{dP_N} \tag{5}$$

and called the Radon-Nikodym derivative of P_{S+N} with respect to P_N [see Halmos (1950), p. 329].

A necessary and sufficient condition that the probability density (5) exist is that P_{S+N} be *absolutely continuous* with respect to P_N in the sense that, for every measurable subset A of Ω,

$$P_N[A] = 0 \text{ implies } P_{S+N}[A] = 0. \tag{6}$$

In order that P_{S+N} be *not* absolutely continuous with respect to P_N, it is necessary and sufficient that there exist a set A such that

$$P_N[A] = 0 \text{ and } P_{S+N}[A] > 0. \tag{7}$$

The probability measures P_N and P_{S+N} are said to be *orthogonal* if there exists a set A such that

$$P_N[A] = 0 \text{ and } P_{S+N}[A] = 1. \tag{8}$$

We can regard (8) as an extreme case of being not absolutely continuous.

The notion of orthogonality derives its importance from detection theory (the theory of testing hypotheses). The simple hypotheses

$$H_0: X(\cdot) = N(\cdot)$$
$$H_1: X(\cdot) = S(\cdot) + N(\cdot)$$

are said to be *perfectly detectable* if there exists a set A such that

$$\begin{aligned} P_N[A] &= \text{prob}\ \{[X(t), t \in T] \in A | H_0\} = 0 \\ P_{S+N}[A] &= \text{prob}\ \{[X(t), t \in T] \in A | H_1\} = 1. \end{aligned} \tag{9}$$

Clearly, the hypotheses H_0 and H_1 are perfectly detectable if and only if P_N and P_{S+N} are orthogonal.

Given the probability measures P_N and P_{S+N}, the following questions arise:

1. Determine whether P_N and P_{S+N} are orthogonal.
2. Determine whether P_{S+N} is absolutely continuous with respect to P_N.
3. Determine the Radon-Nikodym derivative (5) if it exists.

To answer these questions, the natural way to proceed is to approximate the infinite dimensional case by finite dimensional cases. For any finite subset

$$T' = (t_1, \cdots, t_n) \text{ of } T \tag{10}$$

let $P_{N,T'}$ and $P_{S+N,T'}$ denote the probability distributions of $[X(t), t \in T']$ under P_N and P_{S+N}, respectively. *Assume that $P_{S+N,T'}$ is absolutely continuous with respect to $P_{N,T'}$*, with Radon-Nikodym derivative denoted

$$p_{T'} = \frac{dP_{S+N,T'}}{dP_{N,T'}}. \tag{11}$$

The *divergence* between P_{S+N} and P_N on the basis of having observed $[X(t), t \in T']$ is defined by

$$J_{T'} = E_{S+N}(\log p_{T'}) - E_N(\log p_{T'}). \tag{12}$$

Using the theory of martingales, it may be shown that

$$0 \leqslant J_{T'} \leqslant J_{T''} \quad \text{if} \quad T' \subset T''. \tag{13}$$

Consequently, the limit

$$J_T = \lim_{T' \to T} J_{T'} \tag{14}$$

exists and is finite or infinite. Further, it may be shown [see Hajek (1958)] that (*a*) if $J_T < \infty$, then P_{S+N} is absolutely continuous with respect to P_N and

$$p = \frac{dP_{S+N}}{dP_N} = \lim_{T' \to T} p_{T'}; \tag{15}$$

(*b*) if $J_T = \infty$, and both the time series $[N(t), t \in T]$ and $[S(t) + N(t), t \in T]$ are normal, then P_{S+N} and P_N are orthogonal.

We next apply these criteria under the following assumptions.

The noise process $[N(t), t \in T]$ is a normal process with zero means and covariance kernel

$$K(s, t) = E[N(s)\, N(t)], \tag{16}$$

which is positive definite in the sense that for every finite subset $T' = \{t_1, \cdots, t_n\}$ of T the covariance matrix

$$K_{T'} = [K(t_i, t_j)] = \begin{bmatrix} K(t_1, t_1) & \cdots & K(t_1, t_n) \\ \cdot & & \cdot \\ \cdot & & \cdot \\ \cdot & & \cdot \\ K(t_n, t_1) & \cdots & K(t_n, t_n) \end{bmatrix} \tag{17}$$

is nonsingular, with inverse matrix denoted

$$K_{T'}^{-1} = [K^{-1}(t_i, t_j)]. \tag{18}$$

(It should be noted that the assumption of positive definiteness is made only for mathematical convenience in the present exposition; it can be omitted.)

In regard to the signal process, two cases are of most interest:

1. *Sure signal case.* $[S(t), t \in T]$ is a nonrandom function.
2. *Stochastic signal case.* $[S(t), t \in T]$ is a normal time series, independent of the noise process, with zero means and positive definite covariance kernel

$$R(s, t) = E[S(s)\, S(t)]. \tag{19}$$

To employ the criterion (15), we first need to compute the divergence $J_{T'}$, defined by (12). In this section we consider the sure signal case; the stochastic signal case is considered in Section 3.

In the sure signal case

$$\log p_{T'} = (X, S)_{K,T'} - \tfrac{1}{2}(S, S)_{K,T'} \tag{20}$$

where we define for any functions f and g on T

$$(f, g)_{K,T'} = \sum_{s,t \in T'} f(s) K^{-1}(s, t)\, g(t). \tag{21}$$

Consequently

$$J_{T'} = E_{S+N}[(X, S)_{K,T'}] - E_N[(X, S)_{K,T'}] = (S, S)_{K,T'} \tag{22}$$

and

$$J_T < \infty \text{ if and only if } \lim_{T' \to T} (S, S)_{K,T'} < \infty. \tag{23}$$

In words, in the sure signal case, P_{S+N} is absolutely continuous with respect to P_N if and only if $(S, S)_{K,T'}$ approaches a limit as T' tends to T. Fortunately it is possible to characterize those functions $S(\cdot)$ that have this property. To do so, we introduce the notion of a reproducing kernel Hilbert space.

2. REPRODUCING KERNEL HILBERT SPACES

Let $K(s, t)$ be the covariance kernel of a time series $[X(t), t \in T]$. For each t in T, let $K(\cdot, t)$ be the function on T whose value at s in T is equal to $K(s, t)$. It may be shown [see Aronszajn (1950)] that there exists a unique Hilbert space, denoted $H(K; T)$, with the following properties:

1. The members of $H(K; T)$ are real-valued functions on T [if $K(s, t)$ were complex-valued, they would be complex-valued functions].
2. For every t in T

$$K(\cdot, t) \in H(K; T). \tag{I}$$

3. For every t in T and f in $H(K; T)$

$$f(t) = (f, K(\cdot, t))_{K,T}, \tag{II}$$

where the inner product between two functions f and g in $H(K; T)$ is written $(f, g)_{K,T}$.

Example 1. Suppose $T = (1, 2, \ldots, n)$ for some positive integer n and that the covariance kernel K is given by a symmetric positive definite matrix $[K_{ij}]$ with inverse $[K^{ij}]$. The corresponding reproducing kernel space $H(K; T)$

consists of all n-dimensional vectors $\mathbf{f} = (f_1, \cdots, f_n)$ with inner product

$$(\mathbf{f}, \mathbf{g})_{K,T} = \sum_{s,t=1}^{n} f_s K^{st} g_t. \tag{24}$$

To prove (24) we need only to verify that the reproducing property holds for $u = 1, \cdots, n$

$$(\mathbf{f}, K_{\cdot u})_{K,T} = \sum_{s,t=1}^{n} f_s K^{st} K_{tu} = \sum_{s=1}^{n} f_s\, \delta(s, u) = f_u.$$

The inner product may also be written as a ratio of determinants:

$$(\mathbf{f}, \mathbf{g})_{K,T} = - \begin{vmatrix} K_{11} & \cdots & K_{1n} & f_1 \\ \cdot & & \cdot & \cdot \\ \cdot & \cdots & \cdot & \cdot \\ \cdot & & \cdot & \cdot \\ K_{n1} & \cdots & K_{nn} & f_n \\ g_1 & \cdots & g_n & 0 \end{vmatrix} \div \begin{vmatrix} K_{11} & \cdots & K_{1n} \\ \cdot & & \cdot \\ \cdot & & \cdot \\ \cdot & & \cdot \\ K_{n1} & \cdots & K_{nn} \end{vmatrix}. \tag{25}$$

To prove (25), we again need only to verify the reproducing property. When the covariance matrix $\mathbf{K}$ is singular, we may define the corresponding reproducing kernel inner product in terms of the *pseudo-inverse* of the matrix $\mathbf{K}$.

Example 2. Let $T = [t\colon a \leqslant t \leqslant b]$ and let $[N(t), a \leqslant t \leqslant b]$ be the Wiener process; that is, it has independent increments and covariance function

$$K(s, t) = \sigma^2 \min (s, t) \tag{26}$$

for some parameter σ^2. Consider the Hilbert spaces $H(K; T)$ consisting of all functions f on $a \leqslant t \leqslant b$ of the form

$$f(t) = f(a) + \int^t f'(u)\, du \tag{27}$$

for some square integrable measurable function f' on $a \leqslant t \leqslant b$ [which can be called the L_2-derivative of f], with inner product defined by

$$(f, g)_{K,T} = \frac{1}{\sigma^2}\left[\frac{1}{a} f(a)\, g(a) + \int_a^b f'(u)\, g'(u)\, du\right]. \tag{28}$$

If we define

$$\begin{aligned} I_t^{(u)} &= 1 \quad \text{if} \quad a \leqslant u \leqslant t \\ &= 0 \quad \text{if} \quad t < u \leqslant b, \end{aligned} \tag{29}$$

we may rewrite (27):

$$f(t) = f(a) + \int_a^b f'(u)\, I_t(u)\, du.$$

Now the covariance kernel $K(s, t)$ may be represented as

$$K(s, t) = \sigma^2 a + \sigma^2 \int_a^b I_s(u)\, I_t(u)\, du.$$

Therefore, for each t in T, $K(\cdot, t)$ belongs to $H(K; T)$ with L^2 derivative

$$\frac{d}{ds} K(s, t) = \sigma^2 I_t(s).$$

Further,

$$(f, K(\cdot, t))_{K,T} = \frac{1}{\sigma^2}\left[\frac{1}{a} f(a)\sigma^2 a + \int_a^b f'(u)\sigma^2 I_t(u)\, du\right]$$

$$= f(a) + \int_a^t f'(u)\, du = f(t).$$

Thus we see that the reproducing kernel Hilbert space corresponding to the covariance kernel (26) consists of all L_2-differentiable functions on T with inner product given by (28).

The relevance of the theory of reproducing kernel Hilbert spaces to the theory of probability density functionals derives from the following fact: it may be shown (using martingale theory) that

$$\lim_{T' \to T} (S, S)_{K,T'} < \infty \quad \text{if and only if} \quad S \in H(K; T). \tag{30}$$

Further, if $S \in H(K; T)$, then

$$\lim_{T' \to T} (S, S)_{K,T'} = (S, S)_{K,T}. \tag{31}$$

It follows in the sure signal case that P_{S+N} is absolutely continuous with respect to P_N if and only if the signal function $[S(t), t \in T]$ belongs to the reproducing kernel Hilbert space $H(K; T)$ corresponding to the covariance kernel K of the noise process $[X(t), t \in T]$. If $S \in H(K; T)$, then the probability density functional is given by

$$p[X(t), t \in T] = \frac{dP_{S+N}}{dP_N} = \exp\left[(X, S)_{K,T} - \tfrac{1}{2}(S, S)_{K,T}\right] \tag{32}$$

where by $(X, S)_{K,T}$ we mean the limit (in the sense both of convergence with probability one and convergence in quadratic mean)

$$(X, S)_{K,T} = \lim_{T' \to T} (X, S)_{K,T'}. \tag{33}$$

It should be emphasized that although we use inner product notation to write $(X, S)_{K,T}$ this is not a true inner product between two elements in a Hilbert space, since the sample function $[X(t), t \in T]$ does not belong to $H(K)$; that is,

$$\lim_{T' \to T} (X, X)_{K,T'} \text{ is infinite with probability one.} \tag{34}$$

In practice, it will be clear how to define $(X, S)_{K,T}$ by suitably modifying the expression for the inner product between two functions in $H(K)$. Thus for the covariance kernel given by (26), instead of the expression

$$(X, S)_{K,T} = \frac{1}{\sigma^2}\left[\frac{1}{a} X(a)\, S(a) + \int_a^b S'(u)\, X'(u)\, du\right]$$

suggested by (28), we may show that

$$(X, S)_{K,T} = \frac{1}{\sigma^2}\left[\frac{1}{a} X(a)\, S(a) + \int_a^b S'(u)\, dX(u)\right].$$

There is a variety of ways in which one can determine whether a function S belongs to a reproducing kernel Hilbert space $H(K; T)$ and compute the norm $(S, S)_{K,T}$ and the random variable $(X, S)_{K,T}$. These are discussed elsewhere [see Parzen (1961)].

However, certain general principles deserve to be stated at this point.

Roughly speaking, a function $g(\cdot)$ belongs to a reproducing kernel Hilbert space $H(K; T)$ only if it is at least as "smooth" as the functions $K(\cdot, t)$, since every function g in $H(K; T)$ is either a linear combination

$$g(\cdot) = \sum_{i=1}^{n} c_i K(\cdot, t_i)$$

or a limit of such linear combinations. For example, if T is an interval and K is continuous on $T \otimes T$, then every function in $H(K; T)$ is continuous; if K is twice differentiable on $T \otimes T$, then every function in $H(K, T)$ is differentiable.

We are led to the following heuristic conclusion: *In order that a signal not be perfectly detectable in the presence of a noise, it is necessary and sufficient that the signal be as smooth as the noise.* In the case of a sure signal the signal is as smooth as the noise if and only if $S \in H(K; T)$, where K is the covariance kernel of the noise. In the case of stochastic signals the signal is as smooth as the noise if $S \in H(K; T)$ for almost all sample functions of the signal process: a rigorous formulation of this assertion is given in Section 3.

A basic tool in the analytical evaluation of a reproducing kernel inner product is provided by the following theorem.

Integral representation theorem

Let K be a covariance kernel. If (a) a measurable space $(Q, \mathbf{B}, \mu)$ exists and (b) in the Hilbert space of all $\mathbf{B}$-measurable real-valued functions on Q satisfying

$$(f, f)_\mu = \int_Q f^2\, d\mu < \infty \tag{35}$$

there exists a family $[f(t), t \in T]$ of functions satisfying

$$K(s, t) = (f(s), f(t))_\mu = \int_Q f(s)\, f(t)\, d\mu, \tag{36}$$

then the reproducing kernel Hilbert space $H(K; T)$ consists of all functions

g on T, which may be represented as

$$g(t) = \int_Q g^* f(t)\, d\mu, \tag{37}$$

for some unique function g^* in the Hilbert subspace $L[f(t), t \in T]$ of $L_2(Q, \mathbf{B}, \mu)$ spanned by the family of functions $[f(t), t \in T]$. The norm of g is given by

$$(g, g)_{K,T} = (g^*, g^*)_\mu \tag{38}$$

If $[f(t), t \in T]$ spans $L_2(Q, \mathbf{B}, \mu)$, then $X(t)$ may be represented as a stochastic integral with respect to an orthogonal random set function $[Z(B), B \in \mathbf{B}]$ with covariance kernel μ:

$$X(t) = \int_Q f(t)\, dZ \tag{39}$$

$$E[Z(B_1)\, Z(B_2)] = \mu(B_1 B_2). \tag{40}$$

Further

$$(X, g)_{K,T} = \int_Q g^*\, dZ. \tag{41}$$

As an immediate consequence of the integral representation theorem one obtains the following example.

Example 3. *Stationary noise process.* Let $T = [t: -\infty < t < \infty]$ and let $[X(t), -\infty < t < \infty]$ be a stationary time series with spectral density function $f(\omega)$ so that

$$K(s, t) = \int_{-\infty}^{\infty} e^{i\omega(s-t)} f(\omega)\, d\omega. \tag{42}$$

Then $H(K; T)$ consists of all functions g on T of the form

$$g(t) = \int_{-\infty}^{\infty} g^*(\omega) e^{i\omega t} f(\omega)\, d\omega \tag{43}$$

for which the norm

$$\|g\|_{K,T}^2 = \int_{-\infty}^{\infty} |g^*(\omega)|^2 f(\omega)\, d\omega \tag{44}$$

is finite. The corresponding random variable $(X, g)_{K,T}$ can be expressed in terms of the spectral representation of $X(\cdot)$. If

$$X(t) = \int_{-\infty}^{\infty} e^{it\omega}\, dZ(\omega), \tag{45}$$

then

$$(X, g)_{K,T} = \int_{-\infty}^{\infty} g^*(\omega)\, dZ(\omega). \tag{46}$$

Assume that the spectral density function $f(\omega)$ is uniformly bounded. Then $g(t)$, being the Fourier transform of the square integrable function $g^*(\omega) f(\omega)$, is square integrable. Let

$$G(\omega) = \frac{1}{2\pi} \int_{-\infty}^{\infty} e^{-it\omega} g(t)\, dt. \tag{47}$$

Then

$$g^*(\omega) f(\omega) = G(\omega). \tag{48}$$

Assume now that $f(\omega)$ never vanishes. We may then write

$$(g, g)_{K,T} = \int_{-\infty}^{\infty} \left| \frac{G(\omega)}{f(\omega)} \right|^2 f(\omega)\, d\omega = \int_{-\infty}^{\infty} |G(\omega)|^2 \frac{1}{f(\omega)}\, d\omega \tag{49}$$

$$(X, g)_{K,T} = \int_{-\infty}^{\infty} \frac{G(\omega)}{f(\omega)}\, dZ(\omega). \tag{50}$$

To sum up, in the case of a stationary process whose spectral density function is uniformly bounded and never vanishes, the reproducing kernel Hilbert space $H(K; T)$ for $T = (-\infty < t < \infty)$ consists of all space integrable functions $g(t)$ whose Fourier transforms $G(\omega)$ are such that

$$\int_{-\infty}^{\infty} |G(\omega)|^2 \frac{1}{f(\omega)}\, d\omega < \infty. \tag{51}$$

If $f(\omega)$ vanishes, a similar conclusion holds. Let $N = [\omega: f(\omega) = 0]$ and $N^c = [\omega: f(\omega) > 0]$. Then $H(K; T)$ consists of all square integrable functions $g(t)$ whose Fourier transforms $G(\omega)$ vanish on N and such that

$$\int_{N^c} |G(\omega)|^2 \frac{1}{f(\omega)}\, d\omega < \infty.$$

The foregoing results are easily extended to multiple time series $[X_\alpha(t), -\infty < t < \infty, \alpha = 1, 2, \cdots, M]$. Assume that for $\alpha, \beta = 1, 2, \cdots, M$ and $-\infty < s, t < \infty$,

$$K_{\alpha,\beta}(s, t) = E[X_\alpha(s)\, X_\beta(t)] = \int_{-\infty}^{\infty} e^{i\omega(s-t)} f_{\alpha,\beta}(\omega)\, d\omega. \tag{52}$$

Then $H(K; T)$ consists of all functions $g_\alpha(t)$ on

$$T = [(\alpha, t): \alpha = 1, \cdots, M, -\infty < t < \infty], \tag{53}$$

satisfying the condition

$$\sum_{\alpha=1}^{M} \int_{-\infty}^{\infty} g_\alpha^2(t)\, dt < \infty, \tag{54}$$

such that

$$\int_{-\infty}^{\infty} \left[\sum_{\alpha,\beta=1}^{M} G_\alpha(\omega)\, f^{\alpha\beta}(\omega)\, \overline{G_\beta(\omega)} \right] d\omega < \infty, \tag{55}$$

where $\bar{z}$ denotes the complex conjugate of the complex number z,

$$G_\alpha(\omega) = \frac{1}{2\pi} \int_{-\infty}^{\infty} e^{i\omega t}\, g_\alpha(t)\, dt, \tag{56}$$

and $[f^{\alpha\beta}(\omega)]$ is the inverse of the matrix $[f_{\alpha\beta}(\omega)]$. Then $(g, g)_{K,T}$ is given by the expression in (55) and

$$(X, g)_{K,T} = \sum_{\alpha,\beta=1}^{M} \int_{-\infty}^{\infty} \overline{G_\alpha(\omega)}\, f^{\alpha\beta}(\omega)\, dZ_\beta(\omega), \tag{57}$$

where

$$X_\alpha(t) = \int_{-\infty}^{\infty} e^{it\omega}\, dZ_\alpha(\omega). \tag{58}$$

Direct Product Hilbert Spaces

The notion of a direct product space plays an important part in our considerations. Given two function spaces G_1 and G_2, consisting of functions defined on T_1 and T_2, respectively, their direct product space, denoted $G_1 \otimes G_2$, is the Hilbert space completion of the set of functions g on $T_1 \otimes T_2$ of the form

$$g(t_1, t_2) = g_1(t_1)\, g_2(t_2), \tag{59}$$

where $g_1 \in G_1$ and $g_2 \in G_2$. The norm of a function in $G_1 \otimes G_2$ of the form (59) is defined by

$$\|g\|^2_{G_1\otimes G_2} = \|g_1\|^2_{G_1}\|g_2\|^2_{G_2}. \tag{60}$$

The function g defined by (59) is on occasion denoted by $g_1 \otimes g_2$.

It should be noted that if G_1 and G_2 are reproducing kernel Hilbert spaces, with respective reproducing kernels K_1 and K_2 defined on $T \otimes T$, then $G_1 \otimes G_2$ is a reproducing kernel Hilbert space with kernel $K_1 \otimes K_2$, where $K_1 \otimes K_2$ is a function of four real variables defined by

$$K_1 \otimes K_2(s_1, s_2, t_1, t_2) = K_1(s_1, t_1)\, K_2(s_2, t_2) \tag{61}$$

and

$$(g, K_1 \otimes K_2(\cdot, \cdot, t_1, t_2))_{G_1\otimes G_2} = g(t_1, t_2). \tag{62}$$

When $G_1 = G_2 = L_2(T, \mathbf{B}, \mu)$, $G_1 \otimes G_2$ consists of all ($\mathbf{B} \otimes \mathbf{B}$-measurable) functions g on $T \otimes T$ such that

$$\|g\|^2_{G_1\otimes G_2} = \int_T \int_T g^2(s, t)\, \mu(ds)\, \mu(dt) < \infty \tag{63}$$

If G_1 and G_2 are equal to the reproducing kernel Hilbert space consisting of all L_2-differentiable functions on the interval $(t\colon a \leqslant t \leqslant b)$ with norm squared

$$\|g\|^2_{G_1} = \frac{1}{a} g_1^2(a) + \int_a^b |g_1'(t)|^2\, dt, \tag{64}$$

then $G_1 \otimes G_2$ is a reproducing kernel Hilbert space with norm squared

$$\begin{aligned} \|g\|^2_{G_1\otimes G_2} = \frac{1}{a^2} g^2(a, a) &+ \frac{1}{a}\int_a^b \left|\frac{\partial}{\partial s} g(s, a)\right|^2 ds \\ &+ \frac{1}{a}\int_a^b \left|\frac{\partial}{\partial t} g(a, t)\right|^2 dt \\ &+ \int_a^b \int_a^b \left|\frac{\partial}{\partial s}\frac{\partial}{\partial t} g(s, t)\right|^2 ds\, dt. \end{aligned} \tag{65}$$

3. STOCHASTIC SIGNAL CASE

In this section we shall determine conditions for the existence of the probability density functional (6) in the stochastic signal case described before (19).

We shall prove below that P_{S+N} is absolutely continuous with respect to P_N if and only if

$$\|R\|_{H(K)\otimes H(K+R)} < \infty. \tag{66}$$

It may be shown that a sufficient condition for (66) to hold is that

$$\|R\|_{H(K)\otimes H(K)} < \infty. \tag{67}$$

In practice, the condition we shall attempt to verify is (67). Consequently, before proving that (66) is necessary and sufficient for $p = dP_{S+N}/dP_N$ to exist, let us show directly that (67) is a sufficient condition for p to exist and let us obtain an explicit formula for p.

It may be shown that if (67) holds, then the signal process $[S(t), t \in T]$ may be written

$$S(t) = \sum_{\nu=1}^{\infty} \eta_\nu \Phi_\nu(t), \tag{68}$$

where (a) $[\eta_\nu]$ is a sequence of random variables satisfying

$$E(\eta_\alpha, \eta_\beta) = \delta(\alpha, \beta)\lambda_\alpha \tag{69}$$

for a suitable sequence $[\lambda_\nu]$, and (b) $[\Phi_\nu]$ is a sequence of functions in $H(K)$ satisfying

$$(\Phi_\alpha, \Phi_\beta)_{H(K)} = \delta(\alpha, \beta). \tag{70}$$

In fact, $[\lambda_\nu]$ are the eigenvalues and $[\Phi_\nu]$ are the corresponding eigenfunctions of the linear transformation $\mathbf{R}$ on $H(K)$ to itself defined by

$$\mathbf{R}\, h(t) = (h, R(\cdot, t))_{H(K)}. \tag{71}$$

Further

$$\sum_{\nu=1}^{\infty} \lambda_\nu^2 = \|R\|^2_{H(K)\otimes H(K)} < \infty. \tag{72}$$

For $n = 1, 2, \cdots$, let

$$S_n(t) = \sum_{\nu=1}^{n} \eta_\nu \Phi_\nu(t), \quad V_n = (X, \Phi_n)_K. \tag{73}$$

By the developments of Section 1, it follows that P_{S_n+N} is absolutely continuous with respect to P_N with probability density function

$$\begin{aligned} p_n = \frac{dP_{S_n+N}}{dP_N} &= \prod_{\nu=1}^{n} \int_{-\infty}^{\infty} \exp\left(\eta_\nu V_\nu - \tfrac{1}{2}\eta_\nu^2\right) \frac{1}{(2\pi\lambda_\nu)^{1/2}} \exp\left(\frac{-\eta_\nu^2}{2\lambda_\nu}\right) d\eta_\nu \\ &= \prod_{\nu=1}^{n} (1 + \lambda_\nu)^{-1/2} \exp\left(\frac{1}{2} V_\nu^2 \frac{\lambda_\nu}{1 + \lambda_\nu}\right). \end{aligned} \tag{74}$$

By martingale theory it may be shown that (72) implies that the probability density function exists and is given by the limit

$$\frac{dP_{S+N}}{dP_N} = \lim_{n\to\infty} p_n \tag{75}$$

so that

$$\log \frac{dP_{S+N}}{dP_N} = \sum_{\nu=1}^{\infty} \left[-\tfrac{1}{2}\log(1+\lambda_\nu) + \tfrac{1}{2}V_\nu^2 \frac{\lambda_\nu}{1+\lambda_\nu} \right]. \tag{76}$$

If in addition to (72)

$$\sum_{\nu=1}^{\infty} \lambda_\nu < \infty, \tag{77}$$

then the probability density function may be written

$$\log \frac{dP_{S+N}}{dP_N} = -\tfrac{1}{2}\sum_{\nu=1}^{\infty} \log(1+\lambda_\nu) + \tfrac{1}{2}\sum_{\nu=1}^{\infty} V_\nu^2 \frac{\lambda_\nu}{1+\lambda_\nu}. \tag{78}$$

The intuitive meaning of (77) is that almost all sample functions of the signal process $[S(t), t \in T]$ belong to $H(K)$, since from (68)

$$\|S\|_K^2 = \sum_{\nu=1}^{\infty} \eta_\nu^2,$$

$$E[\|S\|_K^2] = \sum_{\nu=1}^{\infty} \lambda_\nu.$$

It appears to establish (77) it would suffice to prove that

$$E[\|S\|_K^2] < \infty.$$

In order to obtain necessary and sufficient conditions that P_{S+N} be absolutely continuous to P_N in the stochastic signal case, let us begin by rephrasing the problem. Let K_1 and K_2 be two positive definite covariance kernels, and let P_i be the probability measure induced on Ω by a normal process $[X(t), t \in T]$ with zero means and covariance kernel K_i. The following questions arise:

1. Determine whether P_1 and P_2 are orthogonal.
2. Determine dP_2/dP_1 if it exists.

We use equations (10) to (15). Let

$$p_{T'} = \frac{dP_{2,T'}}{dP_{1,T'}} = \frac{|K_{2,T'}|^{-1/2} \exp(-\tfrac{1}{2}X^{tr}K_{2,T'}^{-1}X)}{|K_{1,T'}|^{-1/2} \exp(-\tfrac{1}{2}X^{tr}K_{1,T'}^{-1}X)}, \tag{79}$$

$$\begin{aligned} J_{T'} &= E_{P_2}(\log p_{T'}) - E_{P_1}(\log p_{T'}) \\ &= \tfrac{1}{2}\operatorname{trace}(K_{1,T'}^{-1}K_{2,T'} - I - I + K_{2,T'}^{-1}K_{1,T'}). \end{aligned} \tag{80}$$

Amazingly enough, the right-hand side of (80) can be expressed as the norm of a function in the reproducing kernel Hilbert space corresponding to the kernel $K_1 \otimes K_2$, which is a function of four variables (s, s', t, t') defined by

$$K_1 \otimes K_2(s, s', t, t') = K_1(s, t)\, K_2(s', t'). \tag{81}$$

If K_1 and K_2 are nonsingular covariance matrices, we may verify that

$$\text{trace } (K_1 K_2^{-1}) = (K_1, K_1)_{K_1 \otimes K_2}, \tag{82}$$

since

$$\begin{aligned}(K_1, K_1)_{K_1 \otimes K_2} &= \sum_{s,s',t,t'} K_1(s, s')\, K_1^{-1}(s, t)\, K_2^{-1}(s', t')\, K_1(t, t') \\ &= \sum_{s',t,t'} \delta(s', t)\, K_2^{-1}(s', t')\, K_1(t, t') \\ &= \sum_{t,t'} K_2^{-1}(t, t')\, K_1(t, t') \\ &= \text{trace } (K_1 K_2^{-1}).\end{aligned} \tag{83}$$

It may also be proved that

$$\text{trace } I = (K_1, K_2)_{K_1 \otimes K_2}. \tag{84}$$

In this manner we may verify that

$$\text{trace } (K_1 K_2^{-1} + K_2 K_1^{-1} - 2I) = \|K_1 - K_2\|_{K_1 \otimes K_2} \tag{85}$$

$$X^{\text{tr}} K_1^{-1} X - X^{\text{tr}} K_2^{-1} X = (K_2 - K_1, X \otimes X)_{K_1 \otimes K_2}, \tag{86}$$

where $X \otimes X$ is the function on $T \otimes T$ defined by

$$X \otimes X(s, t) = X(s)\, X(t). \tag{87}$$

Using (85) and (86), we may rewrite (79) and (80):

$$p_{T'} = |K_{2,T'}^{-1} K_{1,T'}|^{1/2} \exp [\tfrac{1}{2}(K_2 - K_1, X \otimes X)_{K_1 \otimes K_2, T' \otimes T'}] \tag{88}$$

$$J_{T'} = \tfrac{1}{2}\|K_2 - K_1\|_{K_1 \otimes K_2, T' \otimes T'}. \tag{89}$$

The following conclusions can be immediately inferred:

1. In order that P_1 and P_2 be orthogonal, it is necessary and sufficient that it is *not* so that

$$K_2 - K_1 \quad \text{belongs to } H(K_1 \otimes K_2; T \otimes T). \tag{90}$$

2. If (90) holds, then the Radon-Nikodym derivative exists and is given by the limit (as $T' \to T$) of (88). Formally, we may write

$$\frac{dP_2}{dP_1} = D(K_2^{-1} K_1) \exp [\tfrac{1}{2}(K_2 - K_1, X \otimes X)_{K_1 \otimes K_2, T \otimes T}] \tag{91}$$

if

$$D(K_2^{-1} K_1) = \lim_{T' \to T} |K_{2,T'}^{-1} K_{1,T'}|^{1/2} \tag{92}$$

is assumed to exist.

By using (91), we can sketch a proof of Woodward's theorem on linear transformation of Wiener integrals [Woodward (1961)].

Example 4. To illustrate the use of (67), we consider stationary time series with spectral density functions, so that

$$R(s-t) = \int_{-\infty}^{\infty} e^{i\omega(s-t)} f_S(\omega)\, d\omega,$$

$$K(s-t) = \int_{-\infty}^{\infty} e^{i\omega(s-t)} f_N(\omega)\, d\omega.$$

We now show that a sufficient condition for (67) to hold for any finite interval $T = (t: 0 \leqslant t \leqslant T)$ is that

$$\int_{-\infty}^{\infty} \frac{f_S(\omega)}{f_N(\omega)}\, d\omega < \infty. \tag{93}$$

To prove (93), we write

$$\begin{aligned} \|R\|^2_{K\otimes K, T\otimes T} &= \left\| \int_{-\infty}^{\infty} e^{i\omega s} e^{-i\omega t} f_S(\omega)\, d\omega \right\|^2_{K\otimes K, T\otimes T} \\ &= \int_{-\infty}^{\infty} d\omega_1 \int_{-\infty}^{\infty} d\omega_2 f_S(\omega_1) f_S(\omega_2) (e^{i\omega_1(s-t)}, e^{i\omega_2(s-t)})_{K\otimes K, T\otimes T} \\ &= \int_{-\infty}^{\infty} d\omega_1 \int_{-\infty}^{\infty} d\omega_2 f_S(\omega_1) f_S(\omega_2) |(e^{i\omega_1 s}, e^{i\omega_2 s})_{K,T}|^2 \\ &\leqslant \left[\int_{-\infty}^{\infty} d\omega\, f_S(\omega) \|e^{i\omega s}\|^2_{K,T} \right]^2. \end{aligned}$$

From (49) we may deduce that

$$\frac{1}{T} \|e^{i\omega s}\|^2_{K,T} \leqslant \int_{-\infty}^{\infty} d\lambda [f_N(\lambda)]^{-1} \frac{1}{T} \left| \frac{1}{2\pi} \int_0^T e^{is(\omega-\lambda)}\, ds \right|^2. \tag{94}$$

As T tends to ∞, the right-hand side of (94) tends to

$$[2\pi f_N(\omega)]^{-1} \tag{95}$$

as a limit in mean with respect to the finite measure on $-\infty < \omega < \infty$ with density function $f_S(\omega)$. The desired conclusion may now be inferred.

It might be noted that by using (94) and (95) we can give simple proofs of various extensions to continuous parameter time series of theorems on the asymptotic efficiency of least-squares estimates of regression coefficients given for discrete parameter time series by Grenander and Rosenblatt (1957).

REFERENCES

Aronszajn, N. Theory of reproducing kernels. *Trans. Am. Math. Soc.*, **68** 337–404 (1950).

Grenander, U. and M. Rosenblatt. *Statistical Analysis of Stationary Time Series*, New York: Wiley. Stockholm: Almquist & Wiksell, 1957.

Hajek, J. A property of J-divergence of marginal probability distributions. *Czechoslovak Math. J.*, **8** (83) 460–463 (1958a).

Hajek, J. On a property of normal distribution of any stochastic process (in Russian). *Czechoslovak Math. J.*, **8** (83) 610–618, (1958b). (A translation appears in American Mathematical Society Translations in Probability and Statistics, 1961.)

Hajek, J. On linear estimation theory for an infinite number of observations. *Th. Prob. & Applic.*, **6** 182–193 (1961).

Halmos, P. *Measure Theory.* New York: Van Nostrand, 1950.

Parzen, E. Statistical inference on time series by Hilbert space methods, I. Department of Statistics, Stanford University, Technical Report No. 23, January 2, 1959.

Parzen, E. Regression analysis of continuous parameter time series. *Proc. Fourth Berkeley Symp. Prob. Math. Statist.*, **I**, University of California Press, 1961.

Parzen, E. An approach to time series analysis, *Ann. Math. Statist.*, **32** 951–989 (1961).

Riesz, F., and B. Sz-Nagy. *Functional Analysis.* London: Blackie, 1956.

Woodward, D. A. A general class of linear transformation of Wiener integrals, *Trans. Am. Math. Soc.*, **100** 459–480 (1961).

10

An RKHS Approach to Detection and Estimation Problems—Part II: Gaussian Signal Detection

THOMAS KAILATH, FELLOW, IEEE, AND HOWARD L. WEINERT, MEMBER, IEEE

***Abstract*—The theory of reproducing kernel Hilbert spaces is used to obtain a simple but formal expression for the likelihood ratio (LR) for discriminating between two Gaussian processes with unequal covariances, and to develop a test by which the formal expression can be checked for validity. This LR formula can be evaluated by working separately with each covariance, thus reducing the calculations for the random signal case to those for the simpler known signal problem. In contrast, all previous LR formulas for the unequal covariance problem seem to require calculations involving both covariances simultaneously.**

I. Introduction

We shall describe certain rules for the calculation of likelihood ratios (LR's) in "unequal covariance" Gaussian detection problems, sometimes called "Gaussian signal in Gaussian noise" problems. The literature contains a large number of formulas and procedures for this problem, which is generally considered to be more complicated than the "unequal means, equal covariance" problem, sometimes called the "known signal in Gaussian noise" problem. However, all previous formulas require some calculation involving both covariances simultaneously, e.g., the calculation of simultaneous eigenvalues and eigenfunctions or the solution of certain integral equations [1]–[6]. In contrast, our procedure involves only separate calculations with each covariance, with each calculation being roughly of the same complexity as in the known signal problem.

The basic elements of our procedure are the use of certain formal reproducing kernel Hilbert space (RKHS) expressions, whose validity can then be checked by applying certain "test" procedures. In this sense, the present note generalizes a result, presented in [7] and [8], on the (known signal) unequal-means problem.

It will be useful to begin here with a recapitulation, in a slightly different form, of the result in [8].

Problem I—Unequal Means (Known Signal): Consider the hypotheses

$$H_1: x(t,\omega) = m(t) + n(t,\omega), \qquad t \in I$$

$$H_0: x(t,\omega) = n(t,\omega), \qquad (1)$$

Manuscript received December 17, 1973; revised August 1, 1974. This work was supported in part by the Air Force Office of Scientific Research, AF Systems Command, under Contract AF 44-620-69-C-0101 and in part by the Joint Services Electronics Program under Contract N-00014-67-A-0112-0044.

T. Kailath is with Stanford Electronics Laboratories, Stanford University, Stanford, Calif. 94305.
H. L. Weinert was with Systems Control, Inc., Palo Alto, Calif. He is now with the Department of Electrical Engineering, Johns Hopkins University, Baltimore, Md. 21218.

where ω is the probability-space variable, $m(\cdot)$ is a known nonrandom signal, and $n(\cdot,\omega)$ is a zero-mean Gaussian process with a continuous covariance function

$$En(t,\omega)n(s,\omega) = R(t,s).$$

Suppose that

$$m(\cdot) \in H(R) = \text{the RKHS of } R(\cdot,\cdot).$$

Then the detection problem (1) is nonsingular and the LR can be written

$$\text{LR} = c \cdot \exp U(\omega)$$

where c is a normalizing constant making $E_0[\text{LR}] = 1$ and $U(\omega)$ is a random variable calculated from $x(\cdot,\omega)$ by linear operations[1] and satisfying

$$E_0 x(t,\omega)U(\omega) = m(t), \qquad t \in I. \qquad (2)$$

Here E_0 denotes expectations under hypothesis H_0, and $U(\omega)$ is unique in the usual L_2-sense.

In the literature various methods have been proposed for calculating $U(\omega)$, one of the most common being based on the rule [9]–[11]

$$U(\omega) = \int_I x(t,\omega)a(t)\, dt \qquad (3)$$

where $a(\cdot)$ is the solution of the integral equation

$$\int_I R(t,s)a(s)s = m(t), \qquad t \in I. \qquad (4)$$

This is a useful method, but there are often doubts as to its validity, because the solution $a(\cdot)$ of the Fredholm integral equation of the first kind (4) can be quite ill behaved in the sense that it may contain various "singularity" functions, like the Dirac delta function and its derivatives. Therefore, several authors (see e.g., [9]–[12]) have studied how the use of (3) and (4) may be rigorously justified in a number of cases. The point of the note [7] was that such *a priori* justification could be avoided by merely testing if the proposed $U(\omega)$, found by whatever method, satisfied (2). In particular, in addition to the method (3) and (4), the formula (2) suggests [8] that $U(\omega)$ can often be found as

$$U(\omega) = \langle x(\cdot,\omega), m(\cdot)\rangle_R \qquad (5)$$

where $\langle\cdot,\cdot\rangle_R$ denotes the inner product in the RKHS $H(R)$. The preceding notation is formal because $x(\cdot,\omega)$

[1] That is, $U(\omega)$ is a finite sum of the form $\sum_i c_i x(t_i,\omega)$, or it is a limit in quadratic mean of such sums.

almost surely does not belong to $H(R)$; however, as discussed at some length in [8], in many cases some slight modifications to make the formal expression $\langle x,m \rangle_R$ mathematically well defined (usually just by writing stochastic integrals properly) will yield a random variable $U(\omega)$ that can be tested by seeing if it satisfies (2). It may be noted that RKHS inner products have been explicitly computed for several covariance functions (cf., [8], [13], [14]); some formulas especially suitable for recursive computation have been given in [15].

The reason for this long discussion of the elementary and much studied Problem I is that its solution also underlies that of the following more difficult problem.

Problem II—Unequal Covariances (Random Signal): Consider the hypotheses

$$H_i: \{x(t,\omega),\ t \in I\} \text{ is a zero-mean Gaussian process with continuous covariance function } R_i(t,s), \qquad i = 1,2. \tag{6}$$

We shall assume that the $\{R_i(\cdot,\cdot)\}$ are such that the decision problem is nonsingular, i.e., not capable of resolution with zero probability of error. The conditions for nonsingularity will be described later, but here the point is that it ensures that the LR is well defined. In fact, we shall begin with a somewhat stronger assumption called "strong nonsingularity" [14], which yields a simpler formula for the LR.

We shall prove that with this assumption the LR can be written

$$L[x(\cdot)] = c \cdot \exp - \tfrac{1}{2}U(\omega) \tag{7a}$$

where c is a normalizing constant making $E_2[L] = 1$, and $U(\omega)$ can be any variable calculated from $x(\cdot,\omega)$ by quadratic operations[2] and satisfying the following test.

Let $U(\omega_1,\omega_2)$ be the random variable obtained from $U(\omega)$ by rewriting every quadratic term of the form $x(t,\omega)x(s,\omega)$ as follows:

$$x(t,\omega)x(s,\omega) \to \tfrac{1}{2}x(t,\omega_1)x(s,\omega_2) + \tfrac{1}{2}x(t,\omega_2)x(s,\omega_1). \tag{7b}$$

Note that $U(\omega)$ is the same as $U(\omega,\omega)$. Then we must have the following

$$E_1 U(\omega_1,\omega_2) = 0 = E_2 U(\omega_1,\omega_2) \tag{8a}$$

and

$$E_1\{x(t,\omega_1)E_2[x(s,\omega_2)U(\omega_1,\omega_2)]\} = R_2(t,s) - R_1(t,s), \qquad (t,s) \in I \times I. \tag{8b}$$

Moreover, we shall show that if we can find a symmetric random variable $U(\omega_1,\omega_2)$ that is a "quadratic functional" of $x(\cdot)$ and that obeys (8a) and (8b) and

$$E_1 U(\omega,\omega) < \infty \qquad E_2 U(\omega,\omega) < \infty, \tag{9}$$

then the detection problem is strongly nonsingular and the LR will be given by (7a).

[2] That is, $U(\omega)$ is a finite sum of the form $\sum_{i,j} c_{ij}x(t_i)x(t_j)$, or it is a limit in the mean of such sums.

As in the known signal case, a common method of finding $U(\omega,\omega)$ is via an integral equation [1], [3],

$$U(\omega) = \int_I \int_I x(t,\omega)H(t,s)x(s,\omega)\, dt\, ds \tag{10}$$

where $H(\cdot,\cdot)$ is a solution of the equation

$$\int_I \int_I R_1(t,u)H(u,v)R_2(v,s)\, du\, dv = R_2(t,s) - R_1(t,s), \qquad (t,s) \in I \times I. \tag{11}$$

The solution $H(\cdot,\cdot)$ will usually contain impulsive functions and may not be unique. Therefore, some effort is usually made to directly validate this procedure (see [2], [5], [11], and especially [16] where a generalization of the preceding integral equation is studied). Our test provides an alternative way of doing this, as the interested reader can readily check. Moreover, it also suggests that another form for $U(\omega)$ is

$$U(\omega) = \|x\|_{R_1}^2 - \|x\|_{R_2}^2. \tag{12}$$

This form has the advantage in that we work separately with each covariance and, moreover, that these separate calculations are the same as those necessary for the known signal problem; the difficulty, of course, is that the expression is purely formal especially as each term $\|x\|_{R_i}^2$ is generally infinite with probability one. However, if the problem is (strongly) nonsingular, then we know that the difference (12) must be well behaved; therefore, the infinities must cancel out. In many situations it is easy to see how this cancellation can be arranged, and we can obtain a meaningful expression for $\|x\|_{R_1}^2 - \|x\|_{R_2}^2$ without much difficulty. Even some "trial and error" in manipulating this expression is tolerable, since of course the ultimate test is whether the quantity $U(\omega)$ so obtained satisfies the test (8a) and (8b).

Perhaps the best explanation of this procedure is to consider the following problems, each of which, incidentally, is known to be strongly nonsingular (cf. [14]). However, *a priori* knowledge of this fact is not essential for our solution procedure.

Example 1: Let

$$\left.\begin{aligned} R_1(t,s) &= \tfrac{1}{2}(1 - |t - s|) \\ R_2(t,s) &= 1 + \min(t,s) \end{aligned}\right\} \qquad 0 \le t, s \le T \le 1.$$

Then it is easily verified (see also [8]) that

$$\|x\|_{R_1}^2 = \int_0^T \dot{x}^2(t)\, dt + \frac{[x(0) + x(T)]^2}{2 - T}$$

$$\|x\|_{R_2}^2 = \int_0^T \dot{x}^2(t)\, dt + x^2(0).$$

Note that the $\|x\|_{R_i}^2$ are individually undefined since for $i = 1$ and 2, $\dot{x}$ will contain white noise. However, the difference

$$U(\omega) = \frac{[x(0,\omega) + x(T,\omega)]^2}{2 - T} - x^2(0) \tag{13}$$

is clearly quite meaningful. To check whether we have actually obtained the correct random variable for use in the LR formula (7a), we form

$$U(\omega_1,\omega_2) = \frac{[x(0,\omega_1) + x(T,\omega_1)][x(0,\omega_2) + x(T,\omega_2)]}{2 - T} - x(0,\omega_1)x(0,\omega_2) \quad (14)$$

and see if the conditions (8) and (9) are satisfied. The verification of (8a) is immediate; for (8b), we note that

$$E_2\{x(s,\omega_2)U(\omega_1,\omega_2)\} = \frac{[x(0,\omega_1) + x(T,\omega_1)](2 + s)}{2 - T} - x(0,\omega_1)$$

and, therefore, that

$$\begin{aligned} E_1\{x(t,\omega_1)E_2\{x(s,\omega_2)U(\omega_1,\omega_2)\}\} &= \frac{2 + s}{2 - T}[\tfrac{1}{2}(1 - t) + \tfrac{1}{2}(1 - T + t)] - \frac{1 - t}{2} \\ &= \frac{1}{2} + \frac{t}{2} + \frac{s}{2} \\ &= 1 + \min(t,s) - \tfrac{1}{2}(1 - |t - s|) \\ &= R_2 - R_1. \end{aligned}$$

It can be easily seen that the conditions (9) are met, so that it can be asserted that the problem is strongly nonsingular, without the need for any *a priori* verification.

We turn now to a somewhat more complicated example.

Example 2: Let

$$\left.\begin{aligned} R_1(t,s) &= e^{-1/2|t-s|} \\ R_2(t,s) &= \min(t,s) \end{aligned}\right\} \quad 1 \le t, s \le T.$$

The RKHS norms can be verified to be (also see [8])

$$\|x\|_{R_1}^2 = \int_1^T [\dot{x}(t) + \tfrac{1}{2}x(t)]^2\,dt + x^2(1)$$

$$\|x\|_{R_2}^2 = \int_1^T \dot{x}^2(t)\,dt + x^2(1).$$

The difference is

$$U(\omega) = \int_1^T x(t)\dot{x}(t)\,dt + \frac{1}{4}\int_1^T x^2(t)\,dt \quad (15)$$

and as it stands the right side is not well defined because, under both hypotheses, $\dot{x}(t)$ will contain white noise. We can proceed in several ways to make the right side meaningful. For example, we can choose to interpret the first integral as a so-called "Fisk–Stratonovich" integral

$$\int_1^T x(t)\,dx(t) = \tfrac{1}{2}[x^2(T) - x^2(1)].$$

With this interpretation we have

$$U(\omega) = \frac{1}{2}[x^2(T) - x^2(1)] + \frac{1}{4}\int_1^T x^2(t)\,dt. \quad (16)$$

This random variable is certainly quite meaningful but since other interpretations are possible for the first integral in (15)—as will be seen—we cannot yet be sure whether we have found the proper random variable for use in the LR formula (7a). To check this, we form

$$U(\omega_1,\omega_2) = \frac{1}{2}\left[x(T,\omega_1)x(T,\omega_2) - x(1,\omega_1)x(1,\omega_2) + \frac{1}{2}\int_1^T x(t,\omega_1)x(t,\omega_2)\,dt\right] \quad (17)$$

and find after some algebra that conditions (8) are indeed satisfied. Therefore, $U(\omega)$ as in (16) is in fact the correct random variable. Note also that the conditions (9) are easily met so that we can assert that the problem is strongly nonsingular.

However, as previously mentioned, suppose we had chosen to interpret the first integral in (15) as an Itô integral. With this interpretation, we will obtain a different random variable than (16), namely

$$\frac{[x^2(T) - x^2(1)]}{2} + \frac{1}{4}\int_1^T x^2(t)\,dt - \frac{T - 1}{2} = U_1(\omega), \text{ say.} \quad (18)$$

If we form $U_1(\omega_1,\omega_2)$, we will find that it will satisfy condition (8b) but that it will not satisfy (8a); however, it will be seen that if the term $(T - 1)/2$ is dropped, the resulting random variable will satisfy (8a) and (8b) and, therefore, can be used in the LR formula (7a). In other words, even though in this problem the use of the Itô integral does not directly yield the correct random variable $U(\omega)$, in the process of applying the test we can identify the proper random variable (cf., also the discussion in [7]).

We may expect that similar arguments can be used for more complicated problems. In fact, we have been able to obtain by the preceding methods all explicit LR formulas known to us; moreover, since fairly efficient recursive algorithms[3] are known [15] for computing RKHS norms, the range of applications of the previous method appears to be quite wide.

We turn briefly to certain generalizations before commenting on the relations to previous work.

General Nonsingular Detection

It is known (and will be explained in more detail) that the detection problem can be nonsingular and the LR well defined even if the conditions (9) are not met. In this case, the variable $U(\omega) = U(\omega,\omega)$ will not be well defined, but we shall see that the variable

$$U_0(\omega) = U(\omega,\omega) - E_1 U(\omega,\omega) \quad (19)$$

can be meaningfully interpreted. The LR can now be written

$$L[x(\cdot)] = c_0 \exp -\tfrac{1}{2}U_0(\omega) \quad (20)$$

where c_0 is again a normalizing constant.

[3] More recent results show that the Riccati equations used in [15] can be replaced by the more efficient Chandrasekhar-type equations of [30] whenever the underlying process model is time invariant.

Nonzero Means

Consider the hypotheses

H_i: $x(\cdot)$ has mean-value $m_i(\cdot)$ and covariance R_i, $i = 1,2$.

If this detection problem is nonsingular (again simple RKHS conditions will be given later for this), the LR can be written

$$L[x(\cdot)] = \exp\{U_{m_1}(\omega) - U_{m_2}(\omega) - \tfrac{1}{2}[\|m_1\|_{R_1}^2 - \|m_2\|_{R_2}^2]\} \cdot c_0 \cdot \exp\{-\tfrac{1}{2}U_0(\omega)\} \quad (21)$$

where the $\{U_{m_i}(\omega)\}$ are zero-mean linear functionals of $x(\cdot)$ such that

$$E_i\{x(t,\omega)U_{m_i}(\omega)\} = m_i(t), \qquad t \in I, i = 1,2.$$

Therefore, the $U_{m_i}(\omega)$ can be formally calculated via (3) or (5).

Relation to Previous Results

So much work has been done on the Gaussian detection problem that it would be difficult for any result to be completely new. Nevertheless, our approach, already illustrated in [7], of essentially "guessing" the result by a formal use of RKHS theory and then applying a "test" to check its correctness, does not appear to have been presented before. Results closest to ours appear in the paper [14] of Hájek and the work of Rozanov [6], although several of Rozanov's results were independently developed by us in 1967–1968 (cf., the abstracts of the 1967 International Symposium on Information Theory). Neither Hájek nor Rozanov use the RKHS explicitly. Hájek in [14] introduced the concept of "strong" nonsingularity and gave a LR formula essentially similar to (7a); however, he did not have an explicit way of easily identifying the "correct" random variable $U(\omega)$. Rozanov [6] gave formulas equivalent to (8a) and (8b) and (20), but he apparently did not notice that the random variable $U(\omega)$ could be computed as in (12) by the use of the individual RKHS norms; on the other hand, he essentially reduced the problem of calculating $U(\omega)$ to that of solving a certain integral equation involving both covariances simultaneously. Somewhat similar equations had also been obtained by Kadota [3], [4], as discussed earlier. These integral equation methods tend to obscure the simple relation, implicit in (12), between the known-signal detection problem and the apparently more difficult Gaussian signal problem.

The paper [13] of Parzen was the first to significantly apply RKHS concepts to the Gaussian signal problem. Capon [17], following Parzen [13], gave an explicit LR formula in terms of inner products in the direct-product space of the RKHS $H(R_1)$ and $H(R_2)$; however, he apparently overlooked the possible ambiguity in, and the proper interpretation of, various stochastic integrals that arose in his formulas.

We should also note the paper [18] of Shepp, whose basic results are for the case where one of the processes is a Wiener process. By the use of certain variants of the Wiener process and the "chain rule," these results can be extended to pairs of non-Wiener processes, each of which is "equivalent" to a Wiener process. However, there are pairs for which no such Wiener process can be found, e.g., processes with different means and with covariance function $\max(0, 1 - |t|)$ observed on $[0,T]$, with $T > 1$. We have shown in Part III [31] of this sequence of papers that the proofs of Shepp's results become simpler when a RKHS point of view is used and, moreover, that this point of view enables us to generalize his results with practically no additional effort to the case of two non-Wiener processes. More importantly, Shepp's closed-form expressions are in terms of the solutions to certain integral equations that involve both covariances. Therefore, his formulas, like Kadota's [3], do not exploit the connection with the deterministic signal problem that is revealed by our RKHS formulas.

Finally, we point out that the result in our Example 2 was apparently first found by Striebel [19] and Varberg [20], who obtained it by a careful limiting procedure from the solution to a sampled-data version of the problem.

II. Some Further Properties of RKHS

In [8, part I, section IV] we gave several properties of RKHS. We now present some slightly more sophisticated properties. For various reasons, proofs are given only occasionally. The omitted proofs are either easy or may be found in Aronszajn [21]; the reader may also find it instructive to obtain proofs by using the mapping, discussed at length in Part I, between the RKHS and the space LL_2 of random variables.

A. Dominating Covariances

If R_1 and R_2 are covariance functions, we shall write

$R_1 \ll R_2$, if $R_2 - R_1$ is also a covariance function.

We shall say that

R_2 *dominates* R_1, if $R_1 \ll cR_2$, for some c, $0 < c < \infty$.

We can now state several useful results on the relationships between the RKHS $H(R_1)$ and $H(R_2)$.

Theorem 1: If $R_1 \ll cR_2, 0 < c < \infty$, then $m(\cdot) \in H(R_1)$ implies that $m(\cdot) \in H(R_2)$ (which we shall write[4] as $H(R_1) \subset H(R_2)$) and $\|m\|_{R_2}^2 \le c\|m\|_{R_1}^2$. The converse is also true.

B. Direct Products of RKHS

Given two RKHS's $H(R_1)$ and $H(R_2)$, the direct-product space $H = H(R_1) \otimes H(R_2)$ is defined as the space of functions of the form

$$m(t,s) = \sum_{k=1}^{n} m_{1k}(t)m_{2k}(s), \qquad m_{jk} \in H(R_j)$$

[4] The symbol $\subset$ denotes inclusion in the set-theoretic sense; it does not mean that $H(R_1)$ is a subspace of $H(R_2)$, for then we must have $\|m\|_{R_1}^2 = \|m\|_{R_2}^2$.

and of their completions in the norm

$$\|m(\cdot,*)\|^2_{R_1\otimes R_2} = \sum_{k=1}^{n}\sum_{l=1}^{n} \langle m_{1k},m_{1l}\rangle_{R_1}\langle m_{2k},m_{2l}\rangle_{R_2}.$$

This definition is the obvious modification of the corresponding one for direct products of L_2-spaces, for which

$$L_2[I] \otimes L_2[I]$$

$$= \{\text{space of square-integrable functions } m(\cdot,\cdot) \text{ on } I \times I\}$$

with the norm

$$\|m\|^2_{L_2\otimes L_2} = \int_I \int_I m^2(t,s)\, dt\, ds.$$

A finite-dimensional example will illustrate how $\|\cdot\|_{R_1\otimes R_2}$ can be developed.

Example 3: Let $\boldsymbol{R}_1$ and $\boldsymbol{R}_2$ be two $n \times n$ covariance matrices of the form

$$\boldsymbol{R}_1 = \sum_1^n \lambda_i \boldsymbol{e}_i\boldsymbol{e}_i' \qquad \boldsymbol{R}_2 = \sum_1^n \boldsymbol{e}_i\boldsymbol{e}_i' \tag{22}$$

where the $\{\boldsymbol{e}_i\}$ are the eigenvectors of $\boldsymbol{R}_1$ and $\boldsymbol{R}_2$, i.e., they are the solutions of

$$(\boldsymbol{R}_1 - \lambda_i\boldsymbol{R}_2)\boldsymbol{e}_i = 0, \qquad i = 1,\cdots,n. \tag{23}$$

Then noting that $\boldsymbol{e}_i'\boldsymbol{R}_2^{-1}\boldsymbol{e}_j = \delta_{ij}$ and $\|\boldsymbol{m}\|^2_{R_i} = \boldsymbol{m}'\boldsymbol{R}_i^{-1}\boldsymbol{m}$, we can calculate that

$$\|\boldsymbol{R}_2\|^2_{R_2\otimes R_2} = n \qquad \|\boldsymbol{R}_1\|^2_{R_2\otimes R_2} = \sum_1^n \lambda_i^2$$

$$\|\boldsymbol{R}_1 - \boldsymbol{R}_2\|^2_{R_2\otimes R_2} = \sum_1^n (\lambda_i - 1)^2. \tag{24}$$

The formulas (22)–(24) will be used in Section III to obtain some useful tests for nonsingular detection. Other direct-product spaces and norms can be similarly determined whenever the norms of the $\{H(R_i)\}$ are explicitly known.

We turn now to the basic theorem on direct products of RKHS. However, it will be useful to note first two simple lemmas.

Lemma 1—Series Representations in $H_1 \otimes H_2$: Let $\{f_i(\cdot)\}$ and $\{g_i(\cdot)\}$ be complete orthonormal sets in two Hilbert spaces H_1 and H_2. Then clearly $\{f_i(\cdot)g_j(*)\}$ is a complete orthonormal set in $d_1 \otimes H_2$, and any element in the product space can be written

$$m(t,s) = \sum_1^\infty\sum_1^\infty m_{ij}f_i(t)g_j(s)$$

with

$$\|m\|^2_{H_1\otimes H_2} = \sum_1^\infty\sum_1^\infty |m_{ij}|^2 < \infty. \tag{25}$$

It is easy to see that

$$m_{ij} = \langle f_i(*),\langle g_j(\cdot),m(\cdot,*)\rangle_{H_2}\rangle_{H_1}.$$

Lemma 2: Let $m(\cdot,\cdot) \in H(R_1) \otimes H(R_2)$ and $m_i(\cdot) \in H(R_i)$, $i = 1,2$. Then

$$\langle m(\cdot,*),m_1(\cdot)m_2(*)\rangle_{R_1\otimes R_2} = \langle\langle m(\cdot,*),m_1(\cdot)\rangle_{R_1},m_2(*)\rangle_{R_2}.$$

Theorem 2—Direct Product of RKHS: The direct-product space $H(R_1) \otimes H(R_2)$ of two RKHS's is itself an RKHS with reproducing kernel $R(t_1,t_2,s_1,s_2) = R_1(t_1,s_1)R_2(t_2,s_2)$. That is,

i) $R(\cdot,*,s_1,s_2) \in H(R_1) \otimes H(R_2)$
ii) $\langle m(\cdot,*),R(\cdot,*,s_1,s_2)\rangle_{R_1\otimes R_2} = m(s_1,s_2)$.

The proof follows by direct verification, using the series representation (25) for $m(\cdot,*)$. The importance of Theorem 2 is that it enables us to extend to $H(R_1) \otimes H(R_2)$ the various theorems that have been established for the "one-variable" RKHS $H(R_1)$ or $H(R_2)$.

C. Mappings to Spaces of Random Variables

In [8, part I] we established an isometric isomorphism between the space $LL_2(n)$ and $H(R)$. We recall that LL_2 is the Hilbert space composed of the random variables

$$u_N(\omega) = \sum_1^N c_i x(t_i,\omega), \qquad N < \infty,\ t_i \in I,\ \omega \in \Omega$$

and their limits in the norm

$$\|u_N\|^2 = \int_\Omega u_N^2(\omega)P(d\omega) = E|u_N|^2 = \sum_1^N\sum_1^N c_ic_jR(t_i,t_j)$$

where P is the measure in the probability space $(\Omega,\mathscr{B},P)$ such that

$$Ex(t,\omega)x(s,\omega) = \int_\Omega x(t,\omega)x(s,\omega)P(d\omega) = R(t,s).$$

Then if $H(R)$ is a space of functions on I defined by

$$m(t) = Ex(t)u, \qquad t \in I,\ u \in LL_2$$

and with the norm

$$\|m\|^2_R = E|u|^2,$$

clearly $H(R)$ will also be a Hilbert space. Furthermore, it is an RKHS because

$$\langle R(\cdot,t)m(\cdot)\rangle_R = Ex(t)u = m(t), \qquad t \in I.$$

Similarly consider the product probability space $(\Omega_1 \otimes \Omega_2, \mathscr{B}_1 \otimes \mathscr{B}_2, P_1 \otimes P_2)$ where the measures P_1 and P_2 are induced by covariance functions R_1 and R_2, i.e.,

$$E_i[x(t,\omega)x(s,\omega)] = \int_{\Omega_i} x(t,\omega)x(s,\omega)P_i(d\omega) = R_i(t,s).$$

We can construct a Hilbert space of symmetric random variables of the form

$$U_N(\omega_1,\omega_2) = \sum_{i,j=1}^N c_{ij}x(t_i,\omega_1)x(t_j,\omega_2), \qquad c_{ij} = c_{ji}, \tag{26}$$

and their limits in the norm

$$\begin{aligned}\|U_N(\omega_1,\omega_2)\|^2 &= EU_N^2(\omega_1,\omega_2)\\ &= \int_{\Omega_1}\int_{\Omega_2} U_N^2(\omega_1,\omega_2)P_1(d\omega_1)P(d\omega_2)\\ &= \sum_{i,j,k,l=1}^N c_{ij}c_{kl}R_1(t_i,t_k)R_2(t_j,t_l).\end{aligned} \tag{27}$$

We shall denote this space by $QL_2(x)$, where Q stands for quadratic. By our zero-mean assumptions we also note that

$$EU_N(\omega_1,\omega_2) = 0 = E_1 U_N(\omega_1,\omega_2) = E_2 U_N(\omega_1,\omega_2). \quad (28)$$

Consider the space of functions

$$m(t,s) = Ex(t,\omega_1)x(s,\omega_2)U(\omega_1,\omega_2), \qquad (t,s) \in I \times I,\ U \in QL_2 \quad (29)$$

with norm

$$\|m(\cdot,*)\|^2_{R_1 \otimes R_2} = E|U(\omega_1,\omega_2)|^2. \quad (30)$$

This space is also a Hilbert space. Moreover, since

$$\langle R_1(\cdot,s_1)R_2(*,s_2),m(\cdot,*)\rangle_{R_1 \otimes R_2} = Ex(s_1,\omega_1)x(s_2,\omega_2)U(\omega_1,\omega_2) = m(s_1,s_2)$$

the space of functions (29) can be identified as the RKHS of functions on $I \times I$ with reproducing kernel $R_1(t_1,s_1)R_2 \cdot (t_2,s_2)$ or, equivalently, (cf. Theorem 2) as the direct product of the RKHS $H(R_1)$ and $H(R_2)$. Thus there is an isometric isomorphism between QL_2 and $H(R_1) \otimes H(R_2)$ under the mapping (29).

III. Nonsingularity Conditions and Likelihood Ratio Test

Since the first general results in 1958 of Feldman [22] and Hájek [23] on the conditions for Gaussian hypothesis-testing problems to be nonsingular, a great deal of effort has gone into this problem (cf., the review article of Yaglom [24], which describes most of the work up to 1962). It appears to us that the RKHS approach to this question, as initiated by Parzen [13] and developed in greater detail in the thesis of Oodaira [25] (see also [26]), is the most direct and most general.

It is known, [22], [23] that for Gaussian hypotheses, detection must be either nonsingular or completely singular (i.e., the associated probability measures must be equivalent or singular). We shall give two of the most useful RKHS theorems on nonsingular detection. A summary of RKHS conditions for ninsingularity can be found in Golosov and Tempel'man [27].

Theorem 3—(Parzen [13]): Discrimination between two zero-mean Gaussian processes with covariance functions R_1 and R_2 will be nonsingular, if and only if

$$R_1 - R_2 \in H(R_1) \otimes H(R_2), \text{ i.e., } \|R_1 - R_2\|^2_{R_1 \otimes R_2} < \infty. \quad (31)$$

Inspired by Parzen's theorem, Oodaira derived a modified theorem that is generally easier to apply.

Theorem 4—(Oodaira [25]): The condition (31) in Theorem 3 is equivalent to

i) $\|R_2 - R_1\|^2_{R_2 \otimes R_2} < \infty$,

ii) $H(R_1) = H(R_2)$ (in the set-theoretic sense), or equivalently, there exist two constants $0 < c \le C < \infty$ such that $cR_2 \ll R_1 \ll CR_2$. (The roles of R_1 and R_2 can, of course, be interchanged in this theorem.)

An important corollary is the following.

Corollary 1—Nonzero Means: If the two processes have means $m_k(\cdot)$ and covariances $R_k(\cdot,\cdot)$, $k = 1,2$, then discrimination will be nonsingular if and only if the following additional condition is met:

iii) $m_1 - m_2 \in H(R_1)$ or $H(R_2)$ or $H(R_1 + R_2)$.

Before proving the theorems, it will be useful to express the nonsingularity condition (31) in terms of elements in the space QL_2. The function $R_2 - R_1 \in H(R_1) \otimes H(R_2)$ has an image, say $U(\omega_1,\omega_2)$, in the space QL_2 defined by

$$E\{x(t,\omega_1)x(s,\omega_2)U(\omega_1,\omega_2)\} = R_2(t,s) - R_1(t,s), \qquad (t,s) \in I \times I. \quad (32)$$

The random variable $U(\omega_1,\omega_2)$ is unique up to sets of $P_1 \times P_2$ (product) measure zero and is such that

$$EU^2(\omega_1,\omega_2) = \|R_2 - R_1\|^2_{R_1 \otimes R_2} < \infty. \quad (33)$$

Conversely, if $U(\omega_1,\omega_2) \in QL_2$ satisfies (32), then the isometric relationship between QL_2 and $H(R_1) \otimes H(R_2)$ implies (33). From Theorem 3, we have Corollary 2.

Corollary 2: The detection problem is nonsingular if and only if there exists a (unique) random variable $U(\omega_1,\omega_2) \in QL_2$ that satisfies (32).

This corollary is also proved by Rozanov [6], who does not explicitly use RKHS.

We shall now *outline* one method of proving Theorems 3 and 4 and the basic LR formulas of Section I.

Let $\{T_N\}$, $N = 1,2,\cdots$, be a collection of finite subsets of the original parameter set $[0,T]$,

$$T_N = \{0 = t_1, t_2,\cdots, t_N = T\}, \qquad T_N \subset T_{N+1}$$

whose union is dense in T. On the parameter set T_N, the observation will be the vector

$$\boldsymbol{x} = (x_1,\cdots,x_N) \qquad x_i \triangleq x(t_i), \qquad t_i \in T_N$$

with corresponding covariance matrices, $\boldsymbol{R}_1$ and $\boldsymbol{R}_2$, say, under H_1 and H_2. The LR for $\boldsymbol{x}$ can be written

$$L_N(\boldsymbol{x}) = [\det \boldsymbol{R}_2\boldsymbol{R}_1^{-1}]^{1/2} \exp -\tfrac{1}{2}\boldsymbol{x}'\{\boldsymbol{R}_1^{-1} - \boldsymbol{R}_2^{-1}\}\boldsymbol{x} \quad (34)$$

We shall assume that for any N, the matrices $\boldsymbol{R}_k$ are strictly positive-definite (and thus nonsingular). (If this assumption is not met, the detection problem can be trivially singular.) Therefore, we can find (cf., [28, p. 58]) a nonsingular matrix P such that

$$\boldsymbol{P}\boldsymbol{R}_1\boldsymbol{P}^{-1} = \boldsymbol{\Lambda} = [\lambda_{N,i}\delta_{ij}] \qquad \boldsymbol{P}\boldsymbol{R}_2\boldsymbol{P}^{-1} = \boldsymbol{I}$$

where the $\{\lambda_{N,i}, i = 1,\cdots,N\}$ are the eigenvalues of the matrix equation

$$\boldsymbol{R}_1\boldsymbol{e} = \lambda\boldsymbol{R}_2\boldsymbol{e}. \quad (35)$$

The eigenvalues $\{\lambda_{N,i}\}$ will all be positive, and we shall arrange them in descending order of magnitude

$$\lambda_{N,1} \ge \lambda_{N,2} \cdots \ge \lambda_{N,N} > 0, \qquad \text{for all } N.$$

Furthermore, by the Sturmian separation property (cf. [28, p. 115]), we will have

$$\lambda_{N+1,i+1} \le \lambda_{N,i} \le \lambda_{N+1,i}, \qquad i = 1,\cdots,N. \tag{36}$$

We can now rewrite the LR (34) as

$$2 \ln L_N = -\sum_1^N \tilde{x}_i^2(1 - \lambda_{N,i})\lambda_{N,i}^{-1} - \sum_1^N \ln \lambda_{N,i}, \; \tilde{x} \triangleq Px \tag{37}$$

where the $\{\tilde{x}_i\}$ are independent random variables for we note that

$$E_1[\tilde{x}\tilde{x}'] = PR_1P^{-1} = \Lambda \qquad E_2[\tilde{x}\tilde{x}'] = PR_2P^{-1} = I. \tag{38}$$

Conditions for Singular and Nonsingular Detection

At this point we can proceed in several ways to obtain conditions for singular detection. We shall use a criterion due to Hájek [23], which states that if

$$J_N = E_1[\ln L_N(x)] - E_2[\ln L_N(x)], \tag{39}$$

detection will be nonsingular (singular), if and only if $\lim_{N\to\infty} J_N$ is finite (infinite).

In our problem, it is easy to see that

$$2J_N = \sum_1^N (-2 + \lambda_{N,i}) + \lambda_{N,i}^{-1} = \sum_1^N (\lambda_{N,i} - 1)^2\lambda_{N,i}^{-1}. \tag{40}$$

Therefore, use of Hájek's result will give us the following lemmas.

Lemma 3: Discrimination between zero-mean Gaussian processes with strictly positive-definite covariance functions $R_k(t,s)$, $k = 1,2$, will be nonsingular, if and only if

$$\lim_{N\to\infty} \sum_1^N (\lambda_{N,i} - 1)^2\lambda_{N,i}^{-1} < \infty \tag{41}$$

where the $\{\lambda_{N,i}\}$ are defined by (35); otherwise, discrimination will be singular.

Lemma 4: The condition (41) in Lemma 3 can be replaced by the conditions

$$\lim_{N\to\infty} \sum_1^N (\lambda_{N,i} - 1)^2 < \infty \qquad \lim_{N\to\infty} \lambda_{N,N} > 0. \tag{42}$$

The Sturmian separation property (36) ensures that $\lim \lambda_{N,N}$ is well defined.

Theorems 3 and 4 are just the RKHS forms of Lemmas 3 and 4, as will follow directly from the results on direct products of RKHS and limits of sequences of reproducing kernels.

Proof of Theorem 3: We first note that the finite sum in (41) can be written as an inner product in the direct product of the RKHS $H(R_k)$, $k = 1,2$. In fact, from Example 3 it follows that

$$\sum_1^N (\lambda_{N,i} - 1)^2\lambda_{N,i}^{-1} = \|R_1 - R_2\|^2_{R_1\otimes R_2}.$$

Having recognized this relationship, we now merely have to apply the RKHS approximation theorem ([8, part I, theorem 8]) to complete the proof. It is an easy matter, using the Sturmian separation property of the eigenvalues, to verify that the hypotheses of this theorem are satisfied.

Proof of Theorem 4: We begin by noting that

$$\sum_1^N (\lambda_{N,i} - 1)^2 = \|R_1 - R_2\|^2_{R_2\otimes R_2}$$

from which, again using [8, part I, theorem 8], we obtain condition i) of Theorem 4. For condition ii), we note that the conditions (42) imply the existence of constants c_1,c_2 such that

$$0 < c_1 \le \lambda_{N,i} \le c_2 < \infty, \qquad \text{for any } N \text{ and } i.$$

However, this implies that

$$c_1R_2 \ll R_1 \ll c_2R_2$$

which for continuous covariance functions R_k is equivalent to condition ii) of Theorem 4. This completes the proof.

Likelihood Ratio

When detection is nonsingular, the limit of $\ln L_N(x)$ is well defined, and we can write

$$2 \ln L(x(\cdot)) = \lim_{N\to\infty} 2 \ln L_N(x)$$

$$= \lim \left[-\sum_1^N \tilde{x}_i^2(1 - \lambda_{N,i})\lambda_{N,i}^{-1} - \sum_1^N \ln \lambda_{N,i} \right]. \tag{43}$$

The first term

$$\lim \sum_1^N \tilde{x}_i^2(1 - \lambda_{N,i})\lambda_{N,i}^{-1} = \lim U_N = U, \text{ say}, \tag{44}$$

depends upon the observation $x(\cdot)$, while the second term

$$\lim \sum_1^N \ln \lambda_{N,i} = \lim B_N = B, \text{ say}, \tag{45}$$

is independent of the observation. By standard limiting arguments, it can be shown that U will be well defined, if and only if

$$\lim \sum_1^N |1 - \lambda_{N,i}|\lambda_{N,i}^{-1} < \infty \tag{46}$$

or equivalently, if and only if

$$\lim \sum_1^N |1 - \lambda_{N,i}| < \infty \qquad \lim \lambda_{N,N} > 0. \tag{47}$$

The first condition in (47) is also necessary and sufficient to ensure that the term B is well defined. In fact, note that we can write

$$B = \lim \ln \prod_1^N (1 + \mu_{N,i}), \text{ say, with } \lim \sum_1^N |\mu_{N,i}| < \infty, \tag{48}$$

which can be identified [29, p. 170] as the logarithm of the generalized Fredholm determinant of the operator $\mathscr{R}_1 - I$ on $H(R_2)$.

However, while the conditions (46) or (47) yield nice limits for the quantities U_N and B_N, they are stronger than the necessary and sufficient conditions (41) or (42) for nonsingular detection. In fact, (46) or (47) are just the conditions for what Hájek [14] has termed *strongly nonsingular* detection. While many problems will, in fact, be strongly nonsingular and though it might seem that strong nonsingularity is more meaningful from the point of view of receiver implementation, Rozanov [6], Shepp [18], and others have shown that this distinction is somewhat artificial, as the following simple argument shows. We can rewrite (43) as

$$2 \ln L(x(\cdot)) = -\left[\lim \sum_1^N \{\tilde{x}_i^2(1 - \lambda_{N,i})\lambda_{N,i}^{-1} - (1 - \lambda_{N,i})\}\right] - \left[\lim \sum_1^N \{(\ln \lambda_{N,i}) + (1 - \lambda_{N,i})\}\right] = -[U_0] - [B_0], \text{ say.} \tag{49}$$

Then it is easy to verify that both U_0 and B_0 are well defined, provided that detection is nonsingular, i.e., (41) or (42) are met. Note that we can write

$$B_0 = \lim \ln \prod_1^N (1 + \mu_{N,i})e^{-\mu_{N,i}} \qquad \lim \sum_1^N \mu_{N,i}^2 < \infty, \tag{50}$$

where the limit exists because as $N \to \infty$

$$(1 + \mu_{N,i})e^{-\mu_{N,i}} = (1 + \mu_{N,i})\left(1 - \mu_{N,i} + \frac{\mu_{N,i}^2}{2} + \cdots\right) \to 1 - \frac{\mu_{N,i}^2}{2} + 0(\mu_{N,i}^2),$$

which converges under the conditions (41) or (42) for nonsingular detection. The limit in (50) can be identified [29, p. 171] as the logarithm of the generalized Fredholm–Carleman determinant of the operator $\mathscr{R}_1 - I$ in $H(R_2)$.

RKHS Interpretation of U and U_0

We now note that we can identify U_N and U_{0N} from (46) and (49) as

$$U_N = \|\boldsymbol{x}\|^2_{\boldsymbol{R}_1} - \|\boldsymbol{x}\|^2_{\boldsymbol{R}_2} \qquad U_{0N} = U_N - E_1 U_N.$$

It seems reasonable, therefore, in the infinite-dimensional case to set

$$U = \|x\|^2_{R_1} - \|x\|^2_{R_2} \qquad U_0 = U - E_1 U.$$

The following theorem can now be proved.

Theorem 5—LR Test: Suppose the detection problem is nonsingular. Then a random variable U_0 will be appropriate for use in the LR formula (20), if and only if $U(\omega_1,\omega_2)$ satisfies (8a) and (8b), where $U(\omega_1,\omega_2)$ is the unique random variable obtained from $U_0(\omega)$ by dropping nonquadratic terms and applying the procedure (7b).

Proof: In the finite-dimensional case, the proper random variable clearly is (cf., (44) and (49))

$$U_{0N}(\omega) = U_N(\omega) - E_1 U_N(\omega)$$
$$U_N(\omega) = \boldsymbol{x}'(\boldsymbol{R}_1^{-1} - \boldsymbol{R}_2^{-1})\boldsymbol{x}.$$

Then from (7b)

$$U_N(\omega_1,\omega_2) = \boldsymbol{x}_2'(\boldsymbol{R}_1^{-1} - \boldsymbol{R}_2^{-1})\boldsymbol{x}_1,$$

where $\boldsymbol{x}_i$ is a function of ω_i. Now it is easy to verify that

$$E_1\{E_2\{\boldsymbol{x}_2 U_N(\omega_1,\omega_2)\}\boldsymbol{x}_1'\} = \boldsymbol{R}_2 - \boldsymbol{R}_1$$
$$E_1 U_N(\omega_1,\omega_2) = 0 = E_2 U_N(\omega_1,\omega_2).$$

In other words, using the results of Section II-C, we see that

$$U_N(\omega_1,\omega_2) \in QL_2(\boldsymbol{x}) \sim \boldsymbol{R}_2 - \boldsymbol{R}_1 \in H(\boldsymbol{R}_1) \otimes H(\boldsymbol{R}_2)$$

where the symbol $\sim$ indicates correspondence under the mapping (29). Now applying a theorem of Aronszajn [21, theorem 9-I] (see also [8, theorem 8]) concerning limits of reproducing kernels, we have

$$U(\omega_1,\omega_2) = \lim_{N\to\infty} U_N(\omega_1,\omega_2) \in QL_2(x) \sim R_2 - R_1 \in H(R_1) \otimes H(R_2). \tag{51}$$

The limit is well defined because of the nonsingularity assumption (31) and the isometric isomorphism between $QL_2(x)$ and $H(R_1) \otimes H(R_2)$. In light of the mapping (29) and Fubini's Theorem, it can be seen that (51) implies (8a) and (8b). The converse follows from Corollary 2 and the fact that $U(\omega_1,\omega_2)$ implies a unique $U_0(\omega)$ via

$$U_0(\omega) = U(\omega,\omega) - E_1 U(\omega,\omega).$$

Q.E.D.

Rozanov [6, theorem 7] proves, in a different way, that the proper random variable satisfies (8a) and (8b) but does not give a method of guessing U_0 such as (12) and (14).

IV. Concluding Remarks

We have presented a method for checking the validity of LR formulas for Gaussian processes. The point is that we can calculate the LR's in a number of formal ways, for which a direct justification may be somewhat cumbersome; now we need not worry about this as long as the conjectured expression meets our test. In addition, in this paper we have also given a way of "guessing" the LR, essentially reducing the calculations for the Gaussian (random) signal case to those for the generally simpler known-signal problem.

References

[1] R. Price, "Optimum detection of random signals in noise, with application to scatter multipath communication, I," *Trans. IRE-PGIT*, pp. 125–135, Dec. 1956.
[2] C. R. Rao and V. S. Varadarajan, "Discrimination of Gaussian processes," *Sankhya: Ind. J. Stat.*, ser. A. vol. 25, pp. 303–330, 1963.
[3] T. T. Kadota, "Optimum reception of binary Gaussian signals," *Bell Syst. Tech. J.*, vol. 43, pp. 2767–2810, 1964; vol. 44, pp. 1621–1658, 1965.
[4] ——, "Examples of optimum detection of Gaussian signals and interpretations of white noise," *IEEE Trans. Inform. Theory*, vol. IT-14, pp. 725–734, Sept. 1968.
[5] T. S. Pitcher, "An integral expression for the log likelihood ratio of two Gaussian processes," *SIAM J. Appl. Math.*, pp. 228–233, Mar. 1966.
[6] Yu. A. Rozanov, "Infinite-dimensional Gaussian distributions," *Proc. Steklov Inst. Math., No. 108*, 1968; English translation published by *Amer. Math. Soc.*, 1971.
[7] T. Kailath, "A simple rule for checking the validity of certain detection formulas," *IEEE Trans. Inform. Theory* (Corresp.), vol. IT-13, pp. 144–145, Jan. 1967.
[8] ——, "RKHS approach to detection and estimation problems—

Part I: Deterministic signals in Gaussian noise," *IEEE Trans. Inform. Theory*, vol. IT-17, pp. 530–549, Sept. 1971.
[9] C. Helstrom, *Statistical Theory of Signal Detection*, 2nd. ed. Elmsford, N.Y.: Pergamon, 1968.
[10] H. L. Van Trees, *Detection, Estimation and Modulation Theory*, vol. I. New York: Wiley, 1968.
[11] U. Grenander, "Stochastic processes and statistical inference," *Ark. Mat.*, vol. 1, pp. 195–277, 1950.
[12] T. T. Kadota, "Differentiation of Karhunen-Loève expansion and application to optimum reception of sure signals in noise," *IEEE Trans. Inform. Theory*, vol. IT-13, pp. 255–260, Apr. 1967.
[13] E. Parzen, "Probability density functionals and reproducing kernel hilbert spaces," *Time Series Analysis*, M. Rosenblatt, Ed. New York: Wiley, 1963, ch. 11.
[14] J. Hájek, "On linear statistical problems in stochastic processes," *Czech. Math. J.*, vol. 12, pp. 404–444, 1962.
[15] T. Kailath, R. Geesey, and H. L. Weinert, "Some relations among RKHS norms, Fredholm equations, and innovations representations," *IEEE Trans. Inform. Theory*, vol. IT-18, pp. 341–348, May 1972.
[16] T. T. Kadota, "Generalized optimum receivers of Gaussian signals," *Bell Syst. Tech. J.*, vol. 46, pp. 577–591, 1967; also vol. 46, pp. 883–892, 1967.
[17] J. Capon, "Radon-Nikodym derivatives of stationary Gaussian measures," *Ann. Math. Statist.*, vol. 35, pp. 517–531, 1964.
[18] L. A. Shepp, "Radon-Nikodym derivates of Gaussian measures," *Ann. Math. Statist.*, vol. 37, pp. 321–354, 1966.
[19] C. Striebel, "Densities for stochastic processes," *Ann. Math. Statist.*, vol. 30, pp. 559–567, 1959.
[20] D. E. Varberg, "On equivalence of Gaussian measures," *Pac. J. Math.*, vol. 11, pp. 751–762, 1961.
[21] N. Aronszajn, "Theory of reproducing kernels," *Trans. Amer. Math. Soc.*, vol. 63, pp. 337–404, 1950.
[22] J. Feldman, "Some classes of equivalent Gaussian processes on an interval," *Pac. J. Math.*, vol. 10, pp. 1211–1220, 1960.
[23] J. Hájek, "On a property of normal distributions of any stochastic process," *Czech. Math. J.*, vol. 83, pp. 610–618, 1958.
[24] A. M. Yaglom, "On the equivalence or perpendicularity of two Gaussian probability measures," *Time Series Analysis*, M. Rosenblatt, Ed. New York: Wiley, 1963, ch. 22.
[25] H. Oodaira, "Equivalence and singularity of Gaussian measures," Ph.D. dissertation, Dep. Math., Michigan State Univ., E. Lansing, Mich., Apr. 1963.
[26] G. Kallianpur and H. Oodaira, "The equivalence and singularity of Gaussian measures," in *Time Series Analysis*, M. Rosenblatt, Ed. New York: Wiley, 1963, ch. 19.
[27] Yu. Golosov and A. Tempel'man, "On equivalence of measures corresponding to Gaussian vector-valued function," *Sov. Math.*, vol. 10, pp. 228–232, 1969.
[28] R. E. Bellman, *Matrix Analysis*. New York: McGraw-Hill, 1960.
[29] I. C. Gohberg and M. G. Krein, "Introduction to the theory of linear nonself-adjoint operators," *AMS Translations*, vol. 18, Amer. Math. Soc., Providence, R.I., 1969.
[30] T. Kailath, "Some new algorithms for recursive estimation in constant linear systems," *IEEE Trans. Inform. Theory*, vol. 19, pp. 750–760, Nov. 1973.
[31] T. Kailath and D. Duttweiler, "An RKHS approach to detection and estimation problems—Part III: Generalized Innovations representations and a likelihood-ratio formula," *IEEE Trans. Inform. Theory*, vol. IT-18, pp. 730–745, Nov. 1972.

11

Reprinted from *Sov. Math. Dokl.* **10**:228-232 (1969)

ON EQUIVALENCE OF MEASURES CORRESPONDING TO GAUSSIAN VECTOR-VALUED FUNCTIONS

JU. I. GOLOSOV AND A. A. TEMPEL'MAN

We denote by $\mathcal{H}$ a Hilbert space; $(\cdot, \cdot)$ and $\|\cdot\|$ are the scalar product and the norm in $\mathcal{H}$; $\mathcal{G}(\mathcal{H})$ is the Hilbert space of Hilbert-Schmidt operators with the scalar product $(A, B)_{\mathcal{G}(\mathcal{H})} = \mathrm{Sp}(AB^*)$; $K(\mathcal{H})$ is a set of kernel operators; $\mathcal{G}_+(\mathcal{H})$ and $K_+(\mathcal{H})$ are the corresponding subsets of positive definite operators.

Let $(\Omega, \mathcal{B})$ be a space admitting a measure, and let $\mathcal{P}$ be a probability measure on $\mathcal{B}$. An $\mathcal{H}$-valued random Gaussian variable $\xi(t)$, $t \in T$, determines a mean value $m(t)$ and a correlation function $R(s, t)$: $M(x, \xi(t)) = (x, m(t))$, $M(x, \xi(s) - m(s))\overline{(y, \xi(t) - m(t))} = (x, R(s, t)y)$, $x, y \in \mathcal{H}$; $m(t)$ is an $\mathcal{H}$-valued function; $R(s, t)$ is a $K(\mathcal{H})$-valued positive definite function; $R(s, t) = R(s, t) + m(s) \otimes m(t)$.

1. Let $R(s, t)$, $s, t \in T$, be a $K(\mathcal{H})$-valued positive definite function; let H be a Hilbert space of $\mathcal{H}$-valued functions defined on T; and let $\langle \cdot, \cdot \rangle$ be the scalar product in H.

Definition. A function $R(s, t)$, $s, t \in T$ is said to be a *reproducing kernel of the space* H if for arbitrary $t \in T$, $x \in \mathcal{H}$ we have a) the function $R(\cdot, t)x \in H$; b) $\langle R(\cdot, t)x, m(\cdot)\rangle = (x, m(t))$.

The space H is uniquely determined by the function $R(s, t)$ as the linear span of the functions $m_{t,x}(\cdot) = R(\cdot, t)x$, closed with respect to the scalar product $\langle R(\cdot, s)x, R(\cdot, t)y\rangle = (x, R(s, t)y)$ (see [1]).

Let $R_1(s, t)$ and $R_2(s, t)$, $s, t \in T$, be two $K(\mathcal{H})$-valued kernels; in the tensor product $\mathcal{H} \otimes \mathcal{H}$ we consider the operators: $\Phi(s, u; t, v)(x \otimes y) = R_1(s, t)x \otimes R_2(u, v)y$, $s, u, t, v \in T$, $x, y \in \mathcal{H}$. Then $\Phi(s, u; t, v)$, $(s, u), (t, v) \in T \times T$, is a $K(\mathcal{H} \otimes \mathcal{H})$-valued positive definite function. We write the kernel of $\Phi(s, u; t, v)$ as $R_1(s, t) \otimes R_2(u, v)$. In accordance with our earlier remark, the space $H(R_1 \otimes R_2)$ is spanned on the $(\mathcal{H} \otimes \mathcal{H})$-valued functions of the form $R_1(\cdot, t)x \otimes R_2(\cdot, v)y$. If we set up a mapping between such a function and the corresponding element $R_1(\cdot, t)x \otimes R_2(\cdot, v)y$ belonging to $H(R_1) \otimes H(R_2)$, we have an isomorphism $H(R_1 \otimes R_2) = H(R_1) \otimes H(R_2)$. This fact makes it easy to find $H(R_1 \otimes R_2)$ when we know $H(R_1)$ and $H(R_2)$. To every element of the form $x \otimes y \in \mathcal{H} \otimes \mathcal{H}$ there corresponds an operator $A_{x\otimes y} \in \mathcal{G}(\mathcal{H})$, defined by the formula $A_{x\otimes y}z = (y, z)x$, $z \in \mathcal{H}$. This correspondence can be extended to an isomorphism between the spaces $\mathcal{H} \otimes \mathcal{H}$ and $\mathcal{G}(\mathcal{H})$. We may therefore look on the functions belonging to $H(R_1 \otimes R_2)$ as $\mathcal{G}(\mathcal{H})$-valued operator functions on $T \times T$.

Let $\xi(t)$, $t \in T$, be an $\mathcal{H}$-valued Gaussian function on T, relative to the measure $\mathcal{P}$, with mean value 0 and correlation function $R(s, t)$. We denote by $H(\xi(t), \mathcal{P})$ the linear span of the random variables $(x, \xi(t))$, $x \in \mathcal{H}$, $t \in T$, closed in the sense of mean-square convergence. *We can set up a correspondence between an arbitrary function* $\phi(\cdot) \in H(R)$ *and a random variable* $\gamma \in H(\xi(t), \mathcal{P})$ *such that for arbitrary* $x \in \mathcal{H}$ *we have* $M(x, \xi(t))\gamma = (x, \phi(t))$, $t \in T$; *this correspondence determines an isomorphism between the Hilbert spaces* $H(R)$ *and* $H(\xi(t), \mathcal{P})$; *the random variable* $(x, \xi(t))$ *corresponds to the function* $R(\cdot, t)x$ $(x \in \mathcal{H}, t \in T)$. We write $\gamma = \langle \phi(t), \xi(t)\rangle$. Since the realizations of the random variable $\xi(t)$ do not belong to $H(R)$, we cannot interpret $<\phi(t), \xi(t)>$ literally as a scalar product; in any concrete case, however, we can assign a meaning to this formal expression if we interpret the limit of the quantities that converge to it as the mean-square limit.

We denote by $H^2(\xi(t), \mathcal{P})$ the Hilbert space $H(\zeta(s, t), \mathcal{P})$, where $\zeta(s, t) = \xi(s) \otimes \xi(t) - R(s, t)$ is a random function with values in $\mathcal{H} \otimes \mathcal{H}$ and with the correlation function $\Phi(s, t; u, v)$, where $\Phi(s, t; u, v) \times (x \otimes y) = R(s, u)x \otimes R(t, v)y + R(s, v)y \otimes R(t, u)x$. In accordance with our earlier remark, there exists an isomorphism between the spaces $H^2(\xi(t), \mathcal{P})$ and $H(\Phi)$, under which the random variables $(x \otimes y, \zeta(u, v))$ correspond to the $\mathcal{H} \otimes \mathcal{H}$-valued functions $\Phi(\cdot, \cdot; u, v)(x \otimes y)$, $u, v \in T$.

2. Let $\xi(t)$, $t \in T$, be an $\mathcal{H}$-valued random function on T, and let $\mathcal{B}_\xi$ be the minimal σ-algebra for which all the random variables $\xi(t)$, $t \in T$, are measurable; on $\mathcal{B}_\xi$ suppose given two measures $\mathcal{P}_1$ and $\mathcal{P}_2$, relative to which the $\xi(t)$, $t \in T$, are Gaussian random functions with the characteristics $m_i(t)$ and $R_i(s, t)$, $i = 1, 2$. Theorem 1 reduces the question of equivalence of the measures $\mathcal{P}_1$ and $\mathcal{P}_2$, and the calculation of the Radon-Nikodym derivative $(d\mathcal{P}_2/d\mathcal{P}_1)[\xi(t)]$, to the study of the spaces $H(R_1)$ and $H(R_2)$.

Theorem 1. *The measures $\mathcal{P}_1$ and $\mathcal{P}_2$ are either equivalent or orthogonal. $\mathcal{P}_1$ is equivalent to $\mathcal{P}_2$ if and only if the following conditions hold in one of the combinations* 1)–3), *or* 1) *and* 4), *or* 2) *and* 5), *or* 6):

1) $m_2(t) - m_1(t) \in H(R_1)$;

2) $R_2(\cdot, t)x \in H(R_1)$, *and if* $\langle R_2(\cdot, t)x, \varphi(\cdot)\rangle_1 = 0$ * *for all* $x \in \mathcal{H}$, *then*

3) $R_2(s, t) - R_1(s, t) \in H(R_1 \otimes R_1)$;

4) $R_2(s, t) - R_1(s, t) \in H(R_1 \otimes R_2)$;

5) $R_2(s, t) - m_2(s) \otimes m_2(t) - R_1(s, t) + m_1(s) \otimes m_1(t) \in H(R_1 \otimes R_1)$;

6) $R_2(s, t) - m_2(s) \otimes m_2(t) - R_1(s, t) + m_1(s) \otimes m_1(t) \in H(R_1 \otimes R_2)$.

If the measures $\mathcal{P}_1$ and $\mathcal{P}_2$ are equivalent, then $(d\mathcal{P}_2/d\mathcal{P}_1)[\xi(t)] = D \exp\{½<m_1 - m_2, m_1 - m_2>_1 - <\tilde{\xi}, m_1 - m_2>_1 + ½<A(s, t), \xi(s) \otimes \tilde{\xi}(t) - R_1(s, t)>_{11}\}$; $\tilde{\xi} = \xi - m_2$; $A(s, t) = R_1(s, t) - <R_1(\cdot, s), R_1(\cdot, t)>_2$, i.e. $<f_1, f_2>_1 - <f_1, f_2>_2 = <A(s, t), f_1(s) \otimes f_2(t)>_{11}$, $f_i \in H(R_1)$; $D = \exp\{½ \mathrm{Sp}[\bar{A} + \ln(I - \bar{A})]\}$, *where $\bar{A}$ is the operator in $H(R_1)$ defined by the relations* $<A(\cdot, t)x, f(\cdot)>_1 = (x, (\bar{A}f)(t))$ *for arbitrary $x \in \mathcal{H}$ and $f \in H(R_1)$. The quantities appearing in the expression for* $(d\mathcal{P}_2/d\mathcal{P}_1)[\xi(t)]$ *are defined under the conditions for equivalence.*

This theorem contains as particular cases the criteria for equivalence of the measures, and an expression for the Radon-Nikodym derivative, for measures corresponding to one-dimensional Gaussian functions (see [2, 3, 9, 12]), and for Gaussian measures in Hilbert spaces [6, 7, 11, 13]. In §§3 and 4 we consider certain applications of the method just developed.

3. Let ξ be an $\mathcal{H}$-valued Gaussian random variable with mathematical expectation zero and correlation operator R. Let $R = VV^*$, where $V \in \mathcal{G}(\mathcal{H})$. (We may for instance set $V = V^* = R^{½}$.) We denote by V^{-1} the operator equal to $V^{-1}x = y$ if $Vy = x$, $y \in \overline{V\mathcal{H}}$, and equal to $V^{-1}x = 0$ if $Vx = 0$. We will have $m \in H(R)$ if and only if $m = Vx$, $x \in \mathcal{H}$; $<m_1, m_2> = (V^{-1}m_1, V^{-1}m_2)$. Formally, $<m, \xi> = (V^{-1}m, V^{-1}\xi)$. This expression may be interpreted as follows: Let $e_1, e_2, \cdots$ be a sequence of eigenvectors of the operator R, corresponding to the eigenvalues $\lambda_1, \lambda_2, \cdots$. Then

$$\langle m_1, m_2\rangle = \sum \frac{1}{\lambda_i}(m_1, e_i)(m_2, e_i), \quad \langle m, \xi\rangle = \sum \frac{1}{\lambda_i}(m, e_i)(\xi, e_i)$$

---

* $\langle\cdot, \cdot\rangle_i = \langle\cdot, \cdot\rangle_{H(R_i)}$, $\langle\cdot, \cdot\rangle_{i,j} = \langle\cdot, \cdot\rangle_{H(R_i \otimes R_j)}$.

(the sequence converges in the mean and almost everywhere). The space $H(R_1 \otimes R_2)$ $(R_1, R_2 \in K(\mathcal{H}))$ consists of the operators $K \in \mathcal{G}(\mathcal{H})$ of the form $K = V_1 B V_2^*$, where $B \in \mathcal{G}(\mathcal{H})$, $R_i = V_i V_i^*$, $i = 1, 2$; $\langle K_1, K_2 \rangle_{12} = (B_1, B_2)$, and if $R_1 = R_2 = R$, then $\langle K_1, K_2 \rangle = \Sigma(1/\lambda_i \lambda_j)(K_1 e_i, e_j)(K_2 e_i, e_j)$, $\langle K, \xi \otimes \xi - R \rangle = \Sigma(1/\lambda_i \lambda_j - \delta_{ij})(Ke_i, e_j)(\xi, e_j)(\xi, e_j)$ (the sequence converges in the mean and almost everywhere).

Theorem 2. *Let ξ be an $\mathcal{H}$-valued Gaussian random variable relative to the measures $\mathcal{P}_1$ and $\mathcal{P}_2$. Let m_i and R_i be the corresponding mean values and correlation operators, and let $R_i = V_i V_i^*$. The measures $\mathcal{P}_1$ and $\mathcal{P}_2$ are equivalent on $\mathcal{B}_\xi$ if and only if one of the following alternative combinations of conditions is satisfied:*

1) $m_2 - m_1 = V_1 x$, $x \in \mathcal{H}$;

2) $\overline{R_1 \mathcal{H}} = \overline{R_2 \mathcal{H}}$;

3) $R_2 - R_1 = V_1 B V_1^*$, $B \in \mathcal{G}(\mathcal{H})$; *or* 1) *and*

4) $R_2 - R_1 = V_1 C V_2^* = V_2 C^* V_1^*$, $C \in \mathcal{G}(\mathcal{H})$; *or* 2) *and*

5) $R_2 - m_2 \otimes m_2 - R_1 + m_1 \otimes m_1 = V_1 D V_1^*$, $D \in \mathcal{G}(\mathcal{H})$; *or*

6) $R_2 - m_2 \otimes m_2 - R_1 + m_1 \otimes m_1 = V_1 E V_2^*$, $E \in \mathcal{G}(\mathcal{H})$.

The derivative $(d\mathcal{P}_2/d\mathcal{P}_1)[\xi]$ can be computed by the same formulas as in Theorem 1.

The operator $A = V_1[I - (V_2^{-1}V_1)^* V_2^{-1} V_1] V_1^*$, $\bar{A} = I - V_1(V_2^{-1}V_1)^* V_2^{-1}$ (cf. [6,7,11,13]).

4. Let $T = [a, b]$, $-\infty < a < b < \infty$, $\mathcal{H} = R^n$, let $\Psi(t)$ and $\Phi(t)$ be operator functions, and let the inverses $\Psi^{-1}(t)$ and $\Phi^{-1}(t)$ exist.

Theorem 3. *The operator function*

$$R(s, t) = \Psi(s)\Phi^*(t), s \leqslant t; R(s, t) = \Phi(s)\Psi^*(t), s \geqslant t, \tag{1}$$

is positive definite if and only if:

1) $R(t, t) = \Psi(t)\Phi^*(t) \in K_+(\mathcal{H})$ *for some* $t \in [a, b]$;

2) $\Lambda(t) = \Phi^{-1}(t)\Psi(t) \in K_+(\mathcal{H})$, $t \in [a, b]$;

3) $\Lambda(t) - \Lambda(s) \in K_+(\mathcal{H})$ *for* $t > s$, $s, t \in [a, b]$.

Theorem 4. *An $\mathcal{H}$-valued Gaussian random process $\xi(t)$, $t \in [a, b]$, which is nonsingular and continuous in the mean, is Markovian if and only if its correlation function is representable in the form* (1).

The properties of $\Lambda(t)$ imply that for almost all $t \in [a, b]$ there exists a derivative $\Lambda'(t) \in K_+(\mathcal{H})$. We shall suppose for convenience that all the operators $\Lambda'(t)$ are nonsingular. We introduce the function $M(t)$ and the operator V: $\Lambda'(t) = M(t)M^*(t)$, $\Lambda(a) = VV^*$ (we may, for instance, take $M(t) = M^*(t) = [\Lambda'(t)]^{1/2}$, $V = V^* = [\Lambda(a)]^{1/2}$).

Theorem 5. *Let $\xi(t)$, $t \in [a, b]$ be an R^n-valued process with a correlation function of the form* (1); *then $w(t) = \int_a^t M^{-1}(t)\, d[\Phi^{-1}(t)\xi(t)] + \sqrt{a}\, V^{-1}\Phi^{-1}(a)\xi(a)$ is an R^n-valued Wiener process with the correlation function $I \min(s, t)$. Conversely, $\xi(t) = \Phi(t)\int_a^t M(t)\, dw(t) + \xi(a)$.*

Let us further suppose for convenience that $\xi(a) = 0$. An R^n-valued function $m(t) \in H(R)$ if and only if the function $\Phi^{-1}(t)m(t)$ is absolutely continuous and $f(t) = M^{-1}(t)(\Phi^{-1}(t)m(t))' \in L_2$. If $m_1(t), m_2(t) \in H(R)$, we have $\langle m_1, m_2 \rangle = \int_a^k (f_1(t), f_2(t))\, dt$; $\langle m, \xi \rangle = \int_a^b (M^{-1}(t) \times (\Phi^{-1}(t)m(t))', M^{-1}(t)\, d(\Phi^{-1}(t)\xi(t)))$. Let R_1 and R_2 be two kernels of the form (1). The space $H(R_1 \otimes R_2)$ consists of operator functions $m(s, t)$ such that $\Phi_1^{-1}(s)m(s, t)\Phi_2^{*-1}(t)$ is absolutely continuous and $f(s, t) = M_1^{-1}(s)(\partial^2/\partial s\, \partial t)(\Phi_1^{-1}(s)m(s, t)\Phi_2^{*-1}(t))M_2^{*-1}(t) \in L_2$; $\langle m_1, m_2 \rangle_{1,2} = \int_a^b \int_a^b \mathrm{Sp}[f_1(s, t)f_2^*(s, t)]\, ds\, dt$.

$H^2(\xi(t), \mathcal{P})$ consists of random variables of the form

$$\langle m(s, t), \xi(s) \times \xi(t) - R(s, t)\rangle$$

$$= \int_a^b\int_a^b \mathrm{Sp}\left\{[\Lambda'(s)]^{-1} \frac{\partial^2}{\partial s\,\partial t}(\Phi^{-1}(s)\, m(s, t)\, \Phi^{*-1}(t))\,[\Lambda'(t)]^{-1} d_s d_t (\Phi^{-1}(t)\right.$$
$$\times [\xi(s) \otimes \xi(t)]\, \Phi^{*-1}(s))),$$

where the second stochastic integral is to be understood in the sense of K. Ito; it is defined for arbitrary functions $m(s, t) \in H(R \otimes R)$ (cf. [2, 3, 5]).

Let $\xi(t)$, $t \in [a, b]$ be a nonsingular n-dimensional Gaussian process, continuous in the mean, with the characteristics $m_i(t)$ and $R_i(s, t)$ relative to the measures $\mathcal{P}_i$, $i = 1, 2$ on $\mathcal{B}_\xi$. We write $K_{12}(t) = \Phi_1^{-1}(t)\Phi_2(t)$ and $K_{21}(t) = K_{12}^{-1} = \Phi_2^{-1}(t)\Phi_1(t)$.

Theorem 6. *The measures $\mathcal{P}_1$ and $\mathcal{P}_2$ are equivalent if and only if:*

1) *The vector function $\Phi_1^{-1}(t)(m_2(t) - m_1(t))$ is absolutely continuous and* $M_1^{-1}(t)[\Phi_1^{-1}(t)(m_2(t) - m_1(t))]' \in L_2$;

2) $\Lambda_1'(t)K_{21}^*(t) = K_{12}(t)\Lambda_2'(t)$ *almost everywhere on* $[a, b]$;

3) $K_{12}'(t)$ *and* $K_{21}'(t)$ *exist almost everywhere;*

4) $f(s, t) \in L_2$, *where* $f(s, t) = [M_2^{-1}(t)K_{21}'(t)M_1(s)]^*$ *for* $s \le t$ *and* $f(s, t) = M_1^{-1}(s)K_{12}'(t)M_2(t)$ *for* $s > t$.

The derivative $(d\mathcal{P}_2/d\mathcal{P}_1)[\xi(t)]$ is calculated by the formulas of Theorem 1. *The operator function $A(s, t) = \Phi_1(s)\tilde{A}(s, t)\Phi_1^*(t)$, where*

$$\tilde{A}(s, t) = \int_a^s K_{21}(u)\, K_{12}'(u)\, \Lambda_1(\min(t, u))\, du + \int_a^t \Lambda_1(\min(s, u))\, K_{12}^{*\prime}(u)$$
$$\times K_{21}^*(u)\, du - \int_a^b \Lambda_1(\min(s, u))\, K_{12}^{*\prime}(u)\, K_{21}^*(u)\, (\Lambda_1'(u))^{-1} K_{21}(u)\, K_{12}'(u)$$
$$\times \Lambda_1(\min(t, u))\, du; \quad \ln D = \int_a^b\int_t^b \mathrm{Sp}\, [K_{12}^{*\prime}(u)\, K_{21}^*(u)\, (\Lambda_1'(u))^{-1} K_{21}(u)\, K_{12}'(u)]\, du.$$

Let us consider in particular a random process $\xi(t)$, which relative to the measures $\mathcal{P}_i$, $i = 1, 2$, on $\mathcal{B}_\xi$ is a stationary Gaussian Markov process with $m_i(t) \equiv 0$; $R_i(s, t)m = C_i \exp\{|t - s|Q_i\}C_i$, where $-(Q_i + Q_i^*)$ and C_i are positive definite operators; here $\Psi_i(t) = C_i \exp\{-tQ_i\}$, $\Phi_i(t) = C_i \exp\{tQ_i^*\}$, $\Lambda_i(t) = \exp\{-tQ_i\}\exp\{-tQ_i^*\}$. It is not difficult to show that $\mathcal{P}_1$ and $\mathcal{P}_2$ are equivalent if and only if $C_1(Q_1 + Q_1^*)C_1 = C_2(Q_2 + Q_2^*)C_2$. The measures $\mathcal{P}_1$ and $\mathcal{P}_2$ corresponding to two Gaussian processes with $m_i(t) \equiv 0$, $R_1(s, t) = C\exp\{|t - s|Q\}C$, $R_2(s, t) = B\min(s, t)$, are equivalent if and only if $B = -C(Q + Q^*)C$.

The authors wish to thank A. M. Jaglom for his interest and his valuable counsel.

Institute for Physico-Technical Problems of Energetics
Academy of Sciences of the Lithuanian SSR

Institute of Physics and Mathematics
Academy of Sciences of the Lithuanian SSR

Received 14/JUNE/68

BIBLIOGRAPHY

[1] N. Aronszajn, *Theory of reproducing kernels*, Trans. Amer. Math. Soc. 68 (1950), 337; Russian transl., Matematika 7 (1963), no. 2, 67. MR 14, 479.

[2] Ju. I. Golosov, Dokl. Akad. Nauk SSSR 170 (1966), 242 = Soviet Math. Dokl. 7 (1966), 1162. MR 34 #2081.

[3] ———, *Conditions of equivalence and orthogonality of Gaussian measures, and evaluation of the Radon-Nikodym derivative*, Dissertation, Moscow State University, 1966. (Russian)

[4] Ju. I. Golosov and A. M. Jaglom, *Equivalent Gaussian measures corresponding to generalized random processes*, Abstracts, Internat. Congress of Math., Section on Probability Theory, 1967. (Russian)

[5] T. I. Mirskaja, A. S. Pobedinskaĭte and A. A. Tempel'man, Litovsk. Mat. Sb. 7 (1967), 459.

[6] Ju. A. Rozanov, Dokl. Akad. Nauk SSSR 171 (1966), 1286 = Soviet Math. Dokl. 7 (1966), 1658. MR 34 #8452.

[7] M. G. Šonis, Supplement to G. E. Šilov and Fan Dyk Tin', *Integral, measure and derivative on linear spaces*, "Nauka", Moscow, 1967, pp. 175–189. (Russian) MR 37 #1554.

[8] F. J. Beutler, Ann. Math. Statist. 34 (1963), 424. MR 26 #7033.

[9] J. Capon, IEEE Trans. Information Theory IT 11 (1965), 247. MR 33 #1176.

[10] J. L. Doob, Ann. Math. Statist. 15 (1944), 229. MR 6, 89.

[11] G. Kallianpur and H. Oodaira, "The equivalence and singularity of Gaussian measures," Chapter 19 in *Time series analysis*, Wiley, New York 1963. MR 26 #3163.

[12] E. Parzen, "Probability density functionals and reproducing kernel Hilbert space," Chapter 11 in *Time series analysis*, Wiley, New York 1963. MR 26 #3163.

[13] T. S. Pitcher, SIAM J. Appl. Math. 14 (1966), 228. MR 35 #2379.

Translated by:
A. A. Brown

12

Reprinted by permission from *IEEE Trans. Inform. Theory* **IT-19:**19–28 (1973)

RKHS Approach to Detection and Estimation Problems—Part IV: Non-Gaussian Detection

DONALD L. DUTTWEILER AND THOMAS KAILATH

***Abstract*—We introduce the reproducing-kernel Hilbert space (RKHS) associated with the characteristic functional of a random process and use it to develop a general RKHS theory for non-Gaussian detection. Previously known results for choosing between processes with Gaussian and Poisson statistics are obtained as specializations of this theory.**

I. Introduction

A GAUSSIAN process is completely described by its mean-value function and its covariance function. The covariance, say $R(t,s), t,s \in I \times I$ is a symmetric non-negative definite function and as such has associated with it a special Hilbert space of time functions $H(R)$ that is a reproducing-kernel Hilbert space (*RKHS*) with reproducing kernel $R(t,s)$. This RKHS plays an important role in problems involving the Gaussian process, cf. [1]–[3] and the references therein.

A possibly non-Gaussian process $\{x(t), t \in I\}$ is under certain general conditions completely described by its characteristic functional

$$C(\phi) = E \exp i(x,\phi), \qquad (x,\phi) = \int_I x(t)\phi(t)\, dt \quad (1)$$

when $\phi(\cdot)$ ranges over an appropriate linear space, say Φ, of real functions. To see why this is reasonable, note that if $\phi(\cdot)$ can be concentrated on N discrete points, $C(\phi)$ becomes the Nth-order joint characteristic function that uniquely determines the Nth-order probability distribution function of the process.

A precise statement and detailed verification of this fact and many of the other results in this paper require a rather formidable machinery based on the theory of generalized random processes. We believe, however, that the main ideas can be conveyed in an informal manner and that this approach is useful. It will be adopted in this paper. However, an attempt is made in the Appendixes to give a more careful discussion. In Appendix A we set up the necessary framework to prove in Appendix B that characteristic functionals completely define random processes.

Manuscript received November 22, 1971; revised July 21, 1972.
D. L. Duttweiler is with the Bell Laboratories, Holmdel, N.J. 07733.
T. Kailath is with the Stanford Electronic Laboratories, Stanford, Calif. 94305.

The main fact on which this paper is based is the observation of Hida and Ikeda [4] that the characteristic functional, like the covariance function, is (Hermitian) symmetric and nonnegative definite (i.e., letting asterisks denote complex conjugates,

$$C(\phi - \psi) = C^*(\psi - \phi) \qquad \sum_{j,k=1}^{n} C(\phi_j - \phi_k)u_j u_k^* \geq 0 \quad (2)$$

for any elements $\phi_1(\cdot),\cdots,\phi_n(\cdot)$ and any complex numbers $u_1,\cdots,u_n$) and can, therefore, have associated with it an RKHS, say $H(C)$. We might expect $H(C)$ to play a role in the study of non-Gaussian processes similar to that of $H(R)$ for Gaussian processes.

In this paper we shall use $H(C)$ and certain related spaces to obtain some results for non-Gaussian signal detection. Our main results are essentially the following.

Theorem A: Consider the hypotheses

$$h_1: x(t) = m(t) + n(t)$$
$$h_0: x(t) = n(t), \qquad t \in I \quad (3)$$

where $m(\cdot)$ is a deterministic signal and $n(\cdot)$ is a possibly non-Gaussian process with characteristic functional $C(\phi)$. Then the likelihood ratio exists and has finite variance under hypothesis h_0 if and only if

$$C(\phi) \exp i(m,\phi) \in H(C). \quad (4)$$

Moreover, if (4) holds, the likelihood ratio, say L, can be obtained from the formula

$$L = \langle C(\phi) \exp i(m,\phi), \exp i(x,\phi)\rangle_{H(C)}, \quad (5)$$

where $\langle , \rangle_{H(C)}$ denotes the inner product in $H(C)$.

The results of Theorem A are an easy special case of the following results.

Theorem B: Consider the hypotheses

$$h_j: \{x(t),\ t \in I\} \text{ is a random process with characteristic functional } C_j(\phi), \qquad j = 0,1. \quad (6)$$

Then the likelihood ratio exists and has finite variance under hypothesis h_0 if and only if

$$C_1(\phi) \in H(C_0). \quad (7)$$

Moreover, if (7) holds, the likelihood ratio, say L, can be obtained from the formula

$$L = \langle C_1(\phi), \exp i(x,\phi)\rangle_{H(C_0)}. \quad (8)$$

We immediately remark that (5) and (8) are formal because it turns out that the functional $\exp i(x,\phi)$ almost surely does not belong to $H(C)$. However, we can often rewrite the resulting formulas for L in a mathematically meaningful way and can then verify that L is in fact the likelihood ratio by checking to see if it satisfies (E_0 denotes expectation assuming h_0)

$$E_0[L \exp i(x,\phi)] = C_1(\phi), \qquad \phi \in \Phi. \quad (9)$$

The above procedure is very similar to that used in Parts I–III ([1]–[3]) for the Gaussian detection problem. Some further discussion and an example are given in Section II-E.

The constraint in both these theorems that the likelihood ratio L have finite variance under hypothesis h_0 is nontrivial because all we can assert in general is that L has finite mean, in fact

$$E_0 L = 1. \quad (10)$$

At first one might think that infinite-variance random variables should not be of "physical" interest. However, some reflection shows that this attitude is not consistent with a proper interpretation of the role of mathematical models of physical problems. Without pursuing this point any further here, we merely note that likelihood ratios of interest often have infinite variance. For instance, in the much-studied problem of detecting the presence of a Gaussian process with covariance $K(t,s)$ in a background of white Gaussian noise of unit spectral intensity, we can show (see Example 9) that the likelihood ratio has finite variance if and only if

$$\int_I \int_I K^2(t,s)\, ds\, dt < 1,$$

whereas all that is required for nonsingular detection is that the above integral be finite.

In proving Theorem B we show that any functional of the random process, whether finite-variance or not, that satisfies (9) can be taken as the likelihood ratio. This fact will often help to extend the range of validity of the formulas obtained by using (5) of (8). More deeply, it appears that the RKHS plays a decisive role in determining the properties of large classes of Banach spaces (see, e.g., Schwartz [5] or Kallianpur [6]), and thus we may be able to prove that the finite-variance restriction in Theorems A and B is inconsequential. The potentialities of this embedding of Hilbert spaces in Banach spaces may be of value in some other detection and estimation problems as well.

II. The RKHS Associated With the Characteristic Functional

As in Parts I–III we shall introduce the RKHS via an isometric mapping from a relevant Hilbert space of random variables.

A. Hilbert Spaces of Random Variables

We define the Hilbert space *spanned* by a random process $\{x(t),\ t \in I\}$ as the space say LL_2, consisting of all finite linear combinations

$$\sum_{j=1}^{n} c_j x(t_j)$$

and their limits in the norm $\|u\|_{LL_2}^2 = E|u|^2$. We shall define the Hilbert space *generated* by $\{x(t),\ t \in I\}$ as the space, say NL_2, consisting of all finite products

$$\prod_{j=1}^{n} x(t_j),$$

all linear combinations of these products, and all limits in the norm $\|u\|_{NL_2}^2 = E|u|^2$.[1] To put things in another way, the Hilbert space LL_2 is the closure in the covariance norm of the linear vector space spanned by the $\{x(t), t \in I\}$ while NL_2 is the closure in the same norm of the algebra generated by the $\{x(t), t \in I\}$. More informally, we shall often call LL_2 the "linear" Hilbert space of the process and NL_2 the "nonlinear" Hilbert space of the process.

Both LL_2 and NL_2 have real and complex forms depending on whether or not multiplication by complex constants is permitted. All Hilbert spaces in this paper are to be assumed complex.

An alternate description of the space NL_2 is provided by the following lemma, which will also be useful in associating an RKHS with NL_2.

Lemma 1 (Hida–Ikeda): The set of all finite linear combinations of the random variables $\{\exp i(x,\phi), \phi \in \Phi\}$ is dense in NL_2.

Proof: A proof appears in [7, pp. 76–77].[2]

B. Characteristic Functionals

The exponential function also arises in the definition of the characteristic functional

$$C(\phi) = E \exp i(x,\phi). \tag{11}$$

It is convenient to now give some examples.

Example 1—Gaussian Processes: If $\{x(t), t \in I\}$ is Gaussian with mean $m(t)$ and covariance $R(t,s)$ then

$$(x,\phi) = \int_I x(t)\phi(t)\,dt$$

is Gaussian with mean

$$\int_I m(t)\phi(t)\,dt \triangleq (m,\phi) \tag{12}$$

and covariance

$$\int_I \int_I R(t,s)\phi(t)\phi(s)\,ds\,dt \triangleq (R\phi,\phi). \tag{13}$$

Therefore,

$$C_{GN}(\phi) = \exp\{i(m,\phi) - (\tfrac{1}{2})(R\phi,\phi)\}. \tag{14}$$

Example 2—Poisson White Noise: Another process for which the characteristic functional is easily calculated is Poisson white noise. Poisson white noise has been of recent interest because of its appropriateness in some optical communication and biological system models (see, e.g., [8]–[10]).

Statistically, a random process $\{x(t), t \in I = [0,T]\}$ is Poisson white noise with intensity $\lambda(t)$ [$\lambda(t) \geq 0$ for all $t \in I$ and we shall assume it is locally integrable] if its sample functions consist of unit impulses, integrals of the process over disjoint intervals are statistically independent, and the number $N(a,b)$ of impulses in the interval $[a,b]$ is a Poisson random variable with intensity

$$\int_a^b \lambda(t)\,dt \triangleq \alpha(a,b), \tag{15}$$

that is $\Pr\{N(a,b) = k\} = \exp\{-\alpha(a,b)\}\alpha^k(a,b)/k!$, $k = 0,1,\cdots$. It is not difficult to show (see, e.g., Parzen [11, p. 156]), that the characteristic functional $C_{WPN}(\phi)$ of white Poisson noise is given by

$$C_{WPN}(\phi) = \exp\left\{\int_0^T \lambda(t)(\exp[i\phi(t)] - 1)\,dt\right\}. \tag{16}$$

C. Reproducing Kernel Hilbert Spaces Associated With LL_2 and NL_2

Given the linear space LL_2 spanned by the random process $\{x(t), t \in I\}$, we defined in [1] an isometric Hilbert space $H(R)$ of functions by the mapping

$$J: u \to E[ux(t)] \triangleq f(t) \tag{17a}$$

and norm

$$\|f\|_{H(R)}^2 = E|u|^2. \tag{17b}$$

We note that

$$E[x(s)x(t)] = R(s,t)$$

implies

$$\begin{aligned}\langle f(\cdot), R(\cdot,t)\rangle_{H(R)} &= E[ux(t)]\\ &= f(t),\end{aligned} \tag{18}$$

which shows that as suggested by our notation the space $H(R)$ obtained as in (17) is an RKHS with reproducing kernel $R(t,s)$.

Similarly, given the nonlinear space NL_2 generated by a random process $\{x(t), t \in I\}$, we can define an isometric Hilbert space $H(C)$ of functionals by the mapping

$$J: u \to E[u \exp i(x,\phi)] \triangleq f(\phi) \tag{19a}$$

and the norm

$$\|f\|_{H(C)}^2 = E|u|^2. \tag{19b}$$

We also note that since

$$E[\exp\{i(x,\psi)\}\exp\{-i(x,\phi)\}] = C(\psi - \phi), \tag{20}$$

we have

$$\begin{aligned}\langle f(\cdot), C(\cdot - \phi)\rangle_{H(C)} &= \langle u, \exp\{-i(x,\phi)\}\rangle_{NL_2}\\ &= E[u \exp i(x,\phi)] = f(\phi).\end{aligned} \tag{21}$$

Therefore, the space $H(C)$ of functionals defined as in (19) is an RKHS with reproducing kernel $C(\psi,\phi) \triangleq C(\psi - \phi)$. The space $H(C)$ is a concrete space of well-behaved functionals that is often more convenient to work with than the abstract space of random variables NL_2. The various properties that we described in [1]–[3] for RKHS of functions can be extended, with some obvious changes, to RKHS of functionals.

[1] This norm determines the inner product in NL_2 of two elements u and v as Euv^*. The inner product in LL_2 is the same.

[2] In [7], (x,ϕ) is assumed stationary and Φ is assumed nuclear. The proof as written is valid under the more general conditions of Appendix A.

D. Spanning Functionals

It is often convenient to introduce another kernel and RKHS closely associated with $C(\phi,\psi)$ and $H(C)$. We shall say that a set $\{\mu(\phi), \phi \in \Phi\}$ is a set of *spanning functionals* if for all $\phi \in \Phi$ $\mu(\phi) = \alpha(\phi) \exp i(x,\phi)$, where $\alpha(\phi)$ is a complex constant unequal to zero for any $\phi \in \Phi$. A set of spanning functionals can be used just as the set of exponential functionals was used to define an isometry between NL_2 and a Hilbert space H of functionals. To wit, the mapping

$$J: u \to E[u\mu(\phi)] \triangleq f(\phi) \tag{22a}$$

and norm

$$\|f\|_H^2 = E|u|^2 \tag{22b}$$

define an isometry between NL_2 and H. Let $M(\phi,\psi) = E[\mu(\phi)\mu^*(\psi)] = \alpha(\phi)\alpha^*(\psi)C(\phi,\psi)$. Then

$$\begin{aligned}\langle f(\psi),M(\psi,\phi)\rangle_H &= \langle u,\mu^*(\phi)\rangle_{NL_2}\\ &= E[u\mu(\phi)] = f(\phi),\end{aligned} \tag{23}$$

which shows that $H = H(M)$ is an RKHS with reproducing kernel $M(\phi,\psi)$.

The important property of the set $\{\mu(\phi), \phi \in \Phi\}$ that allows this isometry to be defined is that the set of all finite linear combinations of its elements is dense in NL_2. Thus, it may appear that we have been unnecessarily restrictive in defining spanning functionals and might just as well have defined them as any set of functionals whose finite linear combinations were dense in NL_2. As far as defining isometries is concerned, this observation is correct. However, for the theory of Section III it is convenient to have the more restrictive definition we have made here.

To develop a characterization of $H(M)$ for a particular kernel $M(\phi,\psi)$ (i.e., to develop an explicit and computationally useful description of its elements and norm), it is not necessary to start from the definition (22). Since $H(M)$ is an RKHS with reproducing kernel $M(\phi,\psi)$, it can alternately be defined as the unique Hilbert space H such that

$$M(\cdot,\phi) \in H, \qquad \forall\phi \in \Phi \tag{24a}$$

and

$$\langle f(\cdot),M(\cdot,\phi)\rangle_H = f(\phi), \qquad \forall\phi \in \Phi \tag{24b}$$

and all $f(\cdot) \in H$.

With a clever choice of $\{\mu(\phi), \phi \in \Phi\}$ it is often easier to characterize $H(M)$ than $H(C)$. We note, however, that if a set of spanning functionals can be found for which $H(M)$ can be characterized, $H(C)$ can be characterized also. A function $f(\phi) \in H(C)$ if and only if $\alpha(\phi)f(\phi) \in H(M)$. If $f(\phi)$, $g(\phi) \in H(C)$, then $\langle f(\phi),g(\phi)\rangle_{H(C)} = \langle \alpha(\phi)f(\phi), \alpha(\phi)g(\phi)\rangle_{H(M)}$.

Example 3: Let $\{x(t)\ t \in I\}$ be a Gaussian random process with mean zero and covariance $R(t,s)$. For all $\phi \in \Phi$ define

$$\mu(\phi) = \exp\{i(x,\phi) + (\tfrac{1}{2})(R\phi,\phi)\}. \tag{25}$$

The set $\{\mu(\phi), \phi \in \Phi\}$ is obviously a set of spanning functionals. We have

$$\begin{aligned}M_{GN}(\phi,\psi) &= E[\mu(\phi)\mu^*(\psi)]\\ &= \exp(R\phi,\psi).\end{aligned} \tag{26}$$

Before proceeding with a characterization of $H(M)$ it is necessary to introduce some new notation. We shall let $\otimes^2 H(R)$ denote the direct product (see [2], [6], or [12]) of the RKHS $H(R)$ with itself. For $p \geq 3$, $\otimes^p H(R)$ will denote the direct product of $\otimes^{p-1} H(R)$ and $H(R)$. For notational convenience we shall define $\otimes^1 H(R)$ to equal $H(R)$ itself and $\otimes^0 H(R)$ to be the Hilbert space of complex constants with the inner product

$$\langle \xi_1,\xi_2\rangle_{\substack{0\\ \otimes H(R)}} = \xi_1\xi_2^*.$$

Characterizations of $\otimes^2 H(R)$ for various covariance kernels $R(t,s)$ can be found in [2], [12], or [13]. Usually if $H(R)$ can be characterized, $\otimes^p H(R)$ can be characterized also.

We still need one more notational convention. For any integer p and any function $f_p(t_1,\cdots,t_p) \in \otimes^p H(R)$, $\hat{f}_p(t_1,\cdots,t_p)$ will denote the *symmetric version* of that function, that is,

$$\hat{f}_1(t_1) = f_1(t_1),$$

$$\hat{f}_2(t_1,t_2) = (1/2!)(f_2(t_1,t_2) + f_2(t_2,t_1))$$

and in general

$$\hat{f}_p(t_1,\cdots,t_p) = (1/p!)\sum_{\pi} f_p(t_{\pi(1)},\cdots,t_{\pi(p)}), \tag{27}$$

where the summation ranges over all permutations π of the integers $1,\cdots,p$.

Having introduced this notation it is not difficult to characterize $H(M_{GN})$. The elements of $H(M_{GN})$ are functionals $f(\phi)$ of the form

$$f(\phi) = \sum_{p=0}^{\infty} \int_I \cdots \int_I f_p(t_1,\cdots,t_p) \cdot \phi(t_1)\cdots\phi(t_p)\, dt_1\cdots dt_p, \tag{28}$$

where $f_p(t_1,\cdots,t_p) \in \otimes^p H(R)$ for $p = 0,1,\cdots$ and where

$$\infty > \|f\|^2_{H(M_{GN})} = \sum_{p=0}^{\infty} p!\ \|\hat{f}_p\|^2_{\substack{p\\ \otimes H(R)}}. \tag{29}$$

To establish this characterization of $H(M_{GN})$, it must be shown that $H(M)$ as defined is indeed a Hilbert space and that the two defining RKHS properties (24) of $H(M)$ are satisfied. The details of these verifications can be found in [4] and [14].

Example 3A (White Gaussian Noise): If $\{x(t), t \in I\}$ is white Gaussian noise, the kernel $M_{\text{WGN}}(\phi,\psi)$ is given by

$$M_{\text{WGN}}(\phi,\psi) = \exp \int_I \phi(t)\psi(t)\, dt \tag{30}$$

and $H(M_{\text{WGN}})$ consists of all functions $f(\phi)$ of the form

$$f(\phi) = \sum_{p=0}^{\infty} \int_I \cdots \int_I f_p(t_1,\cdots,t_p) \cdot \phi(t_1)\cdots\phi(t_p)\, dt_1\cdots dt_p, \tag{31}$$

where

$$\infty > \|f(\phi)\|^2_{H(M_{WGN})}$$
$$= \sum_{p=0}^{\infty} p! \int_I \cdots \int_I |\hat{f}_p(t_1,\cdots,t_p)|^2 \, dt_1 \cdots dt_p. \quad (32)$$

Example 4 (Hida–Ikeda): Let $\{x(t), t \in I = [0,T]\}$ be a Poisson white noise process with intensity $\lambda(t)$. For all $\phi \in \Phi$ define

$$\mu(\phi) = \exp\left\{i(x,\phi) - \int_I \lambda(t)(\exp[i\phi(t)] - 1)\, dt\right\}. \quad (33)$$

The set $\{\mu(\phi), \phi \in \Phi\}$ is a set of spanning functionals. We have by simple calculation

$$M_{WPN}(\phi,\psi)$$
$$= \exp\left\{\int_I \lambda(t)(\exp[i\phi(t)] - 1)(\exp[-i\psi(t)] - 1)\, dt\right\}. \quad (34)$$

The elements of $H(M_{WPN})$ are functionals $f(\phi)$ of the form

$$f(\phi) = \sum_{p=0}^{\infty} \int_I \cdots \int_I \lambda(t_1) \cdots \lambda(t_p)(\exp[i\phi(t_1)] - 1)\cdots$$
$$\cdot (\exp[i\phi(t_p)] - 1) f_p(t_1,\cdots,t_p)\, dt_1 \cdots dt_p \quad (35)$$

where

$$\infty > \|f\|^2_{H(M)} = \sum_{p=0}^{\infty} p! \int_I \cdots \int_I \lambda(t_1)\cdots\lambda(t_p)$$
$$\cdot |\hat{f}_p(t_1,\cdots,t_p)|^2\, dt_1 \cdots dt_p. \quad (36)$$

We point out that Hida and Ikeda [4] have also given characterizations of the RKHS associated with general independent increment processes.

E. Congruence Inner Products

The mapping

$$J: u \in NL_2 \to E[u\mu(\phi)] \in H(M) \quad (37)$$

defines an isometry (congruence) and is therefore invertible. Given a function $f(\phi) \in H(M)$, the random variable $J^{-1}(f) \in NL_2$ is given formally by

$$J^{-1}(f) = \langle f(\phi),\mu(\phi)\rangle_{H(M)}. \quad (38)$$

To see the reason for this assertion simply note that

$$\begin{aligned} J(\langle f(\psi),\mu(\psi)\rangle_{H(M)}) &= E[\langle f(\psi),\mu(\psi)\rangle_{H(M)}\mu(\phi)] \\ &= \langle f(\psi),E[\mu(\psi)\mu^*(\phi)]\rangle_{H(M)} \\ &= \langle f(\psi),M(\psi,\phi)\rangle_{H(M)} \\ &= f(\phi). \end{aligned}$$

The inner product in (38) is not a true inner product because $\mu(\phi)$ almost surely will not exist in $H(M)$. Inner products of this type have been called congruence inner products by Parzen [15]. In [1] we explained at some length that the congruence inner product was a well-defined random variable that could always be obtained by a limiting procedure. Thus, here it can be shown that

$$J^{-1}(f) = \lim_{n\to\infty} \sum_{k=1}^{n} \alpha_k \mu^*(\psi_k)$$

if the complex constants α_k and the functions ψ_k are such that $f(\phi)$ is the limit in $H(M)$ of the functionals

$$f_n(\phi) = \sum_{k=1}^{n} \alpha_k M(\phi,\psi_k).$$

However, as noted in [1], often more convenient procedures exist. We showed that in $H(R)$ it often sufficed to replace formal integrals by Wiener integrals. In $H(M)$ we often need to use Itô integrals in order to obtain a meaningful random variable. Note that no matter how a meaningful variable is obtained, its correctness can be directly verified by substituting into the mapping equation (37).

Example 5: Let $\{x(t), t \in I = [0,T]\}$ be Gaussian white noise. As in Example 3, define the spanning functionals $\{\mu(\phi), \phi \in \Phi\}$ by

$$\mu(\phi) = \exp\left\{i\int_0^T x(t)\phi(t)\, dt + \tfrac{1}{2}\int_0^T \phi^2(t)\, dt\right\}. \quad (39)$$

The kernel $M_{WGN}(\phi,\psi)$ and its RKHS $H(M_{WGN})$ are described in Example 3A.

The function $\mu(\phi)$ can be expanded as (see [16])

$$\mu(\phi) = \sum_{p=0}^{\infty} i^p ⨍_0^T ⨍_0^{t_p} \cdots ⨍_0^{t_2} x(t_1)\cdots x(t_p)$$
$$\cdot \phi(t_1)\cdots\phi(t_p)\, dt_1\cdots dt_p \quad (40)$$

where ⨍ denotes at Itô integral (see [16]–[18]). Ignoring the Itô integrals, we have

$$\mu(\phi) = \sum_{p=0}^{\infty} (i^p/p!) \int_0^T \cdots \int_0^T x(t_1)\cdots x(t_p)$$
$$\cdot \phi(t_1)\cdots\phi(t_p)\, dt_1\cdots dt_p \quad (41)$$

and for any function $f(\phi)$ of the form (31),

$$\langle f(\phi),\mu(\phi)\rangle_{H(M_{WGN})} = \sum_{p=0}^{\infty} (-i)^p \int_0^T \cdots \int_0^T \hat{f}_p(t_1,\cdots,t_p)$$
$$\cdot x(t_1)\cdots x(t_p)\, dt_1\cdots dt_p. \quad (42)$$

To make this last equation mathematically meaningful, it is reasonable to try bringing back the Itô integrals, which yields

$$\langle f(\phi),\mu(\phi)\rangle_{H(M_{WGN})}$$
$$= \sum_{p=0}^{\infty} (-i)^p p! ⨍_0^T ⨍_0^{t_p} \cdots ⨍_0^{t_2} \hat{f}_p(t_1,\cdots,t_p)$$
$$\cdot x(t_1)\cdots x(t_p)\, dt_1\cdots dt_p. \quad (43)$$

It can be verified that the mapping J of (37) does indeed transform this latter random variable into $f(\phi)$.

III. Non-Gaussian Detection

One of the most pleasing results of RKHS theory is the following (see [13]): the detection problem

$$h_1: x(t) = m(t) + n(t)$$
$$h_0: x(t) = n(t), \qquad t \in I, \tag{44}$$

where $\{n(t), t \in I\}$ is a Gaussian process with mean zero and covariance $R(t,s)$ is nonsingular if and only if the known signal $m(t)$ exists in $H(R)$. If $m(t) \in H(R)$, the likelihood ratio, say L, is given by

$$L = \exp \{\langle x,m \rangle_{H(R)} - (\tfrac{1}{2})\|m\|_{H(R)}^2\}. \tag{45}$$

Introducing the characteristic functional's RKHS allows a generalization for non-Gaussian detection. Let $\{n(t), t \in I\}$ be a random process with characteristic functional $C(\phi)$ and let $m(t)$ be a known signal. We shall show that the likelihood ratio, say L, for the detection problem

$$h_1: x(t) = m(t) + n(t)$$
$$h_0: x(t) = n(t), \qquad t \in I, \tag{46}$$

exists and has finite variance under hypothesis h_0 if and only if

$$C(\phi) \exp i(m,\phi) \in H(C), \tag{47}$$

in which case it is given by the formula

$$L = \langle C(\phi) \exp i(m,\phi), \exp i(x,\phi) \rangle_{H(C)}. \tag{48}$$

Two points should be emphasized. The first, already made in the Introduction, is that the finite-variance condition is restrictive since some likelihood ratios of interest do have infinite variance. However, equation (48) will often lead one to a valid expression for the likelihood ratio whether or not it has finite variance (see Theorem 2).

The second point is that since this problem is non-Gaussian, the existence of the likelihood ratio does not ensure nonsingular detection, i.e., the impossibility of perfect detection under either hypothesis. A sufficient condition for nonsingular detection that follows easily from the above result is $C(\phi) \exp i(m,\phi) \in H(C)$ and $C(\phi) \exp i(m,-\phi) \in H(C)$.

Actually this detection result is a special case of the following more general theorem, which is an RKHS version of a result apparently first stated (without proof) by Skorokhod [19].

Theorem 1: Consider the hypotheses h_j: $\{x(t), t \in I\}$ is a random process with characteristic functional $C_j(\phi)$, $j = 0,1$. Then the likelihood ratio, say L, exists and has finite variance under hypothesis h_0 if and only if $C_1(\phi) \in H(C_0)$ in which case it can be obtained from the formula

$$L = \langle C_1(\phi), \exp i(x,\phi) \rangle_{H(C_0)}. \tag{49}$$

Proof: A more careful statement and proof of this theorem is given in Appendix C. Here we sketch why the result is plausible.

Suppose L exists and has finite variance under hypothesis h_0. Then $L \in NL_2(h_0)$ and

$$J(L) = E_{h_0}[L \exp i(x,\phi)]$$
$$= E_{h_1}[\exp i(x,\phi)] = C_1(\phi). \tag{50}$$

Therefore $C_1(\phi)$ must exist in $H(C_0)$.

Assume now that $C_1(\phi) \in H(C_0)$. Define

$$u = J^{-1}(C_1(\phi))$$
$$= \langle C_1(\phi), \exp i(x,\phi) \rangle_{H(C_0)}. \tag{51}$$

Then

$$E_{h_0}[u \exp i(x,\phi)] = J(u) = C_1(\phi) \tag{52}$$

and Theorem 2, which follows, establishes that u is indeed the likelihood ratio.

The following theorem, proved in Appendix D, is used in proving Theorem 1 and is of interest in its own right.

Theorem 2: Let h_j and $C_j(\phi), j = 0,1$, be as in Theorem 1. If a random variable u whose mean exists under hypothesis h_0 satisfies for all $\phi \in \Phi$

$$E_{h_0}[u \exp i(x,\phi)] = E_{h_1} \exp i(x,\phi) = C_1(\phi), \tag{53}$$

then the likelihood ratio exists and is equal to u.

A spanning functional version of Theorem 1 is useful in examples.

Corollary 1: Let $\{\mu(\phi), \phi \in \Phi\}$ be a set of spanning functionals and let $\{x(t), t \in I\}$ be a random process known to have statistics associated with either of two hypotheses, h_1 or h_0. Then, the likelihood ratio L exists and has finite variance under hypothesis h_0 if and only if $E_{h_1}\mu(\phi) \in H(M_0)$ in which case it can be obtained from the formula

$$L = \langle E_{h_1}\mu(\phi), \mu(\phi) \rangle_{H(M_0)}. \tag{54}$$

It will of course often be difficult to characterize $H(C_0)$ or $H(M_0)$. We have given characterizations of $H(M_0)$ for Gaussian random processes and Poisson white noise, and these will be used in some simple illustrative examples to follow. Even these simple examples show the complexity apparently inherent in obtaining explicit results for non-Gaussian processes. However, our basic theorems do give a point of departure and more energetic readers may find more examples. For processes in which $H(M_0)$ cannot be characterized, it is possible that Theorem 1 may prove useful in conjunction with approximation techniques (in particular, Weiner's [20] gradient iteration technique) for finding RKHS norms and congruence inner products.

Example 6 (Known Signal in Gaussian Noise): Consider the detection problem

$$h_1: x(t) = m(t) + n(t)$$
$$h_0: x(t) = n(t), \qquad t \in I, \tag{55}$$

where $m(t)$ is a known signal and $\{n(t), t \in I\}$ is Gaussian with mean zero and covariance $R(t,s)$. As noted previously, it is known that this detection problem is nonsingular if and only if $m(t) \in H(R)$ in which case

$$L = \exp \{\langle x,m \rangle_{H(R)} - (\tfrac{1}{2})\|m\|_{H(R)}^2\}. \tag{56}$$

Corollary 1 can be specialized to show that if $m(t) \in H(R)$ detection is nonsingular and L is given by (56). The finite-variance condition it contains precludes using it to show conversely that $m(t)$ must exist in $H(R)$ for detection to be nonsingular.

Assume $m(t) \in H(R)$ and choose the spanning functionals $\{\mu(\phi),\ \phi \in \Phi\}$ as in Example 3. Then

$$E_{h_1}\mu(\phi) = \exp i(m,\phi) = \sum_{p=0}^{\infty} (i)^p(p!)^{-1} \int_I \cdots \int_I m(t_1)\cdots m(t_p) \cdot \phi(t_1)\cdots\phi(t_p)\, dt_1 \cdots dt_p \tag{57}$$

and

$$\|E_{h_1}\mu(\phi)\|^2_{H(M_0)} = \sum_{p=0}^{\infty} p!\, \|i^p(p!)^{-1} m(t_1)\cdots m(t_p)\|^2_{\overset{p}{\otimes} H(R)} = \sum_{p=0}^{\infty} (p!)^{-1} \|m\|_{H(R)}^{2p} = \exp \|m\|_{H(R)}^2. \tag{58}$$

Therefore, detection is nonsingular if $m(t) \in H(R)$. To show that

$$L = \exp \{\langle x,m\rangle_{H(R)} - (\tfrac{1}{2})\|m\|_{H(R)}^2\}, \tag{59}$$

we need only verify that

$$J(\exp \{\langle x,m\rangle_{H(R)} - (\tfrac{1}{2})\|m\|_{H(R)}^2\}) = E_{h_1}\mu(\phi) = \exp i(m,\phi). \tag{60}$$

The verification is not difficult. We shall not give details because it is more interesting to give a constructive, but formal, argument for going from

$$L = J^{-1}(E_{h_1}\mu(\phi)) \tag{61}$$

to (59).

Let $R^{-1}(t,s)$ denote the formal inverse of $R(t,s)$ on $L^2(I)$. In other words $R^{-1}(t,s)$ is the kernel such that

$$\int_I R^{-1}(t,u)R(u,s)\, du = \delta(t - s). \tag{62}$$

Then

$$\begin{aligned} L &= J^{-1}(E_{h_1}\mu(\phi)) = J^{-1}(\exp i(m,\phi)) \\ &= J^{-1}(\exp (R\phi,iR^{-1}m)) \\ &= J^{-1}(M_{GN}(\phi,iR^{-1}m)) = \mu^*(iR^{-1}m) \\ &= \exp \{-i(x,iR^{-1}m) + (\tfrac{1}{2})(RiR^{-1}m,iR^{-1}m)\} \\ &= \exp \{(x,R^{-1}m) - (\tfrac{1}{2})(m,R^{-1}m)\} \\ &= \exp \{\langle x,m\rangle_{H(R)} - (\tfrac{1}{2})\|m\|_{H(R)}^2\}. \end{aligned} \tag{63}$$

The last line of this argument follows from the formal identity (see [15])

$$\langle f,g\rangle_{H(R)} = (R^{-1}f,g). \tag{64}$$

Example 7 (Known Signal in Poisson White Noise): Let the detection problem be the same as that of the previous example except that we now assume the noise to be Poisson white noise of intensity $\lambda(t)$ (see Example 2). It has been proved (see, e.g., [21]) and is intuitively not surprising that there is no nonzero-mean function $m(t)$ such that the likelihood ratio exists. Because of the finite-variance constraint in Corollary 1, it cannot be used to give another proof of this result. However, it can be used to prove the weaker statement that there does not exist a nonzero mean $m(t)$ such that L exists and has finite variance under hypothesis h_0. This may be of interest because the proof of the general result in [21] is rather more difficult to follow.

Suppose to the contrary that there did exist $m(t) \not\equiv 0$ such that L existed in $NL_2(h_0)$. Let $\{\mu(\phi),\ \phi \in \Phi\}$ be the set of spanning functionals described in Example 4. We have

$$E_{h_1}\mu(\phi) = \exp i(m,\phi). \tag{65}$$

By Corollary 1, $\exp i(m,\phi) \in H(M_0)$ and, therefore, it has an expansion of the form

$$\exp i(m,\phi) = \sum_{p=0}^{\infty} F_p(\phi), \tag{66}$$

where

$$F_p(\phi) = \int_I \cdots \int_I \lambda(t_1)\cdots\lambda(t_p)(\exp [i\phi(t_1)] - 1)\cdots (\exp [i\phi(t_p)] \quad 1) f_p(t_1,\cdots,t_p)\, dt_1\cdots dt_p \tag{67}$$

and

$$\infty > \|\exp i(m,\phi)\|^2_{H(M_0)} = \sum_{p=0}^{\infty} p! \int_I \cdots \int_I \lambda(t_1)\cdots\lambda(t_p)|\hat{f}_p(t_1,\cdots,t_p)|^2\, dt_1\cdots dt_p. \tag{68}$$

Since this expansion holds for all $\phi \in \Phi$ and Φ is a linear space, we have (letting ξ denote a dummy variable on the real line

$$i(m,\phi) = \frac{d}{d\xi} \exp i(m,\xi\phi)\Big|_{\xi=0} = \sum_{p=1}^{\infty} \frac{d}{d\xi} F_p(\xi\phi)\Big|_{\xi=0}. \tag{69}$$

But, for all $p \geq 1$ and any $\phi \in \Phi$, it is a straightforward exercise to show that the derivative of $F_p(\xi\phi)$ with respect to ξ evaluated at $\xi = 0$ is zero. Therefore, $i(m,\phi) = 0$ for all ϕ, which implies $m(t) \equiv 0$, a contradiction.

Example 8 (Poisson White Noise Processes With Different Intensities): Consider the hypotheses h_j: $\{x(t),\ t \in I\}$ is Poisson white noise with intensity $\lambda_j(t)$, $(j = 0,1)$, and let $\{\mu(\phi),\ \phi \in \Phi\}$ be the set of spanning functionals described in Example 4 (with $\lambda(t)$ replaced by $\lambda_0(t)$). We have

$$\begin{aligned} E_{h_1}\mu(\phi) &= E_{h_1} \exp \left\{ i(x,\phi) - \int_I \lambda_0(t)(\exp [i\phi(t)] - 1)\, dt \right\} \\ &= \exp \int_I \lambda_0(t)(\exp [i\phi(t)] - 1)(\lambda_1(t)\lambda_0^{-1}(t) - 1)\, dt \\ &= \sum_{p=0}^{\infty} (p!)^{-1} \int_I \cdots \int_I \left[\prod_{q=1}^{p} \lambda_0(t_q)(\exp [i\phi(t_q)] - 1) \cdot (\lambda_1(t_q)\lambda_0^{-1}(t_q) - 1) \right] dt_1\cdots dt_p. \end{aligned} \tag{70}$$

Therefore, $E_{h_1}\mu(\phi) \in H(M_0)$ and the likelihood ratio exists if

$$\|E_{h_1}\mu(\phi)\|^2_{H(M_0)} = \sum_{p=0}^{\infty} (p!)^{-1} \cdot \left(\int_I \lambda_0(t)(\lambda_1(t)\lambda_0^{-1}(t) - 1)^2\, dt\right)^p = \exp \int_I (\lambda_1(t) - \lambda_0(t))^2 \lambda_0^{-1}(t)\, dt. \quad (71)$$

is finite or, equivalently,

$$\int_I (\lambda_1(t) - \lambda_0(t))^2 \lambda_0^{-1}(t)\, dt < \infty. \quad (72)$$

For (72) to be satisfied $\lambda_1(t)$ must be zero whenever $\lambda_0(t)$ is zero, which is an intuitively expected condition.

We have formally from (70)

$$\begin{aligned} E_{h_1}\mu(\phi) &= \exp \int_I \lambda_0(t)(\exp[i\phi(t)] - 1) \cdot (\exp\{-i \cdot i \ln(\lambda_1(t)\lambda_0^{-1}(t))\} - 1)\, dt \\ &= M_0(\phi, i \ln(\lambda_1(t)\lambda_0^{-1}(t))) \\ &= E[\mu(\phi)\mu(-i \ln(\lambda_1(t)\lambda_0^{-1}(t)))] \\ &= J(\mu(-i \ln(\lambda_1(t)\lambda_0^{-1}(t)))). \end{aligned} \quad (73)$$

Therefore,

$$\begin{aligned} L &= \mu(-i \ln(\lambda_1(t)\lambda_0^{-1}(t))) \\ &= \exp\left\{\int_I x(t) \ln(\lambda_1(t)\lambda_0^{-1}(t))\, dt - \int_I \lambda_0(t) \cdot (\lambda_1(t)\lambda_0^{-1}(t) - 1)\, dt\right\} \\ &= \exp \int_I (x(t) \ln(\lambda_1(t)\lambda_0^{-1}(t)) - \lambda_1(t) + \lambda_0(t))\, dt. \end{aligned} \quad (74)$$

This likelihood ratio formula is in agreement with formulas appearing in [22, sec. 4.10] and [8].

Example 9 (Gaussian Processes With Different Covariances): It is possible to use Corollary 1 to show that the likelihood ratio for the problem

$$h_j: \{x(t), t \in I\} \text{ is Gaussian with mean zero and covariance } R_j(t,s), \quad j = 0,1 \quad (75)$$

exists and has finite variance if and only if

$$\|R_1(t,s) - R_0(t,s)\|^2_{\otimes^2 H(R_0)} < 1. \quad (76)$$

Since the demonstration is involved, details will not be given here and the interested reader is referred to [14].

IV. Concluding Remarks

It has been thought the RKHS techniques are restricted to Gaussian processes because Hilbert spaces are linear spaces, and, moreover, RKHS have usually been defined via covariance functions. Hida and Ikeda have shown that since characteristic functionals are nonnegative definite and also provide a complete statistical description, they can be used to generate an RKHS of functionals that takes the place of the RKHS of functions associated with Gaussian processes. We have used this fact to obtain some general results on the existence and calculation of likelihood ratios for non-Gaussian detection problems. Though no new explicit formulas have been obtained, at least so far, there exists some scope for obtaining approximate results.

Of course, it is not only covariance functions and characteristic functionals that are nonnegative definite. In Part V, we shall introduce some other functionals and associated RKHS that are useful in certain parameter estimation problems.

Appendix A

To make sense of the definition

$$C(\phi) = E \exp i \int_I x(t)\phi(t)\, dt \quad (77)$$

it is necessary to specify the set Φ of functions $\phi(\cdot)$ over which $C(\phi)$ is to be defined and to make sure that probability spaces and stochastic integrals are defined in such a way that

$$\int_I x(t)\phi(t)\, dt$$

is well defined and measurable for all $\phi \in \Phi$. The simplest way of meeting these objectives is to change viewpoints and instead of trying to specify $x(\cdot)$ pointwise in time, simply specify integrals of $x(\cdot)$ against a set of testing functions.

Being more precise, let Φ be a normed linear space of real functions on I and let Φ' denote its dual. Define $\beta(\Phi)$ as the smallest σ-field of Φ' sets containing cylinder sets of the form

$$\{\omega \in \Phi': (\omega,\phi_1) \in A_1, \cdots, (\omega,\phi_n) \in A_n\}, \quad (78)$$

where n is an arbitrary finite integer, $A_1, \cdots, A_n$ are Borel sets of the real line, $\phi_1, \cdots, \phi_n$ are functions in Φ, and (ω,ϕ) denotes the value of the functional $\omega \in \Phi'$ evaluated at $\phi \in \Phi$. Making the identification $x = \omega$, a probability measure P on the measurable space $(\Phi', \beta(\Phi))$ defines a random functional $\{(x,\phi), \phi \in \Phi\}$, that can formally be assumed to be generated by a random process $\{x(t), t \in I\}$ through the relation

$$(x,\phi) = \int_I x(t)\phi(t)\, dt, \qquad \phi \in \Phi. \quad (79)$$

We define the characteristic functional $C(\phi)$ of the random functional $\{(x,\phi), \phi \in \Phi\}$ by

$$C(\phi) = E \exp i(x,\phi) = \int \exp\{i(x,\phi)\}\, dP(x), \qquad \phi \in \Phi. \quad (80)$$

Formally, we have

$$C(\phi) = E \exp i \int_I x(t)\phi(t)\, dt. \quad (81)$$

It will not be necessary to restrict the nature of I or Φ. The set I can be chosen as the set of positive integers and Φ as the real Hilbert space of functions

$$\left\{\phi(t): \|\phi(t)\|^2 = \sum_{n=1}^{\infty} \phi^2(n) < \infty\right\}.$$

To study continuous rather than discrete processes, I can be chosen as the positive real line and Φ as the real Hilbert space of functions

$$\left\{\phi(t): \|\phi(t)\|^2 = \int_0^{\infty} \phi^2(t)\,dt < \infty\right\}.$$

Other choices of I and Φ have special merit for particular problems. If I is chosen as the real line and Φ as the space of infinitely differentiable functions with bounded support, it is possible to define rigorously random functionals whose formally associated random processes behave like white noise or even derivatives of white noise (see [23] or [14]).

Appendix B

Lemma 2: Let I, Φ, Φ', and $\beta(\Phi)$ be as in Appendix A and let P_0 and P_1 be two probability measures on $(\Phi',\beta(\Phi))$. If

$$C_0(\phi) = E_0 \exp i(x,\phi) = \int \exp\{i(x,\phi)\}\,dP_0 \tag{82}$$

equals

$$C_1(\phi) = E_1 \exp i(x,\phi) = \int \exp\{i(x,\phi)\}\,dP_1 \tag{83}$$

for all $\phi \in \Phi$, then $P_1 = P_0$.

Proof: For any finite integer n and any functions $\phi_1,\cdots,\phi_n \in \Phi$, the restrictions of P_0 and P_1 to the smallest σ-field measuring the random variables $\{(x,\phi_j), j = 1,\cdots,n\}$ must be identical since finite-dimensional distributions are completely defined by characteristic functions. By [17, p. 605], P_0 and P_1 must be identical.

We emphasize that no assumptions about the nature of Φ have been needed. In particular, we have not needed to assume that Φ is a nuclear space (see [7] or [23]). All characteristic functionals are easily shown to be continuous, to be nonnegative definite, and to evaluate to one at $\phi(t) \equiv 0$. Assuming Φ to be nuclear is necessary only if one wants to assume that all continuous and nonnegative-definite functionals evaluating to one at $\phi(t) \equiv 0$ have associated probability measures on $(\Phi',\beta(\Phi))$ (the Bockner–Minlos theorem, see [7] or [23]).

Appendix C

Theorem 1: Let I, Φ, Φ', and $\beta(\Phi)$ be as defined in Appendix A and let P_0 and P_1 be two probability measures on $(\Phi',\beta(\Phi))$. Define

$$C_j(\phi) = \int \exp\{i(x,\phi)\}\,dP_j, \qquad j = 0,1. \tag{84}$$

Then, $P_1 \ll P_0$ (P_1 is absolutely continuous with respect to P_0) and the Radon–Nikodym derivative of P_1 with respect to P_0, dP_1/dP_0, exists in $NL_2(P_0)$ (the space of finite variance random variables on $(\Phi',\beta(\phi),P_0)$) if and only if $C_1(\phi) \in H(C_0)$ in which case it can be obtained from the formula

$$dP_1/dP_0 = J^{-1}(C_1(\Phi)) \tag{85}$$

where J^{-1} is the inverse of the isometry

$$J: u \in NL_2(P_0) \to \int u \exp\{i(x,\phi)\}\,dP_0 \in H(C_0).$$

Proof: If $P_1 \ll P_0$ and $dP_1/dP_0 \in NL_2(P_0)$, then

$$\begin{aligned} J(dP_1/dP_0) &= \int (dP_1/dP_0) \exp\{i(x,\phi)\}\,dP_0 \\ &= \int \exp\{i(x,\phi)\}\,dP_1 \\ &= C_1(\phi) \end{aligned} \tag{86}$$

and thus $C_1(\phi)$ must exist in $H(C_0)$.

Assume now that $C_1(\phi) \in H(C_0)$. Let $u = J^{-1}(C_1(\phi))$. We have for all $\phi \in \Phi$

$$\int u \exp\{i(x,\phi)\}\,dP_0 = J(u) = C_1(\phi) \tag{87}$$

and by Theorem 2 the proof is complete.

Appendix D

Theorem 2: If a random variable u whose mean exists (P_0) satisfies for all $\phi \in \Phi$

$$\int u \exp\{i(x,\phi)\}\,dP_0 = \int \exp\{i(x,\phi)\}\,dP_1, \tag{88}$$

then $P_1 \ll P_0$ and $dP_1/dP_0 = u$.

Proof: Taking complex conjugates of both sides of (88) we obtain

$$\int u^* \exp\{i(x,-\phi)\}\,dP_0 = \int \exp\{i(x,-\phi)\}\,dP_1, \qquad \phi \in \Phi, \tag{89}$$

or since $\phi \in \Phi$ if and only if $-\phi \in \Phi$,

$$\int u^* \exp\{i(x,\phi)\}\,dP_0 = \int \exp\{i(x,\phi)\}\,dP_1, \qquad \phi \in \Phi. \tag{90}$$

Equations (88) and (90) imply

$$\int (u - u^*) \exp i(x,\phi)\,dP_0 = 0, \qquad \phi \in \Phi, \tag{91}$$

and therefore (see [7, pp. 76–77]—the arguments as given are valid for Φ not nuclear, (x,ϕ) not stationary, and $u - u^*$ having only an existing mean and not necessarily finite variance)

$$u - u^* = 0 \tag{92}$$

or in other words u is real valued.

Let

$$u^+(x) = \begin{cases} u(x), & u(x) \geq 0 \\ 0, & u(x) < 0 \end{cases} \tag{93}$$

and

$$u^-(x) = \begin{cases} 0, & u(x) \geq 0 \\ -u(x), & u(x) < 0. \end{cases} \tag{94}$$

Since

$$\int u\, dP_0 = \int u \exp\{i(x,0)\}\, dP_0$$

$$= \int \exp\{i(x,0)\}\, dP_1$$

$$= \int dP_1 = 1, \tag{95}$$

the constant

$$c = \int u^+\, dP_0 \tag{96}$$

is greater than zero.

Define the probability measures P^+ and P^- on $(\Phi',\beta(\Phi))$ by

$$P^+(A) = c^{-1} \int_A u^+\, dP_0, \qquad A \in \beta(\Phi), \tag{97}$$

and

$$P^-(A) = c^{-1} \left(\int_A u^-\, dP_0 + P_1(A) \right), \qquad A \in \beta(\Phi). \tag{98}$$

For all $\phi \in \Phi$ we have

$$\int \exp\{i(x,\phi)\}\, dP_1$$

$$= \int u \exp\{i(x,\phi)\}\, dP_0$$

$$= \int u^+ \exp\{i(x,\phi)\}\, dP_0 - \int u^- \exp\{i(x,\phi)\}\, dP_0. \tag{99}$$

Rearrangement gives

$$\int \exp\{i(x,\phi)\}\, dP^- = \int \exp\{i(x,\phi)\}\, dP^+. \tag{100}$$

By Appendix B, Lemma 1, P^- and P^+ must be identical. Therefore, for all $A \in \beta(\Phi)$

$$c^{-1} \int_A u^+\, dP_0 = c^{-1} \left(\int_A u^-\, dP_0 + P_1(A) \right). \tag{101}$$

Multiplying by c and rearranging, we have

$$\int_A u\, dP_0 = P_1(A), \tag{102}$$

which shows $P_1 \ll P_0$ and $dP_1/dP_0 = u$.

REFERENCES

[1] T. Kailath, "RKHS approach to detection and estimation problems—Part I: Deterministic signals in Gaussian noise," *IEEE Trans. Inform. Theory*, vol. IT-17, pp. 530–549, Sept. 1971.

[2] ——, "An RKHS approach to detection and estimation problems —Part II: Simultaneous diagonalization and Gaussian signals," submitted to *IEEE Trans. Inform. Theory*.

[3] T. Kailath and D. L. Duttweiler, "RKHS approach to detection and estimation problems—Part III: Generalized innovations representations and a likelihood-ratio formula," *IEEE Trans. Inform. Theory*, vol. IT-18, pp. 730–745, Nov. 1972.

[4] T. Hida and N. Ikeda, "Analysis on Hilbert space with reproducing kernel arising from multiple Wiener integral," in *Proc. 5th Berkeley Symp. Math. Stat. and Prob.*, vol. 2, pt. 1. Berkeley, Calif.: Univ. California Press, 1967.

[5] L. Schwartz, *Radon Measures on Arbitrary Topological Spaces*. Bombay, India: Tata Inst. Fundamental Res. Also, preprint prepared by the Dep. Math., Univ. Maryland, College Park.

[6] G. Kallianpur, "The role of reproducing-kernel Hilbert spaces in the study of Gaussian processes," *Advances in Probability and Other Topics*, vol. 2, P. Ney, Ed. New York: Dekker, 1970.

[7] T. Hida, *Stationary Stochastic Processes*. Princeton, N.J.: Princeton Univ. Press, 1970.

[8] I. Bar-David, "Communication under the Poisson regime," *IEEE Trans. Inform. Theory*, vol. IT-15, pp. 31–37, Jan. 1969.

[9] B. Reiffen and H. Sherman, "An optimum demodulator for Poisson processes: Photon source detectors," *Proc. IEEE*, vol. 51, pp. 1316–1320, Oct. 1963.

[10] D. L. Snyder, "Filtering and detection for doubly stochastic Poisson processes," *IEEE Trans. Inform. Theory*, vol. IT-18, pp. 91–102, Jan. 1972.

[11] E. Parzen, *Stochastic Processes*. San Francisco: Holden-Day, 1962.

[12] N. Aronszajn, "Theory of reproducing kernels," *Trans. Amer. Math. Soc.*, vol. 63, pp. 337–404, May 1950.

[13] J. Capon, "Radon-Nikodym derivatives of stationary Gaussian measures," *Ann. Math. Statist.*, vol. 35, pp. 517–531, June 1964.

[14] D. Duttweiler, "Reproducing-kernel Hilbert space techniques for detection and estimation problems," Ph.D. dissertation, Dep. Elec. Eng., Stanford Univ., Stanford, Calif., June 1970.

[15] a. E. Parzen, "Statistical inference on time series by Hilbert space methods," Appl. Math. Statist. Lab., Stanford Univ., Stanford, Calif., Tech. Rep. 23, 1959. This report and others by Parzen on RKHS are reprinted in E. Parzen, *Time Series Analysis Papers*. San Francisco: Holden-Day, 1967.
b. ——, "Statistical inference on time series by RKHS methods, II," in *Proc. 12th Biennial Canadian Math. Congr.*, R. Pyke, Ed. Providence, R.I.: Amer. Math. Soc., 1969, pp. 1–37.

[16] P. A. Frost, "Estimation in continuous-time nonlinear systems," Ph.D. dissertation, Dep. Elec. Eng., Stanford Univ., Stanford, Calif., June 1968.

[17] J. L. Doob, *Stochastic Processes*. New York: Wiley, 1953.

[18] K. Itô, *Lectures on Stochastic Processes*. Bombay, India: Tata Inst. Fundamental Res., 1961.

[19] A. V. Skorokhod, "Constructive methods of specifying stochastic processes," *Russian Math. Surveys*, vol. 20, pp. 63–83, May/June 1965.

[20] H. J. Weiner, "The gradient iteration in time series," *J. Soc. Indust. Appl. Math.*, vol. 13, pp. 1096–1101, Dec. 1965.

[21] A. V. Skorokhod, "Limit theorems for stochastic processes with independent increments," *Theory Prob. Appl.*, vol. 2, pp. 138–171, 1957.

[22] U. Grenander, "Stochastic processes and statistical inference," *Ark. Math.*, vol. 1, pp. 195–227, Jan. 1950.

[23] I. M. Gelfand and N. Ya. Vilenkin, *Generalized Functions*, vol. 4. New York: Academic Press, 1964.

13

Reprinted from pages 329–354 of *2nd Int. Symp. Inform. Theory Proc.*, B. N. Petrov and F. Csaki, eds., Budapest, 1973, 451p.

ON LINEAR REGRESSION ESTIMATES

A. A. TEMPEL'MAN
USSR

The paper studies the estimates of the unknown "regression coefficients" a_i contained in the mathematical expectation $m(t) = \sum_{i=1}^{n} a_i \varphi_i(t)$ of the random function $\xi(t)$, $t \in T$ (T is an abstract set). In Section 1 necessary notions of the reproducing kernels are given. In Section 2 certain properties of Hilbert spaces generated by random functions are examined, especially their connections with the corresponding kernels. Section 3 deals with the best unbiased regression estimates; Theorem 3.1 is an extension of the well-known result of Hájek [7] and Parzen [13]–[15]; it includes the situations in which some or all regression coefficients can be estimated without error. In Sections 4 and 5, the arbitrary "simple" unbiased linear regression estimates are investigated, i.e. the estimates of the type $\hat{m}(t) = \sum_{i=1}^{n} \hat{a}_i \varphi_i(t)$; for that purpose, wide use is made of Rozanov's method of constructing unbiased linear regression estimates by kernels (positively defined functions) ([17], [18]). In Section 4 it is shown that any simple unbiased linear regression estimate is defined by some kernel; this makes it possible to give an efficiency criterion for simple estimates and to introduce the notion of "one-type" simple estimates. In Section 5 the convergence is proved for one-type regression estimates when the scope of observations is expanded; a general criterion for consistency of regression estimates in terms of their defining kernels is also established. These results generalize the well-known Rozanov's [17], [18] theorems on consistency and convergence of unbiased linear regression estimates for stationary stochastic processes. By way of examples of application of these results, the conditions of convergence and consistency of the least-squares estimates are found for certain types of stationary and non-stationary stochastic processes and fields. The findings of Section 5 bear a close relation to the classical ergodic theorem, being its generalization in several directions. The major results were published in a communication by the author [21].

§ 1. HILBERT SPACES WITH REPRODUCING KERNELS

1. Definition. Let H be a Hilbert space of real-valued functions defined on a certain set T; the scalar product of two functions ψ and φ from H will be denoted by $\langle\psi, \varphi\rangle_H$. Let $R(.,.)$ be a real-valued function of two variables on the set T.

Definition 1.1. If (a) $R(.,t) \in H$ for any fixed $t \in T$ and (b) $\langle R(.,t), \varphi(.)\rangle_H = \varphi(t)$ for any $\varphi \in H$ and $t \in T$, then $R(s,t)$ is called *the producing kernel* of the Hilbert space H; we shall also say that H is *the Hilbert space with the reproducing kernel* R and denote it as $H(R)$ or $H(R,T)$.

This termonology is justified by the fact that the Hilbert space $H(R)$ and its reproducing kernel R determine each other in a unique manner. The Hilbert space of functions on the set T possesses a reproducing kernel, iff for any $t \in T$ the "point" functional $f_t(\varphi) = \varphi(t)$ on H is continuous. Hence, any subspace M of the space $H(R)$ also possesses a reproducing kernel; that kernel $K_M^R(s,t)$ is given by the expression: $K_M^R(.,t) = \Pi_M^{H(R)} R(.,t)$ where $\Pi_M{}^{H(R)}$ is the projector in the space $H(R)$ onto the subspace M. For the function $R(s,t)$, $s, t \in T$ to be the reproducing kernel of some Hilbert space of functions defined on T, it is necessary and sufficient to be positively definite; this means that $\sum_{i,j=1}^{k} \alpha_i \alpha_j R(t_i, t_j) \geq 0$ for any $t_1, \ldots t_k \in T$ and for any real numbers $\alpha_1, \ldots \alpha_k$ $(1 \leq k \leq \infty)$.* The Hilbert space $H(R)$ is generated by the family of functions $\{R(.,t), t \in T\}$. These statements are proved in a paper by N. Aronszajn [2].

Examples of Hilbert spaces with reproducing kernels may be found in [2], [12], [13]–[15].

2. The sum of reproducing kernels. Let $K_i(s,t)$, $i \in \overline{1, n}$, be kernels on the set T; obviously, their sum $K = \sum_{i=1}^{n} K_i$ is also a kernel. The connection between the Hilbert spaces $H(K)$ and $H(K_i)$ is established by the following theorem (see [2], § 6).

Theorem 1.1. 1) The Hilbert space $H(K)$ consists of the functions of the form:

$$\varphi(t) = \sum_{i-1}^{n} \varphi_i(t), \qquad \varphi_i(t) \in H(K_i); \tag{1.1}$$

2) $$\|\varphi\|_{H(K)}^2 = \min_{\sum_{i=1}^{n} \varphi_i = \varphi} \left\{ \sum_{i=1}^{n} \|\varphi_i\|_{H(K_i)}^2 \right\};$$

3) $$H(K) = \sum_{i=1}^{n} \oplus H(K_i), \text{ iff } H(K_i) \cap H(K_j) = \{0\} \text{ at } i \neq j.$$

* Such functions will be referred to as *kernels* further on.

Consider a simple special case, which is important for the further reasoning. Let $R(s, t)$ be a kernel, $\vartheta(\,.\,)$ a function on T and $\Theta(s, t) = \vartheta(s)\vartheta(t)$. Evidently, $\Theta(s, t)$ is a kernel and $H(\Theta)$ consists of functions of the form $\psi_a(t) = a\vartheta(t)$ $(-\infty < a < \infty)$. Let us consider the kernel $K(s, t) = R(s, t) + \Theta(s, t)$. If $\vartheta \notin H(R)$, then, evidently, $H(K) = H(R) \oplus H(\Theta)$. If, however, $\vartheta \in H(R)$ then, obviously, the sets of functions $H(R)$ and $H(K)$ coincide; all the representations of (1.1) for the function ϑ have the form: $\vartheta(t) = (1 - a)\vartheta(t) + a\vartheta(t)$ and hence

$$\| \vartheta \|^2_{H(K)} = \min_{-\infty<a<\infty} [(1-a)^2 \| \vartheta \|^2_{H(R)} + a^2] = \frac{\| \vartheta \|^2_{H(R)}}{1 + \| \vartheta \|^2_{H(R)}} < 1 .$$

The argument of this paragraph is summarized by the following lemma.

Lemma 1.1. Let $K(s, t) = R(s, t) + \vartheta(s)\vartheta(t)$.

Then:

1) if $\vartheta \notin H(R)$, then $H(K) = H(R) \oplus H(\Theta)$, where $\Theta(s, t) = \vartheta(s)\vartheta(t)$; $\| \vartheta \|_{H(K)} = \| \vartheta \|_{H(\Theta)} = 1$;

2) if $\vartheta \in H(R)$, then $H(K)$ and $H(R)$ coincide as sets of functions; $\| \vartheta \|_{H(K)} < 1$.

3. Restrictions of reproducing kernels. Let $T_\alpha \subset T$. Obviously, the restriction $R^\alpha(s, t)$ of the kernel $R(s, t)$ onto the set T_α is also a kernel. Let $H_\alpha(R)$ be the subspace in $H(R)$ generated by the family of functions $\{R(\,.\,, t), t \in T_\alpha\}$.

Theorem 1.2. The function $\varphi^\alpha(t), t \in T_\alpha$, belongs to $H(R^\alpha)$, iff it can be continued to a function $\varphi(t)$, $t \in T$, belonging to $H(R)$. The function $\Pi^{H(R)}_{H_\alpha(R)}\varphi$ is the same for all continuations φ of the function φ^α. The correspondence $\varphi^\alpha \leftrightarrow \Pi^{H(R)}_{H_\alpha(R)}\varphi$ is an isometric isomorphism between $H(R^\alpha)$ and $H_\alpha(R)$.

This statement is proved in § 5 of [2].

Let A be a direction and $\{T_\alpha, \alpha \in A\}$ a net of subsets of the set T; φ^α denotes the restriction of the function φ onto T_α.

Theorem 1.3. If the net $\{T_\alpha, \alpha \in A\}$ is increasing (i.e. $T_{\alpha_1} \subseteq T_{\alpha_2}$ if $\alpha_1 < \alpha_2$) and $\bigcup_{\alpha \in A} T_\alpha = T$), then for any ψ, $\varphi \in H(R)$

$$\lim_{\alpha \in A} \langle \psi^\alpha, \varphi^\alpha \rangle_{H(R^\alpha)} = \langle \psi, \varphi \rangle_{H(R)} .$$

This is contained in the general theorem of Aronszajn on monotone kernel sequences (see [2], § 9, Theorem 1). It is also a corollary of the theorem on the convergence of monotone sequences of projectors (see e.g. [1], § 38): by virtue of Theorem 1.2

$$\langle \psi^\alpha, \varphi^\alpha \rangle_{H(R^\alpha)} = \langle \Pi^{H(R)}_{H_\alpha(R)} \psi, \Pi^{H(R)}_{H_\alpha(R)} \varphi \rangle_{H(R)} .$$

4. Majorization, domination and orthogonality of kernels. In this subsection a number of important types of relations between kernels is discussed.

Definition 1.2. We shall say that a kernel R *is majorized by a kernel* B (notation: $R \lhd B$), if the function $B - R$ is a kernel;

— a kernel R *is bounded in relation to a kernel* B, if $R \lhd kB$ at a certain non-negative k;

— a kernel R *is dominated by a kernel* B (notation: $R \ll B$), if $R(t, t) = \lim_{\Lambda \in U_R^B} \Lambda(t, t)$, where $U_R^B = \{\Lambda : \Lambda \lhd R, \Lambda \lhd kB\}$ is a direction of kernels with the relation of order $\lhd$);

— kernels R and B *are reciprocally orthogonal* (notation: $R \perp B$), if no kernel Λ exists for which $\Lambda \lhd R$ and $\Lambda \lhd B$.

Example 1.1. Let $T = \mathfrak{R}^r$ (the r-dimensional Euclidean space), R and B be stationary kernels, i.e.

$$R(s, t) = \int_{R^r} e^{i(\lambda,(t-s))} F_R(d\lambda) , \qquad B(s, t) = \int_{R^r} e^{i(\lambda,(t-s))} F_B(d\lambda) ,$$

where F_R and F_B are finite Borel measures on $\mathfrak{R}^r$. Then the relations (a) $R \perp B$, (b) $R \ll B$, (c) $R \lhd kB$ are equivalent to (a) $F_R \perp F_B$ (measures F_R and F_B are orthogonal), (b) $F_R \ll F_B$ (F_R is absolutely continuous relative to F_B), (c) $F_R \ll F_B$ and $\frac{dF_R}{dF_B}(\lambda) \leq kF_B$ almost everywhere, respectively. Let us also note that the space $H(R)$ consists of all the functions of the form: $\varphi(t) = \int_{\mathfrak{R}^r} e^{i(\lambda,t)} h(\lambda) F(d\lambda)$ with $\int_{\mathfrak{R}^r} | h |^2 F(d\lambda) < \infty$.

We shall write $R \oplus B$ instead of $R + B$, if $R \perp B$. If $R \lhd kB$, then $H_B(R)$ denotes the closure in $H(B)$ of the linear variety $H(R)$.

The following theorem connects these relations between the kernels R and B with some relations between the spaces $H(R)$ and $H(B)$.

Theorem 1.4. (a) $R \lhd kB$, iff for the sets $H(R) \subseteq H(B)$;

(b) $R \ll B$, iff the set $H(R) \cap H(B)$ is dense in the space $H(R)$;

(c) $R \perp B$, iff $H(R) \cap H(B) = \{0\}$, i.e. iff $H(R + B) = H(R) \oplus H(B)$;

(d) if $R \lhd kB$ and $B \ll R$, then $H(B) = H_B(R)$.

The statement (a) of this theorem is due to Aronszajn [2], the remaining statements are easily deducible from the results of [2].

Let us consider kernels R and B. Let $R(. , t) \in H(B)$ at all $t \in T$. Then we can define the operator L_R^B converting every function φ from $H(B)$ into the function $L_R^B \varphi$ given by the expression:

$$(L_R^B\varphi)(t) = < R(., t), \varphi(.) >_{H(B)}. \qquad (1.2)$$

Theorem 1.5. $R(. , t) \in H(B)$ at all $t \in T$ and the operator L_R^B is a bounded linear operator in the space $H(B)$, iff $R \lhd kB$. In that case

a) L_R^B is a positive Hermitian operator in $H(B)$ and its norm

$$|| L_R^B || = k_0 = \min\{k : R \lhd kB\};$$

b) L_R^B is a continuous linear operator from $H(B)$ into $H(R)$ with the norm $|| L_R^B || = \sqrt{k_0}$;

c) the Hermitian operator $(L_R^B)^{1/2}$ in the space $H(B)$ defined by the relation $L_R^B = (L_R^B)^{1/2}(L_R^B)^{1/2}$ maps the subspace $H_B(R)_0$ onto the set $H(R)$ in a one-to-one manner; it is an isometry from the space $H_B(R)_0$ onto the space $H(R)$;

d) $L_R^B \varphi = 0$ iff $\varphi \in H(B) \ominus H_R(B)$.

The proofs of all statements of Theorem 1.5 can be found in Aronszajn's paper [2].

§ 2. HILBERT SPACES OF RANDOM VARIABLES

1. Basic definitions. Let $(\Omega, \mathfrak{B})$ be a measurable space. We shall call a *random variable* (r.v.) any measurable real-valued function $\xi = \xi(\omega)$ on Ω. A family of random variables $\xi(t)$, $t \in T$ is called *a random function* (r.f.) on the set T. Random functions on the r-dimensional real Euclidean space $\mathfrak{R}^r$ or on the r-dimensional integer lattice Z^r are called *random fields*; in the one-dimensional case, special terminology is also used: a r.f. on the real axis $\mathfrak{R}$ is called a *random process*, and on Z a *random sequence*.

If a probability measure P is given on $(\Omega, \mathfrak{B})$, we can consider in relation to it all the various characteristics of the r.v. ξ, including the *expectation* $E_p\xi$ and the *variance* $D_p\xi$. Further $\xi(t)$ is a fixed r.f. everywhere and *we shall consider only such measures P on $\mathfrak{B}$* for which $E_p[\xi(t)]^2 < \infty$ *for any* $t \in T$. In relation to such a measure for the r.f. $\xi(t)$ *its expectation* $m(t) = E_p\xi(t)$, *correlation function* $K(s, t) = E_p\xi(s)\xi(t)$ and *proper correlation function* $R(s, t) = E_p(\xi(s) - m(s))(\xi(t) - m(t))$ are defined. Obviously, $K(s, t) = R(s, t) + m(s)m(t)$. It is well known that $K(s, t)$ and $R(s, t)$ are kernels in the sense of § 1.

Along with r.f. $\xi(t)$ we shall often consider the r.f. $\xi_0(t) = \xi(t) - m(t)$, obtained by eliminating the determinate "trend" $m(t) = E_p\xi(t)$. The characteristics and properties of the r.f. $\xi_0(t)$ will be called the *proper* characteristics and properties of r.f. $\xi(t)$; a case in point is the proper correlation function $R(s, t)$ introduced above.

One can relate with any measure P the real Hilbert space of random variables $H_p(\xi)$ with the scalar product $\langle \eta, \zeta \rangle_{H_p(\xi)} = E_p\eta\zeta$, generated by the random variables $\xi(t)$, $t \in T$. More exactly, the elements of the space $H_p(\xi)$ are classes of P-equivalent r.v.; in the subsequent discussion we use the same notation both for such a class and any of its representatives. The convergence of a sequence of elements η_n to the element η in the space $H_p(\xi)$ is the mean square convergence of r.v. η_n to r.v. η in relation to the measure $P: \lim_{n\to\infty} E_p(\eta_n - \eta)^2 = 0$; we shall write this as follows: $(P) \lim_{n\to\infty} \eta_n = \eta$.

Let $\mathcal{P}_K$ (resp. $\mathcal{P}_{R,m}$) be the class of all probability measures in relation to which the r.f. $\xi(t)$ has the correlation kernel K (resp. the proper correlation kernel R and the expectation $m(\,.)$). If P and $Q \in \mathcal{P}_K$, then the correspondence $\xi(t) \leftrightarrow \xi(t)$ establishes the natural isometric isomorphism between the spaces $H_p(\xi)$ and $H_Q(\xi)$. We shall often use the same notations for elements of different spaces H_p, $P \in \mathcal{P}_K$, corresponding to each other under this isomorphism.

2. *Canonical isometry.* In this subsection an arbitrary measure $P \in \mathcal{P}_K$ is fixed.

Definition 2.1. Let B be an arbitrary kernel on T; the mapping $\varkappa$ of the space $H(B)$ onto $H_p(\xi)$ will be called *canonical,* if $\varkappa\big(B(\cdot, t)\big) = \xi(t)$ for any $t \in T$.

Underlying all applications of the theory of reproducing kernels to random functions is the following fact, established by M. Loève.

Theorem 2.1. There exists a unique continuous linear canonical mapping i of the space $H(K)$ into $H_p(\xi)$; this mapping is an isometry from $H(K)$ onto $H_p(\xi)$.

Following Parzen [14], let: $i(\varphi) = \langle \xi(.), \varphi(.)\rangle_{H(K)}$. If the set T is finite, then, with probability 1, the realizations of $\xi(t)$ belong to the space $H(K)$; then $\langle \xi, \varphi \rangle_{H(K)}$ coincides with the scalar product of the functions $\xi(.)$ and $\varphi(.)$ in $H(K)$. If the set T is infinite, the realizations of $\xi(t)$, excepting a few very special situations, do not belong to $H(K)$; in that case $\langle \xi, \varphi \rangle_{H(K)}$ is merely a symbol. Nevertheless, this notation is convenient from the heuristic point of view: a sense can be given to the formally written expression $\langle \xi, \varphi \rangle_{H(K)}$, by interpreting all linear operations in it in "P-stochastic" terms, i.e. as limits in the mean square in relation to the measure P; then $\langle \xi, \varphi \rangle_{H(K)} = i(\varphi)$. Thus, the notation $\langle \xi, \varphi \rangle_{H(K)}$ suggests explicit expressions for the canonical isometry i. To some extent this useful heuristic rule is explained by Theorem 2.3 given below, which shows that the r.v. $\langle \xi, \varphi \rangle_{H(K)}$ is the P-stochastic limit of the net of r.v. $\langle \xi^\alpha, \varphi^\alpha \rangle_{H(K^\alpha)}$ constructed by the restrictions of the functions ξ, φ and K onto finite sets $T_\alpha \subset T$ $(\bigcup_{\alpha \in A} T_\alpha = T)$.

Theorem 2.2. ([7], [13]–[15]). If $E_p\xi(t) \equiv m(t)$ then for any $\varphi \in H(K)$

$$E_p\langle \xi, \varphi \rangle_{H(K)} = \langle m, \varphi \rangle_{H(K)}. \tag{2.1}$$

Let us formulate a simple but useful assertion on the connection between the spaces $H_p(\xi^\alpha)$ and $H_p(\xi)$.

Theorem 2.3. Let $T_\alpha \subset T$ and ξ^α, φ^α, K^α be the restrictions of the functions ξ, φ, and K onto T_K. The Hilbert space $H_p(\xi^\alpha)$ is a subspace of $H_p(\xi)$; then

$$\langle \xi^\alpha, \varphi^\alpha \rangle_{H(K^\alpha)} = \langle \xi, \Pi^{H(K)}_{H\alpha(K)} \varphi \rangle_{H(K)}. \tag{2.2}$$

Now let us give a stochastic analogue of Theorem 1.3.

Theorem 2.4. Let $\{T_\alpha, \alpha \in A\}$ be an increasing net of sets, $\bigcup_{\alpha \in A} T_\alpha = T$ and $P \in \mathcal{P}_K$; then, whatever the function $\varphi \in H(K)$, there exists the limit

$$(P) \lim_{\alpha \in A} \langle \xi^\alpha, \varphi^\alpha \rangle_{H(K^\alpha)} = \langle \xi, \varphi \rangle_{H(K)}$$

(the equality holds with P-probability 1).

This statement is a simple corollary of Theorem 2.3 and of the theorem on convergence of monotone sequences of projectors.

3. P-scalars in $H_p(\xi)$. We shall call a *P-scalar* a random variable γ, P-almost everywhere equals to a real constant. For the further discussion it is essential to establish a criterion permitting to judge the presence or absence of non-zero P-scalars in $H_p(\xi)$ i.e. scalars $\gamma \neq 0$ P-almost everywhere. We shall presently see that this question is connected with the relation between the mean value $m(\,.)$ of the r.f. $\xi(t)$ and its proper correlation kernel R. Let us denote by $\mathfrak{R}_p$ the one-dimensional Hilbert space consisting of classes of P-equivalent P-scalars. As previously, K is the "complete" correlation kernel of r.f. $\xi(t)$. It is easy to deduce the following lemma from Theorem 2.2.

Lemma 2.1. The functional $l(\gamma) = E_p(\gamma)$ is a continuous linear functional on $H_p(\xi)$; its norm $\| l \| = \| m \|_{M(K)}$.

Additional information about the norm of the functional l is given by

Lemma 2.2. 1) $\| l \| \leq 1$; 2) $\| l \| < 1$, iff $H_p(\xi) \cap \mathfrak{R}_p = \{0\}$.

Proof. The first statement follows from the relations:

$$| l(\gamma) | = | E_p\gamma | \leq (E_p\,\gamma^2)^{1/2} = \| \gamma \|_{H_p(\xi)}\,. \tag{2.3}$$

The second statement of the lemma follows from the fact that the equality in (2.3) is true only if $\mathfrak{D}_p\gamma = E_p\gamma^2 - (E_p\gamma)^2 = 0$, that is when $\gamma \in \mathfrak{R}_p$.

The criterion being sought for follows from Lemmas 2.1, 2.2 and 1.1:

Theorem 2.5. The necessary and sufficient condition for non-zero P-scalars to be absent in $H_p(\xi)$ is that the expectation $m(\,.) \in H(R)$.

4. Spaces $H_p^*(\xi)$ and $H_p^\varphi(\xi)$. Suppose that $m(\,.) \in H(R)$, then, as we have just seen, there are no non-zero scalars in the space $H_p(\xi)$ and, hence, it is possible to introduce in it the new scalar product: $\langle \eta, \zeta \rangle^* = E_p(\eta - E_p\eta) \times$ $\times (\zeta - E_p\zeta)$. The space $H_p(\xi)$ with this scalar product will be denoted by $H_p^*(\xi)$. It can be easily seen that $H_p^*(\xi)$ is a Hilbert space and for any $\eta \in H_p(\xi)$

$$(1 - \| l \|^2)^{1/2} \| \eta \|_{H_p(\xi)} \leq \| \eta \|_{H_p^*(\xi)} \leq \| \eta \|_{H_p(\xi)} \tag{2.4}$$

where l is the functional considered in subsection 3.

Let us remind of the notation: $\xi_0(t) = \xi(t) - m(t)$. It is easy to prove the following statement.

Lemma 2.3. If $m \in H(R)$, then the correspondence $\gamma \leftrightarrow \gamma - E_p\gamma$ is an isometric isomorphism between the Hilbert spaces $H_p^*(\xi)$ and $H_p(\xi_0)$; $\xi(t) \leftrightarrow \xi_0(t)$ for every $t \in T$. We designate:

$$\langle \xi, \varphi \rangle_{H(R)} = \langle \xi_0, \varphi \rangle_{H(R)} + \langle m, \varphi \rangle_{H(R)}\,. \tag{2.5}$$

The following well known proposition can be inferred from Theorem 2.1 and Lemma 2.3.

Theorem 2.6. If $m \in M(R)$, the mapping $\varphi \xrightarrow{i_p^*} \langle \xi, \varphi \rangle_{H(R)}$ is the unique canonical isometry from $H(R)$ onto $H_p^*(\xi)$.

If $m \notin H(\mathcal{R})$, the bilinear form $\langle \eta, \zeta \rangle^* = E\zeta(\eta - E_p\eta)(\zeta - E_p\zeta)$ is no longer a scalar product: $\langle \eta, \eta \rangle^* = 0$ for any $\eta \in \mathcal{R}_p$. Therefore a generalized construction is given. Let us consider a function $\psi \in H(R)$ and denote: $\vartheta(t) = m(t) - \psi(t)$, $\xi_\psi(t) = \xi_0(t) + \psi(t)$; $\xi_\vartheta(t) = \xi_0(t) + \vartheta(t)$. Let P_ϑ be such a measure on $\mathcal{B}$ that the P_ϑ-distribution of the r.f. $\xi(t)$ coincides with the P-distribution of the r.f. $\xi_\vartheta(t)$. It can be easily proved that the bilinear form

$$\langle \eta, \zeta \rangle^{(\psi)} = E_p(\eta - E_p\, \eta)\,(\zeta - E_p\, \zeta) + E_{p\vartheta}\eta \cdot E_{p\vartheta}\, \zeta$$

possesses all the properties of a scalar product. The space $H_p(\xi)$ with this scalar product is complete; we shall denote it by $H_p^\psi(\xi)$. The spaces $H_p(\xi)$ and $H_p^*(\xi)$ belong to the family of spaces $\{H_p^\psi(\xi), \psi \in H(R)\}$; namely: $H_p(\xi) = H_p^0(\xi)$ and $H_p^*(\xi) = H_p^m(\xi)$ if $m \in H(R)$. It is easy to see the truth of the following statement.

Theorem 2.7. If $m(\,.) \notin H(\mathcal{R})$, then $H_p^\psi(\xi) = H_p^*(\xi_\psi) \oplus \mathcal{R}_p$.

The following obvious corollary of Eq. (2.4) and Theorem 2.7 is valid both for $m \in H(\mathcal{R})$ and for $m \notin H(\mathcal{R})$.

Theorem 2.8. The identical mapping $H_p^\psi(\xi) \leftrightarrow H_p(\xi)$ is continuous in both directions.

5. M-majorants of a kernel. Let $R(s, t)$, $s, t \in T$, be a kernel; M be an n-dimensional space of real-valued functions on T; $\varphi_1, \ldots, \varphi_u$ — a finite set of functions from M; $B = \{b_{ij}\}_1^u$ — a positively definite matrix (it may be degenerate). The function

$$B(s, t) = R(s, t) + \sum_{i,j=1}^{u} b_{ij}\, \varphi_i(s)\, \varphi_j(t) \tag{2.6}$$

is a kernel.

Definition 2.2. Every kernel of the form (2.6) possessing the property $M \subseteq H(B)$ will be called an *M-majorant* of the kernel R.

Let $M^{(\varphi)}$ be the span of the functions $\varphi_1, \ldots, \varphi_u$. It can be easily seen that if the matrix $\{b_{ij}\}$ is nondegenerate and $[M \cap H(R)] + M^{(\varphi)} = M$, then the kernel B defined by (2.6) is an M-majorant of R. If $M \subset H(R)$, the kernel R is an M-majorant of itself.

Let us fix an M-majorant $\tilde{R}$ of R and consider M as a subspace of $H(\tilde{R})$. Let $M_\psi = M \cap H(R)$, $M_\vartheta = M \ominus M_\psi$ and let $M_\vartheta = H(\Theta)$, i.e. Θ is the reproducing kernel of the space M_ϑ. Evidently, $M = M\psi \oplus M_\vartheta$ and we have the following decomposition of any function $m \in M$:

$$m(t) = \psi(t) + \vartheta(t), \quad \text{where} \quad \psi = \Pi^M_{M_\psi} m, \ \vartheta = \Pi^M_{M\vartheta}\, m.$$

This decomposition is uniquely defined by the M-majorant $\tilde{R}$. Thus, for every $m \in M$, $\tilde{R}$ defines a unique space $H_p^\psi(\xi)$.

6. Embeddings. Let P and Q be probability measures on $\mathcal{B}$; as previously, we assume that $E_p[\xi(t)]^2 < \infty$ and $E_Q[\xi(t)]^2 < \infty$ for any $t \in T$.

Definition 2.3. An element $\gamma' \in H_p(\xi)$ is called *the linear image of the element* $\gamma \in H_p(\xi)$, if $(P) \lim_{n\to\infty} \sum_{i=1}^{k_n} a_i^{(n)}\xi(t_i^{(n)}) = \gamma'$ provided $(Q) \lim_{n\to\infty} \sum_{i=1}^{k_n} a_i^{(n)}\xi(t_i^{(n)}) = \gamma$ (here $a_i^{(n)} \in \mathcal{R}$), $t_i^{(n)} \in T$).

Evidently, an element $\gamma \in H_Q(\xi)$ can possess only one linear image in $H_p(\xi)$.

Definition 2.4. The Hilbert space $H_Q(\xi)$ is *embeddable* into the Hilbert space $H_p(\xi)$, if any element $\gamma \in H_Q(\xi)$ has a linear image γ' in $H_p(\xi)$; the operator $A: \gamma \to \gamma'$ will be called the *embedding* of the Hilbert space $H_Q(\xi)$ into $H_p(\xi)$.

Note 2.1. If an element $\gamma \in H_Q(\xi)$ possesses a linear image γ' in $H_p(\xi)$, then it also has a linear image γ'' in the Hilbert space $\frac{H_{p+Q}}{2}(\xi)$; the class γ'' consists of random variables belonging both to the class γ' and to the class γ. Hence, any r.v. from γ can be re-defined on a set of Q-measure zero so that it will belong to the class γ'. This re-definition can be achieved, according to Definition 2.3, by interpreting all Q-stochastic linear operations, defining the r.v. from γ, as P-stochastic linear operations. For instance, if $T = [a, b]$ and $\gamma = (Q) \int_a^b \xi(t)\varphi(t)dt$ (Q-stochastic integral) possesses a linear image γ' in the space $H_p(\xi)$, then there also exists the P-stochastic integral $(P)_p \int_a^b (t)\varphi(t)dt = \gamma'$.

The following simple lemma characterizes the embedding operators.

Lemma 2.4. An operator A from $H_Q(\xi)$ into $H_p(\xi)$ is an embedding if: 1) A is a continuous linear operator from $H_Q(\xi)$ into $H_p(\xi)$; 2) $A\xi(t) = \xi(t)$ for any $t \in T$.

Let R and B be kernels on T; M be a finite-dimensional space of functions on T, $M \subset H(B)$; $\tilde{R}$ is an M-majorant of the kernel R. We consider the decomposition $M = M_\varphi \oplus M_\vartheta$ (in $H(\tilde{R})$) connected with $\tilde{R}$ and use the notations of the previous subsections. If $R \lhd kB$, we can regard the bounded linear operators L_R^B and L_θ^B, defined in subsection 4 of § 1, from $H(B)$ into $H(R)$ and from $H(B)$ into $M_\vartheta = H(\Theta)$, respectively.

Theorem 2.9. If $R \lhd kB$ and $m \in H(B)$, every space $H_Q(\xi)$ with $Q \in \mathscr{P}_{B,m}$ is embeddable into any space $H_p(\xi)$ with $P \in \mathscr{P}_{R,m}$. If $M \subset H(B)$ then for every $m \in M$ the embedding A_R^B is given* by

$$A_R^B\langle \xi, \varphi\rangle_{H(B)} = \langle \xi_\varphi, L_R^B\varphi\rangle_{(H)R} + \langle \vartheta, L_\theta^B\varphi\rangle_{H(\theta)}. \qquad (2.7)$$

Proof. We have: $L_R^B B(.\,,t), = R(.\,,t), L_\theta^B B(.\,,t) = \theta(.\,,t)$.

Hence, $A_R^B \xi(t) = A_R^B\langle \xi(\,.\,), \quad B(\,.\,,t)\rangle_{H(B)} = \langle \xi(\,.\,), \quad R(\,.\,,t)\rangle_{H(R)} +$

$$\langle \vartheta(\,.\,), \Theta(.\,,t)\rangle_{H(\theta)} = \xi_\varphi(t) + \vartheta(t) = \xi(t)\,.$$

* Up to the natural isomorphism.

The linearity of the operator A_R^B is obvious; it is also easy to prove that it is bounded:

$$||A_R^B\langle \xi, \varphi\rangle_{H(B)}||^{2\Psi}_{H_p(\xi)} = ||\langle \xi_\Psi, L_R^B \varphi\rangle_{H(R)}||^2_{H_p^*(\xi_\Psi)} + |\langle \vartheta, L_\theta^B \varphi\rangle_{H(B)}|^2 =$$

$$= || L_R^B\varphi ||^2_{H(R)} + |\langle \vartheta, L_\theta^B\varphi\rangle_{H(\theta)}|^2 \leq k \,|| \varphi ||^2_{H(B)} + k_\theta \,|| \vartheta ||^2_{H(\theta)} \,|| \varphi ||^2_{H(B)} =$$

$$= (k, + k_\theta \,|| \vartheta ||^2_{H(\theta)}) \,|| \langle \xi, \varphi\rangle_{H(B)}||^2_{H_Q^*(\xi)},$$

where k_θ is a constant such that $\Theta \lhd k_\theta B$; it remains to make use of Theorem 2.8 and Lemma 2.4.

The following theorem is the converse of Theorem 2.9.

Theorem 2.10. Let R and B be kernels on T. If for some function $m \in H(B)$ and for some measures $Q \in \mathscr{P}_{B,m}$ and $P \in \mathscr{P}_{R,m}$ the space $H_Q(\xi)$ is embeddable into the space $H_p(\xi)$, then $R \lhd kB$ for some $k < \infty$.

Proof. Let us consider the random variables $\sum_{i=1}^n a_i\xi(t_i)$, where a_i are real numbers and $t_i \in T$. We have:

$$\sum_{i,j=1}^n a_i a_j R(t_i, t_j) = E_p\left[\sum_{i=1}^n \left(a_i\,\xi(t_i) - a_i\,m(t_i)\right)\right]^2 \leq$$

$$\leq \left\|\sum_{i=1}^n a_i\,\xi(t_i)\right\|^2_{H_p^\Psi(\xi)} \leq ||A||^2 \left\|\sum_{i=1}^n a_i\,\xi(t_i)\right\|^2_{H\ (\xi)} = ||A||^2 \sum_{i,j=1}^n a_i a_j B(t_i, t_j)$$

where $||A||$ is the norm of the embedding A as an operator from $H_Q^*(\xi)$ into $H_p^\Psi(\xi)$. The theorem is proved.

The following properties of the embedding of A_R^B valid for any $P \in \mathscr{P}_{R,m}$, $Q \in \mathscr{P}_{B,m}$, $m \in M$ are inferrable from Theorem 2.9.

Corollary 2.1. The element $A_R^B\langle \xi, \varphi\rangle_{H(B)}$ is a P-scalar iff $L_R^B\varphi = 0$, i.e. iff $\varphi \in H(B) \ominus H_B(R)$.

Corollary 2.2. If $M \subseteq H(R)$, then $A_R^B\langle \xi, \varphi\rangle_{H(B)} = \langle \xi, \mathfrak{L}_R^B\varphi\rangle_{H(R)}$

Corollary 2.3. $A_R^B = A_R^{\tilde{R}} A_{\tilde{R}}^B$

Corollary 2.4. R.v. $A_R^{\tilde{R}}\langle \xi, \varphi\rangle_{H(\tilde{R})}$ is a P-scalar, iff $\varphi \in H(\theta) = H(\tilde{R}) \ominus H(R)$.

Note 2.2. According to subsection 2 of this section, the symbol $\langle \xi, \varphi\rangle_{H(B)}$ denotes the class of Q-equivalent r.v., which is an element in $H_Q(\xi)$ at a fixed measure $Q \in \mathscr{P}_{B,m}$. This symbol will often be used also in the further discussion to designate the linear image of that element in any space $H_p(\xi)$. In that case, our formal "Q-stochastic scalar product" $\langle \xi, \varphi\rangle_{H(B)}$ must be re-defined into the "P-stochastic scalar product", in conformity with Note 2.1. We have already proceeded in that manner more than once. For instance, we used the same symbol $\langle \xi, \varphi\rangle_{H(R)}$ to designate the element $i_p(\varphi)$ in $H_p(\xi)$ and $i_Q(\varphi)$ in $H_Q(\xi)$, when $P \neq Q$ and $P, Q \in \mathscr{P}_R$. Further, the same symbol $\langle \xi, \varphi\rangle_{H(R)}$ was also used to designate the element $i_p^*(\varphi)$ in the case when $P \in \mathscr{P}_{R,m}$ and $m \in H(R)$; this is well justified, since at $P \in \mathscr{P}_{R,m}$,

$Q_0 \in \mathscr{P}_{R,0}$ and $Q \in \mathscr{P}_R$ each of the spaces $H_p^*(\xi)$, $H_{Q_0}(\xi)$ and $H_Q(\xi)$ is embeddable into the remaining ones and for any function $\varphi \in H(R)$ the elements $i_p^*(\varphi)$, $i_{Q_0}(\varphi)$ and $i_Q(\varphi)$ are the linear images of one another.

We shall need further the following property of "unbiasedness" of the embedding A_R^B.

Theorem 2.11. Under the conditions of Theorem 2.9 for any function $\varphi \in H(B)$

$$E_p A_R^B \langle \xi, \varphi \rangle_{H(B)} = E_Q \langle \xi, \varphi \rangle_{H(B)} = \langle m, \varphi \rangle_{H(B)}.$$

Proof. By virtue of Eq. (2.7), Theorem 2.2 and the assertions (c) and (d) of Theorem 1.5

$$E_p A_R^B \langle \xi, \varphi \rangle_{H(B)} = E_p \langle \xi_\psi, L_R^B \Pi_{H_B(R)}^{H(B)} \varphi \rangle_{H(R)} + \langle \vartheta, L_\Theta^B \Pi_{H(\Theta)}^{H(B)} \varphi \rangle_{H(\Theta)} =$$

$$= \langle \psi, L_R^B \Pi_{H(R)}^{H(B)} \varphi \rangle_{H(R)} + \langle \vartheta, L_\Theta^B \Pi_{H(\Theta)}^{H(B)} \varphi \rangle_{H(\Theta)} =$$

$$= \langle \psi, \Pi_{H_B(R)}^{H(B)} \varphi \rangle_{H(B)} + \langle \vartheta, \Pi_{H(\Theta)}^{H(B)} \varphi \rangle_{H(B)} = \langle \psi, \varphi \rangle_{H(B)} + \langle \vartheta, \varphi \rangle_{H(B)} = \langle m, \varphi \rangle_{H(B)}.$$

We conclude this section by a simple example illustrating Note 2.2, Theorem 2.7 and Corollaries 2.1 and 2.4.

Example 2.1. Let $\xi(t)$ be a random process on the segment $[0, 1]$ with $m(t) \equiv 1$ and $R(s, t) = \min(s, t)$. We assume $\tilde{R}(s, t) = \min(s, t) + 1$. The space $H(R)$ consists of absolutely continuous functions φ such that $\varphi'(\,.\,) \in \in \mathfrak{L}^2[0, 1]$ and $\varphi(0) = 0$. Evidently, $H(\tilde{R}) = H(R) \oplus \mathfrak{R}$ and $H(\tilde{R})$ consists of absolutely continuous functions φ, the derivatives of which $\varphi' \in \mathfrak{L}^2[0, 1]$. If $\psi, \varphi \in H(R)$, then $\langle \psi, \varphi \rangle_{H(R)} = \int_0^1 \psi'(t)\, \varphi'(t)\, dt$, $\langle \xi, \varphi \rangle_{H(R)} = \int_0^1 \varphi'(t)\, d\xi(t)$. If $\psi, \varphi \in H(\tilde{R})$, then $\langle \psi, \varphi \rangle_{H(\tilde{R})} = \int_0^1 \psi'(t)\, \varphi'(t)\, dt + \psi(0)\, \varphi(0)$; $\langle \xi, \varphi \rangle_{H(\tilde{R})} = = \int_0^1 \varphi'(t)\, d\xi(t) + \xi(0)\, \varphi(0)$. At $\varphi \equiv m \equiv 1$ the symbol $\langle \xi, \varphi \rangle_{H(R)}$ is not defined, and $\langle \xi, \varphi \rangle_{H(\tilde{R})} = \xi(0)$. Further, $D_p \xi(0) > 0$ at $P \in \mathscr{P}_{\tilde{R},m}$ (since $m \in H(\tilde{R})$ there are no non-zero P-scalars in $H_p(\xi)$); if, however, $P \in \mathscr{P}_{R,m}$, then $\xi(0) = 1$ with P-probability 1.

§ 3. BEST UNBIASED LINEAR REGRESSION ESTIMATES

1. Problem statement. Unbiased linear regression estimates. Let $\xi(t), t \in T$ be a r. f. with known proper correlation function R; let it be also known that the mathematical expectation of r.f. $\xi(t)$ is $m(t) = \sum_{i=1}^{n} a_i\, \varphi_i(t)$, where $n < \infty$ and $\varphi_1, \ldots, \varphi_n$ are known functions on T, and a_i unknown constants; in other words, it is known that $m(\,.\,) \in M$ the finite-dimensional space spanned by the functions $\varphi_1, \ldots, \varphi_n$; without loss of generality, we can assume that the functions $\varphi_1, \ldots, \varphi_n$ are linearly independent. It is necessary to estimate the function $m(t)$, or, which is the same, the coefficients a_i on the base of an observed realization of the r.f. $\xi(t)$.

Let us recall that $\mathscr{P}_{R,m}$ is the family of all probability measures on $\mathscr{B}$ in relation to which the proper correlation kernel of $\xi(t)$ is $R(s,t)$ and $E_p\xi(t) = m(t)$; $\mathscr{P}_{R,M} = \bigcup_{m \in M} \mathscr{P}_{R,m}$.

Definition 3.1. A r.f. $\hat{m}(t)$ is called a *linear regression estimate* relative to the family $\mathscr{P}_{R,M}$ based on the realization of $\xi(t)$ on T (in abbreviate form $\mathscr{P}_{R,M}$-l.e.) if: 1) for any $t \in T$ and $P \in \mathscr{P}_{R,M}$ the r.v. $\hat{m}(t) \in H_p(\xi)$ and 2) for any pair of measures $P, Q \in \mathscr{P}_{R,M}$ and for every $t \in T$ the elements $\hat{m}(t)$ in $H_p(\xi)$ and in $H_Q(\xi)$ are linear images of each other.

Condition 2) contains the natural requirement of "operational independence" of the estimate $\hat{m}(t)$ of the unknown parameter $m(t)$.

Definition 3.2. A $\mathscr{P}_{R,M}$-l.e. $m(t)$ is called an *unbiased* $\mathscr{P}_{R,M}$-l.e. ($\mathscr{P}_{R,M}$ u.l.e.) if $E_p\hat{m}(t) = E_p\xi(t)$ on T for every $P \in \mathscr{P}_{R,M}$.

Definition 3.3. A $\mathscr{P}_{R,M}$-u.l.e. $\hat{m}(t)$ is called the *best* $\mathscr{P}_{R,M}$-u.l.e. if $D_p\hat{m}(t) \leq D\hat{n}(t)$ for all $t \in T$ and $P \in \mathscr{P}_{R,M}$ whatever the $\mathscr{P}_{R,M}$-u.l.e. $\hat{n}(t)$ is.

2. *Estimate* $\hat{m}_{R,M}(t)$. Let $P \in \mathscr{P}_{R,M}$ and $\tilde{R}$ be a M-majorant of the kernel R. Consider in $H_p(\xi)$ the elements*

$$\hat{m}_{R,M}(t) = \langle \xi(\,.\,),\ K_M^{\tilde{R}}(\,.\,,t)\rangle_{H(\tilde{R})}, \quad t \in T. \tag{3.1}$$

Recall that the function $K_M^{\tilde{R}}(\,.\,,t) = \Pi_M^{H(\tilde{R})}\tilde{R}(\,.\,,t)$ is the reproducing kernel of the subspace M in $H(\tilde{R})$. First of all let us show that the element $\hat{m}_{R,M}(t)$ is completely defined by the kernel R and the space M.

Lemma 3.1. Equation (3.1) defines one and the same element $\hat{m}_{R,M}(t)$ in $H_p(\xi)$ irrespective of the choice of the M-majorant $\tilde{R}$; in particular, if $M \subseteq H(R)$, then $\hat{m}_{R,M}(t) = \langle \xi(\,.\,),\ K_M^R(\,.\,,t)\rangle_{H(R)}$.

Proof. Let $\tilde{R}_1 = R + \Phi_1$ and $\tilde{R}_2 = R + \Phi_2$ be two M-majorants of R. Let $M_\Psi = M \cap H(R)$ and let R_0 be the reproducing kernel of the subspace $H(R) \ominus M_\Psi$. We have $R = R_0 \oplus K_{M_\Psi}^R$, $\tilde{R}_i = R_0 \oplus (K_{M_\Psi}^R + \varphi_i)$, $(i = 1, 2)$ and using Theorem 1.1 we obtain $H(\tilde{R}_1) = H(\tilde{R}_2) = H(R_0) + M$ (as linear spaces) and $\tilde{R}_i = R_0 \oplus K_M^{\tilde{R}_i}$. Hence, $\tilde{R}_1 \lhd k\tilde{R}_2$ and according to Theorem 2.9 for every $m \in M$ the spaces $H_Q(\xi)$, $Q \in \mathscr{P}_{\tilde{R}_1,m}$, are embeddable into any space $H_p(\xi)$, $P \in \mathscr{P}_{\tilde{R}_2,m}$. For every $t \in T$ we have

$$\begin{aligned} L_{\tilde{R}_1}^{\tilde{R}_2} K_M^{\tilde{R}_2}(s,t) &= \langle \tilde{R}_1(\,.\,,s),\ K_M^{\tilde{R}_1}(\,.\,,t)\rangle_{H(\tilde{R}_1)} = \\ &= \langle R_0(\,.\,,s) + K_M^{\tilde{R}_1}(\,.\,,s),\ K_M^{\tilde{R}_2}(\,.\,,t)\rangle_{H(\tilde{R}_1)} = \\ &= \langle K_M^{\tilde{R}_1}(\,.\,,s),\ K_M^{\tilde{R}_2}(\,.\,,t)\rangle_{H(\tilde{R}_1)} = K_M^{\tilde{R}_1}(s,t) \end{aligned}$$

* See Note 2.2 on the notations used.

and thus, $A^{\tilde{R}_2}_{\tilde{R}_1}\langle\xi(\,.\,), K^{\tilde{R}_2}_{M}(\,.\,,t)\rangle_{H(\tilde{R}_2)} = \langle\xi(\,.\,), K^{\tilde{R}_1}_{M}(\,.\,,t)\rangle_{H(\tilde{R}_1)}$. Since $A^{\tilde{R}_2}_{R} = A^{\tilde{R}_1}_{R} A^{\tilde{R}_2}_{\tilde{R}_1}$, the theorem is proved.

Evidently the element $\hat{m}_{R,M}(t)$ in $H_p(\xi)$, $P \in \mathscr{P}_{R,M}$, is the linear image of the element $\hat{m}_{\tilde{R},M}(t) \in H_Q(\xi)$ where $Q \in \mathscr{P}_{R,M}$.

Let $\{\varphi_1, \ldots, \varphi_n\}$ be a basis in M; it may be easily verified that $K^{\tilde{R}}_{M}(s,t) = \sum_{i,j=1}^{n} d_{ij}\varphi_i(s)\varphi_j(t)$ where $\{d_{ij}\}_1^n = \mathscr{G}^{-1}$ and $\mathscr{G} = \{\langle\varphi_i, \varphi_j\rangle_{H(\tilde{R})}\}_1^n$ is the Gram matrix of the basis $\{\varphi_1, \ldots, \varphi_n\}$. Hence

$$\hat{m}_{R,M}(t) = \sum_{j=1}^{n} \hat{a}_j \varphi_j(t) \tag{3.2}$$

where

$$\hat{a}_j = \sum_{i=1}^{n} d_{ij}\langle\xi, \varphi_i\rangle_{H(\tilde{R})}; \tag{3.3}$$

in other words, the coefficients $\hat{a}_i$ are the unique solution of the "normal" system of equations:

$$\sum_{j=1}^{n}\langle\varphi_i, \varphi_j\rangle_{H(\tilde{R})}\hat{a}_j = \langle\xi, \varphi_i\rangle_{H(\tilde{R})}, \, i = 1, \ldots, n. \tag{3.4}$$

If the basis $\{\varphi_i\}$ is orthonormal, then $\hat{a}_i = \langle\xi, \varphi_i\rangle_{H(\tilde{R})}$.

Let $M = M_\Psi \oplus M_\vartheta$ be a decomposition of the subspace M corresponding to $\tilde{R}$ (see subsection 5 of § 2), i.e. $M_\Psi = M \cap H(R)$ and $M_\vartheta = H(\theta) = M \ominus M_\Psi$; $\tilde{R}_1 = R \oplus \theta$. Evidently, $\tilde{R}_1$ is also an M-majorant of R and $L^{\tilde{R}_1}_{R}\varphi = \varphi$ for any $\varphi \in M_\Psi$. Let $\{\psi_1, \ldots, \psi_k\}$ and $\{\vartheta_1, \ldots, \vartheta_l\}$ be some orthonormal bases in $H(K^{R}_{M\Psi})$ and in M_ϑ, respectively. Any function $m \in M$ is representable as:

$$m(t) = \sum_{i=1}^{k} b_i\psi_i(t) + \sum_{i=1}^{l} c_i\vartheta_i(t) = \psi(t) + \vartheta(t), \tag{3.5}$$

where $\psi = \sum_{i=1}^{k} b_i\psi_i = \Pi^{H(\tilde{R}_1)}_{M_\Psi} m$ and $\vartheta = \sum_{i=1}^{l} c_i\vartheta_i = \Pi^{H(\tilde{R}_1)}_{M_\vartheta} m$.

Formulas (2.7), (3.2) and (3.3) assume the appearance:

$$\hat{m}_{R,M}(t) = \sum_{i=1}^{k}\hat{b}_i\psi_i(t) + \sum_{i=1}^{l}\hat{c}_i\vartheta_i(t) \tag{3.6}$$

where

$$\hat{b}_i = \langle\xi, \psi_i\rangle_{H(\tilde{R}_1)} = \langle\xi_\Psi, \psi_i\rangle_{H(R)}; \tag{3.7'}$$

$$\hat{c}_i = \langle\xi, \vartheta_i\rangle_{H(\tilde{R}_1)} = \langle m, \vartheta_i\rangle_{H(\tilde{R}_1)} = c_i; \tag{3.7''}$$

evidently, $D\hat{b}_i > 0$ and $D\hat{c}_i = 0$.

Formulas (3.2)–(3.3) permit to consider $\hat{m}_{R,M}(t)$ as a r.f. on T, defined for almost all $\omega \in \Omega$. Obviously, this r.f. is a $\mathcal{S}_{R,M'}$-u.l.e.; moreover, $\hat{E}b_i = E\langle \xi, \psi_i \rangle_{H(\tilde{R})} = \langle m, \psi_i \rangle_{H(\tilde{R})} = b_i$ (Theorem 2.11) and $\hat{c}_i = c$. Hence, $\hat{m}_{R,M}(t)$ is a $\mathcal{S}_{R,M}$-u.l.e. Hájek [7] and Parzen [13]–[15] proved that if $M \subseteq H(R)$ the r.f. $\hat{m}_{R,M}(t)$ is the unique* best $\mathcal{S}_{R,M}$-u.l.e.

In the general case it follows that the r.f. $\hat{\psi}(t) = \hat{m}_{R,M}(t) - \vartheta(t) = \hat{m}_{R,M\Psi}(t)$ is the best $\mathcal{S}_{R,M\Psi}$-u.l.e., and $D_p\hat{\psi}(t) = D_p\hat{m}_{R,M}(t)$. If $\hat{m}_1(t)$ is an arbitrary $\mathcal{S}_{R,M}$-u.l.e., then $\hat{\psi}_1(t) = \hat{m}_1(t) - \vartheta(t)$ is $\mathcal{S}_{R,M\Psi}$-u.l.e. and $D_p\hat{m}_1(t) = D_p\hat{\psi}_1(t)$. This implies the following generalization of the above results of Hájek and Parzen:

Theorem 3.1. The r.f. $\hat{m}_{R,M}(t)$ is the unique best $\mathcal{S}_{R,M}$-u.l.e. based on the realization of the r.f. $\xi(t)$ on T.

From formulas (3.6)–(3.7) the criterion of the possibility of error-free estimation of the regression coefficients a_i immediately follows, which is due in the case $n = 1$ to Ylvisaker [23] and in general to Rozanov [16]:

Theorem 3.2. $D_p\hat{m}_{R,M}(t) = 0$ at any measure $P \in \mathcal{S}_{R,M}$ iff $M \cap H(R) = \{0\}$.

A simple example is given to illustrate Lemma 3.1 and Theorem 3.2.

Example 3.1. Consider a random process $\xi(t)$, $t \in [0, 1]$, with the proper correlation kernel $R(s, t) = \min(s, t)$ and the expectation $m(t) = a_1\varphi_1(t) + a_2\varphi_2(t)$ where a_1 and a_2 are unknown constants (we choose the functions φ_1 and φ_2 below). A function $\varphi \in H(R)$ iff it is absolutely continuous, $\varphi(0) = 0$ and $\varphi' \in \mathcal{S}^2[0, 1]$; $\langle \psi, \varphi \rangle_{H(R)} = \int_0^1 \psi'(t)\,\varphi'(t)\,\mathrm{d}t(R)$; $\langle \xi, \varphi \rangle_{H,R} = \int_0^1 \varphi'(t)\,\mathrm{d}\xi(t)$. Three typical cases are considered:

1) $\varphi_1(t) = t$, $\varphi_2(t) = t^2$; in this case $M \subset H(R)$, $k = 2$, $l = 0$. One can take $\psi_1(t) = t$, $\psi_2(t) = \sqrt{3}(t - t^2)$; then $\hat{b}_1 = \langle \xi, \psi_1 \rangle_{H(R)} = \xi(1)$; $\hat{b}_2 = \langle \xi, \psi_2 \rangle_{H(R)} = \sqrt{3}\left(\xi(1) - 2\int_0^1 t\,\mathrm{d}\,\xi(t)\right)$.

2) $\varphi_1(t) = t + 1$, $\varphi_2(t) \equiv 1$; here $\varphi_1, \varphi_2 \notin H(R)$, $k = l = 1$ and $M_\Psi = \{f(\,.\,) : f(t) \equiv ct,\ -\infty < c < \infty\}$. We assume: $\tilde{R}(s, t) = R(s, t) + 1$, $\psi_1(t) = t$, $\vartheta_1(t) \equiv 1$. Then $M_\vartheta = \mathfrak{R}$ (real axis), $\langle \psi, \varphi \rangle_{H(R)} = \int_0^1 \psi'(t)\,\varphi'(t)\,\mathrm{d}t + \psi(0)\,\varphi(0)$; $\langle \xi, \varphi \rangle_{H(\tilde{R})} = \int_0^1 \varphi'(t)\,\mathrm{d}\xi + \xi(0)\,\varphi(0)$. Thus, $\hat{b}_1 = \langle \xi, \psi_1 \rangle_{H(\tilde{R})} = \xi(1) - \xi(0)$, $\hat{c}_1 = \langle \xi, \vartheta_1 \rangle_{H(\tilde{R})} = \xi(0) \equiv 1$ and $\hat{m}_{R,M}(t) = (\xi(1) - \xi(0)t) + \xi(0)$. We can also assume $\tilde{R}(s, t) = R(s, t) + (1 + s)(1 + t)$. Then $\langle \psi, \varphi \rangle_{H(\tilde{R})} = \int_0^1 (\psi(t) - \psi(0)\,t)'\,(\varphi(t) - \varphi(0)\,t)'\,\mathrm{d}t + \psi(0)\,\varphi(0)$; $\langle \xi, \varphi \rangle_{H(\tilde{R})} = \int_0^1 (\varphi(t) - \varphi(0)\,t)'\,\mathrm{d}(\xi(t) - \xi(0)\,t) + \varphi(0)\,\xi(0)$; $\hat{b}_1 = \langle \xi, \varphi_1 \rangle_{H(\tilde{R})} = \xi(1) - 2\xi(0)$;

* Up to P-equivalency.

$\hat{c}_1 = \xi(0)$ and again $\hat{m}_{R,M}(t) = (\xi(1) - \xi(0)t) + \xi(0)$ in accordance with Lemma 3.1. Note that $D_p\hat{m}_{R,M}(t) = t > 0$ at $t > 0$, although φ_1 and $\varphi_2 \notin H(R)$.

3) $\varphi_1(t) \equiv 1$; $\varphi_2(t) = t^{-1}$; in this case $H(R) \cap M = \{0\}$ $(k = 0,\ l = 2)$ and the function $m(t)$ can be estimated without error: $\hat{a}_2 = \lim_{t\to 0} t\xi(t) = a_2$ and $\hat{a}_1 = \lim_{t\to 0} (\xi(t) - a_2 t^{-1}) = a_1$.

§ 4. SIMPLE REGRESSION ESTIMATES AND DEFINING KERNELS

1. Estimates defined by kernels. In the previous section we assumed the proper correlation kernel R of the r.f. $\xi(t)$ to be known and constructed the best $\mathcal{P}_{R,M}$-u.l.e. $\hat{m}_{R,M}(t)$. But the concrete calculation of this estimate is often rather difficult; it becomes, however, quite impossible if the kernel R is unknown. In such cases other estimates instead of $\hat{m}_{R,M}(t)$ should be used. If a kernel B is known to be in some sense "near" to the kernel R, it is natural to make use of the estimate $\hat{m}_{B,M}(t)$ (practically, statistical estimates of the function R are often used as such kernels). This gives rise to two questions: (1) under what conditions is the estimate $\hat{m}_{B,M}(t)$ a $\mathcal{P}_{R,M}$-u.l.e.? (2) which $\mathcal{P}_{R,M}$-u.l. estimates can be obtained in the above way using all possible kernels B? In this subsection we shall discuss the first question, postponing the second one till subsection 2.

Let us consider a $\mathcal{P}_{R,M}$-u.l.e. $\hat{m}(t)$ and a kernel B.

Definition 4.1. We shall say that the $\mathcal{P}_{R,M}$-u.l.e. $\hat{m}(t)$ *is defined by the kernel B*, if for any $t \in T$ and $m \in M$ the element $\hat{m}(t) \in H_p(\xi)$ $(P \in \mathcal{P}_{R,M})$ is the linear image of the element $\hat{m}_{B,M}(t) \in H_Q(\xi)$ $(Q \in \mathcal{P}_{B,M})$; in this case we shall also say that B is a *defining kernel* (of the estimate $\hat{m}(t)$) or that the r.f. $\hat{m}_{B,M}(t)$ is a $\mathcal{P}_{R,M}$-u.l.e.

The notion of an estimate defined by a kernel is close to the notion of a pseudo-best estimate introduced by Rozanov [17], [18].

Evidently, for every measure $P \in \mathcal{P}_{R,M}$ the estimate $\hat{m}(t)$ is defined by the kernel B uniquely up to P-equivalency.

Note 4.1. Assuming that $P \ll Q$ the estimate $\hat{m}(t)$ defined by the kernel B could be characterized by the equality $\hat{m}(t) = \hat{m}_{B,M}(t)$ P-almost everywhere. However, in many important cases $P \perp Q$ (see, e.g. subsections 3 and 4 of this section) and then, since $\hat{m}_{B,M}(t)$ is determined up to Q-equivalency, *any* r.f. $m(t)$ satisfies the equality given above.

In accordance with Note 2.2 we shall further on denote the $\mathcal{P}_{R,M}$-u.l.e. defined by the kernel B by $\hat{m}_{B,M}(t)$ too.

We have already touched upon defining kernels; Lemma 3.1 shows that the best $\mathcal{P}_{R,M}$-u.l.e. $\hat{m}_{R,M}(t)$ is defined not only by the kernel R, but also by any M-majorant $\tilde{R}$ of the kernel R.

Theorem 4.1. The condition $R \lhd kB$ is necessary and sufficient for the estimate $\hat{m}_{B,M}(t)$ of any finite-dimensional space M to be a $\mathcal{P}_{R,M}$-u.l.e.

Proof. The sufficiency of the condition $R \lhd kB$ follows from Theorems 2.9 and 2.11.

N e c e s s i t y. Let us consider some measures $P_0 \in \mathscr{P}_{R,0}$ and $Q_0 \in \mathscr{P}_{B,0}$. Let $\varphi \in H(B)$ and M_φ denote the one dimensional space generated by the function φ. Evidently, $P_0 \in \mathscr{P}_{R,M_\varphi}$, $Q_0 \in \mathscr{P}_{R,M_\varphi}$ whatever the function $\varphi \in H(B)$ is. Further, $\hat{m}_{B,M_\varphi}(t) = \dfrac{\langle \xi, \varphi\rangle_{H(B)}}{\|\varphi\|^2_{H(B)}}\varphi(t)$. If for any function $\varphi \in H(B)$ the r.f. $\hat{m}_{B,M_\varphi}(t)$ is a $\mathscr{P}_{R,M_\varphi}$-u.l.e., then all the elements $\langle \xi, \varphi\rangle_{H(B)}$ of the space $H_{Q_0}(\xi)$ have linear images in $H_{P_0}(\xi)$, i.e. $H_{P_0}(\xi)$ is embeddable into $H_{P_0}(\xi)$. It remains to make use of Theorem 2.10.

2. *Simple unbiased linear regression estimates.* In this subsection we answer the second question: what is the class of $\mathscr{P}_{R,M}$-u.l.e. defined by kernels?

Definition 4.2. A $\mathscr{P}_{R,M}$-u.l.e. $m(t)$ will be referred to as *simple*, if for any measure $p \in \mathscr{P}_{R,M}$ with P-probability 1 $\hat{m}(t) = \sum_{i=1}^{n} \eta_i \varphi_i(t)$, where $\eta_i \in H_p(\xi)$, $\varphi_i \in M$.

Evidently, if $E_p\, \xi(t) = \sum_{i=1}^{n} a_i \varphi_i(t)$ and the functions $\varphi_1, \ldots, \varphi_n$ are linearly independent, then $E_p \eta_i = a_i$, i.e. η_i are unbiased linear estimates of the coefficients a_i. An example of a simple $\mathscr{P}_{R,M}$-u.l.e. is the estimate $\hat{m}_{R,M}(t)$.

Theorem 4.2. A $\mathscr{P}_{R,M}$-u.l.e. is simple iff it is defined by some kernel B such that $R \lhd kB$.

Proof. The s u f f i c i e n c y of this condition is evident. Let us prove its n e c e s s i t y. Let $\hat{m}(t) = \sum_{i=1}^{n} \hat{a}_i \varphi_i(t)$ be a simple $\mathscr{P}_{R,M}$-u.l.e. Without loss of generality, one can assume that $M \subset H(R)$ and that $\{\varphi_1, \ldots, \varphi_n\}$ is an orthonormal basis in M. Let $P \in \mathscr{P}_{R,M}$ and $\psi_i \in H(R)$ be a function in $H(R)$ such that $\hat{a}_i = \langle \xi, \psi_i\rangle_{H(R)}$ $(i = 1, \ldots, n)$. Using Theorem 2.9 it can be seen that the functions ψ_i are defined by that property in a unique fashion independent of the choice of the measure $P \in \mathscr{P}_{R,M}$. We designate: M_1 the subspace of $H(R)$ generated by the functions $\psi_1, \ldots, \psi_n$; R_0 the reproducing kernel of the subspace $H(R) \ominus M_1$ and $B(s, t) = R_0(s, t) + \sum_{i=1}^{n} \varphi_i(s)\varphi_i(t)$. Let us show that the kernel $B(s, t)$ defines the estimate $\hat{m}(t)$. Since this estimate is unbiased, for all $i \in \overline{1, n}$

$$a_i = E_p \hat{a}_i = \langle m, \psi_i\rangle_{H(R)}, \tag{4.1}$$

if $E_p\, \xi(t) = m(t) = \sum_{i=1}^{n} a_i \varphi_i(t)$. For $m = \varphi_i$ Eq. (4.1) gives

$$\langle \varphi_j, \psi_i\rangle_{H(R)} = \delta_{ij} \qquad (i, j = 1, \ldots, n). \tag{4.2}$$

Hence, all $\varphi_i \notin H(R_0)$, $H(B) = H(R_0) \oplus M$ and the sets $H(R)$ and $H(B)$ coincide; thus, $R \lhd kB$ and $B \lhd k_1 R$. Let us consider the operators L_R^B

and L_B^R (cf. subsection 4 of § 1). Since these operators are continuous and $L_R^B B(\,.\,,t) = R(\,.\,,t)$ and $L_B^R R(\,.\,,t) = B(\,.\,,t)$ for all $t \in T$, they are mutually inverse. We have:

$$(L_B^R \psi_i)(t) = \langle B(\,.\,,t), \psi_i(\,.\,)\rangle_{H(R)} = \Big\langle \sum_{k=1}^{n} \varphi_k(\,.\,)\, \varphi_k(t), \psi_i(\,.\,)\Big\rangle_{H(R)} = \varphi_i(t)$$

and, hence, $L_R^B \varphi_i = \psi_i$. This fact implies that $\hat{m}_{B,M}(t)$ is a $\mathcal{S}_{R,M}$-u.l.e. and

$$\hat{m}_{B,M}(t) = \sum_{i=1}^{n} \langle \xi, \varphi_i \rangle_{H(B)} \varphi_i(t) = \sum_{i=1}^{n} \langle \xi, L_R^B \varphi_i \rangle_{H(R)} \varphi_i(t) =$$

$$= \sum_{i=1}^{n} \langle \xi, \psi_i \rangle_{H(R)} \varphi_i(t) = \sum_{i=1}^{n} \hat{a}_i \varphi_i(t) = \hat{m}(t)\,.$$

The theorem is proved.

Note 4.2. The defining kernel B constructed during the proof of Theorem 4.1 is far from unique. Indeed, if $B_1 \perp B$ then the kernel $B \oplus B_1$ also defines the estimate $\hat{m}(t)$.

Since any kernel B with the property $R \lhd kB$ defines a certain $\mathcal{S}_{R,M}$-u.l.e., we can regard the choice of a defining kernel as the choice of a "rule" for building estimates. This permits to introduce the notion of "one-type" estimates or estimates constructed by the same "rule". The latter fact is especially important, if we wish to formulate the relation of "similarity" of two estimates $\hat{m}_1$ and $\hat{m}_2$, based on realizations of the r.f. $\xi(t)$ on different subsets T_1 and T_2 or estimates constructed for different "model" spaces M_1 and M_2.

Definition 4.3. Let us consider a family of $\mathcal{S}_{R\alpha,M\alpha}$-estimates $\{\hat{m}_\alpha(t), t \in T\}$ based on restrictions of the r.f. $\xi(t)$ on the subsets $T_\alpha \subset T$; the estimates $\hat{m}_\alpha(t)$ will be called *one-type* if they have a common defining kernel.

In the next subsection we shall examine by way of example a well known class of one-type estimates.

3. *Least-squares estimates and δ-kernels.* Let T be a finite set. Let us connect with every simple estimate $\hat{m}(t) = \sum_{i=1}^{n} \gamma_i \varphi_i(t)$ a random variable

$$l(\gamma_1, \ldots, \gamma_n) = \sum_{t \in T} [\hat{m}(t) - \xi(t)]^2 = \sum_{t \in T} \left(\sum_{i=1}^{n} \gamma_i \varphi_i(t) - \xi(t) \right)^2 \tag{4.3}$$

characterizing the "closeness" of the estimate $\hat{m}(t)$ to the r.f. $\xi(t)$. The least-square rule consists in choosing such an estimate $\hat{m}(t)$ for which the value $l(\gamma_1, \ldots, \gamma_n)$ is the least with probability 1. As known, the coefficients $\gamma_1, \ldots, \gamma_n$ of such an estimate must satisfy the system of equations

$$\frac{\partial l(\gamma_1, \ldots, \gamma_n)}{\partial \gamma_i} = 0 \qquad (i \in \overline{1, n}),$$

i.e. the system

$$\sum_{k=1}^{n} \left(\sum_{t \in T} \varphi_k(t)\, \varphi_i(t) \right) \gamma_k = \sum_{t \in T} \xi(t)\, \varphi_i(t)\,; \qquad i = 1, \ldots, n\,. \tag{4.4}$$

That system is compatible, if the functions φ_k are linearly independent. Its solution leads to a simple u.l.e. which is called the *least-squares estimate* (l.s.e.). Consider now the Kronecker kernel δ on T: $\delta(s, t) = 1$ if $s = t$ and $\delta(s,t) = 0$ if $s \neq t$. In this case $\langle \psi, \varphi \rangle_{H(\delta)} = \sum_{t \in T} \psi(t)\, \varphi(t)$, $\langle \xi, \varphi \rangle_{H(\delta)} = \sum_{t \in T} \xi(t)\, \varphi(t)$ and the system of Eqs. (4.4) coincides with the normal system for the estimate $\hat{m}_{\delta,M}(t)$ (see subsection 2 of § 3). Thus, the least-squares estimates are defined by the kernel δ. The above assertions also hold in case if the set T is countable; it is necessary only that the series $\sum_{t \in T} \varphi_k(t)\, \varphi_i(t)$ and $\sum_{t \in T} \xi(t)\, \varphi_i(t)$ in Eqs. (4.4) be convergent. This is the case if $R \lhd k\delta$ and $\varphi_k \in H(\delta), k \in \overline{1, n}$. Thus, in this case l.s.e. are defined by the "Kronecker kernel" δ.

Finally, suppose that T is a measurable set in the r-dimensional space $\mathfrak{R}^r$. In this case summation in the expressions $\sum_{t \in T} \varphi_k(t)\, \varphi_i(t)$ and $\sum_{t \in T} \xi(t)\, \varphi_i(t)$ is replaced by integration with respect to Lebesgue measure. These integrals can be considered as $\langle \varphi_k, \varphi_i \rangle_{\mathfrak{L}^2(T)}$ and $\langle \xi, \varphi_i \rangle_{\mathfrak{L}^2(T)}$, where $\mathfrak{L}^2(T)$ is the Hilbert space of all functions φ with $\| \varphi \|_{\mathfrak{L}^2(T)} = \left(\int_T [\varphi(s)]^2 \, ds \right)^{1/2} < \infty$. This Hilbert space does not possess a reproducing kernel but it does within the theory of generalized functions. Namely, every function $\psi \in \mathfrak{L}^2(T)$ can be considered as a (regular) generalized function $f(u) = \int_T \psi(t)\, u(t) dt$, $u \in D$ (D is the space of the basic functions — see, e.g., [3]) and as a space of generalized functions $\mathfrak{L}^2(T)$ has the reproducing kernel $\delta(u, v) = \int_T u(t)\, v(t)\, dt$, $u, v \in D$ (the "Dirac δ-kernel") which is the correlation function of the "white noise" (see, e.g. [4]). If R is a continuous kernel on T, the relation $R \lhd k\delta$ is fulfilled on T if $\int_T [R(s, t)]^2 \, ds < \infty$ for almost all $t \in T$ and the operator L_R^δ: $(L_R^\delta \psi)(t) = \int_T [R(s, t)]^2 \psi(s)\, ds$ is a continuous linear operator in $\mathfrak{L}_T^2$. This is, e.g., the case when $\sup_{t \in T} \int_T | R(s, t) | \, ds < \infty$ or when R is a Hilbert–Schmidt kernel, i.e. when $\int_T \int_T [R(s, t)] ds\, dt < \infty$. If R is a stationary kernel (see Example 1.1) the relation $R \lhd k\delta$ is fulfilled if the spectral measure F_R of R possesses density f_R and $f_R \leq k$ a.e. (with respect to Lebesgue measure). Now we can see easily that the l.s.e. is determined by the Dirac δ-kernel. If $R \lhd k\delta$ and $M \subset H(\delta) = \mathfrak{L}^2(T)$ the integrals $\int_T \varphi_i \varphi_j \, dt$ and $(P) \int_T \xi \varphi_j \, dt$ exist and so the l.s.e. is well defined. Note that the $\mathcal{S}_{R,M}$-u.l.e. $\hat{m}_{\delta,M}(t)$ is determined as well in case $\varphi_i \notin \mathfrak{L}^2(T)$, when the least-squares approach is not valid. Let us also note that $P \perp Q$ if P and Q are Gaussian measures with respect to the (generalized) r.f. $\xi(t)$ and $P \in \mathcal{P}_{R,M}, Q \in \mathcal{P}_{\delta,M}$.

4. Kernels defining the best unbiased linear estimate. This subsection proposes a criterion of "efficiency" of a simple $\mathcal{P}_{R,M}$-u.l.e. in terms of its defining kernel.

Lemma 4.1. If $R \lhd kB$ and $M \subseteq H(R)$, then $\Pi_M^{H(R)} L_R^B K_M^B(\cdot\, , t) \equiv K_M^R(\cdot\, , t)$.

Proof. Evidently, the function $\Lambda(\,.\,, t) = \Pi_M^{H(R)} L_R^B K_M^B(\,.\,, t) \in M$ at any $t \in T$. Further, at any $\varphi \in M$ and $t \in T$ $\langle \Lambda(\,.\,, t), \varphi(\,.\,)\rangle_{H(R)} = \langle L_R^B K_M^B(\,.\,, t), \varphi(\,.\,)\rangle_{H(R)} = \langle K_M^B(\,.\,, t), \varphi(\,.\,)\rangle_{H(R)} = \varphi(t)$. Thus, the function $\Lambda(s, t)$ is the reproducing kernel $K_M^R(s, t)$ of the subspace M of $H(R)$, q.e.d.

Theorem 4.3. Let $R \lhd kB$ and $M \cap H(B)$. The $\mathcal{P}_{R,M}$-u.l.e. $\hat{m}_{B,M}(t) \equiv \equiv \hat{m}_{R,M}(t)$ on T, iff the space $M \cap H(B)$ is invariant in relation to the operator L_R^B (in this case $L_R^B(M \cap H(B)) = M \cap H(R)$).

Proof. Let $\tilde{R}$ be an M-majorant of R. Recall that $\hat{m}_{R,M}(t) \equiv \langle \xi(\,.\,), K_M^R(\,.\,, t)\rangle_{H(\tilde{R})}$ and $\hat{m}_{B,M}(t) = \langle \xi(\,.\,), K_M^{\tilde{B}}(\,.\,, t)\rangle_{H(B)} = \langle \xi(\,.\,), L_{\tilde{R}}^{B} K_M^B(\,.\,,t)\rangle_{H(\tilde{R})}$. Evidently, $\hat{m}_{B,M}(t) \equiv \hat{m}_{R,M}(t)$ iff $L_R^{\tilde{B}} K_M^{\tilde{B}}(\,.\,, t) \equiv K_M^{\tilde{R}}(\,.\,, t)$. Since the families of functions $\{K_M^{\tilde{R}}(\,.\,, t), t \in T\}$ and $\{L_{\tilde{R}}^{\tilde{B}} K_M^{\tilde{B}}(\,.\,, t), t \in T\}$ generate the space M and $L_{\tilde{R}}^{\tilde{B}} M$, respectively, this condition, by virtue of Lemma 4.1, is equivalent to the condition $L_{\tilde{R}}^{\tilde{B}} M = M$. The latter one in its term is equivalent to $L_R^B(M \cap H(B)) = H(R) \cap M$. The theorem is proved.

The case when the sample is of finite size ($T = \{1, \ldots, s\}$) and $\hat{m}_{B,M}(t) = = \hat{m}_{\delta,M}(t)$ (l.s.e.) was earlier treated by another method by Kruskal [11].

Corollary 4.1. Let $R(s, t) = \sum_{i=1}^{r} \oplus\, l_i R_i(s, t)$, where R_i are kernels and l_i are non-negative numbers, and let $M \cap H(R) = \sum_{i=1}^{r} [M \cap H(R_i)]$. Then the $\mathcal{P}_{R,M}$-u.l.e. $\hat{m}_{R,M}$ is defined by any kernel $B(s, t) = \sum_{i=1}^{r} \oplus\, \lambda_i R_i(s, t)$ with positive λ_i.

This statement immediately follows from Theorem 4.3, since $L_R^B \varphi = \dfrac{l_i}{\lambda_i} \varphi$ if $\varphi \in H(R_i)$.

In particular if $M \subseteq H(R)$ and $B(s, t) = \lambda R(s, t) (\lambda > 0)$ then $\hat{m}_{B,M}(t) = = \hat{m}_{R,M}(t)$. In connection with the discussion in Note 4.1 let us remark that if $H_p(\xi)$ is infinite-dimensional and $\lambda \neq 1$ then the Gaussian measures $P \in \mathcal{P}_{R,M}$ and $Q \in \mathcal{P}_{B,M}$ are mutually orthogonal.

We do not dwell here on the question of the asymptotic efficiency of u.l. estimates, referring the reader to [6], [8]–[10], [18]–[20] and the literature cited there.

5. Error-free simple estimates. Let us now establish a criterion of the simple $\mathcal{P}_{R,M}$-u.l.e. $\hat{m}_{B,M}(t)$ (i.e. the estimate defined by the kernel B) being error-free, i.e. $D_p \hat{m}_{B,M}(t) \equiv 0$ for any measure $P \in \mathcal{P}_{R,M}$ or, in other words, with

P-probability 1 $\hat{m}_{B,M}(t) \equiv E_p\xi(t)$. From Corollary 2.1 the following result can be easily deduced.

Theorem 4.4. Let $R \lhd kB$. $\mathcal{P}_{R,M}$-u.l.e. $\hat{m}_{B,M}(t)$ is error-free iff $M \cap H(B) \subseteq$ $\subseteq H(B) \ominus H_B(R)$, i.e. iff $L_R^B(M \cap H(B)) = \{0\}$.

In the special case of $B = R$ this theorem turns into Corollary 4.1. Note that if $M \subset H(B)$ and $M \not\subset H(R) \cup (H(B) \ominus H_B(R))$, then the estimate $\hat{m}_{B,M}$ is not error-free despite the fact that the best $\mathcal{P}_{R,M}$-u.l.e. $\hat{m}_{R,M}(t)$ is error-free; thus, in that case the estimate $\hat{m}_{B,M}(t)$ is of an evidently poorer quality than $\hat{m}_{R,M}(t)$.

This criterion presupposes the availability of exhaustive information concerning the kernel R; there are no grounds to believe that the statistician will normally dispose of such information. Therefore, the following criterion is of interest as well, which can be inferred from Theorems 4.4 and 1.4

Corollary 4.2. Let $R \lhd kB$. If $M \cap H(B) = \{0\}$, then the estimate $\hat{m}_{B,M}(t)$ is error-free; if $B \ll R$ and the estimate $\hat{m}_{B,M}(t)$ is error-free, then $M \cap$ $\cap H(B) = \{0\}$.

§ 5. CONVERGENCE AND CONSISTENCY OF SIMPLE ESTIMATES

1. General theorems. This section deals with the behaviour of simple regression estimates under an increasing scope of observations. It is natural for that purpose to confine the discussion to the case when the method of estimate building remains constant, i.e. to the case of one-type estimates. We shall presently see that with an increasing scope of observations one-type estimates converge to the estimate of the same type based on the maximal set of observations.

Let A be a direction and $\{T_\alpha, \alpha \in A\}$ a net of subsets of the set T, while $T_{\alpha_1} \subset T_{\alpha_2}$ if $\alpha_1 < \alpha_2$ and $\bigcup_{\alpha \in A} T_\alpha = T$. We designate: $\xi^\alpha(t)^2$, $m_\alpha(t)$, $R^\alpha(s, t)$. restriction onto the set T_α of the functions $\xi(t)$, $m(t)$ and $R(s,t)$; M is a randomi finite-dimensional space of functions $m(\,.\,)$ on T; M^α is the space consistng of the restrictions onto T_α of the functions $m \in M$; and $\hat{m}^\alpha_{R,M}(t)$ is the best P_{R^α,M^α}-u.l.e. for the random function $\xi^\alpha(t)$, $t \in T_\alpha$.

Theorem 5.1. Let $R \lhd kB$. Then:

1) the estimates $\hat{m}^\alpha_{B,M}(t)$ are $\mathcal{P}_{R^\alpha,M^\alpha}$-u.l.e. based on the realization of $\xi(t)$ on T_α;

2) if $P \in \mathcal{P}_{R,M}$, then at any $t \in T$ there exists $(P) \lim_{\alpha \in a^\pi} \hat{m}^\alpha_{B,M}(t)$;

3) $(P) \lim_{\alpha \in A} \hat{m}^\alpha_{B,M}(t) \equiv \hat{m}_{B,M}(t)$ with P-probability 1.

Proof. The first statement immediately follows from Theorem 4.1. . The remaining statements need only to be proved for the case $M \not\subset H(B)$, since for $M \not\subset H(B)$ one can consider, instead of the kernel B on T, any of its M-majorants. Further, since for $Q \in \mathcal{P}_{B,M}$ the space $H_Q(\xi)$ is embeddable into the space $H_p(\xi)$ (Theorem 2.9), it is sufficient to prove the mean square convergence of the estimates $\hat{m}^\alpha_{B,M}(t)$ to the estimate $\hat{m}_{B,M}(t)$ with respect to a measure $Q \in \mathcal{P}_{B,M}$. Let $\varphi_1, \ldots, \varphi_n$ be a basis in M. According to Theorem 1.3

$$\lim_{\alpha \in A} \langle \varphi_i^\alpha, \varphi_j^\alpha \rangle_{H(B^\alpha)} = \langle \varphi_i, \varphi_j \rangle_{H(B)} ; \tag{5.1}$$

Theorems 1.3 and 2.4 imply that

$$(Q) \lim_{\alpha \in A} \langle \xi^\alpha, \varphi_i^\alpha \rangle_{H(B^\alpha)} = \langle \xi, \varphi_i \rangle_{H(B)} . \tag{5.2}$$

Hence, first of all, the determinants of the Gram matrices $\mathcal{G}^\alpha = \{\langle \varphi_i^\alpha, \varphi_j^\alpha \rangle_{H(B^\alpha)}\}$ converge to the determinant of the Gram matrix $\mathcal{G} = \{\langle \varphi_i, \varphi_j \rangle_{H(B)}\}$, and, therefore, under sufficiently large α, the restrictions $\varphi_1^\alpha, \ldots, \varphi_n^\alpha$ are also linearly independent and form a basis in M^α. The estimates $\hat{m}_{B,M}(t)$ and $\hat{m}^\alpha_{B,M}(t)$ then have the appearance:

$$\hat{m}_{B,M}(t) = \sum_{i=1}^{n} \hat{a}_i \varphi_i(t) , \tag{5.3}$$

$$\hat{m}^\alpha_{B,M}(t) = \sum_{i=1}^{n} \hat{a}_i^\alpha \varphi_i(t) ,$$

while

$$\hat{a}_i = \sum_{j=1}^{n} d_{ij} \langle \xi, \varphi \rangle_{H(B)} , \tag{5.4}$$

$$\hat{a}_i^\alpha = \sum_{j=i}^{n} d_{ij}^\alpha \langle \xi^\alpha, \varphi^\alpha \rangle_{H(B^\alpha)} ,$$

where $\{d_{ij}^\alpha\}$ and $\{d_{ij}\}$ are matrices inverse to the Gram matrices $\mathcal{G}^\alpha$ and $\mathcal{G}$ (see subsection 2, § 3). It is easy to obtain $\lim_{\alpha \in A} d_{ij}^\alpha = d_{ij}$ for all $i, j \in \overline{1, n}$ from Eq. (5.1); hence and from Eqs. (5.2) and (5.4) we obtain for any $i \in \overline{1,n}$ $(Q) \lim_{\alpha \in A} \hat{a}_i^\alpha = \hat{a}_i$. It remains to apply formulas (5.3).

The following inverse theorem is also true.

Theorem 5.2. If for any finite-dimensional space of functions $M \subset H(B)$ 1) any estimate $\hat{m}^\alpha_{B,M}(t)$ is $\mathcal{P}_{R^\alpha, M^\alpha}$-u.l.e. and 2) with respect to any measure $P \in \mathcal{P}_{R,M}$ for all $t \in T$ there exists $(P) \lim_{\alpha \in A} \hat{m}^\alpha_{B,M}(t)$ then $R \lhd kB$ with some positive number $k < \infty$.

Proof. In keeping with Theorem 4.1, from condition 1) follows the existence of such constants k_α, for which $R^\alpha \lhd k_\alpha B^\alpha$. These constants can be chosen in such a manner that $\| L^{B^\alpha}_{R^\alpha} \| = \sqrt{k_\alpha}$ (here $L^{B^\alpha}_{R^\alpha}$ is the operator from $H(B^\alpha)$ into $H(R^\alpha)$ defined in subsection 4, § 1). According to Theorem 1.2 the mapping $\varphi^\alpha \xrightarrow{\tilde{i}_{\alpha,B}} \Pi^{H(B)}_{H_\alpha(B)} \varphi$ is an isometry from $H(B^\alpha)$ onto $H_\alpha(B)$. Consider the linear operator L_α from $H(B)$ into $H(R)$, defined by the expression $L_\alpha = \tilde{i}_{\alpha,R} L^{B^\alpha}_{R^\alpha} \tilde{i}^{-1}_{\alpha,B} \Pi^{H(B)}_{H_\alpha(B)}$. Evidently, $\| L_\alpha \| = \| L^{B^\alpha}_{R^\alpha} \| = \sqrt{k_\alpha}$. Let us now fix a function $\varphi \in H(B) \setminus \{0\}$ and a measure $P \in \mathcal{P}_{R,\varphi}$. Let M be the one-dimensional space generated by the function φ. Then $\hat{m}^{\alpha}_{B,M}(t) = (\|\varphi^{\alpha}\|^{-2}_{H(B^{\alpha})}\langle\xi^{\alpha}_{c}, L^{B^{\alpha}}_{R^{\alpha}}\varphi^{\alpha}\rangle_{H(R^{\alpha})} + 1)\,\varphi(t)$. According to Theorem 1.3, $\lim_{\alpha\in A}\|\varphi^{\alpha}\|_{H(B^{\alpha})} = \|\varphi\|_{H(B)}$; according to Theorem 2.3, $\langle\xi^{\alpha}_{0}, L^{B^{\alpha}}_{R^{\alpha}}\varphi^{\alpha}\rangle_{H(R^{\alpha})} = \langle\xi_0, \tilde{i}_{\alpha,R}L^{B^{\alpha}}_{R^{\alpha}}\varphi^{\alpha}\rangle_{H(R)} = \langle\xi_0, L_{\alpha}\varphi\rangle_{H(R)}$; now from the condition of the theorem it is easily deducible that in $H(R)$ there exists $\lim_{\alpha\in A} L_{\alpha}\varphi$. In view of the arbitrary nature of the choice of the function φ, this means that the operators L_{α} strongly converge, and, therefore, $\sup_{\alpha\in A} k_{\alpha} = \sup_{\alpha\in A}\|L_{\alpha}\|^2 < \infty$. Evidently, $R \lhd kB$ with $k = \sup_{\alpha\in A} k_{\alpha}$, q.e.d.

Definition 5.1. A net of $\mathscr{P}_{R^{\alpha},M^{\alpha}}$-u.l.e. estimates $\{\hat{m}_{\alpha}(t),\ \alpha\in A\}$ based on realizations of the r.f. $\xi(t)$ on T_{α} will be called $\mathscr{P}_{R,M}$-consistent, if for any measure $P\in\mathscr{P}_{R,M}$ there exists the limit $(P)\lim_{\alpha\in A}\hat{m}_{\alpha}(t)$ and the equality $(P)\lim_{\alpha\in A}\hat{m}_{\alpha}(t) \equiv E_p\,\xi(t)$ holds with P-probability 1.

The following theorem can be inferred from Theorems 5.1 and 4.4 and from Corollary 4.2.

Theorem 5.3 Let $R\lhd kB$ for certain $0<k<\infty$. The condition $M\cap H(B)\subseteq H(B)\ominus H_B(R)$ (or $L^B_R\, M\cap H(B) = \{0\}$) is necessary and sufficient for $\mathscr{P}_{R,M}$-consistency of the net of estimates $\{\hat{m}^{\alpha}_{B,M}(t), \alpha\in A\}$. Thus if $M\cap H(B) = \{0\}$, then the sequence of estimates $\hat{m}^{\alpha}_{B,M}(t)$ is $\mathscr{P}_{R,M}$-consistent; if $B\ll R$ and the sequence $\hat{m}^{\alpha}_{B,M}(t)$ is $\mathscr{P}_{R,M}$-consistent, then $M\cap H(B) = \{0\}$.

If $T\subset\mathscr{R}^r$ and the kernels R and B are stationary we can reformulate Theorems 5.1 and 5.2 paraphrasing the relations $R\lhd kB$, $R\ll B$, $\varphi\in H(B)$ in spectral terms (see Example 1.1). We obtain then a proposition close to a theorem of Rozanov (see [8], [17], [18]).

2. *Convergence and consistency of the best unbiased linear estimates.* From Theorems 5.1 and 5.3 the following result due to Rozanov [16] is deducible.

Theorem 5.4. 1) If $P\in\mathscr{P}_{R,M}$, then there exists the limit $(P)\lim_{\alpha\in A}\hat{m}^{\alpha}_{R,M}(t) = \hat{m}_{R,M}(t)$.

2) The sequence $\hat{m}^{\alpha}_{R,M}(t)$ is $\mathscr{P}_{R,M}$-consistent if $M\cap H(R) = \{0\}$.

3. *Convergence and consistency of least-squares estimates.* In this subsection we apply the theorems of subsection 1 to the estimates $\hat{m}^{\alpha}_{\delta,M}(t)$, defined in the discrete case by the Kronecker kernel, and in the continuous case by the Dirac kernel and coincident at $M\subset H(\delta)$ and $R\lhd k\delta$ with the least-square estimates (cf. subsection 3 of § 4). To derive Theorems 5.5 and 5.7 given below one just has to paraphrase conditions $R\lhd k\delta$, $\delta\ll R$ and $M\cap H(\delta) = \{0\}$. In Theorems 5.5, 5.6 and in Corollaries 5.1 and 5.2, T and $T_{\alpha}(\alpha\in A)$ are measurable sets in $\mathscr{R}^r$, $T = \bigcup_{\alpha\in A} T_{\alpha}$ and $T_{\alpha_1}\subset T_{\alpha_2}$ if $\alpha_1<\alpha_2$; M is a finite-dimensional space of locally integrable functions on T and $\{f : f\in M, f = 0 \text{ a.e.}\} = \{0\}$; $\xi(t)$ is properly continuous in a mean square random field on R^r with the correlation function $R(s,t)$.

Theorem 5.5. In order that for every M and for any $t \in T$ $\lim_{\alpha \in A} \hat{m}^{\alpha}_{\delta,M}(t) = \hat{m}_{\delta,M}(t)$ should exist it is necessary and sufficient that $\int_T [R(s,t)]^2 \, ds < \infty$ for almost all $t \in T$ and the operator $L^{\delta}_R : (L^{\delta}_R \varphi)(t) = \int_T R(s,t)\varphi(s) \, ds$ be a continuous linear operator in $\mathfrak{L}^2(T)$.* In this case the net of estimates $\{\hat{m}^{\alpha}_{\delta,M}, \alpha \in A\}$ is consistent iff for every $\varphi \in M$ either $\int_T [\varphi(t)]^2 \, dt = \infty$ or $(L^{\delta}_R \varphi)(t) \equiv 0$.

Let χ_T be the indicator of T. If $\varphi \in \mathfrak{L}^2(T)$, $\tilde{\varphi}_T$ will denote the Fourier–Plancherel transform of $\varphi_T = \varphi\chi_T$.

Corollary 5.1. Let $\xi(t)$ be a properly homogeneous continuous mean square random field on $\mathfrak{R}^r$ (see for example [22]); then 1) in order that for any M on every T the limit $\lim_{\alpha \in A} \hat{m}^{\alpha}_{\delta,M}(t) \equiv \hat{m}_{\delta,M}(t)$ should exist, it is necessary and sufficient for the field $\xi(t)$ to possess a bounded proper spectral density $f(\lambda)$; in this case: 2) the estimates $\hat{m}^{\alpha}_{\delta,M}(t)$ are consistent on T iff for every $\varphi \in M$ either $\int_T [\varphi(t)]^2 \, dt = \infty$ or $\tilde{\varphi}_T(\lambda) = 0$ a.e. with respect to the proper spectral measure of the field $\xi(t)$. Thus the condition $\int_T [\varphi(t)]^2 \, dt = \infty$ is sufficient for consistency; it is necessary if $f(\lambda) > 0$ a.e. with respect to Lebesgue measure.

A similar result has been earlier obtained by Kholevo [9].

The boundedness condition of the spectral density can be removed by giving up the arbitrary nature of the choice of the space M and the sets T_{α}. The following theorem is a step in this direction. Let $\{\varphi_1, \ldots, \varphi_n\}$ be a basis in M, and for any $\alpha \in A$ the restrictions $\varphi^{\alpha}_1, \ldots, \varphi^{\alpha}_n$ be also lineary independent and $\int_{T_{\alpha}} [\varphi^{\alpha}_i(t)]^2 \, dt < \infty$. We designate by $\mathfrak{D}^{\alpha} = \{d^{\alpha}_{ij}\}^n_1$ the matrix inverse to the Gram matrix $\mathfrak{G}^{\alpha} = \{\int_{T_{\alpha}} \varphi^{\alpha}_i \varphi^{\alpha}_j \, dt\}^n_1$.

Theorem 5.6. Let $\xi(t)$ be a properly homogeneous continuous in mean square random field on $\mathfrak{R}^r$ with absolutely continuous proper spectrum. Suppose that $\sup_{\alpha \in A} \{|d^{\alpha}_{ij}| \int_{T_{\alpha}} |\varphi_j(t)| \, dt\} < \infty$ for all $i, j \in \overline{1, n}$. Then 1) for every $\alpha \in A$ the l.s.e. $\hat{m}_{\alpha}(t)$ based on the realization of $\xi(t)$ on T_{α} is well-defined; 2) there exists $\lim_{\alpha \in A} \hat{m}_{\alpha}(t)$ for any $t \in T$; moreover, statement 2) of corollary 5.1 is true.

Proof. Since $\varphi^{\alpha}_i \in \mathfrak{L}^1(T_{\alpha}) \cap \mathfrak{L}^2(T_{\alpha})$ $(i \in \overline{1, n}, \alpha \in A)$ there exist the integrals $\int_{T_{\alpha}} \varphi_i(t)\varphi_j(t,) \, dt$ and $\int_{T_{\alpha}} \xi(t)\varphi_i(t) \, dt$ and hence statement 1) is proved. Further,

* Convenient sufficient conditions: $\sup_{t \in T} \int_T |R(s,t)| \, ds < \infty$ and $\int_T \int_T [B(s,t)]^2 \, ds \, dt < \infty$.

$\xi(t) = \int_{R^r} e^{i(\lambda,t)} Z(d\lambda) + m(t)$ where $Z(\cdot)$ is the proper spectral random measure of the field $\xi(t)$. Let $f_k(\cdot)$ $(k = 1, 2, \ldots)$ be an increasing sequence of positive measurable functions such that $\lim_{k\to\infty} f_k(\lambda) = f(\lambda)$. Let us consider the random fields

$$\xi_k(t) = \int_{\mathfrak{R}^r} e^{i(\lambda,t)} \sqrt{\frac{f_k(\lambda)}{f(\lambda)}}\, Z(d\lambda) + m(t)$$

and

$$\Delta_k(t) = \int_{\mathfrak{R}^r} e^{i(\lambda,t)} \left(1 - \sqrt{\frac{f_k(\lambda)}{f(\lambda)}}\right) Z(d\lambda) = \xi(t) - \xi_k(t)\,.$$

We denote: $R_{\Delta k}(s, t)$ is the correlation kernel of the field $\Delta_k(t)$; $\hat{m}^{\alpha,k}_{\delta,M}(t)$ and $\hat{m}^k_{\delta,M}(t)$ are $\mathcal{S}_{R,M}$-u.l. estimates based on the realizations of the fields $\xi^\alpha_k(t)$ and $\xi_k(t)$ respectively, defined by the kernel δ. Obviously, $R_{\Delta_k}(s, t) = \int_{\mathfrak{R}^r} e^{i(\lambda, t-s)} (\sqrt{f(\lambda)} - \sqrt{f_k(\lambda)})^2\, d\lambda$ and thus $|R_{\Delta_k}(s, t)| \leq R_{\Delta_k}(0, 0) = \int_{\mathfrak{R}^r} [\sqrt{f(\lambda)} - \sqrt{f_k(\lambda)}]^2 d\lambda$. Hence

$$\lim_{k\to\infty} R_{\Delta_k}(0, 0) = 0\,. \tag{5.5}$$

Further, $\hat{m}_\alpha(t) = \sum_{i=1}^n \hat{a}^\alpha_i \varphi_i(t)$ and $\hat{m}^{\alpha,k}_{\delta,M}(t) = \sum_{i=1}^n \hat{a}^{\alpha,k}_i \varphi_i(t)$, where $\hat{a}^\alpha_i = \sum_{i,j=1}^n d^\alpha_{ij} \times \int_{T_\alpha} \xi(t)\, \varphi_i(t)\, \mathrm{d}t$ and $a^{\alpha,k}_i = \sum_{j=1}^n d^\alpha_{ij} \int_{T_\alpha} \xi_k(t)\, \varphi_j(t)\, \mathrm{d}t$.

We have

$$E(\hat{a}^\alpha_i - \hat{a}^{\alpha,k}_i)^2 = \sum_{j,l=1}^n d^\alpha_{ij}\, d^\alpha_{il} \int_{T_\alpha}\int_{T_\alpha} R_{\Delta_k}(s, t)\, \varphi_j(s)\, \varphi_l(t) \mathrm{d}s\, \mathrm{d}t \leq$$

$$\leq R_{\Delta_k}(0, 0) \left(\sum_{j=1}^n |\mathrm{d}_{ij}| \int_{T_\alpha}^\alpha |\varphi_j(t)|\, \mathrm{d}t\right)^2. \tag{5.6}$$

From relations (5.5) and (5.6) and the condition of the theorem it follows that $E(\hat{a}^{\alpha,k}_i - \hat{a}^\alpha_i)^2 \to 0$ uniformly with respect to $\alpha \in A$ when $k \to \infty$. According to Corollary 5.1, there exists $\lim_{\alpha\in A} \hat{a}^{\alpha,k}_i = \hat{a}^k_i$. The latter two limit relations imply that equal (with probability 1) limits $\lim_{k\to\infty} \hat{a}^k_i$ and $\lim_{\alpha\in A} \hat{a}^\alpha$ exist. Hence it is easy to derive the assertion of the theorem.

This theorem assumes a particularly simple appearance when the space M is one-dimensional, i.e. when $m(t) = a\varphi(t)$, when φ is a known locally integrable function, and a is an unknown constant. We have

Corollary 5.2. Let $\xi(t), t \in T$, be a properly homogeneous continuous mean square random field with absolutely continuous proper spectrum and $m(t) = a\varphi(t)$ $(\varphi(t) \not\equiv 0)$. Suppose that $\sup_{\alpha\in A} \frac{\int_{T_\alpha} |\varphi(t)|\, dt}{\int_{T_\alpha} [\varphi(t)]^2\, dt} < \infty$ and $\int_{T_\alpha} [\varphi(t)]^2 dt < \infty$ for all $\alpha \in A$.

Then

1) there exists $\hat{\xi} = \lim \left(\int_{T_\alpha} [\varphi(t)]^2\, dt \right)^{-1} \int_{T_\alpha} \xi(t)\, \varphi(t)\, dt$;

2) if $\int_T [\varphi(t)]^2\, dt = \infty$ then $\hat{\xi} = a$ with probability 1; iff $\int_T [\varphi(t)]^2 dt < \infty$ or $f(\lambda)\, \tilde{\varphi}_T(\lambda) = 0$ a.e.

This is a generalized version of the classical statistical ergodic theorem for homogeneous random fields with absolutely continuous spectrum. If $\varphi(t) \equiv 1$, Corollary 5.2 asserts that for such fields, for any increasing net of measurable sets $\{T_\alpha, \alpha \in A\}$ $\hat{\xi} = \lim_{\alpha\in A} [\mu(T_\alpha)]^{-1} \int_{T_\alpha} \xi(t)\, dt$ exists (μ denotes the Lebesgue measure on $\mathcal{R}^r$); $\hat{\xi} = E\xi(0)$ with probability 1 iff either $\lim_{\alpha\in A} \mu(T_\alpha) = \infty$ or the Fourier transform $\tilde{\chi}_T(\lambda)$ of the indicator $\chi_T(t)$ equals 0 a.e. with respect to the spectral measure of the field.

Finally, consider the case of arbitrary discrete observations: $T = \{t_1, t_2, \ldots\}$, $T_k = \{t_1, \ldots, t_k\}$, $k = 1, 2, \ldots$, where t_i are elements of an arbitrary set S. Theorems 5.1 and 5.2 imply

Theorem 5.7. In order that for any finite-dimensional space M for any $t \in T$ the limit $\lim_{k\to\infty} \hat{m}^k_{\delta,M}(t)$ exists it is necessary and sufficient that the sequence of the norms of the matrices $R^{(k)} = \{R(t_i, t_j)\}_1^k$ are bounded,* if, moreover, $\sum_{i=1}^{\infty} [\varphi(t_i)]^2 = \infty$ for $\varphi \in M \setminus \{0\}$, then the sequence $\hat{m}^k_{\delta,M}(t)$ is consistent; if the matrices $R^{(k)}$ are non-degenerate, this condition is also necessary for the consistency of l.s.e.

Theorems 5.5 and 5.6 and Corollary 5.1 can be easily transferred to properly homogeneous random fields on the integer lattice Z^r and to generalized properly homogeneous random fields on $\mathcal{R}^r$.

The apparatus of operator reproducing kernels (cf. [5]) makes it possible without any substantial modifications of statements and proofs, to transfer all the results presented in this paper onto vector-valued random functions and, in particular, to examine the estimates of regression coefficients based on several realizations of a r.f. $\xi(t)$.

* Convenient sufficient conditions: $\sup_{i\le j<\infty} \sum_{i=1}^{\infty} |R(t_i, t_j)| < \infty$ and $\sum_{i,j=1}^{\infty} [R(t_i, t_j)]^2 < \infty$.

REFERENCES

1. Akhiezer, N. I. and Glazman, I. M., *Teoriya lineinykh operatorov v gil'bertovom prostranstve* (Theory of linear operators in Hilbert space). Nauka, Moscow, 1966.
2. Aronszajn, N., Theory of reproducing kernels. *Trans. Amer. Math. Soc.* **68** **3** (1950) 337–404.
3. Gel'fand, I. M. and Shilov, G. E., *Obobschennye funktsii i ikh primenenie* (Generalized functions and their application). Fizmatgiz, Moscow 1958.
4. Gel'fand, I. M. and Vilenkin, N. Y., *Nekotorye primenenija garmonicheskogo analiza, Osnashchennye gil'bertovy prostranstva* (Some applications of harmonic analysis, Equipped Hilbert spaces). Fizmatgiz, Moscow 1961.
5. Golosov, Yu. I. and Tempel'man, A. A., Ob ekvivalentnosti mer, sootvetstvuyushchikh gaussovskim vektornoznachnym funktsiyam (On equivalence of measures corresponding to Gaussian vector-valued functions). *DAN SSSR* **184** **6** (1969) 1271–1274.
6. Grenander, U. and Rosenblatt, M., *Statistical analysis of stationary time series.* Uppsala, 1956.
7. Hájek, J., On statistical problems in stochastic processes. *Czechoslovakian Mathematical Journal* **12** **87** (1962) 404–444.
8. Ibragimov, I. A. and Rozanov, Yu. A., *Gaussovskie sluchainye protsessy* (Gaussian random processes). Nauka, Moscow 1970.
9. Kholevo, A. S., Ob otsenkakh koeffitsientovreg ressii (On estimation of regression coefficients). *Teoriya veroyatnostei i ee primeneniya* **14** **1** (1969) 78–101.
10. Kholevo, A. S., Ob asimptoticheskoi effektivnosti psevdonailutshykh otsenok (On asymptotic efficiency of pseudo-best estimators). *Teoriya veroyatn. i ee primen.* 16 **3** (1971) 524–534.
11. Kruskal, W., When are Gauss-Markov and least squares estimators identical? A coordinate-free approach. *Ann. Math. Statist.* **39** **1** (1968) 70–75.
12. Mirskaya, T. I., Pabedinskaite, A. S. and Tempel'man, A. A., Gil'bertovy prostranstva nekotorykh vosproizvodvashchikh yader i ekvivalentnost' gaussovskikh mer (Hilbert spaces of some reproducing kernels and the equivalence of Gaussian measures). *Litovsk. matem. sborn.* **7** **3** (1967) 459–469.
13. Parzen, E., Statistical inference on time series by Hilbert space methods, I. Department of Statistics, Stanford University, *Techn. Report*, No. **23**, 1959.
14. Parzen, E., Regression analysis of continuous parameter time series. *Proc. IVth Berkeley Symp.* **1** (1961) 479–489.
15. Parzen, E., An approach to time series analysis. *Ann. Math. Statist.* **32** (1961) 951–989.
16. Rozanov, Yu. A., *Gaussovskie beskonechnomernye raspredeleniya* (Gaussian infinite-dimensional distributions). Nauka, Moscow 1968.
17. Rozanov, Yu. A., O novom klasse statisticheskikh otsenok (On a new class of statistical estimates). *Sovetsko-Yaponskii simpozium po teorii veroyatnostei*, Khabarovsk, Aug. 1969, Novosibirsk, AN SSSR, 1969, 231–238.
18. Rozanov, Yu. A., *On a new class of estimates. Multivariate analysis.* Academic Press, New York, 1969, Vol. 2, 437–441.
19. Rozanov, Yu. A., Kozlov, M. V., Ob asimptoticheski effektivnom otsenivatanii koeffitsientov regressii (On asymptotically effective estimation of regression coefficients). *DAN SSSR*, **188** **1** (1969) 37–40.
20. Striebel, Ch. T., Efficient estimation of a regression parameter for certain second order processes. *Ann. Math. Statist.* **32** **4** (1961) 1299—1313.
21. Tempel'man, A. A., O lineinykh otsenkakh regressii (On linear regression estimates). *DAN SSSR* **191** **4** (1970) 772–775.
22. Yaglom, A. M., Nekotorye klassy sluchainykh polei v n-mernom prostranstve, rodstvennye statsionarnym sluchainym protsessam (Some classes of random fields in n-dimensional space, related with stationary random processess). *Teor. veroyatn. i ee primen.* **2** 3 (1957) 292–338.
23. Ylvisaker, N. D., On linear estimation for regression problems on time series. *Ann. Math. Statist.* **33** **3** (1962) 1077–1084.

14

Reprinted by permission from *IEEE Trans. Inform. Theory* **IT-19:**29–37 (1973)

RKHS Approach to Detection and Estimation Problems—Part V: Parameter Estimation

DONALD L. DUTTWEILER AND THOMAS KAILATH

Abstract—**Using reproducing-kernel Hilbert space (RKHS) techniques, we obtain new results for three different parameter estimation problems. The new results are 1) an explicit formula for the minimum-variance unbiased estimate of the arrival time of a step function in white Gaussian noise, 2) a new interpretation of the Bhattacharyya bounds on the variance of an unbiased estimate of a function of regression coefficients, and 3) a concise formula for the Cramér–Rao bound on the variance of an unbiased estimate of a parameter determining the covariance of a zero-mean Gaussian process.**

I. Introduction

In THE preceding parts of this series [1]–[4], we used reproducing-kernel Hilbert space (RKHS) techniques to study various detection and least squares estimation problems; here such techniques are applied to three different parameter estimation problems.

Parzen [5] was the first to realize the applicability of RKHS methods to parameter estimation problems. He used the RKHS generated by the expectation of a product of likelihood ratios to reformulate in concise RKHS terms Barankin's [6] theory of minimum-variance unbiased estimation.

We review Parzen's work in Section III and then apply his results in the following section to the much studied problem of estimating the arrival time of a step function in white Gaussian noise. We obtain explicit formulas for the minimum-variance unbiased estimate and its variance.

The second problem studied (Section V) is of the regression type. Here, the observed random process is

$$x(t) = \theta_1 m_1(t) + \cdots + \theta_n m_n(t) + n(t), \qquad t \in I, \quad (1)$$

where $n(\cdot)$ is a zero-mean Gaussian process with covariance $R(t,s)$, the known functions $m_1(\cdot),\cdots,m_n(\cdot)$ are elements of $H(R)$, and the unknown regression coefficients take values from the real line. From an observation of $x(\cdot)$ an estimate of some possibly nonlinear function $f(\boldsymbol{\theta}) = f(\theta_1,\cdots,\theta_n)$ is desired. By applying Parzen's theory, we obtain a concise formula for the variance of the minimum-variance unbiased estimate of $f(\boldsymbol{\theta})$, and using a natural orthogonal decomposition of the fundamental RKHS, give a new interpretation of the Bhattacharyya bounds [7].

This regression problem has been extensively studied, especially in the case where $f(\boldsymbol{\theta})$ is linear and we do not obtain any computationally new formulas. Perhaps the major appeal of our results on this problem is their compactness and the directness of their derivations by RKHS techniques.

In the final problem, an observed random process $\{x(t),\ t \in I\}$ is assumed to be zero-mean Gaussian with a covariance function drawn from the set $\{R_\theta(t,s),\ \theta \in \Theta\}$ where the indexing set Θ is a subset of the real line. From an observation of this process an estimate of θ or, more generally, a function $f(\theta)$ is desired. In Section VI we show that the Cramér–Rao (CR) bound on the variance of an unbiased estimate of $f(\theta)$ is given by

$$\mathrm{CR} = 2\dot{f}^2(\theta_0)\|\dot{R}_{\theta_0}(t,s)\|^{-2}_{\otimes^2 H(R_{\theta_0})}, \quad (2)$$

where θ_0 is the true parameter value, the dot denotes (as it will throughout this paper) differentiation with respect to θ, the Hilbert space $\otimes^2 H(R_{\theta_0})$ is the direct product of the RKHS $H(R_{\theta_0})$ with itself, and $H(R_{\theta_0})$ is the RKHS associated with $R_{\theta_0}(t,s)$, $t, s \in I$.

Equation (2) generalizes previous results of Levin [8] and Sakrison [9]. Levin found an asymptotic formula for the Cramér–Rao bound on the variance of an unbiased estimate of a parameter determining the covariance of a stationary Gaussian random process. Sakrison obtained an integral-equation version of (2), but his analysis required the observed process to have a white-noise component. Our formula does not require stationarity or a white-noise component in the observed process, and is applicable for discrete as well as continuous observation intervals.

Often it will be difficult to explicitly evaluate expression (2) because $\otimes^2 H(R_{\theta_0})$ cannot be easily characterized. (We call an explicit and computationally useful description of the elements and norm of a Hilbert space a *characterization*). It should be noted, however, that since $\otimes^2 H(R_{\theta_0})$ is itself an RKHS with reproducing kernel $R_{\theta_0}(\boldsymbol{t},\boldsymbol{s}) = R_{\theta_0}(t_1,s_1)R_{\theta_0}(t_2,s_2)$, any of the several computational procedures for approximating RKHS norms (see e.g., [10]) can be used to approximate (2) to any desired accuracy.

We assume that readers have some knowledge of basic RKHS theory as presented in [1]. Apart from this requirement, Sections I–V are self-contained. For the last section, a prior reading of Part IV is desirable since the mapping discussed at length therein is reintroduced.

II. Unbiased Parameter Estimation

The theory of minimum-variance unbiased estimation will be briefly reviewed, primarily to introduce our notation. Additional details may be found in [11] and [12], among many other references.

The general parameter estimation problem is, given a

Manuscript received December 10, 1971; revised July 21, 1972.
D. L. Duttweiler is with the Bell Laboratories, Holmdel, N.J. 17733.
T. Kailath is with the Stanford Electronic Laboratories, Stanford, Calif. 94305.

random process $\{x(t), t \in I\}$, the statistics of which depend on an unknown nonrandom parameter $\theta \in \Theta$, to estimate some function $f(\theta)$.[1] The set I of observation times may be any subset of the real line; in particular, it may be discrete or continuous. The set Θ is often also a subset of the real line, but this assumption is not needed here.

A function u of the observed process $\{x(t), t \in I\}$ is said to be an *unbiased estimate* of $f(\theta)$ if

$$E_\theta u = f(\theta), \qquad \theta \in \Theta \tag{3}$$

where E_θ denotes the expectation assuming θ to be the true parameter value. The function $f(\theta)$ is said to be *estimable* at $\theta_0 \in \Theta$ if there exists an unbiased estimate u of $f(\theta)$ having finite variance at θ_0, that is, an unbiased estimate u such that $E_{\theta_0}[u - f(\theta_0)]^2 < \infty$. An unbiased estimate $u_{\theta_0}^*$ of a function $f(\theta)$ estimable at θ_0 is said to be a *minimum-variance unbiased estimate* (MVUE) at θ_0 if

$$E_{\theta_0}[u_{\theta_0}^* - f(\theta_0)]^2 \le E_{\theta_0}[u - f(\theta_0)]^2 \tag{4}$$

for all other unbiased estimates u. For any given θ_0 an estimable function always has a unique MVUE. However, for any pair θ_0, θ_0' of parameters, $u_{\theta_0}^*$ is not necessarily equal to $u_{\theta_0'}^*$. If $u_{\theta_0}^* = u_{\theta_0'}^*$ for all $\theta_0' \in \Theta$, the MVUE is said to be *uniform*.

III. An RKHS Approach to MVUE

Throughout this section and those remaining, dependencies on the true parameter θ_0 will be suppressed. All unsubscripted expectations are to be assumed to be taken with θ_0 as the true parameter. We shall say estimable rather than estimable at θ_0 and MVUE rather than MVUE at θ_0.

For all $\theta \in \Theta$ define $\rho(\theta)$ as the likelihood ratio for the hypotheses h_1: $\{x(t), t \in I\}$ has statistics determined by the parameter θ and h_0: $\{x(t), t \in I\}$ has statistics determined by the parameter θ_0. We assume that $\rho(\theta)$ exists and that $E\rho^2(\theta) < \infty$ for all $\theta \in \Theta$. These assumptions are made explicitly or implicitly in almost all studies of MVUE.

Let $LL_2[\rho(\theta), \theta \in \Theta]$ denote the real Hilbert space of random variables that are either finite linear combinations of the random variables $\{\rho(\theta), \theta \in \Theta\}$ or else are limits of such random variables under the norm

$$\|u\|^2_{LL_2[\rho(\theta),\theta\in\Theta]} = Eu^2. \tag{5}$$

Since the set of finite linear combinations of the random variables $\{\rho(\theta), \theta \in \Theta\}$ is dense in $LL_2[\rho(\theta), \theta \in \Theta]$, we can define (as done in this series several times by now) an isometric Hilbert space H by the mapping

$$J: u \in LL_2[\rho(\theta), \theta \in \Theta] \to E[u\rho(\theta)] \triangleq g(\theta) \in H \tag{6a}$$

[1] Let $\mathscr{R}^I$ denote the space of real-valued functions defined on I and let $\beta(\mathscr{R}^I)$ denote the smallest σ-field containing sets of the form $\{x(\cdot) \in \mathscr{R}^I: x(t_1) \le a_1, \cdots, x(t_n) \le a_n\}$. A more precise statement is that the probability measure P on $(\mathscr{R}^I, \beta(\mathscr{R}^I))$ defining the statistics of $\{x(t), t \in I\}$ is only known to be drawn from a set $\{P_\theta, \theta \in \Theta\}$.

and the norm

$$\|g\|_H^2 = \|u\|^2_{LL_2[\rho(\theta),\theta\in\Theta]} = Eu^2. \tag{6b}$$

Define the kernel $G(\theta_1,\theta_2)$ by

$$G(\theta_1,\theta_2) = E[\rho(\theta_1)\rho(\theta_2)], \qquad \theta_1, \theta_2 \in \Theta. \tag{7}$$

Since

$$\langle g(\theta_1), G(\theta_1,\theta_2)\rangle_H = \langle u, \rho(\theta_2)\rangle_{LL_2[\rho(\theta),\theta\in\Theta]} = g(\theta_2) \tag{8}$$

for all $g(\theta_1) \in H$ and all $\theta_2 \in \Theta$, we see that the Hilbert space H is an RKHS with reproducing kernel $G(\theta_1,\theta_2)$ and can alternately be defined as the unique real Hilbert space $H(G)$ such that

$$G(\cdot,\theta) \in H(G), \qquad \forall\theta \in \Theta \tag{9a}$$

and

$$\langle g(\cdot), G(\cdot,\theta)\rangle_{H(G)} = g(\theta), \qquad \forall\theta \in \Theta, \qquad \forall g(\cdot) \in H(G). \tag{9b}$$

As usual (see [1]) we have the formal relationship

$$J^{-1}(g) = \langle g(\theta), \rho(\theta)\rangle_{H(G)}. \tag{10}$$

Theorem 1 (Parzen): A function $f(\theta)$ is estimable if and only if $f(\theta) \in H(G)$, in which case its MVUE, say u^*, is given by

$$u^* = \langle f(\theta), \rho(\theta)\rangle_{H(G)} \tag{11}$$

and has variance

$$\|f(\theta)\|^2_{H(G)} - f^2(\theta_0). \tag{12}$$

Making the weak assumption that Θ is either countable or a separable metric space, we have by the approximation theorem (see, e.g. [5, theorem 6E])

$$\|f(\theta)\|^2_{H(G;\Theta)} = \sup \|f(\theta)\|^2_{H(G;\Theta')}, \tag{13}$$

where the supremum is to be taken over all finite-dimensional subsets Θ' of Θ that include θ_0, and $H(G;\Theta')$ is the finite-dimensional RKHS determined by $G(\theta_1,\theta_2)$ restricted to $\Theta' \times \Theta'$. Combining (12) and (13), we obtain

$$\text{var } u^* = \sup \{\|f(\theta)\|^2_{H(G;\Theta')} - f^2(\theta_0)\}, \tag{14}$$

which is Barankin's [6] main result. Since by Theorem 1 the expression within brackets on the right side of (14) is also the variance of the estimate of least variance in the class of estimates unbiased only over Θ', Parzen's theory provides an alternate and simpler derivation of one of the results recently reported by Glave [13, sec. III].

IV. Estimating Arrival Time

Assume that a random process $x(\cdot)$ is of the form $x(t) = \alpha 1(t - \theta) + n(t)$, $t \in I = [0,T]$, where $n(\cdot)$ is white Gaussian noise, α is a known constant, $1(t)$ is a unit step function at the origin, and $\theta \in \Theta = [0,L]$ is an unknown parameter that is to be estimated. Without loss of generality (scaling is possible through α), it can be assumed that the white Gaussian noise has unit (two-sided) spectral intensity. The reasonable condition $T > L$ is also assumed.

To carry out Parzen's procedure for finding the MVUE of θ and its variance, we must find $\rho(\theta)$, evaluate $G(\theta_1,\theta_2)$, and characterize $H(G)$. The first two steps are easy. It is well known (see, e.g. [11]) that the likelihood ratio $\rho(\theta)$ is equal to

$$\exp\left\{\int_0^T [x(t) - \alpha 1(t-\theta_0)][\alpha 1(t-\theta) - \alpha 1(t-\theta_0)]\,dt - \tfrac{1}{2}\int_0^T [\alpha 1(t-\theta) - \alpha 1(t-\theta_0)]^2\,dt\right\}, \quad (15)$$

and it is straightforward to calculate

$$G(\theta_1,\theta_2) = \begin{cases} \exp\{\alpha^2 \cdot \min(|\theta_1 - \theta_0|,|\theta_2 - \theta_0|)\}, & (\theta_1 - \theta_0)(\theta_2 - \theta_0) \geq 0 \\ 1, & \text{otherwise.} \end{cases} \quad (16)$$

The difficult step in applying Parzen's theory is often in characterizing $H(G)$. Fortunately, we have been able to make this characterization for the above kernel. The elements of $H(G)$ are functions $g(\theta)$ that are absolutely continuous and such that[2]

$$\begin{aligned}\infty &> \|g(\theta)\|^2_{H(G)} \\ &= g^2(\theta_0) + \int_0^L \dot{g}^2(\theta)\alpha^{-2}\exp\{-\alpha^2|\theta - \theta_0|\}\,d\theta. \quad (17)\end{aligned}$$

To verify this characterization, it is necessary only to show that $H(G)$ as defined by (17) is a Hilbert space and that the two RKHS properties (9a) and (9b) are satisfied. The interesting step is to verify (9b). We shall make this verification for $\theta \in [\theta_0,L]$. For $\theta \in [0,\theta_0]$ a similar argument holds. We have (letting u denote a dummy integration variable)

$$\begin{aligned}\langle g(\cdot),G(\cdot,\theta)\rangle_{H(G)} &= g(\theta_0)G(\theta_0,\theta) \\ &\quad + \int_0^{\theta_0} \dot{g}(u)\left(\frac{d}{du}1\right)\alpha^{-2} \\ &\quad \exp\{-\alpha^2(\theta_0 - u)\}\,du \\ &\quad + \int_{\theta_0}^{\theta} \dot{g}(u)\left(\frac{d}{du}\exp\{\alpha^2(u-\theta_0)\}\right)\alpha^{-2} \\ &\quad \cdot \exp\{\alpha^{-2}(u-\theta_0)\}\,du \\ &\quad + \int_{\theta}^{L} \dot{g}(u)\left(\frac{d}{du}\exp\{\alpha^2(\theta-\theta_0)\}\right)\alpha^{-2} \\ &\quad \cdot \exp\{-\alpha^2(u-\theta_0)\}\,du \\ &= g(\theta_0) + \int_{\theta_0}^{\theta} \dot{g}(u)\alpha^2\exp\{\alpha^2(u-\theta_0)\}\alpha^{-2} \\ &\quad \cdot \exp\{-\alpha^2(u-\theta_0)\}\,du \\ &= g(\theta_0) + \int_{\theta_0}^{\theta}\dot{g}(u)\,du = g(\theta). \quad (18)\end{aligned}$$

Using this characterization and Theorem 1, we obtain for an unbiased estimate u of $f(\theta) = \theta$

$$\begin{aligned}\operatorname{var}[u] &\geq \|f(\theta)\|_{H(G)} - f^2(\theta_0) \\ &= \|\theta\|^2_{H(G)} - {\theta_0}^2 \\ &= \alpha^{-4}(2 - \exp\{-\alpha^2\theta_0\} - \exp\{-\alpha^2(L-\theta_0)\}). \quad (19)\end{aligned}$$

The MVUE, say u^*, is given formally (the inner product in the first line is a congruence inner product) by

$$\begin{aligned}u^* &= \langle\theta,\rho(\theta)\rangle_{H(G)} \\ &= \theta_0\rho(\theta_0) + \int_0^L 1\cdot\dot{\rho}(\theta)\alpha^{-2}\exp\{-\alpha^2|\theta-\theta_0|\}\,d\theta \\ &= \theta_0 + \int_0^{\theta_0}\dot{\rho}(\theta)\alpha^{-2}\exp\{\alpha^2(\theta-\theta_0)\}\,d\theta \\ &\quad + \int_{\theta_0}^{L}\dot{\rho}(\theta)\alpha^{-2}\exp\{\alpha^2(\theta_0-\theta)\}\,d\theta. \quad (20)\end{aligned}$$

However, the function $\dot{\rho}(\theta)$ is not well behaved. In [14] rigorous meaning is given to (20) by using the Itô calculus. A simpler way is just to integrate by parts, which yields

$$\begin{aligned}u^* &= \theta_0 + \rho(L)\alpha^{-2}\exp\{\alpha^2(\theta_0 - L)\} \\ &\quad - \rho(0)\alpha^{-2}\exp\{-\alpha^2\theta_0\} \\ &\quad - \int_0^{\theta_0}\rho(\theta)\exp\{\alpha^2(\theta-\theta_0)\,d\theta\} \\ &\quad + \int_{\theta_0}^{L}\rho(\theta)\exp\{\alpha^2(\theta_0-\theta)\}\,d\theta. \quad (21)\end{aligned}$$

Since $\rho(\theta)$ is with probability one a continuous function of θ, the integrals in (21) are well defined. Furthermore, it is easy to verify that with u^* as given by (21),

$$E[u^*\rho(\theta)] = f(0) = 0 \quad (22)$$

and thus that our formal arguments have led us to a meaningful and correct expression for u^*.

The expression (21) for u^* is of course too cumbersome to be of anything but theoretical interest. We have given it since it was so easy to derive, once the bound on the variance of an unbiased estimate had been obtained.

The above arguments extend easily to the problem of estimating the arrival time of a rectangular pulse in white Gaussian noise if the rather restrictive condition that the duration of the pulse is longer than the *a priori* range of possible arrival times is imposed. Let

$$x(t) = \alpha r(t-\theta) + n(t), \qquad t \in I = [0,T], \quad (23)$$

where $n(\cdot)$, α, and θ are as before, but

$$r(t) = \begin{cases} 1, & t \in [0,S] \\ 0, & \text{otherwise.} \end{cases} \quad (24)$$

With this notation, the constraint that the duration of the pulse be longer than the *a priori* range of arrival times is simply $S \geq L$. The reasonable condition $T \geq S + L$ is also assumed to be met.

[2] Since through the parallelogram identity inner products are specified by norms, we need only define norms when making RKHS characterizations.

Proceeding as before it is easy to show that

$$G(\theta_1,\theta_2) = \begin{cases} \exp\{2\alpha^2 \cdot \min(|\theta_1 - \theta_0|,|\theta_2 - \theta_0|), & (\theta_1 - \theta_0)(\theta_2 - \theta_0) \geq 0 \\ 1, & \text{otherwise.} \end{cases} \quad (25)$$

$$\operatorname{var} u^* = 4^{-1}\alpha^{-4}(2 - \exp\{-2\alpha^2\theta_0\} - \exp\{-2\alpha^2(L - \theta_0)\}) \quad (26)$$

and that

$$u^* = \theta_0 - \rho(0)\alpha^{-2}\exp\{-2\alpha^2\theta_0\} + \rho(L)\alpha^{-2}\exp\{-2\alpha^2(L - \theta_0)\} - \int_0^{\theta_0}\rho(\theta)\exp\{2\alpha^2(\theta - \theta_0)\} + \int_{\theta_0}^{L}\rho(\theta)\exp\{2\alpha^2(\theta_0 - \theta)\}\,d\theta. \quad (27)$$

The condition $S \geq L$ is needed to give $G(\theta_1,\theta_2)$ the above simple form. Without it we are unable to characterize $H(G)$ and obtain explicit solutions via Theorem 1.

There are few results to compare with ours in either of these two arrival-time problems. For both problems the Cramér–Rao bound is trivial, being equal to zero. For the rectangular-pulse problem Swerling [15] obtained a lower bound on the variance. He was the first to use Barankin's theory for engineering problems. His bound also has the virtue of being valid both for $S < L$ and $S > L$. However, it is an asymptotic bound and, as stated by Ziv and Zakai [16], the details of his derivation are not completely clear. Ziv and Zakai [16] show that (their methods do not require $S \geq L$, but the following specialization of one of their formulas does)

$$\hat{\varepsilon}^2 \geq \begin{cases} 0.37\alpha^{-4}, & L\alpha^2 \geq 6.48 \\ \alpha^{-4}(L\alpha^2/2)^2 \operatorname{erfc}\{L\alpha^2/2\}, & \text{otherwise} \end{cases} \quad (28)$$

where

$$\hat{\varepsilon}^2 = \sup_{\theta_0 \in [0,L]} \operatorname{var}[u^*_{\theta_0}] \quad (29)$$

and

$$\operatorname{erfc}\xi = \int_{\xi}^{\infty}(2\pi)^{-1/2}\exp\{-u^2/2\}\,du. \quad (30)$$

From (26) we have

$$\hat{\varepsilon}^2 = 2^{-1}\alpha^{-4}(1 - \exp\{-\alpha^2 L\}), \quad (31)$$

which can be verified to be consistent with (28). The method of [16] is quite ingenious and can be adapted to a wide variety of problems.

V. Unbiased Estimation of Functions of Regression Coefficients

In this section the observed random process is assumed to be of the form

$$x(t) = \sum_{j=1}^{n}\theta(j)m_j(t) + n(t), \qquad t \in I, \quad (32)$$

where $n(\cdot)$ is Gaussian with mean zero and covariance $R(t,s)$, the unknown function $\theta(j), j = 1,\cdots,n$, is an element of n-dimensional Euchlidean space E^n, and the n known functions $m_1(t),\cdots,m_n(t)$ are linearly independent elements of $H(R)$. The function $f(\theta)$ to be estimated is not restricted; in particular, it need not be linear.

The true parameter θ_0 is assumed to be zero; that is, it is assumed that $\theta_0(j) = 0, j = 1,\cdots,n$. We can make this assumption without loss of generality, for if $\theta_0 \neq 0$ and the function $f(\theta)$ is to be estimated, we can instead solve the equivalent problem of estimating $f'(\theta')$ from an observation of $\{x'(t), t \in I\}$ where

$$x'(t) = x(t) - \sum_{j=1}^{n}\theta_0(j)m_j(t), \quad (33)$$

$$\theta' = \theta - \theta_0, \quad (34)$$

and

$$f'(\theta') = f(\theta' - \theta_0) - f(\theta_0). \quad (35)$$

To apply Theorem 1, $\rho(\theta)$ and $G(\theta_1,\theta_2)$ must be found and $H(G)$ characterized. For this problem none of these steps is difficult. The likelihood ratio $\rho(\theta)$ is given by (see [5] or [1])

$$\rho(\theta) = \exp\left\{\left\langle x(t), \sum_{j=1}^{n}\theta(j)m_j(t)\right\rangle_{H(R)} - 1/2\left\|\sum_{j=1}^{n}\theta(j)m_j(t)\right\|^2_{H(R)}\right\}. \quad (36)$$

Using the fact that when $\theta = \theta_0 = 0$, the processes $x(\cdot)$ and $n(\cdot)$ are identical and the random variable

$$\left\langle x(t), \sum_{j=1}^{n}[\theta_1(j) + \theta_2(j)]m_j(t)\right\rangle_{H(R)}$$

is Gaussian with mean zero and covariance

$$\left\|\sum_{j=1}^{n}[\theta_1(j) + \theta_2(j)]m_j(t)\right\|^2_{H(R)}, \quad (37)$$

it is easy to see that

$$G(\theta_1,\theta_2) = \exp\left\{\sum_{j,k=1}^{n}\theta_1(j)\theta_2(k)\langle m_j(t),m_k(t)\rangle_{H(R)}\right\}. \quad (38)$$

Define the kernel $A(j,k), j,k = 1,\cdots,n$, as the inverse of the kernel

$$\langle m_j(t),m_k(t)\rangle_{H(R)}, \qquad j,k = 1,\cdots,n; \quad (39)$$

that is, define $A(j,k)$ as the unique kernel satisfying for all $j,k = 1,\cdots,n$

$$\sum_{l=1}^{n}A(j,l)\langle m_l(t),m_k(t)\rangle_{H(R)} = \delta_{jk}, \quad (40)$$

where δ_{jk} is the Kronecker delta function. The existence and uniqueness of $A(j,k)$ follows from the assumption that the $m_j(t)$ are linearly independent in $H(R)$. Then, we have (see [5] or [1])

$$G(\theta_1,\theta_2) = \exp\langle\theta_1,\theta_2\rangle_{H(A)}, \quad (41)$$

where $H(A)$ is the finite-dimensional RKHS generated by the kernel $A(j,k)$.

By introducing the kernel $A(j,k)$ in this somewhat artificial manner, we have manipulated $G(\theta_1,\theta_2)$ into the same functional form as the kernel $M_{WGN}(\phi,\psi)$ whose RKHS we already characterized in [4]. Thus, letting $\otimes^p H(A)$ denote the pth direct product of $H(A)$ with itself and carets $\wedge$ denote symmetric versions of functions (concepts to be more fully explained), we have the following characterization of $H(G)$: a function $g(\theta)$ belongs to $H(G)$ if and only if it has the decomposition

$$g(\theta) = \sum_{p=0}^{\infty} \langle g_p(j_1,\cdots,j_p),\theta(j_1)\cdots\theta(j_p)\rangle_{\otimes^p H(A)}, \quad (42)$$

where

$$\infty > \|g\|^2_{H(G)} = \sum_{p=0}^{\infty} p!\, \|\hat{g}_p\|^2_{\otimes^p H(A)}. \quad (43)$$

For $p = 2,3,\cdots$ the pth direct product of $H(A)$ with itself, $\otimes^p H(A)$, can (there are many equivalent ways) be defined as the unique Hilbert space of functions with p arguments ranging over the first n integers such that for any functions $g_1(j),\cdots,g_p(j) \in H(A)$ and any function $g_p(j_1,\cdots,j_p) \in \otimes^p H(A)$, the product $g_1(j_1)\cdots g_p(j_p)$ exists in $\otimes^p H(A)$ and

$$\langle g_p(j_1,\cdots,j_p),g_1(j_1)\cdots g_p(j_p)\rangle_{\otimes^p H(A)} = \langle\cdots\langle g_p(j_1,\cdots,j_p),g_1(j_1)\rangle_{H(A)},\cdots,g_p(j_p)\rangle_{H(A)}. \quad (44)$$

The direct product $\otimes^p H(A)$ is itself an RKHS with reproducing kernel

$$A(\boldsymbol{j},\boldsymbol{k}) = A(j_1,k_1)\cdots A(j_p,j_p) \quad (45)$$

(see, e.g., [17]). For notational convenience we also defined $\otimes^1 H(A)$ as the RKHS $H(A)$ itself and $\otimes^0 H(A)$ as the Hilbert space of real constants with inner product

$$\langle\xi_1,\xi_2\rangle_{\otimes^0 H(A)} = \xi_1\xi_2. \quad (46)$$

The symmetrization $\hat{g}_p(j_1,\cdots,j_p)$ of a function $g_p(j_1,\cdots,j_p)$ is defined by

$$\hat{g}_p(j_1,\cdots,j_p) = (1/p!)\sum_{\pi} g(j_{\pi(1)},\cdots,j_{\pi(p)}) \quad (47)$$

when the summation ranges over all permutations π of the integers $1,\cdots,p$. For example,

$$\hat{g}_1(j_1) = g_1(j_1) \quad (48)$$

and

$$\hat{g}_2(j_1,j_2) = \tfrac{1}{2}[g_2(j_1,j_2) + g_2(j_2,j_1)]. \quad (49)$$

Combining the characterization (42) and (43) with Theorem 1, we obtain the following theorem.

Theorem 2: For the regression problem of this section a function $f(\theta)$ is estimable if and only if it can be decomposed as

$$f(\theta) = \sum_{p=0}^{\infty} \langle f_p(j_1,\cdots,j_p),\theta(j_1)\cdots\theta(j_p)\rangle_{\otimes^p H(A)}, \quad (50)$$

where

$$\infty > \sum_{p=0}^{\infty} p!\, \|\hat{f}_p\|^2_{\otimes^p H(A)}. \quad (51)$$

If $f(\theta)$ is estimable, then its MVUE, say u^*, has variance

$$\sum_{p=1}^{\infty} p!\, \|\hat{f}_p\|^2_{\otimes^p H(A)}. \quad (52)$$

The infinite sum (67) provides a sequence of monotonically increasing lower bounds on the variance of an unbiased estimate u of an estimable function $f(\theta)$. We have for $k \geq 1$

$$\text{var}\,[u] \geq \text{var}\,[u^*] \geq \sum_{p=1}^{k} p!\, \|\hat{f}_p\|^2_{\otimes^p H(A)}. \quad (53)$$

It is notationally cumbersome but otherwise straightforward to verify (see [14]) that (53) is the kth Bhattacharyya bound (see [7] or [12]). Thus, for this particular problem (regression in Gaussian noise) we can interpret the Bhattacharyya bounds as arising from a natural orthogonal decomposition of a basic underlying Hilbert space.

VI. Estimating a Parameter Determining the Covariance of a Gaussian Random Process

From an observation of a zero-mean Gaussian random process $\{x(t),\ t \in I\}$ with a covariance drawn from the set $\{R_\theta(t,s),\ \theta \in \Theta\}$ an estimate of a function $f(\theta)$ (often the identity function) is desired. The set I is assumed to be a subset of the real line and may be discrete or continuous. For convenience, the indexing set Θ is also assumed to be a subset of the real line. The generalization for Θ a subset of n-dimensional Euclidean space should be obvious.

For this problem we are unable to characterize $H(G)$. Somewhat cumbersome expressions for $G(\theta_1,\theta_2)$ are available ([18] or [14, appendix D]), which could be used with Theorem 1 and RKHS approximation procedures to approximate the MVUE and its variance to any desired accuracy, but to obtain explicit analytical results we must lower our objectives. In this section we calculate the Cramér–Rao bound on the variance of the MVUE.

Instead of the RKHS $H(G)$, the RKHS of Part IV will play the important role. Before turning to the derivation of the Cramér–Rao bound we shall reintroduce this RKHS.

Let NL_2 denote the real Hilbert space of real-valued finite-variance (assuming θ_0 to be the true parameter) nonlinear functionals of the random process $\{x(t),\ t \in I\}$[3] with inner product

$$\langle u,v\rangle_{NL_2} = E_{\theta_0}[uv], \quad (54)$$

and let $H(R)$ denote the RKHS of functions on I generated by $R_{\theta_0}(t,s), t,s \in I$. [$H(R_{\theta_0})$ is perhaps a more consistent notation, but because of the extensive use that will be made of this RKHS it is convenient to have a more compact notation.] For all $\phi \in H(R)$ define

$$\mu(\phi) = \exp\{\langle x,\phi\rangle_{H(R)} - \tfrac{1}{2}\|\phi\|^2_{H(R)}\}. \quad (55)$$

It can be shown (see [19]) that the set of all finite linear combinations of the random variables $\{\mu(\phi),\ \phi \in H(R)\}$ is dense in NL_2. Intuitively, this fact is not surprising since

[3] More precisely, the elements of NL_2 are finite-variance random variables on the probability space $(\mathscr{R}^I,\beta(\mathscr{R}^I),P_{\theta_0})$.

with

$$\phi(\cdot) = \sum_{j=1}^{n} \lambda_j R_{\theta_0}(t_j,\cdot) \tag{56}$$

the function $\mu(\phi)$ equals

$$\exp\left\{\sum_{j=1}^{n} \lambda_j x(t_j) - \tfrac{1}{2}\sum_{j,k=1}^{n} \lambda_j\lambda_k R_{\theta_0}(t_j,t_k)\right\}. \tag{57}$$

Thus (see [1]), we can define an isometric Hilbert space H by the mapping

$$J: u \in NL_2 \to E_{\theta_0}[u\mu(\phi)] \triangleq g(\phi) \in H \tag{58a}$$

and norm

$$\|g\|_H^2 = Eu^2. \tag{58b}$$

Let

$$M(\phi,\psi) = E[\mu(\phi)\mu(\psi)] = \exp\langle\phi,\psi\rangle_{H(R)}. \tag{59}$$

Since

$$\langle g(\cdot),M(\cdot,\phi)\rangle_H = \langle u,\mu(\phi)\rangle_{NL_2} = E[u\mu(\phi)] = g(\phi), \tag{60}$$

$H = H(M)$ is an RKHS with reproducing kernel $M(\phi,\psi)$ and can alternately be described as the unique Hilbert space of functionals such that

$$M(\cdot,\phi) \in H(M), \qquad \forall\phi \in H(R) \tag{61}$$

and

$$\langle g(\cdot),M(\cdot,\phi)\rangle_{H(M)} = g(\phi), \qquad \forall g(\cdot) \in H(M),\ \forall\phi \in H(R). \tag{62}$$

In the notation of Section V we can characterize $H(M)$ as the space of functionals $g(\phi)$ decomposable as

$$g(\phi) = \sum_{p=0}^{\infty} \langle g_p(t_1,\cdots,t_p),\phi(t_1)\cdots\phi(t_p)\rangle_{\otimes^p H(R)}, \tag{63}$$

where $g_p(t_1,\cdots,t_p) \in \otimes^p H(R)$ for $p = 0,1,\cdots$ and

$$\infty > \|g(\phi)\|_{H(M)}^2 = \sum_{p=0}^{\infty} p!\ \|\hat{g}_p\|^2_{\otimes^2 H(R)}. \tag{64}$$

Returning to the main derivation, the Cramér–Rao bound, say CR, is given by (see, e.g. [11, p. 79])

$$\text{CR} = \dot{f}^2(\theta_0)/E_{\theta_0}\left(\frac{d}{d\theta}\ln P[x(\cdot)\mid\theta]\bigg|_{\theta=\theta_0}\right)^2, \tag{65}$$

where the dot denotes differentiation with respect to θ and $P[x(\cdot)\mid\theta]$ denotes the probability density function when θ is the true parameter. [This density function has only formal meaning if I is continuous. There are a number of steps in the derivation to follow that obviously lack rigor. A rigorous version of the result is presented as a theorem and proof in the Appendix.] We shall show that the denominator of this expression, denoted by D for convenience, can be written

$$D = \tfrac{1}{2}\|\dot{R}_{\theta_0}(t_1 s)\|^2_{\otimes^2 H(R)}. \tag{66}$$

First let

$$\rho(\theta) = P[x(\cdot)/\theta]/P[x(\cdot)/\theta_0] = \text{likelihood ratio}. \tag{67}$$

Then

$$D = E_{\theta_0}\left(\frac{\dot{P}[x(\cdot)\mid\theta_0]}{P[x(\cdot)\mid\theta_0]}\right)^2 = E_{\theta_0}\dot{\rho}^2(\theta_0) = \|\dot{\rho}(\theta_0)\|^2_{NL_2}. \tag{68}$$

Since NL_2 and $N(M)$ are isometric under the linear mapping J,

$$D = \|J[\dot{\rho}(\theta_0)]\|^2_{H(M)} = \left\|\frac{d}{d\theta}J[\rho(\theta)]\bigg|_{\theta=\theta_0}\right\|^2_{H(M)}. \tag{69}$$

Evaluating $J[\rho(\theta)]$ is not difficult. Remembering that $\rho(\theta)$ is a likelihood ratio, we have

$$J[\rho(\theta)] = E_{\theta_0}[\rho(\theta)\mu(\phi)] = E_\theta\mu(\phi) = E_\theta\exp\{\langle x,\phi\rangle_{H(R)} - \tfrac{1}{2}\|\phi\|^2_{H(R)}\} \tag{70}$$

If $\{x(t),\ t \in I\}$ is Gaussian with mean zero and covariance $R_\theta(t,s)$, the random variable $\langle x,\phi\rangle_{H(R)}$ is Gaussian with mean zero and covariance

$$\begin{aligned}E_\theta[\langle x,\phi\rangle_{H(R)}\langle x,\phi\rangle_{H(R)}] &= E_\theta[\langle x(t),\phi(t)\rangle_{H(R)}\langle x(s),\phi(s)\rangle_{H(R)}] \\ &= E_\theta\langle x(t)x(s),\phi(t)\phi(s)\rangle_{\otimes^2 H(R)} \\ &= \langle R_\theta(t,s),\phi(t)\phi(s)\rangle_{\otimes^2 H(R)}.\end{aligned} \tag{71}$$

Therefore,

$$\begin{aligned}J[\rho(\theta)] &= \exp\{\tfrac{1}{2}\langle R_\theta(t,s),\phi(t)\phi(s)\rangle_{\otimes^2 H(R)} - \tfrac{1}{2}\|\phi\|^2_{H(R)}\} \\ &= \exp\{\tfrac{1}{2}\langle R_\theta(t,s),\phi(t)\phi(s)\rangle_{\otimes^2 H(R)} - \tfrac{1}{2}\langle R_{\theta_0}(t,s),\phi(t)\phi(s)\rangle_{\otimes^2 H(R)}\} \\ &= \exp\{\tfrac{1}{2}\langle K_\theta(t,s),\phi(t)\phi(s)\rangle_{\otimes^2 H(R)}\},\end{aligned} \tag{72}$$

where

$$K_\theta(t,s) = R_\theta(t,s) - R_{\theta_0}(t,s). \tag{73}$$

Continuing formally,

$$\frac{d}{d\theta}J[\rho(\theta)] = \frac{d}{d\theta}\exp\{\tfrac{1}{2}\langle K_\theta(t,s),\phi(t)\phi(s)\rangle_{\otimes^2 H(R)}\} = \tfrac{1}{2}\langle\dot{K}_\theta(t,s),\phi(t)\phi(s)\rangle_{\otimes^2 H(R)}\,J[\rho(\theta)] \tag{74}$$

and

$$\frac{d}{d\theta}J[\rho(\theta)]\bigg|_{\theta=\theta_0} = \tfrac{1}{2}\langle\dot{K}_{\theta_0}(t,s),\phi(t)\phi(s)\rangle_{\otimes^2 H(R)} \tag{75}$$

Therefore, from (69) we have

$$\begin{aligned}D &= \|\tfrac{1}{2}\langle\dot{K}_{\theta_0}(t,s),\phi(t)\phi(s)\rangle_{\otimes^2 H(R)}\|^2_{H(M)} \\ &= 2!\ \|\tfrac{1}{2}\hat{\dot{K}}_{\theta_0}(t,s)\|^2_{\otimes^2 H(R)} \\ &= \tfrac{1}{2}\|\dot{K}_{\theta_0}(t,s)\|^2_{\otimes^2 H(R)} = \tfrac{1}{2}\|\dot{R}_{\theta_0}(t,s)\|^2_{\otimes^2 H(R)}\end{aligned} \tag{76}$$

and the derivation is complete.

Example 1: Suppose that a zero-mean Gaussian process $\{x(t), t \in [0,T]\}$ has a covariance drawn from the set $(\alpha/2\theta)\exp\{-\theta|t-s|\}$, $0 < \theta < \infty$,[4] where α is a known constant and that an estimate of θ is desired. We have $\Theta = (0,\infty)$, $I = [0,T]$, $f(\theta) = \theta$, and

$$R_\theta(t,s) = (\alpha/2\theta)\exp\{-\theta|t-s|\}. \tag{77}$$

The norm in $H(R)$ is given by (see, e.g. [1])

$$\|\phi\|^2_{H(R)} = (1/\alpha)\left(\int_0^T (\theta_0\phi(t) + \dot{\phi}(t)^2\, dt + 2\theta_0\phi^2(0)\right) \tag{78}$$

and the norm of a function $\phi(t,s) \in \overset{2}{\otimes} H(R)$ is given by (see [21])

$$\alpha^2\|\phi(t,s)\|^2_{\overset{2}{\otimes} H(R)} = \int_0^T\int_0^T \left(\frac{\partial}{\partial t}\frac{\partial}{\partial s}\phi(t,s)\right)^2 ds\, dt + \phi^2(0,0) + \int_0^T \left(\frac{\partial}{\partial t}\phi(t,0)\right)^2 dt + \int_0^T \left(\frac{\partial}{\partial s}\phi(0,s)\right)^2 ds. \tag{79}$$

We have

$$\dot{R}_{\theta_0}(t,s) = -((\alpha/2\theta)|t-s| + \alpha/(2\theta_0^2))\exp\{-\theta_0|t-s|\} \tag{80}$$

and after some straightforward but tedious calculations using the characterization (79), we find

$$\|\dot{R}_{\theta_0}(t,s)\|^2_{\overset{2}{\otimes} H(R)} = T/\theta_0 + 1/\theta_0^2. \tag{81}$$

Therefore, any unbiased estimate of θ must have a variance greater than or equal to

$$\begin{aligned} CR &= 2(T\theta_0^{-1} + \theta_0^{-2})^{-1} \\ &= (2\theta_0/T)(1 + \theta_0^{-1}T^{-1})^{-1}. \end{aligned} \tag{82}$$

The bound (82) is consistent with the expectation that as T becomes larger better estimates of θ should be possible.

Levin [8] obtained an asymptotic formula for the Cramér–Rao bound for the special case of stationary covariances, showing formally that if a stationary zero-mean Gaussian process with a covariance known to be from the set $\{R_\theta(t-s), \theta \in \Theta\}$ is observed over the time interval $[0,T]$ where T is large, the Cramér–Rao bound on the variance of an unbiased estimate of θ is approximately

$$2\left(T\int_{-\infty}^{\infty} \dot{S}_{\theta_0}^2(f)S_{\theta_0}^{-2}(f)\, df\right)^{-1}, \tag{83}$$

where

$$S_\theta(f) = \int_{-\infty}^{\infty} R_\theta(t)\exp\{-j2\pi ft\}\, dt. \tag{84}$$

[4] It is necessary to have this type of dependence on θ to make the estimation nontrivial. In particular, if the set of *a priori* possible covariances is $\{\alpha\exp\{-\theta|t-s|\}, 0 < \theta < \infty\}$, the parameter θ is uniquely determinable (see [20]).

More precisely, letting CR (T) denote the Cramér–Rao bound when the observation interval is $[0,T]$, Levin's result is

$$\lim_{T\to\infty} T\cdot \mathrm{CR}(T) = 2\left(\int_{-\infty}^{\infty} \dot{S}_{\theta_0}^2(f)S_{\theta_0}^{-2}(f)\, df\right)^{-1}. \tag{85}$$

For the problem in Example 1, we have from (82)

$$\lim_{T\to\infty} T\cdot \mathrm{CR}(T) = 2\theta_0, \tag{86}$$

which is also the result that would be obtained from (85).

Appendix

Theorem 3: Let Θ and I be subsets of the real line and for all $\theta \in \Theta$ let P_θ be a probability measure defining a Gaussian random process with mean zero and covariance $R_\theta(t,s)$ on the measure space $(\mathscr{R}^I, \beta(\mathscr{R}^I))$. Let $\theta_0 \in \Theta$ be the true parameter and $f(\theta)$, a function defined on θ that is to be estimated. Define $H(R)$ as the RKHS generated by $R_{\theta_0}(t,s)$ and

$$K_\theta(t,s) = R_\theta(t,s) - R_{\theta_0}(t,s), \qquad \theta \in \Theta. \tag{87}$$

Assume that

i) there exists $\varepsilon > 0$ such that $|\theta - \theta_0| < \varepsilon$ implies $\theta \in \Theta$ and $\|K_\theta(t,s)\|_{\overset{2}{\otimes} H(R)} < \infty$,

ii) $\dot{f}(\theta_0)$ exists,

and

iii) $(\partial/\partial\theta_1)(\partial/\partial\theta_2)\langle K_{\theta_1}(t,s), K_{\theta_2}(t,s)\rangle_{\overset{2}{\otimes} H(R)}$ exists at $\theta_1 = \theta_2 = \theta_0$.

Then, if u is any function measurable on $(\mathscr{R}^I, B(\mathscr{R}^I))$ and such that

$$\int u\, dP_\theta \triangleq E_\theta u = f(\theta) \tag{88}$$

for all $\theta \in \Theta$, the variance of u at θ_0, that is, $E_{\theta_0}(u - f(\theta_0))^2$, is at least

$$2\dot{f}^2(\theta_0)\|\dot{R}_{\theta_0}(t,s)\|^{-2}_{\overset{2}{\otimes} H(R)}. \tag{89}$$

Proof: Remembering $K_{\theta_0}(t,s) = 0$ and using conditions i) and iii), we have

$$\infty > \lim h^{-2}\|K_{\theta_0+h}(t,s)\|^2_{\overset{2}{\otimes} H(R)} = \frac{\partial}{\partial\theta_1}\frac{\partial}{\partial\theta_2}\langle K_{\theta_1}(t,s), K_{\theta_2}(t,s)\rangle_{\overset{2}{\otimes} H(R)}\Big|_{\theta_1=\theta_2=\theta_0} \tag{90}$$

Therefore there exists a constant δ such that $0 < \delta \le \varepsilon$ and for all $h \in (-\delta,\delta)$

$$\|K_{\theta_0+h}(t,s)\|^2_{\overset{2}{\otimes} H(R)} < \tfrac{1}{2}. \tag{91}$$

For all $\theta \in (\theta_0 - \delta, \theta_0 + \delta)$ let $\rho(\theta)$ denote the Radon–Nikodym derivative of the probability measure P_θ with respect to P_{θ_0}. This derivative must exist and be of finite mean-square at θ_0, that is, it must exist and satisfy

$$E_{\theta_0}\rho^2(\theta) = \int \rho^2(\theta)\, dP_{\theta_0} < \infty \tag{92}$$

(see [14, lemma 5.2.1]).

It is simple to verify that

$$f(\theta) - f(\theta_0) = E_{\theta_0}[(u - f(\theta_0))(\rho(\theta) - 1)] = \int (u - f(\theta_0))(\rho(\theta) - 1)\, dP_{\theta_0}. \quad (93)$$

By the Schwarz inequality, we have

$$(f(\theta) - f(\theta_0))^2 \leq \int (u - f(\theta_0))^2\, dP_{\theta_0} \cdot \int (\rho(\theta) - 1)^2\, dP_{\theta_0} \quad (94)$$

or after rearrangement

$$\text{var}\,[u] \triangleq \int (u - f(\theta_0))^2\, dP_{\theta_0} \geq (f(\theta) - f(\theta_0))^2 \left(\int (\rho(\theta) - 1)^2\, dP_{\theta_0}\right)^{-1}. \quad (95)$$

Since (100) holds for all $\theta = \theta_0 + h$ with $|h| < \delta$ and $\dot{f}(\theta)$ exists, the proof will be complete if it can be shown that

$$\lim_{h \to 0} h^{-2} D(\theta_0 + h) = \tfrac{1}{2} \|\dot{R}_{\theta_0}(t,s)\|^2_{2\,\otimes H(R)}, \quad (96)$$

where

$$D(\theta) = \int (\rho(\theta) - 1)^2\, dP_{\theta_0} = \int \rho^2(\theta)\, dP_{\theta_0} - 1. \quad (97)$$

Let NL_2, $\{\mu(\phi), \phi \in H(R)\}$, $M(\phi,\psi)$, $H(M)$, and J be as in Section VI. Since NL_2 and $H(M)$ are isometric under J, we have for any $\theta \in (\theta_0 - \delta, \theta_0 + \delta)$

$$D(\theta) = \|\rho(\theta)\|^2_{NL_2} - 1 = \|J(\rho(\theta))\|_{H(M)}{}^2 - 1. \quad (98)$$

But,

$$\begin{aligned} J(\rho(\theta)) &= E_{\theta_0}[\rho(\theta)\mu(\phi)] = E_\theta \mu(\phi) \\ &= \exp\{\tfrac{1}{2}\langle K_\theta(t,s), \phi(t)\phi(s)\rangle_{2\,\otimes H(R)}\} \\ &= \sum_{n=0}^{\infty} (1/n!) 2^{-n} \langle K_\theta(t,s), \phi(t)\phi(s)\rangle^n_{2\,\otimes H(R)} \\ &= \sum_{\substack{p=0 \\ p\,\text{even}}}^{\infty} [(p/2)!]^{-1} 2^{-p/2} \langle K_\theta(t_1,t_2) \cdots K_\theta(t_{p-1},t_p), \\ &\qquad \cdot \phi(t_1) \cdots \phi(t_p)\rangle_{p\,\otimes H(R)}. \end{aligned} \quad (99)$$

Therefore, letting $f_p = K_\theta(t_1,t_2) \cdots K_\theta(t_{p-1},t_p)$ the norm of $J(p(\theta))$ in $H(M)$ equals

$$\sum_{\substack{p=0 \\ p\,\text{even}}}^{\infty} p! [(p/2)!]^{-2} 2^{-p} \|\hat{f}_p\|^2_{p\,\otimes H(R)}$$

or, equivalently,

$$\sum_{n=0}^{\infty} (2n)!\,(n!)^{-2} 2^{-2n} \|\hat{f}_{2n}\|^2_{2n\,\otimes H(R)} \quad (100)$$

and

$$D(\theta) = \sum_{n=1}^{\infty} (2n)!\,(n!)^{-2} 2^{-2n} \|\hat{f}_{2n}\|^2_{2n\,\otimes H(R)} \quad (101)$$

Let $S(\theta)$ equal the summation (106) minus its first term. Using the facts that

$$(2n!)(n!)^{-2} 2^{-2n} \leq 1, \qquad \forall n \quad (102)$$

and the norm of the symmetrization of a function is no greater than the norm of that function, we have for $\theta \in (\theta_0 - \delta, \theta_0 + \delta)$

$$\begin{aligned} S(\theta) &\leq \sum_{n=2}^{\infty} \|K_\theta(t_1,t_2) \cdots K_\theta(t_{2n-1},t_{2n})\|^2_{2n\,\otimes H(R)} \\ &= \sum_{n=2}^{\infty} \|K_\theta(t,s)\|^{2n}_{2\,\otimes H(R)} \\ &= \|K_\theta(t,s)\|^4_{2\,\otimes H(R)} \sum_{n=0}^{\infty} \|K_\theta(t,s)\|^{2n}_{2\,\otimes H(R)} \\ &\leq \|K_\theta(t,s)\|^4_{2\,\otimes H(R)} \sum_{n=0}^{\infty} (\tfrac{1}{2})^n = 2\|K_\theta(t,s)\|^4_{2\,\otimes H(R)}. \end{aligned} \quad (103)$$

Since

$$\lim_{h \to 0} h^{-2} \|K_{\theta_0+h}(t,s)\|^2_{2\,\otimes H(R)}$$

exists,

$$\lim_{h \to 0} h^{-2} S(\theta_0 + h) = 0. \quad (104)$$

Therefore

$$\begin{aligned} \lim_{h \to 0} h^{-2} D(\theta_0 + h) &= \lim_{h \to 0} h^{-2} 2!\,(1!)^{-2} 2^{-2} \|K_{\theta_0+h}(t_1,t_2)\|^2_{2\,\otimes H(R)} \\ &= \lim_{h \to 0} \tfrac{1}{2} h^{-2} \|K_{\theta_0+h}(t_1,t_2)\|^2_{2\,\otimes H(R)}. \end{aligned} \quad (105)$$

Conditions i) and iii) and the equivalence of norm and pointwise convergence in reproducing-kernel Hilbert spaces guarantee that this limit exists and equals

$$\tfrac{1}{2}\|\dot{K}_{\theta_0}(t,s)\|^2_{2\,\otimes H(R)} = \tfrac{1}{2}\|\dot{R}_{\theta_0}(t,s)\|^2_{2\,\otimes H(R)}. \quad (106)$$

References

[1] T. Kailath, "RKHS approach to detection and estimation problems—Part I: Deterministic signals in Gaussian noise," *IEEE Trans. Inform. Theory*, vol. IT-17, pp. 530–549, Sept. 1971.

[2] ——, "An RKHS approach to detection and estimation problems —Part II: Simultaneous diagonalization and Gaussian signals," submitted to *IEEE Trans. Inform. Theory*.

[3] D. L. Duttweiler and T. Kailath, "RKHS approach to detection and estimation problems—Part III: Generalized innovations representations and a likelihood-ratio formula," *IEEE Trans. Inform. Theory*, vol. IT-18, pp. 730–745, Nov. 1972.

[4] ——, "RKHS approach to detection and estimation problems—Part IV: Non-Gaussian detection," *IEEE Trans. Inform. Theory*, this issue, pp. 19–28.

[5] A.E. Parzen, "Statistical inference on time series by Hilbert space methods," Appl. Math. Statist. Lab., Stanford Univ., Stanford, Calif., Tech. Rep. 23, 1959. This report and others by Parzen on RKHS are reprinted in E. Parzen, *Time Series Analysis Papers*. San Francisco: Holden-Day, 1967.
b. ——, "Statistical inference on time series by RKHS methods, II," *Proc. 12th Biennial Canadian Math. Congress*, R. Pyke, Ed. Providence, R.I.: Amer. Math. Soc., 1969, pp. 1–37.

[6] E. W. Barankin, "Locally best unbiased estimates," *Ann. Math. Statist.*, vol. 20, pp. 477–501, 1949.

[7] A. Bhattacharyya, "On some analogues of the amount of information and their use in statistical estimation," *Sankhya Ind. J. Stat.*, vol. 8, pp. 1–14, 201–218, 315–328, 1946.

[8] M. J. Levin, "Power spectrum parameter estimation," *IEEE Trans. Inform. Theory*, vol. IT-11, pp. 100–107, Jan. 1965.

[9] D. J. Sakrison, "Efficient recursive estimation; Applications to estimating the parameters of a covariance function," *Int. J. Eng. Sci.*, vol. 3, pp. 461–481, 1965.

[10] H. J. Weiner, "The gradient iteration in time series," *J. Soc. Ind. Appl. Math.*, vol. 13, pp. 1096–1101, Dec. 1965.

[11] H. L. Van Trees, *Detection, Estimation, and Modulation Theory, Part I.* New York: Wiley, 1968.

[12] C. R. Rao, *Advanced Statistical Methods in Biometric Research.* New York: Wiley, 1952.

[13] F. E. Glave, "A new look at the Barankin lower bound," *IEEE Trans. Inform. Theory*, vol. IT-18, pp. 349–356, May 1972.

[14] D. L. Duttweiler, "Reproducing-kernel Hilbert space techniques for detection and estimation problems," Ph.D. dissertation. Dep. Elec. Eng., Stanford Univ., Stanford, Calif., June 1970.

[15] P. Swerling, "Parameter estimation for waveforms in additive Gaussian noise," *J. SIAM*, vol. 7, pp. 152–166, June 1959.

[16] J. Ziv and M. Zakai, "Some lower bounds on signal parameter

estimation," *IEEE Trans. Inform. Theory*, vol. IT-15, pp. 386–391, May 1969.
[17] N. Aronszajn, "Theory of reproducing kernels," *Trans. Amer. Math. Soc.*, vol. 68, pp. 337–404, May 1950.
[18] A. B. Baggeroer, "Barankin bounds on the variance of estimates of the parameters of a Gaussian random process," M.I.T. Res. Lab. Electron., Quart. Prog. Rep. 92, Jan. 15, 1969.
[19] T. Hida, *Stationary Stochastic Processes*. Princeton, N.J.: Princeton Univ. Press, 1970.
[20] M. I. Schwartz, "On the detection of known binary signals in Gaussian noise of exponential covariance," *IEEE Trans. Inform. Theory*, vol. IT-11, pp. 330–335, July 1965.
[21] J. Capon, "Radon–Nikodym derivatives of stationary Gaussian measures," *Ann. Math. Statist.*, vol. 35, pp. 517–531, June 1964.

Part IV

APPLICATIONS TO STOCHASTIC PROCESSES

Editor's Comments on Papers 15 Through 20

Various statistical applications of reproducing kernel Hilbert spaces were documented in Parts II and III. In Part IV applications to some of the more difficult aspects of stochastic processes will be considered.

The first two papers deal with the analysis of nonlinear functionals of stochastic processes. In Paper 15, Hida and Ikeda study the congruence relation between the nonlinear span of an independent-increment process and the RKHS determined by its characteristic functional. These results are used to derive orthogonal expansions of nonlinear functionals of such processes. In Paper 16, Kallianpur shows how to express the nonlinear span of a Gaussian process as the direct sum of tensor products of its linear span. Similar results had been obtained independently by Neveu [1]. Other presentations and extensions of the results of these papers can be found in works by Hida [2, Ch. 4 and 6; 3, Ch. 4], Kallianpur [4,Ch. 6; 5], Segall and Kailath [6], and Huang and Cambanis [7]. There are obvious implications here for nonlinear estimation. See Parzen [8] for a formal solution to the general nonlinear least-squares estimation problem. Some restricted nonlinear estimation problems have been studied using Kallianpur's results by Huang and Cambanis [9, 10] and Marcus et al. [11].

Another important area in which RKHS have been applied is that of canonical, or innovations, representations for Gaussian processes. The history of this area to 1973 is excellently surveyed by Kailath [12]. The first RKHS application was by Hida [13] in connection with the existence of canonical representations. Hida's paper has been reprinted along with several other classical papers in this area in a volume by Ephremides and Thomas [14]. Further work can be found in papers by Siraya [15], Hitsuda [16], and Mosca [17]. Paper 17 by Kailath and Duttweiler and Paper 18 by Kallianpur and Oodaira present results, obtained independently and almost simultaneously, on generalized innovations representations in which the innovations process need not be white noise. Paper 17 contains an interesting discussion of the notion of causality in a RKHS, along with many useful examples. Both papers also give likelihood-ratio formulas for discriminating between Gaussian processes. Further generalizations and refinements can be found in work by Kallianpur [18], Saeks [19], and Bromley and Kallianpur [20].

The last two papers in Part IV deal with Markovian properties of multidimensional Gaussian processes (random fields). In Paper 19, Molchan gives a RKHS development of some early results of H. P. McKean on multidimensional Brownian motion. Similar work was carried out independently by Assouad [21]. In Paper 20, Pitt gives RKHS conditions for more general Gaussian fields to be Markovian. Russek [22] has applied Pitt's results to solutions of stochastic boundary value problems. Pitt's later work is referenced by Adler [23, Appendix], who gives a brief discussion of this research area. Other relevant references include Ciesielski [24], Molchan [25], Kallianpur and Mandrekar [26], Okabe [27], Kotani [28], and Kotani and Okabe [29].

A few other results deserve mention here. Ylvisaker [30] gives RKHS conditions under which a function can be the mean value of a stochastic process with given correlation function. Corresponding conditions on the cross-covariance function of two processes with given covariances are derived by Siraya [31]. Molchan and Golosov [32] characterize the RKHS of processes with certain types of power spectra.

REFERENCES

1. J. Neveu, *Gaussian Random Processes* (in French), University of Montreal Press, Montreal, 1968, 224p.
2. T. Hida, *Stationary Stochastic Processes,* Princeton University Press, Princeton, N. J., 1970, 161p.
3. T. Hida, *Brownian Motion,* Springer-Verlag, New York, 1980, 325p.
4. G. Kallianpur, *Stochastic Filtering Theory,* Springer-Verlag, New York, 1980, 316p.
5. G. Kallianpur, Homogeneous Chaos Expansions, in *Statistical Models and Turbulence,* M. Rosenblatt and C. Van Atta, eds., Springer-Verlag, Berlin, 1972, pp. 230–254.
6. A. Segall and T. Kailath, Orthogonal Functionals of Independent-Increment Processes, *IEEE Trans. Inform. Theory* **IT-22:**287–298 (1976).
7. S. T. Huang and S. Cambanis, Stochastic and Multiple Wiener Integrals for Gaussian Processes, *Ann. Probab.* **6:**585–614 (1978).
8. E. Parzen, Extraction and Detection Problems and Reproducing Kernel Hilbert Spaces, *J. Soc. Ind. Appl. Math. Control* **1:**35–62 (1962).
9. S. T. Huang and S. Cambanis, Spherically Invariant Processes: Their Nonlinear Structure, Discrimination, and Estimation, *J. Multivar. Anal.* **9:**59–83 (1979).
10. S. T. Huang and S. Cambanis, On the Representation of Nonlinear Systems with Gaussian Inputs, *Stochastics* **2:**173–189 (1979).
11. S. I. Marcus, S. K. Mitter, and D. Ocone, Finite-Dimensional Nonlinear Estimation for a Class of Systems in Continuous and Discrete Time, in *Analysis and Optimisation of Stochastic Systems,* O. L. R. Jacobs, M. H. A. Davis, M. A. H. Dempster, C. J. Harris, and P. C. Parks, eds., Academic Press, London, 1980, pp. 387–406.
12. T. Kailath, A View of Three Decades of Linear Filtering Theory, *IEEE Trans. Inform. Theory* **IT-20:**146–181 (1974).
13. T. Hida, Canonical Representations of Gaussian Processes and their Applications, *Kyoto Univ. Coll. Sci. Mem.* **A33:**109–155 (1960).
14. A. Ephremides and J. B. Thomas, eds., *Random Processes: Multiplicity Theory and Canonical Decompositions,* Benchmark Papers in Electrical Engineering and Computer Science, vol. 2, Dowden, Hutchinson, & Ross, Stroudsburg, Pa., 1973, 408p.
15. T. N. Siraya, On Canonical Representations for Stochastic Processes of Multiplicities One and Two, *Theory Probab. Appl.* **18:**153–158 (1973).
16. M. Hitsuda, Multiplicity of Some Classes of Gaussian Processes, *Nagoya Math. J.* **52:**39–46 (1973).
17. E. Mosca, Weaker Conditions for Innovations Informational Equivalence in the Independent Gaussian Case, *IEEE Trans. Autom. Control* **AC-24:**63–69 (1979); **AC-25:**607–609 (1980).
18. G. Kallianpur, Canonical Representations of Equivalent Gaussian Processes, in *Stochastic Processes and Related Topics,* M. L. Puri, ed., Academic Press, New York, 1975, pp. 195–221.
19. R. Saeks, Reproducing Kernel Resolution Space and its Applications, *J. Franklin Inst.* **302:**331–355 (1976).
20. C. Bromley and G. Kallianpur, Gaussian Random Fields, *Appl. Math. Optimiz.* **6:**361–376 (1980).
21. P. Assouad, Study of a Reproducing Space Associated with Brownian

Motion with Time Parameter in R^n (in French), *Acad. Sci. (Paris) C. R.* **269:**36-37 (1969).
22. A. Russek, Gaussian n-Markovian Processes and Stochastic Boundary Value Problems, *Z. Wahrscheinlichkeitstheorie* **53:**117-122 (1980).
23. R. J. Adler, *The Geometry of Random Fields,* Wiley, Chichester, England, 1981, 280p.
24. Z. Ciesielski, On Lévy's Brownian Motion with Several-Dimensional Time, in *Probability-Winter School,* Z. Ciesielski, K. Urbanik, and W. A. Woyczynski, eds., Springer-Verlag, Berlin, 1975, pp. 29-56.
25. G. M. Molchan, Characterization of Gaussian Fields with Markovian Property, *Sov. Math. Dokl.* **12:**563-567 (1971).
26. G. Kallianpur and V. Mandrekar, The Markov Property for Generalized Gaussian Random Fields, *Inst. Fourier, Grenoble Ann.* **24:**143-167 (1974).
27. Y. Okabe, On a Markovian Property of Gaussian Processes, in *Proceedings of the Second Japan-USSR Symposium on Probability Theory,* G. Maruyama and Ju. V. Prokhorov, eds., Springer-Verlag, Berlin, 1973, pp. 340-354.
28. S. Kotani, On a Markov Property for Stationary Gaussian Processes with a Multidimensional Parameter, in *Proceedings of the Second Japan-USSR Symposium on Probability Theory,* G. Maruyama and Ju. V. Prokhorov, eds., Springer-Verlag, Berlin, 1973, pp. 239-250.
29. S. Kotani and Y. Okabe, On a Markovian Property of Stationary Gaussian Processes with a Multi-Dimensional Parameter, in *Hyperfunctions and Pseudo-Differential Equations,* H. Komatsu, ed., Springer-Verlag, Berlin, 1973, pp. 153-163.
30. D. Ylvisaker, A Generalization of a Theorem of Balakrishnan, *Ann. Math. Stat.* **32:**1337-1339 (1961).
31. T. N. Siraya, On Subordinate Processes, *Theory Probab. Appl.* **22:**129-133 (1977).
32. G. M. Molchan and Ju. I. Golosov, Gaussian Stationary Processes with Asymptotic Power Spectrum, *Sov. Math. Dokl.* **10:**134-137 (1969).

15

Reprinted by permission from pages 117-143 of *Proceedings of the Fifth Berkeley Symposium on Mathematical Statistics and Probability*, vol. 2, L. M. Le Cam and J. Neyman, eds., University of California Press, Berkeley and Los Angeles, 1967, 447p.

ANALYSIS ON HILBERT SPACE WITH REPRODUCING KERNEL ARISING FROM MULTIPLE WIENER INTEGRAL

TAKEYUKI HIDA
NAGOYA UNIVERSITY
and
NOBUYUKI IKEDA
OSAKA UNIVERSITY

1. Introduction

The multiple Wiener integral with respect to an additive process with stationary independent increments plays a fundamental role in the study of the flow derived from that additive process and also in the study of nonlinear prediction theory. Many results on the multiple Wiener integral have been obtained by N. Wiener [14], [15], K. Itô [5], [6], and S. Kakutani [8] by various techniques. The main purpose of our paper is to give an approach to the study of the multiple Wiener integral using reproducing kernel Hilbert space theory.

We will be interested in stationary processes whose sample functions are elements in E^* which is the dual of some nuclear pre-Hilbert function space E. For such processes we introduce a definition of *stationary process* which is convenient for our discussions. This definition, given in detail by section 2, definition 2.1, is a triple $\mathbf{P} = (E^*, \mu, \{T_t\})$, where μ is a probability measure on E^* and $\{T_t\}$ is a flow on the measure space (E^*, μ) derived from shift transformations which shift the arguments of the functions of E.

In order to facilitate the discussion of the Hilbert space $L_2 = L_2(E^*, \mu)$, we shall introduce a transformation τ defined by the following formula:

$$(1.1) \qquad (\tau\varphi)(\xi) = \int_{E^*} e^{i\langle x,\xi\rangle}\varphi(x)\mu(dx) \qquad \text{for} \quad \varphi \in L_2,$$

where $\langle \cdot, \cdot \rangle$ denotes the bilinear form of $x \in E^*$ and $\xi \in E$. This transformation τ from L_2 to the space of functionals on E is analogous to the ordinary Fourier transform. By formula (1.1) and a requirement that τ should be a unitary transformation, $\mathfrak{F} \equiv \tau(L_2(E^*, \mu))$ has to be a Hilbert space with reproducing kernel $C(\xi - \eta)$, $(\xi, \eta) \in E \times E$, where C is the *characteristic functional* of the measure μ defined by

$$(1.2) \qquad C(\xi) = \int_{E^*} e^{i\langle x,\xi\rangle}\mu(dx), \qquad \xi \in E.$$

The first task in section 2 will be to establish the explicit correspondence between L_2 and $\mathfrak{F}$.

Another concept which we shall introduce in section 2 is a group $\mathbf{G}(\mathbf{P})$ associated with a stationary process $\mathbf{P}$. Consider the set of all linear transformations $\{g\}$ on E satisfying the conditions that

$$(1.3)\qquad \begin{aligned} &\text{(i)}\quad C(g\xi) = C(\xi) \quad \text{for every} \quad \xi \in E, \quad \text{and} \\ &\text{(ii)}\quad \text{that } g \text{ be a homeomorphism on } E. \end{aligned}$$

Obviously the collection $\mathbf{G}(\mathbf{P})$ of all such g's forms a group with respect to the operation $(g_1g_2)\xi = g_1(g_2\xi)$. The collection $\mathbf{G}(\mathbf{P})$ includes not only shift transformations S_h, h real, defined by

$$(1.4)\qquad (S_h\xi)(t) = \xi(t - h),$$

but also some other transformations depending on the form of the characteristic functional.

An interesting subclass of stationary processes is the class of processes with *independent values at every point* (Gelfand and Vilenkin [4]). Sections 4–6 will be devoted to discussions of some typical such processes. Roughly speaking they are the stationary processes obtained by subtracting the mean functions from the derivatives with respect to time of additive processes with independent stationary increments. In these cases the independence at every point can be illustrated rather clearly in the space $\mathfrak{F}$ by using a direct product decomposition of it in the sense of J. von Neumann [10]. Furthermore, because of the particular form of the characteristic functional, we can easily get an infinite direct sum decomposition of $\mathfrak{F}$:

$$(1.5)\qquad \mathfrak{F} = \sum_{n=0}^{\infty} \oplus \mathfrak{F}_n.$$

Each $\mathfrak{F}_n$ appearing in the last expression is invariant under every V_g, $g \in \mathbf{G}(\mathbf{P})$ defined by

$$(1.6)\qquad (V_gf)(\xi) = f(g\xi), \qquad f \in \mathfrak{F},$$

that is, $V_g(\mathfrak{F}_n) \subset \mathfrak{F}_n$ for every $g \in \mathbf{G}(\mathbf{P})$. With the aid of these two different kinds of decompositions, we shall investigate $\mathfrak{F}$ and discuss certain applications.

In section 5, we shall consider *Gaussian white noise,* although many of the results are already known. To us, it is the most fundamental example of a stationary process. Here the subspace $\mathfrak{F}_n$ corresponds to the multiple Wiener integral introduced by K. Itô [5] and also to Wiener's homogeneous chaos of degree n. Kakutani [7] expressed the subspace of $L_2(E^*, \mu)$ corresponding to $\mathfrak{F}_n$ in terms of the product of Hermite polynomials in L_2. These results are important in the determination of the spectrum of the flow $\{T_t\}$ of Gaussian white noise. It should be noted that the L_2 space for this process enjoys properties similar to those of $L_2(S^n, \sigma_n)$ over an n-dimensional sphere S^n with the uniform probability measure σ_n. As is mentioned by H. Yosizawa (oral communication), the multiple Wiener integral plays the role of spherical harmonics in

$L_2(S^n, \sigma_n)$; for example, $\{\mathfrak{F}_n, V_g, g \in \mathbf{G}(\mathbf{P})\}$ is an irreducible representation of the group $\mathbf{G}(\mathbf{P})$. This suggests that one should consider certain aspects of the theory which are analogous to the analysis on the finite dimensional sphere. This is done by H. Yosizawa and others (oral communication).

We will also discuss the Hilbert space $\mathfrak{F}$ arising from *Poisson white noise* which is another fundamental example of a stationary process (section 6).

Finally we will show that our approach is applicable to the study of generalized white noise and even to an arbitrary sequence of independent identically distributed random variables. Further discussions, such as the detailed proofs of the theorems stated in this paper and certain of the applications of this work, will appear elsewhere.

We would like to express our deep thanks to Professor J. W. Van Ness for his help in preparing the manuscript.

2. Definitions and preparatory considerations

Before defining the term stationary process let us first introduce some notation. Let E be a real nuclear pre-Hilbert space. Denote the inner product by $\langle \cdot, \cdot \rangle$; it determines the norm $\|\cdot\|$. Let H be the completion of E in the norm $\|\cdot\|$ and E^* be the dual of E. Then by the usual identification $H^* = H$ for Hilbert spaces, we have the following relation: $E \subset H \subset E^*$.

Let $\mathfrak{B} = \mathfrak{B}(E^*)$ be the σ-algebra generated by all cylinder sets in E^*. If $C(\xi)$, $\xi \in E$, is a continuous positive definite functional with $C(0) = 1$, then there exists a unique probability measure μ on the measurable space $(E^*, \mathfrak{B})$ such that

$$C(\xi) = \int_{E^*} \exp\,[i\langle x, \xi\rangle]\mu(dx), \qquad \xi \in E \tag{2.1}$$

(cf. Gelfand and Vilenkin [4]).

In what follows we shall deal only with the case in which E is a subset of $R^{\mathbf{T}}$, where R is the field of real numbers and $\mathbf{T}$ is the additive group of real numbers or one of its subgroups. Every element of E then has a coordinate representation $\xi = (\xi(t), t \in \mathbf{T})$. For every h, we consider the point transformation S_h defined by (1.4). Whenever we are concerned with the S_h's, we always assume that E is invariant under all of them. For each S_t we define a transformation T_t on E^* as follows:

$$T_t\colon T_t x \in E^*, \quad t \in \mathbf{T} \ \text{ with } \ \langle T_t x, \xi\rangle = \langle x, S_t \xi\rangle \quad \text{for every } \xi. \tag{2.2}$$

Obviously $\{T_t\colon t \in \mathbf{T}\}$ forms a group satisfying

$$T_t T_s = T_s T_t = T_{t+s}, \qquad s, t \in \mathbf{T}, \tag{2.3}$$

$$T_0 = I \quad \text{(identity)}. \tag{2.4}$$

The group $\{T_t; t \in \mathbf{T}\}$ can be considered as a transformation group acting on E^*.

Let $\mathfrak{B}(\mathbf{T})$ be the topological Borel field of $\mathbf{T}$.

Definition 2.1. *The transformation group* $\{T_t, t \in \mathbf{T}\}$ *is called a group of shift transformations if* $T_t x = f(x, t)$ *is measurable with respect to* $\mathfrak{B} \times \mathfrak{B}(\mathbf{T})$. *The* *triple* $\mathbf{P} = (E^*, \mu, \{T_t\})$ *is called a stationary process if* μ *is invariant under shift transformations* T_t, $t \in \mathbf{T}$.

DEFINITION 2.2. *The functional* $C(\xi)$, $\xi \in E$, *defined by* (2.1) *for a stationary process* $\mathbf{P} = (E^*, \mu, \{T_t\})$, *is called* the characteristic functional *of* $\mathbf{P}$.

Having introduced these definitions, we begin our investigation of stationary processes. For convenience, we assume the following throughout the remainder of the paper.

ASSUMPTION 2.1. *There exists a system* $\{\xi_n^\circ\}_{n=1}^\infty$, $\xi_n^\circ \in E$, *which forms a complete orthonormal system in* H.

For the measure space, $(E^*, \mathcal{B}, \mu)$, associated with a stationary process, we can form the Hilbert space $L_2 = L_2(E^*, \mu)$ of all square summable complex-valued functions with the inner product

$$\langle\langle \varphi, \psi \rangle\rangle = \int_{E^*} \varphi(x)\overline{\psi(x)}\mu(dx), \qquad \varphi, \psi \in L_2. \tag{2.5}$$

LEMMA 2.1. *The closed linear subspace of* L_2 *spanned by* $\{e^{i\langle x,\xi\rangle}, \xi \in E\}$ *coincides with the whole space* L_2.

This can be proved by using the uniqueness theorem for Fourier inverse transform (for detailed proof, see Prohorov [13]).

The next lemma is due to Aronszajn ([1], part I, section 2).

LEMMA 2.2. *For any stationary process* $\mathbf{P} = (E^*, \mu, \{T_t\})$ *there always exists a smallest Hilbert space* $\mathfrak{F} = \mathfrak{F}(E, C)$ *of functionals on* E *with reproducing kernel* $C(\xi - \eta)$, $(\xi, \eta) \in E \times E$, *where* $C(\xi)$, $\xi \in E$, *is the characteristic functional of* $\mathbf{P}$.

Let us denote the inner product in $\mathfrak{F}$ by $(\cdot, \cdot)$. We now state some of the properties of $\mathfrak{F}$ obtained by Aronszajn:

(i) for any fixed $\xi \in E$, $C(\cdot - \xi) \in \mathfrak{F}$;

(ii) $(f(\cdot), C(\cdot - \xi)) = f(\xi)$ for any $f \in \mathfrak{F}$;

(iii) $\mathfrak{F}$ is spanned by $\{C(\cdot - \xi), \xi \in E\}$.

From these properties and lemma 2.1 we can prove the following theorem.

THEOREM 2.1. *The transformation* τ *defined by*

$$(\tau\varphi)(\xi) = \int_{E^*} \varphi(x)e^{i\langle x,\xi\rangle}\mu(dx) \tag{2.6}$$

is a unitary operator from L_2 *onto* $\mathfrak{F}$.

In fact, the relation

$$\tau\left(\sum_{j=1}^n a_j e^{-i\langle x,\xi_j\rangle}\right)(\cdot) = \sum_{j=1}^n a_j C(\cdot - \xi_j) \tag{2.7}$$

shows that τ preserves norm since this relation can be extended to the entire space.

Note that

$$(\tau 1)(\cdot) = C(\cdot) \tag{2.8}$$

and define

$$\begin{aligned} L_2^* &= \{\varphi; \varphi \in L_2, \langle\langle \varphi, 1 \rangle\rangle = 0\}, \\ \mathfrak{F}^* &= \{f; f \in \mathfrak{F}, (f(\cdot), C(\cdot)) = 0\}. \end{aligned} \tag{2.9}$$

Then τ restricted to L_2^* is still a unitary operator from L_2^* onto $\mathfrak{F}^*$. On the other hand, if we define U_t, $t \in \mathbf{T}$, by

$$(2.10) \qquad U_t\colon\ (U_t\varphi)(x) = \varphi(T_t x) \in L_2 \qquad \text{for} \quad \varphi \in L_2,$$

then $\{U_t, t \in \mathbf{T}\}$ forms a group of unitary operators on L_2 satisfying

$$(2.11) \qquad \begin{aligned} &U_t U_s = U_s U_t = U_{t+s}, \qquad s, t \in \mathbf{T},\\ &U_0 = I \quad \text{(identity)},\\ &U_t \text{ is strongly continuous.}\end{aligned}$$

Moreover, we can see that $\{\tilde{U}_t; \tilde{U}_t = \tau \cdot U_t \tau^{-1}, t \in \mathbf{T}\}$ is also a group of unitary operators acting on $\mathfrak{F}$. For simplicity we also use the symbol U_t for $\tilde{U}_t$. Further, we write U_t even when U_t (or $\tilde{U}_t$) is restricted to L_2^* (or $\mathfrak{F}^*$).

In connection with $\mathbf{G(P)}$ we can consider a group $\mathbf{G^*(P)}$ of linear transformations g^* acting on E^* as follows:

$$(2.12) \qquad \mathbf{G^*(P)} = \{g^*; g^*x \in E^* \text{ for every } x \in E^*, \langle g^*x, \xi\rangle = \langle x, g\xi\rangle \text{ holds for every } x \in E^* \text{ with } g \in \mathbf{G(P)}\}.$$

From the definition we can easily prove lemma 2.3.

LEMMA 2.3. *The collection* $\mathbf{G^*(P)}$ *is a group with respect to the operation*

$$(2.13) \qquad (g_1^* g_2^*)x = g_1^*(g_2^* x).$$

Also,

$$(2.14) \qquad (g^*)^{-1} = (g^{-1})^*.$$

REMARK 2.1. Detailed discussions concerning $\mathbf{G(P)}$ and $\mathbf{G^*(P)}$ will appear elsewhere. For the related topics on such groups we would like to refer to M. G. Kreĭn [9].

By (1.3) (i), it can be proved that every $g^* \in \mathbf{G^*(P)}$ is measure preserving; that is, $\mu(d(g^*x)) = \mu(dx)$. Hence, by the usual method, we can define a unitary operator V_{g^*} acting on $L_2(E^*, \mu)$ by

$$(2.15) \qquad (V_{g^*}\varphi)(x) = \varphi(g^*x), \qquad g^* \in \mathbf{G^*(P)}, \qquad \varphi \in L_2(E^*, \mu).$$

Similarly, we define

$$(2.16) \qquad (V_g f)(\xi) = f(g\xi), \qquad g \in \mathbf{G(P)}, \qquad f \in \mathfrak{F}.$$

Obviously, $\{V_{g^*}; g^* \in \mathbf{G^*(P)}\}$ and $\{V_g; g \in \mathbf{G(P)}\}$ form groups of unitary operators on $L_2(E^*, \mu)$ and $\mathfrak{F}$, respectively.

We now have the following relation between V_{g^*} and V_g,

$$(2.17) \qquad (\tau(V_{g^*}\varphi))(\xi) = V_{g^{-1}}(\tau\varphi)(\xi), \qquad \xi \in E,$$

which is proved by the equations

$$(2.18) \qquad \begin{aligned}\int_{E^*} e^{i\langle x,\xi\rangle}\varphi(g^*x)\mu(dx) &= \int_{E^*} e^{i\langle g^{*-1}x,\xi\rangle}\varphi(x)\mu(dg^{*-1}x)\\ &= \int_{E^*} e^{i\langle x, g^{-1}\xi\rangle}\varphi(x)\mu(dx) = (\tau\varphi)(g^{-1}\xi).\end{aligned}$$

Another important concept relating to stationary processes is the purely nondeterministic property. Let $\mathbf{T}_t$ be a set of the form

(2.19) $$\mathbf{T}_t = \{s; s \in \mathbf{T}, s \leq t\}$$

and $\mathcal{B}_t$ be the smallest Borel field generated by all cylinder sets of the form

(2.20)
$$\{x; (\langle x, \xi_1\rangle, \cdots, \langle x, \xi_n\rangle) \in B^n, \xi_k \in E, \mathrm{supp}\,(\xi_k) \subset \mathbf{T}_t, 1 \leq k \leq n, B^n \in \mathcal{B}(R^n)\}.$$

The subspaces of L_2, $L_2(t)$ and $L_2^*(t)$ are defined by

(2.21)
$$L_2(t) = \{\varphi; \varphi \in L_2, \varphi \text{ is } \mathcal{B}_t\text{-measurable}\}, \qquad t \in \mathbf{T},$$
$$L_2^*(t) = \{\varphi; \varphi \in L_2(t), \langle\langle \varphi, 1\rangle\rangle = 0\}, \qquad t \in \mathbf{T}.$$

Corresponding to $L_2(t)$ and $L_2^*(t)$, we can define

(2.22)
$$\mathfrak{F}(t) = \mathfrak{S}\{C(\cdot - \xi); \xi \in E, \mathrm{supp}\,(\xi) \subset \mathbf{T}_t\}$$
$$\mathfrak{F}^*(t) = \{f; f \in \mathfrak{F}(t), (f(\cdot), C(\cdot)) = 0\},$$

where $\mathfrak{S}\{\ \}$ denotes the subspace spanned by elements written in the bracket. Then we can easily prove the following proposition.

PROPOSITION 2.1. *For every* $t \in \mathbf{T}$,

(2.23) $$L_2(t) \cong \mathfrak{F}(t), \qquad \text{(isomorphic)}$$

and

(2.24) $$L_2^*(t) \cong \mathfrak{F}^*(t), \qquad \text{(isomorphic)}$$

under the transformation τ *restricted to* $L_2(t)$ *and* $L_2^*(t)$, *respectively.*

DEFINITION 2.5. *If*

(2.25) $$L_2^*(-\infty) \equiv \bigcap_{s \in \mathbf{T}} L_2^*(s) = \{0\}$$

holds, then $\mathbf{P} = (E^*, \mu, \{T_t\})$ *is called purely nondeterministic.*

This definition was given by M. Nisio [12] for the case where E^* is an ordinary function space. By definition and proposition 2.1, $\mathbf{P}$ is purely nondeterministic if and only if

(2.26) $$\mathfrak{F}^*(-\infty) = \bigcap_{s \in \mathbf{T}} \mathfrak{F}^*(s) = \{0\}$$

holds.

We are now in a position to develop certain basic concepts relative to stationary processes. We would like to emphasize the importance of a stationary process with independent values at every point.

DEFINITION 2.6 (Gelfand and Vilenkin [4]). *A stationary process* $\mathbf{P} = (E^*, \mu, \{T_t\})$ *will be called a process with independent values at every point if its characteristic functional* $C(\xi)$, $\xi \in E$, *satisfies*

(2.27) $$C(\xi_1 + \xi_2) = C(\xi_1)C(\xi_2), \quad \text{whenever} \quad \mathrm{supp}\,(\xi_1) \cap \mathrm{supp}\,(\xi_2) = \emptyset.$$

If E is the space $\mathcal{K}$ of C^∞-functions with compact supports introduced by L. Schwartz, this definition coincides with that of Gelfand and Vilenkin. If E

is the space s of rapidly decreasing sequences, then we have a sequence of independent random variables with the same distribution.

PROPOSITION 2.2. *If* **P** *is a stationary process with independent values at every point, then it is purely nondeterministic.*

PROOF. For $f \in \mathfrak{F}^*(t)$ there exists a sequence $\{f_n\}$ such that $\text{l.i.m.}_{n\to\infty} f_n = f$ and

$$(2.28) \qquad f_n(\cdot) = \sum_{k=1}^{N_n} a_k^{(n)} C(\cdot - \xi_k^{(n)}), \qquad \xi_k^{(n)} \in E, \qquad \text{supp}\,(\xi_k^{(n)}) \subset \mathbf{T}_t.$$

Since **P** is a stationary process with independent values at every point, we have

$$(2.29) \qquad f_n(\xi) = \sum_{k=1}^{N_n} a_k^{(n)} C(\xi - \xi_k^{(n)}) = C(\xi) \sum_{k=1}^{N_n} a_k^{(n)} C(-\xi_k^{(n)}) = C(\xi) f_n(0)$$

for any ξ with $\text{supp}\,(\xi) \subset \mathbf{T}_t^c$. However, $f_n(\xi)$ has to be zero since

$$(2.30) \qquad f_n(0) = (f_n(\cdot), C(\cdot - 0)) = 0 \qquad \text{for } f_n \in \mathfrak{F}^*.$$

Thus, $f(\xi) = 0$. If $f \in \bigcap_{t\in\mathbf{T}} \mathfrak{F}^*(t)$, then $f(\xi) = 0$ for every ξ with $\text{supp}\,(\xi) \subset \mathbf{T}_t^c$ for every t. Hence, we have $f(\cdot) = 0$.

Now we can proceed to the analysis of the $L_2(E^*, \mu)$ space arising from a stationary process **P** with independent values at every point. First we discuss polynomials on E^*. The function expressed in the form

$$(2.31) \qquad \varphi(x) = P(\langle x, \xi_1\rangle, \cdots, \langle x, \xi_n\rangle), \qquad \xi_1, \cdots, \xi_n \in E, \qquad x \in E^*,$$

where P is a polynomial of n variables with complex coefficients, is called a *polynomial* on E^*. If P is of degree p, we say that $\varphi(x)$ is p-th *degree polynomial*, and if, in particular, P is *homogeneous*, we say the same of φ.

Throughout the remainder of this section we shall assume the following.

ASSUMPTION 2.2. *The following conditions hold:*

(i) $\int_{E^*} |\langle x, \xi\rangle|^p \mu(dx) < \infty$ *for* $\xi \in E$ *and every integer* $p \geq 1$,

(ii) $\int_{E^*} \langle x, \xi\rangle \mu(dx) = 0$ *for every* $\xi \in E$.

With these assumptions we see that the set M of all polynomials on E^* forms a linear manifold of L_2. Consequently, $\tau(M) = \{\tau(\varphi); \varphi \in M\}$ is defined and $\tau(M) \subset \mathfrak{F}$.

DEFINITION 2.7. *An operator* D_ξ, $\xi \in E$, *is defined by*

$$(2.32) \qquad (D_\xi f)(\cdot) = \text{l.i.m.}_{\epsilon\to 0} \frac{1}{\epsilon} [f(\cdot + \epsilon\xi) - f(\cdot)]$$

if the limit exists, D_ξ *is called a* differential operator, *and its domain is denoted by* $\mathfrak{D}(D_\xi)$. *We define* $\mathfrak{D}$ *as* $\bigcap_{\xi\in E} \mathfrak{D}(D_\xi)$.

LEMMA 2.4. *If* **P** *satisfies assumption* 2.2, *we have the following:*

(i) $C(\cdot - \xi) \in \mathfrak{D}$ *for every* $\xi \in E$, *and* $\prod_{j=1}^n D_{\xi_j} C(\cdot - \xi)$ *belongs to* $\mathfrak{D}$ *for any* n *and* $\xi_1, \cdots, \xi_n \in E$;

(ii) $\tau(M) \subset \mathfrak{D}$;

(iii) *For any* $\xi_1, \cdots, \xi_n \in E$ *and any choice of positive integers* $k_1, \cdots, k_n$, *we have*

$$\tau^{-1}\left\{\left(\prod_{j=1}^{n} D_{\xi_j}^{k_j}\right) C(\cdot)\right\}(x) = (i)\sum k_j \prod_{j=1}^{n} \langle x, \xi_j\rangle^{k_j}, \qquad x \in E^*; \tag{2.33}$$

(iv) *$(i)^{-1}D_\xi$ is a self-adjoint operator, the domain of which includes $\{C(\cdot - \xi); \xi \in E\} \cup \tau(M)$;*

(v) *for any $f \in \tau(M)$ and $\xi_1, \xi_2 \in E$, we have*

$$D_{\xi_1}D_{\xi_2}f(\cdot) = D_{\xi_2}D_{\xi_1}f(\cdot). \tag{2.34}$$

PROOF. By assumption 2.2,

$$\underset{\epsilon\to 0}{\text{l.i.m.}} \left\{\frac{1}{\epsilon}\,(e^{i\epsilon\langle x,\xi\rangle} - 1)e^{i\langle x,\eta\rangle} - i\langle x, \xi\rangle e^{i\langle x,\eta\rangle}\right\} = 0. \tag{2.35}$$

Using τ, the above relation proves that

$$\underset{\epsilon\to 0}{\text{l.i.m.}}\ \frac{1}{\epsilon}\,\{C(\cdot + \epsilon\xi + \eta) + C(\cdot + \eta)\} \tag{2.36}$$

exists and is equal to $\tau\{i\langle x, \xi\rangle e^{i\langle x,\eta\rangle}\}$.

In a similar way, we can prove the second assertion using assumption 2.2. (ii). By assumption 2.2, $\exp\{i \sum_{j=1}^{n} t_j\langle x, \xi_j\rangle\}$ is differentiable infinitely many times (in L_2-norm) with respect to $t_1, \cdots, t_{n-1}$ and t_n, and we have

$$(i)^{-\sum k_j}\left(\partial^{\sum k_j}/\partial t_1^{k_1}\cdots\partial t_n^{k_n}\right) \exp\left\{i \sum_{j=1}^{n} t_j\langle x, \xi_j\rangle\right\}\Big|_{t_1=t_2=\cdots=t_n=0} = \prod_{j=1}^{n} \langle x, \xi_j\rangle^{k_j}. \tag{2.37}$$

The right-hand side belongs to L_2, and mapping by τ, we have (2.33).

For assertion (iv), if $f \in \tau(M)$, we have

$$\begin{aligned}(D_\xi f)(\eta) - (D_\xi f(\cdot), C(\cdot - \eta)) &= \left(\underset{\epsilon\to 0}{\text{l.i.m.}}\ \frac{1}{\epsilon}\,[f(\cdot + \epsilon\xi) - f(\cdot)], C(\cdot - \eta)\right)\\ &= \lim_{\epsilon\to 0} \frac{1}{\epsilon}\,\{f(\eta + \epsilon\xi) - f(\eta)\},\\ (f(\cdot), D_\xi C(\cdot - \eta)) &= \lim_{\epsilon\to 0} \frac{1}{\epsilon}\,(f(\eta - \epsilon\xi) - f(\eta)) = -(D_\xi f)(\eta),\end{aligned} \tag{2.38}$$

which prove that $(i)^{-1}D_\xi$ is self-adjoint.

For (v) consider $(f(\eta + \epsilon\xi_1 + \epsilon\xi_2) - f(\eta + \epsilon\xi_2) - f(\eta + \epsilon\xi_1) + f(\eta))/\epsilon^2$. Arguments like the above prove that D_{ξ_1} and D_{ξ_2} are commutative.

N. Wiener [15] discussed the following decomposition of $\mathfrak{F}(L_2)$ for the case of Gaussian white noise. Consider a system K of elements of $\mathfrak{F}$ defined by

$$\begin{aligned}K_n &= \left\{\prod_{j=1}^{n} D_{\xi_{k_j}} C(\cdot); k_1, \cdots, k_n = 1, 2, \cdots\right\}, \qquad n \geq 1,\\ K_0 &= \{C(\cdot)\}, \quad \text{and} \quad K = \bigcup_{n=0}^{\infty} K_n\end{aligned} \tag{2.39}$$

where $\{\xi_n^{0)}\}_{n=1}^{\infty}$ is the system appearing in assumption 2.1. Since the system K_n

forms a countable set, we can arrange all the elements in linear order. We shall denote them by $\{g_k^{(n)}(\cdot)\}$. Let **P** be a stationary process satisfying assumption 2.2 and let $\mathfrak{F}_0$ be the one-dimensional space spanned by $C(\cdot)$, that is, $\mathfrak{F}_0 = K_0$. Put $f_k^{(1)} = g_k^{(1)}$ and define

$$\mathfrak{F}_1 = \mathfrak{S}\{f_k^{(1)}; k = 1, 2, \cdots\}. \tag{2.40}$$

Obviously, $\mathfrak{F}_0$ and $\mathfrak{F}_1$ are mutually orthogonal. Suppose that $\{\mathfrak{F}_j\}_{j=0}^{n-1}$ are defined and mutually orthogonal, consider

$$f_k^{(n)} = P_{\left(\sum_{j=0}^{n-1} \oplus \mathfrak{F}_j\right)^{\perp}} g_k^{(n)}, \qquad k = 1, 2, \cdots, \tag{2.41}$$

where $P_{(\cdot)}$ denotes the projection on $(\cdot)$. Then $\mathfrak{F}_n$ is defined by

$$\mathfrak{F}_n = \mathfrak{S}\{f_k^{(n)}; k = 1, 2, \cdots\}, \tag{2.42}$$

and it is orthogonal to $\sum_{j=0}^{n-1} \oplus \mathfrak{F}_j$. This procedure can be continued until there are no more elements $g_k^{(n+1)}$ not belonging to $\sum_{j=0}^{n} \oplus \mathfrak{F}_j$. Finally

$$\overline{K} \text{ (= the closure of } K \text{ in } \mathfrak{F}) = \sum_{j\geq 0} \oplus \mathfrak{F}_j. \tag{2.43}$$

DEFINITION 2.8. *The direct sum decomposition* (2.43) *is called Wiener's direct sum decomposition.*

THEOREM 2.2. *Let* **P** *be a stationary process satisfying assumptions* 2.1 *and* 2.2. *If*

$$\sum_{n=0}^{\infty} g_\xi^{(n)}(\cdot)/n!, \qquad g_\xi^{(n)}(\cdot) = (D_\xi)^n C(\cdot), \tag{2.44}$$

converges for every ξ, *then Wiener's direct sum decomposition satisfies the following properties:*

$$\text{(i)} \quad \sum_{n=0}^{\infty} \oplus \mathfrak{F}_n = \mathfrak{F}, \qquad \text{(ii)} \quad U_t(\mathfrak{F}_n) = \mathfrak{F}_n. \tag{2.45}$$

PROOF. By assumption,

$$\sum_{n=0}^{\infty} (i)^n \langle x, \xi\rangle^n/n! = \tau^{-1}\left(\sum_{n=0}^{\infty} g_\xi^{(n)}(\cdot)/n!\right), \qquad (g_\xi^{(0)}(\cdot) = C(\cdot)), \tag{2.46}$$

converges and the sum is equal to $\exp\{i\langle x, \xi\rangle\}$. Hence,

$$C(\cdot - \xi) = \sum_{n=0}^{\infty} \frac{g_\xi^{(n)}(\cdot)}{n}. \tag{2.47}$$

On the other hand, by the construction of the $\mathfrak{F}_n$'s we can prove

$$g_\xi^{(n)} \in \sum_{j=0}^{\infty} \oplus \mathfrak{F}_j. \tag{2.48}$$

Therefore, $C(\cdot - \xi) \in \sum_{j=0}^{\infty} \oplus \mathfrak{F}_j$ which proves

$$\mathfrak{F} \subset \sum_{n=0}^{\infty} \oplus \mathfrak{F}_n (\subset \mathfrak{F}), \tag{2.49}$$

since $\{C(\cdot - \xi)\}$ spans the entire $\mathfrak{F}$.

The second assertion is easily proved noting that

(2.50) $$U_t g_\xi^{(n)}(\cdot) = g_{S_t\xi}^{(n)}(\cdot)$$

and

(2.51) $$\mathfrak{S}(U_t K_n) = \mathfrak{S}(K_n).$$

Let us return to the group **G(P)**. If

(2.52) $$V_g(\mathfrak{F}_n) \subset \mathfrak{F}_n \qquad \text{for every} \quad g \in \mathbf{G(P)},$$

we call the decomposition $\mathfrak{F} = \sum_{n=0}^{\infty} \oplus \mathfrak{F}_n$ *invariant* with respect to **G(P)**. This concept is important in connection with the Wiener's direct sum decomposition. We shall discuss this topic in the later sections (4–6).

3. Orthogonal polynomials and reproducing kernels

From now on we shall deal with the decomposition of $L_2(E^*, \mu)$ and $\mathfrak{F}(E, C)$ associated with a stationary process with independent values at every point. First we consider, in this section, the simple case where E is a finite dimensional space. We can find a relation between the space with reproducing kernel and Rodrigues' formula for classical orthogonal polynomials. Such considerations will aid us in considering the case where E is an infinite dimensional nuclear space and will be preparation for later discussions.

Let ν be a probability measure (distribution) on R^1 and $\tilde{C}$ be its Fourier-Stieltjes transform (characteristic function); that is,

(3.1) $$\tilde{C}(\lambda) = \int_{R^1} e^{i\lambda x}\nu(dx), \qquad \lambda \in R^1.$$

Appealing to Aronszajn's results [1] stated in lemma 2.2, we obtain the smallest Hilbert space $\tilde{\mathfrak{F}} = \tilde{\mathfrak{F}}(R^1, \tilde{C})$, the reproducing kernel of which is $\tilde{C}(\lambda - \mu)$, $\lambda, \mu \in R^1$. By theorem 2.1, there exists an isomorphism $\tilde{\tau}$ which maps $\tilde{L}_2 = L_2(\nu; R^1) = \{f; \int_{R^1} |f(x)|^2\nu(dx)\}$ onto $\tilde{\mathfrak{F}}$ in the following way:

(3.2) $$(\tilde{\tau}f)(\lambda) = \int_{R^1} e^{i\lambda x}f(x)\nu(dx).$$

We shall examine this isomorphism $\tilde{\tau}$ in detail in the following examples. It is more interesting to discuss the analysis on $\tilde{\mathfrak{F}}$ rather than on $\tilde{L}_2$, since, for one thing, the development of functions belonging to $\tilde{L}_2$ in terms of orthogonal polynomials turns out to be the power series expansion in $\tilde{\mathfrak{F}}$.

3.1. *Gaussian distribution.* Consider the case where

(3.3) $$\nu(dx) = \nu(x; \sigma^2)\,dx = (2\pi\sigma^2)^{-1/2} \exp\left\{-\frac{x^2}{2\sigma^2}\right\} dx,$$

then

(3.4) $$\tilde{C}(\lambda, \sigma^2) = \int_{R^1} e^{i\lambda x}\nu(x; \sigma^2)\,dx = \exp\left(-\frac{\sigma^2}{2}\lambda^2\right).$$

Choose Hermite polynomials

$$(3.5)\qquad H_n(x;\sigma^2) = \frac{(-1)^n\sigma^{2n}}{n!}\frac{1}{\nu(x;\sigma^2)}\frac{d^n}{dx^n}\nu(x;\sigma^2), \qquad n = 1, 2, \cdots$$

(Rodrigues' formula), which form a complete orthonormal system in $\tilde{L}_2$. The isomorphism $\tilde{\tau}$ maps $H_n(x)$ to the n-th degree monomial of λ times $\tilde{C}$. In fact,

$$(3.6)\qquad (\tilde{\tau}H_n(\cdot\,;\sigma^2))(\lambda) = \sigma_n\lambda^n\tilde{C}(\lambda:\sigma^2),\ \sigma_n = \frac{\sigma^{2n}i^n}{n!}.$$

The proof of the formula (3.6) is as follows:

$$(3.7)\qquad (\tilde{\tau}H_n(\cdot,\sigma^2))(\lambda)$$

$$= \frac{(-1)^n\sigma^{2n}}{n!}\int_{R^1}\{\exp(i\lambda x)\}\left\{\nu(x;\sigma^2)^{-1}\frac{d^n}{dx^n}\nu(x;\sigma^2)\right\}\nu(x;\sigma^2)\,dx$$

$$= \frac{(-1)^n\sigma^{2n}}{n!}\int_{R^1}\exp(i\lambda x)\left\{\frac{d^n}{dx^n}\nu(x;\sigma^2)\right\}dx$$

$$= \frac{\sigma^{2n}}{n!}i^n\lambda^n\tilde{C}(\lambda;\sigma^2).$$

More generally, we have

$$(3.8)\qquad \left\{\tilde{\tau}\left(\sum_{n=0}^{\infty}a_nH_n(\cdot\,;\sigma^2)\right)\right\}(\lambda) = \left(\sum_{n=0}^{\infty}a_n\sigma_n\lambda^n\right)\tilde{C}(\lambda:\sigma^2).$$

3.2. *Poisson distribution.* Let $\nu(dx)$ be given by

$$(3.9)\qquad \nu(dx) = \nu(x,c)\delta_{S_c}(dx) = \frac{c^{x+c}}{\Gamma(x+c+1)}e^{-c}, \qquad x\in S_c,$$

where $S_c = \{-c, 1-c, 2-c, \cdots\}$. We obtain orthogonal polynomials with respect to the measure $\nu(x,c)\delta_{S_c}(dx)$, which are called *generalized Charlier polynomials,* by the following generalized Rodrigues' formula (cf. Bateman and others ([2], p. 222, and p. 227)):

$$(3.10)\qquad p_n(x,c) = (-c)^n(\nu(x,c))^{-1}\Delta_x^n\,\nu(x-n,c) = L_n^{x+c-n}(c)n!, \qquad x\in S_c,$$

where Δ_x^n denotes the n-th order difference operator acting on functions of x. The relation

$$(3.11)\qquad \tilde{C}(\lambda:c) = \int_{R^1}e^{i\lambda x}\nu(x;c)\delta_{S_c}(dx) = \sum_{x\in S_c}e^{i\lambda x}\nu(x,c)$$

$$= \exp\{e^{i\lambda c} - 1 - i\lambda c\}$$

is easily obtained. Now put

$$(3.12)\qquad Q_n(x,c) = \frac{1}{n!c^n}P_n(x,c);$$

then we get the orthogonality relation for Q_n:

$$(3.13)\qquad \sum_{x\in S_c}Q_n(x,c)Q_m(x,c)\nu(x,c) = \delta_{n,m}, \qquad n,m = 1, 2, \cdots.$$

Every P_n, of course, belongs to $L^2(\nu, R^1)$, and it is transformed by $\tilde{\tau}$ into

$$(3.14) \qquad (\tilde{\tau}P_n(\cdot, c))(\lambda) = c^n(e^{i\lambda} - 1)^n\tilde{C}(\lambda: c).$$

This is proved as follows:

$$(3.15) \qquad \begin{aligned}(\tilde{\tau}P_n(\cdot, c))(\lambda) &= \int_{R^1} e^{i\lambda x}(-c)^n((\nu(x, c))^{-1}\Delta_x^n\nu(x - n, c))\nu(x, c)\delta_{S_c}(dx) \\ &= (-c)^n(-1)^n \sum_{x \in S_c} (\Delta_x^n e^{i\lambda x})\nu(x, c) \\ &= c^n \sum_{x \in S_c} \sum_{m=0}^{n} (-1)^{n-m} \binom{n}{m} e^{im\lambda}e^{ix\lambda}\nu(x, c) \\ &= c^n(e^{i\lambda} - 1)^n\tilde{C}(\lambda: c).\end{aligned}$$

Note that the last expression is of the monomial form of $(e^{i\lambda} - 1)$ times $\tilde{C}$.

4. Stationary process with independent values at every point

This section is devoted to the study of the general theory for a certain class of stationary processes with independent values at every point. Let $\mathbf{P} = (E^*, \mu, \{T_t\})$ be a stationary process, where $\mathbf{T}$ is a set of real numbers and E is the function space $\mathcal{S}$ in the sense of L. Schwartz, and let its characteristic function be given by

$$(4.1) \qquad \begin{aligned} C(\xi) &= \exp \int_{-\infty}^{\infty} \alpha(\xi(t))\, dt, \\ \alpha(x) &= (-\sigma^2x^2)/2 + \int_{-\infty}^{\infty} \left(e^{ixu} - 1 - \frac{ixu}{1 + u^2}\right) \frac{1 + u^2}{u^2}\, d\beta(u).\end{aligned}$$

Here $0 \leq \sigma^2 < \infty$ and $d\beta(u)$ is a bounded measure on $(-\infty, \infty)$ such that $d\beta(\{0\}) = 0$. Obviously, $\mathbf{P}$ satisfies (2.27); that is, it is a stationary process with independent values at every point.

For the moment let us turn our attention from the flow $\{U_t, t \text{ real}\}$ to the direct sum decomposition of $\mathfrak{F} = \mathfrak{F}(\mathcal{S}, C)$ mentioned in section 1. Define

$$(4.2) \qquad \begin{aligned} K_0(\xi, \eta) &= \exp\left(\int \alpha(\xi(t))\, dt\right) \exp\left(\int \alpha(-\eta(t))\, dt\right) = C(\xi)C(-\eta), \\ K_1(\xi, \eta) &= \int \alpha(\xi(t) - \eta(t))\, dt - \int \alpha(\xi(t))\, dt - \int \alpha(-\eta(t))\, dt, \\ K_p(\xi, \eta) &= \frac{1}{p!}(K_1(\xi, \eta))^p, \qquad p \geq 2, \\ k_p(\xi, \eta) &= \frac{1}{p!} K_0(\xi, \eta)(K_1(\xi, \eta))^p, \qquad p \geq 0, \qquad \xi, \eta \in \mathcal{S}.\end{aligned}$$

Note that $C(\xi - \eta) = \sum_{p=0}^{\infty} k_p(\xi, \eta)$. We then prove the following lemma using the fact that $\alpha(x - y)$ is conditionally positive definite (cf. Gelfand and Vilenkin [4], chapter 3).

LEMMA 4.1. *The functionals* $K_0(\xi, \eta)$, $K_p(\xi, \eta)$, *and* $k_p(\xi, \eta)$, $p = 0, 1, \cdots$, $(\xi, \eta) \in \mathcal{S} \times \mathcal{S}$, *are all positive definite and continuous.*

Again appealing to the Aronszajn's theorem (lemma 2.2), we obtain the Hilbert spaces $\mathfrak{F}$, $\mathfrak{F}_p$ and $\mathfrak{F}_p$, $p = 0, 1, 2, \cdots$ with reproducing kernels C, K_p, and k_p respectively. Consider subspaces $\mathfrak{F}_p$ and $\mathfrak{F}_p$. We use the symbol $\otimes^*$ to express the direct product of subspaces in the sense of Aronszajn [1]. Hereafter we use subscripts to distinguish the various norms.

LEMMA 4.2. *The space $\mathfrak{F}_p$ is the class of all restrictions of functionals belonging to $\mathfrak{F}_1 \otimes^* \mathfrak{F}_1 \otimes^* \cdots \otimes^* \mathfrak{F}_1$ (p times) to the diagonal set $\mathcal{S}_p = \{(\xi, \cdots, \xi); \xi \in \mathcal{S}\}$. The norm $\|\cdot\|_{\mathfrak{F}_p}$ in $\mathfrak{F}_p$ can be expressed in the form*

$$(4.3) \qquad \|f\|_{\mathfrak{F}_p} = \inf_{\substack{f' \in \mathfrak{F}_1 \otimes^* \cdots \otimes^* \mathfrak{F}_1 \\ f = f'/\mathcal{S}_p}} \|f'\|_{\mathfrak{F}_1 \otimes^* \cdots \otimes^* \mathfrak{F}_1}$$

where $f'/\mathcal{S}_p$ denotes the restriction of f' to $\mathcal{S}_p$.

PROOF. By the definition of K_p and by Aronszajn ([1], section 8, theorem II) the assertion is easily proved.

LEMMA 4.3. *The space $\mathfrak{F}_p$ is the class of all restrictions of functionals belonging to $\mathfrak{F}_0 \otimes^* \mathfrak{F}_p$ to the diagonal set $\mathcal{S}_2 = \{(\xi, \xi); \xi \in \mathcal{S}\}$. The norm $\|\cdot\|_{\mathfrak{F}_p}$ can be expressed in the form*

$$(4.4) \qquad \|f\|_{\mathfrak{F}_p} = \inf_{\substack{f' \in \mathfrak{F}_0 \otimes^* \mathfrak{F}_p \\ f = f'/\mathcal{S}_2}} \|f'\|_{\mathfrak{F}_0 \otimes^* \mathfrak{F}_p}.$$

LEMMA 4.4. *The space $\mathfrak{F}_p$, $p = 1, 2, \cdots$, are mutually orthogonal subspaces of $\mathfrak{F}$.*

PROOF. Put

$$(4.5) \qquad K_{p,q}(\xi, \eta) = K_p(\xi, \eta) + K_q(\xi, \eta), \qquad p \neq q.$$

Then the Hilbert space $\mathfrak{F}_{p,q}$ with reproducing kernel $K_{p,q}(\xi, \eta)$ is expressible in the form

$$(4.6) \qquad \mathfrak{F}_{p,q} = \mathfrak{F}_p \oplus \mathfrak{F}_q.$$

To prove this assertion we first show that

$$(4.7) \qquad \mathfrak{F}_p \cap \mathfrak{F}_q = \{0\}.$$

Suppose $p > q$; then

$$(4.8) \qquad K_p(\xi, \eta) = K_q(\xi, \eta) \frac{q!(p-q)!}{p!} K_{p-q}(\xi, \eta).$$

Consequently, $\mathfrak{F}_p$ is the class of all restrictions of functionals belonging to $\mathfrak{F}_q \otimes^* \mathfrak{F}_{p-q}$ to the diagonal set $\mathcal{S}_2$. Now suppose $f \in \mathfrak{F}_p \cap \mathfrak{F}_q$ and let $\{f_k^{(q)}\}$ be a complete orthonormal system in $\mathfrak{F}_q$. Since $f \in \mathfrak{F}_p$, it can be expressed in the form

$$(4.9) \qquad f(\xi) = \sum_{k=1}^{\infty} g_k(\xi) f_k^{(q)}(\xi), \qquad g \in \mathfrak{F}_{p-q}$$

(remark attached to theorem II, Aronszajn ([1], p. 361)). But by assumption, f belongs to $\mathfrak{F}_q$. Consequently, all the g_k's must be zero, which implies $f = 0$.

Let us recall the discussion of Aronszajn ([1], part I, section 6). By (4.7), if $f_p \in \mathfrak{F}_p$, $f_q \in \mathfrak{F}_q$, then

$$\|f_p + f_q\|^2_{\mathfrak{F}_{p,q}} = \|f_p\|^2_{\mathfrak{F}_p} + \|f_q\|^2_{\mathfrak{F}_q},$$
$$(4.10)\qquad \|f_p - f_q\|^2_{\mathfrak{F}_{p,q}} = \|f_p\|^2_{\mathfrak{F}_p} + \|-f_q\|^2_{\mathfrak{F}_q} = \|f_p\|^2_{\mathfrak{F}_p} + \|f_q\|^2_{\mathfrak{F}_q},$$

which imply $\mathrm{Re}\,(f_p, f_q) = 0$. Similarly, we have $\mathrm{Im}\,(f_p, f_q) = 0$. Thus we have proved (4.6) and the lemma.

Further, by the proof of lemma 4.4, we can show the following. If $\mathfrak{F}_{p,q}$ is the reproducing kernel Hilbert space with kernel $k_p(\xi, \eta) + k_q(\xi, \eta)$, then

$$(4.11)\qquad \mathfrak{F}_{p,q} = \mathfrak{F}_p \oplus \mathfrak{F}_q, \qquad p \neq q,$$

and if $f \in \mathfrak{F}_{p,q}$, then

$$(4.12)\qquad (f(\cdot), k_p(\cdot, \xi)) \equiv f_p(\xi)$$

is the projection of f on $\mathfrak{F}_p$.

LEMMA 4.5. *The kernel $K_p(\xi, \eta)$ and $k_p(\xi, \eta)$, $p \geq 0$, are $\mathbf{G}(\mathbf{P})$-invariant; that is,*

$$(4.13)\qquad K_p(g\xi, g\eta) = K_p(\xi, \eta), \qquad k_p(g\xi, g\eta) = k_p(\xi, \eta)$$

for every $g \in \mathbf{G}(\mathbf{P})$.

PROOF. It is sufficient to prove that K_0 and K_1 are $\mathbf{G}(\mathbf{P})$-invariant. For K_0 this is easily proved by (4.2) and the definition of $\mathbf{G}(\mathbf{P})$. Concerning K_1, we have

$$(4.14)\qquad K_1(g\xi, g\eta) = \int_{-\infty}^{\infty} \alpha((g\xi)(t) - (g\eta)(t))\,dt - \int_{-\infty}^{\infty} \alpha((g\xi)(t))\,dt$$
$$- \int_{-\infty}^{\infty} \alpha(-(g\eta)(t))\,dt = \int_{-\infty}^{\infty} \alpha(g(\xi - \eta)(t))\,dt - \int_{-\infty}^{\infty} \alpha((g\xi)(t))\,dt$$
$$- \int \alpha(-(g\eta)(t))\,dt = K_1(\xi, \eta),$$

since every $g \in \mathbf{G}(\mathbf{P})$ keeps the integral $\int_{-\infty}^{\infty} \alpha(\xi(t))\,dt$ invariant.

Now we shall state one of our main results.

THEOREM 4.1. *The space $\mathfrak{F}$ has the direct sum decomposition*

$$(4.15)\qquad \mathfrak{F} = \sum_{p=0}^{\infty} \oplus \mathfrak{F}_p,$$

and it is $\mathbf{G}(\mathbf{P})$-invariant. The kernel $k_p(\cdot, \xi)$ is a projection operator in the following sense:

$$(4.16)\qquad (f(\cdot), k_p(\cdot, \xi)) \equiv f_p(\xi)$$

is the projection of f on $\mathfrak{F}_p$.

PROOF. By lemma 4.4,

$$(4.17)\qquad \sum_{p=0}^{n} k_p(\xi, \eta), \qquad (\xi, \eta) \in \mathbb{S} \times \mathbb{S},$$

will be the reproducing kernel of the subspace $\sum_{p=0}^{n} \oplus \mathfrak{F}_p$. Noting that

$$(4.18)\qquad C(\xi, \eta) = \sum_{p=0}^{\infty} k_p(\xi, \eta),$$

and

$$\left\| \sum_{p=n}^{m} k_p(\cdot, \eta) \right\|_{\mathfrak{F}}^2 = \sum_{p=n}^{m} k_p(\eta, \eta), \tag{4.19}$$

we conclude

$$\mathfrak{F} = \sum_{p=0}^{\infty} \oplus \mathfrak{F}_p \tag{4.20}$$

(cf. Aronszajn [1], part I, section 9). The **G(P)**-invariantness of $\mathfrak{F}_p$ comes from the definition of $\mathfrak{F}_p$ and lemma 4.5. By (4.12) and the above discussions, we have the last assertion.

Coming back to L_2 space, we have the following decomposition:

$$L_2 = \sum_{p=0}^{\infty} \oplus L_2^{(p)} \qquad \text{with} \quad \tau(L_2^{(p)}) = \mathfrak{F}_p. \tag{4.21}$$

5. Gaussian white noise

In the following three sections we shall discuss some typical stationary processes with independent values at every point. First we deal with Gaussian white noise, the characteristic functional of which is

$$C(\xi) = \exp\left\{-\tfrac{1}{2}\int_{-\infty}^{\infty} \xi(t)^2\, dt\right\}, \qquad \xi \in \mathcal{S}' \tag{5.1}$$

namely the particular case where $\alpha(x) = -\frac{1}{2}x^2$ in the formula (4.1). Consequently, $K_p(\xi, \eta)$ is of the form

$$K_p(\xi, \eta) = \frac{1}{p!}\langle \xi, \eta\rangle^p = \frac{1}{p!}\left(\int_{-\infty}^{\infty} \xi(t)\eta(t)\, dt\right)^p. \tag{5.2}$$

Now put

$$\hat{L}_2(R^p) = \{F; F \in L_2(R^p),\ F \text{ is symmetric}\},$$

$$\hat{F}(t_1, \cdots, t_p) = \frac{1}{p!}\sum_{\pi} F(t_{\pi(1)}, \cdots, t_{\pi(p)}), \qquad \text{(symmetrization)} \tag{5.3}$$

where π denotes the permutation of integers $1, 2, \cdots, p$. Define $I_p^*(\xi; F)$ by

$$I_p^*(\xi; F) = \int_{-\infty}^{\infty}\!\!\cdots\!\int \xi(t_1)\cdots\xi(t_p)F(t_1, \cdots, t_p)\, dt_1, \cdots, dt_p, \qquad \xi \in \mathcal{S}, \tag{5.4}$$

then we have

$$I_p^*(\xi; F) = I_p^*(\xi; \hat{F}), \qquad \text{for every} \quad \xi \in \mathcal{S} \quad \text{and} \quad F \in L_2(R^p). \tag{5.5}$$

THEOREM 5.1. *For Gaussian white noise we have the following properties:*

(i) $\mathfrak{F}_p = \{f(\cdot); f(\xi) = I_p^*(\xi; F),\ F \in L_2(R^p)\}$,

(ii) $(I_p^*(\cdot\,; F), I_p^*(\cdot\,; G))_{\mathfrak{F}_p} = p! \int_{-\infty}^{\infty}\!\!\cdots\!\int \hat{F}(t_1, \cdots, t_p)\overline{G(t_1, \cdots, t_p)}\, dt_1\cdots dt_p.$

PROOF. Define $\tilde{L}_2(R^p)$ and $\mathfrak{F}_p$ by

$$\tilde{L}_2(R^p) = \left\{F; F(t_1, \cdots, t_p) = \frac{1}{p!}\sum_{k=1}^{n} a_k\xi_k(t_1)\cdots\xi_k(t_p),\right.$$
$$\left. a_k \text{ complex},\ \xi_1, \cdots, \xi_n \in \mathcal{S}\right\} \tag{5.6}$$

and

(5.7) $$\mathfrak{F}_p = \{f(\cdot); f(\xi) = I_p^*(\xi; F),\ F \in \tilde{L}_2(R^p)\}.$$

Then we can prove $\mathfrak{F}_p \subseteq \mathfrak{F}_p$. If F and G are elements of $\tilde{L}_2(R^p)$ of the form

(5.8) $$F = \frac{1}{p!}\sum_{k=1}^{n} a_k\xi_k(t_1)\cdots\xi_k(t_p), \qquad G = \frac{1}{p!}\sum_{k=1}^{n} b_k\eta_k(t_1)\cdots\eta_k(t_p),$$

where a_k's and b_k's are complex numbers and $\xi_k,\ \eta_k \in \mathcal{S}$, we have

(5.9) $$\begin{aligned}(I_p^*(\cdot\,; F), I_p^*(\cdot\,; G))_{\mathfrak{F}_p} &= \sum_{k=1}^{n}\sum_{j=1}^{n} a_k\bar{b}_j \frac{1}{p!}\int_{-\infty}^{\infty}\cdots\int \xi_k(t_1)\cdots\xi_k(t_p)\eta_j(t_1)\cdots\eta_j(t_p)\,dt_1\cdots dt_p\\ &= p!\int_{-\infty}^{\infty}\cdots\int F(t_1,\cdots,t_p)\overline{G(t_1,\cdots,t_p)}\,dt_1\cdots dt_p.\end{aligned}$$

Since $\tilde{L}_2(R^p)$ is dense in $\hat{L}_2(R^p)$, we can prove that $\mathfrak{F}_p$ is also dense in $\mathfrak{F}_p$. Indeed,

(5.10) $$\mathfrak{F}_p = \{f(\cdot); f(\xi) = I_p^*(\xi; F),\ F \in \hat{L}_2(R^p)\}.$$

Thus, by (5.5), we get (i).

The second assertion is easily verified using (5.5) and (5.9).

Take a complete orthonormal system $\{\mathring{\xi}_j\}_{j=1}^{\infty}$ in $L_2(R^1)$ such that all the $\mathring{\xi}_j$'s belong to $\mathcal{S}$ (cf. assumption 2.1).

COROLLARY 5.1 (M. G. Kreĭn [9]). *Define the functional*

(5.11) $$\Phi_{(k_1,\cdots,k_q)}^{(j_1,\cdots,j_q)}(\xi) = \frac{\sqrt{p!}}{\sqrt{k_1!\cdots k_q!}}\prod_{n=1}^{q}\langle\xi, \mathring{\xi}_{j_n}\rangle^{k_n};$$

then

(5.12) $$\left\{\Phi_{(k_1,\cdots,k_q)}^{(j_1,\cdots,j_q)} : \begin{array}{ll} j_1,\cdots,j_q & \text{different positive integers,}\\ k_1,\cdots,k_q & \text{different positive integers such that } \sum_1^q k_j = p\end{array}\right\}$$

forms a complete orthonormal system in $\mathfrak{F}_p$.

PROOF. The set of functionals on $\mathcal{S}\times\cdots\times\mathcal{S}$ (p times) of the form

(5.13) $$\left\{\prod_{n=1}^{p}\langle\xi, \mathring{\xi}_{j_n}\rangle;\ j_1,\cdots,j_p = 1, 2, \cdots\right\}$$

is a complete orthonormal system in $\mathfrak{F}_1 \otimes^* \cdots \otimes^* \mathfrak{F}_1$ (p times).

On the other hand, we have

(5.14) $$\begin{aligned}&(\Phi_{(k_1,\cdots,k_q)}^{(j_1,\cdots,j_q)}, \Phi_{(k_1',\cdots,k_q')}^{(j_1',\cdots,j_q')})_{\mathfrak{F}_p}\\ &\quad= p!\frac{\sqrt{p!}}{\sqrt{k_1!\cdots k_q!}}\frac{\sqrt{p!}}{\sqrt{k_1'!\cdots k_q'!}}\left(\frac{1}{p!}\right)^2\\ &\qquad\sum_{\pi}\sum_{\pi'}\int_{-\infty}^{\infty}\cdots\int \mathring{\xi}_{j_1}(t_{\pi(1)})\cdots\mathring{\xi}_{j_1}(t_{\pi(k_1)})\mathring{\xi}_{j_2}(t_{\pi(k_1+1)})\cdots\\ &\qquad\times\mathring{\xi}_{j_2}(t_{\pi(k_1+k_2)})\cdots\mathring{\xi}_{j_q}(t_{\pi(p)})\cdots\mathring{\xi}_{j_1'}(t_{\pi'(k_1')})\cdots\mathring{\xi}_{j_q'}(t_{\pi'(p)})\,dt_1\cdots dt_p.\end{aligned}$$

The right-hand side vanishes if $((j_1, k_1), \cdots, (j_q, k_q)) \neq ((j_1', k_1'), \cdots, (j_q', k_q'))$ (as sets) and is equal to

$$\frac{1}{k_1! \cdots k_q!}\left[\sum_{\pi^{(k_1)}}\left(\int \mathring{\xi}_{j_1}(t)^2\, dt\right)^{k_1}\right]\cdots\left[\sum_{\pi^{(k_q)}}\left(\int \mathring{\xi}_{j_q}(t)^2\, dt\right)^{k_q}\right] = 1 \tag{5.15}$$

otherwise, where $\pi^{(k)}$ denotes the permutation of k integers. Moreover, if $q \neq q'$, then (5.14) obviously vanishes. Thus we have proved the corollary.

REMARK 5.1. The above result has already been proved by M. G. Kreĭn ([9], section 4), although it is stated in a somewhat different form.

COROLLARY 5.2. *If $H_n(x; 1)$ denotes the Hermite polynomial defined by* (3.5) *with $\sigma = 1$, then*

$$\tau^{-1}\{\Phi^{(j_1,\cdots,j_q)}_{(k_1,\cdots,k_q)}(\cdot)C(\cdot)\}(x) = \sqrt{p!}\sqrt{k_1!\cdots k_q!}(i)^{-p}\prod_{m=1}^{q} H_{k_m}(\langle x, \mathring{\xi}_{j_m}\rangle; 1). \tag{5.16}$$

PROOF. The formula

$$\begin{aligned}\int_{S^*} e^{i\langle x,\xi\rangle}\prod_{m=1}^{q} H_{k_m}(\langle x, \mathring{\xi}_{j_m}\rangle; 1)\mu(dx) \\ = \int_{S^*}\prod_{m=1}^{q}[H_{k_m}(\langle x, \mathring{\xi}_{j_m}\rangle; 1)e^{i\langle x,\mathring{\xi}_{j_m}\rangle\langle\xi,\mathring{\xi}_{j_m}\rangle}]\mu(dx)\int_{S^*} e^{i\sum_{l\notin(j_1,\cdots,j_q)}\langle x,\mathring{\xi}_l\rangle\langle\xi,\mathring{\xi}_l\rangle}\mu(dx) \\ = \prod_{m=1}^{q}\int_{-\infty}^{\infty} e^{i\langle\xi,\mathring{\xi}_{j_m}\rangle x}H_{k_m}(x; 1)\frac{1}{\sqrt{2\pi}}e^{-\frac{x^2}{2}}\, dx\, C\left(\sum_{l\notin(j_1,\cdots,j_q)}\langle\xi, \mathring{\xi}_l\rangle\mathring{\xi}_l\right)\end{aligned} \tag{5.17}$$

becomes

$$\frac{(i)^p}{k_1!\cdots k_q!}\prod_{m=1}^{q}\langle\xi, \mathring{\xi}_{j_m}\rangle^{k_m}C(\xi). \tag{5.18}$$

This proves (5.16).

REMARK 5.2. From theorem 4.2 and the above result, we get the orthogonal development of the elements of L_2 due to Cameron and Martin [3].

In the above discussion we use an important property of Gaussian white noise, that is, the equivalence of independence and orthogonality. For other cases discussed here, the multiple Wiener integral due to K. Itô [6] plays an important role.

Let $\{I_j\}_{j=1}^{n}$ be a finite partition of $\mathbf{T}$. Then we have

$$C(\xi) = \prod_{j=1}^{n} C(\xi\chi_{I_j}), \qquad \xi \in \mathcal{S}, \tag{5.19}$$

where χ_{I_j} is the indicator function of I_j. Note that $C(\xi\chi_{I_j})$ has meaning even though $\xi\chi_I$ may not be in $\mathcal{S}$.

Now if we consider the restriction of $C(\xi)$ to $\mathcal{K}(\bar{I}_j)$; then

$$C_{I_j}(\xi) = C(\xi), \qquad \xi \in \mathcal{K}(\bar{I}_j), \tag{5.20}$$

is a continuous positive definite functional. Therefore, we can follow exactly the same arguments as we did for $C(\xi)$. Let us use the symbols $\mathfrak{F}(I_j)$, $\hat{\mathfrak{F}}_p(I_j)$, and $\mathfrak{F}_p(I_j)$ to denote the Hilbert spaces corresponding to $\mathfrak{F}$, $\hat{\mathfrak{F}}_p$, and $\mathfrak{F}_p$ defined for $C(\xi)$. Then we have

(5.21) $$\mathfrak{F} = \prod_{j=1}^{n} \otimes^* \mathfrak{F}(I_j)$$

by the formula (5.19). We can also prove

(5.22) $$\prod_{j=1}^{n} \otimes^* \mathfrak{F}(I_j) \cong \prod_{j=1}^{n} \otimes \mathfrak{F}(I_j), \qquad \text{(isomorphic)}$$

by J. von Neumann's theory [10].

Let $\mathcal{B}(I_j)$ be the smallest Borel field generated by sets of the form

(5.23) $$\{x; \langle x, \xi\rangle \in B\} \ \xi \in \mathcal{K}(\bar{I}), \ B \text{ is a one-dimensional Borel set,}$$

and let $L_2(I_j)$ be the Hilbert space defined by

(5.24) $$L_2(I_j) = \{\varphi; \varphi \in L_2, \varphi \text{ is } \mathcal{B}(I_j)\text{-measurable}\}.$$

Then by (5.21),

(5.25) $$L_2 = \prod_{j=1}^{n} \otimes^* L_2(I_j).$$

Because of the particular form of $C(\xi)$, we can prove that

(5.26) $$\underset{q\to\infty}{\text{l.i.m.}} \langle x, \xi_q\rangle$$

exists if ξ_q tends to χ_{I_j} in $L_2(R^1)$ as $q \to \infty$. We denote the above limit by $\langle x, \chi_{I_j}\rangle$.

We are now in a position to define the multiple Wiener integral of K. Itô. Let $F(t_1, \cdots, t_p)$ be a special elementary function (see K. Itô [5], p. 160) defined as follows:

(5.27) $$F(t_1, \cdots, t_p) = \begin{cases} a_{i_1,\cdots,i_p}, & \text{for } (t_1, \cdots, t_p) \in T_{i_1} \times \cdots \times T_{i_p}, \\ 0, & \text{otherwise,} \end{cases}$$

where the T_i's are mutually disjoint finite intervals. For such F, $I_p(x; F)$ is defined by

(5.28) $$I_p(x; F) = \sum_{i_1,\cdots,i_p} a_{i_1,\cdots,i_p} \prod_{j=1}^{p} \langle x, \chi_{T_{i_j}}\rangle.$$

This function satisfies the following properties (5.29)–(5.32): for any two special elementary functions F and G,

(5.29) $$I_p(x; F + G) = I_p(x; F) + I_p(x; G),$$

(5.30) $$I_p(x; F) = I_p(x; \hat{F}),$$

where $\hat{F}$ is the symmetrization of F;

(5.31) $$I_p(x; F) \in L_2 \quad \text{for any } p \text{ and any special elementary function } F,$$

and

(5.32) $$\begin{aligned} &\langle\langle I_p(x; F), I_p(x; G)\rangle\rangle = p!(\hat{F}, \hat{G})_{L_2(R^p)}, \\ &\langle\langle I_p(x; F), I_q(x; G)\rangle\rangle = 0, \qquad \text{if } p \neq q. \end{aligned}$$

The map I_p can be extended to a bounded linear operator from $L_2(R^p)$ to L_2, which will be denoted by the same symbol I_p. The integral $I_p(x; F)$ is called *the multiple Wiener integral*. It is essentially the same as that of K. Itô except that we can consider complex $L_2(R^p)$ functions as integrands.

THEOREM 5.2. *For every $F \in L_2(R^p)$, we have*

$$I_p(x; F) = (i)^p \tau^{-1}\{I_p^*(\cdot\,; F)C(\cdot)\}(x). \tag{5.33}$$

PROOF. If F is a special elementary function, (5.33) is obvious by the definition of I^* and τ. In fact, if F is defined by (5.27),

$$I_p^*(\xi; F)C(\xi) = \sum_{i_1, \cdots, i_p} a_{i_1, \cdots, i_p} \prod_{j=1}^{p} \langle \xi, \chi_{T_{i_j}} \rangle. \tag{5.34}$$

Hence, we have

$$\tau^{-1}\{I_p^*(\cdot\,; F)C(\cdot)\}(x) = \sum_{i_1, \cdots, i_p} a_{i_1, \cdots, i_p} \prod_{j=1}^{p} [\tau^{-1}\{\langle \cdot, \chi_{T_{i_j}} \rangle C(\cdot)\}]. \tag{5.35}$$

Since $\tau^{-1}\{(\langle \cdot, \chi_{T_i} \rangle C(\cdot)\}(x) = (i)^{-1}\langle x, \chi_{T_i} \rangle$, the above formula is equal to

$$(i)^{-p} \sum_{i_1, \cdots, i_p} a_{i_1, \cdots, i_p} \prod_{j=1}^{p} \langle x, \chi_{T_{i_j}} \rangle. \tag{5.36}$$

Such a relation can be extended to the case of general F.

S. Kakutani [7] also gave a direct sum decomposition of L_2 using the addition formula for Hermite polynomials. It is known that Kakutani's decomposition is the same as that obtained by using multiple Wiener integrals. Conversely, this addition formula can be illustrated by using the decomposition of $\mathfrak{F}$. This was shown by N. Kôno (private communication) in the following way.

Let I, I_1, and I_2 be finite intervals such that $I = I_1 + I_2$; then

$$\frac{1}{n!}\langle \cdot, \chi_I \rangle^n C(\cdot) = \sum_{k=0}^{n} \frac{1}{k!}\langle \cdot, \chi_{I_1} \rangle^k C(\cdot\chi_{I_1}) \frac{1}{(n-k)!}\langle \cdot, \chi_{I_2} \rangle^{n-k} C(\cdot\chi_{I_2}) C(\cdot\chi_{I^c}). \tag{5.37}$$

Noting that

$$\frac{1}{k!}\langle \cdot, \chi_{I_j} \rangle^k C(\cdot\chi_{I_j}) \in \mathfrak{F}(I_j), \qquad j = 1, 2, \tag{5.38}$$
$$C(\cdot\chi_{I^c}) \in \mathfrak{F}(I^c)$$

and that (5.25) holds, we have

$$\tau^{-1}\left\{\frac{1}{k!}\langle \cdot, \chi_{I_1} \rangle^k C(\cdot\chi_{I_1}) \frac{1}{(n-k)!}\langle \cdot, \chi_{I_2} \rangle^{n-k} C(\cdot\chi_{I_2}) C(\cdot\chi_{I^c})\right\}(x) \tag{5.39}$$
$$= \tau_{I_1}^{-1}\left\{\frac{1}{k!}\langle \cdot, \chi_{I_1} \rangle^k C(\cdot\chi_{I_1})\right\}(x)$$
$$\cdot\, \tau_{I_2}^{-1}\left\{\frac{1}{(n-k)!}\langle \cdot, \chi_{I_2} \rangle^{n-k} C(\cdot\chi_{I_2})\right\}(x) \cdot \tau_{I^c}^{-1}\{C(\cdot\chi_{I^c})\}(x),$$

where τ_I denotes the mapping from $L_2(I)$ to $\mathfrak{F}(I)$ which is similar to τ. Here each factor of the right-hand side is expressed in the form

$$\tau_{I_1}^{-1}\left\{\frac{1}{k!}\langle \cdot, \chi_{I_1} \rangle^k C(\cdot\chi_{I_1})\right\}(x) = (i)^{-k} H_k(\langle x, \chi_{I_1} \rangle; |I_1|) \in L_2(I_1), \tag{5.40}$$

$$\tau_{I_2}^{-1}\left\{\frac{1}{(n-k)!}\langle \cdot, \chi_{I_2} \rangle^{n-k} C(\cdot\chi_{I_2})\right\}(x) \tag{5.41}$$
$$= (i)^{-n+k} H_{n-k}(\langle x, \chi_{I_2} \rangle; |I_2|) \in L_2(I_2),$$

(5.42) $$\tau_{I^c}^{-1}\{C(\cdot\chi_{I^c})\}(x) = 1 \in L_2(I^c),$$

where $|I|$ denotes the length of the interval I. On the other hand, since

(5.43) $$\tau^{-1}\left(\frac{1}{n!}\langle\cdot, \chi_I\rangle^n C(\cdot)\right)(x) = H_n(\langle x, \chi_I\rangle; |I|),$$

we get

(5.44) $$H_n(\langle x, \chi_I\rangle; |I|) = \sum_{k=0}^{n} H_k(\langle x, \chi_{I_1}\rangle; |I_1|)H_{n-k}(\langle x, \chi_{I_2}\rangle; |I_2|).$$

Therefore,

(5.45) $$H_n(x + y; |I_1| + |I_2|) = \sum_{k=0}^{n} H_k(x; |I_1|)H_{n-k}(y; |I_2|)$$

for almost all $(x, y) \in R^2$ with respect to the Gaussian measure,

(5.46) $$\frac{1}{\sqrt{2\pi|I_1|\,|I_2|}} \exp\left[-\frac{x^2}{2|I_1|} + \frac{y^2}{2|I_2|}\right] dx\,dy.$$

Since $H_n(x; \sigma^2)$ is a continuous function of x, (5.45) is true for all $(x, y) \in R^2$. Indeed, (5.45) is the addition formula obtained by S. Kakutani [7].

Let us further note that N. Kōno has shown that (5.45) can also be proved by using the *Gauss transform* defined by

(5.47) $$\tilde{\varphi}(y) = \int_{S^*} \varphi(x + iy)\mu(dx), \qquad \varphi \in L_2(S^*, \mu), \qquad y \in S^*.$$

This transformation is well-defined for polynomials. Since the transformation is bounded and linear, and since polynomials form a dense set in $L_2(S^*, \mu)$, we can extend (5.47) to all of $L_2(S^*, \mu)$.

Let $\check{\varphi}$ be the Gauss inverse transform of φ. We can then introduce an operation $\circ$ from $L_2(S^*, \mu) \times L_2(S^*, \mu)$ to $L_2(S^*, \mu)$:

(5.48) $$(\varphi \circ \psi)(y) = \widetilde{(\check{\varphi} \circ \check{\psi})}(y) \qquad \text{for} \quad \varphi, \psi \in L_2(S^*, \mu) \quad \text{and} \quad y \in S^*.$$

By simple computations we can prove the following: if $\varphi(x) = H_n(\langle x, \xi_1\rangle, 1)$ and $\psi(x) = H_m(\langle x, \xi_2\rangle, 1)$,

(5.49) $$(\varphi \circ \psi)(y) = \begin{cases} c_{n,m}H_{n+m}(\langle y, \xi\rangle, 1) & \text{for} \quad \xi_1 = \xi_2 = \xi, \\ c'_{n,m}\varphi(y)\psi(y) & \text{for} \quad \langle \xi_1, \xi_2\rangle = 0. \end{cases}$$

More generally, we can prove that if $\varphi \in L_2^{(n)}$ and $\psi \in L_2^{(m)}$, then

(5.50) $$\varphi \circ \psi \in L_2^{(n+m)}.$$

This operation becomes simpler when it is considered in $\mathfrak{F}$. We shall use the same symbol $\circ$ to express the corresponding operation, namely,

(5.51) $$f \circ g = \tau((\tau^{-1}f) \circ (\tau^{-1}g)) \qquad \text{for} \quad f, g \in \mathfrak{F}.$$

Recalling that the $\mathfrak{F}_n$ appearing in Wiener's direct sum decomposition of $\mathfrak{F}$ is $\tau(L_2^{(n)})$, we have the following proposition.

PROPOSITION 6.1. *The spaces* $\{\mathfrak{F}_n\}_{n=0}^{\infty}$ *form a graded ring with respect to the operation* $\circ$. *(For definition, see Zariski and Samuel* [17], p. 150).

6. Poisson white noise

In this section we shall deal with Poisson white noise, which is another typical stationary process with independent values at every point. Our goal is to find the explicit expressions for $\mathfrak{F}$, $\hat{\mathfrak{F}}_p$, and $\mathfrak{F}_p$ and also to look for relations between the multiple Wiener integrals and the Charier polynomials. Since Poisson white noise enjoys many properties similar to those of Gaussian white noise, we shall sometimes skip the detailed proofs except when there is an interesting difference from Gaussian case.

The characteristic functional of Poisson white noise $\mathbf{P}$ is given by

$$C(\xi) = \exp\left\{\int_{-\infty}^{\infty} (e^{i\xi(t)} - 1 - i\xi(t))\, dt\right\}, \qquad \xi \in \mathcal{S}, \tag{6.1}$$

that is, $\alpha(x) = (e^{ix} - 1 - ix)$ in the expression (4.1). Hence, $K_p(\xi, \eta)$ is expressed in the form

$$K_p(\xi, \eta) = \frac{1}{p!}\left(\int_{-\infty}^{\infty} P(\xi(t))P(\eta(t))\, dt\right)^p, \qquad \xi, \eta \in \mathcal{S}, \tag{6.2}$$

where $P(x) = e^{ix} - 1$. For $F \in L_2(R^p)$ we define $J_p^*(\xi; F)$ by

$$J_p^*(\xi; F) = \int \overset{\infty}{\underset{-\infty}{\cdots}} \int P(\xi(t_1))\cdots P(\xi(t_p))F(t_1, \cdots, t_p)\, dt_1\cdots dt_p. \tag{6.3}$$

Obviously,

$$J_p^*(\xi; F) = J_p^*(\xi; \hat{F}), \qquad \xi \in \mathcal{S}, \tag{6.4}$$

still holds (cf. (5.30)).

THEOREM 6.1. *For Poisson white noise, we have*

(i) $\hat{\mathfrak{F}}_p = \{f(\cdot); f(\xi) = J_p^*(\xi; F),\ F \in L_2(R^p)\}$

(ii) $(J_p^*(\cdot\,; F), J_p^*(\cdot\,; G))_{\mathfrak{F}_p} = p! \int \overset{\infty}{\underset{-\infty}{\cdots}} \int \hat{F}(t_1, \cdots, t_p)\overline{\hat{G}(t_1, \cdots, t_p)}\, dt_1\cdots dt_p$

for any $F, G \in L_2(R^p)$.

PROOF. The proof is nearly the same as that of theorem 5.1. Thus, we shall just point out the necessary changes. The spaces $\tilde{L}_2(R^p)$ and $\hat{\mathfrak{F}}_p$ have to be defined in the following way:

$$\tilde{L}_2(R^p) = \left\{F; F(t_1, \cdots, t_p) = \frac{1}{p!}\sum_{k=1}^{n} a_k P(\xi_k(t_1))\cdots P(\xi_k(t_p)); \quad a_k \text{ complex},\ \xi_k \in \mathcal{S}\right\},$$

$$\hat{\mathfrak{F}}_p = \{f(\cdot); f(\xi) = J_p^*(\xi; F), f \in \tilde{L}_2(R^p)\}. \tag{6.5}$$

If we prove that $\tilde{L}_2(R^p)$ is dense in $\hat{L}_2(R^p)$, then the rest of the proof is exactly the same as that of theorem 5.1. To do this, note that the totality of all linear combinations of functions such as $\chi_{T_1}(t_1)\cdots\chi_{T_p}(t_p)$ with disjoint finite intervals $\{T_j\}_{j=1}^{p}$ is dense in $L_2(R^p)$, and also note that the fact that

$$\int \overset{\infty}{\underset{-\infty}{\cdots}} \int \sum_k a_k P(\xi_k(t_1))\cdots P(\xi_k(t_1))\chi_{T_1}(t_1)\cdots\chi_{T_p}(t_p)\, dt_1\cdots dt_p = 0 \tag{6.6}$$

for any choice of $\{a_k\}$ and ξ_k's in $\mathcal{S}$ implies that

$$\chi_{T_1}(t_1)\cdots\chi_{T_p}(t_p) = 0, \qquad \text{a.e.} \tag{6.7}$$

We can therefore prove that $\tilde{L}_2(R^p)$ is dense in $\hat{L}_2(R^p)$.

The direct product decomposition of $\mathfrak{F}$ and L_2 is the same as in section 5. For any finite partition $\{I_j\}_{j=1}^n$ of $\mathbf{T}$,

$$C(\xi) = \prod_{j=1}^{n} C(\xi\chi_{I_j}), \qquad \xi \in \mathcal{S}, \tag{6.8}$$

still holds. Therefore we have, using the same notation,

$$\mathfrak{F} = \prod_{j=1}^{n} \otimes^* \mathfrak{F}(I_j), \tag{6.9}$$

$$L_2 = \prod_{j=1}^{n} \otimes^* L_2(I_j). \tag{6.10}$$

Moreover, we can define the multiple Wiener integral with respect to Poisson white noise similarly. First note that $\langle x, \chi_I\rangle$ is defined as an element of L_2. If F is a special elementary function given by (5.27), then $J_p(x; F)$ is defined by

$$J_p(x; F) = \sum_{i_1,\cdots,i_p} a_{i_1,\cdots,i_p} \prod_{j=1}^{p} \langle x, \chi_{T_{i_j}}\rangle. \tag{6.11}$$

The map J_p can be extended to a bounded linear operator from $L_2(R^p)$ to L_2 as was done in section 5 (cf. K. Itô [6], section 3).

THEOREM 6.2. *For every* $F \in L_2(R^p)$,

$$J_p(x; F) = \tau^{-1}\{J_p^*(\cdot\,; F)C(\cdot)\}(x). \tag{6.12}$$

PROOF. This proof is also the same as that of theorem 5.2, except for the following relation:

$$\tau(\langle x, \chi_{T_1}\rangle\cdots\langle x, \chi_{T_p}\rangle)(\xi) = \prod_{j=1}^{p} \int_{T_j} P(\xi(t_j))\, dt_j C(\xi). \tag{6.13}$$

From the last theorem we can show that

$$\mathfrak{F} = \sum_{p=0}^{\infty} \oplus \mathfrak{F}_p \tag{6.14}$$

is nothing but Wiener's direct sum decomposition. This fact can also be proved using a certain addition formula for a one-parameter family of generalized Charier polynomials: let $\nu(x, c)$ be given by

$$\nu(x, c) = \frac{c^{x+c}}{\Gamma(x + c + 1)} e^{-c}, \qquad x = -c, 1 - c, 2 - c, \cdots, \tag{6.15}$$

and let

$$P_n(x, c) = n!P_n^*(x, c) = (-c)^n(\nu(x, c))^{-1}\Delta_x^n\nu(x - n, c), \tag{6.16}$$

where Δ_x^n is a difference operator of order n; then the formula is

$$P^*(x + y, c) = \sum_{k=0}^{n} P_k^*(x, c_1)P_{n-k}^*(y, c_2), \qquad c_1, c_2 > 0,\ c = c_1 + c_2. \tag{6.17}$$

7. Concluding remarks

The theorems given in sections 5 and 6 extend to generalized white noise. Furthermore, we shall show that a sequence of independent identically distributed random variables can be dealt with in our scheme. We do not take up detailed discussions but summarize some of their properties.

7.1. *Generalized white noise.* We now discuss the stationary process with the characteristic functional

$$(7.1)\qquad \begin{aligned} C(\xi) &= \exp\left\{\int_{-\infty}^{\infty} \alpha(\xi(t))\,dt\right\}, \qquad \xi \in \mathcal{S}, \\ \alpha(x) &= \int_{-\infty}^{\infty} \left(e^{ixu} - 1 - \frac{ixu}{1+u^2}\right)\frac{1+u^2}{u^2}\,d\beta(u), \end{aligned}$$

which is the one obtained from (4.1) by eliminating the Gaussian part $-(\sigma^2/2)x^2$. Then $K_p(\xi, \eta)$ is expressible as follows:

$$(7.2)\qquad K_p(\xi, \eta) = \frac{1}{p!}\left(\iint_{-\infty}^{\infty} P(\xi(t)u)P(\eta(t)u)\,d\nu(t, u)\right)$$

where $d\nu(t, u) = dt\,d\beta(u)$. We introduce the following notations:

$$D_p = R^{2p}, \qquad dm = u^2\,d\nu, \qquad dm_p = dm \times \cdots \times dm \ (p \text{ times}),$$

$$(7.3)\qquad L_2(D_p; m_p) = \left\{F;\ F \text{ is square summable with respect to } \frac{1}{p!}\,dm_p\right\},$$

$$\begin{aligned} \hat{L}_2(D_p; m_p) = \{F;\ &F \in L_2(D_p; m_p),\ F((t_1, u_1), \cdots, (t_p, u_p)) \\ &= F((t_{\pi(1)}, u_{\pi(1)}), \cdots, (t_{\pi(p)}, u_{\pi(p)})) \text{ for any permutation } \pi\}. \end{aligned}$$

Define $M_p^*(\xi; F)$ by

$$(7.4)\qquad \begin{aligned} M_p^*(\xi; F) = \int_{-\infty}^{\infty}\cdots\int P(\xi(t_1)u_1)\cdots P(\xi(t_p)u_p)F((t_1, u_1), \cdots, (t_p, u_p)) \\ \times u_1\cdots u_p\,d\nu(t_1, u_1)\cdots d\nu(t_p, u_p) \end{aligned}$$

using the same technique as in sections 5 and 6. Then we have

$$(7.5)\qquad M_p^*(\xi; F) = M_p^*(\xi; \hat{F}), \qquad \xi \in \mathcal{S},$$

where

$$(7.6)\qquad \hat{F}((t_1, u_1), \cdots, (t_p, u_p)) = \frac{1}{p!}\sum_{\pi} F((t_{\pi(1)}, u_{\pi(1)}), \cdots, (t_{\pi(p)}, u_{\pi(p)})).$$

For generalized white noise with characteristic functional (7.1), we have the following results:

$$\text{(i)}\qquad \begin{aligned} \mathfrak{F}_p &= \{f(\cdot); f(\xi) = M_p^*(\xi; F),\ F \in L_2(D_p, m_p)\} \\ M_p^*(\xi; F) &= M_p^*(\xi; \hat{F}); \end{aligned}$$

$$\text{(ii)}\qquad \begin{aligned} (M_p^*(\cdot\,; F), M_p^*(\cdot\,; G))_{\mathfrak{F}_p} = p!\iint_{-\infty}^{\infty} \hat{F}((t_1, u_1), \cdots, (t_p, u_p)) \\ \times \overline{\hat{G}((t_1, u_1), \cdots, (t_p, u_p))}\,dm_p((t_1, u_1), \cdots, (t_p, u_p)). \end{aligned}$$

Let us emphasize some of the important differences from Gaussian or Poisson white noise. First we cannot expect that the decomposition $\mathfrak{F} = \sum_{p=0}^{\infty} \oplus \mathfrak{F}_p$, where $\mathfrak{F}_p$ corresponds to the $\mathfrak{H}_p$ appearing in (i), will be the Wiener's direct sum decomposition. However, $\tau^{-1}(\mathfrak{F}_n)$ coincides with the multiple Wiener integral introduced by K. Itô [6]. The next remarkable thing concerns the direct product decomposition.

Let $\{I_j\}_{j=1}^{n}$ and $\{J_k\}_{k=1}^{m}$ be finite partitions of $\mathbf{T}$ and R^1, respectively, and define

$$(7.7) \qquad C(\xi; I_j \times J_k) = \exp\left\{\int_{I_j}\int_{J_k}\left(e^{i\xi(t)u} - 1 - \frac{i\xi(t)u}{1+u^2}\right)\frac{1+u^2}{u^2}\, d\beta(u)\, dt\right\}.$$

Recall $C(\xi; I_j \times J_k)$ defines the subspaces $\mathfrak{F}(I_j \times J_k)$, $\mathfrak{F}_p(I_j \times J_k)$ and $\mathfrak{H}_p(I_j \times J_k)$. Since

$$(7.8) \qquad C(\xi) = \prod_{j=1}^{n}\prod_{k=1}^{m} C(\xi; I_j \times J_k),$$

we have

$$(7.9) \qquad \mathfrak{F} = \prod_{j=1}^{n}\prod_{k=1}^{m} \otimes^* \mathfrak{F}(I_j \times J_k).$$

Now we note a connection with K. Itô's multiple Wiener integral. It seems to be difficult to start in the same way as in sections 5 and 6 by introducing $\langle x, \chi_I \rangle$ in L_2. However, if we consider $\mathfrak{F}$, we can proceed by defining for finite intervals I and J,

$$(7.10) \qquad M^*(\xi; I \times J) = \int_I \int_J P(\xi(t)u)u\, d\nu(t, u),$$

$$(7.11) \qquad M(x; I \times J) = \tau^{-1}(M^*(\cdot\,; I \times J)C(\cdot)).$$

$M(x; \cdot)$ can be considered as a random measure as in K. Itô ([6], section 3) and using it, we can define the multiple Wiener integral $I_p(F)$. Let us denote it by $M_p(x; F)$. Then, for every $F \in L_2(D_p, m_p)$ we can easily prove that

$$(7.12) \qquad \tau^{-1}(M_p^*(\cdot\,; F)C(\cdot)) = M_p(x; F).$$

Rather than discuss the group $\mathbf{G(P)}$ in detail, we shall just give a simple example.

EXAMPLE. Consider the case where

$$(7.13) \qquad \alpha(x) = |x|^\theta, \qquad 0 < \theta < 2.$$

This corresponds to the symmetric stable distribution with exponent θ. A transformation g on E belongs to $\mathbf{G(P)}$, that is,

$$(7.14) \qquad C(g\xi) = C(\xi), \qquad \xi \in E$$

if and only if

$$(7.15) \qquad \int_{-\infty}^{\infty} |(g\xi)(t)|^\theta\, dt = \int_{-\infty}^{\infty} |\xi(t)|^\theta\, dt.$$

Then g,

$$(7.16) \qquad (g\xi)(t) = c\xi(c^\theta t), \qquad c > 0,$$

is an example satisfying (7.15).

7.2. *A sequence of independent random variables.* Consider a stationary process $\mathbf{P} = (E^*, \mu, \{T_t\})$ with independent values at every point, where $E = s = (\xi = \{\xi^k\}_{k=-\infty}^{\infty}; \xi^k$ real) is the space of rapidly decreasing sequences and $\mathbf{T}$ is the additive group of integers. A system of independent identically distributed random variables arises in the following way.

Take a sequence $\{\xi_n\}_{n=-\infty}^{\infty}$ of sequences,

$$\xi_n = \{\xi_n^k\}_{k=-\infty}^{\infty} \in s, \ \xi_n^k = \delta_{n,k}. \tag{7.17}$$

Since the ξ_n's have disjoint supports, the $\langle x, \xi_n\rangle \equiv X_n(x)$, $-\infty < n < \infty$, are mutually independent.

Further, we have

$$U_t X_n(x) = X_n(T_t x) = X_{n+t}(x). \tag{7.18}$$

In view of the above, $T = T_1$ is called *a Bernoulli automorphism* and $\mathbf{P}$ is called *a stationary process with a Bernoulli automorphism T.*

If $\tilde{C}(z)$ is the characteristic function of $X_n(x)$, that is,

$$\tilde{C}(z) = \int_{s^*} e^{izX_n(x)}\mu(dx), \tag{7.19}$$

then the characteristic functional C of $\mathbf{P}$ is expressible in the form

$$C(\xi) = \prod_{k=-\infty}^{\infty} \tilde{C}(\xi^k), \qquad \xi = \{\xi^k\}_{k=-\infty}^{\infty} \in s. \tag{7.20}$$

We can now form the Hilbert space $\mathfrak{F} = \mathfrak{F}(s, C)$ with reproducing kernel C given by (7.20).

The direct product decomposition and the direct sum decomposition of $\mathfrak{F}$ can be done in section 4. We would like to mention two particular cases of stationary processes with Bernoulli automorphisms.

(a) *The Gaussian case.* Let $\mathbf{P} = (s^*, \mu, \{T_t\})$ be the stationary process with a Bernoulli automorphism. Suppose that the characteristic functional of P is given by

$$C(\xi) = \exp\{-\tfrac{1}{2}\|\xi\|^2\}, \qquad \xi \in s \tag{7.21}$$

where $\|\xi\|^2 = \sum_{k=-\infty}^{\infty} (\xi^k)^2$. In this case, the subspace $\mathfrak{F}_p$ of $\mathfrak{F}(s, C)$ turns out to be the following:

$$\mathfrak{F}_p = \Big\{f(\cdot); f(\eta) = \sum_{j_1,\cdots,j_p=-\infty}^{\infty} a(j_1, \cdots, j_p)\eta^{j_1}\cdots\eta^{j_p}C(\eta), \quad \eta = \{\eta^n\}_{n=-\infty}^{\infty} \in s, \ a(j_1, \cdots, j_p) \in \hat{l}_2(R^p)\Big\}, \tag{7.22}$$

where $\hat{l}_2(R^p)$ is defined by

$$\hat{l}_2(R^p) = \Big\{a(j_1, \cdots, j_p); \sum_{j_1,\cdots,j_p=-\infty}^{\infty} |a(j_1, \cdots, j_p)|^2 < \infty, \ a(j_1, \cdots, j_p) \text{ is symmetric with respect to } j_k\text{'s}\Big\}. \tag{7.23}$$

If the $f_k(\cdot)$, $k = 1, 2$, in $\mathfrak{F}$ are given by

$$f_k(\eta) = \sum_{j_1,\cdots,j_p=-\infty}^{\infty} a_k(j_1, \cdots, j_p)\eta^{j_1}\cdots\eta^{j_p}C(\eta), \qquad k = 1, 2, \tag{7.24}$$

then we have the following:

$$(f_1, f_2)_{\mathfrak{F}} = p! \sum_{j_1,\cdots,j_p=-\infty}^{\infty} a_1(j_1, \cdots, j_p)\overline{a_2(j_1, \cdots, j_p)}. \tag{7.25}$$

Actually $\mathfrak{F} = \sum_{p=0}^{\infty} \oplus \mathfrak{F}_p$ is the Wiener's direct sum decomposition. Thus, the subspace $L_2^{(p)}$ of $L_2(\mathcal{S}^*, \mu)$, corresponding to $\mathfrak{F}_p$, is expressed as

$$L_2^{(p)} = \mathfrak{S}\left\{\prod_{k=1}^{n} H_{p_k}(X_{q_k}(x); 1); \{q_k\} \text{ different}, \sum_{1}^{n} p_k = p\right\}. \tag{7.26}$$

(b) *Poisson case.* Consider the stationary process $\mathbf{P} = (\mathcal{S}^*, \mu, \{T_t\})$ whose characteristic functional is

$$C(\xi) = \exp\left\{\sum_{j=-\infty}^{\infty} (e^{i\xi^j} - 1 - i\xi^j)\right\}, \qquad \xi = \{\xi^j\}_{j=-\infty}^{\infty} \in \mathcal{S}. \tag{7.27}$$

Of course, $\mathbf{P}$ is a stationary process with a Bernoulli automorphism. The interesting thing is that $L_2^{(p)}$ is spanned by elements of the form

$$\sum_{p_1,\cdots,p_n} a_{p_1,\cdots,p_n} \prod_{k=1}^{n} Q_{p_k}(X_{q_k}(x); 1), \sum_{k=1}^{n} p_k = p, \tag{7.28}$$

where Q_n is the function defined by (3.12). Note that the Q_n's form a complete orthonormal system in $L_2(S_1, d\nu(x, 1))$ (for notation, see (3.9)).

We can also prove that

$$\mathfrak{F}_p = \left\{f(\cdot); f(\eta) = \sum_{j_1,\cdots,j_p=-\infty}^{\infty} a(j_1, \cdots, j_p) \prod_{k=1}^{p} (e^{i\eta^{j_k}} - 1)C(\eta), \right. \tag{7.29}$$

$$\left. a(j_1, \cdots, j_p) \in \hat{\ell}_2(R^p)\right\}.$$

Although the expression of $f(\cdot)$ in (7.29) is quite different from that in (7.22), we still have the same formula for the inner product, that is, (7.25).

REFERENCES

[1] N. ARONSZAJN, "Theory of reproducing kernels," *Trans. Amer. Math. Soc.*, Vol. 68 (1950), pp. 337–404.

[2] H. BATEMAN *et al.*, *Higher Transcendental Functions*, Vol. 2, New York, McGraw-Hill, 1953.

[3] R. H. CAMERON and W. T. MARTIN, "The orthogonal development of non-linear functionals in a series of Fourier-Hermite functionals," *Ann. of Math.*, Vol. 48 (1947), pp. 385–392.

[4] I. M. GELFAND and N. YA. VILENKIN, *Generalized Functions*, IV, New York, Academic Press, 1964.

[5] K. ITÔ, "Multiple Wiener integral," *J. Math. Soc. Japan*, Vol. 3 (1951), pp. 157–169.

[6] ———, "Spectral type of the shift transformation of differential process with stationary increments," *Trans. Amer. Math. Soc.*, Vol. 81 (1956), pp. 253–263.

[7] S. KAKUTANI, "Determination of the flow of Brownian motion," *Proc. Nat. Acad. Sci.*, Vol. 36 (1950), pp. 319–323.

[8] ———, "Spectral analysis of stationary Gaussian processes," *Proceedings of the Fourth Berkeley Symposium on Mathematical Statistics and Probability*, Berkeley and Los Angeles, University of California Press, 1961, Vol. 2, pp. 239–247.

[9] M. G. KREĬN, "Hermitian-positive kernels on homogeneous spaces," I and II, *Amer. Math. Soc. Transl.*, Ser. 2, Vol. 34 (1963), pp. 69–164. (*Ukrain. Mat. Ž.*, Vol. 1 (1949), pp. 64–98; Vol. 2 (1950), pp. 10–59.)

[10] J. VON NEUMANN, "On infinite direct product," *Compositio Math.*, Vol. 6 (1939), pp. 1–77.

[11] ———, "On rings of operators," *Ann. of Math.*, Vol. 50 (1949), pp. 401–485.

[12] M. NISIO, "Remarks on the canonical representation of strictly stationary processes," *J. Math. Kyoto Univ.*, Vol. 1 (1961), pp. 129–146.

[13] YU. V. PROHOROV, "The method of characteristic functionals," *Proceedings of the Fourth Berkeley Symposium on Mathematical Statistics and Probability*, Berkeley and Los Angeles, University of California Press, 1961, Vol. 2, pp. 403–419.

[14] N. WIENER, "The homogeneous chaos," *Amer. J. Math.*, Vol. 60 (1938), pp. 897–936.

[15] ———, *Nonlinear Problems in Random Theory*, Technology press of M.I.T., Wiley, 1958.

[16] N. WIENER and A. WINTNER, "The discrete chaos," *Amer. J. Math.*, Vol. 65 (1943), pp. 279–298.

[17] O. ZARISKI and P. SAMUEL, *Commutative Algebra*, II, Princeton, Van Nostrand, 1960.

16

Reprinted from pages 49, 51–83 of *Advances in Probability and Related Topics*, vol. 2, P. Ney, ed., Marcel Dekker, New York, 1970, 248p.

The Role of Reproducing Kernel Hilbert Spaces in the Study of Gaussian Processes*

G. Kallianpur

SCHOOL OF MATHEMATICS
INSTITUTE OF TECHNOLOGY
UNIVERSITY OF MINNESOTA
MINNEAPOLIS, MINNESOTA

* *This research was supported by NSF Grant GP-11888.*

1. Introduction

The main theme of this article is the relationship that exists between a Gaussian stochastic process and its reproducing kernel Hilbert space (RKHS).

Let $\{x_t, t \in T\}$ be a real Gaussian stochastic process defined on a probability space $(\Omega_x, \bar{\mathscr{A}}(\Omega_x), P_0)$ where $\bar{\mathscr{A}}(\Omega_x)$ is taken to be the completion with respect to P_0 of $\mathscr{A}(\Omega_x)$, the minimal σ-field with respect to which all the random variables $x_t(\omega)$ $(t \in T)$ are measurable. The mean function $E[x(t)]$ will be assumed to be zero and the covariance $E[x(t)x(s)]$ will be denoted by $R(t, s)$. We shall assume that the infinite set T is a complete, separable metric space. It is well-known (see [15]) that R determines a Hilbert space $H(R)$, called the RKHS of R (or of the process), which has the following properties: $H(R)$ consists of real functions f on T such that

$$R(\cdot, t) \in H(R), \tag{1.1}$$

$$(f, R(\cdot, t)) = f(t) \tag{1.2}$$

for every t in T. Because of the assumptions on T, $H(R)$ is also separable and infinite-dimensional. Let $L^2(\Omega_x, \bar{\mathscr{A}}(\Omega_x), P_0)$ or, briefly, $L^2(\Omega_x, P_0)$ be the Hilbert space of square integrable random variables on Ω_x (with the usual identification of functions differing on null sets), and let $L_1(X)$ be the closed linear subspace of $L^2(\Omega_x, P_0)$ spanned by the finite, real linear combinations $\sum_{i=1}^n c_i x_{t_i}$. The existence of a congruence (i.e., isometric isomorphism) between $L_1(X)$ and $H(R)$ is by now too well-known to require special comment. This congruence has played an important part in the investigation of linear estimation and prediction problems in time-series analysis.

The major part of this work is taken up with the study of the homogeneous chaos of a Gaussian process in which the congruence between the corresponding tensor products of $L_1(X)$ and $H(R)$ is exploited. Homogeneous chaos, or, equivalently, the Cameron-Martin decomposition of $L^2(\Omega_x, P_0)$, was originally obtained for Wiener space. Section 4 of this article (Sections 2 and 3 contain preparatory and auxiliary results on tensor and symmetric-tensor products of Hilbert spaces) presents a general and abstract version of this theory. Apart from its theoretical interest, such an extension of nonlinear techniques to other Gaussian processes is of considerable value in problems of nonlinear estimation and filtering. However, the ideas developed in Section 4 can hardly be called new; they are essentially due to I. E. Segal [16, 17], K. Ito [7], and S. Kakutani [8] and are also related to more recent work of T. Hida and N. Ikeda [5]. Our work is in the spirit of Kakutani's paper, in which all the basic ideas are outlined.

While the results of Sections 2–4 were being given in a course, a similar, somewhat earlier, effort by J. Neveu came to the author's attention [14]. Neveu also goes into questions not treated here, such as the equivalence and singularity of Gaussian measures. In Section 5 we study another aspect of the relationship between the Gaussian measure μ defined on a suitable linear space and its RKHS $H(R)$, in particular, the fact that for a large class of processes, $H(R)$ is a subset of the domain space of μ, has μ-measure zero, but nonetheless serves to determine (through its Hilbert space structure) many important properties of the process. This point is further emphasized in Sections 6 and 7, which contain generalizations of theorems about the Wiener measure proved by Cameron and Graves [3]. Most of the results of these sections are either given without proof or the proofs are only briefly sketched, since they are discussed in two of the author's (as yet unpublished) recent papers [9, 10]. Their inclusion has seemed appropriate to us because, logically, they come within the scope of the present study.

Many of the results, particularly those of Section 5, are closely connected with the abstract Wiener spaces of Gross [4], and their proofs are based on ideas introduced there.

2. Tensor Products and Symmetric Tensor Products of Hilbert Spaces

Let H_1 and H_2 be real separable Hilbert spaces with inner products $(\, , \,)_1$ and $(\, , \,)_2$. We shall assume that the reader is familiar with the definition of the tensor- (or direct-) product Hilbert space of H_1 and H_2, written $H_1 \otimes H_2$. (See, e.g., K. Maurin [13].) Here we shall briefly recapitulate the results on tensor products of Hilbert spaces which will be useful to us later, as well as dwell in greater detail on the notion of the symmetric tensor product.

For any two elements h_1, h_2 ($h_i \in H_i$), let $h_i \otimes h_2$ denote their tensor product, which is an element of $H_1 \otimes H_2$. Let (,) denote the inner product for $H_1 \otimes H_2$.

Lemma 2.1. (i) If h_i, g_i are elements of H_i then

$$(h_1 \otimes h_2, g_1 \otimes g_2) = (h_1, g_1)_1 (h_2, g_2)_2.$$

(ii) The set of all finite linear combinations of the form

$$\sum_{k=1}^{n} c_k (h_1{}^k \otimes h_2{}^k),$$

where the c_k are real, $h_1{}^k \in H_1$, and $h_2{}^k \in H_2$, are dense in $H_1 \otimes H_2$.

(iii) Let $\{f_\alpha\}$ be a set that spans H_1, i.e., such that H_1 is the closed linear manifold spanned by the f_α. Similarly, let $\{g_\beta\}$ be a set that spans H_2. Then the set of elements $\{f_\alpha \otimes g_\beta\}$ spans $H_1 \otimes H_2$.

(iv) Let $\{e_i\}_1{}^\infty$ be a complete orthonormal system (CONS) in H_1 and $\{f_j\}_1{}^\infty$ be a CONS in H_2. Then $\{e_i \otimes f_j,\ i, j = 1, 2, \ldots\}$ is a CONS in $H_1 \otimes H_2$.

As is well-known, statement (iv) can be (and often is, see Maurin [11]) used to provide a definition of $H_1 \otimes H_2$. Both the definition of the tensor product and Lemma 2.1 generalize in an obvious way to the case of any finite number of Hilbert spaces H_i ($i = 1, \ldots, p$). The tensor-product Hilbert space in that case is written $H_1 \otimes \ldots \otimes H_p$. When all the spaces are the same, say, H, we introduce the shorter notation $\otimes^p H$.

Let $h_1 \otimes \ldots \otimes h_p$ be an element of $\otimes^p H$. Define

$$\sigma(h_1 \otimes \cdots \otimes h_p) = \frac{1}{p!} \sum_{\pi} h_{\pi_1} \otimes \cdots \otimes h_{\pi_p} \tag{2.1}$$

where $\pi = (\pi_1, \ldots, \pi_p)$ is a permutation of the integers $(1, \ldots, p)$.

Definition 2.1. The symmetric tensor-product Hilbert space $\sigma[\otimes^p H]$ is the closed linear subspace of $\otimes^p H$ generated by elements of the form

$$\sum_{k=1}^{n} c_k \sigma(h_1{}^k \otimes \cdots \otimes h_p{}^k).$$

In fact, it can be shown that σ defined by (2.1) can be extended to define a projection operator on $\otimes^p H$ whose range is $\sigma[\otimes^p H]$. We shall follow these ideas more closely for $H = H(R)$, the RKHS of the Gaussian process.

First we introduce the following notation. Write $T^p = Tx \ldots_p xT$ and let $\otimes^p R$ denote the following covariance function on $T^p \times T^p$:

$$\overset{p}{\otimes} R(s_1, \ldots, s_p; t_1, \ldots, t_p) = R(s_1, t_1) \ldots R(s_p, t_p) \qquad (2.2)$$

for $(s_1, \ldots, s_p)$, $(t_1, \ldots, t_p)$ in T^p. Letting $H(\otimes^p R)$ be the RKHS generated by $\otimes^p R$, we have the following lemma:

Lemma 2.2. $\otimes^p H(R) \cong H(\otimes^p R)$.

Proof. By Lemma 2.1 (ii), the set of elements

$$\{R(\cdot, t_1) \otimes \ldots \otimes R(\cdot, t_p), (t_1, \ldots, t_p) \in T^p\}$$

spans $\otimes^p H(R)$. Also, by the definition of a RKHS, the set

$$\{(\otimes^p R)(\cdot\,; t_1, \ldots, t_p), (t_1, \ldots, t_p) \in T^p\}$$

spans $H(\otimes^p R)$. From (i) of Lemma 2.1 we have

$$\begin{aligned}
&(R(\cdot, s_1) \otimes \cdots \otimes R(\cdot, s_p), R(\cdot, t_1) \otimes \cdots \otimes R(\cdot, t_p)) \\
&= \prod_{i-1}^{p} (R(\cdot, s_i), R(\cdot, t_i)) \\
&= \prod_{i-1}^{p} R(s_i, t_i) \\
&= ((\overset{p}{\otimes} R)(\cdot\,; s_1, \ldots, s_p), \overset{p}{\otimes} R(\cdot\,; t_1, \ldots, t_p)). \qquad (2.3)
\end{aligned}$$

The inner products appearing at the extreme ends of the relation (2.3) refer, as is obvious from the context, to the Hilbert spaces $\otimes^p H(R)$ and $H(\otimes^p R)$, respectively. In fact, throughout this paper, unless the context makes it necessary, we shall use the same symbol for the inner product of different Hilbert spaces. (The same applies also to the norm.) From (2.3) the correspondence

$$R(\cdot, t_1) \otimes \cdots \otimes R(\cdot, t_p) \leftrightarrow (\overset{p}{\otimes} R)(\cdot\,; t_1, \ldots, t_p) \qquad (2.4)$$

can be extended to a congruence (i.e., an isometric isomorphism) between $\otimes^p H(R)$ and $H(\otimes^p R)$. In what follows we write K in place of $\otimes^p R$.

Lemma 2.3. Let $\mathscr{P}: H(K) \to H(K)$ be defined by

$$(\mathscr{P}F)(t_1, \ldots, t_p) = \frac{1}{p!} \sum_{\pi} F_\pi(t_1, \ldots, t_p), \tag{2.5}$$

where π is a permutation of $(1, \ldots, p)$ and

$$F_\pi(t_1, \ldots, t_p) = F(t_{\pi_1}, \ldots, t_{\pi_p}), \qquad F \in H(K). \tag{2.6}$$

Then $\mathscr{P}$ is an orthogonal projection operator on $H(K)$ whose range S is the closed linear subspace of $H(K)$ consisting of all symmetric functions in $H(K)$.

Proof. If $F \in H(K)$, there exists a sequence $\{\sum_{k=1}^n c_k K(\cdot\,; u_1{}^k, \ldots, u_p{}^k)\}$ which converges to F in $H(K)$-norm. Taking inner products with $K(\cdot\,; t_{\pi_1}, \ldots, t_{\pi_p})$ gives

$$\begin{aligned}
F(t_{\pi_1}, \ldots, t_{\pi_p}) &= \lim_{n\to\infty} \sum_{k=1}^{n} c_k K(u_1{}^k, \ldots, u_p{}^k; t_{\pi_1}, \ldots, t_{\pi_p}) \\
&= \lim_{n\to\infty} \sum_{k=1}^{n} c_k \prod_{i=1}^{p} R(u_i{}^k, t_{\pi_i}) \\
&= \lim_{n\to\infty} \sum_{k=1}^{n} c_k R(u_{\alpha_1}^k, t_1) \ldots R(u_{\alpha_p}^k, t_p) \\
&= \lim_{n\to\infty} \sum_{k=1}^{n} c_k K(u_{\alpha_1}^k, \ldots, u_{\alpha_p}^k; t_1, \ldots, t_p) \\
&= \lim_{n\to\infty} (\zeta^{(n)}, K(\cdot\,; t_1, \ldots, t_p))_{H(K)}
\end{aligned} \tag{2.7}$$

where $\alpha = (\alpha_1, \ldots, \alpha_p)$ is a permutation of $\pi = (\pi_1, \ldots, \pi_p)$ and hence of $(1, \ldots, p)$, and $\zeta^{(n)} \in H(K)$ is given by

$$\zeta^{(n)} = \sum_{k=1}^{n} c_k K(\cdot\,; u_{\alpha_1}^k, \ldots, u_{\alpha_p}^k). \tag{2.8}$$

From (2.8) it is easy to verify that

$$\| \zeta^{(n)} \|^2_{H(K)} \to \| F \|^2_{H(K)} \tag{2.9}$$

as $n \to \infty$. It is also easily checked that

$$\lim_{n\to\infty} (\zeta^{(n)}, \phi) \tag{2.10}$$

exists if ϕ is a finite linear combination of the elements $K(\cdot\,; t_1, \ldots, t_p)$. Since the set of elements $\{K(\cdot\,; t_1, \ldots, t_p), (t_1, \ldots, t_p) \in T^p\}$ spans $H(K)$, then (2.10) and (2.9) (boundedness of the sequence of norms $\| \zeta^{(n)} \|$) together imply that $\zeta^{(n)}$ converges weakly to an element G in $H(K)$. Hence, for all $(t_1, \ldots, t_p) \in T^p$,

$$\begin{aligned} G(t_1, \ldots, t_p) &= (G, K(\cdot\,; t_1, \ldots, t_p))_{H(K)} \\ &= \lim_{n\to\infty} (\zeta^{(n)}, K(\cdot\,; t_1, \ldots, t_p))_{H(K)} \\ &= F(t_{\pi_1}, \ldots, t_{\pi_p}) \\ &= F_\pi(t_1, \ldots, t_p) \end{aligned}$$

from (2.7). We thus conclude $F_\pi = G$. Furthermore, for any $h \in H(K)$,

$$| (F_\pi, h)_{H(K)} | = \lim_{n\to\infty} | (\zeta^{(n)}, h)_{H(K)} | \leq \| F \|_{H(K)} \cdot \| h \|_{H(K)},$$

from which it follows that $\| F_\pi \|_{H(K)} \leq \| F \|_{H(K)}$. Repeating the argument but working with the inverse permutation π^{-1} we obtain the reverse inequality. Hence

$$\| F_\pi \|_{H(K)} = \| F \|_{H(K)}. \tag{2.11}$$

Turning now to σ, it is clear from its definition that it is linear. Also from (2.5) and (2.11),

$$\| \mathscr{P} F \|_{H(K)} \leq \frac{1}{p!} \sum_{\pi} \| F_\pi \|_{H(K)} = \| F \|_{H(K)}. \tag{2.12}$$

Let F, G be elements in $H(K)$. Then, defining $\zeta_\pi^{(n)}$ exactly as $\zeta^{(n)}$ was defined above, we have

$$
\begin{aligned}
(\mathscr{P}F, G)_{H(K)} &= \frac{1}{p!}\sum_{\pi}(F_{\pi}, G)_{H(K)} \\
&= \frac{1}{p!}\sum_{\pi}\lim_{n\to\infty}(\zeta_{\pi}^{(n)}, G)_{H(K)} \\
&= \frac{1}{p!}\sum_{k} c_k\left\{\frac{1}{p!}\sum_{\alpha} G(u_{\alpha_1}^{k}, \ldots, u_{\alpha_p}^{k})\right\} \\
&= \lim_{n\to\infty}\sum_{k=1}^{n} c_k(\mathscr{P}G)(u_1{}^k, \ldots, u_p{}^k) \\
&= \lim_{n\to\infty}\left(\sum_{k=1}^{n} c_k K(\cdot\,; u_1{}^k, \ldots, u_p{}^k), \mathscr{P}G\right)_{H(K)} \\
&= (F, \mathscr{P}G)_{H(K)}. \qquad (2.13)
\end{aligned}
$$

(In the above steps, the permutation $\alpha = \pi^{-1}$.) Formulas (2.12) and (2.13) show that the linear operator $\mathscr{P}$ is bounded and self-adjoint. Since, obviously, $\mathscr{P}^2 = \mathscr{P}$, it follows that $\mathscr{P}$ is an orthogonal projection as asserted in the lemma. The identification of the range S of $\mathscr{P}$ is also immediate and the lemma is proved.

Let

$$K_{\mathscr{P}}(\cdot\,; t_1, \ldots, t_p) = \mathscr{P}K(\cdot\,; t_1, \ldots, t_p). \qquad (2.14)$$

Theorem 2.1

$$\sigma[\overset{p}{\otimes} H(R)] \cong S = H(K_{\mathscr{P}}). \qquad (2.15)$$

Proof. It follows from Lemma 2.3 that for every $(t_1, \ldots, t_p) \in T^p$, $\mathscr{P}K(\cdot\,; t_1, \ldots, t_p)$ belongs to S and it is easy to see that the set $\{\mathscr{P}K(\cdot\,; t_1, \ldots, t_p)\}$ spans the subspace S. Furthermore, $K_{\mathscr{P}}$ is a covariance function (i.e. positive semidefinite) on $T \times T$. For any $h \in S$,

$$
\begin{aligned}
(h, K_{\mathscr{P}}(\cdot\,; t_1, \ldots, t_p))_{H(K)} &= (\mathscr{P}h, K(\cdot\,; t_1, \ldots, t_p))_{H(K)} \\
&= h(t_1, \ldots, t_p), \qquad (2.16)
\end{aligned}
$$

from (2.14) and the fact that $\mathscr{P}h = h$. Hence

$$S = H(K_{\mathscr{P}}), \qquad (2.17)$$

the inner product in $H(K_{\mathscr{P}})$ being the same as the $H(K)$-inner-product. To show that $\sigma[\otimes^p H(R)] \cong S$, observe that the set of elements $\{\sigma[R(\cdot, t_1) \otimes \ldots \otimes R(\cdot, t_p)]\}$, where σ is defined in (2.1), spans $\sigma[\otimes^p H(R)]$. Letting φ denote the congruence relation of Lemma 2.2, it can be seen from (2.4) that

$$\begin{aligned} &\varphi(\sigma[R(\cdot, t_1) \otimes \cdots \otimes R(\cdot, t_p)]) \\ &= \varphi\Big(\frac{1}{p!}\sum_{\pi} R(\cdot, t_{\pi_1}) \otimes \cdots \otimes R(\cdot, t_{\pi_p})\Big) \\ &= \frac{1}{p!}\sum_{\pi} \overset{p}{\otimes} R(\cdot\,; t_{\pi_1}, \ldots, t_{\pi_p}) \\ &= \mathscr{P}K(\cdot\,; t_1, \ldots, t_p) = K_{\mathscr{P}}(\cdot\,; t_1, \ldots, t_p). \end{aligned} \tag{2.18}$$

Formula (2.18) and the fact that $\{K_{\mathscr{P}}(\cdot\,; t_1, \ldots, t_p)\}$ spans S complete the proof of the theorem.

Remark. It is clear from the above proof that the projection operators σ of (2.1) and $\mathscr{P}$ of (2.5) are connected by the relation

$$\sigma = \varphi^{-1}\mathscr{P}\varphi. \tag{2.19}$$

From the congruence of $L_1(X)$ and $H(R)$, Lemma 2.2, and Theorem 2.1, we deduce

Corollary 2.1

$$\overset{p}{\otimes} L_1(X) \cong H(\overset{p}{\otimes} R) \tag{2.20}$$

and

$$\sigma[\overset{p}{\otimes} L_1(X)] \cong \sigma[\overset{p}{\otimes} H(R)] \cong H(K). \tag{2.21}$$

We shall denote the congruence relations between $\sigma[\otimes^p L_1(X)]$ and either of the other two Hilbert spaces by the same symbol, ψ_p.

3. CONS in $\sigma[\otimes^p H(R)]$ and $\sigma[\otimes^p L_1(X)]$

We shall assume throughout that $\{e_i\}_1^\infty$ is a fixed (but arbitrary) CONS in $H(R)$ and that $\{\xi_i\}_1^\infty$ are the random variables over Ω_x such that ξ_i is the element of $L_1(X)$ which corresponds to e_i in $H(R)$ under the isomorphism ψ_1 between the two Hilbert spaces. Let $p \geq 1$ be any integer. From Lemma 2.1 (iv) (extended to p-fold tensor products), it is clear that the elements $e_{i_1} \otimes \ldots \otimes e_{i_p}$, as $i_1, \ldots, i_p$ range independently from 1 to infinity, form a CONS for $\otimes^p H(R)$. It is more convenient to arrange the elements of this CONS in the following manner. Let $\lambda_1, \ldots, \lambda_r$ be the distinct integers in the sequence $(i_1, \ldots, i_p)$ with λ_1 occurring n_1 times, λ_2 occurring n_2 times, etc. Then $n_i > 0$ and $n_1 + \cdots + n_r = p$. The following lemma gives a description of CONS in $\sigma(\otimes^p H(R))$ and $\sigma(\otimes^p L_1(X))$ in terms of $\{e_i\}$ and $\{\xi_i\}$. Define

$$e_{\lambda_1,\ldots,\lambda_r}^{n_1,\ldots,n_r} = \frac{\sqrt{p!}}{(n_1! \ldots n_r!)^{\frac{1}{2}}} \sigma(e_{i_1} \otimes \cdots \otimes e_{i_p}). \tag{3.1}$$

As defined earlier,

$$\sigma(e_{i_1} \otimes \cdots \otimes e_{i_p}) = \frac{1}{p!} \sum_{(j)\sim(i)} e_{j_1} \otimes \cdots \otimes e_{j_p}, \tag{3.2}$$

where (j) is a permutation $(j_1, \ldots, j_p)$ of $(i) = (i_1, \ldots, i_p)$. We define the elements $\xi_{\lambda_1,\ldots,\lambda_r}^{n_1,\ldots,n_r}$ in a similar manner.

Lemma 3.1. (i) The elements $\{e_{\lambda_1,\ldots,\lambda_r}^{n_1,\ldots,n_r}, \sum_1^r n_i = p, \lambda_2 = 1, 2, \ldots, n_i > 0\}$ form a CONS in $\sigma(\otimes^p H(R))$;

(ii) The elements $\{\xi_{\lambda_1,\ldots,\lambda_r}^{n_1,\ldots,n_r}, \sum_1^r n_i = p, \lambda_i = 1, 2, \ldots, n_i > 0\}$ form a CONS in $\sigma(\otimes^p L_1(X))$.

Proof. (i) There are $N = p!/(n_1! \ldots n_r!)$ distinct permutations of $(i) = (i_1, \ldots, i_p)$. Denoting by $C_1, \ldots, C_N$ the classes representing the distinct permutations, we have

$$\begin{aligned} \| \sum_{(j)\sim(i)} e_{j_1} \otimes \cdots \otimes e_{j_p} \|^2 &= \| \sum_{l=1}^{N} \sum_{(j)\in C_l} e_{j_1} \otimes \cdots \otimes e_{j_p} \|^2 \\ &= \sum_{l=1}^{N} (n_1! \cdots n_r!)^2 = p!(n_1! \cdots n_r!). \end{aligned} \tag{3.3}$$

Formulas (3.1–3.3) give

$$\| e_{\lambda_1,\ldots,\lambda_r}^{n_1,\ldots,n_r} \| = 1. \tag{3.4}$$

Also, if $(i) = (i_1, \ldots, i_p) \neq (k_1, \ldots, k_p) = k$, then

$$\left(\sum_{(j)\sim(i)} e_{j_1} \otimes \cdots \otimes e_{j_p}, \sum_{(m)\sim(k)} e_{m_1} \otimes \cdots \otimes e_{m_p} \right)$$

is clearly zero, so that it follows from (3.1) that the set $\{e^{n_1,\ldots,n_r}_{\lambda_1,\ldots,\lambda_r}\}$ is orthonormal. Let $h \in \sigma[\bigotimes^p H(R)]$ be such that for all $(n_1, \ldots, n_r)$, $(\lambda_1, \ldots, \lambda_r)$,

$$(h, e^{n_1,\ldots,n_1}_{\lambda_1,\ldots,\lambda_r}) = 0. \tag{3.5}$$

Again from (3.1) and the definition of $\sigma[\bigotimes^p H(R)]$, the left-hand side of (3.5) equals

$$(\sigma h, e_{i_1} \otimes \cdots \otimes e_{i_p}) = (h, e_{i_1} \otimes \cdots \otimes e_{i_p}).$$

Hence

$$(h, e_{i_1} \otimes \cdots \otimes e_{i_p}) = 0 \tag{3.6}$$

for all sequences $(i_1, \ldots, i_p)$ of positive integers. Equation (3.6) implies that $h = 0$ and (i) is proved. The proof of (ii) follows from (i) upon using the congruence ψ_p established in Corollary 2.1.

4. Homogeneous Chaos

First we shall define the following linear subspaces of $L^2(\Omega_x, \mathscr{A}(\Omega_x), P_0)$. Let $G_0 = \{1\}$ be the closed linear subspace spanned by the constant random variables, and for $p \geq 1$ let $\hat{G}$ be the linear subspace of all polynomials in $\{\xi_i\}_1^\infty$ of degree not exceeding p. Let us write $G_p = \hat{G}_p \Theta \hat{G}_{p-1}$, the set of all polynomials in $\hat{G}_p$ orthogonal to every polynomial in $\hat{G}_{p-1}$. Finally, let $\bar{G}_p$ be the closed linear subspace spanned by G_p. In all of this we shall set $\bar{G}_0 = \hat{G}_0 = G_0$. Following Kakutani [8] we state

Definition 4.1. The subspace $\bar{G}_p$ of $L^2(\Omega_x, \mathscr{A}(\Omega_x), P_0)$ is called to pth homogeneous chaos. G_p is called the pth polynomial chaos.

To simplify the notation in the next lemma we shall write a polynomial θ_p of degree at most p in the variables $u_1, \ldots, u_r$ as

$$\theta_p[u_1, \ldots, u_r] = \sum a_{m_1,\ldots,m_r} u_1^{m_1} \ldots u_r^{m_r} \tag{4.1}$$

where the m_i are nonnegative integers and

$$a_{m_1,\ldots,m_r} = 0 \quad \text{if} \quad m_1 + \cdots + m_r > p. \tag{4.2}$$

Let $h_n(x)$ denote the nth normalized Hermite polynomial in x, i.e.,

$$h_n(x) = (-1)^n \frac{1}{\sqrt{n!}} e^{x^2/2}\left(\frac{d}{dx}\right)^n e^{-x^2/2}. \tag{4.3}$$

Lemma 4.1. A random variable γ belongs to G_p ($p \geq 1$) if and only if it is of the form

$$\gamma(\omega) = \sum a_{m_1,\ldots,m_r} h_{m_1}[\xi_{\lambda_1}(\omega)] \ldots h_{m_r}[\xi_{\lambda_r}(\omega)] \tag{4.4}$$

for some choice of integers $\lambda_1, \ldots, \lambda_r$. In the expression on the right-hand side of (4.4), the summation is over $m_i \geq 0$, r and $\lambda_1, \ldots, \lambda_r$ being fixed, $h_n(x)$ is the nth normalized Hermite polynomial given in (4.3), and the coefficients $a_{m_1,\ldots,m_r}$ satisfy

$$a_{m_1,\ldots,m_r} = 0 \quad \text{if} \quad m_1 + \cdots + m_r \neq p. \tag{4.5}$$

Proof. Suppose $\gamma \in G_p$. Then since $\gamma \in \hat{G}_p$, γ is a polynomial θ_p of degree less than or equal to p in some r of the random variables $\{\xi_i\}_1^\infty$, say $\xi_{\lambda_1}, \ldots, \xi_{\lambda_r}$, i.e.,

$$\theta_p(\xi_{\lambda_1}, \ldots, \xi_{\lambda_r}) = \sum c_{m_1,\ldots,m_r} \xi_{\lambda_1}^{m_1} \ldots \xi_{\lambda_r}^{m_r} \tag{4.6}$$

where $m_i \geq 0$ and $m_1 + \cdots + m_r \leq p$. Clearly, the polynomial $\theta_p(x_1, \ldots, x_r)$ in the r real variables $x_1, \ldots, x_r$ belongs to $L^2[(-\infty, \infty)^r, \nu_r]$, where

$$\nu_r(dx_1, \ldots, dx_r) = (2\pi)^{-r/2} e^{-\frac{1}{2}(x_1^2+\ldots+x_r^2)} dx_1 \ldots dx_r. \tag{4.7}$$

Since the space $L^2[(-\infty, \infty)^r, \nu_r]$ is spanned by the complete orthonormal family $\{h_{n_1}(x_1) \ldots h_{n_r}(x_r), n_i \geq 0\}$, it follows that

$$\theta_p(x_1, \ldots, x_r) = \sum_{\substack{n_i=0 \\ (i=1,\ldots,r)}}^{\infty} a_{n_1,\ldots,n_r} h_{n_1}(x_1) \ldots h_{n_r}(x_r). \tag{4.8}$$

From (4.6) and (4.8) it follows that

$$a_{n_1,\ldots,n_r} = \sum_{\substack{m_i \geq 0 \\ m_1+\cdots+m_r \leq p}} c_{m_1,\ldots,m_r} \prod_{i=1}^{r} \left(\frac{1}{\sqrt{2\pi}} \int_{-\infty}^{\infty} e^{-\frac{1}{2}x_i^2} x_i^{m_i} h_{n_i}(x_i) dx_i\right). \tag{4.9}$$

If $n_1 + \cdots + n_r > p$, there is at least one m_i $(i = 1, \ldots, r)$ for which $m_i < n_i$ so that

$$\frac{1}{\sqrt{2\pi}} \int_{-\infty}^{\infty} e^{-\frac{1}{2}x_i^2} x_i^{m_i} h_{n_i}(x_i) dx_i = 0.$$

Hence every term in the sum on the right-hand side of (4.9) vanishes and we obtain

$$a_{n_1, \ldots, n_r} = 0 \quad \text{whenever } n_1 + \cdots + n_r > p. \tag{4.10}$$

On the other hand, if $n_1 + \cdots + n_r < p$,

$$\begin{aligned} a_{n_1, \ldots, n_r} &= E[\theta_p(\xi_{\lambda_1}, \ldots, \xi_{\lambda_r}) h_{n_1}(\xi_{\lambda_1}) \ldots h_{n_r}(\xi_{\lambda_r})] \\ &= 0, \end{aligned} \tag{4.11}$$

since θ_p is orthogonal to all polynomials in $\hat{G}_{p-1}$ and $h_{n_1}(\xi_{\lambda_1}) \ldots h_{n_r}(\xi_{\lambda_r})$ is a polynomial of degree $n_1 + \cdots + n_r < p - 1$. Formulas (4.10) and (4.11) prove (4.4) and (4.5). Conversely, if γ is a random variable of the form (4.4) where (4.5) holds, then clearly $\gamma \in \hat{G}_p$ and γ is orthogonal to $\hat{G}_{p-1}$ and hence belongs to G_p. This completes the proof of the Lemma.

The remaining theorems in this section comprise our general discussion of homogeneous chaos for Gaussian processes. The central theorem (Theorem 4.3) gives the homogeneous chaos decomposition of the L^2-space of a Gaussian process. The other theorems establish congruences between several Hilbert spaces of interest. We have tried to present them in a form which brings out clearly the part played by the RKHS of the process and exploits the "duality" between the tensor products of the space of random variables and the corresponding tensor products of the RKHS.

The classical result of Cameron and Martin for the Wiener space, as well as Ito's version of the same in terms of multiple Wiener integrals, follows as an immediate consequence of Theorem 4.3. It will be seen later that some work of Hida and Ikeda and the homogeneous chaos decomposition for the abstract Wiener spaces of Gross are also important particular cases of the main theorem of this section. The proofs given below are short because they are based on ideas and results already introduced in Sections 2 and 3.

Let H_p $(p = 0, 1, 2, \dots)$ be Hilbert spaces with inner products $(\ ,\ )_p$. We shall denote by $\sum_{p\geq 0} \oplus H_p$ their orthogonal (external) direct-sum Hilbert space. (See [1] for definitions and other details.)

Theorem 4.1

$$\sum_{p\geq 0} \oplus\, \sigma[\overset{p}{\otimes} L_1(X)] \overset{\psi}{\cong} \sum_{p\geq 0} \oplus\, \sigma[\overset{p}{\otimes} H(R)]. \tag{4.12}$$

Furthermore, let u be an element of the Hilbert space on the left-hand side of (4.12) and let

$$\psi(u) = f. \tag{4.13}$$

Then

$$u = \sum_{p\geq 0} \sum_{\substack{n_1+\cdots+n_r=p\\ n_i\geq 0}} \sum_{\lambda_1,\dots,\lambda_r} (f_p, e^{n_1,\dots,n_r}_{\lambda_1,\dots,\lambda_r})_p \xi^{n_1,\dots,n_r}_{\lambda_1,\dots,\lambda_r} \tag{4.14}$$

where

$$f_p \in \sigma[\overset{p}{\otimes} H(R)] \quad (p = 0, 1, \dots) \tag{4.15}$$

are uniquely determined and

$$\sum_{p\geq 0} \sum_{\substack{n_1+\cdots+n_r=p\\ n_i\geq 0}} \sum_{\lambda_1,\dots,\lambda_r} (f_p, e^{n_1,\dots,n_r}_{\lambda_1,\dots,\lambda_r})_p^2 < \infty. \tag{4.16}$$

Proof. Formula (4.12) follows from (2.21) of Corollary 2.1. Let $u \in \sum_{p\geq 0} \oplus\, \sigma[\otimes^p L_1(X)]$. Then

$$u = \sum_{p\geq 0} u_p \tag{4.17}$$

where

$$u_p \in \sigma[\overset{p}{\otimes} L_1(X)] \tag{4.18}$$

are uniquely determined by u,

$$u_p \perp u_q \quad (p \neq q), \tag{4.19}$$

and

$$\sum_{p\geq 0} \| u_p \|_p^2 < \infty. \tag{4.20}$$

(Note that (4.19) means that u_p and u_q are orthogonal as elements of the Hilbert space $\sum_{p\geq 0} \oplus\, \sigma[\otimes^p L_1(X)]$.)

From Lemma 3.1 (ii) it follows that

$$u_p = \sum_{\substack{n_1+\cdots+n_r=p \\ n_i\geq 0}} \sum_{\lambda_1,\ldots,\lambda_r} (u_p, \xi^{n_1,\ldots,n_r}_{\lambda_1,\ldots,\lambda_r})_p \xi^{n_1,\ldots,n_r}_{\lambda_1,\ldots,\lambda_r}. \tag{4.21}$$

Let

$$f_p = \psi_p(u_p) \tag{4.22}$$

where ψ_p is the congruence introduced in Corollary 2.1. Clearly f_p satisfies (4.15). Also from (2.21) and Lemma 3.1 (i),

$$(u_p, \xi^{n_1,\ldots,n_r}_{\lambda_1,\ldots,\lambda_r})_p = (f_p, e^{n_1,\ldots,n_r}_{\lambda_1,\ldots,\lambda_r})_p. \tag{4.23}$$

The remaining assertions of the theorem now follow from (4.17–4.23).

Lemma 4.2. For each $p \geq 1$,

$$\sigma[\overset{p}{\otimes} L_1(X)] \overset{J_p}{\cong} \bar{G}_p. \tag{4.24}$$

Proof. Recalling the definition given in Section 3 of $\xi^{n_1,\ldots,n_r}_{\lambda_1,\ldots,\lambda_r}$ in terms of the ξ_i, define J_p first for $\xi^{n_1,\ldots,n_r}_{\lambda_1,\ldots,\lambda_r}$ by

$$J_p(\xi^{n_1,\ldots,n_r}_{\lambda_1,\ldots,\lambda_r}) = h_{n_1}(\xi_{\lambda_1}) \ldots h_{n_r}(\xi_{\lambda_r}). \tag{4.25}$$

For finite linear combinations of $\xi^{n_1,\ldots,n_r}_{\lambda_1,\ldots,\lambda_r}$ define J_p by extending (4.25) in the usual way. It then follows from Lemma 4.1 that the J_p thus obtained is an isometric isomorphism between the linear manifold spanned by the $\xi^{n_1,\ldots,n_r}_{\lambda_1,\ldots,\lambda_r}$ and G_p. J_p then extends to become a congruence between the closed linear manifolds respectively spanned by $\{\xi^{n_1,\ldots,n_r}_{\lambda_1,\ldots,\lambda_r}\}$ and $\{h_{n_1}(\xi_{\lambda_1}) \ldots h_{n_r}(\xi_{\lambda_r})\}$. This completes the proof of (4.24).

The following result can easily be proved by applying Lemma 4.2 and using the definition of direct sum. The proof is omitted.

Theorem 4.2

$$\sum_{p\geq 0} \oplus \sigma[\overset{p}{\otimes} L_1(X)] \overset{J}{\cong} \sum_{p\geq 0} \oplus \bar{G}_p. \tag{4.26}$$

Let us recall that since $\{\xi_i\}_1{}^\infty$ is a CONS in $L_1(X)$, and $\{e_i\}$ with

$e_i = \psi_1(\xi_i)$ is a CONS in $H(R)$, we have

$$x(t, \omega) = \sum_{i=1}^{\infty} \xi_i(\omega) e_i(t) \tag{4.27}$$

with the series converging a.s. for each fixed t.

From (4.27) and since each ξ_i is $\bar{\mathscr{A}}(\Omega_x)$-measurable, it follows that

$$\bar{\mathscr{A}}(\xi) = \bar{\mathscr{A}}(\Omega_x) \tag{4.28}$$

where $\bar{\mathscr{A}}(\xi)$ denotes the completion of the σ-field generated by the random variables $\{\xi_i(\omega)\}_1^{\infty}$.

Theorem 4.3

$$L^2(\Omega_x, \bar{\mathscr{A}}(\Omega_x), P_0) = \sum_{p \geq 0} \oplus \bar{G}_p \cong \sum_{p \geq 0} \oplus \sigma[\overset{p}{\otimes} L_1(X)]. \tag{4.29}$$

For any $u \in L^2(\Omega_x, \bar{\mathscr{A}}(\Omega_x), P_0)$, the ($L^2$-convergent) expansion

$$u = \sum_{p \geq 0} \sum_{n_1 + \cdots + n_r = p} \sum_{\lambda_1, \ldots, \lambda_r} (F_p, e_{\lambda_1, \ldots, \lambda_r}^{n_1, \ldots, n_r})_p h_{n_1}(\xi_{\lambda_1}) \ldots h_{n_r}(\xi_{\lambda_r}) \tag{4.30}$$

holds, where

$$F_p, e_{\lambda_1, \ldots, \lambda_r}^{n_1, \ldots, n_r} \in \sigma[\overset{p}{\otimes} H(R)]. \tag{4.31}$$

Proof. The second assertion in (4.29) has already been shown in Theorem 4.2. To prove the first part, let $u \in L^2(\Omega_x, \bar{\mathscr{A}}(\Omega_x)P_0)$ be such that

$$u \perp \sum_{p \geq 0} \oplus \bar{G}_p. \tag{4.32}$$

Then for every integer N,

$$E\{u h_{n_1}(\xi_1) \cdots h_{n_N}(\xi_N)\} = 0$$

for all $n_1, \ldots, n_N \geq 0$. Hence, taking conditional expectations we have

$$E\{E(u \mid \xi_1, \ldots, \xi_N) h_{n_1}(\xi_1) \ldots h_{n_N}(\xi_N)\} = 0 \tag{4.33}$$

for all nonnegative integers $n_1, \ldots, n_N$.

Since $E(u \mid \xi_1, \ldots, \xi_N) \in L^2(\Omega_x, \mathscr{A}(\xi_1, \ldots, \xi_N), P_0)$, it follows from (4.33) that

$$E(u \mid \xi_1, \ldots, \xi_N) = 0 \quad \text{a.s.} \tag{4.34}$$

N being arbitrary, making $N \to \infty$ in (4.34) a standard martingale theorem gives

$$E(u \mid \mathscr{A}(\xi)) = 0 \quad \text{a.s.,} \tag{4.35}$$

which implies

$$u = 0 \quad \text{a.s.} \tag{4.36}$$

since (4.28) holds. Thus (4.29) is proved. The relation (4.30) can be easily shown to follow from Theorems 4.1 and 4.2.

Let $\bar{A}$ be the class of all infinite sequences $\{a^{n_1,\ldots,n_r}_{\lambda_1,\ldots,\lambda_r}\}$ such that

$$\sum_{p\geq 0} \sum_{\substack{n_1+\cdots+n_r=p\\ n_i\geq 0}} \sum_{\lambda_1,\ldots,\lambda_r} (a^{n_1,\ldots,n_r}_{\lambda_1,\ldots,\lambda_r})^2 < \infty. \tag{4.37}$$

Define the real-valued function ϕ on $H(R)$ by

$$\phi(m) = \sum_{p\geq 0} \sum_{\substack{n_1+\cdots+n_r=p\\ n_i\geq 0}} \sum_{\lambda_1,\ldots,\lambda_r} a^{n_1,\ldots,n_r}_{\lambda_1,\ldots,\lambda_r} \prod_{i=1}^{r} \frac{(m, e_{\lambda_i})^{n_i}}{\sqrt{n_i!}} \tag{4.38}$$

($m \in H(R)$), where $\{a^{n_1,\ldots,n_r}_{\lambda_1,\ldots,\lambda_r}\}$ belongs to $\bar{A}$. It is not hard to verify that the correspondence

$$\phi \sim \{a^{n_1,\ldots,n_r}_{\lambda_1,\ldots,\lambda_r}\} \tag{4.39}$$

is one-to-one by showing that $\phi(m) = 0$ for all m in $H(R)$ implies that all the coefficients $a^{n_1,\ldots,n_r}_{\lambda_1,\ldots,\lambda_r}$ vanish. Let $\mathscr{F}$ be the class of all functions ϕ on $H(R)$ given by (4.38). If $\psi \in \mathscr{F}$ and

$$\psi \sim \{b^{n_1,\ldots,n_r}_{\lambda_1,\ldots,\lambda_r}\} \tag{4.40}$$

and if we define [,] by

$$[\phi, \psi] = \sum_{p\geq 0} \sum_{\substack{n_1+\cdots+n_r=p\\ n_i\geq 0}} \sum_{\lambda_1,\ldots,\lambda_r} a^{n_1,\ldots,n_r}_{\lambda_1,\ldots,\lambda_r} b^{n_1,\ldots,n_r}_{\lambda_1,\ldots,\lambda_r} \tag{4.41}$$

(the series converges by Schwarz's inequality and (4.37)), then it is easy to see that [,] defines an inner product in $\mathscr{F}$. In his thesis [12], R. LePage has proved the following result.

Theorem 4.4. $\mathscr{F}$ with the inner product [,] is a RKHS with kernel Γ_R on $H(R) \times H(R)$ given by

$$\Gamma_R(m, m_0) = \exp\{(m, m_0)\} \quad (m, m_0 \in H(R)). \tag{4.42}$$

We give here only an indication of the proof. First, completeness is proved from (4.41) in the usual way. After some computation it is verified that

$$\Gamma_R(m, m_0) = \sum_{p\geq 0} \sum_{\substack{n_1+\cdots+n_r=p \\ n_i\geq 0}} \sum_{\lambda_1,\ldots,\lambda_r} \prod_{i=1}^{r} \frac{(m, e_{\lambda_i})^{n_i}(m_0, e_{\lambda_i})^{n_i}}{n_i!}. \tag{4.43}$$

Setting

$$\gamma_{\lambda_1,\ldots,\lambda_r}^{n_1,\ldots,n_r} = \prod_{i=1}^{r} \frac{(m_0, e_{\lambda_i})^{n_i}}{\sqrt{n_i!}},$$

it follows immediately from (4.43) that the sequence $\{\gamma_{\lambda_1,\ldots,\lambda_r}^{n_1,\ldots,n_r}\} \in \bar{A}$ and that $\Gamma_R(\cdot, m_0) \in \mathscr{F}$ for every $m_0 \in H(R)$ since

$$\Gamma_R(m, m_0) = \sum_{p\geq 0} \sum_{\substack{n_1+\cdots+n_r=p \\ n_i\geq 0}} \sum_{\lambda_1,\ldots,\lambda_r} \gamma_{\lambda_1,\ldots,\lambda_r}^{n_1,\ldots,n_r} \prod_{i=1}^{r} \frac{(m, e_{\lambda_i})^{n_i}}{\sqrt{n_i!}}.$$

Finally, if $\phi \in \mathscr{F}$, from (4.41) we have for every $m_0 \in H(R)$ the reproducing property

$$\begin{aligned}[\phi, \Gamma_R(\cdot, m_0)] &= \sum_{p\geq 0} \sum_{\substack{n_1+\cdots+n_r=p \\ n_i\geq 0}} \sum_{\lambda_1,\ldots,\lambda_r} a_{\lambda_1,\ldots,\lambda_r}^{n_1,\ldots,n_r} \prod_{i=1}^{r} \frac{(m_0, e_{\lambda_i})^{n_i}}{\sqrt{n_i!}} \\ &= \phi(m_0). \end{aligned} \tag{4.44}$$

We shall denote the RKHS of the above theorem by $H(\Gamma_R)$. For fixed $m \in H(R)$, introduce the abbreviation $m^{\otimes p}$ for $m \otimes \ldots {}_p\!\otimes m$. Then the series $\sum_{p\geq 0} (m^{\oplus p}/\sqrt{p!})$ defines an element of $\sum_{p\geq 0} \oplus [\otimes^p H(R)]$ which we denote by $\exp[\hat{\otimes}\, m]$.

Lemma 4.3. For every $m \in H(R)$,

$$\exp[\hat{\otimes}\, m] \in \sum_{p\geq 0} \oplus\, \sigma[\overset{p}{\otimes} H(R)]. \tag{4.45}$$

Furthermore, if $m_0 \in H(R)$

$$(\exp[\hat{\otimes}\, m], \exp[\hat{\otimes}\, m_0]) = \exp\{(m, m_0)\}. \tag{4.46}$$

Proof. Clearly, since $m^{\otimes p} \in \sigma[\otimes^p H(R)]$, (4.45) is proved if we show that

$$\sum_{p\geq 0}^{\infty} \frac{1}{p!} \| m^{\otimes p} \|_p^2 < \infty. \tag{4.47}$$

Now

$$\frac{1}{p!} \| m^{\otimes p} \|_p^2 = \sum_{\substack{n_1+\cdots+n_r=p \\ n_i\geq 0}} \sum_{\lambda_1,\ldots,\lambda_r} \left(\frac{m^{\otimes p}}{\sqrt{p!}}, e^{n_1,\ldots,n_r}_{\lambda_1,\ldots,\lambda_r} \right)_p^2,$$

and from (3.1),

$$\begin{aligned}
\left(\frac{m^{\otimes p}}{\sqrt{p!}}, e^{n_1,\ldots,n_r}_{\lambda_1,\ldots,\lambda_r} \right)_p &= \left(\frac{m^{\otimes p}}{\sqrt{p!}}, \frac{\sqrt{p!}}{\sqrt{n_1!\ldots n_r!}} \sigma(e_{\lambda_1}^{\otimes n_1} \otimes \cdots \otimes e_{\lambda_r}^{\otimes n_r}) \right)_p \\
&= \left(\prod_{i=1}^{r} n_i! \right)^{-\frac{1}{2}} (\sigma(m^{\otimes p}), e_{\lambda_1}^{\otimes n_1} \otimes \cdots \otimes e_{\lambda_r}^{\otimes n_r})_p \\
&= \left(\prod_{i=1}^{r} n_i! \right)^{-\frac{1}{2}} (m^{\otimes p}, e_{\lambda_1}^{\otimes n_1} \otimes \cdots \otimes e_{\lambda_r}^{\otimes n_r})_p \\
&= \left(\prod_{i=1}^{r} n_i! \right)^{-\frac{1}{2}} (m, e_{\lambda_1})^{n_1} \cdots (m, e_{\lambda_r})^{n_r}.
\end{aligned} \tag{4.48}$$

Hence

$$\begin{aligned}
\sum_{p\geq 0}^{\infty} \frac{1}{p!} \| m^{\otimes p} \|_p^2 &= \sum_{p\geq 0}^{\infty} \sum_{\substack{n_1+\cdots+n_r=p \\ n_i\geq 0}} \sum_{\lambda_1,\ldots,\lambda_r} \prod_{i=1}^{r} \frac{(m, e_{\lambda_i})^{2n_i}}{n_i!} \\
&= \exp\{(m, m)\} < \infty
\end{aligned}$$

and (4.45) is proved. From (4.48) we obtain

$$\begin{aligned}
&(\exp[\hat{\otimes}\, m], \exp[\hat{\otimes}\, m_0] \\
&= \sum_{p\geq 0} \sum_{\substack{n_1+\cdots+n_r=p \\ n_i\geq 0}} \sum_{\lambda_1,\ldots,\lambda_r} \prod_{i=1} \frac{(m, e_{\lambda_i})^{n_i} (m_0, e_{\lambda_i})^{n_i}}{n_i!} \\
&= \exp\{(m, m_0)\}
\end{aligned}$$

and the lemma is proved.

Theorem 4.5

$$H(\Gamma_R) \cong \sum_{p\geq 0} \oplus\, \sigma[\overset{p}{\otimes} H(R)]. \tag{4.49}$$

Proof. We shall first show that the family

$$\{\exp[\hat{\otimes}\, m], m \in H(R)\} \quad \text{spans} \quad \sum_{p\geq 0} \oplus\, \sigma[\overset{p}{\otimes}\, H(R)].$$

Let

$$f = \sum_{p\geq 0} \sum_{\substack{n_1+\cdots+n_r=p \\ n_i\geq 0}} \sum_{\lambda_1,\ldots,\lambda_r} a^{n_1,\ldots,n_r}_{\lambda_1,\ldots,\lambda_r} e^{n_1,\ldots,n_r}_{\lambda_1,\ldots,\lambda_r} \tag{4.50}$$

be such that

$$(f, \exp[\hat{\otimes}\, m]) = 0 \quad \text{for all} \quad m \in H(R). \tag{4.51}$$

The left-hand side of (4.51) equals

$$\sum_{p\geq 0} \sum_{\substack{n_1+\cdots+n_r=p \\ n_i\geq 0}} \sum_{\lambda_1,\ldots,\lambda_r} a^{n_1,\ldots,n_r}_{\lambda_1,\ldots,\lambda_r} \prod_{i=1}^{r} \frac{(m, e_{\lambda_i})^{n_i}}{\sqrt{n_i!}}.$$

Hence (4.51) and the remark concerning the one-to-one-ness of the correspondence (4.39) yields $f = 0$. Finally from (4.46) we have for all $m, m_0 \in H(R)$,

$$(\exp[\hat{\otimes}\, m], \exp[\hat{\otimes}\, m_0]) = [\Gamma_R(\cdot, m), \Gamma_R(\cdot, m_0)]. \tag{4.52}$$

The conclusion of the theorem now follows by standard arguments upon noting the fact that $\{\exp[\hat{\otimes}\, m], m \in H(R)\}$ spans the Hilbert space on the right-hand side of (4.49) and that $\{\Gamma_R(\cdot, m), m \in H(R)\}$ spans $H(\Gamma_R)$.

Theorems 4.1, 4.2, 4.3, and 4.5 are the main results of this section which investigate the structure of the L^2-space of a Gaussian process. We shall now make a few comments on the classical theorem of Cameron and Martin [2] (together with the version using multiple Wiener integrals due to Ito [6]).

Suppose that $(\Omega_x, \bar{\mathscr{A}}(\Omega_x), P_0) = (C, \bar{\mathscr{A}}(C), \mu_w)$ where the latter is Wiener space, i.e., C is the Banach space of real continuous functions $x(t)$ on $[0, 1]$ such that $x(0) = 0$, $\mathscr{A}(C)$ is the σ-field of Borel sets in C, $\bar{\mathscr{A}}(C)$ its completion with respect to the Wiener measure μ_w. The Cameron-Martin theorem and Ito's version of it are given below (Theorems 4.1 and 4.2 [6]). The notation has been changed slightly to conform to ours. We shall write $I = [0, 1]$, $I^p = [0, 1]^p$, $L^2(I)$

$= L^2(I$, Lebesgue) and $\hat{L}^2(I^p)$ is the closed linear subspace of $L^2(I^p)$ consisting of all real symmetric functions.

Theorem 4.6. Let $\{\varphi_i(t)\}_1^\infty$ be a CONS in $L^2(I)$. Then any $L^2(C, \mu_w)$-functional u has the development

$$u(x) = \sum_{p\geq 0} \sum_{\substack{n_1+\cdots+n_r=p\\ n_i\geq 0}} \sum_{\lambda_1,\ldots,\lambda_r} a^{n_1,\ldots,n_r}_{\lambda_1,\ldots,\lambda_r} \prod_{i=1}^{r} h_{n_i}\left[\int_0^1 \varphi_{\lambda_i}(t)\, dx(t)\right]. \tag{4.53}$$

Also

$$u = \sum_{p\geq 0} I_p(f_p) = \sum_{p\geq 0} I_p(\tilde{f}_p) \tag{4.54}$$

where $f_p \in L^2(I^p)$ is given by

$$f_p(t_1, \ldots, t_p) = \sum_{\substack{n_1+\cdots+n_r=p\\ n_i\geq 0}} \sum_{\lambda_1,\ldots,\lambda_r} a^{n_1,\ldots,n_r}_{\lambda_1,\ldots,\lambda_r}$$
$$\cdot\, \varphi_{\lambda_1}(t_1) \cdots \varphi_{\lambda_1}(t_{n_1})\varphi_{\lambda_2}(t_{n_1+1}) \cdots \varphi_{\lambda_2}(t_{n_1+n_2})$$
$$\cdot\, x \cdots \varphi_{\lambda_r}(t_{n_1+\cdots+n_{r-1}+1}) \cdots \varphi_{\lambda_r}(t_p), \tag{4.55}$$

$I_p(f_p)$ is the multiple Wiener integral of degree p and $\tilde{f}_p$ is the symmetrization of f_p. (See (2.5) and (2.6) of Section 2.)

It is obvious that Theorem 4.3 is applicable to $L^2(C, \mathscr{A}(C), \mu w)$. From (4.30) we may write $u = \sum_{p\geq 0} u_p$, where

$$u_p = \sum_{\substack{n_1+\cdots+n_r=p\\ n_i\geq 0}} \sum_{\lambda_1,\ldots,\lambda_r} (F_p, e^{n_1,\ldots,n_r}_{\lambda_1,\ldots,\lambda_r})_p \prod_{i=1}^{r} h_{n_i}(\xi_{\lambda_i}), \tag{4.56}$$

where $F_p \in \sigma[\otimes^p H(R)]$ is uniquely determined by u_p. Since for the standard Wiener process, $R(t, s) = \min(t, s)$ $(t, s \in I)$, it is well-known that $H(R)$ consists of absolutely continuous functions $F \in L^2(I)$ with square integrable derivatives, and if $F(t) = \int_0^t f(u)du$, $G(t) = \int_0^t g(u)du$, then $(F, G)_{H(R)} = (f, g)_{L^2(I)}$. Hence, if $\{\varphi_i\}_1^\infty$ is a CONS in L^2(I) then $\{e_i\}_1^\infty$, where $e_i(t) = \int_0^t \varphi_i(u)du$, is a CONS in $H(R)$. It is easy to see that the correspondence θ_1 given by $\theta_1\varphi_i = e_i$ $(i = 1, 2, \ldots)$ (and defined by linearity for finite linear combinations) extends to a congruence between $L^2(I)$ and $H(R)$. It follows that for any integer p,

$$L^2(I^p) \simeq \bigotimes^p L^2(I) \simeq \bigotimes^p H(R) \tag{4.57}$$

and

$$\hat{L}^2(I^p) \cong \sigma[\overset{p}{\otimes} L^2(I)] \cong \sigma[\overset{p}{\otimes} H(R)]. \tag{4.58}$$

Under the congruence defined by (4.58), the image in $\hat{L}^2(I^p)$ of $e^{n_1 \cdots n_r}_{\lambda_1 \cdots \lambda_r}$ is $\varphi^{n_1 \cdots n_r}_{\lambda_1 \cdots \lambda_r}$ where $\{\varphi^{n_1 \cdots n_r}_{\lambda_1 \cdots \lambda_r}\}$ is the CONS in $\hat{L}^2(I^p)$ obtained from the sequence

$$\{\varphi_{\lambda_1}(t_1) \cdots \varphi_{\lambda_1}(t_{n_1})\varphi_{\lambda_2}(t_{n_1+1}) \cdots \varphi_{\lambda_2}(t_{n_1+n_2}) \cdots \varphi_{\lambda_r}(t_{n_1+\cdots+n_{r-1}+1}) \cdots \varphi_{\lambda_r}(t_p)\}$$

(here $(\lambda_1, \ldots, \lambda_r)$ runs over all finite sequences of distinct integers and $n_i \geq 0$ such that $n_1 + \cdots + n_r = p$), which is a CONS in $L^2(I^p)$. Also let the image in $\hat{L}^2(I^p)$ of F_p be $\tilde{f}_p$. It then follows that

$$\begin{aligned}(F_p, e^{n_1 \ldots n_r}_{\lambda_1 \ldots \lambda_r})_p &= (\tilde{f}_p, \varphi^{n_1 \ldots n_r}_{\lambda_1 \ldots \lambda_r})_{L^2(I^p)} \\ &= a^{n_1 \ldots n_r}_{\lambda_1 \ldots \lambda_r}, \quad \text{say.}\end{aligned} \tag{4.59}$$

The random variables ξ_i in $L_1(X)$ are given by the Wiener integrals

$$\xi_i(x) = \int_0^1 \varphi_i(t)\, dx(t). \tag{4.60}$$

Substituting from (4.59) and (4.60) in (4.56) and using (4.30), we obtain the Cameron-Martin expansion (4.53) of Theorem 4.6. From the way it is defined, it also follows that $\tilde{f}_p$ is the symmetrization of the function f_p defined by (4.55).

From the congruence ψ_p established in Corollary 2.1 between $\sigma[\otimes^p L_1(X)]$ and $\sigma[\otimes^p H(R)]$ and from (4.58) we have

$$\hat{L}^2(I^p) \cong \sigma[\overset{p}{\otimes} L_1(X)]. \tag{4.61}$$

Assertion (4.61) suggests that it is appropriate to define a multiple Wiener integral of degree p ($p \geq 1$) to be any element of $\sigma[\otimes^p L_1(X)]$.

For every function $\tilde{f}_p$ in $\hat{L}^2(I^p)$ ($p \geq 1$) by (4.61) there is a unique element η_p in $\sigma[\otimes^p L_1(X)]$, and hence by the congruence (4.24) established in Lemma 4.2, there exists a unique random variable $u_p(x)$ in $\bar{G}_p$ which gives a "concrete" realization of the multiple Wiener integral of degree p, and we write

$$I_p(\tilde{f}_p) = u_p(x). \tag{4.62}$$

I_0 (constant function) is defined to be the same constant function. From (4.53), $u_p(x)$ has the expansion

$$u_p(x) = \sum_{\substack{n_1+\cdots+n_r=p \\ n_i \geq 0}} \sum_{\lambda_1,\ldots,\lambda_r} a_{\lambda_1,\ldots,\lambda_r}^{n_1,\ldots,n_r} \prod_{i=1}^{r} h_{n_i}\left[\int_0^1 \varphi_{\lambda_i}(t)\, dx(t)\right]. \tag{4.63}$$

Assertions (4.62) and (4.63) together give the familiar (see [6]) expression for the multiple Wiener integral of $\tilde{f}_p$. It is seen that

$$I_p\colon\ \hat{L}^2(I^p) \to \bar{G}_p \tag{4.64}$$

is a one-to-one linear mapping, and with a little effort its usual properties can be verified:

$$\| I_p(\tilde{f}_p) \|_p = \sqrt{p!}\ \| \tilde{f}_p \|_{L^2(I^p)}. \tag{4.65}$$

The above discussion of multiple Wiener integrals and the derivation of the Cameron-Martin expansion holds also for a Gaussian generalized white noise process and can be deduced from our Theorem 4.3. This deduction of the Hida-Ikeda theorem [5] will be referred to later.

The somewhat abstract definition of multiple Wiener integrals as elements of suitable symmetric tensor products of Hilbert spaces goes back to ideas of Ito [6], if not to Segal [16], and has been known by Kakutani, who briefly mentions it in his Berkeley Symposium paper [8]. What we have done here is merely present these ideas in some detail in the general context of the theorems of this section.

5. Gaussian Processes in Linear Spaces and Their RKH Spaces

The importance of the RKHS $H(R)$ and its symmetric tensor products in the discussion of the homogeneous chaos of a Gaussian process has, we hope, already been made clear in the results of the last section. In these theorems the probability space Ω_x on which the random variables of the process are defined was not assumed to possess any particular structure or to be endowed with a topology. $H(R)$, however, is always a linear space (in fact, a Hilbert space) of functions on T. When Ω_x is a function space or a linear topological space of functions on T, it happens in the most interesting cases that $H(R)$

is a subset of Ω_x. This is so in the two most extensively studied classes of Gaussian processes: (*a*) abstract Wiener processes and (b) Gaussian measures on duals of nuclear spaces. The former is a class of Gaussian measures on separable Banach spaces recently introduced by L. Gross [4]. Typical among these are Gaussian measures defined on $C(T)$ where $C(T)$ is the Banach space (with sup norm) of real-valued functions on a compact metric space T. The Wiener process is, of course, the most celebrated example of these processes and the source from which all later research has flowed in this area. The latter (viz. (b)) are the so-called generalized processes, which include the stationary Gaussian white noise process (see [5]). For these processes it is possible to learn more about the measure from its RKHS. It is a curious fact that in so far as the measure is concerned, the topology of Ω_x is intrinsically of less importance than the relationship of Ω_x to $H(R)$. This point has been noted in [4]. One the other hand, $H(R)$ itself is, measure-theoretically speaking, a negligible set, a fact which seems to be well-known and accepted, although a general proof of such a statement does not seem to be available in the literature.

We investigate these questions in the present section. It will be assumed throughout that the covariance function is continuous and the mean functions of all Gaussian processes are zero. The following probabilistic set-up is general enough for many purposes. As before, we assume that T is a complete, separable metric space and set $\Omega_x = X$, where X satisfies the following conditions:

$$X \text{ is a real linear space of functions } x(t) \text{ on } T; \tag{5.1}$$

$$X \supset H(R). \tag{5.2}$$

Let $\mathscr{A}(X)$ be the σ-field generated by cylinder sets of the form

$$E = \{x \in X;\ [x(t_1), \ldots, x(t_n)] \in B\} \tag{5.3}$$

(B an n-dimensional Borel set), P_0 a Gaussian measure with covariance R, and $\bar{\mathscr{A}}(X)$ the completion of $\mathscr{A}(X)$ with respect to P_0.

Remark. Theorem 4.3 applies to $L^2(X, \bar{\mathscr{A}}(X), P_0)$.

Lemma 5.1. Let $T = \Phi$, a locally convex, real, linear topological space and let $X = \Phi^*$, the topological dual to Φ. Suppose μ

is a Gaussian measure on $(\Phi^*, \mathscr{A}(\Phi^*))$ whose covariance functional R is given by

$$R(\varphi, \psi) = \int_{\Phi^*} \langle x, \varphi\rangle\langle x, \psi\rangle\mu(dx) \tag{5.4}$$

where $\varphi, \psi \in \Phi$ and $\langle x, \varphi\rangle$ denotes the value of x at φ. Assume that

$$R(\varphi, \varphi) = 0 \quad \text{implies that} \quad \varphi = 0. \tag{5.5}$$

Then

$$H(R) \subset \Phi^*. \tag{5.6}$$

Proof. The details of the proof can be easily filled in. It is enough to observe that, by definition, $H(R)$ consists of linear functionals on Φ which can be shown to be continuous in the topology of Φ. Thus we have (5.6).

Let Φ be a nuclear space and μ a Gaussian measure on Φ^*, with covariance given by

$$R(\varphi, \psi) = \int_{-\infty}^{\infty} \varphi(t)\psi(t)\, dt. \tag{5.7}$$

In this case (Φ^*, μ) is called a white noise process. It is a stationary process in the sense described in [5]. It is easy to show then that

$$H(R) \cong L^2(\mathbb{R}^1, \text{Leb.}), \qquad (\mathbb{R}^1 = (-\infty, \infty)). \tag{5.8}$$

The homogeneous chaos and the Cameron-Martin expansion for the generalized Gaussian white noise process now follow as a consequence of Theorem 4.3. Furthermore, if we use Lemma 5.1 and the congruence (5.8), the definition of multiple Wiener integrals and the ensuing discussion given at the end of Section 4 apply also to the white noise process. This has been done in [5].

We now turn to the definition of an abstract Wiener space. It will be necessary to introduce some related definitions. We shall assume that the reader is familiar with the notion of cylinder-set measure on any locally convex, linear topological space L. (See, e.g., [4].) An equivalent concept is that of the weak distribution in L. Let L^* be the topological dual of L.

Definition 5.1. A weak distribution on L is an equivalence class of linear maps F from L^* to the linear space $M(\Omega, \mathscr{A}(\Omega), P)$

of random variables on some probability space $(\Omega, \mathscr{A}(\Omega), P)$ (the choice of which depends on F).

Definition 5.2. If $L = H$ is a separable Hilbert space, F is called a canonical normal distribution (or simply a canonical distribution) on H if to each h in H^* the real random variable $F(h)$ is normally distributed with mean 0 and variance $\| h \|^2$. (Here $\| h \|$ denotes H^*-norm). From this point on, F will always be a representative of the canonical distribution.

If f is a *tame* function on L, i.e., a function of the form

$$f(x) = \varphi[\langle y_1, x\rangle, \ldots, \langle y_n, x\rangle] \tag{5.9}$$

where $y_1, \ldots, y_n \in L^*$, φ is a Baire function of n real variables, and $\langle y, x\rangle$ is the value of y at x, then

$$\tilde{f}(\omega) = \varphi[F(y_1)(\omega), \ldots, F(y_n)(\omega)] \quad (\omega \in \Omega) \tag{5.10}$$

is a random variable on Ω having the same distribution as f under the weak distribution. We shall use Gross's notation to denote by $\tilde{f}$ the random variable (as described above) corresponding to any tame function on L.

Let H be a separable Hilbert space and let $\mathscr{F}$ be the family of all finite-dimensional projections on H. Note that $\| Px \|_1$ is a tame function if $P \in \mathscr{F}$.

Definition 5.3. A norm or seminorm $\| x \|_1$ on H is said to be measurable if for every $\varepsilon > 0$ there exists a projection $P_\varepsilon \in \mathscr{F}$ such that

$$\text{Prob}(\| Px \|_1^\sim > \varepsilon) < \varepsilon \tag{5.11}$$

for all $P \perp P_\varepsilon$ $(P \in \mathscr{F})$.

Let H be a separable Hilbert space, F the canonical distribution on H, and $\| x \|_1$ a measurable norm on H. Let B be the Banach space which is the completion of H with respect to $\| x \|_1$. Then we may identify B^*, the dual of B, with a subset of H^* and write

$$B^* \subset H^* = H \subset B. \tag{5.12}$$

Furthermore, the canonical distribution on H induces a weak distribution on B if we restrict the map F to B^*. Let μ be the cylinder-set measure on B determined by this weak distribution.

Gross has shown that if $\| x \|_1$ is a measurable norm on H and μ the cylinder-set measure on B induced by the canonical distribution on H, then μ extends to a countably additive Gaussian measure $(B, \mathscr{A}(B))$.

Definition 5.4. $(B, \bar{\mathscr{A}}(B), \mu)$ is called an abstract Wiener space or process.

Lemma 5.1 does not cover abstract Wiener spaces since, in general, B is not the dual of a Banach space. But condition (5.2) has been independently established in Corollary 1, ref. [10] and is stated here as

Lemma 5.2. If R is the covariance of μ, then

$$H(R) = H \subset B. \tag{5.13}$$

We are now in a position to state and prove a result, involving minimal assumptions, concerning the measure-theoretic negligibility of $H(R)$.

Theorem 5.1. Let m be the Gaussian probability measure on $(\mathbb{R}^T, \mathscr{A}(\mathbb{R}^T))$ determined by the consistent family of finite-dimensional normal distributions defined by the covariance R. If m^* denotes the outer measure corresponding to m, then

$$m^*(H(R)) = 0. \tag{5.14}$$

Proof. First, we observe a simple fact which suggests the definition of abstract Wiener space. Let H be an arbitrary, separable (infinite-dimensional) Hilbert space and let ν be the cylinder-set measure corresponding to the canonical weak distribution on H. Then ν is not countably additive. Secondly, it is clear that we may take as a representative of the canonical distribution

$$F : H^* \to M(\mathbb{R}^T, \mathscr{A}(\mathbb{R}^T), m) \tag{5.15}$$

given by

$$F[R(\cdot, t)](\omega) = \omega(t) \quad (t \in T, \omega \in \mathbb{R}^T). \tag{5.16}$$

It is easy to verify that the family of finite-dimensional distributions associated with the cylinder-set measure ν_R corresponding to F

coincides with the finite-dimensional family determined by the covariance R. From the first remark, it thus follows that the finite-dimensional family of normal distributions defined by R cannot determine a countably additive probability measure on $\mathscr{A}(H(R))$, the family of Borel sets of the Hilbert space $H(R)$. A somewhat picturesque way of expressing this is: for every covariance R, the corresponding finite-dimensional normal family is a white noise process in $H(R)$. To conclude from this that (5.14) holds, we show that the assumption $m^*(H(R)) > 0$ leads to a contradiction. Actually, we shall assume that

$$m^*(H(R)) = 1; \tag{5.17}$$

since it is not too difficult to reduce the general case to (5.17) (e.g., by using Problem 2*f*, page 71, in P. Halmos' *Measure Theory*). Let $\hat{\mathscr{A}}(H(R))$ be the σ-field of subsets of $H(R)$ of the form

$$A = H(R) \cap \tilde{A} \tag{5.18}$$

where $\tilde{A} \in \mathscr{A}(R^T)$.

Suppose A in (5.18) is a cylinder set

$$A = \{x \in H(R) : [x(t_1), \ldots, x(t_n)] \in B\},$$

B being an n-dimensional Borel set. But then

$$A = \{x \in H(R) : [(R(\cdot, t_1), x), \ldots, (R(\cdot, t_n), t_n), x)] \in B\}$$

where $R(\cdot, t_1), \ldots, R(\cdot, t_n)$ are regarded as elements of $H(R)$, and we identify $H^*(R)$ with $H(R)$. Hence $A \in \mathscr{A}(H(R))$, which implies that $\hat{\mathscr{A}}(H(R)) \subset \mathscr{A}(H(R))$. Conversely, $\mathscr{A}(H(R))$ being the family of Borel sets of the separable Hilbert space, $H(R)$ is also the minimal σ-field with respect to which all the continuous linear functionals are measurable. Let $h \in H(R)$. Then there exists a sequence h_n of finite linear combinations $\sum c_i^n R(\cdot, t_i^n)$ such that $\| h - h_n \| \to 0$. Hence for every x in $H(R)$,

$$(h, x) = \lim_{n\to\infty} (h_n, x) = \lim_{n\to\infty} \sum c_i^n x(t_i^n). \tag{5.19}$$

Equations (5.19) says that (h, x) is the pointwise limit of a sequence of functions of x which are measurable with respect to $\hat{\mathscr{A}}(H(R))$. It follows that (h, x) is $\hat{\mathscr{A}}(H(R))$-measurable, from which we have $\mathscr{A}(H(R)) \subset \hat{\mathscr{A}}(H(R))$. Hence

$$\hat{\mathscr{A}}(H(R)) = \mathscr{A}(H(R)). \tag{5.20}$$

Now define μ on $\hat{\mathscr{A}}(H(R))$ using (5.18) as follows:

$$\mu(A) = m(\tilde{A}). \tag{5.21}$$

From (5.17), (5.20), and (5.21) we see that μ is a well-defined, countably additive Gaussian measure on the family $\mathscr{A}(H(R))$ of the infinite-dimensional Hilbert space $H(R)$. Furthermore, the cylinder-set measure determined by μ coincides with that, *viz.*, ν_R, defined by the canonical distribution in $H(R)$. Since it has already been shown above that ν_R cannot be countably additive, we arrive at a contradiction. The theorem is thus proved.

We conclude the discussion of this section by stating some results proved in [10]. The proofs, which are not short and are not reproduced here, make use of the properties of measurable norms established by Gross [4]. The first result may be regarded as providing a necessary and sufficient criterion in terms of $H(R)$ for a separable Gaussian process $x(t)$ $(0 \leq t \leq 1)$ to have (a.s.) continuous sample paths.

Theorem 5.2. Let R be a continuous covariance on $[0, 1] \times [0, 1]$. Then the canonical normal distribution on $H(R)$ extends to a Gaussian (countably additive) measure on $\overline{H(R)}$, the closure of $H(R)$ in $C[0, 1]$ if and only if $\| x \|_1 = \sup_{0\leq t\leq 1} | x(t) |$ is a measurable norm on $H(R)$.

Let T be a compact metric space and let $C = C(T)$ be the Banach space of real-valued continuous functions with norm $\| x \|_1 = \sup_{t\in T} | x(t) |$. Let C_0 be any closed linear subspace of C (in particular, C_0 could be C itself) and $\mathscr{A}(C_0)$ the family of Borel subsets of C_0.

Theorem 5.3. Let μ be a Gaussian measure on $(C_0, \mathscr{A}(C_0))$ with covariance R. Then the following assertions are true.

$$H(R) \subset C_0; \tag{5.22}$$

if $\overline{H(R)}$ denotes the closure of $H(R)$ in C_0, then

$$\overline{H(R)} = \text{support } (\mu). \tag{5.23}$$

For $x_0 \in C_0$,

$$\mu\{x \in C_0 \colon \| x - x_0 \|_1 < \varepsilon\} > 0 \tag{5.24}$$

for every $\varepsilon > 0$ if and only if

$$x_0 \in \overline{H(R)}. \tag{5.25}$$

Theorem 5.4. Let B be an arbitrary separable Banach space and let μ be a Gaussian measure on $(B, \mathscr{A}(B))$ with continuous covariance functional. Then there exists a Hilbert space H congruent to the RKHS of μ such that

$$H \subset B \tag{5.26}$$

and

$$\bar{H} = \text{support } (\mu), \tag{5.27}$$

$\bar{H}$ being the closure of H in B.

6. A Zero-One Law

The results in this section and the next are generalizations of similar results obtained by Cameron and Graves for Wiener processes [3]. The proofs of these authors rely heavily on "Fourier-Hermite" analysis in Wiener space, whereas the proofs of the theorems presented here (which have been given in detail in [9]) bring to light yet another interesting relationship between a Gaussian measure and its RKHS. We revert to the probability space $(X, \bar{\mathscr{A}}(X), P_0)$ described by (5.1) and (5.2) at the beginning of the previous section.

Definition 6.1. A subset M of X is called an r-module if $x_1, x_2 \in M$ and r_1, r_2 are any rationals, then $r_1 x_1 + r_2 x_2 \in M$.

Theorem 6.1. If $M \in \bar{\mathscr{A}}(X)$ is an r-module which has positive P_0-measure, then

$$M \supset H(R). \tag{6.1}$$

The conclusion of this theorem is surprising and interesting in view of Theorem 5.1. Theorem 6.1 plays a key role in the derivation of the following zero-one law.

Theorem 6.2. If $M \in \bar{\mathscr{A}}(X)$ is an r-module, then

$$P_0(M) = 0 \quad \text{or} \quad 1. \tag{6.2}$$

The above results contain as special cases similar properties for abstract Wiener spaces and for Gaussian measures on $(\Phi^*, \mathscr{A}(\Phi^*))$ where Φ is nuclear.

7. Essentially Additive Functionals on Gaussian Probability Spaces

Let $(X, \bar{\mathscr{A}}(X), P_0)$ be the probability space of Section 6 and let F be any random variable on it. F is said to be a Gaussian additive functional if

$$F(x_1 + x_2) = F(x_1) + F(x_2) \tag{7.1}$$

for every x_1, x_2 in X. F is said to be essentially additive if there is a measurable additive functional F_0 such that

$$F(x) = F_0(x) \quad \text{a.s.} \quad P_0. \tag{7.2}$$

Theorem 7.1. Let $G \in L^2(X, \bar{\mathscr{A}}(X)P_0)$ be such that $E(G) = 0$ and

$$G(x + re_j) = G(x) + rG(e_j) \tag{7.3}$$

for each $x \in X$, all rational numbers r and for $j = 1, 2, \ldots,$ where $\{e_j\}_1^\infty$ is a fixed CONS in $H(R)$. Then, in the expansion for G given in Theorem 4.3 (or Theorem 4.1), only the term for $p = 1$ remains, all other terms vanish.

The proof of this result, which we indicate here, makes use of some facts about Gaussian processes taken from [9] and [15]. For convenience of reference we state them as a lemma. Let τ_m be the translation through a fixed element m of X, i.e., $\tau_m x = x + m \, (x \in X)$; and let $P_m = P_0 \tau_m^{-1}$. Then P_m is a Gaussian measure with mean function m and covariance R. We use the notation $P_m \equiv P_0$ to mean that the two measures are absolutely continuous with respect to each other.

Lemma 7.1. Let $m \in H(R)$. Then

$$P_m \equiv P_0 \quad \text{relative to} \quad \mathscr{A}(X). \tag{7.4}$$

Writing $\varrho_m(x) = ((dP_m)/(dP_0))(x)$, we have

$$\varrho_m(x) = \exp\left\{u_m(x) - \frac{1}{2} \| m \|^2\right\} \tag{7.5}$$

where $u_m(x)$ is the element in $L_1(X)$ which corresponds to m under the congruence between $L_1(X)$ and $H(R)$.

$$\varrho_m \in L^2(X, \bar{\mathscr{A}}(X), P_0) \tag{7.6}$$

$$(\varrho_{m'}, \varrho_{m_2})_{L^2(X, P_0)} = \exp\{(m', m_2)_{H(R)}\} \quad (m', m_2 \in H(R)). \tag{7.7}$$

$$\varrho\colon \quad H(R) \to L^2(X, \bar{\mathscr{A}}(X), P_0) \tag{7.8}$$

defined by (7.5) is continuous;

$$\text{The family } \{\varrho_m, m \in H(R)\} \text{ spans } L^2(X, \bar{\mathscr{A}}(X), P_0); \tag{7.9}$$

$F \in L^2(X, \bar{\mathscr{A}}(X), P_0)$ is almost surely a constant if and only if for all m in $H(R)$,

$$F \perp (\varrho_m - \varrho_0). \tag{7.10}$$

Proof of Theorem 7.1. We give only an outline of the proof. Setting

$$\hat{G}(m) = \int_X G(x)\varrho_m(x)P_0(dx) \tag{7.11}$$

and applying (7.3), (7.9), and (7.10) we show that

$$\hat{G}(m + m_0) = \hat{G}(m) + \hat{G}(m_0) \tag{7.12}$$

for every m and m_0 in $H(R)$ and that

$$\hat{G}(m) \text{ is a continuous function on } H(R). \tag{7.13}$$

Hence

$$\hat{G} \in H^*(R). \tag{7.14}$$

By the definition of $H(\Gamma_R)$ given in Section 4, it follows that

$$\hat{G} \in H(\Gamma_R), \tag{7.15}$$

so that

$$\hat{G}(m) = [\hat{G}(\cdot), \Gamma_R(\cdot, m)] \quad (m \in H(R)). \tag{7.16}$$

From Theorems 4.3 and 4.5 we know that there is a congruence, say, ψ, between $L^2(X, \bar{\mathscr{A}}(X), P_0)$ and $H(\Gamma_R)$. Hence from (7.16),

$$\hat{G}(m) = (\psi^{-1}\hat{G}, \varrho_m)_{L^2(X,P_0)}. \tag{7.17}$$

From (7.11) and (7.17) it follows that for all $m \in H(R)$,

$$(G, \varrho_m)_{L^2(X,P_0)} = (\psi^{-1}\hat{G}, \varrho_m)_{L^2(X,P_0)}. \tag{7.18}$$

Hence

$$G = \psi^{-1}\hat{G}. \tag{7.19}$$

From (7.19), (7.14), and the natural isomorphism between $H^*(R)$ and $H(R)$ under which an element $\hat{m}$ of $H(R)$ corresponds to $\hat{G}$, it follows that

$$G(x) = (\hat{m}, x)^{\sim} \quad \text{a.s.} \tag{7.20}$$

where, following Gross's notation, we write $(\hat{m}, x)^{\sim}$ to denote the element of $L_1(X)$ which corresponds to $\hat{m}$ in $H(R)$. This proves the assertion of Theorem 7.1. The following result is an immediate consequence.

Theorem 7.2. Let G be an essentially additive, measurable functional belonging to $L^2(X, \bar{\mathscr{A}}(X), P_0)$ such that $EG = 0$. Then there exists an $\hat{m}$ in $H(R)$ such that

$$G(x) = (\hat{m}, x)^{\sim} \quad \text{a.s.} \tag{7.21}$$

If we take $(X, \bar{\mathscr{A}}(X), P_0)$ to be an abstract Wiener space $(B, \bar{\mathscr{A}}(B), \mu)$, then the following somewhat stronger result can be proved.

Theorem 7.3. Let G be an essentially additive, measurable functional on the abstract Wiener space $(B, \bar{\mathscr{A}}(B), \mu)$ whose associated RKHS is H. Then there exists a unique element $g \in H$ such that

$$G(x) = (g, x)^{\sim} \quad \text{for a.e.} \quad x \text{ in } B. \tag{7.22}$$

The above result has also been obtained directly from the argument of [3] by Kuelbs in recent paper [11].

References

1. A. I. Plessner, Fundamental ideas in the theory of Hermitian operators, *Uspehi Mat. Nauk.* **1** (1946) No. 1, pp. 192–216.
2. Cameron, R. H., and Martin, W. T., The orthogonal development of non-linear functionals in series of Fourier-Hermite functionals, *Ann. Math.* **48** (1947).
3. Cameron, R. H., and Graves, R. E., Additive functionals on a space of continuous functions, *Trans. Am. Math. Soc.* **70** (1951).
4. Gross, L., Abstract Wiener spaces, *Proc. 5th Berkeley Symp.*, Vol. II, Pt. 1 (1965).
5. Hida, T., and Ikeda, N., Analysis on Hilbert space with reproducing kernel arising from multiple Wiener integral, *Proc. 5th Berkeley Symp.*, Vol. II, Pt. 1 (1965).
6. Ito, K., Multiple Wiener integral, *J. Math. Soc. Japan* **3** (1951).
7. Ito, K., Spectral type of the shift transformation of differential processes with stationary increments, *Trans. Am. Math. Soc.* **81** (1956).
8. Kakutani, S., Spectral analysis of stationary Gaussian processes, *Proc. 4th Berkeley Symp.* Vol. 2 (1961).
9. Kallianpur, G., Zero-one laws for Gaussian processes, to appear in *Trans. Am. Math. Soc.*
10. Kallianpur, G., Abstract Wiener processes and their reproducing kernel Hilbert spaces, to appear in *Zeits. für Wahrscheinlichkeitstheorie.*
11. Kuelbs, J., Abstract Wiener spaces and applications to analysis, to appear in the *Pacific Journal of Mathematics.*
12. LePage, R., An isometry related to unbiased estimation, Chapter V, Unpublished Thesis, U. of Minnesota (1967).
13. Maurin, K., Methods of Hilbert spaces, Monografie Mat. **45**, Polish Scientific Publishers, Warsaw, 1967.
14. Neveu, J., Processus alèatoires Gaussiens, Sèminaire de mathèmatiques supèrieures (1968).
15. Parzen, E., Statistical inference on time series by Hilbert space methods, (Tech. Rep. No. 23) I. Department of Statistics, Stanford U. (1959).
16. Segal, I. E., Tensor algebras over Hilbert spaces, *Trans. Am. Math. Soc.* **81** (1956).
17. Segal, I. E., Distributions in Hilbert space and canonical systems of operators, *Trans. Am. Math. Soc.* **88** (1958).

17

An RKHS Approach to Detection and Estimation Problems—Part III: Generalized Innovations Representations and a Likelihood-Ratio Formula

THOMAS KAILATH, FELLOW, IEEE, AND D. DUTTWEILER, MEMBER, IEEE

Abstract—**The concept of a white Gaussian noise (WGN) innovations process has been used in a number of detection and estimation problems. However, there is fundamentally no special reason why WGN should be preferred over any other process, say, for example, an *n*th-order stationary autoregressive process. In this paper, we show that by working with the proper metric, any Gaussian process can be used as the innovations process. The proper metric is that of the associated reproducing kernel Hilbert space. This is not unexpected, but what is unexpected is that in this metric some basic concepts, like that of a causal operator and the distinction between a causal and a Volterra operator, have to be carefully reexamined and defined more precisely and more generally.**

It is shown that if the problem of deciding between two Gaussian processes is nonsingular, then there exists a causal (properly defined) and causally invertible transformation between them. Thus either process can be regarded as a generalized innovations process. As an application, it is shown that the likelihood ratio (LR) for two arbitrary Gaussian processes can, when it exists, be written in the same form as the LR for a known signal in colored Gaussian noise. This generalizes a similar result obtained earlier for white noise.

The methods of Gohberg and Krein, as specialized to reproducing kernel spaces, are heavily used.

I. Introduction

In SEVERAL previous papers, see, e.g., [1]–[3], and the references therein, we have shown the usefulness of representing a Gaussian random process as the response of a causal and causally invertible linear filter to a white Gaussian noise (WGN) input. The point of this so-called "innovations representation" is that in many problems WGN, because of its delta-function covariance, is very easy to work with. In particular, it is usually much simpler to handle than the original process. However, there may be problems in which some other process is "simpler" and more natural. For example, in working with stationary processes with rational spectral densities, the so-called Ornstein–Uhlenbeck process, with spectrum

$$S(\omega) = \frac{1}{\alpha^2 + \omega^2}, \qquad -\infty < \omega < \infty \tag{1a}$$

or covariance function

$$R(t) = (2\alpha)^{-1} \exp[-\alpha|t|] \tag{1b}$$

may be easier to work with than WGN or its integrated version, the Wiener process. In this paper we shall solve a generalized innovations representation problem: given processes $\{x(t), t \in I\}$ and $\{x_0(t), t \in I\}$ with covariance functions R and R_0 we shall show how to relate x and x_0 by a causal and causally invertible linear transformation. Our solution will rely heavily on some reproducing kernel Hilbert space (RKHS) concepts and on some results of Gohberg and Krein [4] on operator factorization.

As one application of these generalized innovations representations we shall derive a generalization of the "estimator-correlator" type of likelihood ratio (LR) formula that was obtained in [2] for discrimination between a Gaussian random process and white Gaussian noise; here we shall be able to remove the restriction that one of the processes be white. We feel that, as in the white-noise case,

Manuscript received August 31, 1971; revised March 20, 1972. This work was supported in part by the Applied Mathematics Division, Air Force Office of Scientific Research under Contract AF 44-620-69-C-0101 and in part by the Joint Services Electronics Program under Contract N-00014-67-A-0112-0044.

T. Kailath is with the Department of Electrical Engineering, Stanford University, Stanford, Calif. 94305.
D. Duttweiler is with the Bell Telephone Laboratories, Holmdel, N.J.

the structural information on the random processes provided by their generalized innovations representations will lead to other applications as well.

In this connection we should remark that one method of relating x and x_0 is via innovations representations of x and x_0 in terms of WGN or simple variants of WGN. When such WGN representations are known, this route is probably the simplest in a computational sense, especially as several efficient numerical algorithms are available for determining WGN innovations representations [1]–[3]. However, there are processes for which such representations may not be known or may not exist and therefore, at least for theoretical and aesthetic reasons, it is of interest to discuss the problem of generalized innovations representations.

After these cautionary remarks, we can return to this generalized problem of relating two Gaussian processes x and x_0. For convenience of discussion, we shall assume that the basic process is x_0 and we shall state our results in terms of its RKHS $H(R_0)$. However, it will be clear how to recast the statements in terms of x and $H(R)$. When $H(R_0)$ satisfies a rather strong condition, gaplessness, the generalization of white-noise results is easier. This condition cannot be completely defined without introducing considerable detail. For now we will merely note that $H(R_0)$ is not gapless if I is a discrete set, or if I is an interval and the whitening filter for the covariance $R_0(t,s)$, $(t,s) \in I$, needs initial-condition random variables. We shall assume gaplessness in this introductory section and treat the more general case in later sections.

Our main result will be the following theorem, which generalizes similar results described in [1] and [2] for white noise. It is actually a special case of certain more general results that will be established in Section III.

Theorem 1: Assume $H(R_0)$ is gapless and that $R(t,s) = R_0(t,s) + K(t,s)$ where (in operator notation) $I + K > 0$ and K is Hilbert-Schmidt. Then the following is true.

a) The Wiener–Hopf equation $h + \{hK\}_+ - \{K\}_+ = 0$, where $\{A\}_+$ equals the causal part of the operator A, has a unique causal and Hilbert-Schmidt solution h.

b) $I - h$ is invertible and its inverse is of the form $I + k$ where k is causal and Hilbert-Schmidt.

c) If $x_0(\cdot)$ has covariance $R_0(t,s)$, the process $x(\cdot)$ defined by

$$x = (I + k)x_0$$

has covariance $R_0(t,s) + K(t,s) = R(t,s)$ and is causally and causally invertibly related to $x_0(\cdot)$.

d) Equivalently, $I + k$ "factors" the covariance $I + K$, i.e.,

$$(I + K) = (I + k)(I + k^*),$$

where k^* represents the adjoint of the operator k.

It is to be emphasized that the operators I, K, h, and k in this theorem (and all operators to be considered in the remainder of this section) are operators on $H(R_0)$. The operator K is related to the kernel $K(t,s)$ appearing in the decomposition $R(t,s) = R_0(t,s) + K(t,s)$ by

$$(K\phi)(t) = \langle K(t,s),\phi(s)\rangle_{H(R_0)}.$$

Note that the defining kernel for the identity operator I in $H(R_0)$ is $R_0(t,s)$ (i.e., $(I\phi)(t) = \phi(t) = \langle R_0(t,s),\phi(s)\rangle_{H(R_0)}$) and not the delta function as it is in $L^2[0,1]$.

In Section III we shall also present a slightly more general version of the following representation theorem, which should be compared with [2, theorem 2].

Theorem 2: Assume $H(R_0)$ is gapless. A zero-mean random process $x(\cdot)$ has a covariance of the form $R_0(t,s) + K(t,s)$, where K is Hilbert-Schmidt and $I + K > 0$, if and only if it is expressible as

$$x(t) = z(t) + n(t), \qquad t \in I,$$

where

a) $n(\cdot)$ has zero mean and covariance $R_0(t,s)$;
b) $E[z(\cdot)] \equiv 0$;
c) $\operatorname{tr}\{E[z(t)z(s)]\} < \infty$;
d) $E[z(t) \mid n(\tau), \tau \in I] = E[z(t) \mid n(\tau), \tau \in I \text{ and } \tau \le t]$.

Theorem 1 can now be given [5] a generalized innovations interpretation. Let

$$x(t) = z(t) + n(t),$$

where $z(\cdot)$ and $n(\cdot)$ satisfy the conditions of Theorem 2 and let

$$K(t,s) = E[z(t)z(s) + z(t)n(s) + z(s)n(t)].$$

By Theorem 2, K will be Hilbert-Schmidt and $I + K$ will be strictly positive. Then the generalized innovations process can be defined by

$$x_0(t) = x(t) - \hat{z}(t),$$

where

$$\hat{z}(t) = (hx)(t)$$

and h is the unique Hilbert-Schmidt solution of the Wiener–Hopf equation

$$h + \{hK\}_+ - \{K\}_+ = 0.$$

We note that, in contrast to [1]–[3] where $n(\cdot)$ was white noise, $\hat{z}(\cdot)$ is not the causal linear least squares estimate of $z(t)$. It is a causal estimate, but the sense in which it is optimum depends on $R_0(t,s)$. For example, if $R_0(t,s) = t \wedge s$,

$$\hat{z}(t) \triangleq (hx)(t) = \int_0^t \hat{\dot{z}}(s)\, ds,$$

where $\hat{\dot{z}}(s)$ is the causal linear least squares estimate of $\dot{z}(s)$.

Before going on, let us observe that in this generalized nonwhite-noise representation problem, certain concepts like causality, the causal component of an operator, and the difference between causal and Volterra operators, have to be carefully defined. A direct extension from the white-noise case can lead to errors. For instance, in the white-noise problem, a causal filter (operator) has an impulse response (kernel) that satisfies the "causality" constraint

$$h(t,s) = 0, \qquad s > t. \tag{2}$$

However, when the Ornstein–Uhlenbeck process is used as the basic process, it turns out that the impulse response (kernel) of a causal filter (operator) must obey

$$h(t,s) = h(t,t)e^{-\alpha(s-t)}, \qquad s > t. \tag{3}$$

Similarly, if the Wiener process is used as the basic process, a causal impulse response must obey

$$h(t,s) = h(t,t), \qquad s > t. \tag{4}$$

We shall see in Section II how these requirements arise from the formal definition of causality, but some heuristic words of explanation may be useful. To begin with, a causal filter should be one that gives zero weight to the future, i.e., to inputs occurring at times s greater than the present time t. It would seem that this is precisely what the usual constraint (2) ensures. However, some reflection shows that in general there can be two components of the future, a completely "predictable" future and a completely "unpredictable" future. A causal filter should give zero weight to the unpredictable future and unit weight to the predictable future. For example, having observed a Wiener process up to time t, the predictable part of the future can be taken as the least squares prediction of the future given the past. As is well known, the best prediction is equal to the most recently observed value; the weighting $h(t,t)$ for $s > t$ picks up this predictable component of the future. Similarly for the Ornstein–Uhlenbeck process, the best prediction is, in an obvious notation,

$$\hat{x}(t + \lambda) = e^{-\alpha\lambda}x(t), \qquad \lambda > 0.$$

The condition (3) on the impulse response reflects the unit-weighting of this predicted part of the future. Of course, for white noise, the future is completely uncertain and the whole future is given zero weight, as in (2).

One might wonder why the predictable part is defined in terms of least squares prediction. There is perhaps no completely satisfactory answer, but the basic reason is that the least squares criterion is intimately related to the geometry of the Hilbert-space framework in which linear Gaussian problems are usually best studied. The basic space is the Hilbert space of random variables with the covariance norm. However, it is often convenient to work with isometric spaces of functions on the real line. For white noise, the proper space is $L_2(I)$, the space of square-integrable functions on the interval I. For other processes, a good space is often the RKHS associated with the covariance function of the process. The white-noise case can be regarded either as a limiting case of a pseudo-RKHS associated with the delta covariance, or more rigorously as an RKHS of functionals [6].

It is well known in operator theory that certain concepts have to be distinguished in L_2 and non-L_2 spaces. For example, while in L_2 the concepts of a causal and a Volterra operator agree, this may not be so in an arbitrary Hilbert space. For example, in the space R^n, a causal operator is represented by a lower triangular matrix; however, this operator will also be Volterra only if all diagonal entries in the matrix are zero.

In Section II we shall show how to define causal operators in an RKHS. The concepts of projection chains and operator integrals will be introduced and used to define the operation of extracting causal components. In Section III we shall give Gohberg and Krein's operator integral solution to the generalized covariance-factorization problem and also give an alternate solution based on a Wiener–Hopf equation. Our discussions will rely heavily on the work of Gohberg and Krein. We should note that their work is couched in terms of a general Hilbert space, not necessarily an RKHS. However, all their explicit examples are in terms of the space L_2. There are several other explicit examples of RKHS with many interesting features that have aided us in following the abstract presentation of Gohberg and Krein [4]. Moreover, for applications to detection theory, the reproducing kernel Hilbert spaces will be the most appropriate.

We should also note that generalized causal operators in our sense (i.e., essentially the sense of Gohberg and Krein) have been introduced in several recent papers on generalized system and stability theory, e.g., Porter [7], [8], Saeks [9], [10], and Willems [11], [12], among others. However, again in all these papers the only explicit examples are essentially for the spaces $L_2[I]$, where causal operators are very familiar. The simple but unexpected examples we give of nontrivially different causal operators in various RKHS may perhaps be helpful in gaining a better appreciation of the generalizations envisaged in [7]–[12].

In Section IV we present a likelihood ratio formula for two arbitrary Gaussian processes. Readers most interested in this application can turn directly to that section to gain further motivation for the rather abstract discussions of Sections II and III.

II. Some Operators in $H(R)$

A kernel $A(t,s)$, $(t,s) \in I$, defines a bounded linear operator A by the relation

$$(Af)(t) = \langle A(t,s), f(s)\rangle_{H(R)}$$

if it is sufficiently well behaved. An obvious necessary condition on $A(t,s)$ for such a relation to be possible is that $A(t,\cdot) \in H(R)$ for all $t \in I$. A sufficient condition will be given in Lemma 2. Necessary and sufficient conditions can be formulated (see Aronszajn [14, pp. 373–374]), but we shall not need them here.

The following lemma guarantees that every bounded linear operator has a defining kernel belonging to $H(R)$. An analogous lemma does not exist for L_2 spaces.

Lemma 1: Let A be a bounded linear operator on $H(R)$. Define the vector $r_t(\cdot) \in H(R)$ by $r_t(\cdot) = R(t,\cdot)$ and define the kernel $A(t,s)$, $(t,s) \in I$, by

$$A(t,s) = (A^* r_t)(s),$$

where A^* is the adjoint of the operator A. Then, for all $f(\cdot) \in H(R)$ and all $t \in I$

$$(Af)(t) = \langle A(t,s), f(s)\rangle_{H(R)}.$$

Proof: We have

$$\langle A(t,s),f(s)\rangle_{H(R)}$$
$$= \langle (A^*r_t)(s),f(s)\rangle_{H(R)} = \langle r_t(s),(Af)(s)\rangle_{H(R)}$$
$$= \langle R(t,s),(Af)(s)\rangle_{H(R)} = (Af)(t).$$

As already tacitly done in the introduction and in the preceding, we use the same symbol for an operator and its defining kernel. This convention will greatly simplify later discussions and should cause no confusion. The one exception will be the use of the symbol I for the identity operator even though its defining kernel is $R(t,s)$.

It is easy to show that the defining kernel $A^*(t,s)$ for the operator adjoint to a bounded linear operator A is equal to $A(s,t)$. Sums and products of operators are defined in the usual way.

The next lemma, in addition to asserting that Hilbert–Schmidt operators correspond to kernels in $H(R) \otimes H(R)$, the direct product of $H(R)$ with itself (see [14]), also gives a sufficient condition for a kernel to define a bounded linear operator.

Lemma 2: If A is a Hilbert–Schmidt operator on $H(R)$, then its defining kernel exists in $H(R) \otimes H(R)$ and the Hilbert–Schmidt norm of A, $\|A\|$, equals $\|A(t,s)\|_{H(R)\otimes H(R)}$. If a kernel $A(t,s)$ exists in $H(R) \otimes H(R)$, then it defines a Hilbert–Schmidt operator A and $\|A\| = \|A(t,s)\|_{H(R)\otimes H(R)}$.

A proof of this lemma is given in [14].

A. Causal Operators

For all $\lambda \in I$ let I_λ denote the set $\{t \in I : t \le \lambda\}$. We shall define an operator A in $H(R)$ to be causal if for all $\lambda \in I$

$$f_1(t) = f_2(t), \qquad t \in I_\lambda$$

implies

$$(Af_1)(t) = (Af_2)(t), \qquad t \in I_\lambda.$$

Equivalently, A is causal if for all $\lambda \in I$

$$f(t) = 0, \qquad t \in I_\lambda$$

implies

$$(Af)(t) = 0, \qquad t \in I_\lambda.$$

This definition of causality is the same as the usual definition of causality of an operator in $L_2[0,1]$. An operator A in $L_2[0,1]$ is causal if and only if its defining kernel $A(t,s)$ satisfies $A(t,s) = 0$ (a.e.) for $s > t$. In $H(R)$ the characterization of kernels defining causal operators is not as easy, as is shown by the next two examples. In all examples

$$I = [0,T].$$

Example 1—Wiener Process $R_1(t,s) = \min(t,s) = t \wedge s$: An operator A is related to its defining kernel $A(t,s)$ through the relation

$$(Af)(t) = \langle A(t,s),f(s)\rangle_{H(R_1)} = \int_0^T \left(\frac{\partial}{\partial s} A(t,s)\right) \dot{f}(s)\, ds.$$

If the operator A is to be causal, then we must have $\partial A(t,s)/\partial s = 0$ for $t < s \le T$ or, equivalently, $A(t,s) = A(t,t)$ for $t < s \le T$.

Example 2—Ornstein–Uhlenbeck Process $R_2(t,s) = (2\alpha)^{-1} \cdot \exp[-\alpha|t-s|]$: An argument similar to the preceding shows that an operator A is causal if and only if

$$A(t,s) = A(t,t)e^{-\alpha(s-t)}, \qquad t < s \le T.$$

B. Projection Operators and Truncation

It is convenient to develop another definition of causal operators in terms of projection operators. For all $\lambda \in I$ let $H_\lambda(R)$ denote the subspace of $H(R)$ spanned by $\{R(t,\cdot)\}_{t\in I_\lambda}$ and let $H_\lambda^\perp(R)$ denote the subspace of $H(R)$ that is orthogonal to $H_\lambda(R)$. By the projection theorem every vector $f \in H(R)$ has a unique decomposition

$$f = f_\lambda + f_\lambda^\perp,$$

where $f_\lambda \in H_\lambda(R)$ and $f_\lambda^\perp \in H_\lambda^\perp(R)$. Define the operators P_λ and $P_\lambda^\perp$ by

$$P_\lambda f = f_\lambda$$

and

$$P_\lambda^\perp f = f_\lambda^\perp = f - f_\lambda = (I - P_\lambda)f.$$

P_λ and $P_\lambda^\perp$ are called *projection operators* since they project a vector onto a linear subspace. $(P_\lambda f)(\cdot)$ will be called the *truncation* in $H(R)$ of $f(\cdot)$ at λ. P_λ and $P_\lambda^\perp$ are both bounded self-adjoint linear operators with unity norm.

In an L_2 Hilbert space

$$(P_\lambda f)(t) = \begin{cases} f(t), & t \le \lambda \\ 0, & t > \lambda \end{cases}$$
$$= f(t)\mathbf{1}(\lambda - t), \qquad \text{say.}$$

In an RKHS the truncation of a function is not as easily specified.

The following lemmas, whose easy proofs are given in [5], and theorem will be quite useful in determining the truncation $(P_\lambda f)(\cdot)$ of a function $f(\cdot)$ in $H(R)$.

Lemma 3: Let $f \in H(R)$ and $\lambda \in I$. Then, $(P_\lambda f)(t) = f(t)$ for all $t \in I_\lambda$.

Lemma 4: Let f_1 and $f_2 \in H(R)$ and let $\lambda \in I$. If $f_1(t) = f_2(t)$ for all $t \in I_\lambda$, then $P_\lambda f_1 = P_\lambda f_2$.

Lemma 5: Let $f,g \in H(R)$ and $\lambda \in I$. If $P_\lambda f \ne g$ and $g(t) = f(t)$ for all $t \in I_\lambda$, then $\|g\|_{H(R)} > \|P_\lambda f\|_{H(R)}$.

Theorem 3 (Restriction Theorem): Let $f \in H(R)$ and $\lambda \in I$. If $g \in H(R)$ satisfies a) $g(t) = f(t)$ for all $t \in I_\lambda$, and b) $\|g\|_{H(R)}^2 \le \|h\|_{H(R)}^2$ for any function $h \in H(R)$ such that $h(t) = g(t)$, $t \in I_\lambda$, then $g = P_\lambda f$.

Proof: This theorem follows directly from Lemmas 3 and 5. The reader may find it useful to compare this proof with the proof of [6, theorem 7].

We will give some examples of the application of this theorem. The covariances in the last two examples are not of much interest in their own right, but will be useful later to illustrate a concept.

Example 1 (cont'd.)—Wiener Process: The norm of a function $f \in H(R_1)$ is given by (see, e.g., [6])

$$\|f\|_{H(R_1)}^2 = \int_0^T \dot{f}^2(t)\, dt.$$

For any $\lambda \in I$, $(P_\lambda f)(t)$ must equal $f(t)$ for $t \in I_\lambda = [0,\lambda]$. The minimum-norm condition requires $d/dt(P_\lambda f)(t) = 0$

for $t \in (\lambda,T]$. Therefore

$$(P_\lambda f)(t) = \begin{cases} f(t), & 0 \le t \le \lambda \\ f(\lambda), & \lambda < t \le T \end{cases}$$
$$= f(t \wedge \lambda).$$

Example 2 (cont'd.)—Ornstein–Uhlenbeck Process: The norm of a function $f \in H(R_2)$ is given by (see, e.g., [6])

$$\|f\|_{H(R_2)}^2 = \int_0^T (\dot{f}(t) + \alpha f(t))^2 \, dt + 2\alpha f^2(0).$$

To satisfy the minimum-norm requirement, we must have for any $\lambda \in I$

$$\frac{d}{dt}(P_\lambda f) + \alpha(P_\lambda f)(t) = 0, \qquad t \in (\lambda,T].$$

Since functions in $H(R_2)$ are absolutely continuous,

$$(P_\lambda f)(\lambda+) = (P_\lambda f)(\lambda) = f(\lambda).$$

Solving the differential equation with this initial condition and remembering that $(P_\lambda f)(t)$ must equal $f(t)$ for $t \in [0,\lambda]$, we find

$$(P_\lambda f)(t) = \begin{cases} f(t), & 0 \le t \le \lambda \\ f(\lambda)e^{-\alpha(t-\lambda)}, & \lambda < t \le T \end{cases}$$
$$= f(t \wedge \lambda)e^{-\alpha(t-\lambda)1(t-\lambda)}.$$

Example 3—Triangular Covariance $R_3(t,s) = 1 - |t - s|$, $0 \le t, s \le T \le 1$: The norm is given by [14]

$$\|f\|_{H(R_3)}^2 = \frac{1}{2}\frac{(f(0) + f(T))^2}{2 - T} + \tfrac{1}{2}\int_0^T \dot{f}^2(t) \, dt.$$

The calculus of variations must be used to find $P_\lambda f$. The final result (see [5, ch. 3] for the algebraic details) is

$$(P_\lambda f)(t) = \begin{cases} f(t), & 0 \le t \le \lambda \\ f(\lambda) - \dfrac{t - \lambda}{2 - \lambda}(f(0) + f(\lambda)), & \lambda < t \le T \end{cases}.$$

Example 4—Free Wiener Process: $R_4 = 1 + t \wedge s$: Using essentially the same arguments as in Example 3, it is easy to show that for all $\lambda \in [0,1]$

$$(P_\lambda f)(t) = f(t \wedge \lambda).$$

Example 5—Free Doubly Integrated Wiener Process: Let

$$R_5(t,s) = 1 + \int_0^t \int_0^s \left(1 + \int_0^u \int_0^v (1 + \xi \wedge \tau) \, d\xi \, d\tau\right) du \, dv.$$

A function $f(t)$ exists in $H(R_5)$ if its second derivative exists and is absolutely continuous and

$$\|f\|_{H(R_5)}^2 = \int_0^T \dddot{f}^2(t) \, dt + \ddot{f}^2(0) + \dot{f}^2(0) + f^2(0) < \infty.$$

We can show that

$$(P_0 f)(t) = f(0), \qquad 0 \le t \le T,$$

and for $\lambda \in (0,T]$

$$(P_\lambda f)(t) = \begin{cases} f(t), & 0 < t \le \lambda \\ f(\lambda) + (t - \lambda)\dot{f}(\lambda) + \dfrac{(t - \lambda)^2}{2}\ddot{f}(\lambda), & \lambda < t \le T \end{cases}.$$

C. Further Discussion of Causal Operators

The following lemma gives an alternative definition of causal operators.

Lemma 6: A bounded linear operator A is causal if and only if

$$P_\lambda A = P_\lambda A P_\lambda, \qquad \forall \lambda \in I.$$

Proof: Assume A is causal and let $\lambda \in I$ and $f \in H(R)$. By Lemma 3 we have

$$(f - P_\lambda f)(t) = 0, \qquad t \in I_\lambda.$$

Since A is causal,

$$(A(f - P_\lambda f))(t) = 0, \qquad t \in I_\lambda.$$

Using Lemma 4 we have

$$P_\lambda A(f - P_\lambda f) = 0$$

or since f was arbitrary,

$$P_\lambda A = P_\lambda A P_\lambda.$$

Assume now that $P_\lambda A = P_\lambda A P_\lambda$ for all $\lambda \in I$. Let $f \in H(R)$ and satisfy $f(t) = 0$, $t \in I_\lambda$. We have for all $t \in I_\lambda$

$$(Af)(t) = (P_\lambda Af)(t) \qquad \text{(Lemma 3)}$$
$$= (P_\lambda A P_\lambda f)(t)$$
$$= 0 \qquad \text{(Lemma 4)}.$$

There are two ways in which the concept of an anticausal operator on $L_2[0,T]$ might be abstracted for operators on $H(R)$. The first, which is perhaps the abstraction most natural for engineers, is to define an operator A to be anticausal if and only if for all $\lambda \in I$

$$f_1(t) = f_2(t), \; t \in I - I_\lambda = \{t : t \in I, t > \lambda\}$$

implies

$$(Af_1)(t) = (Af_2)(t), \qquad t \in I - I_\lambda$$

The other possible definition, which is the one that we will use, is that A is anticausal if and only if A^* is causal. Unfortunately, although these definitions are equivalent on $L_2(I)$, they are not in general on $H(R)$. The underlying reason for the difference is that in general the space that $P_\lambda^\perp$ projects onto is not equal to the space spanned by $\{R(t_i), t \in I - I_\lambda\}$, although it is for the special case $R(t,s) = \delta(t - s)$.

Having defined H to be anticausal if and only if A^* is causal, we note that equivalently A is anticausal if and only if for all $\lambda \in I$

$$P_\lambda^\perp A = P_\lambda^\perp A P_\lambda^\perp.$$

By substituting $P_\lambda^\perp = I - P_\lambda$ into this equation, still another definition of an anticausal operator is obtained: A is anticausal if and only if

$$AP_\lambda = P_\lambda A P_\lambda, \qquad \forall \lambda \in I.$$

We note that products (and, trivially, sums) of causal operators are causal. This fact is obvious from the original definition of causal operators. It is interesting to give a proof based on the projection-operator definition of causality. If A_1 and A_2 are any two causal operators, we have for any $\lambda \in I$

$$P_\lambda A_1 A_2 = P_\lambda A_1 P_\lambda A_2 = P_\lambda A_1 P_\lambda A_2 P_\lambda = P_\lambda A_1 A_2 P_\lambda.$$

D. Chains of Projection Operators

Let Q_1 and Q_2 be any two projection operators on a Hilbert space $\mathscr{H}$. Q_1 is said to be less than or equal to Q_2 ($Q_1 \leq Q_2$) if

$$Q_1 = Q_1 Q_2 = Q_2 Q_1$$

and strictly less than Q_2 ($Q_1 < Q_2$) if $Q_1 \leq Q_2$ and $Q_1 \neq Q_2$. Equivalently, $Q_1 \leq Q_2$ ($Q_1 < Q_2$) if the linear subspace onto which Q_1 projects is (properly) contained in the linear subspace onto which Q_2 projects.

Let $\mathscr{P} = \{P_\lambda\}_{\lambda \in I}$ and let $P^{(1)}$ and $P^{(2)}$ be any two distinct members of $\mathscr{P}$. Clearly, $P^{(1)} < P^{(2)}$ or $P^{(2)} < P^{(1)}$. Not all sets of projection operators have such a property. Sets having such a property are said to be *ordered*. The set of projection operators $\mathscr{P}^\perp = \{P_\lambda^\perp\}_{\lambda \in I}$ is also ordered.

Denote by $\bar{\mathscr{P}}$ the set of operators that are elements of $\mathscr{P}$ or else are strong limits of operators from $\mathscr{P}$. $\bar{\mathscr{P}}$ will be closed in the sense of strong convergence. In other words, limits of strongly converging operators from $\bar{\mathscr{P}}$ will themselves be in $\bar{\mathscr{P}}$. It can be shown that the elements of $\bar{\mathscr{P}}$ are projection operators and that $\bar{\mathscr{P}}$ is ordered.

$\bar{\mathscr{P}}$ need not contain the identity and null operators. We will let $\bar{\mathscr{P}}_B = \bar{\mathscr{P}} \cup \{0\} \cup \{I\}$. A set of projection operators such as $\bar{\mathscr{P}}_B$ that contains the identity and null vectors is said to be bordered. A set of projection operators that is ordered, closed in the sense of strong convergence, and bordered is a *chain*.

A pair of operators (P^-, P^+) from $\bar{\mathscr{P}}_B$ forms a *gap* if $P^+ > P^-$ and there does not exist another operator $P \in \bar{\mathscr{P}}_B$ such that $P^- < P < P^+$. $\bar{\mathscr{P}}_B$ (or $H(R_0)$) will be said to be gapless if no two operators from $\bar{\mathscr{P}}_B$ form a gap. By the linear subspace onto which a gap projects we will mean the linear subspace onto which the projection operator $\Delta P = P^+ - P^-$ projects. The *dimension of a gap* is the dimension of the linear subspace onto which it projects. Two gaps of a chain must project on to orthogonal subspaces. Since we have assumed that $H(R_0)$ is separable, the chain $\bar{\mathscr{P}}_B$ must have a countable number of gaps and each gap must be of countable dimension. Such chains can always be imbedded in a larger chain having only one-dimensional gaps. For instance, if the chain $\bar{\mathscr{P}}_B$ has one gap (P^-, P^+) of dimension N with $2 \leq N \leq \infty$, the chain $\bar{\mathscr{P}}_B \cup \{P^{(n)}\}_{n=1}^{N-1}$ where

$$P^{(n)}(t,s) = P^-(t,s) + \sum_{k=1}^{n} e_k(t) e_k(s)$$

and $\{e_k\}_{k=1}^N$ is any orthonormal set of vectors spanning the linear subspace of the gap (P^-, P^+) will have only one-dimensional gaps. A chain having only one-dimensional gaps is a *maximal chain*. We will let $\mathscr{P}_M$ denote any maximal chain containing $\bar{\mathscr{P}}_B$.

The following examples should help clarify these ideas.

Example 1 (cont'd.)—Wiener Process: The operators in the set $\mathscr{P} = \{P_\lambda\}_{\lambda \in I}$ were described earlier. Strong limits of operators in the set $\mathscr{P}$ are themselves in $\mathscr{P}$, so $\bar{\mathscr{P}} = \mathscr{P}$. Since $P_0 = 0$ and $P_T = I$, $\bar{\mathscr{P}}$ is already bordered and $\bar{\mathscr{P}}_B = \bar{\mathscr{P}} = \mathscr{P}$. Finally, $\bar{\mathscr{P}}_B$ has no gaps and therefore $\mathscr{P}_M = \bar{\mathscr{P}}_B = \bar{\mathscr{P}} = \mathscr{P}$.

Example 2 (cont'd.)—Ornstein–Uhlenbeck Process: For this example $\bar{\mathscr{P}} = \mathscr{P}$. However, the null operator is not an element of $\bar{\mathscr{P}}$ so $\bar{\mathscr{P}}_B = \bar{\mathscr{P}} \cup \{0\}$. There is one gap in the chain $\bar{\mathscr{P}}_B$. It is $(0, P_0)$. Since this gap is one dimensional, $\mathscr{P}_M = \bar{\mathscr{P}}_B$.

Example 3 (cont'd.)—Triangular Covariance: As in Example 2, $\bar{\mathscr{P}} = \mathscr{P}$, $\bar{\mathscr{P}}_B = \bar{\mathscr{P}} \cup \{0\}$, $(0, P_0)$ is the one gap of $\bar{\mathscr{P}}_B$, and $\mathscr{P}_M = \bar{\mathscr{P}}_B$.

Example 4 (cont'd.)—Free Wiener Process: As in Example 2, $\bar{\mathscr{P}} = \mathscr{P}$, $\bar{\mathscr{P}}_B = \bar{\mathscr{P}} \cup \{0\}$, $(0, P_0)$ is the one gap of $\bar{\mathscr{P}}_B$, and $\mathscr{P}_M = \bar{\mathscr{P}}_B$.

Example 5 (cont'd.)—Free Doubly Integrated Wiener Process: The strong limit of the operators $P_1, P_{1/2}, P_{1/3}, \cdots$, which we will denote by P_{0++}, is described by

$$(P_{0++} f)(t) = f(0) + t\dot{f}(0) + \frac{t^2}{2} \ddot{f}(0).$$

Since P_{0++} is not in $\mathscr{P}$ and all other limits of sequences of operators from $\mathscr{P}$ are in $\mathscr{P}$, $\bar{\mathscr{P}} = \mathscr{P} \cup \{P_{0++}\}$. The null operator does not exist in $\bar{\mathscr{P}}$ so $\bar{\mathscr{P}}_B = \bar{\mathscr{P}} \cup \{0\}$. The chain $\bar{\mathscr{P}}_B$ has two gaps, $(0, P_0)$ and (P_0, P_{0++}). The gap $(0, P_0)$ is one dimensional, but (P_0, P_{0++}) is two dimensional. An orthonormal pair of functions in and spanning the projection space of (P_0, P_{0++}) is $e_1(t) = t$ and $e_2(t) = t^2/2$. Define

$$P_{0+}(t,s) = P_0(t,s) + e_1(t)e_1(s) = P_0(t,s) + ts.$$

For an arbitrary vector $f \in H(R_5)$ we have

$$(P_{0+} f)(t) = f(0) + t\dot{f}(0).$$

$\mathscr{P}_M$ can be chosen as $\bar{\mathscr{P}}_B \cup \{P_{0+}\}$. A different choice of orthonormal functions would lead to a different chain $\mathscr{P}_M$.

We will say that a bounded linear operator A on $H(R)$ is causal with respect to $\mathscr{P}_M$ if $PA = PAP$ for all $P \in \mathscr{P}_M$ and anticausal with respect to $\mathscr{P}_M$ if $AP = PAP$ for all $P \in \mathscr{P}_M$. Since $\mathscr{P} \subset \mathscr{P}_M$, an operator that is causal with respect to $\mathscr{P}_M$ is causal, and an operator anticausal with respect to $\mathscr{P}_M$ is anticausal.

We may note that a chain $\mathscr{P}_M$ such that $AP = PAP$ for all $P \in P_M$ is called an *eigenchain* of the operator A. This name is used because $AP = PAP$ implies that the range of P is an *invariant* subspace of the operator A. In this language, Lemma 6 states that A is causal if and only if $\{P_\lambda^\perp\}$ is an eigenchain of A, or equivalently, the subspaces $\{H_\lambda^\perp(R)\}$ are invariant subspaces of A. It may be noted (by using Lemmas 3 and 4) that

$$H_\lambda^\perp(R) = \{f \in H(R) : f(t) = 0,\ t \in I_\lambda\},$$

which shows that the present definition of causality is the same as the one first given (cf. Section II-A).

E. Operator Integrals

Following Gohberg and Krein [4], we will describe a number of different types of operator integrals.

A partition $\mathscr{D}$ of the chain $\mathscr{P}_M$ is a finite set of operators from $\mathscr{P}_M$ containing the null and identity operators. A partition $\mathscr{D}'$ is a refinement of $\mathscr{D}$ if $\mathscr{D} \subset \mathscr{D}'$.

Let $F(P)$ denote an operator function that assigns to each $P \in \mathscr{P}_M$ a bounded linear operator $F(P)$. The integral

$$\int F(P)\, dP$$

will be said to exist and equal a bounded linear operator A if there exists for any $\varepsilon > 0$ a partition $\mathscr{D}_\varepsilon$ such that if $\mathscr{D} = \{0 = P^{(0)} < P^{(1)} < \cdots < P^{(n)} = I\}$ is any refinement of $\mathscr{D}_\varepsilon$ and the projectors $Q_j \in \mathscr{P}_M$ are chosen arbitrarily under the constraint $P^{(k-1)} \leq Q_k \leq P^{(k)}$, $k = 1,\cdots,n$,

$$\left| A - \sum_{k=1}^{n} F(Q_k)(P^{(k)} - P^{(k-1)}) \right| < \varepsilon.$$

The integral

$$\int dPF(P)$$

will be defined in terms of the convergence of sums of the form

$$\sum_{k=1}^{n} (P^{(k)} - P^{(k-1)})F(Q_k).$$

Clearly, if $\int F(P)\, dP$ exists, then $\int dPF^*(P)$ exists and $\int dPF^*(P) = (\int F(P)\, dP)^*$.

If $\mathscr{P}_M$ has gaps, a necessary condition for the convergence of $\int F(P)\, dP$ is

$$(F(P^+) - F(P^-))(P^+ - P^-) = 0 \qquad (5)$$

for all gaps (P^+,P^-) of $\mathscr{P}_M$. (No matter how $\mathscr{D}_\varepsilon$ is chosen, $\mathscr{D}$ can be chosen with $P^+ = P^{(k)}$ and $P^- = P^{(k-1)}$ for some k. Since Q_k can be chosen as either P^+ or P^-, (5) must be true.) Condition (5) is a strong condition when $\mathscr{P}_M$ has gaps and it is useful to define operator integrals not requiring it for their existence. The integral

$$\int_m F(P)\, dP$$

is defined in terms of the convergence of sums of the form

$$\sum_{k=1}^{n} F(P^{(k-1)})(P^{(k)} - P^{(k-1)})$$

and the integral

$$\int_M F(P)\, dP$$

in terms of the convergence of sums of the form

$$\sum_{k=1}^{n} F(P^{(k)})(P^{(k)} - P^{(k-1)}).$$

Whenever $\int F(P)\, dP$ exists, $\int_m F(P)\, dP$ and $\int_m F(P)\, dP$ exist and equal $\int F(P)\, dP$. The condition (5) that is necessary for the existence of $\int F(P)\, dP$ is by no means necessary for the existence of $\int_M F(P)\, DP$ and $\int_M F(P)\, dP$.

Integrals of the form $\int_m dPF(P)$ and $\int_M dPF(P)$ are defined in the obvious way. Analogous statements to those made in the preceding paragraph about the existence of integrals of the form $\int F(P)\, dP$, $\int_m F(P)\, dP$, and $\int_M F(P)\, dP$ can be made for $\int dPF(P)$, $\int_m dPF(P)$, and $\int_M dPF(P)$.

We can now define the causal and anticausal component-extracting operations $\{\ \}_+$ and $\{\ \}_-$. We shall assume first that $\mathscr{P}_M$ is gapless. Define for any Hilbert–Schmidt operator A

$$\{A\}_+ = \int dPAP$$

and

$$\{A\}_- = \int PA\, dP.$$

Both of these integrals can be shown to exist. (The validity of this statement depends on both the assumption that $\mathscr{P}_M$ is gapless and the assumption that A is Hilbert–Schmidt.)

The following lemma shows that $\{\ \}_+$ and $\{\ \}_-$ as previously defined do indeed have the properties intuitively expected of them.

Lemma 7: Let $\mathscr{P}_M$ be gapless, A be a Hilbert–Schmidt operator, and $\{A\}_+$ and $\{A\}_-$ be defined as previously. Then, a) $A = \{A\}_+ + \{A\}_-$; b) if A is causal with respect to $\mathscr{P}_M$, $\{A\}_+ = A$ and $\{A\}_- = 0$; and c) if A is anticausal with respect to $\mathscr{P}_M$, $\{A\}_- = A$ and $\{A\}_+ = 0$.

Example 1 (cont'd.)—Wiener Process: $\mathscr{P} = \bar{\mathscr{P}} = \bar{\mathscr{P}}_B = \mathscr{P}_M$ and $\mathscr{P}_M$ is gapless.

Let A be an arbitrary Hilbert–Schmidt operator with defining kernel $A(t,s)$. It is easily verified that for all $\lambda \in I$, $P_\lambda(t,s) = t \wedge s \wedge \lambda$. We have for any $\lambda \in I$,

$$\begin{aligned}(AP_\lambda)(t,s) &= \langle A(t,\tau), P_\lambda(\tau,s)\rangle_{H(R_1)} \\ &= \int_0^T \left(\frac{\partial}{\partial\tau} A(t,\tau)\right)\left(\frac{\partial}{\partial\tau}(\tau \wedge s \wedge \lambda)\right) d\tau \\ &= A(t, s \wedge \lambda)\end{aligned}$$

and for all $\lambda,\xi \in I$,

$$\begin{aligned}(P_\xi A P_\lambda)(t,s) &= \langle P_\xi(t,\tau), (AP_\lambda)(\tau,s)\rangle_{H(R_1)} \\ &= A(t \wedge \xi, s \wedge \lambda).\end{aligned}$$

The operator $\{A\}_+$ is the limit, as the partition $\{0 = \lambda_0 < \lambda_1 < \cdots < \lambda_n = T\}$ becomes finer, of operators of the form

$$\sum_{k=1}^{n} (P_{\lambda_k} - P_{\lambda_{k-1}})AP_{\lambda_k}.$$

We have for all $(t,s) \in I$,

$$\{A\}_+(t,s) = \lim \sum_{k=1}^{n} A(t \wedge \lambda_k, s \wedge \lambda_k) - A(t \wedge \lambda_{k-1}, s \wedge \lambda_k)$$
$$= \lim \sum_{\substack{k=1 \\ k:\lambda_k < +}}^{n} A(\lambda_k, s \wedge \lambda_k) - A(\lambda_{k-1}, s \wedge \lambda_k)$$
$$= \int_0^t A_1(\lambda, s \wedge \lambda)\, d\lambda,$$

where $A_1(t,s) = \partial A(t,s)/\partial t$.

When $\mathscr{P}_M$ is not gapless, the integrals defining $\{A\}_+$ and $\{A\}_-$ will not exist for all Hilbert–Schmidt operators. Instead, we must define

$$\{A\}_{+,m} = \int_m dPAP, \{A\}_{+,M} = \int_M dPAP$$
$$\{A\}_{-,m} = \int_m PA\, dP, \text{ and } \{A\}_{-,M} = \int_M dPAP.$$

All these integrals will exist if A is Hilbert–Schmidt. The following lemma summarizes a number of relations among $\{A\}_{\pm,m}, \{A\}_{\pm,M}$.

Lemma 8: Let A be a Hilbert–Schmidt operator and $\{A\}_{+,m}$, $\{A\}_{+,M}$, $\{A\}_{-,m}$, and $\{A\}_{-,M}$ be defined as previously. Then the following is true.

a)
$$A = \{A\}_{+,M} + \{A\}_{-,m}$$
$$= \{A\}_{+,m} + \{A\}_{-,M}.$$

b)
$$\{A\}_{+,M} - \{A\}_{+,m} = \sum_{j=1}^{N_G} (P_j^+ - P_j^-)A(P_j^+ - P_j^-)$$

and

$$\{A\}_{-,M} - \{A\}_{-,m} = \sum_{j=1}^{N_G} (P_j^+ - P_j^-)A(P_j^+ - P_j^-),$$

where $1 \le N_G \le \infty$ and $(P_1^-,P_1^+),\cdots,(P_{N_G}^-,P_{N_G}^+)$ are the gaps of $\mathscr{P}_M$.

c) If A is causal with respect to $\mathscr{P}_M$, $\{A\}_{+,M} = A$ and $\{A\}_{-,m} = 0$.

d) If A is anticausal with respect to $\mathscr{P}_M$, $\{A\}_{-,M} = A$ and $\{A\}_{+,m} = 0$.

Example 4 (cont'd.)—Free Wiener Process: Let A be any Hilbert–Schmidt operator in $H(R_4)$. Proceeding as previously it is possible to show

$$\{A\}_{+,m}(t,s) = \int_0^t A_1(\lambda, s \wedge \lambda)\, d\lambda.$$

Using property b) of Lemma 8 we have

$$\{A\}_{+,M} = \{A\}_{+,m} + (P_0 - 0)A(P_0 - 0)$$
$$= \{A\}_{+,m} + P_0AP_0$$

and

$$\{A\}_{+,M}(t,s) = \int_0^t A_1(\lambda, s \wedge \lambda)\, d\lambda + A(0,0).$$

III. Generalized Covariance Factorization

In this section we shall solve the generalized covariance factorization problem described in Section I. Divorced from its stochastic setting, the generalized covariance factorization problem is to find a causal and causally invertible bounded linear operator A such that

$$I + K = AA^*.$$

The operator $I + K$, being strictly positive and self-adjoint, will possess a strictly positive self-adjoint inverse (see Riesz and Nagy [15, p. 265]). Defining H by

$$(I + K)^{-1} = I - H$$

we have

$$I = (I + K)(I - H) = I + K - H - KH$$

or

$$0 = H + KH - K. \tag{6}$$

But also,

$$I = (I - H)(I + K) = I + K - H - HK$$

or

$$0 = H + HK - K, \tag{7}$$

and comparison of (6) and (7) shows that H and K commute. Rewriting (6) as

$$H = K(I - H)$$

and remembering that the product of a Hilbert–Schmidt operator and a bounded linear operator is Hilbert–Schmidt, it is apparent that the operator H must be Hilbert–Schmidt.

It is convenient to let $\mathscr{G}_2(\mathscr{P}_M)$ denote the set of operators A that are Hilbert–Schmidt, are causal with respect to $\mathscr{P}_M$, and satisfy

$$(P^+ - P^-)A(P^+ - P^-) = 0$$

for all gaps (P^-,P^+) of $\mathscr{P}_M$. The following lemma and theorem, which are due to Gohberg and Krein [4, p. 478], essentially solve the generalized covariance factorization problem.

Lemma 9: If $A \in \mathscr{G}_2(\mathscr{P}_M)$, $(I + A)^{-1}$ exists and is of the form $I + B$ where $B \in \mathscr{G}_2(\mathscr{P}_M)$.

Theorem 4: Let K be a self-adjoint Hilbert–Schmidt operator such that $I + K > 0$ and let $I - H = (I + K)^{-1}$. Define

$$I - H_P = (I + PKP)^{-1}, \qquad P \in \mathscr{P}_M$$

and

$$Z = \sum_j (P_j^+ - P_j^-)H_{P_j^+}(P_j^+ - P_j^-),$$

where $(P_1^-,P_1^+),\cdots$ are the gaps of $\mathscr{P}_M$. Then, a) the integral

$$\tilde{h} = \int_m dPKP(I - H_P) \tag{8}$$

exists and $\tilde{h} \in \mathscr{G}_2(\mathscr{P}_M)$ and b)

$$I - H = (I - \tilde{h})^*(I - Z)(I - \tilde{h}).$$

We will first show how these results solve the generalized

covariance factorization problem when $\mathscr{P}_M$ is gapless. Letting $h = \tilde{h}$ and noting that $Z = 0$ if $\mathscr{P}_M$ is gapless, we have

$$I - H = (I - h)^*(I - h).$$

Since $h \in \mathscr{G}_2(\mathscr{P}_M)$, $(I - h)^{-1}$ exists and is of the form $I + k$ with k existing in $\mathscr{G}_2(\mathscr{P}_M)$. We have

$$\begin{aligned} I + K &= (I - H)^{-1} \\ &= ((I - h)^*(I - h))^{-1} \\ &= (I - h)^{-1}(I - h^*)^{-1} \\ &= (I + k)(I + k)^*. \end{aligned}$$

The filter $I + k$ factors the covariance $I + K$. We note that it is not necessary to invert $I - h$ to find k. The relation

$$-k^* = h + hK - K, \tag{9}$$

which follows from

$$\begin{aligned} (I + k)^* &= (I + k)^{-1}(I + K) \\ &= (I - h)(I + K), \end{aligned}$$

can be used to find k.

A. Alternative Method Based on Wiener–Hopf Equations

In $L_2[0,1]$ it is known that h can be found through a Wiener–Hopf equation and an operator integral does not need to be evaluated. The following theorem generalizes this technique. We note that the theorem is related to a famous lemma on factorization in normed rings [4, p. 168].

Theorem 5: Assume $\mathscr{P}_M$ is gapless and let h be as previously defined.

Then h is the unique operator in $\mathscr{G}_2(\mathscr{P}_M)$ satisfying

$$h + \{hK\}_+ - \{K\}_+ = 0.$$

Proof: Extracting causal components of both sides of (9), we have

$$\{-k^*\}_+ = \{h\}_+ + \{hK\}_+ - \{K\}_+.$$

Since h is causal with respect to $\mathscr{P}_M$, $\{h\}_+ = h$, and since k is causal with respect to $\mathscr{P}_M$, k^* is anticausal with respect to $\mathscr{P}_M$ and $\{-k^*\}_+ = 0$. We have

$$0 = h + \{hK\}_+ - \{K\}_+$$

showing that h satisfies the Wiener–Hopf equation.

To prove uniqueness, assume that two unequal operators h_1 and h_2 both in $\mathscr{G}_2(\mathscr{P}_M)$ satisfy the Wiener–Hopf equation. Define for $j = 1,2$

$$\begin{aligned} -k_j^* &= h_j + h_jK - K \\ I - g_j &= (I + k_j)^{-1} \\ I + f_j &= (I - h_j)^{-1}. \end{aligned}$$

The operators k_1, k_2, g_1, g_2, f_1, and f_2 must all exist in $\mathscr{G}_2(\mathscr{P}_M)$. We have for $j = 1,2$

$$(I + k_j)^* = (I - h_j)(I + K)$$

or after premultiplication by $(I - g_j)^*$ and postmultiplication by $I - H$

$$I - H = (I - g_j)^*(I - h_j).$$

Therefore,

$$(I - g_1)^*(I - h_1) = (I - g_2)^*(I - h_2)$$

or

$$(I + k_2)^*(I - g_1)^* = (I - h_2)(I + f_1) \triangleq I + A.$$

The operator A is Hilbert–Schmidt and both causal and anticausal with respect to $\mathscr{P}_M$. But then, it must be the null operator, for we have from properties b) and c) of Lemma 7 that $\{A\}_- = 0$ and $\{A\}_+ = 0$. Since $A = 0$,

$$(I - h_2)(I + f_1) = I$$

or

$$(I - h_2) = (I + f_1)^{-1} = I - h_1.$$

We shall give an example to illustrate the computations involved in solving a Wiener–Hopf equation on an RKHS.

Example 6: Let $I = [0,T]$, $R_0(t,s) = t \wedge s = R_1(t,s)$ in our earlier notation. Also let $K(t,s) = \alpha ts$ where $\alpha T > -1$. If $\alpha < 0$, the covariance $R_1(t,s) + K(t,s) = t \wedge s + \alpha ts$ is the covariance of a Wiener process pinned to zero at $-1/\alpha$. The RKHS $H(R_1)$ and the operators P_λ, $\lambda \in [0,T]$, have been described earlier in Example 1. As noted there, $\mathscr{P} = \bar{\mathscr{P}} = \mathscr{P}_B = \mathscr{P}_M$, and $\mathscr{P}_M$ is gapless for covariance $R_1(t,s)$. It is easy to show that the condition $\alpha T > -1$ is both necessary and sufficient for $I + K > 0$. K is Hilbert–Schmidt since

$$\begin{aligned} \|K\|^2 &= \|K(t,s)\|^2_{H(R_1)\otimes H(R_1)} \\ &= \int_0^T \int_0^T \left(\frac{\partial}{\partial t}\frac{\partial}{\partial s}\alpha ts\right)^2 ds\, dt \\ &= \alpha^2 T^2 < \infty. \end{aligned}$$

A formula for the causal component $\{A\}_+$ of an arbitrary Hilbert–Schmidt operator in $H(R_1)$ was given earlier. Using the evaluations

$$K_1(t,s) = \alpha s$$

and

$$\begin{aligned} (hK)_1(t,s) &= \frac{\partial}{\partial t}(hK)(t,s) = \frac{\partial}{\partial t}\langle h(t,\tau),K(\tau,s)\rangle_{H(R_1)} \\ &= \frac{\partial}{\partial t}\int_0^T \left(\frac{\partial}{\partial \tau} h(t,\tau)\right) \alpha s\, d\tau \\ &= \frac{\partial}{\partial t}\alpha s h(t,T) = \alpha s h_1(t,T) \end{aligned}$$

we have

$$\{K\}_+(t,s) = \int_0^t \alpha(s \wedge \lambda)\, d\lambda$$

$$\{hK\}_+(t,s) = \int_0^t \alpha(s \wedge \lambda)h_1(\lambda,T)\, d\lambda.$$

Substituting in the Wiener–Hopf equation $h + \{hK\}_+ - \{K\}_+ = 0$, we have

$$h(t,s) + \int_0^t \alpha(s \wedge \lambda)h_1(\lambda,T)\,d\lambda - \int_0^t \alpha(s \wedge \lambda)\,d\lambda = 0. \tag{10}$$

Differentiation with respect to t shows that

$$h_1(t,s) + \alpha(s \wedge t)h_1(t,T) - \alpha(s \wedge t) = 0$$

and letting $s = T$ gives

$$h_1(t,T) + \alpha t h_1(t,T) - \alpha t = 0$$

or

$$h_1(t,T) = \frac{\alpha t}{1 + \alpha t}.$$

Finally, substituting into (10), we have

$$h(t,s) = \int_0^t \alpha(s \wedge \lambda)\left(1 - \frac{\alpha\lambda}{1 + \alpha\lambda}\right) d\lambda = \int_0^t \frac{\alpha(s \wedge \lambda)}{1 + \alpha\lambda}\,d\lambda.$$

The preceding example is also solved by Kailath and Geesey [16], who do it by reduction to the once-differentiated problem with $R_0(t,s) = \delta(t - s)$ and $K(t,s) = \alpha$. Their approach is considerably simpler computationally. This example has been included to illustrate the preceding theory only. We do not wish to stress Theorem 5 as a computational tool; it is of more interest for its generality.

We will now consider the case where $\mathscr{P}_m$ has gaps. In this case the operator Z defined in Theorem 4 will not be the null operator. It is easy to show from its defining equation that Z is self-adjoint, Hilbert–Schmidt, and both causal and anticausal with respect to $\mathscr{P}_M$. This latter property implies Z commutes with all operators $P \in \mathscr{P}_M$, for we have for any $P \in \mathscr{P}_M$

$$PZ = PZP = ZP.$$

The operator $I - Z$ must be strictly positive, for if it were not, there would exist a nonzero function $f \in H(R_0)$ such that $\langle f,(I - Z)f\rangle_{H(R_0)} \leq 0$, and then $g = (I - \tilde{h})^{-1}f$ would satisfy $\langle g,(I - H)g\rangle_{H(R_0)} \leq 0$, violating the strict positiveness of $I - H$. A square root of $I - Z$ is any bounded linear operator that when multiplied by itself equals $I - Z$. $I - Z$ will in general possess many square roots. However, it will have a unique strictly positive self-adjoint square root (see Riesz and Nagy [15, p. 265]). Moreover, this particular square root, which we will denote by $I - W$, commutes with all operators commuting with $I - Z$. In particular, we will have for any $P \in \mathscr{P}_M$

$$(I - W)P = P(I - W).$$

It can also be proved [5] that W is a Hilbert–Schmidt operator.

Since $I - W$ is self-adjoint and strictly positive, it will have an inverse, which we will denote by $I + V$, that is self-adjoint and strictly positive. The operator V will be Hilbert–Schmidt. For any $P \in \mathscr{P}_M$

$$(I + V)P = (I + V)P(I - W)(I + V) = (I + V)(I - W)P(I + V) = P(I + V),$$

which shows that $I + V$ commutes with all $P \in \mathscr{P}_M$.

Let

$$I + \tilde{k} = (I - \tilde{h})^{-1} \qquad (I - h) = (I - W)(I - \tilde{h}),$$

and

$$I + k = (I - h)^{-1} = (I - \tilde{h})^{-1}(I - W)^{-1} = (I + \tilde{k})(I + V).$$

The operators h and k will be Hilbert–Schmidt and causal with respect to $\mathscr{P}_M$. We have

$$I - H = (I - \tilde{h})^*(I - Z)(I - \tilde{h}) = (I - \tilde{h})^*(I - W)^*(I - W)(I - \tilde{h}) = (I - h)^*(I - h)$$

and

$$I + K = (I - H)^{-1} = (I + k)(I + k)^*.$$

We have proved the following theorem.

Theorem 6: Let K, Z, and $\tilde{h}$ be as defined in Theorem 4 and let $I - W$ be the unique strictly positive self-adjoint square root of $I - Z$. Define the operators h and k by

$$I - h = (I - W)(I - \tilde{h})$$

and

$$I + k = (I - h)^{-1}.$$

Then, the filter defined by the operator $I + k$ is causal and causally invertible and provides a factorization for the covariance $R_0(t,s) + K(t,s)$.

We note that, as in the gapless case, it is not necessary to invert $I - h$ to find k. The operator k can be found from the equation

$$-k^* = h + hK - K.$$

When $\mathscr{P}_M$ has gaps, the operator h can no longer be defined as the solution to a Wiener–Hopf equation. However, $\tilde{h}$, to which it is simply related, is the solution of a Wiener–Hopf equation.

Theorem 7: Let $\tilde{h}$ be as defined in Theorem 4. Then $\tilde{h}$ is the unique operator in $\mathscr{G}_2(\mathscr{P}_M)$ satisfying

$$\tilde{h} + \{\tilde{h}K\}_{+,m} - \{K\}_{+,m} = 0.$$

Proof: We have

$$I - H = (I - \tilde{h})^*(I - Z)(I - \tilde{h}).$$

Postmultiplication by $I + K$ and then premultiplication by first $(I + \tilde{k})^* = (I - \tilde{h}^*)^{-1}$ and then $I + D \triangleq (I - Z)^{-1}$ shows

$$(I + D)(I + \tilde{k})^* = (I - \tilde{h})(I + K)$$

or

$$-D - D\tilde{k}^* - \tilde{k}^* = \tilde{h} + \tilde{h}K - K. \tag{11}$$

The operators D, $D\tilde{k}^*$, and $\tilde{k}^*$ are all Hilbert–Schmidt and anticausal with respect to $\mathscr{P}_M$. Therefore,

$$\{D\}_{+,m} = \{D\tilde{k}^*\}_{+,m} = \{\tilde{k}^*\}_{+,m} = 0.$$

We have

$$\{\tilde{h}\}_{+,m} = \{\tilde{h}\}_{+,M} - \sum_j (P_j^+ - P_j^-)\tilde{h}(P_j^+ - P_j^-),$$

where $(P_1^-,P_1^+),(P_2^-,P_2^+),\cdots$ are the gaps of $\mathscr{P}_M$. Since $\tilde{h} \in \mathscr{G}_2(\mathscr{P}_M)$, $\sum_j (P_j^+ - P_j^-)\tilde{h}(P_j^+ - P_j^-) = 0$, and since $\tilde{h}$ is causal, $\{\tilde{h}\}_{+,M} = \tilde{h}$. Performing the operation $\{\ \ \}_{+,m}$ on both sides of (11) and using the preceding relations, we obtain

$$\tilde{h} + \{\tilde{h}K\}_{+,m} - \{K\}_{+,m} = 0.$$

The proof of uniqueness is similar to the proof of uniqueness in Theorem 5.

Example 7: Let $I = [0,T]$ where $T < 2$, $R_0(t,s) = 1 + t \wedge s = R_4(t,s)$ in our earlier notation and $R_0(t,s) + K(t,s) = \frac{1}{2}(1 + |t - s|)$. $H(R_4)$ and the operators P_λ, $\lambda \in [0,T]$ are described in Example 4. As noted there $\bar{\mathscr{P}} = \mathscr{P}$ and $\mathscr{P}_M = \bar{\mathscr{P}}_B = \mathscr{P} \cup \{0\}$. $(0,P_0)$ is the one gap of $\mathscr{P}_M$. It is easy to show that K is Hilbert-Schmidt and that $T < 2$ is both necessary and sufficient for $I + K > 0$. As noted earlier the causal component $\{A\}_{+,m}$ of any Hilbert-Schmidt operator A in $H(R_4)$ is given by the formula

$$\{A\}_{+,m}(t,s) = \int_0^t A_1(\lambda, s \wedge \lambda)\, d\lambda.$$

We have

$$K(t,s) = \tfrac{1}{2}(1 - |t - s|) - 1 - t \wedge s = -\tfrac{1}{2}(1 + t + s)$$

$$K_1(t,s) = -\tfrac{1}{2}$$

and

$$(\tilde{h}K)_1(t,s) - \frac{\partial}{\partial t}\langle \tilde{h}(t,\tau),K(\tau,s)\rangle_{H(R_0)} = \frac{\partial}{\partial t}\left(\tilde{h}(t,0)(-\tfrac{1}{2}(1 + s)) + \int_0^T \left(\frac{\partial}{\partial \tau}\tilde{h}(t,\tau)\right)\left(-\frac{1}{2}\right) d\tau\right) = \tfrac{1}{2}s\tilde{h}_1(t,0) - \tfrac{1}{2}\tilde{h}_1(t,T).$$

Therefore,

$$\{K\}_{+,m}(t,s) = -\tfrac{1}{2}t,$$

and

$$\{\tilde{h}K\}_{+,m}(t,s) = \int_0^t (-\tfrac{1}{2}(s \wedge \lambda)\tilde{h}_1(\lambda,0) - \tfrac{1}{2}\tilde{h}_1(\lambda,T))\, d\lambda.$$

Substituting in the Wiener–Hopf equation $\tilde{h} + \{\tilde{h}K\}_{+,m} - \{K\}_{+,m} = 0$, we have

$$\tilde{h}(t,s) - \tfrac{1}{2}\int_0^t ((s \wedge \lambda)\tilde{h}_1(\lambda,0) + \tilde{h}_1(\lambda,T))\, d\lambda + \tfrac{1}{2}t = 0 \tag{12}$$

or after differentiating with respect to t

$$\tilde{h}_1(t,s) - \tfrac{1}{2}((s \wedge t)\tilde{h}_1(t,0) + \tilde{h}_1(t,T)) + \tfrac{1}{2} = 0. \tag{13}$$

Substituting $s = 0$ and $s = T$ in (13), we obtain two equations that can be solved simultaneously to find $\tilde{h}_1(t,0)$ and $\tilde{h}_1(t,T)$. These values can then be substituted into (12) to yield

$$\tilde{h}(t,s) = -\int_0^t \frac{(s \wedge \lambda) + 2}{2 - \lambda}\, d\lambda.$$

To find $h(t,s)$ we must first find $Z(t,s)$ and $W(t,s)$. The one gap of $\mathscr{P}_M$ is $(0,P_0)$. We have $(P_0KP_0)(t,s) = -\frac{1}{2}$. The equation

$$H_{P_0} + H_{P_0}P_0KP_0 - P_0KP_0 = 0, \tag{14}$$

which follows from $I - H_{P_0} = (I + P_0KP_0)^{-1}$, is easily solved to find $H_{P_0}(t,s) = -1$. Therefore,

$$Z(t,s) = ((P_0 - 0)H_{P_0}(P_0 - 0))(t,s) = H_{P_0}(t,s) = -1 \tag{15}$$

$$(I - Z)(t,s) = 2 + t \wedge s \qquad (I - W)(t,s) = \sqrt{2} + t \wedge s \tag{16}$$

and

$$W(t,s) = 1 + t \wedge s - \sqrt{2} - t \wedge s = 1 - \sqrt{2}. \tag{17}$$

Finally, $(I - W)(I - \tilde{h}) = I - h$ implies $h = \tilde{h} + W - W\tilde{h}$, and using (16) and (17) in this equation, we find

$$h(t,s) = 1 - \sqrt{2} - \int_0^t \frac{(s \wedge \lambda) + 2}{2 - \lambda}\, d\lambda.$$

The problem of the preceding example can be reduced to the problem of Example 6 by subtracting initial-condition random variables (see Kailath and Geesey [16]). This procedure is of course simpler; again we emphasize that the results of this paper are primarily of interest for their theoretical completeness.

The solutions of Wiener–Hopf equations on $L_2[0,1]$ can be related to the solutions of matrix Riccati differential equations (see Kailath [17]) when the kernel $K(t,s)$ is separable. It seems that a generalization of this theory might be possible for RKHS Wiener–Hopf equations. Such a theory might introduce in a natural way the theory of reduced-order Riccati differential equations [3]. We have been unable so far to obtain such a generalization.

To conclude this section, we shall quote from [5] a representation theorem that generalizes slightly Theorem 2.

Theorem 8: If a zero mean-random process $\{x(t), t \in I\}$ has a covariance of the form $R_0(t,s) + K(t,s)$ where K is Hilbert-Schmidt and $I + K > 0$, then it is expressible as

$$x(t) = z(t) + n(t), \qquad t \in I,$$

where

a) $n(\cdot)$ has zero mean and covariance $R_0(t,s)$;
b) $E[z(\cdot)] = 0$;
c) tr $\{E[z(t)z(s)]\} < \infty$;
d) $E[z(t) \mid n(\tau), \tau \in I] = E[z(t) \mid n(\tau), \tau \in I_t]$ for all $t \in I$.

If $\mathscr{P} = \mathscr{P}_M$ and $\mathscr{P}_M$ has no gaps, the converse holds;

that is, a random process $x(\cdot)$ defined by

$$x(t) = z(t) + n(t),$$

where $z(\cdot)$ and $n(\cdot)$ obey a), b), c), and d) has a covariance of the form $R_0(t,s) + K(t,s)$ where K is Hilbert–Schmidt and $I + K > 0$.

IV. Discrimination Between Gaussian Processes of Unequal Covariance

Consider the discrimination problem 1) $h_1 : \{x(t), t \in I\}$ is a zero-mean second-order Gaussian process with covariance $R_0(t,s) + K(t,s)$, $(t,s) \in I$, versus 2) $h_0 : \{x(t), t \in I\}$ is a zero-mean second-order Gaussian process with covariance $R_0(t,s)$, $(t,s) \in I$. We are assuming zero means, but this is not essential and can be easily generalized. We shall present a likelihood ratio formula that generalizes one derived in [2] for $x(\cdot)$ being a Wiener process under h_0.

This detection problem is known to be nonsingular (that is, the probability measures P_1 and P_0 induced on the measure space $(\mathscr{R}^I, B(\mathscr{R}^I))$ by the $x(\cdot)$ process under hypotheses 1 and 0, respectively, are equivalent) if and only if K is Hilbert–Schmidt and $I + K > 0$ (see Kallianpur and Oodaira [13]). Since the operator K is self-adjoint and Hilbert–Schmidt, the kernel $K(t,s)$ will have a diagonal expansion of the form

$$K(t,s) = \sum_{j=1}^{\infty} \lambda_j e_j(t) e_j(s), \tag{18}$$

where

$$\sum_{j=1}^{\infty} \lambda_j^2 = \|K\|^2 < \infty$$

and the set of eigenvectors $\{e_j\}_{j=1}^{\infty}$ is a complete orthonormal set (CONS). $R_0(t,s)$ will have the expansion

$$R_0(t,s) = \sum_{j=1}^{\infty} e_j(t) e_j(s) \tag{19}$$

and therefore $R_0(t,s) + K(t,s)$ will have the expansion

$$R_0(t,s) + K(t,s) = \sum_{j=1}^{\infty} (1 + \lambda_j) e_j(t) e_j(s). \tag{20}$$

These infinite sums converge in $H(R_0)$ for each fixed s or t and also pointwise for all t,s.

The condition $I + K > 0$ implies

$$1 + \lambda_j > 0, \qquad j = 1,2,\cdots.$$

By using the diagonal expansions (19)–(20), it can be shown (see, e.g., Root [18]) that the likelihood ratio dP_1/dP_0 for our problem is given by[1]

$$L = \lim_{N\to\infty} \prod_{j=1}^{N} \frac{1}{\sqrt{1 + \lambda_j}} \exp\left\{\frac{1}{2} x_j^2 \frac{\lambda_j}{1 + \lambda_j}\right\}, \tag{21}$$

where the $\{x_j\}$ are random variables that can be written symbolically as

$$x_j = \langle x(t), e_j(t)\rangle_{H(R_0)}.$$

We should remark that the inner product $\langle x(t), e_j(t)\rangle_{H(R_0)}$ cannot be defined rigorously as a true inner product nor as a congruence inner product since under one of the hypotheses $x(\cdot)$ does not have covariance $R_0(t,s)$. See the Appendix for the proper definition.

We will start with the formula (21) and show that another formula for the likelihood ratio is

$$\delta_W(-1) \exp\left\{\langle\!\langle hx, x\rangle_{H(R_0)} - \tfrac{1}{2}\|hx\|^2_{H(R_0)}\right\}, \tag{22}$$

where h and W are the operators defined in Section III, $\delta_W(\lambda)$ is the Fredholm–Carleman determinant of W, and $\langle\!\langle\ ,\ \rangle_{H(R_0)}$ is an inner product generalizing the concept of an Itô integral.

When $H(R_0)$ is gapless, $\delta_W(-1) = 1$. This was the case treated in [2], where we assumed $R_0(t,s) = t \wedge s$. Our general arguments here will parallel those of [2].

A. Itô Inner Products

Let us first define the Itô inner product. If A is any Hilbert–Schmidt operator and $\{\phi_j\}_{j=1}^{\infty}$, any CONS in $H(R_0)$, the defining kernel $A(t,s)$ [which exists in $H(R_0) \otimes H(R_0)$] will have the expansion, convergent in $H(R_0) \otimes H(R_0)$,

$$A(t,s) = \sum_{j,k=1}^{\infty} A_{jk} \phi_j(t) \phi_k(s),$$

where

$$A_{jk} = \langle \phi_j, A\phi_k\rangle_{H(R_0)} \qquad \sum_{j,k=1}^{\infty} A_{jk}^2 = \|A\|^2 < \infty.$$

The Itô inner product $\langle\!\langle Ax, x\rangle_{H(R_0)}$ will be defined by

$$\langle\!\langle Ax, x\rangle_{H(R_0)} = \lim_{n\to\infty} \sum_{j,k=1}^{n} A_{jk}(x_j x_k - \delta_{jk}), \tag{23}$$

where

$$x_j = \langle x(t), \phi_j(t)\rangle_{H(R_0)}$$

and the inner product must be defined as in the Appendix. For this definition of $\langle\!\langle Ax, x\rangle_{H(R_0)}$ to be acceptable, we must verify that under hypothesis h_0 the limit (23) exists and is independent of the particular choice made for the CONS $\{\phi_j\}_{j=1}^{\infty}$. (The existence and nondependence on basis under h_1 will then follow by the equivalence of the probability measures P_1 and P_0.) We shall only describe the verification that the limit exists. The other verification is straightforward, but tedious.

To show existence of the limit (23) under h_0, let

$$\langle\!\langle A_n x, x\rangle_{H(R_0)} \triangleq \sum_{j,k=1}^{n} A_{jk}(x_j x_k - \delta_{jk})$$

and let $\mathscr{F}_n$ denote the smallest Borel field on which the random variables $x_1, \cdots, x_n$ are measurable. $\{\langle\!\langle A_n x, x\rangle_{H(R_0)}, \mathscr{F}_n\}_{n=1}^{\infty}$ will be a martingale. Also, for any n

[1] We shall use "lim" to denote limits with probability one and "l.i.m." to denote limits in quadratic mean.

$$E[\langle\!\!\langle A_n x,x\rangle_{H(R_0)}]^2$$

$$= E\left[\sum_{j,k,l,m}^{n} A_{jk}A_{lm}(x_jx_k - \delta_{jk})(x_lx_m - \delta_{lm})\right]$$

$$= \sum_{j,k,l,m}^{n} A_{jk}A_{lm}(\delta_{jl}\delta_{km} + \delta_{jm}\delta_{kl})$$

$$= \sum_{j,k=1}^{n} A_{jk}^2 + A_{jk}A_{kj}$$

$$\leq 2\|A\|^2.$$

Therefore, the sequence

$$E[\langle\!\!\langle A_1 x,x\rangle_{H(R_0)}]^2, E[\langle\!\!\langle A_2 x,x\rangle_{H(R_0)}]^2, \cdots$$

is uniformly bounded, and by the martingale convergence theorem of Doob [19, p. 319], the limit of the random variables $\langle\!\!\langle A_n x,x\rangle_{H(R_0)}$ will exist with probability one.

We have chosen to call $\langle\!\!\langle Ax,x\rangle_{H(R_0)}$ an "Itô inner product" since when $I = [0,T]$, $R_0(t,s) = t \wedge s$, and A is a causal operator on $H(R_0)$, $\langle\!\!\langle Ax,x\rangle_{H(R_0)}$ can be shown to equal

$$\int_0^T\!\!\!\!\!-\; \int_0^t \left(\frac{\partial}{\partial t}\frac{\partial}{\partial s} A(t,s)\right) \dot{x}(s)\dot{x}(t)\, ds\, dt, \tag{24}$$

where we have used the dash to indicate that the outer integral is an Itô stochastic integral. Since A is causal (24) can be rewritten as

$$\int_0^T\!\!\!\!\!-\; \int_0^T \left(\frac{\partial}{\partial t}\frac{\partial}{\partial s} A(t,s)\right) \dot{x}(s)\dot{x}(t)\, ds\, dt. \tag{25}$$

Notice that the symbolism $\langle\!\!\langle Ax,x\rangle_{H(R_0)}$ suggests (25).

Returning to our discrimination problem, we have from (21)

$$L^2 = \lim_{n\to\infty} \prod_{j=1}^{n} \frac{1}{1+\lambda_j} \exp\left\{\frac{\lambda_j x_j^2}{1+\lambda_j}\right\}$$

$$= \lim_{n\to\infty} \prod_{j=1}^{n} \left(1 - \frac{\lambda_j}{1+\lambda_j}\right) \exp\left\{\frac{\lambda_j}{1+\lambda_j}\right\}$$

$$\cdot \exp\left\{\frac{\lambda_j}{1+\lambda_j}(x_j^2 - 1)\right\}.$$

The operator H defined by $I - H = (I + K)^{-1}$ will have the diagonal expansion

$$H(t,s) = \sum_{j=1}^{\infty} \frac{\lambda_j}{1+\lambda_j} e_j(t)e_j(s).$$

For any Hilbert–Schmidt self-adjoint operator A with eigenvalues $\mu_1,\mu_2,\cdots$, the Fredholm–Carleman determinant of A, $\delta_A(\mu)$, is defined by

$$\delta_A(\mu) = \prod_{j=1}^{\infty} (1 + \mu\mu_j) \exp\{-\mu\mu_j\}. \tag{26}$$

The square summability of the eigenvalues ensures $\delta_A(\mu)$ is finite. Recognizing the eigenvalues of H as $\{\lambda_j/(1+\lambda_j)\}_{j=1}^{\infty}$, it is apparent that

$$\lim_{n\to\infty} \prod_{j=1}^{n} \left(1 - \frac{\lambda_j}{1+\lambda_j}\right) \exp\left\{\frac{\lambda_j}{1+\lambda_j}\right\} = \delta_H(-1)$$

and

$$\lim_{n\to\infty} \prod_{j=1}^{n} \exp\left\{\frac{\lambda_j}{1+\lambda_j}(x_j^2 - 1)\right\} = \exp\{\langle\!\!\langle Hx,x\rangle_{H(R_0)}\}.$$

Thus, we have

$$L^2 = \delta_H(-1) \exp\{\langle\!\!\langle Hx,x\rangle_{H(R_0)}\}$$

or

$$L = \sqrt{\delta_H(-1)} \exp\{\tfrac{1}{2}\langle\!\!\langle Hx,x\rangle_{H(R_0)}\}. \tag{27}$$

This formula for the likelihood ratio is of some interest in its own right. It was derived in [2] for the special case of $I = [0,T]$ and $R_0(t,s) = t \wedge s$ and shown to be equivalent to a formula due to Shepp [20].

The causal Hilbert–Schmidt operator h, introduced in Section III, satisfies

$$I - H = (I - h)^*(I - h)$$

or, equivalently,

$$H = h + h^* - h^*h.$$

By the linearity (with respect to operators) of the definition of the Itô inner product, we have

$$\langle\!\!\langle Hx,x\rangle_{H(R_0)} = \langle\!\!\langle hx,x\rangle_{H(R_0)} + \langle\!\!\langle h^*x,x\rangle_{H(R_0)} - \langle\!\!\langle hh^*x,x\rangle_{H(R_0)}. \tag{28}$$

By the symmetry in the Itô product definition,

$$\langle\!\!\langle h^*x,x\rangle_{H(R_0)} = \langle\!\!\langle hx,x\rangle_{H(R_0)}. \tag{29}$$

Letting

$$(h^*h)_{jk} = \langle e_j, h^*he_k\rangle_{H(R_0)}$$

and

$$h_{jk} = \langle e_j, he_k\rangle_{H(R_0)}$$

we have

$$\langle\!\!\langle h^*hx,x\rangle_{H(R_0)} = \lim_{n\to\infty} \sum_{j,k=1}^{n} (h^*h)_{jk}(x_jx_k - \delta_{jk})$$

$$= \lim_{n\to\infty} \sum_{j,k=1}^{n} \left(\sum_{l=1}^{\infty} h_{lj}h_{lk}\right)(x_jx_k - \delta_{jk})$$

$$= \|hx\|_{H(R_0)}^2 - \|h\|^2, \tag{30}$$

where

$$(hx)(t) = \langle h(t,s), x(s)\rangle_{H(R_0)}$$

and is an element of $H(R_0)$ with probability one. Using (28)–(30) in (27), we have

$$L = \sqrt{\delta_H(-1)} \exp\{\langle\!\!\langle hx,x\rangle_{H(R_0)} - \tfrac{1}{2}\|hx\|_{H(R_0)}^2 + \tfrac{1}{2}\|h\|^2\}. \tag{31}$$

Gohberg and Krein [21, p. 169] show that if for a particular number λ, Hilbert–Schmidt operators A, B, and C satisfy

$$I + \lambda A = (I + \lambda B)(I + \lambda C)$$

then

$$\delta_A(\lambda) = \delta_B(\lambda)\delta_C(\lambda) \exp\{-\lambda^2 \operatorname{tr}\{BC\}\}.$$

An alternate proof of this result is given by Kailath [2, appendix I]. Since $I - H = (I - h)^*(I - h)$,[2]

$$\delta_H(-1) = \delta_{h^*}(-1)\delta_h(-1) \exp\{-\text{tr}\{h^*h\}\}. \quad (32)$$

Letting W and $\tilde{h}$ be defined as in Section III, we have

$$I - h = (I - W)(I - \tilde{h}).$$

Therefore,

$$\delta_h(-1) = \delta_W(-1)\delta_{\tilde{h}}(-1) \exp\{-\text{tr}\{W\tilde{h}\}\}.$$

Similarly,

$$\delta_{h^*}(-1) = \delta_{\tilde{h}}(-1)\delta_W(-1) \exp\{\text{tr}\{\tilde{h}^*W\}\}.$$

Another result of Gohberg and Krein's that will be useful is that all operators that are in $\mathcal{G}_2(\mathcal{P}_M)$ themselves or have adjoints that are in $\mathcal{G}_2(\mathcal{P}_M)$ have traces that are 0 and Fredholm–Carleman determinants that equal 1. Since $\tilde{h} \in \mathcal{G}_2(\mathcal{P}_M)$ and $W^*\tilde{h} = W\tilde{h} \in \mathcal{G}_2(\mathcal{P}_M)$,

$$\delta_h(-1) = \delta_{h^*}(-1) = \delta_W(-1). \quad (33)$$

Using (32) and (33) in (31), we have finally

$$L = \delta_W(-1) \exp\{\langle\!\langle hx,x\rangle_{H(R_0)} - \tfrac{1}{2}\|hx\|^2_{H(R_0)}\}. \quad (34)$$

If $\mathcal{P}_M$ is gapless, W will equal 0 and $\delta_W(-1)$ will equal 1. Since for the pseudo-RKHS of white noise, $L_2[0,T]$, $\mathcal{P}_M$ is gapless, the white-noise analog of formula (34) does not contain a multiplying constant. This makes the appearance of the nonunity (when $\mathcal{P}_M$ has gaps) multiplying constant $\delta_W(-1)$ in the likelihood ratio (34) somewhat unexpected. The first of the following three examples, which is one-dimensional, may give some insight into the appearance of this multiplying constant; also see the remarks and the last reference in Section V.

Example 8—Density for Two Gaussian Variates: Let $I = \{0\}$, $R_0(0,0) = 1$, and $K(0,0) + R_0(0,0) = \sigma^2$. The elements of $H(R_0)$ are constants and the inner product of any two elements is their product. The condition $I + K > 0$ implies $\sigma^2 > 0$.

The likelihood ratio for this problem is, of course, just the ratio of the density of a normal random variable with zero mean and variance σ^2 to the density of a normal random variable with zero mean and unity variance. We have

$$L = \frac{(2\pi\sigma^2)^{-1/2} \exp\{-x^2(0)/2\sigma^2\}}{(2\pi)^{-1/2} \exp\{-x^2(0)/2\}}$$
$$= \sigma^{-1} \exp\{(x^2(0)/2)(1 - \sigma^{-2})\}. \quad (35)$$

Though it had at first seemed surprising to us, we shall show that this expression can also have the form (34).

For this example $\mathcal{P} = \bar{\mathcal{P}} = \{\mathcal{P}_0\} = \{I\}$ and $\mathcal{P}_M = \bar{\mathcal{P}}_B = \{0\} \cup \{I\}$. The one gap of $\mathcal{P}_M$ is $(0,I)$. The defining kernels for the operators h and W are easily found to be

$$h(0,0) = 1 - \gamma = W(0,0),$$

where, for notational convenience, we define

$$\gamma = 1/\sigma.$$

The RKHS $H(R_0)$ is one dimensional and the function $\{\phi_1(0) = 1\}$ forms a CONS. We have

$$h_{11} = \langle\phi_1,h\phi_1\rangle_{H(R_0)} = 1 - \gamma$$

$$x_1 = \langle x,\phi_1\rangle_{H(R_0)} = x(0)$$

$$\langle\!\langle hx,x\rangle_{H(R_0)} = \lim_{n\to\infty} \sum_{j,k=1}^{n} h_{jk}(x_jx_k - \delta_{jk})$$
$$= (1 - \gamma)(x^2(0) - 1).$$

The operator W has one eigenvalue, $1 - \gamma$, and its Fredholm–Carleman determinant is $\gamma \exp\{1 - \gamma\}$. The norm $\|hx\|^2_{H(R_0)}$ is easily evaluated as $(1 - \gamma)^2x^2(0)$. Therefore, (34) yields

$$L = \gamma \exp\{1 - \gamma\} \exp\{(1 - \gamma)(x^2(0) - 1)$$
$$- (1 - \gamma)^2x^2(0)/2\}$$
$$= \sigma^{-1} \exp\{x^2(0)(1 - \sigma^{-2})/2\},$$

which agrees with (35).

Example 9—Pinned Wiener Process: Let I, $R_0(t,s)$, and $K(t,s)$ be as in Example 6, where the operator h was found. We have

$$(hx)(t) = \langle h(t,s),x(s)\rangle_{H(R_0)}$$
$$= \int_0^T \left(\frac{\partial}{\partial s}\int_0^t \frac{\alpha(s \wedge \lambda)}{1 + \alpha\lambda}\,d\lambda\right) \dot{x}(s)\,ds$$
$$= \int_0^t \int_s^t \frac{\alpha}{1 + \alpha\lambda}\,d\lambda \dot{x}(s)\,ds$$
$$= \int_0^t \int_0^\lambda \frac{\alpha}{1 + \alpha\lambda}\,\dot{x}(s)\,ds\,d\lambda = \int_0^t \frac{\alpha x(\lambda)}{1 + \alpha\lambda}\,d\lambda$$

$$\|hx\|^2_{H(R_0)} = \int_0^T \left(\frac{\partial}{\partial t}(hx)(t)\right)^2 dt = \int_0^T \left(\frac{\alpha x(t)}{1 + \alpha t}\right)^2 dt$$

$$\langle\!\langle hx,x\rangle_{H(R_0)} = \int\!\!\!\!-_0^T \frac{\alpha x(t)}{1 + \alpha t}\,\dot{x}(t)\,dt = \int\!\!\!\!-_0^T \frac{\alpha x(t)\,dx(t)}{1 + \alpha t}.$$

Furthermore, since $\mathcal{P}_M$ is gapless, $W = 0$, so that (34) yields

$$L = \exp\left\{\int\!\!\!\!-_0^T \frac{\alpha x(t)}{1 + \alpha t}\,dx(t) - \frac{1}{2}\int_0^T \left(\frac{\alpha x(t)}{1 + \alpha t}\right)^2 dt\right\}.$$

This formula is also obtained by Kailath [2], who shows its equivalence to the formula

$$L = \frac{1}{\sqrt{1 - T}} \exp\left\{-\frac{x^2(T)}{2(1 - T)}\right\},$$

first obtained by Varberg [23].

Example 10—Triangular Covariance: Let T, $R_0(t,s)$, and $K(t,s)$ be as in Example 7, where the operators h and W were found. We have

[2] We have not defined eigenvalues, Fredholm–Carleman determinants, or the trace operation for operators such as h that are not self-adjoint. By working on a complex Hilbert space it is possible to do so (see [21], [22]). We shall not comment further on this problem since our final result (34) does not contain the trace or Fredholm–Carleman determinant of a nonself-adjoint operator.

$$(hx)(t) = \langle h(t,s),x(s)\rangle_{H(R_0)}$$
$$= h(t,0)x(0) + \int_0^T \left(\frac{\partial}{\partial s} h(t,s)\right) \dot{x}(s)\, ds$$
$$= \left(1 - \sqrt{2} - \int_0^t \frac{2}{2-\lambda}\, d\lambda\right) x(0)$$
$$- \int_0^t \int_s^t \frac{1}{2-\lambda}\, d\lambda \dot{x}(s)\, ds$$
$$= (1 - \sqrt{2})x(0) - \int_0^t \frac{x(\lambda) + x(0)}{2-\lambda}\, d\lambda$$
$$\|hx\|^2_{H(R_0)} = (hx)^2(0) + \int_0^T \left(\frac{\partial}{\partial t}(hx)(t)\right)^2 dt$$
$$= (1 - \sqrt{2})^2 x^2(0) + \int_0^T \left(\frac{x(t) + x(0)}{2-t}\right)^2 dt$$
$$\measuredangle hx,x\rangle_{H(R_0)} = (1 - \sqrt{2})(x^2(0) - 1)$$
$$- \rlap{\,-}\int_0^T \frac{x(t) + x(0)}{2-t} \dot{x}(t)\, dt.$$

The operator W has only one eigenvalue, $1 - \sqrt{2}$. We have

$$\delta_W(-1) = (1 + (-1)(1 - \sqrt{2})) \exp\{-(-1)(1 - \sqrt{2})\}$$
$$= \sqrt{2} \exp\{1 - \sqrt{2}\}.$$

Therefore, (34) yields

$$L = \sqrt{2} \exp\{1 - \sqrt{2}\}$$
$$\cdot \exp\left\{(1 - \sqrt{2})(x^2(0) - 1) - \rlap{\,-}\int_0^T \frac{x(t) + x(0)}{2-t}\, dx(t)\right\}$$
$$\cdot \exp\left\{-\tfrac{1}{2}\left((1 - \sqrt{2})^2 x^2(0) + \int_0^T \left(\frac{x(t) + x(0)}{2-t}\right)^2 dt\right)\right\}$$
$$= \sqrt{2} \exp\left\{-\tfrac{1}{2}x^2(0) - \rlap{\,-}\int_0^T \frac{x(t) + x(0)}{2-t}\, dx(t)\right.$$
$$\left. - \frac{1}{2}\int_0^T \frac{x(t) + x(0)}{2-t}\right\}. \tag{36}$$

Shepp [20] and Hájek [24] also solve this problem and obtain the likelihood ratio for it as

$$\frac{2}{\sqrt{2-T}} \exp\left\{\frac{x^2(0)}{2} - \frac{(x(0) + x(T))^2}{2(2-T)}\right\}. \tag{37}$$

The equivalence of (36) and (37) can be established by noting that both formulas agree for $T = 0$ and that the Itô differentials of both are identical (see the verification of equivalence of the likelihood ratio formulas (33) and (34) in Kailath [2]).

V. Concluding Remarks

In Section IV we have used purely Hilbert-space arguments to derive the LR formula (34), which generalized the earlier formula of [2] for the case where one of the processes was a Wiener process. However, in [2] we had given an alternative derivation, similar to one of Hitsuda [25], that was almost entirely based on the theory of martingales of a Wiener process. Our present generalization suggests that we should seek a similar derivation here; though we have not carried out all the details, it seems that such a proof can be obtained by starting with a general formula of Doléans-Dade [26].

In the Introduction we noted some recent work in network theory that used generalized causality concepts as discussed in Section II; we may remark that our identification of this definition with the ideas of Gohberg and Krein makes possible the generalization of many important network-theoretical results, e.g., the Hilbert transform relations between real and imaginary parts (cf. [4, ch. III]).

After this paper had been submitted, Kallianpur sent us a preprint [27] in which he and Oodaira obtain some results similar to those of Sections III and IV. In particular, they also establish Theorem 4 and establish the likelihood ratio formula (34) for the special case where $x_0(\cdot)$ is a free Wiener process. No discussion is given of the generalized Wiener–Hopf theory. Reference [27] was motivated by a comment in our paper on the white-noise case [2] that the work of Gohberg and Krein could be used to extend the results to nonwhite processes; incidentally, we had carried out [5] the extension before the appearance in print of [2].

Acknowledgment

The first steps to the results of this paper were taken with R. Geesey. Of course, the work progressed very rapidly with the discovery of the deep and beautiful studies of Gohberg and Krein; for the inadvertent discovery in May 1968 of their books, T. Kailath is indebted to G. Wallenstein and his insistence on browsing in a Leningrad bookstore.

Appendix

We wish to define an inner product of the form $\langle x(t),f(t)\rangle_{H(R_0)}$ in such a way that it has meaning for both a process $x(\cdot)$ with covariance $R_0(t,s)$ and a process $x(\cdot)$ with covariance $R_0(t,s) + K(t,s)$, where the operator K is Hilbert–Schmidt and the operator $I + K$ is strictly positive.

If $x(\cdot)$ has covariance $R_0(t,s)$, $\langle x(t),f(t)\rangle_{H(R_0)}$ is defined as a congruence inner product, that is, as the unique random variable u in $LL_2(x)$ such that

$$E[ux(\cdot)] = f(\cdot).$$

Since $u \in LL_2(x)$, it is expressible as a limit in quadratic mean of finite linear combinations of the random variables $\{x(t)\}_{t\in T}$. Since the convergence is in quadratic mean, there will exist a subsequence that will converge almost surely. Let

$$\sum_{j=1}^{N_1} \xi_j^{(1)} x(t_j^{(1)}),\ \sum_{j=1}^{N_2} \xi_j^{(2)} x(t_j^{(2)}), \cdots$$

be any such sequence and define

$$\langle x(t),f(t)\rangle_{H(R_0)} = \text{l.i.m.} \sum_{j=1}^{N_n} \xi_j^{(n)} x(t_j^{(n)}). \tag{A1}$$

Clearly, this definition of $\langle x(t),f(t)\rangle_{H(R_0)}$ is consistent with the congruence inner product definition when $x(\cdot)$ has covariance

$R_0(t,s)$. Definition (A1) also has meaning when $x(\cdot)$ has covariance $R_0(t,s) + K(t,s)$ because the probability measures associated with zero-mean Gaussian processes with covariances $R_0(t,s)$ and $R_0(t,s) + K(t,s)$ are equivalent (mutually absolutely continuous): almost sure properties are preserved under absolutely continuous changes of measure.

References

[1] T. Kailath, "The innovations approach to detection and estimation theory," *Proc. IEEE*, vol. 58, pp. 680–695, May 1970.

[2] ——, "Likelihood ratios for Gaussian processes," *IEEE Trans. Inform. Theory*, vol. IT-16, pp. 276–288, May 1970.

[3] T. Kailath and R. A. Geesey, "An innovations approach to least squares estimation—Part IV: Recursive estimation given lumped covariance functions," *IEEE Trans. Automat. Contr.*, vol. AC-16, pp. 720-727, Dec. 1971.

[4] I. Gohberg and M. Krein, "Theory and applications of Volterra operators in Hilbert space," *Transl. Math. Mono.*, vol. 24, Amer. Math. Soc., Providence, R.I., 1970.

[5] D. Duttweiler, "Reproducing kernel Hilbert space techniques for detection and estimation problems," Ph.D. dissertation, Dep. Elec. Eng., Stanford Univ., Stanford, Calif., June 1970.

[6] T. Kailath, "RKHS approach to detection and estimation problems—Part I: Deterministic signals in Gaussian noise," *IEEE Trans. Inform. Theory*, vol. 17, pp. 530–549, Sept. 1971.

[7] W. A. Porter, "Some circuit theory concepts revisited," *Int. J. Contr.*, vol. 12, pp. 433–448, 1970.

[8] ——, "A basic optimization problem in linear systems," *Math. Syst. Theory*, vol. 5, pp. 20–44, 1971.

[9] R. Saeks, "Causality in Hilbert space," *SIAM Rev.*, vol. 12, pp. 357-383, 1970.

[10] ——, "State in Hilbert space," *SIAM Rev.*, to be published; also Univ. Notre Dame, Notre Dame, Ind., Tech. Memo. EE 6912a, 1969.

[11] J. C. Willems, "Stability, instability, invertibility and causality," *SIAM J. Contr.*, vol. 7, pp. 645–671, 1969.

[12] J. C. Willems, "The generation of Lyapunov functions for input-output stable systems," *SIAM J. Contr.*, vol. 9, pp. 105–134, 1971.

[13] G. Kallianpur and H. Oodaira, "The equivalence and singularity of Gaussian measures," in *Proc. Symp. Time Series Analysis*, M. Rosenblatt, Ed. New York: Wiley, 1963.

[14] N. Aronszajn, "Theory of reproducing kernels," *Trans. Amer. Math. Soc.*, vol. 63, pp. 337–404, May 1950.

[15] F. Riesz and B. S. Nagy, *Functional Analysis*. New York: Ungar, 1955.

[16] T. Kailath and R. Geesey, "Covariance factorization—An explication via examples," in *Proc. 2nd Asilomar Conf. Systems and Circuits*, Monterey, Calif., Nov. 1968.

[17] T. Kailath, "Fredholm resolvents, Wiener–Hopf equations, and Riccati differential equations," *IEEE Trans. Inform. Theory*, vol. IT-15, pp. 665–672, Nov. 1969.

[18] W. L. Root, "Singular Gaussian measures in detection theory," in *Proc. Symp. Time Series Analysis*, M. Rosenblatt, Ed. New York: Wiley, 1963.

[19] J. L. Doob, *Stochastic Processes*. New York: Wiley, 1953.

[20] L. A. Shepp, "Radon–Nikodym derivatives of Gaussian measures," *Ann. Math. Statist.*, vol. 37, pp. 321–354, 1966.

[21] I. Gohberg and M. G. Krein, "On the factorization of operators in Hilbert space," *Acta Sci. Math.* (in Hungarian), vol. 25, pp. 90–123, 1964; also translated in *Amer. Math. Soc. Trans.*, vol. 51, 1966.

[22] I. M. Gelfand and N. Ya. Vilenkin, *Generalized Functions*, vol. 4. New York: Academic Press, 1964.

[23] D. E. Varberg, "On equivalence of Gaussian measures," *Pac. J. Math.*, vol. 11, pp. 751–762, 1961.

[24] J. Hájek, "Linear statistical problems," *Czech. Math. J.*, vol. 87, pp. 404–444, Dec. 1962.

[25] M. Hitsuda, "Representation of Gaussian processes equivalent to Wiener process," *Osaka J. Math.*, vol. 5, pp. 299–312, Dec. 1968.

[26] C. Doléans-Dade, "Quelques applications de la formule de changement de variables pour les semimartingales," *Z. Wahr. verw. Geb.*, vol. 16, pp. 181–194, 1970.

[27] G. Kallianpur and H. Oodaira, "Nonanticipative representations of equivalent Gaussian processes," submitted for publication.

18

Reprinted from *Ann. Probab.* **1**:104–122 (1973)

NON-ANTICIPATIVE REPRESENTATIONS OF EQUIVALENT GAUSSIAN PROCESSES[1]

BY G. KALLIANPUR AND H. OODAIRA

University of Minnesota

Given two equivalent Gaussian processes the notion of a *non-anticipative* representation of one of the processes with respect to the other is defined. The main theorem establishes the existence of such a representation under very general conditions. The result is applied to derive such representations explicitly in two important cases where one of the processes is (i) a Wiener process, and (ii) a N-ple Gaussian Markov process. Radon-Nikodym derivatives are also discussed.

1. Introduction. Let $(X(t), P)$ and $(X(t), Q)$, $(0 \leqq t \leqq 1)$ be equivalent Gaussian processes given on some space $(\Omega, \mathscr{A})$ which are quadratic mean continuous, have zero mean functions and covariance functions Γ_P and Γ_Q. The term "equivalent" is here used in the sense that the probability measures P and Q are mutually absolutely continuous with respect to the σ-field $\mathscr{B}$ generated by the random variables $(X(t))$.

By a representation of $(X(t), P)$ in terms of $(X(t), Q)$ we mean a family of random variables $(Y(t))$ $(0 \leqq t \leqq 1)$ on Ω such that

$$(Y(t), Q) \tag{1.1}$$

is a quadratic mean continuous Gaussian process with zero mean and covariance Γ_P, and for each t

$$Y(t) \in L(X; 1) \tag{1.2}$$

where $L(X; 1)$ is the linear space of the process $(X(t), Q)$. (The precise definitions of terms and notation used here are given in Sections 3 and 4.) When $(X(t), Q)$ is a standard Wiener process the following representation for all $(X(t), P)$ equivalent to it was given by Shepp (1966).

$$Y(t) = X(t) - \int_0^t \left[\int_0^1 M(s, u)\, dX(u)\right] ds \tag{1.3}$$

where $M(s, u)$ is a square integrable kernel related to Γ_P. From the point of view of statistical or engineering applications a drawback of this representation is that in order to find $Y(t)$ from (1.3) we need to know all the values of $X(s)$, $(0 \leqq s \leqq 1)$. It is desirable to obtain a representation which involves only the values $(X(s), 0 \leqq s \leqq t)$, i.e. the "present and past" but not the "future" values $X(s)$ $(s > t)$. Just such a representation has recently been given by Hitsuda (1968). Such a representation will be called non-anticipative (see Section 4).

Received September 13, 1971; revised May 23, 1972.

[1] This work was supported in part by NSF Grant GP-1188-8.

Key words and phrases. Equivalent Gaussian processes, representation, Volterra operators, factorization.

The main purpose of this paper is to establish the existence of non-anticipative representations in the most general case. This is done in Section 4 and is based on the fundamental work of Gohberg and Krein on the factorization of operators of the form $I - T$ where T is Hilbert-Schmidt ([2] Chapter IV). In Sections 5 and 6 we apply the general theorem to derive non-anticipative representations when $(X(t), Q)$ is, respectively, the non-standard Wiener process (thus incidentally obtaining an alternative proof of Hitsuda's result) and a Gaussian N-ple Markov process. The problem of the Radon-Nikodym derivative is considered in Section 7. The possibility of using Gohberg and Krein's ideas in this connection was raised by Kailath (1970) who has formally derived Hitsuda's result using the result of [2]. (See comments at the end of Section 5.)

The following remarks form the starting point of our investigations which, we hope, put our problem in its most general setting. Starting from a given representation $(Y(t), Q)$ and using the necessary and sufficient conditions for equivalence obtained in Kallianpur and Oodaira ([5], 1963), (see Theorem 4.2 below), it is easy to verify that (1.1) and (1.2) imply the existence of a bounded linear operator $\tilde{F}$ on $L(X; 1)$ with the following properties: For every t

$$Y(t) = \tilde{F}X(t)\,, \tag{1.4}$$

$$\tilde{S} = \tilde{F}^*\tilde{F}\,, \tag{1.5}$$

$\tilde{S}$ being the operator in $L(X; 1)$ corresponding to the operator S of Theorem 4.2, so that $\tilde{S} = I - \tilde{T}$ where $\tilde{T}$ is Hilbert-Schmidt and 1 is not a point of the spectrum of $\tilde{T}$. Conversely, if $\tilde{F}$ is an operator on $L(X; 1)$ satisfying (1.5) then $(Y(t), Q)$ where $Y(t)$ is defined by (1.4), is a representation of $(X(t), P)$ in the sense of (1.1) and (1.2). Thus every representation of $(X(t), P)$ uniquely corresponds (the uniqueness is easily seen) to a factorization of S (or $\tilde{S}$) of the type (1.5). What we intend to do is to pick out the particular factorization that corresponds to the non-anticipative representation. That such a factorization exists is not obvious a priori and that is where Gohberg and Krein's theory of special factorization enters into the picture in a natural fashion. Before considering it in Section 4 we discuss some basic ideas and results in the next section.

2. Factorization of self-adjoint positive invertible operators. Let H be a separable Hilbert space. A family of orthoprojectors $\pi = \{P\}$ is called a chain if for any distinct $P_1, P_2 \in \pi$, either $P_1 < P_2$ or $P_2 < P_1$, where $P_1 < P_2$ means $P_1H \subset P_2H$, i.e., $P_1P_2 = P_2P_1 = P_1$. We shall write $P_1 \leqq P_2$ if either $P_1 < P_2$ or $P_1 = P_2$. A chain π is said to be bordered if $\pi \ni 0, I$. The closure of a chain π is the set of all operators which are the strong limits of sequences in π. The closure of a chain is again a chain and if a chain coincides with its closure, it is said to be closed. A pair (P^-, P^+) of orthoprojectors in a closed chain π with $P^- < P^+$ is called a gap of π if for any $P \in \pi$ either $P \leqq P^-$ or $P \geqq P^+$, and the dimension of $P^+ - P^-$, i.e., $\dim [P^+H \ominus P^-H]$, is called the dimension of the gap (P^-, P^+). A chain is said to be maximal if it cannot be enlarged, or, equivalently, if it is bordered, closed and its gaps (if any) are one-dimensional. A chain π is called an eigenchain of a bounded linear operator A on H if $PAP = AP$ for all $P \in \pi$.

Let π be a closed chain. A partition ζ of π is a chain consisting of a finite number of elements $\{P_0 < P_1 < \cdots < P_n\}$ of π such that $P_0 = \min_{P \in \pi} P$ and $P_n = \max_{P \in \pi} P$. Let $F(P)$ be an operator function defined on π and having as its values bounded linear operators on H. For a partition $\zeta = \{P_0 < P_1 < \cdots < P_n\}$ of π, define

$$S(\zeta) = \textstyle\sum_{j=1}^{n} F(P_{j-1})\Delta P_j , \qquad \Delta P_j = P_j - P_{j-1} .$$

An operator A is called the limit in norm of $S(\zeta)$, denoted by

$$A = (m) \textstyle\int_\pi F(P)\, dP , \tag{2.1}$$

if for any $\varepsilon > 0$ there exists a partition $\zeta(\varepsilon)$ of π such that, for every partition $\zeta \supset \zeta(\varepsilon)$, $\|S(\zeta) - A\| < \varepsilon$. If the limit of $S(\zeta)$ exists, we shall say that the integral (2.1) converges. The integral

$$B = (m) \textstyle\int_\pi dPF(P)$$

is defined analogously.

The dual $\pi^\perp$ of a chain π is a chain consisting of all orthoprojectors of the form $P^\perp = I - P$, $P \in \pi$. If π is an eigenchain of an operator A, then the dual chain $\pi^\perp$ is an eigenchain of the adjoint operator A^*.

By a special factorization of an operator A along a chain π we mean the representation of A in the form

$$A = (I + X_+)D(I + X_-) , \tag{2.2}$$

where X_+ and X_- are Volterra operators (i.e., completely continuous operators with the one-point spectrum $\lambda = 0$) having π and $\pi^\perp$ as eigenchains respectively, D commutes with all $P \in \pi$, and $D - I$ is completely continuous.

The factors $I + X_+$, $I + X_-$ are invertible, and if A is invertible, so is the factor D. If an invertible operator A admits a special factorization relative to a maximal chain π, then the factorization is unique, and from the uniqueness it follows that if a self-adjoint invertible operator $A = A^*$ has such a factorization, then $X_+^* = X_-$ and $D^* = D$.

The following theorem is a special case of Theorems 6.1 and 6.2, Chapter IV, Gohberg–Krein [2]. We denote by $\mathscr{S}_2$ the class of all Hilbert-Schmidt operators on H.

THEOREM 2.1. *Let π be a maximal chain. Then, for every operator $T \in \mathscr{S}_2$ such that each of the operators $I - PTP$, $P \in \pi$, is invertible, the integrals*

$$\begin{aligned} X_+ &= (m) \textstyle\int_\pi (I - PTP)^{-1} PT\, dP , \\ X_- &= (m) \textstyle\int_\pi dPTP(I - PTP)^{-1} \end{aligned} \tag{2.3}$$

converge in norm, and the operator $A = (I - T)^{-1}$ has a special factorization (2.2) *along π with X_+, X_-, $D - I \in \mathscr{S}_2$ and*

$$D = I + \textstyle\sum_j (P_j^+ - P_j^-)[(I - P_j^+ T P_j^+)^{-1} - I](P_j^+ - P_j^-) , \tag{2.4}$$

where $\{(P_j^-, P_j^+)\}$ is the set of all gaps in the chain π.

For the convenience of the reader it is perhaps worth pointing out that the deduction of this result from the above-mentioned theorems of Gohberg and Krein is based on the fact that if $T \in \mathcal{S}_2$ then the integral $(m) \int_\pi PT\,dP$ converges in uniform norm (in fact, in Hilbert-Schmidt norm) and belongs to $\mathcal{S}_2$. The verification is simple and is a part of the proof of Theorem 10.1, Chapter I of [2].

LEMMA 2.1. *Let $T \in \mathcal{S}_2$. If $I - T$ is self-adjoint, positive and invertible, then, for any orthoprojector P, $I - PTP$ is invertible.*

The proof is immediate as can be seen from the following inequality.

$$\begin{aligned}\langle (I - PTP)f, f\rangle &= \langle (I - P)f, f\rangle + \langle (I - T)Pf, Pf\rangle \\ &= \|(I - P)f\|^2 + \|(I - T)^{\frac{1}{2}}Pf\|^2 \\ &\geqq \|(I - P)f\|^2 + c^2\|Pf\|^2 \geqq c_1^2\|f\|^2 ,\end{aligned}$$

where c is some positive constant, $\langle \cdot, \cdot \rangle$ and $\|\cdot\|$ are the inner product and the norm of H, and $c_1^2 = \min(1, c^2)$.

LEMMA 2.2. *If a self-adjoint, positive, invertible operator A has a special factorization* (2.2), *then the factor D is self-adjoint, positive and invertible.*

PROOF. We need only to prove the positive definiteness of D. Since $I + X_+$, $I + X_-$ are invertible,

$$\langle Df, f\rangle = \langle (I + X_+)^{-1}A(I + X_-)^{-1}f, f\rangle .$$

Set $(I + X_-)^{-1}f = g$. Then

$$\begin{aligned}\langle Df, f\rangle &= \langle (I + X_+)^{-1}Ag, (I + X_-)g\rangle \\ &= \langle (I + X_-)^*(I + X_+)^{-1}Ag, g\rangle \\ &= \langle (I + X_+)(I + X_+)^{-1}Ag, g\rangle \\ &= \langle Ag, g\rangle \\ &\geqq 0 .\end{aligned}$$

LEMMA 2.3. *If V is a Volterra operator $\in \mathcal{S}_2$ with π as an eigenchain, then the operator $W = (I + V)^{-1} - I$ is also Volterra $\in \mathcal{S}_2$ and has π as an eigenchain.*

PROOF. That $W \in \mathcal{S}_2$ follows immediately from the relation

$$W + V + VW = 0 . \tag{2.5}$$

Since V and $I + W$ are permutable, we have (cf. [6] page 426)

$$r_{V+VW} \leqq r_V \cdot r_{I+W}$$

where r_A denotes the spectral radius of A, i.e., the radius of the smallest closed disk centered at 0 which contains all the spectrum of A. By assumption $r_V = 0$, so $r_{V+VW} = 0$, i.e., $V + VW$ is Volterra. Hence, from (2.5), W is Volterra. Since $I + W$ is the resolvent at 1 of the operator $-V$,

$$W = \sum_{n=1}^{\infty} (-1)^n V^n ,$$

the right-hand side converging in norm. It is readily verified that, for $P \in \pi$, $PV^nP = V^nP$ for any n. Hence we have $PWP = WP$.

From Theorem 2.1 and Lemmas 2.1–2.3 we have the following theorem.

THEOREM 2.2. *Let $S = I - T$ with $T \in \mathscr{S}_2$ be a self-adjoint positive and invertible operator. Then S and S^{-1} have the factorizations along any maximal chain $\pi = \{P\}$*

$$S = (I + W_-)D^{-1}(I + W_+)$$
$$S^{-1} = (I + X_+)D(I + X_-) ,$$

where

(a) *W_+, W_-, X_+, X_- are Volterra operators $\in \mathscr{S}_2$, X_+, X_- are given by* (2.3), *the integral converging in norm, and $I + W_+ = (I + X_+)^{-1}$, $I + W_- = (I + X_-)^{-1}$,*

(b) *W_+, X_+ have π and W_-, X_- have $\pi^\perp$ as eigenchains,*

(c) *$W_+^* = W_-$ and $X_+^* = X_-$,*

(d) *D is a self-adjoint, positive and invertible operator given by* (2.4), *and*

(e) *$D - I \in \mathscr{S}_2$, $DP = PD$ for all $P \in \pi$.*

3. Chains of orthoprojectors associated with a Gaussian process. Let $\{X(t), t \in [a, b]\}$ be a Gaussian process defined on a probability space $(\Omega, \mathscr{A}, Q)$ with $E_Q X(t) \equiv 0$ and covariance function $\Gamma_Q(s, t)$, where $[a, b]$ is taken to be either a finite closed or an infinite interval. For the sake of simplicity we assume that $[0, 1] \subset [a, b]$. Throughout the paper we make the following assumption:

(3.1) $\{X(t)\}$ is continuous in quadratic mean (q.m.) .

For $0 \leqq t \leqq 1$, let $L(X; t)$ be the closed linear subspace spanned by $X\{(\tau), 0 \leqq \tau \leqq t\}$ of $L^2(\Omega, \mathscr{A}, Q)$, and let $\tilde{P}(t)$ be the orthoprojector defined on $L(X; 1)$ with range $L(X; t)$. We are interested in a maximal chain containing the chain $\{\tilde{P}(t), 0 \leqq t \leqq 1\}$ (or $\{\tilde{P}(t), 0 < t \leqq 1\}$).

Let

$$L(X; t+) = \textstyle\bigcap_{s>t} L(X; s)$$

and

$L(X; t -) =$ the smallest closed linear space containing all $L(X; s)$, $s < t$.

Obviously $L(X; 0) \subset L(X; 0+)$ and $L(X; t -) \subset L(X; t) \subset L(X; t+)$ for $t > 0$. It is also easy to verify that $L(X; t-) = L(X; t)$ for all $0 < t \leqq 1$. Since by the assumption (3.1) the Hilbert space $L(X; 1)$ is separable, the set of discontinuities $D = \{t \in [0, 1] \mid L(X; t) \neq L(X; t+)\}$ is at most countable. Let $\tilde{P}(t_j+)$ be the orthoprojector with range $L(X; t_j+)$ for $t_j \in D$. The closure of the chain $\{\tilde{P}(t), 0 \leqq t \leqq 1\}$ consists of $\{\tilde{P}(t), 0 < t \leqq 1\}$ and $\{\tilde{P}(t_j+), t_j \in D\}$. If $D \neq \emptyset$, it has gaps $(\tilde{P}(t_j), \tilde{P}(t_j+))$, $t_j \in D$. If the dimension of gap $(\tilde{P}(t_j), \tilde{P}(t_j+))$ is $n_j > 1$, we write the space $(\tilde{P}(t_j+) - \tilde{P}(t_j))L(X; 1)$ as the orthogonal sum of one-dimensional subspaces $L(j, i)$:

$$(\tilde{P}(t_j+) - \tilde{P}(t_j))L(X, 1) = \textstyle\sum_{i=1}^{n_j} \oplus L(j, i) .$$

Let $\tilde{Q}(j, k)$ be the orthoprojector with range $\sum_{i=1}^{k} \oplus L(j, i)$. Now consider the family of orthoprojectors $\tilde{\pi}$ consisting of 0, $\{\tilde{P}(t), 0 \leqq t \leqq 1\}$, $\{\tilde{P}(t_j+), t_j \in D\}$ and $\{\tilde{P}(t_j) + \tilde{Q}(j, k), k = 1, \cdots, n_j - 1, t_j \in D\}$. It is clear that $\tilde{\pi}$ is a chain and maximal.

REMARK 1. A maximal chain containing the chain $\{\tilde{P}(t), 0 \leqq t \leqq 1\}$ is obviously not unique, in general. If the dimension of gap $(\tilde{P}(t_j), \tilde{P}(t_j+))$ is > 1, we may take different orthoprojectors $\tilde{Q}(j, k)$.

REMARK 2. The gap $(\tilde{P}(0), \tilde{P}(0+))$, if it exists and $\tilde{P}(0) \neq 0$, is special. Instead of filling in the gap $(\tilde{P}(0), \tilde{P}(0+))$, we may insert any set of orthoprojectors $\{\tilde{Q}(j), j = 0, 1, \cdots, n\}$ such that $\tilde{Q}(0) = 0$, $\tilde{Q}(n) = \tilde{P}(0+)$ and $\dim(\tilde{Q}(j) - \tilde{Q}(j-1)) = 1$. The maximal chain thus obtained will suffice for our purposes. In other words, we need only a maximal chain $\tilde{\pi}$ containing $\{\tilde{P}(t), 0 < t \leqq 1\}$.

As we shall see the space $L(X; 0+)$ is of particular interest. It can be trivial, can be n-dimensional ($1 \leqq n < \infty$), or even infinite dimensional.

EXAMPLE 1. If $\{X(t), 0 \leqq t \leqq 1\}$ is a Wiener process, then $L(X; t+) = L(X; t)$ for all $t \in [0, 1]$, i.e., $D = \varnothing$. If $X(0) = 0$, i.e., if $\{X(t)\}$ is the standard Wiener process, the chain $\tilde{\pi} = \{\tilde{P}(t), 0 \leqq t \leqq 1\}$ is maximal. However, if $X(0) \neq 0$, the chain $\tilde{\pi} = \{0, \tilde{P}(t), 0 \leqq t \leqq 1\}$ is maximal and has a one-dimensional gap $(0, \tilde{P}(0) = \tilde{P}(0+))$.

EXAMPLE 2. Let $\{X(t), a \leqq t \leqq b\}$ be an N-ple Gaussian Markov process, where $a < 0$, $b \geqq 1$ (see Section 6). Then $L(X; t+) = L(X; t)$ for all $t \in (0, t]$ and the space $L(X; 0+)$ is N-dimensional.

Let ψ denote the congruence (isometric isomorphism) from $L(X; 1)$ onto the reproducing kernel Hilbert space (RKHS) $H \equiv H(\Gamma_Q)$ with reproducing kernel $\Gamma_Q(s, t)$, $0 \leqq s, t \leqq 1$, such that $\psi X(t) = \Gamma_Q(\cdot, t)$, $0 \leqq t \leqq 1$. We note the following relation between subspaces of $L(X; 1)$ and of H.

Let

$$F(t) = \{f \in H \mid f(s) = 0, 0 \leqq s \leqq t\}$$

and

$$M(t) = H \ominus F(t)\,, \quad \text{the orthogonal complement of} \quad F(t)\,.$$

LEMMA 3.1. $\psi[L(X; t)] = M(t)$.

PROOF. Let $f \in F(t)$ and $\xi = \psi^{-1}f$. Then, for all $s \in [0, t]$,

$$0 = f(s) = \langle f(\cdot), \Gamma(\cdot, s)\rangle = (\xi, X(s))\,,$$

where $\langle \cdot, \cdot \rangle$, $(\cdot, \cdot)$ denote respectively the inner products of H and $L(X; 1)$. Hence $\xi \perp L(X; t)$, and thus $f \in F(t)$ if and only if $\psi^{-1}f \in L(X; 1) \ominus L(X; t)$. This is equivalent to the assertion of the lemma.

Let $P(t)$ denote the orthoprojector on H with range $M(t)$. Then $P(t) = \psi\tilde{P}(t)\psi^{-1}$. If $\tilde{\pi} = \{\tilde{P}\}$ is a maximal chain containing $\{\tilde{P}(t), 0 \leqq t \leqq 1\}$, then, obviously, the chain $\pi = \{\psi\tilde{P}\psi^{-1}\}$ is maximal and contains $\{P(t), 0 < t \leqq 1\}$. We shall consistently use the following notation. If A is any linear operator on H, $\tilde{A}$ denotes the operator on $L(X; 1)$ given by $\tilde{A} = \psi^{-1}A\psi$.

4. Non-anticipative representations of equivalent Gaussian processes. Let $\mathscr{B}$ denote the σ-field generated by the random variables $\{X(t), 0 \leqq t \leqq 1\}$. Let P be another probability measure on $(\Omega, \mathscr{B})$ such that $\{X(t), 0 \leqq t \leqq 1, P\}$ is a q.m. continuous Gaussian process with $E_P X(t) = 0$ and covariance function $\Gamma_P(s, t)$. ($\{X(t), 0 \leqq t \leqq 1, Q\}$ is, by assumption, a q.m. continuous Gaussian process with $E_Q X(t) = 0$ and covariance function $\Gamma_Q(s, t)$.) Assume that P and Q are equivalent, i.e., mutually absolutely continuous relative to $\mathscr{B}$.

A non-anticipative representation of a Gaussian process with respect to another is defined as follows.

DEFINITION. The process $\{X(t), 0 \leqq t \leqq 1, P\}$ has a non-anticipative representation with respect to $\{X(t), 0 \leqq t \leqq 1, Q\}$ if there is a Gaussian process $\{Y(t), 0 \leqq t \leqq 1, Q\}$, having zero mean and Γ_P for its covariance, with the following property:

(4.1) $$Y(t) \in L(X; t) \quad \text{for each} \quad t \in [0, 1] .$$

REMARK. (4.1) implies $L(Y; t) \subset L(X; t)$ for $t \in [0, 1]$.

We shall now prove the following main theorem.

THEOREM 4.1. *Every Gaussian process $\{X(t), 0 \leqq t \leqq 1, P\}$ which is equivalent to a given Gaussian process $\{X(t), 0 \leqq t \leqq 1, Q\}$ has a non-anticipative representation with respect to $\{X(t), 0 \leqq t \leqq 1, Q\}$. The processes are assumed to be* q.m. *continuous.*

The proof is based on Theorem 2.2. and the following necessary and sufficient conditions for equivalence of P and Q (cf. [5]).

THEOREM 4.2. *Gaussian measures P and Q are equivalent if and only if Γ_P defines an operator S on the* RKHS *$H(\Gamma_Q)$ with the following properties*:

(a) $\Gamma_P(\cdot, t) = S\Gamma_Q(\cdot, t)$ *for* $0 \leqq t \leqq 1$,

(b) *S is a bounded, self-adjoint, positive operator,*

(c) $T = I - S \in \mathscr{S}_2$,

(d) $1 \notin \sigma(T)$, *the spectrum of T.*

PROOF OF THEOREM 4.1. Consider a maximal chain π in $H(\Gamma_Q)$ described in the preceding section. Applying Theorem 2.2 to the operator S defined in Theorem 4.2, we have

$$S = (I + W_-)\Delta(I + W_+) ,$$

where $\Delta = D^{-1}$. The operator Δ is self-adjoint and positive. Since D commutes with all $P \in \pi$, we have $\Delta^{\frac{1}{2}}P = P\Delta^{\frac{1}{2}}$ for all $P \in \pi$. If we write

$$F = \Delta^{\frac{1}{2}}(I + W_+)$$

then

$$S = F^*F ,$$

because $F^* = (I + W_+^*)\Delta^{\frac{1}{2}} = (I + W_-)\Delta^{\frac{1}{2}}$.

Consider now the operator $\tilde{F}$ on $L(X; 1)$ corresponding to F:

(4.2) $$\tilde{F} = \tilde{\Delta}^{\frac{1}{2}}(I + \tilde{W}_+) .$$

Define

$$Y(t) = \tilde{F}X(t) , \qquad t \in [0, 1] .$$

Since $\tilde{F}$ is a linear operator on $L(X; 1)$, $\{Y(t), 0 \leqq t \leqq 1, Q\}$ is Gaussian. Furthermore, $E_Q Y(t) \equiv 0$ and

$$\begin{aligned} E_Q Y(s)Y(t) &= (\tilde{F}X(s), \tilde{F}X(t)) \\ &= \langle F\Gamma_Q(\cdot, s), F\Gamma_Q(\cdot, t)\rangle \\ &= \langle S\Gamma_Q(\cdot, s), \Gamma_Q(\cdot, t)\rangle \\ &= \langle \Gamma_P(\cdot, s), \Gamma_Q(\cdot, t)\rangle \\ &= \Gamma_P(s, t) . \end{aligned}$$

Thus $\{Y(t), 0 \leqq t \leqq 1, Q\}$ is a Gaussian process with zero mean and covariance function $\Gamma_P(s, t)$. We have also for each $t \in [0, 1]$,

$$\begin{aligned} Y(t) &= \tilde{\Delta}^{\frac{1}{2}}(I + \tilde{W}_+)X(t) \\ &= \tilde{\Delta}^{\frac{1}{2}}(I + \tilde{W}_+)\tilde{P}(t)X(t) \\ &= \tilde{\Delta}^{\frac{1}{2}}\tilde{P}(t)(I + \tilde{W}_+)\tilde{P}(t)X(t) \\ &= \tilde{P}(t)\tilde{\Delta}^{\frac{1}{2}}(I + \tilde{W}_+)\tilde{P}(t)X(t) \\ &= \tilde{P}(t)Y(t) . \end{aligned}$$

This shows that $Y(t) \in L(X; t)$. The proof of the theorem is complete.

We have shown that the non-anticipative representation $Y(t)$ is given by

(4.3) $$Y(t) = \tilde{\Delta}^{\frac{1}{2}}(I + \tilde{W}_+)X(t) .$$

We are indebted to Yu. A. Rozanov for the remark that the representation (4.3) can be cast in the form

(4.4) $$Y(t) = X(t) + \tilde{G}X(t)$$

where $\tilde{G}$ is a Hilbert-Schmidt operator in the space $L(X; 1)$ such that for each t

(4.5) $$\tilde{G}L(X; t) \subseteqq L(X; t) .$$

This is easily deduced from Theorems 2.2 and 4.1 as follows:
Write $\tilde{G} = \tilde{F} - I$. From (4.2) we have

(4.6) $$\tilde{G} = \tilde{G}_1 + \tilde{G}_2$$

where $\tilde{G}_1 = \tilde{\Delta}^{\frac{1}{2}} - I$ and $\tilde{G}_2 = \tilde{\Delta}^{\frac{1}{2}}\tilde{W}_+$.
Now

$$\tilde{G}_1 = \tilde{D}^{-\frac{1}{2}}(I - \tilde{D}^{\frac{1}{2}}) = \tilde{D}^{-\frac{1}{2}}(I + \tilde{D}^{\frac{1}{2}})^{-1}(I - \tilde{D}) .$$

Hence it follows that $\tilde{G}_1$ is Hilbert-Schmidt (i.e. belongs to $\mathscr{S}_2$), self-adjoint and that $\tilde{G}_1\tilde{P} = \tilde{P}\tilde{G}_1$ for all $\tilde{P} \in \tilde{\pi}$. Since W_+ belongs to $\mathscr{S}_2$ and has $\tilde{\pi}$ as an eigenchain and since $\tilde{\Delta}^{\frac{1}{2}}$ is a self-adjoint operator with $\tilde{\pi}$ as an eigenchain we conclude that $\tilde{G}_2 = \tilde{\Delta}^{\frac{1}{2}}\tilde{W}_+$ belongs to $\mathscr{S}_2$ and has $\tilde{\pi}$ as an eigenchain. This proves the assertions (4.4) and (4.5).

In fact, we can strengthen the above remark and show that (4.6) yields a decomposition of the non-anticipative Hilbert-Schmidt operator $\tilde{G}$ into the sum of two such operators $\tilde{G}_1$ and $\tilde{G}_2$ *where* $\tilde{G}_2$ *is Volterra*. That $\tilde{G}_2$ is Volterra follows from the fact that the Volterra operator $\tilde{W}_+$ and the bounded self-adjoint operator $\Delta^{\frac{1}{2}}$ have the maximal chain $\tilde{\pi}$ for a common eigenchain. For if $(\tilde{P}^-, \tilde{P}^+)$ is any gap of $\tilde{\pi}$ we have

$$(\tilde{P}^+ - \tilde{P}^-)\tilde{\Delta}^{\frac{1}{2}}\tilde{W}_+(\tilde{P}^+ - \tilde{P}^-) = \tilde{\Delta}^{\frac{1}{2}}(\tilde{P}^+ - \tilde{P}^-)\tilde{W}_+(\tilde{P}^+ - \tilde{P}^-) = 0 .$$

From Theorem 5.1 of Chapter I of [2] it follows that $\tilde{\Delta}^{\frac{1}{2}}\tilde{W}_+$ has a triangular representation and is therefore, of course, a Volterra operator. The triangular representation is given by

$$\tilde{\Delta}^{\frac{1}{2}}\tilde{W}_+ = \int_{\tilde{\pi}} \tilde{P}\tilde{H}\,d\tilde{P} \tag{4.7}$$

where

$$\tilde{H} = (\tilde{\Delta}^{\frac{1}{2}}\tilde{W}_+) - (\tilde{\Delta}^{\frac{1}{2}}\tilde{W}_+)^* = \tilde{\Delta}^{\frac{1}{2}}\tilde{W}_+ - \tilde{W}_-\tilde{\Delta}^{\frac{1}{2}} . \tag{4.8}$$

Let us recall that $\tilde{\pi}$ is a maximal chain containing the orthoprojectors $\tilde{P}(t)$ $(0 \leqq t \leqq 1)$ defined in Section 3. Since $\tilde{P}(t)X(t) = X(t)$ it follows from (4.7) that

$$\tilde{G}_2X(t) = \int_{\tilde{\pi}} \tilde{P}\tilde{H}\,d\tilde{P}X(t) = \int_{\tilde{\pi}} \tilde{P}\tilde{H}\,d\tilde{P}\tilde{P}(t)X(t) = \int_{\tilde{\pi}_t} \tilde{P}\tilde{H}\,d\tilde{P}X(t) \tag{4.9}$$

where $\tilde{\pi}_t$ is the chain $\{\tilde{P} \in \tilde{\pi} : \tilde{P} \leqq \tilde{P}(t)\}$. The last step in (4.9) is easily verified from the definition of the operator integral.

We thus arrive at an alternate and perhaps more interesting version of Theorem 4.1.

THEOREM 4.3. *Every Gaussian process* $\{X(t), 0 \leqq t \leqq 1, P\}$ *(satisfying the conditions stated at the beginning of the section) which is equivalent to a given Gaussian process* $\{X(t), 0 \leqq t \leqq 1, Q\}$ *has a non-anticipative representation* $Y(t)$ *given by*

$$Y(t) = X(t) + (\tilde{\Delta}^{\frac{1}{2}} - I)X(t) + \int_{\tilde{\pi}_t} \tilde{P}\tilde{H}\,d\tilde{P}X(t) . \tag{4.10}$$

If we know more about the nature of the $(X(t), Q)$ process it is reasonable to expect a more "concrete" representation for the operator $\tilde{G}$, leading to an explicit evaluation of $\tilde{\Delta}^{\frac{1}{2}}$ and the replacement of the last term in (4.10) by an expression involving conventional stochastic integrals. We investigate this question in the following sections and obtain explicit forms for $Y(t)$ for some important special cases.

5. Non-anticipative representation of a Gaussian process equivalent to a Wiener process. Suppose that $\{X(t), 0 \leqq t \leqq 1, Q\}$ is a Wiener process with $EX(t) = 0$, $E(X(t) - X(0))^2 = t$ and $EX^2(0) = \sigma^2 \geqq 0$. Then

$$\begin{aligned}\Gamma_Q(s, t) &= EX(s)X(t) \\ &= \sigma^2 + \min(s, t) \\ &= \int_0^1 \chi(s, u)\chi(t, u)\mu(d\mu) ,\end{aligned} \tag{5.1}$$

where

$$\chi(t, u) = 1 \quad \text{if} \quad 0 \leqq u \leqq t$$
$$= 0 \quad \text{if} \quad t < u \leqq 1$$

and the measure μ assigns point mass σ^2 at $u = 0$ and is Lebesgue measure over $(0, 1]$. (5.1) implies that there is an isometric isomorphism from $H(\Gamma_Q)$ onto $L^2([0, 1], \mu)$ which sends $\Gamma_Q(\cdot, t)$ to $\chi(t, u)$, and any element $f \in H(\Gamma_Q)$ is represented in the form

$$f(t) = \int_0^t \hat{f}(u)\mu(du) = \hat{f}(0) + \int_0^t \hat{f}(u)\, du$$

with $\hat{f} \in L^2([0, 1], \mu)$. Correspondingly, there is an isometric isomorphism θ from $L(X; 1)$ onto $L^2([0, 1], \mu)$ such that $\theta X(t) = \chi(t, u)$ and, for any $\xi \in L(X; 1)$, we have

$$\xi = \hat{f}(0) + \int_0^1 \hat{f}(u)\, dB(u)\,, \tag{5.2}$$

where $B(u) = X(u) - X(0)$ and $\hat{f} = \theta\xi$.

It is easy to see that

$$\theta[L(X; t)] = \{\hat{f} \in L^2([0, 1], \mu) \,|\, \hat{f}(s) = 0 \text{ a.e. } \mu \text{ for } t < s \leqq 1\}\,.$$

The chain $\pi = \{0, P(t) = \theta\tilde{P}(t)\theta^{-1}, 0 \leqq t \leqq 1\}$ is maximal and $P(t) \in \pi$ is characterized by

$$P(t)\hat{f}(u) = \hat{f}(u)\chi(t, u)\,. \tag{5.3}$$

We shall denote by A the operator on $L^2([0, 1], \mu)$ corresponding to an operator $\tilde{A}$ on $L(X; 1)$ (or A on $H(\Gamma_Q)$), $L^2([0, 1], \mu)$ being a representation of $H(\Gamma_Q)$.

LEMMA 5.1. *Let $K(u, v)$ be the kernel in $L^2([0, 1] \times [0, 1], \mu \times \mu)$ corresponding to a Hilbert-Schmidt operator K on $L^2([0, 1], \mu)$. If the chain π is an eigenchain for K, then*

$$K(u, v) = 0 \quad \text{a.e.} \quad \mu \times \mu \quad \textit{for} \quad u > v$$

and if, in addition, K is Volterra and $\sigma^2 > 0$,

$$K(0, 0) = 0\,.$$

PROOF. Since, by assumption,

$$KP(t)\hat{f} = P(t)KP(t)\hat{f}\,, \qquad 0 \leqq t \leqq 1\,,$$

for $\hat{f} \in L^2([0, 1], \mu)$, it follows from (5.3) that

$$\int_0^t K(u, v)\hat{f}(v)\mu(dv) = 0 \quad \text{a.e.} \quad \mu \quad \text{for} \quad u > t\,. \tag{5.4}$$

Hence we have

$$\int_0^1\int_0^1 K(u, v)(1 - \chi(t, u))\chi(t, v)\hat{g}(u)\hat{f}(v)\mu(d\mu)\mu(dv) = 0$$

for all $0 \leqq t \leqq 1$ and for any $\hat{f}, \hat{g} \in L^2([0, 1], \mu)$. Let $\hat{f}(u) = \chi((a, b], u) = \chi(b, u) - \chi(a, u)$ and $\hat{g}(v) = \chi((c, d], v) = \chi(d, v) - \chi(c, v)$. Then, for either $b \leqq c$ or $a \leqq d$,

$$\int_0^1\int_0^1 (1 - \chi(u, v))K(u, v)\chi((a, b], u)\chi((c, d], v)\mu(d\mu)\mu(dv) = 0\,.$$

Since the family $\{\chi((a, b], u)\chi((c, d], v)$ with $b \leqq c$ or $a \geqq d\}$ spans $L^2([0, 1] \times [0, 1]$, Lebesgue measure), we have

$$(1 - \chi(u, v))K(u, v) = 0 \quad \text{a.e.} \quad \mu \times \mu \qquad \begin{aligned} &\text{for} \quad u, v \geqq 0 \quad \text{if} \quad \sigma^2 = 0 \\ &\text{for} \quad u, v > 0 \quad \text{if} \quad \sigma^2 > 0 , \end{aligned}$$

i.e.,

$$K(u, v) = 0 \quad \text{a.e.} \quad \mu \qquad \begin{aligned} &\text{for} \quad u > v \geqq 0 \quad \text{if} \quad \sigma^2 = 0 \\ &\text{for} \quad u > v > 0 \quad \text{if} \quad \sigma^2 > 0 . \end{aligned}$$

In the case $\sigma^2 > 0$, setting $t = 0$ in (5.4), we immediately obtain

$$K(u, 0) = 0 \quad \text{a.e.} \quad \mu \qquad \text{for} \quad u > 0 .$$

Finally, if $\sigma^2 > 0$, we have, for $\hat{f}(v) = \chi(0, v)$,

$$K\hat{f} = K(0, 0)\sigma^2\hat{f} ,$$

and hence, if K is Volterra, $K(0, 0) = 0$, for otherwise $K(0, 0)\sigma^2$ would be a nonzero eigenvalue. The proof is complete.

LEMMA 5.2. *Let $T(u, v)$ be the $L^2([0, 1] \times [0, 1], \mu \times \mu)$ kernel corresponding to the Hilbert-Schmidt operator $T = I - S$. Then for $\hat{f} \in L^2([0, 1], \mu)$,*

$$\text{(5.5)} \qquad \begin{aligned} \Delta^{\frac{1}{2}}\hat{f}(u) &= [1 - \sigma^2 T(0, 0)]^{\frac{1}{2}}\hat{f}(0) \qquad && \textit{with } 1 - \sigma^2 T(0, 0) > 0 \textit{ for } u = 0 \\ &= \hat{f}(u) && \textit{for } 0 < u \leqq 1 . \end{aligned}$$

PROOF. The only possible gap is $(0, P(0) = P(0+))$, and so $D = I + P(0)[(I - P(0)TP(0))^{-1} - I]P(0)$. Direct verification shows that $\Delta = D^{-1} = I - P(0)TP(0)$. From (5.2) it follows that

$$\begin{aligned} \Delta\hat{f}(u) &= [1 - \sigma^2 T(0, 0)]\hat{f}(0) \qquad && \text{if} \quad u = 0 \\ &= \hat{f}(u) && \text{if} \quad 0 < u \leqq 1 . \end{aligned}$$

Hence we have (5.5). If $\sigma^2 > 0$, then $1 - \sigma^2 T(0, 0) > 0$, because $\Delta^{\frac{1}{2}}$ is positive.

Consider now

$$\theta Y(t) = \Delta^{\frac{1}{2}}(I + W_+)\theta X(t) = \Delta^{\frac{1}{2}}(I + W_+)\chi(t, \cdot) .$$

Applying Lemma 5.1 to the Volterra operator W_+, we have

$$\begin{aligned} (I + W_+)\chi(t, \cdot) &= \chi(t, \cdot) + \textstyle\int_0^1 W_+(\cdot, v)\chi(t, v)\mu(dv) \\ &= \chi(t, \cdot) + \textstyle\int_0^t W_+(\cdot, v)\, dv , \end{aligned}$$

where $W_+(u, v)$ is the Volterra kernel (i.e., $W_+(u, v) = 0$ a.e. μ for $u > v$ and $W_+(0, 0) = 0$ if $\sigma^2 > 0$) in $L^2([0, 1] \times [0, 1], \mu \times \mu)$ corresponding to W_+. Hence, by Lemma 5.2,

$$\begin{aligned} \theta Y(t)(u) &= [1 - \sigma^2 T(0, 0)]^{\frac{1}{2}}\{1 + \textstyle\int_0^t W_+(0, v)\, dv\} \qquad && \text{for} \quad u = 0 \\ &= \chi(t, u) + \textstyle\int_0^t W_+(u, v)\, dv && \text{for} \quad 0 < u \leqq 1 . \end{aligned}$$

Taking $\xi = Y(t)$ in (5.2), we thus obtain the non-anticipative representation

$$\begin{aligned} Y(t) &= [1 - \sigma^2 T(0, 0)]^{\frac{1}{2}}\{1 + \textstyle\int_0^t W_+(0, v)\, dv\}X(0) \\ &\quad + \textstyle\int_0^1 \chi(t, u)\, dB(u) + \int_0^1 \int_0^t W_+(u, v)\, dv\, dB(u) \\ &= [1 - \sigma^2 T(0, 0)]^{\frac{1}{2}}\{1 + \textstyle\int_0^t W_+(0, v)\, dv\}X(0) \\ &\quad + B(t) + \textstyle\int_0^t \{\int_0^v W_+(u, v)\, dB(u)\}\, dv , \end{aligned} \tag{5.6}$$

where $W_+(u, v)$ is a Volterra kernel in $L^2([0, 1] \times [0, 1])$ and $1 - \sigma^2 T(0, 0) > 0$.

The non-anticipative representation of Gaussian process equivalent to the standard Wiener process. If, in particular, $X(0) = 0$, i.e., if $\{X(t), 0 \leqq t \leqq 1, Q\}$ is the standard Wiener process $\{B(t), 0 \leqq t \leqq 1, Q\}$, then (5.6) becomes

$$Y(t) = X(t) + \textstyle\int_0^t \{\int_0^v W_+(u, v)\, dX(u)\}\, dv .$$

This formula has been obtained by Hitsuda (1968) using martingale theory and Girsanov's theorem, and also formally by Kailath (1970) using Gohberg–Krein's results.

6. Non-anticipative representation of Gaussian process equivalent to an N-ple Gaussian-Markov process. Suppose that the process $\{X(t), a \leqq t \leqq b, Q\}$ $(a < 0 < 1 \leqq b)$ with $E_Q X(t) \equiv 0$ has the representation of the form

$$X(t) = \textstyle\int_a^t F(t, u)\, dB(u) ,$$

where $\{B(u), a \leqq u \leqq b\}$ is the standard Wiener process,

$$F(t, u) = \textstyle\sum_{j=1}^N f_j(t) g_j(u)$$

and $\{f_j(t)\}$, $\{g_j(u)\}$ satisfy the following conditions.

(a) $f_j(t) \in C^{N-1}[a, b]$, $j = 1, \cdots, N$, and $\det(f_j(t_i)) \neq 0$ for any choice of N distinct indices $\{t_i\}$,

(b) $g_j(u) \in L^2([a, b] \cap (-\infty, t])$ for all $t < \infty$, $j = 1, \cdots, N$, and $\{g_j\}$ is linearly independent,

(c) $\partial^k/\partial t^k F(t, u)|_{t=u} = 0$, $k = 0, 1, \cdots, N-2$, $\partial^{N-1}/\partial t^{N-1} F(t, u)|_{t=u} \neq 0$ on $[a, b]$.

Then, for $0 \leqq t \leqq 1$, we have

$$X(t) = \textstyle\sum_{j=1}^N f_j(t)\eta_j + \int_0^1 \chi(t, u) F(t, u)\, dB(u) . \tag{6.1}$$

where

$$\eta_j = \textstyle\int_a^0 g_j(u)\, dB(u) , \qquad j = 1, 2, \cdots, N .$$

Let C be the covariance matrix of $\eta_j, j = 1, 2, \cdots, N$, and let H_N be the N-dimensional space consisting of column vectors α with inner product $\langle \alpha, \beta \rangle = \alpha^* C \beta$, $*$ denoting the transpose. From the assumptions (a)—(c) it follows that the family of functions $\{\chi(\tau, u)F(\tau, u), 0 \leqq \tau \leqq t\}$ spans the space $L^2[0, t]$ for each $0 \leqq t \leqq 1$. Then (6.1) implies that there is an isometric isomorphism θ from $L(X; 1)$ onto the space $H_N \oplus L^2[0, 1]$ such that

$$\theta X(t) = \{(f_j(t)); \chi(t, u)F(t, u)\} \in H_N \oplus L^2[0, 1]$$

and any element $\xi \in L(X; 1)$ has the representation

$$\xi = \sum_{j=1}^{N} \alpha_j \eta_j + \int_0^1 \varphi(u)\, dB(u) , \tag{6.2}$$

where $\alpha = (\alpha_1, \cdots, \alpha_N)^* \in H_N$, $\varphi \in L^2[0, 1]$ and $\theta\xi = \{a; \varphi\}$. Just as in the preceding section, the operator on $H_N \oplus L^2[0, 1]$ corresponding to an operator $\tilde{A}$ on $L(X; 1)$ will be denoted by A.

Since $\theta[L(X; t)] = H_N \oplus L^2[0, t]$ for $0 < t \leqq 1$, we have $L(X; t+) = L(X; t)$ for all $0 < t \leqq 1$, and $\theta[L(X; 0+)] = H_N$. Thus the chain $\tilde{\pi} = \{0, \tilde{P}(0+), \tilde{P}(t), 0 < t \leqq 1\}$ is closed. It has an N-dimensional gap $(0, \tilde{P}(0+))$. Let H_j, $j = 0, 1, \cdots, N$, denote the subspaces of H_N consisting of vectors of the form $(\alpha_1, \cdots, \alpha_j, 0, \cdots, 0)^*$. Define the orthoprojectors Q_j, $j = 0, 1, \cdots, N$, on $H_N \oplus L^2[0, 1]$ with range H_j. Then the chain $\pi = \{0 = Q_0, Q_1, \cdots, Q_{N-1}, Q_N = P(0+), P(t), 0 < t \leqq 1\}$, where $P(t) = \theta\tilde{P}(t)\theta^{-1}$, is maximal, all gaps (Q_{j-1}, Q_j) being one-dimensional. Q_j, $P(t)$ are characterized by

$$\begin{aligned} Q_j[H_N \oplus L^2[0, 1]] &= H_j , \qquad j = 0, 1, \cdots, N , \\ P(t)[H_N \oplus L^2[0, 1]] &= H_N \oplus L^2[0, t] . \end{aligned} \tag{6.3}$$

It is convenient to use the following matrix form for linear operators K on $H_N \oplus L^2[0, 1]$. We shall write

$$K = \begin{bmatrix} K_{11} & K_{12} \\ K_{21} & K_{22} \end{bmatrix} ,$$

where K_{11} is an operator on H_N, i.e., an $N \times N$ matrix, K_{12} an operator from $L^2[0, 1]$ into H_N, which may be represented in the form of a column vector $(K(i, v))$, K_{21} an operator from H_N into $L^2[0, 1]$, which may be written as a row vector $(K(u, j))$, and K_{22} an operator on $L^2[0, 1]$. If we write an element $h = \{h_1; h_2\} \in H_N \oplus L^2[0, 1]$, $h_1 \in H_N$, $h_2 \in L^2[0, 1]$, as a column vector, then Kh is obtained by the usual multiplication rule.

LEMMA 6.1. *Let K be a Hilbert-Schmidt Volterra operator on $H_N \oplus L^2[0, 1]$ having π as an eigenchain. Then,*

(i) $K_{11} = (k_{ij})$ *is a Volterra matrix*, i.e. $k_{ij} = 0$ *for* $i \geqq j$,
(ii) $K_{21} = (K(u, j)) = 0$ a.e.
(iii) $K_{22}(u, v) = 0$ a.e. *for* $u > v$,

where $K_{22}(u, v)$ is the $L^2([0, 1] \times [0, 1])$ kernel corresponding to K_{22}.

PROOF. By assumption,

$$Q_n K Q_n = K Q_n , \qquad n = 0, 1, 2, \cdots, N , \tag{6.4}$$

and

$$P(t) K P(t) = K P(t) , \qquad 0 < t \leqq 1 . \tag{6.5}$$

Let $h = \{\alpha; \varphi\} \in H_N \oplus L^2[0, 1]$, where $\alpha = (\alpha_1, \cdots, \alpha_N)^* \in H_N$ and $\varphi \in L^2[0, 1]$. Then, from (6.3) and (6.4),

$$Q_N K Q_N h = \{K_{11}\alpha; 0\} = K Q_N h = \{K_{11}\alpha; K_{21}\alpha\} .$$

Hence $K_{21} = 0$, which is (ii). Also

$$KQ_n h = \{(\sum_{j=1}^n k_{ij}\alpha_j); \sum_{j=1}^n K(u, j)\alpha_j\}$$
$$= \{(\sum_{j=1}^n k_{ij}\alpha_j); 0\}$$

and

$$Q_n KQ_n h = \{(\sum_{j=1}^n k_{1j}\alpha_j, \cdots, \sum_{j=1}^n k_{nj}\alpha_j, 0, \cdots, 0)^*; 0\}.$$

Thus $\sum_{j=1}^n k_{ij}\alpha_j = 0$ for $i > n$. Taking $\alpha = e_j = (\delta_{ij})$, $j = 1, \cdots, N$, we have $k_{ij} = 0$ for $i > j$. We have also $k_{jj} = 0$ for all j, for otherwise e_j would be an eigenvector with non-zero eigenvalue k_{jj}. (iii) follows by the same argument as in the proof of Lemma 5.1, using (6.5) and taking $h = \{0; \varphi\}$.

From the lemma we see that the operator W_+ is of the form

$$W_+ = \begin{bmatrix} W_{11} & W_{12} \\ 0 & W_{22} \end{bmatrix}$$

where W_{11} is a Volterra matrix (w_{jk}) with $w_{jk} = 0$ for $j \geqq k$ and the kernel $W_{22}(u, v)$ of W_{22} is a Volterra kernel.

Now the operator $\Delta^{\frac{1}{2}}$ is given in the following form.

LEMMA 6.2.

$$\Delta^{\frac{1}{2}} = \sum_{j=1}^N (d_j/d_{j-1})^{\frac{1}{2}}(Q_j - Q_{j-1}) + (I - Q_N),$$

where $d_0 = 1$, $d_j > 0$, $j = 1, 2, \cdots, N$, *are the principal minors of the matrix* $(I - T_{11})$, T_{11} *being the* $N \times N$ *matrix as the component of the operator* T.

PROOF.

$$D = I + \sum_{j=1}^N (Q_j - Q_{j-1})[(I - Q_j TQ_j)^{-1} - I](Q_j - Q_{j-1})$$
$$= I - Q_N + \sum_{j=1}^N (Q_j - Q_{j-1})(I - Q_j TQ_j)^{-1}(Q_j - Q_{j-1}).$$

Now

$$I - Q_j TQ_j = I - Q_j + Q_j(I(j) - T(j))Q_j,$$

where $I(j)$ is the $j \times j$ identity matrix and $T(j) = Q_j TQ_j$ regarded as a $j \times j$ matrix. Hence we see that

$$(I - Q_j TQ_j)^{-1} = I - Q_j + Q_j(I(j) - T(j))^{-1}Q_j$$

and

$$D = I - Q_N + \sum_{j=1}^N (Q_j - Q_{j-1})(I(j) - T(j))^{-1}(Q_j - Q_{j-1}).$$

Regarding now $Q_j - Q_{j-1}$ as a $j \times j$ matrix with (j, j)th element $= 1$ and all other elements $= 0$, we have

$$D = I - Q_N + \sum_{j=1}^N (d_{j-1}/d_j)(Q_j - Q_{j-1}).$$

Therefore

$$\Delta = D^{-1} = \sum_{j=1}^N (d_j/d_{j-1})(Q_j - Q_{j-1}) + I - Q_N$$

and

$$\Delta^{\frac{1}{2}} = \sum_{j=1}^{N} (d_j/d_{j-1})^{\frac{1}{2}}(Q_j - Q_{j-1}) + I - Q_N .$$

That $d_j > 0, j = 1, 2, \cdots, N$, follows from the positive definiteness of $S = I - T$.

In the matrix form $\Delta^{\frac{1}{2}}$ can be represented as

$$\Delta^{\frac{1}{2}} = \begin{bmatrix} \Delta_{11} & 0 \\ 0 & I \end{bmatrix}$$

where Δ_{11} is the diagonal matrix with diagonal elements $(d_j/d_{j-1})^{\frac{1}{2}}$.

Thus we have

$$\begin{aligned} \Delta^{\frac{1}{2}}(I + W_+)\{(f_j(t)); \chi(t, u)F(t, u)\} \\ = \{(d_j/d_{j-1})^{\frac{1}{2}}[f_j(t) + \textstyle\sum_{k=j+1}^{N} w_{jk} f_k(t) + \int_0^t W_{12}(j, v)F(t, v)\, dv]; \\ \chi(t, u)F(t, u) + \textstyle\int_0^1 W_{22}(u, v)\chi(t, v)F(t, v)\, dv\} \end{aligned}$$

and hence, from (6.2),

$$\begin{aligned} Y(t) &= \tilde{\Delta}^{\frac{1}{2}}(I + \tilde{W}_+)X(t) \\ &= \textstyle\sum_{j=1}^{N} c_j(t)\eta_j + \int_0^t F(t, u)\, dB(u) \\ &\quad + \textstyle\int_0^t \{\int_0^v W_{22}(u, v)\, dB(u)\}F(t, v)\, dv , \end{aligned} \tag{6.6}$$

where

$$c_j(t) = (d_j/d_{j-1})^{\frac{1}{2}}\{f_j(t) + \textstyle\sum_{k=j+1}^{N} w_{jk} f_k(t) + \int_0^t W_{12}(j, v)F(t, v)\, dv\} ,$$
$$W_{12}(j, v) \in L^2[0, 1],\ j = 1, 2, \cdots, N ,$$

and $W_{22}(u, v)$ is a Volterra kernel in $L^2([0, 1] \times [0, 1])$.

REMARK. (6.6) may be written in the form

$$\begin{aligned} Y(t) &= \textstyle\sum_{j=1}^{N} b_j(t)X^{(j-1)}(0) + \int_0^t F(t, u)\, dB(u) \\ &\quad + \textstyle\int_0^t \{\int_0^v W_{22}(u, v)\, dB(u)\}F(t, v)\, dv , \end{aligned} \tag{6.7}$$

where $X^{(j)}(0)$ are the derivatives of $X(t)$ at $t = 0$. (6.7) may be obtained directly using a different maximal chain containing $\{P(t), 0 < t \leqq 1\}$.

7. Radon-Nikodym derivatives. Hitsuda [3] showed also that if $\{X(t), 0 \leqq t \leqq 1, Q\}$ is the standard Wiener process and if a Gaussian measure P is equivalent to Q, then the Radon–Nikodym (RN) derivative dP/dQ is given by

$$dP/dQ = \exp\{\textstyle\int_0^1 (\int_0^v k(u, v)\, dX(u))\, dX(v) - \frac{1}{2}\int_0^1 (\int_0^v k(u, v)\, dX(u))^2\, dv\} \tag{7.1}$$

where $k(u, v)$ is a Volterra kernel in $L^2([0, 1] \times [0, 1])$. In [4] Kailath derived the above form of RN derivative from Shepp's result [8] using a certain identity for Carleman–Fredholm determinants. In this section we shall obtain a similar form of RN derivatives for the case considered in Section 5.

First we note that, for any equivalent Gaussian measures P and Q corresponding to general Gaussian processes $\{X(t), 0 \leqq t \leqq 1, P\}$ and $\{X(t), 0 \leqq t \leqq 1, Q\}$ with $E_P X(t) = E_Q X(t) = 0$ and covariance functions Γ_P, Γ_Q, the RN derivative

dP/dQ is given by

$$dP/dQ = \lim_n \prod_{j=1}^n (1-\lambda_j)^{-\frac{1}{2}} \exp\left\{-\tfrac{1}{2}\sum_{j=1}^n \frac{\lambda_j}{1-\lambda_j} X_j^2\right\}, \tag{7.2}$$

where $\{\lambda_j\}$ are the eigenvalues of the operator $T = I - S$ on the RKHS $H(\Gamma_Q)$ defined in Theorem 4.2 and $X_j = \psi^{-1}\varphi_j$, φ_j denoting the eigenfunction corresponding to λ_j (cf. e.g. [7]).

Suppose now that $\{X(t), 0 \leqq t \leqq 1, Q\}$ is a Wiener process (see Section 5). Then, because of the isometric isomorphism between $H(\Gamma_Q)$ and $L^2([0,1],\mu)$, $\{\lambda_j\}$ and $\{\varphi_j\}$ can be taken to be the eigenvalues and eigenfunctions of the operator $T \in \mathscr{S}_2$ on $L^2([0,1],\mu)$, and

$$X_j = \varphi_j(0)X(0) + \int_0^1 \varphi_j(u)\,dB(u)\,.$$

Let U be the Fredholm resolvent operator of T at 1, i.e.,

$$(I+U)(I-T) = I$$

and let $U(u,v)$ be the corresponding kernel. (We shall denote by $A(u,v)$ the $L^2([0,1]\times[0,1],\mu\times\mu)$ kernel corresponding to an operator $A\in\mathscr{S}_2$ on $L^2([0,1],\mu)$.) Then the kernel $U(u,v)$ has the expansion

$$U(u,v) = \sum_{j=1}^\infty \frac{\lambda_j}{1-\lambda_j}\varphi_j(u)\varphi_j(v)\,.$$

We define the double integral $\int_0^1\int_0^1 H(u,v)\,dX(u)\,dX(v)$ for any symmetric kernel $H(u,v)\in L^2([0,1]\times[0,1],\mu\times\mu)$ by

$$\begin{aligned}\int_0^1\int_0^1 & H(u,v)\,dX(u)\,dX(v)\\ &= H(0,0)[X^2(0)-\sigma^2] + X(0)\int_0^1 H(u,0)\,dB(u)\\ &\quad + X(0)\int_0^1 H(0,v)\,dB(v) + \int_0^1\int_0^1 H(u,v)\,dB(u)\,dB(v)\,,\end{aligned}$$

where the last term is the usual Itô's double Wiener integral. This double integral is quite similar to the usual one and we have

$$\lim_n \sum_{j=1}^n \frac{\lambda_j}{1-\lambda_j}(X_j^2-1) = \int_0^1\int_0^1 U(u,v)\,dX(u)\,dX(v)\,.$$

Let $\delta_A(\lambda)$ denote the Carleman–Fredholm determinant of an operator $A\in\mathscr{S}_2$, i.e.,

$$\delta_A(\lambda) = \prod_{j=1}^\infty [(1-\lambda\lambda_j(A))\exp(\lambda\lambda_j(A))]\,,$$

where $\lambda_j(A)$ are the eigenvalues of A. If A is Volterra, then, by definition, $\delta_A(\lambda) = 1$. Since

$$\begin{aligned}\lim_n \prod_{j=1}^n (1-\lambda_j)^{-\frac{1}{2}} & \exp\left\{-\tfrac{1}{2}\sum_{j=1}^n \frac{\lambda_j}{1-\lambda_j}\right\}\\ &= \lim_n \prod_{j=1}^n \left[\left(1+\frac{\lambda_j}{1-\lambda_j}\right)\exp\left\{-\frac{\lambda_j}{1-\lambda_j}\right\}\right]^{\frac{1}{2}}\\ &= [\delta_U(-1)]^{\frac{1}{2}}\,,\end{aligned}$$

(7.2) can be written in the form

$$dP/dQ = [\delta_U(-1)]^{\frac{1}{2}} \exp\{-\tfrac{1}{2} \textstyle\int_0^1 \int_0^1 U(u, v)\, dX(u)\, dX(v)\} . \tag{7.3}$$

We have, by Theorem 2.2,

$$I + U = S^{-1} = (I + X_+)D(I + X_-) .$$

Define the operator V by

$$I + V = (I + X_+)(I + X_-) .$$

Then the operator

$$U - V = (I + X_+)(D - I)(I + X_-)$$

is 0 or one-dimensional according as $X(0) = 0$ or $X(0) \neq 0$, because

$$D - I = P(0)[(I - P(0)TP(0))^{-1} - I]P(0)$$

and

$$U - V = P(0)(I + X_+)(D - I)(I + X_-)P(0) .$$

We apply the following relation to U and V (see [1] page 172): if $A, B \in \mathscr{S}_2$ and $A - B$ is nuclear, then, for λ such that $(I - \lambda A)^{-1}$ exists,

$$\delta_A(\lambda) = \delta_B(\lambda)[D_{B/A}(\lambda)]^{-1} \exp\{\lambda \operatorname{tr}(A - B)\} ,$$

where $D_{B/A}(\lambda)$ is the perturbation determinant of A by $B - A$, i.e.,

$$D_{B/A}(\lambda) = \textstyle\prod_{j=1}^{\infty} (1 - \nu_j)$$

and $\{\nu_j\}$ denote the eigenvalues of the operator $\lambda(B - A)(I - \lambda A)^{-1}$. Then we have

$$\delta_U(-1) = \delta_V(-1)[D_{V/U}(-1)]^{-1} \exp\{-\operatorname{tr}(U - V)\} .$$

Since the eigenvalue of the one-dimensional operator $U - V$ (for the case $X(0) \neq 0$) is $\sigma^2 T(0, 0)/(1 - \sigma^2 T(0, 0))$, which is easily found by applying Lemma 5.1 to X_+ and $X_- = X_+{}^*$,

$$\begin{aligned} \operatorname{tr}(U - V) &= \textstyle\int_0^1 (U - V)(u, u)\mu(du) \\ &= (U - V)(0, 0) \cdot \sigma^2 \\ &= \frac{\sigma^2 T(0, 0)}{1 - \sigma^2 T(0, 0)} , \end{aligned}$$

and the eigenvalue of the one-dimensional operator $(U - V)S$ is $\sigma^2 T(0, 0)$, and hence

$$D_{V/U}(-1) = 1 - \sigma^2 T(0, 0) .$$

Furthermore, from the identity (see [1] page 169): for $A, B \in \mathscr{S}_2$ and $I - C = (I - A)(I - B)$,

$$\delta_C(1) \exp\{\operatorname{tr}(AB)\} = \delta_A(1)\delta_B(1) ,$$

it follows that

$$\delta_V(-1) = \exp\{-\operatorname{tr}(X_+ X_-)\}\,.$$

Thus we have

$$[\delta_U(-1)]^{\frac{1}{2}} = (1 - \sigma^2 T(0,0))^{-\frac{1}{2}} \exp\left\{-\frac{1}{2}\frac{\sigma^2 T(0,0)}{1-\sigma^2 T(0,0)}\right\} \times \exp\{-\tfrac{1}{2}\operatorname{tr}(X_+ X_-)\}\,. \tag{7.4}$$

Now consider

$$\begin{aligned}\textstyle\int_0^1\int_0^1 U(u,v)\,dX(u)\,dX(v) &= \textstyle\int_0^1\int_0^1 (U-V)(u,v)\,dX(u)\,dX(v)\\ &\quad + \textstyle\int_0^1\int_0^1 (X_+ + X_-)(u,v)\,dX(u)\,dX(v)\\ &\quad + \textstyle\int_0^1\int_0^1 (X_+ X_-)(u,v)\,dX(u)\,dX(v)\,.\end{aligned}$$

Since $(U - V)(u, v) = 0$ a.e. $\mu \times \mu$ for $(u, v) \neq (0, 0)$, the first term is

$$(U - V)(0,0)[X^2(0) - \sigma^2] = \frac{T(0,0)}{1-\sigma^2 T(0,0)}X^2(0) - \frac{\sigma^2 T(0,0)}{1-\sigma^2 T(0,0)}$$

The second term can be written as the iterated integral

$$2\textstyle\int_0^1\{\int_0^v X_+(u,v)\,dX(u)\}\,dX(v)$$

using properties of stochastic integrals and noting that $X_+(0, 0) = X_-(0, 0) = 0$ if $\sigma^2 > 0$. The third term is equal to

$$\textstyle\int_0^1\{\int_0^v X_+(u,v)\,dX(u)\}^2\mu(dv) - \operatorname{tr}(X_+X_-)\,,$$

which follows easily from the definition of the double integral. Therefore we have

$$\begin{aligned}&\exp\{-\tfrac{1}{2}\textstyle\int_0^1\int_0^1 U(u,v)\,dX(u)\,dX(v)\}\\ &\quad= \exp\left\{-\frac{1}{2}\frac{T(0,0)}{1-\sigma^2 T(0,0)}X^2(0)\right\}\exp\left\{\frac{1}{2}\frac{\sigma^2 T(0,0)}{1-\sigma^2 T(0,0)}\right\}\\ &\qquad\times \exp[-\textstyle\int_0^1\{\int_0^v X_+(u,v)\,dX(u)\}\,dX(v)\\ &\qquad -\tfrac{1}{2}\textstyle\int_0^1\{\int_0^v X_+(u,v)\,dX(u)\}^2\mu(dv)]\exp\{\tfrac{1}{2}\operatorname{tr}(X_+X_-)\}\,.\end{aligned} \tag{7.5}$$

Substituting (7.4) and (7.5) in (7.3), we obtain the following form for the RN derivative.

$$\begin{aligned}dP/dQ &= (1-\sigma^2 T(0,0))^{-\frac{1}{2}}\exp\left\{-\frac{1}{2}\frac{T(0,0)}{1-\sigma^2 T(0,0)}X^2(0)\right\}\\ &\quad\times \exp[-\textstyle\int_0^1\{\int_0^v X_+(u,v)\,dX(u)\}\,dX(v)\\ &\quad - \tfrac{1}{2}\textstyle\int_0^1\{\int_0^v X_+(u,v)\,dX(u)\}^2\mu(dv)]\,.\end{aligned}$$

In [8] Shepp has derived the RN derivative for the more general case when $X(t)$ is a so-called "free" Wiener process. Our method is different and, we feel, can be used to deduce his result in its full generality.

Note added. After this paper had been submitted the authors learned from Professor T. Kailath that he and Dr. D. Duttweiler have also obtained Theorem 4.1 of this paper. They also derive a general likelihood ratio formula and consider several applications from the engineering standpoint.

REFERENCES

[1] GOHBERG, I. C. and KREIN, M. G. (1969). *Introduction to Linear Nonself-Adjoint Operators* (Eng. trans.). American Mathematical Society, Providence.

[2] GOHBERG, I. C. and KREIN, M. G. (1970). *Theory and Applications of Volterra Operators in Hilbert Space* (Eng. trans.). American Mathematical Society, Providence.

[3] HITSUDA, M. (1968). Representation of Gaussian processes equivalent to Wiener process. *Osaka J. Math.* **5** 299–312.

[4] KAILATH, T. (1970). Likelihood ratios of Gaussian processes. *IEEE Trans. Information Theory* **IT-16** 276–288.

[5] KALLIANPUR, G. and OODAIRA, H. (1963). The equivalence and singularity of Gaussian measures. *Proceedings of Symposium on Time Series Analysis* Wiley, New York, 279–291.

[6] RIESZ, F. and SZ.-NAGY, B. (1955). *Functional Analysis.* Ungar, New York.

[7] ROZANOV, YU. A. (1971). Infinite-dimensional Gaussian distributions (Eng. trans.). American Mathematical Society, Providence.

[8] SHEPP, L. A. (1966). Radon-Nikodym derivatives of Gaussian measures. *Ann. Math. Statist.* **37** 321–354.

[9] KAILATH, T. and DUTTWEILER, D. (1972). An RKHS approach to detection and estimation problems, Part III: Generalized innovations representations and a likelihood-ratio formula. To appear.

19

ON SOME PROBLEMS CONCERNING BROWNIAN MOTION IN LÉVY'S SENSE

G. M. MOLCHAN

(*Translated by H. C. Folguera*)

P. Lévy defined Brownian motion $x(t)$ with multi-dimensional time as a random function of the point t in n-dimensional Euclidean space E^n, which possesses the properties: a) every finite-dimensional distribution for the variable $x(t) - x(0)$ is a Gaussian distribution; b) $\mathbf{E}|x(t) - x(\tau)|^2 = |t - \tau|$, $|t| = \sqrt{t_1^2 + \cdots + t_n^2}$, and $\mathbf{E}x(t) = 0$. We shall consider a Lévy field in an odd-dimensional space E^n with the condition $x(0) = 0$. In this case the Brownian motion is completely determined by its correlation function

$$R(t, \tau) = \tfrac{1}{2}(|t| + |\tau| - |t - \tau|). \tag{1}$$

In the first part of the article we study the Hilbert space $H(x, G)$ of random variables with norm $\|\eta\|^2 = \mathbf{E}\eta^2$, generated by the closure of the linear hull of the variables $x(t)$, where t varies in the domain G. The description of the space $H(x, G)$ is given in terms of the isomorphic model of the functional Hilbert space $H(R, G) = \{m(t), t \in G\}$ with reproducing kernel $R(t, \tau)$, $(t, \tau) \in G \times G$ (see [8], [10]). The isomorphism of the spaces $H(x, G)$ and $H(R, G)$ is provided by the unitary mapping $\varphi : \eta \leftrightarrow m(t) = \mathbf{E}x(t)\eta = (x(t), \eta)$, i.e., the scalar product of two elements $m_i(t)$ in $H(R, G)$ can be defined as $\langle m_1, m_2\rangle_{H(R,G)} = (\varphi^{-1}m, \cdot \varphi^{-1}m_2)$. The description of the space $H(R, G)$ solves essentially the problem concerning the equivalence of the measures in the space of functions $\{f(t), t \in G\}$, which correspond to Brownian Lévy motion and to an arbitrary Gaussian field $y(t)$ in the domain G with mean value $a(t)$ and correlation function $b(t, \tau) = \mathbf{E}[y(t) - a(t)] \times [y(\tau) - a(\tau)]$, $t, \tau \in G$. That is, the mean value $a(t)$ must be an element of $H(R, G)$, $R(t, \tau) - b(t, \tau)$ must be an element of the tensor product $H(R, G) \times H(R, G)$ and the equation $\langle b(t, \tau), \varphi(\tau)\rangle_{H(R,G)} = 0$ must have only the null solution in $H(R, G)$ (see [2], [3] and [10]).

The study of the space $H(R, G)$ has independent interest, because it is closely related with the poly-harmonic operator $\Delta^{(n+1)/2}$ (Δ is the Laplacian operator). It is possible to give a visual probabilistic meaning to the Green's function of such an operator for the domain G, which allows one to construct it for domains with arbitrarily non-smooth boundaries (subsection 2.5.).

The infinitesimal boundary σ-algebra of the domain G (see [4]) is described in Section 3 in terms of the functions of $H(R, E^n)$, and from there the results obtained by McKean on Markovian properties of the Lévy fields follow.

In conclusion one problem of integral geometry is considered which is closely related with the representation of the Lévy field in Chentsov's form [11].

1. Description of the Functional Space $H(R, E^n)$

Lévy's field admits of a non-degenerate spectral representation

$$x(t) = \frac{1}{\sqrt{\Omega_{n+1}}} \int_{E^n} \frac{e^{i(t,\lambda)} - 1}{|\lambda|^{(n+1)/2}} b(d^n\lambda), \tag{2}$$

in the whole space, where $b(d^n\lambda)$ *is the Gaussian random measure of elementary volume* $d^n\lambda = d\lambda_1\, d\lambda_2 \cdots d\lambda_n$ *in* E^n: $\mathbf{E}b(d^n\lambda) = 0$, $\mathbf{E}b(\Delta^n\lambda_1)\overline{b(\Delta^n\lambda_2)} = \text{meas}(\Delta^n\lambda_1 \cap \Delta^n\lambda_2$; Ωp *is the area of the unit sphere in* E^p.

The integral (2) is understood in the mean square sense, and the non-degeneracy of the representation means the coincidence of the closure of the linear hull of $\{x(t), t \in E^n\}$ and of $\{b(d^n\lambda), \lambda \in E^n\}$ in the metric $\|\eta\|^2 = \mathbf{E}\eta^2$.

This statement is a consequence of general theorems on spectral representation of locally homogeneous and locally isotropic fields in E^n, obtained by A. M. Yaglom in his article [12, Theorems 6 and 13]. However, the proof can be given directly starting from the representation of the correlation function $R(t, \tau)$ in the form:

$$R(t, \tau) = \frac{1}{\Omega_{n+1}} \int_{E^n} [e^{i(t,\lambda)} - 1][e^{-i(\tau,\lambda)} - 1] \frac{d^n\lambda}{|\lambda|^{n+1}}. \tag{3}$$

Corollary 1.1. *The function* $m(t)$ *belongs to the space* $H(R, E^n)$ *if and only if* $m(t)$ *admits a representation of the form*

$$m(t) = \frac{1}{\sqrt{\Omega_{n+1}}} \int_{E^n} \frac{e^{i(t,\lambda)} - 1}{|\lambda|^{(n+1)/2}} f(\lambda) d^n\lambda, \tag{4}$$

where $f(\lambda) \in L^2(E^n)$.

PROOF. The closure of the random variables $\{b(d^n\lambda), \lambda \in E^n\}$ coincides with all the random variables of the form $\eta = \int f(\lambda) b(d^n\lambda)$, where $f \in L^2(E^n)$ (see, for instance, [5]). Because of the representation (2) and its non-degeneracy, we see that an arbitrary element of the space $H(R, E^n)$ has the form

$$(x(t), \eta) = (\Omega_{n+1})^{-1/2} \int_{E^n} [e^{i(t,\lambda)} - 1]|\lambda|^{-(n+1)/2} f(\lambda)\, d^n\lambda.$$

Corollary 1.2. *All the elements of the space* $H(R, E^n)$ *have generalized derivatives in Sovolev's sense, of order* $(n + 1)/2$, *which are square integrable in* E^n, *i.e.*, $m(t) \in W_2^{(n+1)/2}(E^n)$; *and, therefore,* $m(t)$ *has all the intermediate derivatives, belonging to* L^2 *in each bounded domain* G, $(m^{(k)}(t) \in L^2_{\text{loc}}(E^n)$, $k < (n - 1)/2)$, *and the functions* $m(t)$ *themselves are Hölder continuous, with exponent* $\frac{1}{2}$ *and vanish at* $t = 0$.

PROOF. For a finite differentiable function φ and the operator

$$\partial^k = \frac{\partial^k}{\partial x_1^{k_1} \cdots \partial x_n^{k_n}}, \qquad k = k_1 + \cdots + k_n = (n + 1)/2,$$

we have

$$(\partial^k\varphi, m(t))_{L^2} = \frac{1}{\sqrt{\Omega_{n+1}}} \int \hat{\varphi}(\lambda) \frac{(i\lambda_1)^{k_1} \cdots (i\lambda_n)^{k_n}}{|\lambda|^k} f(\lambda)\, d^n\lambda = (\varphi(\lambda), \psi(\lambda))_{L_2},$$

where $\psi(\lambda)$ is the Fourier transform of the function const. $\dfrac{(i\lambda_1)^{k_1} \cdots (i\lambda_n)^{k_n}}{|\lambda|^k} \cdot f(\lambda)$ in L^2, since

$$\left| \frac{(i\lambda_1)^{k_1} \cdots (i\lambda_n)^{k_n}}{|\lambda|^k} \right| < 1.$$

(From here on $\hat{\varphi}(\lambda)$ is the Fourier transform of the function $\varphi(t)$.) The second part of the statement follows from the imbedding theorem [7]. The last statement can be obtained more easily directly from the definition of $m(t) = (x(t), \eta) \in H(R, E^n)$; indeed,

$$|m(t) - m(\tau)| \leqq \|\eta\| \, \|x(t) - x(\tau)\| = \|\eta\| \, |t - \tau|^{1/2}$$

Theorem 1. *The scalar product in $H(R, E^n)$ has the form*

$$\langle m_1, m_2 \rangle = \begin{cases} \dfrac{\Omega_{n+1}}{(2\pi)^n} (\Delta^k m_1(t), \quad \Delta^k m_2(t))_{L^2} & \text{for} \quad n = 4k - 1, \\ \dfrac{\Omega_{n+1}}{(2\pi)^n} \displaystyle\sum_{i=1}^{n} \left(\frac{\partial}{\partial t_i} \Delta^k m_1, \frac{\partial}{\partial t_i} \Delta^k m_2 \right)_{L^2} & \text{for} \quad n = 4k + 1. \end{cases} \tag{5}$$

PROOF. Let $m_i(t)$ correspond to the random variables $\eta_i = \int f_i(\lambda) b(d^n\lambda)$. We shall prove the theorem for the case $n = 4k + 1$. We have

$$(\eta_1, \eta_2) = (f_1, f_2)_{L^2} = \sum_{i=1}^{n} \left(\frac{\lambda_i}{|\lambda|} f_1, \frac{\lambda_i}{|\lambda|} f_2 \right) = \frac{1}{(2\pi)^n} \sum_{i=1}^{n} \left(\widehat{\frac{\lambda_i f_1}{|\lambda|}}, \widehat{\frac{\lambda_i f_2}{|\lambda|}} \right)$$

$$= \frac{\Omega_{n+1}}{(2\pi)^n} \sum_{i=1}^{n} \left(\frac{\partial}{\partial t_i} \Delta^k m_1, \frac{\partial}{\partial t_i} \Delta^k m_2 \right)_{L^2},$$

since from (4),

$$\frac{\partial}{\partial t_k} \Delta^k m = \int_{E^n} \frac{i\lambda_k}{|\lambda|} f(\lambda) d^n\lambda.$$

Theorem 2. *The function $m(t)$ belongs to $H(R, E^n)$ if and only if $m(t)$ is a Hölder continuous solution of the problem*

$$\Delta^{[(n+1)/4]} m(t) = f, \qquad m(0) = 0, \tag{6}$$

in E^n, with exponent $\frac{1}{2}$, where $f \in L^2(E^n)$ for $n = 4k - 1$ and $f \in W_2^1(E^n) \cap L^2_{\text{loc}}$ for $n = 4k + 1$. The solution of problem (6) *is always unique.*

PROOF. The necessity of the conditions of the theorem has been proved above. We shall show the sufficiency, for example, for $n = 4k - 1$. The reproducing property of the kernel $R(t, \tau)$ allows us to write immediately a particular solution of problem (6),

$$m(t) = \frac{\Omega_{n+1}}{(2\pi)^n} (\Delta^k R(t, \cdot), f(\cdot))_{L^2}.$$

We shall show that the homogeneous problem (6) has only the zero solution. Clearly, it suffices to prove that a solution $m_0(t)$ of the homogeneous problem is a weak solution of $\Delta^k m_0(t) = 0$ in all of the space E^n. Then, because of the inequality $|m_0(t)| < c|t|^{1/2}$, $m_0(t) \equiv m_0(0)$ (see [6, p. 166], i.e., $m_0(t) \equiv 0$. Let $h(t)$ be an infinitely differentiable function such that

$$h(t) = \begin{cases} 0 & \text{for} \quad |t| > 1, \\ 1 & \text{for} \quad |t| < \frac{1}{2}, \end{cases}$$

and $|h(t)| < 1$. Then for an arbitrary finite regular function φ we have

$$(m_0(t), \Delta^k \varphi) = (m, \Delta^k[\varphi \cdot h(t/\varepsilon)]) + (m_0, \Delta^k[\varphi \cdot (1 - h(t/\varepsilon)]).$$

The second term is equal to zero, because $\varphi \cdot (1 - h(t/\varepsilon))$ is a finite regular function, concentrated outside a $\frac{1}{2}\varepsilon$-neighborhood of zero, and

$$\left| \int_{|t|<\varepsilon} m \cdot \Delta^k[\varphi \cdot h(t/\varepsilon)] \, dt \right| \leqq C \max_{|t|<\varepsilon} m(t) \max_{\substack{0 \leqq l \leqq 2k \\ -\infty < t < \infty}} |\varphi^{(l)}(t)| \max_{0 \leqq l \leqq 2k} \int |h^{(l)}(t/\varepsilon)| d^n t$$

$$\leqq C \max_{|t|<\varepsilon} m(t) \cdot \|\varphi\|_{C^{2k}} \cdot \|h\|_{C^{2k}} \cdot \varepsilon^{n-2k},$$

i.e., $(m_0(t), \Delta^k \varphi) = 0$, which it was required to prove.

2. Description of the Space $H(R, G)$

Let us consider an arbitrary domain G in E^n and the space of random variables $H(x, G)$ or the isomorphic functional space $H(R, G) = \{m(t), t \in G\}$, reproduced by the kernel $R(t, \tau)$ in the domain G.

2.1. *Every function $m(t) \in H(R, G)$ admits of a unique extension $\tilde{m}(t)$ to all of the space E^n such that $\tilde{m}(t) \in H(R, E^n)$ and $\|m(t)\|_{H(R,G)} = \|\tilde{m}(t)\|_{H(R,E^n)}$.*

In fact, let $\check{x}(t)$ be the projection of $x(t)$, $t \notin G$, on $H(x, G)$. Then, for the function $m(t)$ which corresponds to the random variable $\eta \in H(x, G)$, $\tilde{m}(t) = (\check{x}(t), \eta) = (x(t), \eta) \in H(R, E^n)$ is the sought extension. It is unique because the imbedding $H(x, G) \subset H(R, E^n)$ is unitary.

2.2. *As a consequence of subsection 2.1, we obtain that every element $m(t)$ of $H(R, G)$ belongs to $W_2^{(n+1)/2}(G)$, and its extension $\tilde{m}(t) \in W_2^{(n+1)/2}(E^n)$. In the case of a domain G with smooth boundary, it follows additionally from the imbedding theorem that $\tilde{m}(t)$ is a continuous extension of $m(t)$ which preserves the continuity in L^2 of all the generalized derivatives*

$$\frac{\partial^k}{\partial t_1^{k_1} \cdots \partial t_n^{k_n}} m(t), \quad k = k_1 + \cdots + k_n \leqq (n-1)/2,$$

on the boundary in the following sense: if we shift an arbitrary piece of the boundary $S_1 \subset \Gamma$ of the domain G by a vector Δ in such a way that $S_1 + \Delta$ remains in $E^n - G$, then $\int_{S_1} |\tilde{m}^{(k)}(t + \Delta) - m^{(k)}(t)|^2 dS \to 0$ if $|\Delta| \to 0$.

2.3. *The extension of the element $m(t)$ of $H(R, G)$ is an $(n+1)/2$-harmonic function outside $G \cup \{0\}$, i.e., $\Delta^{(n+1)/2}\tilde{m}(t) = 0$ for $t \notin G \cup \{0\}$. Therefore, in the case of a domain G with smooth boundary, $\tilde{m}(t)$ is the solution of the Dirichlet problem for the equation $\Delta^{(n+1)/2}u = 0$ in the domain $E^n - (G \cup \{0\})$ with boundary data $m^{(l)}(t)|_\Gamma \in L^2_{\text{loc}}(\Gamma)$, $0 \leqq l \leqq (n-1)/2$, and $m(0) = 0$ if $0 \notin G$, in the class of functions $W_2^{(n+1)/2}$. The solution takes on the boundary data in the sense of subsection 2.2.*

PROOF. We shall show that $\tilde{m}(t)$ is an $(n+1)/2$-harmonic function outside G. For the function $R(t, \cdot)$, $t \in G$, which corresponds to the variable $x(t) \in H(x, G)$ the statement is true. The simplest way to see it is by using the representation (3) which shows that

$$\Delta^{(n+1)/2} R_t(\tau) = \frac{(-1)^{(n+1)/2}}{\Omega_{n+1}} [\delta(t - \tau) - \delta(\tau)], \tag{7}$$

where $\delta(\cdot)$ is the generalized Dirac function. But the system of functions $\{R_t(\tau), t \in G\}$ is complete in $H(R, G)$, and therefore for every function $m(t)$ there exists a sequence $m_k(t) \in H(R, G)$ which possesses an $(n+1)/2$-harmonic extension outside $G \cup \{0\}$ and which converges towards $m(t)$ in $H(R, G)$, and, therefore, also in $H(R, E^n)$. For an arbitrary infinitely differentiable function φ with compact support outside $G \cup \{0\}$ we shall have

$$\begin{aligned}(m(t), \Delta^{(n+1)/2}\varphi) &= (\Delta^{(n+1)/4}\tilde{m}(t), \quad \Delta^{(n+1)/4}\varphi) \\ &= \lim_{k\to\infty} (\Delta^{(n+1)/4}\tilde{m}_k(t), \quad \Delta^{(n+1)/4}\varphi) = \lim_{k\to\infty} (\Delta^{(n+1)/2}\tilde{m}_k(t), \varphi) = 0.\end{aligned}$$

From here it follows [6], that $\tilde{m}(t)$ is infinitely differentiable and an $(n+1)/2$-harmonic function outside $G \cup \{0\}$.

2.4. *If the domain G is compact, the extensions of the elements of $H(R, G)$ satisfy at ∞ the inequality*

$$\left| \frac{\partial^k \tilde{m}(t)}{\partial t_1^{k_1} \cdots \partial t_n^{k_n}} \right| < \frac{c}{|t|^k}, \qquad 0 \leqq k \leqq n. \tag{8}$$

The estimate (8) was obtained by Browder (see [4]) for functions $\tilde{m}(t) \in W_2^{(n+1)/2}(E^n)$ which are $(n+1)/2$-harmonic outside the sphere $|t| = r$. Obviously, these conditions are fulfilled for $m(t) \in H(R, G)$.

Theorem 3. *The function $m(t)$ belongs to the space $H(R, G)$, where G is a bounded domain with smooth boundary, if and only if $m(t) \in W_2^{(n+1)/2}(G)$, and there exists an $(n+1)/2$-harmonic extension of $m(t)$ outside $G \cup \{0\}$ in the class of functions $W_2^{(n+1)/2}(E^n)$ with $m(0) = 0$, or $m(t)$ must* *be the restriction to G of a continuous solution of the following problem*:

$$\Delta^{[(n+1)/4]}m(t) = f, \qquad m(0) = 0, \qquad |m(t)| < \text{const.},$$

where $f \in L^2(E^n)(f \in W_2^1(E^n) \cap L^2_{\text{loc}})$, *for* $n = 4k - 1, (n = 4k + 1)$, *and* $\Delta^{[(n+1)/4]}m(t) = 0$ *outside* $G \cup \{0\}$.

The scalar product in $H(R, G)$ *is given by formula* (5), *where the* $m_i(t)$ *are the* $(n+1)/2$-*harmonic extensions of the elements of* $H(R, G)$ *to all of the space* E^n.

PROOF. The necessity of the conditions of the theorem follows from subsections 2.1–2.4 and Theorem 2. We shall prove the sufficiency of these conditions. Let the function $m(t), t \in G$, be such that its $(n+1)/2$-harmonic extension to $E^n - G$ $\tilde{m}(t) \in W_2^{(n+1)/2}(E^n)$ and $\tilde{m}(0) = 0$. Because of subsection 2.4, the function $\tilde{m}(t)$ is bounded and, by Theorem 2, $\tilde{m}(t) \in H(R, E^n)$. Let $\tilde{m}(t)$ correspond to a certain random variable $\eta \in H(x, E^n)$, let $\check{\eta}$ be the projection of η on $H(x, G)$ and $\check{m}(t)$ be its image in $H(R, E^n)$. The functions $\tilde{m}(t)$ and $\check{m}(t)$ coincide on G and they are $(n+1)/2$-harmonic outside $G \cup \{0\}$. Because of the theorem on uniqueness for the interior and exterior Dirichlet problem [4], [7] and since, according to subsection 2.2, $\tilde{m}(t)$ and $\check{m}(t)$ take on the same boundary values, they necessarily coincide also outside G. The proof is concluded.

2.5. *The polyharmonic extension of a function* $m(t) \in H(R, G)$ *to the domain* $E^n - (G \cup \{0\})$ *can be obtained if the Green's function for the operator* $\Delta^{(n+1)/2}$ *in the domain* $E^n - (G \cup \{0\})$ *is known. By the Green function we mean the solution of the equation*

$$\Delta^{(n+1)/2}g_\tau(t) = \delta(t - \tau), \qquad t, \tau \in E^n - G, \tag{9}$$

with zero boundary data

$$\frac{\partial^k}{\partial t_1^{k_1} \cdots \partial t_n^{k_n}} g_\tau(t)|_\Gamma = 0, \qquad 0 \leqq k \leqq (n-1)/2,$$

in the class of functions $W_2^{(n+1)/2}(E^n - G)$. *If the domain G is simply connected and has smooth boundary* Γ, *then the* $(n+1)/2$-*harmonic extension of* $m(t)$ *is given by the Green's formula*

$$\tilde{m}(t) = \int_\Gamma \sum_{i=0}^{(n-1)/2} (\partial^+ \Delta^i m(t) \Delta^{(n-1)/2 - i} g_\tau(t) - \Delta^{(n-1)/2 - i} m(t) \partial^+ \Delta^i g_\tau(t))\, dS. \tag{10}$$

(Here ∂^+ is the derivative in the direction of the exterior normal.)

We shall show, that the Green's function in the domain $E^n - G$, *where G is an arbitrary domain which contains* 0, *always exists in the following sense*: *the function* $g_\tau(t)$ *satisfies the equation* (9) *and, when extended as zero on G, belongs to the space* $W_2^{(n+1)/2}(E^n)$.

Obviously such a definition coincides with the definition given above if G is a bounded domain with smooth boundary. This follows from the continuity in L^2 of all derivatives up to the $(n-1)/2$-th of the function $m(t) \in W_2^{(n+1)/2}(E^n)$ on the boundary Γ, in the sense of subsection 2.2.

PROOF. We decompose the random variable $x(\tau)$ into the component $\hat{x}(\tau)$ which is orthogonal to the space $H(x, G)$ and into the projection $\check{x}(\tau)$ onto this space. The image $\check{x}(\tau)$ in $H(R, E^n): \varphi\check{x}_\tau = \mathbf{E}\check{x}(\tau)x(t)$ is an $(n+1)/2$-harmonic function of t outside G (subsection 2.3.) for $\tau \notin G$, and therefore,

$$\Delta^{(n+1)/2}(x(t), \check{x}(\tau)) = \Delta_t^{(n+1)/2}R(t, \tau) = \frac{(-1)^{(n+1)/2}}{\Omega_{n+1}} \cdot \delta(t - \tau)$$

when $\tau, t \notin G$. By construction, in the domain G, $\mathbf{E}\hat{x}(\tau)x(t) \equiv 0$. Therefore, the function $(-1)^{(n+1)/2}\Omega_{n+1}\mathbf{E}\hat{x}(\tau)x(t) = g(t, \tau)$ is the sought for Green's function, since $g_\tau(t) \in H(R, E^n)$ and this means that $g_\tau(t) \in W_2^{(n+1)/2}$.

2.6. Let us calculate the scalar product in $H(R, G)$, if $G = \{t : |t| \leqq r\}$ is a sphere in the three-dimensional space E^3.

Theorem 4. *The scalar product in* $H(R, G)$ *has the form*

$$\begin{aligned} \langle m_1, m_2 \rangle = {} & \frac{1}{8\pi} \int_{|t| \leqq r} \Delta m_1(t)\, \Delta m_2(t)\, d^3 t \\ & - \frac{1}{2r} \iint_{\Gamma^- \times \Gamma^-} \frac{\partial}{\partial |t|} \left\{ |t| m_1(t) \frac{\partial}{\partial r} \left(|\tau^*| \frac{|\tau^*|^2 - |t|^2}{|\tau^* - t|^3} - r \right) m_2(\tau)|\tau| \right\} \frac{\partial}{\partial |t|} dO_t\, dO_\tau. \end{aligned} \tag{11}$$

The last integral is understood as a limit

$$\lim_{\varepsilon,\delta\to 0}\int_{|t|=r-\varepsilon}\int_{|\tau|=r-\delta}\frac{\partial}{\partial|t|}\{\cdot\}\frac{\partial}{\partial|\tau|}\,dO_t\,dO_\tau.$$

The point $\tau^* = (r^2/|\tau^2|)\tau$ *is symmetric to* τ *with respect to the sphere* $|t| = r$, dO *is the unitary Haar measure on the sphere.*

PROOF. By Theorem 3,

$$\langle m_1, m_2\rangle = \frac{1}{8\pi}(\Delta m_1, \Delta m_2)_{L^2(G)} + \frac{1}{8\pi}(\Delta\tilde{m}_1, \Delta\tilde{m}_2)_{L^2(E^n-G)},$$

where $\tilde{m}_i$ are biharmonic extensions of m_i outside the sphere $|t| = r$. This extension is given by the formula

$$\tilde{m}(t) = \left[1 - \frac{1}{2}\left(\frac{r}{|t|} + \frac{|t|}{r}\right)\right]\int_\Gamma \frac{\partial|\tau|m(\tau)}{\partial|\tau|}\,dO_\tau + \frac{|t|^2 - r^2}{2r}\int_\Gamma \frac{\partial}{\partial|\tau|}\frac{|\tau|m(\tau)}{|t-\tau|^3}\,dO_\tau,$$

which can be obtained easily by constructing the Green's function for the sphere (see [4]). This gives us $\Delta\tilde{m}(t)$. In order to simplify the computations we expand $\tilde{m}(t)$ on the sphere of radius $|t|$ in spherical functions $Y_l^{(k)}(t)$:

$$\tilde{m}(t) = \sum_{(k,l)} \tilde{m}_l^{(k)}(|t|)Y_l^{(k)}(t), \qquad \tilde{m}_l^{(k)}(|t|) = \int_{|\tau|=|t|}\tilde{m}(\tau)Y_l^{(k)}(\tau)\,dO_\tau.$$

By using the identity

(12) $$\sum_{(k,l)} Y_l^{(k)}(t)Y_l^{(k)}(\tau)\left(\frac{|t|}{|\tau|}\right)^n = |\tau|\cdot\frac{|\tau|^2 - |t|^2}{|\tau - t|^3}, \qquad |t| < |\tau|,$$

we find

$$\Delta\tilde{m}(t) = \sum_{l=1}^{\infty}\sum_{k=-l}^{l}\frac{d}{d|\tau|}[|\tau|^{l+1}m_l^k(|\tau|)](r)\cdot\frac{Y_l^k(t)}{|t|^{l+1}}\frac{(1-2l)}{r}$$

and since

$$\int_{|t|>r} Y_l^k(t)Y_{l'}^{k'}(t)|t|^{-2l-2}\,d^3t = \frac{4\pi}{2l-1}r^{-2n+1}\delta_k^{k'}\delta_l^{l'}, \qquad n \geqq 1,$$

where δ_p^k is the Kronecker symbol,

$$\begin{aligned}\frac{1}{8\pi}(\Delta\tilde{m}_1, \Delta\tilde{m}_2)_{L^2(\bar{G})} &= -\frac{1}{2r}\sum_{(k,l)l\geqq 1}\frac{d}{d|\tau|}[|\tau|^{l+1}m_{1l}^k(|\tau|)](r)\cdot\frac{d}{d|t|}[|t|^{l+1}m_{2l}^k(|t|)](r)\,\frac{d}{dr}r^{-2n+1}\\ &= \lim_{\substack{|t|\to r\\|\tau|\to r}}\frac{(-1)}{2r}\iint_{\Gamma\times\Gamma}\frac{\partial^3}{\partial|\tau|\,\partial|t|\,\partial r}\\ &\quad\times\left[m_1(\tau)m_2(t)\sum_{\substack{(l,k)\\l\geqq 1}} Y_l^k(t)Y_l^k(\tau)\frac{|t|^l|\tau|^l}{r^{2l}}|t|\,|\tau|r\right]dO_t\,dO_\tau.\end{aligned}$$

Again using identity (12), we obtain (11).

3. Markovian Properties of Lévy Fields

In this section we shall obtain from Theorem 3 certain known results obtained by McKean on Markovian properties of Lévy's fields in odd-dimensional spaces.

Let G be a simply connected domain, which contains 0, with smooth boundary Γ. Following [4], we call $H_- = \cap_{k\geqq 1} H(x, G_k)$ the past of the field $x(t)$ (here G_k is the open $1/k$-neighborhood of G), $H_+ = H(x, E^n - G)$ is the future; $\partial^+H = \cap_{k\geqq 1} H(x, G_k - G)$ the space of germs; $\partial H = \cap_{k\geqq 1} H(x, G \cap (E - G)_k)$ is the present of the field $x(t)$. Since $x(t)$ is a Gaussian field, the Hilbert spaces H_-, H_+, ∂^+H, ∂H can be identified with the corresponding minimal σ-algebras $\mathfrak{B}_-$, $\mathfrak{B}_+$, $\partial^+\mathfrak{B}$, $\partial\mathfrak{B}$ with respect to which the considered random variables are measurable.

3.1. *For a Lévy field in an odd-dimensional space there exists a non-degenerate σ-algebra $\mathfrak{B} \subset \mathfrak{B}_-(\mathfrak{B} \neq \mathfrak{B}_-)$, which separates the future and the past of the field, i.e., under the condition $\dot{\mathfrak{B}}$ the events $B_- \in \mathfrak{B}_-$ and $B_+ \in \mathfrak{B}_+$ are independent. The minimal separation algebra coincides with the algebras $\partial\mathfrak{B} = \partial^+\mathfrak{B}$, which can be identified with the $(n+1)/2$-harmonic functions outside $\Gamma \cup \{0\}$ which are continuous in E^n, equal to zero at 0 and belong to the class $W_2^{(n+1)/2}(E^n)$.*

Proof. From Theorem 3 it follows that $\partial^+ H$ and ∂H consist of classes of functions which belong to $H(R, E^n)$ and are $(n+1)/2$-harmonic outside $\Gamma \cup \{0\}$. It is easy to see that each such function belongs to $H(R, G_k - G)$, and to $H(R, G \cap \bar{G}_k)$. This implies the coincidence of $\partial^+ H$ and ∂H.

Let $\eta_+ + \eta_- = \eta \in H^+$ be the expansion of a random variable η into $\eta_- \in H_-$ and $\eta_+ \perp H_-$. Then, by Theorem 3, its image in $H(R, E^n)$: $\varphi\eta = m(t)$ is an $(n+1)/2$-harmonic function in $G - \{0\}$, and $\varphi\eta_- = m_-(t)$ is an $(n+1)/2$-harmonic function outside G; moreover, $m(t) = m_-(t)$ on G. Therefore, $m_-(t) \in \partial^- H$ or, which is the same, $H_+ \ominus H_- \subset \partial^- H$. From the inclusion $\partial^- H \subset H_- \cap H$ follows the coincidence of the spaces $H_+ \ominus H_-$ and $\partial^- H$, and hence follows the separation property of the algebra $\partial^-\mathfrak{B}$ and its minimality, because $\partial^- H$ consists only of projections of the future onto the past. $\partial^- H$ is not all the past H_-; for instance, $\varphi x(t) = R_t(\tau) \in H_-$ if $t \in G - (\Gamma \cup \{0\})$, but it is not a polyharmonic function in $G - \{0\}$, and therefore, it does not belong to ∂H.

3.2. The functions $m(t) \in \partial H$ have generalized Sobolev derivatives $\partial^k m(t) \in L^2(\Gamma)$, $0 \leqq k \leqq (n-1)/2$, on Γ and they are uniquely determined by these derivatives. In order to do this it is necessary to solve two boundary value problems for the equation $\Delta^{(n+1)/2}u = 0$ in the domains G and $E^n - G$ in the class of functions $W_2^{(n+1)/2}$ and with boundary data: $u(0) = 0$, $\partial^k u|_\Gamma = \partial^k m|_\Gamma$, $0 \leqq k \leqq (n-1)/2$. Therefore, the algebra $\partial\mathfrak{B}$ can be identified with the Dirichlet boundary data on Γ for the operator $\Delta^{(n+1)/2}$. In terms of random variables, this means that the algebra $\partial\mathfrak{B}$ is determined by the generalized derivatives of the field in Γ up to the $(n+1)/2$-th (see [4]).

3.3. *The infinitesimal σ-algebra $\mathfrak{B}_-\{0\}$ of the field is trivial.*

Proof. By Theorem 3, $H_-\{0\}$ consists of all functions which are $(n+1)/2$-harmonic outside $\{0\}$, $m(0) = 0$, $|m(t)| <$ const. and $\Delta^{[(n+1)/4]}m(t) \in L^2(E^n)$ $(W_2^1(E^n))$, for $n = 4k-1$ $(n = 4k+1)$. Since we have $(\Delta^{(n+1)/2}m(t), \varphi) = 0$ over the finite smooth functions φ with support outside $\{0\}$, $\Delta^{(n+1)/2}m(t)$, as a generalized function on the space of the finite infinitely differentiable functions is equal to $\sum_{k=0}^N c_k\delta^{(k)}(t)$ [6]. (For $n = 4k+1$, $\Delta^k \sum_{i=1}^n (\partial/\partial t_i) f_i = \sum_{k=0}^N c_k \delta^{(k)}(t)$, where $f_i = (\partial/\partial t_i)\Delta^k m \in L^2(E^n)$.) By applying the Fourier transform to both sides, we obtain

$$\sum_{i=1}^n \frac{\lambda_i}{|\lambda|}\hat{f}_i(\lambda) = \frac{\mathbf{P}_N(\lambda)}{|\lambda|^{2k+1}}.$$

Since the left-hand side of this equality belongs to $L^2(E^n)$, also $\mathbf{P}_N(\lambda)|\lambda|^{-2k-1} \in L^2(E^n)$, which is only possible if $\mathbf{P}_N(\lambda) \equiv 0$. Thus, the function $m(t)$ is polyharmonic in the whole space, and bounded; therefore, $m(t) \equiv m(0) = 0$.

4. One Problem of Integral Geometry

While studying the space of functions $H(R, E^n)$, the spectral representation of the Lévy field proposed by Chentsov in his paper [11] can be used as a starting point:

$$x(t) = c(n)\int_{(0t)} b(dr \times d0), \tag{13}$$

where $(0, t)$ is the sphere of diameter $\vec{0t}$; $b(dr \times d0)$ is a Gaussian random measure with independent values, i.e., $\mathbf{E}b(dr \times d0) = 0$, $\mathbf{E}|b(dr \times d0)|^2 = dr \times d0$ and $c^2(n) = 2(n-1)\int_0^{\pi/2} \sin^{n-2}\theta\, d\theta$.

4.1. *The representation* (13) *is non-degenerate. Therefore, the function $m(t)$ belongs to the space $H(R, E^n)$ if and only if $m(t)$ admits a representation of the form*

$$m(t) = c(n)\int_{(0t)} f(\lambda)\, d|\lambda|\, d0_\lambda, \qquad f \in L^2(dr \times d0). \tag{14}$$

Proof. Let $f \in L^2(dr \times d0)$ be a function such that the integral (14) of f over every sphere $(0, t)$ is equal to zero. Let us perform an inversion $t \to t^* = t/|t|^2$; then the sphere $(0, t)$ goes over into the half-space $\{(t, \tau) \geqq |t|\}$ and f is a function whose integrals over the half-spaces are equal to zero. Therefore, $f = 0$ (see the Radon problem in [8]), i.e., the representation (13) is non-degenerate. The second part of the statement can be obtained similarly to subsection 1.1.

4.2. Let us consider the function $f(\lambda) \in L^2(G, dr \times d0)$ defined in the domain $G = \{\lambda : |\lambda| \leqq r\}$ and let us assume that the integrals (14) of $f(\lambda)$ over the spheres $(0, t)$ contained in the domain of definition of the function $f(\lambda)$ are known. Let us formulate the problem of reconstructing f by means of its integrals $m(t)$. If $r = \infty$, i.e., $G = E^n$, the problem of reconstructing f after the inversion $t \to t/|t|^2$ goes over into the classical Radon problem on reconstruction of a function if its integrals over the half-spaces are known. The solution of the Radon problem is always unique. In the case $r < \infty$ the problem will have infinitely many solutions, which is equivalent to the degeneracy of the representation (13) in the sphere $|t| \leqq r$ (see [4]).

Theorem 5: *The function $m(t)$ in the domain $G = \{t : |t| \leqq 1\}$ is representable in the form of an integral, over the spheres $(0, t) \subset G$, of a function f in $L^2(G, dr \times d0)$ if and only if $m(t) \in H(R, G)$. Among the solutions $f(t), t \in G$, of equation (14), there exists a unique solution which possesses minimal norm in $L^2(G, dr \times d0)$. This is the only solution that, extended as zero outside G, corresponds to a function $m(t), t \in E^n$, in (14), which possesses the properties: $m(t) \in W_2^{(n+1)/2}(E^n)$; moreover, $m(t)$ is an $(n + 1)/2$-harmonic function outside G.*

PROOF. If $m(t) \in H(R, G)$, then $m(t)$ admits the representation (14) with a function f concentrated in the sphere G. In fact, by the isomorphism $\varphi : H(x, G) \leftrightarrow H(R, G)$ the function $m(t)$ corresponds to the random variable $\eta = \varphi^{-1}m(t)$. From subsection 4.1, $\eta = \int f(t)b(dr \times d0)$, where the function f is concentrated in G (since the subset $x(t), |t| < 1$, which is dense in $H(x, G)$ has such a representation) and $\|\eta\| = \|f\|_{L^2(dr \times d0)}$. Therefore, $m(t) = \int_{(0,t)} f \, dr \, d0$, where f is concentrated in G. Conversely, let $m(t)$ be representable in that form. We define a random variable $\eta = \int f(\lambda)b(dr \times d0)$, belonging to $H(x, E^n)$, and we shall obtain its projection $\check{\eta}$ on $H(x, G)$. The function $\varphi\check{\eta} \in H(R, G)$, where $\varphi\check{\eta} = (x(t), \check{\eta}) = (x(t), \eta) = m(t)$ for $|t| < 1$, i.e., $m(t) \in H(R, G)$. We remark that $\|f\|_{L^2(G, dr \times d0)} = \|\eta\| \leqq \|\check{\eta}\| = \|m(t)\|_{H(R,G)}$, where the projection η on $H(x, G)$ does not depend on f and is equal to $\varphi^{-1}m(t)$. From the uniqueness of the representation of the random variable $\eta \in H(x, E^n)$ in the form $\eta = \int fb(dr \times d0)$, it follows that the function f, which corresponds in this representation to $\varphi^{-1}m(t)$, is the unique solution of (14) with minimum norm in $L^2(dr \times d0)$, equal to $\|m(t)\|_{H(R,G)}$. Finally, Theorem 3 implies that, for the function f corresponding to $\varphi^{-1}m(t)$, the integrals (14) satisfy the equation $\Delta^{(n+1)/2}m(t) = 0$ for $|t| > 1$.

Thus, in order to find the function f from its integrals $m(t), |t| < 1$, under the condition of minimality of the norm of f in $L^2(dr \times d0)$, it is necessary to solve the Dirichlet problem for the equation $\Delta^{(n+1)/2}\tilde{m}(t) = 0$ in the domain $|t| > 1$, with boundary data

$$\left. \frac{\partial^k \tilde{m}}{\partial t_1^{k_1} \cdots \partial t_n^{k_n}} \right|_\Gamma = \left. \frac{\partial^k m}{\partial t_1^{k_1} \cdots \partial t_n^{k_n}} \right|_\Gamma, \qquad 0 \leqq k \leqq (n-1)/2,$$

in the class of functions $W_2^{(n+1)/2}$. The function $\tilde{m}(t)$ will immediately define the integrals (14) of f; then, by the inversion $t \to t/|t|^2$, the problem is reduced to the Radon problem.

The author expresses his deep gratitude to Ya. G. Sinai for his interest and guidance, and to Yu. I. Golosov for helpful conversations.

Received by the editors
May 24, 1966

REFERENCES

[1] P. LÉVY, *Processus Stochastique et Mouvement Brownien*, Paris, Gauthier-Villars, 1948.
[2] YU. A. ROZANOV, *On the densities of Gaussian distributions*, Theory Prob. Applications, 11 (1966), pp. 170–179.
[3] YU. I. GOLOSOV, *On one method for the calculation of the Radon-Nikodym derivative of two Gaussian measures*, Dokl. Akad. Nauk SSSR, 170 (1966), pp. 11–13.
[4] H. P. MCKEAN, JR., *Brownian motion with a several-dimensional time*, Theory Prob. Applications, 8 (1963), pp. 335–354.
[5] J. L. DOOB, *Stochastic Processes*, Wiley and Sons, Inc., New York, 1953.

[6] I. M. Guel'fand and G. E. Shilov, *Generalized Functions*, Vol. 1–2, Fizmatgiz, 1958. (In Russian.)

[7] S. L. Sobolev, *Some Applications of Functional Analysis to Mathematical Physics*, Novosibirsk, Izd-vo SO AN SSSR, 1962.

[8] N. Aronszajn, *Theory of reproducing kernels*, Trans. Amer. Math. Soc., 68 (1950), pp. 337–404.

[9] I. M. Gel'fand, M. I. Graev and N. Ya. Vilenkin, *Generalized Functions*, Vol. 5, Fizmatgiz, 1962. (In Russian.)

[10] J. Hajek, *On linear statistical problems in stochastic processes*, Czechoslov. Math. J., 12 (1962), pp. 404–444.

[11] N. N. Chentsov, *Multi-parametric Lévy Brownian motion for several parameters and generalized white noise*, Theory Prob. Applications, 2 (1957), pp. 265–266.

[12] A. M. Yaglom, *Some classes of random fields in the n-dimensional space, related to stationary random processes*, Theory Prob. Applications, 2 (1957), pp. 273–320.

20

Reprinted from *Arch. Ration. Mech. Anal.* **43:**367–391 (1971)

A Markov Property for Gaussian Processes with a Multidimensional Parameter

LOREN D. PITT

Communicated by M. KAC

Introduction

In this paper we discuss a class of Gaussian processes $\{X_t\colon t \in T\}$ for which the time domain T is an n-dimensional space rather than the usual real interval. Of special interest to us are those processes which we call Markovian. These are intuitively described by requiring that for any smooth surface Γ which separates T into complementary domains, what happens inside is independent of what happens outside when conditioned by knowledge of X_t on (and near) Γ.

The discussion centers on the reproducing kernel Hilbert space associated with the covariance function of the process. Those spaces associated with Markov processes are characterized by a locality condition on their inner products that requires functions with disjoint support to be orthogonal. The discussion of these matters is very general and together with the introduction of local reproducing kernel spaces appears in Part I.

With appropriate additional assumptions the inner product in the reproducing kernel space can be identified as coming from a non-negative Dirichlet form

$$\langle u, v\rangle = \sum_{\alpha,\beta} \int_T a_{\alpha\beta} D^\alpha u \overline{D^\beta v}.$$

Moreover, if this form is elliptic the associated process $\{X_t\}$ is found to have certain generalized normal derivatives on the surface and the "boundary" σ-field is seen to be generated by these normal derivatives. We call such a process Markovian of finite order.

It is then shown that the least squares prediction problem for the process is intimately related to the Dirichlet problem for an associated elliptic operator. Formulas which solve the Dirichlet problem may be reinterpreted to give a solution of the prediction problem. These matters are discussed in Part II. Those readers familiar with the work of McKEAN [8] and MOLCHAN [9] on Brownian motion with a multidimensional time parameter will easily recognize this as an extension of their work.

In Part III we have done some spectral theory for stationary Gaussian Markov processes on R_n. This material is not complete but does contain the characterization of the spectral densities of Markov processes of finite order as being inverses of non-negative elliptic polynomials. Further work in this direction will appear in [11].

We have attempted to make this paper accessible to an audience of probabilists that is neither used to processes with a several-dimensional time nor familiar with Dirichlet problems for higher order operators. For this reason we have not stressed formal completeness and generality, but we have included such items as the example in Section 1 of a Markov process whose time parameter runs over the circle. This example is the simplest possible and shows in the absence of all technical problems the identity of the least squares prediction problem for the process and the Dirichlet problem for the associated differential operator.

This work was begun at Rockefeller University during the year 1969–70. The ideas of H. P. MCKEAN, Jr. appear throughout this paper, and with gratitude I acknowledge his influence.

1. An Example: A Stationary Markov Process on the Circle

Let $\{X_t\colon -\infty<t<\infty\}$ be a real continuous stationary Gaussian process that is periodic with period 2π and has zero expectation. We consider $\{X_t\}$ to be a process defined for t on the circle $C=R_1 (\mathrm{mod}\, 2\pi)$, and call $\{X_t\}$ *simply Markovian* if*:

For each interval $I=[\alpha, \beta]$ on the circle C, the "past" $=\{X_t\colon t\in I\}$ *and the* "future" $=\{X_t\colon t\notin I\}$ *are conditionally independent given the* "present" $=\{X_\alpha, X_\beta\}$.

Since the process is Gaussian this is equivalent to the condition that for $t\notin I$,

$$E\{X_t \mid X_s\colon s\in I\}=E\{X_t \mid X_\alpha, X_\beta\}.$$

Our process $\{X_t\}$ is described by the periodic covariance function $R(t-s) = EX_tX_s$. Covariance functions of Markov processes are easily characterized.

Proposition 1.1. *A non-constant covariance function $R(t-s)$ on C corresponds to a Markov process if and only if R is the Green's function of a second order operator*

$$Lf=\alpha f-\gamma f''; \qquad \alpha, \gamma>0^{\dagger} \tag{1.1}$$

on C.

Note. Our proof makes no use of the reproducing kernel spaces introduced later. The relationship with the Dirichlet problem for L and the fact that as operators $R=L^{-1}$ are, however, still to be stressed.

Proof. First suppose that L is given and R is the Green's function for L. For any proper sub-interval $I=[\alpha, \beta]$ of C, the space of functions on I satisfying $Lf=0$ has dimension two and is spanned by the functions $R(s-\alpha)$ and $R(s-\beta)$. For $t\notin I$ we have $LR(\cdot-t)=0$ on I, and thus $R(s-t)$ satisfies a relation

$$R(s-t)=A(t)\,R(s-\alpha)+B(t)\,R(s-\beta); \qquad s\in I, \ t\in I \tag{1.2}$$

---

* This definition is a natural extension of the classical definition for processes with time domain $T=[0, \infty)$. In both cases the topological present is a point set separating the time domain into complementary domains.

† The restrictions on α and γ are consequences of the fact that R is positive definite.

where the functions A and B are continuous. But $R(s-t)=EX_tX_s$ so that (1.2) implies that $X_t-A(t)X_\alpha-B(t)X_\beta$ and X_s are uncorrelated for $s\in I$. Since the process is Gaussian, this implies independence.

Equivalently we have for $t\notin I$

$$E\{X_t|X_s\colon s\in I\}=A(t)X_\alpha+B(t)X_\beta. \tag{1.3}$$

Taking conditional expectations on both sides given X_α and X_β, it follows that

$$E\{X_t|X_s\colon s\in I\}=E\{X_t|X_\alpha, X_\beta\} \tag{1.4}$$

and thus $\{X_t\}$ is Markovian.

Conversely we note that the steps leading from (1.2) to (1.4) are reversible, and we conclude that if $\{X_t\}$ is Markovian then functions A and B must exist, which satisfy (1.2). Differentiating* $R(s-t)$ twice with respect to s, equation (1.2) shows that as functions of t each of the three functions $R(s-t)$, $R'(s-t)$ and $R''(s-t)$ is a linear combination of $A(t)$ and $B(t)$. $R(t)$ thus satisfies an equation

$$\lambda R(t)+\mu R'(t)+R''(t)=0 \quad \text{on } (0,2\pi). \tag{1.5}$$

Moreover, R is even about π since X_t has period 2π and the odd and even parts of (1.5) must vanish separately. We conclude that

$$\lambda R(t)+R''(t)=0 \quad \text{on } (0,2\pi). \tag{1.6}$$

Without loss of generality we may assume that $R(0)=EX_t^2=1$. With this normalization the unique solution of (1.6), which is even about π, is

$$R(t)=\begin{cases}\dfrac{\operatorname{Cosh}[(t-\pi)\sqrt{-\lambda/\nu}]}{\operatorname{Cosh}(\pi\sqrt{-\lambda/\nu})} & \text{if } \lambda\cdot\nu<0\\[2ex] \dfrac{\operatorname{Cos}[(t-\pi)\sqrt{\lambda/\nu}]}{\operatorname{Cos}(\pi\sqrt{\lambda/\nu})} & \text{if } \lambda\cdot\nu>0.\end{cases} \tag{1.7}$$

The first is a Green's function as required, while the second is easily eliminated as a possibility.

In fact, we must have $|R(t)|\leqq 1$, and for $\lambda\cdot\nu>0$ we have $R(\pi)=1/\cos(\pi\sqrt{\lambda/\nu})$. The only possibility is that $R(\pi)=\pm 1$, but this would imply $X_{\pi+h}=X_h$. This clearly cannot happen for a non-constant Markov process and the proof is complete.

Before leaving this example we make two further observations. First, note that equation (1.2) implies $LA(t)=LB(t)=0$. It then follows from (1.3) that the function $u(t)=E\{X_t|X_s\colon s\in I\}$ satisfies

$$\begin{aligned} &Lu(t)=0 && \text{for } t\notin I\\ &u(\alpha)=X_\alpha, && u(\beta)=X_\beta. \end{aligned} \tag{1.8}$$

That is, the prediction of X_t given $\{X_s\colon s\in I\}$ is the solution of the Dirichlet problem $Lu(t)=0$ on $C-I$ with the boundary data $u(\alpha)=X_\alpha$ and $u(\beta)=X_\beta$.

* $R(t)$ is easily shown to be smooth on the interval $(0, 2\pi)$ by considering the smooth convolutions $\int R(s-t)\phi(t)dt=R(s-\alpha)\int A(t)\phi(t)dt+R(s-\beta)\int B(t)\phi(t)dt$, where ϕ is a smooth function vanishing on I.

Lastly, we comment that an obvious generalization of our definition of a Markov process is available. We call $\{X_t: t\in C\}$ Markovian of order p if the sample paths have $p-1$ continuous derivatives and if for each interval I, the past $=\{X_t: t\in I\}$ and the future $=\{X_t: t\notin I\}$ are conditionally independent when given the present $=\{X_\alpha, X'_\alpha, \ldots, X_\alpha^{(p-1)}, X_\beta, \ldots, X_\beta^{(p-1)}\}$. Our proposition and its proof then have easy extensions to the effect that $\{X_t\}$ is Markovian of order p if and only if $R(t)$ is the Green's function of an operator of order $2p$.

This result is well-known when $T=R_1$ instead of the circle. The Fourier version states that a stationary Gaussian process is Markov of order p if and only if its spectral density is the inverse of a polynomial of degree $2p$; see LEVINSON & MCKEAN [7]. The reader may also wish to compare this example with the work of DOOB [4], HIDA [5] and DOLPH & WOODBURY [3].

I. General Theory

2. Splitting Fields and the Markov Property

We begin by fixing some notations and reviewing the concept of splitting fields introduced by MCKEAN [8].

A fixed probability space (Ω, Σ, P) is given. The σ-field Σ is assumed to be complete, and a field $\mathscr{F}$ is always to be understood as a sub σ-field of Σ which contains all null sets of Σ. The field generated by a collection $\{f_\alpha: \alpha\in A\}$ of random variables will be denoted by $\sigma\{f_\alpha: \alpha\in A\}$.

We will use HUNT's excellent operator formalism ([6], p. 44) and write $\mathscr{F}f$ for the conditional expectation of the random variable f given the field $\mathscr{F}$. This notation is especially appropriate when f is square integrable; $\mathscr{F}f$ is then simply the orthogonal projection of f onto the space $L^2(\Omega, \mathscr{F}, P)$.

Let $\mathscr{F}$ and $\mathscr{G}$ be two subfields of Σ and let $\mathscr{S}$ be a subfield of $\mathscr{F}$. We say that *$\mathscr{F}$ and $\mathscr{G}$ split over $\mathscr{S}$ or that $\mathscr{S}$ is a splitting field of $\mathscr{F}$ and $\mathscr{G}$ if $\mathscr{F}$ and $\mathscr{G}$ are conditionally independent given $\mathscr{S}$*, that is, if

$$\mathscr{S}(fg)=\mathscr{S}(f)\cdot\mathscr{S}(g) \tag{2.1}$$

for all bounded f and g with $\sigma\{f\}\subset\mathscr{F}$ and $\sigma\{g\}\subset\mathscr{G}$.

Splitting fields are easily characterized by the following lemma, which surprisingly seems to be new.

Lemma 2.1. *A field $\mathscr{S}\subset\mathscr{F}$ is a splitting field if and only if $\mathscr{S}$ contains the field*

$$\mathscr{S}_0=\sigma\{\mathscr{F}g: g \text{ bounded and } \mathscr{G}\text{-measurable}\}. \tag{2.2}$$

Proof. If $\mathscr{S}_0\subset\mathscr{S}\subset\mathscr{F}$ we will show that (2.1) holds. In fact, since $\mathscr{S}\subset\mathscr{F}$ we have $\mathscr{S}(f\cdot g)=\mathscr{S}\mathscr{F}(f\cdot g)=\mathscr{S}(f\cdot\mathscr{F}g)$. But $\mathscr{S}_0\subset\mathscr{S}$ implies that $\mathscr{S}g=\mathscr{F}g$ and thus $f\cdot\mathscr{F}g=f\cdot\mathscr{S}g$. Hence $\mathscr{S}(f\cdot g)=\mathscr{S}(f\cdot\mathscr{S}g)=\mathscr{S}(f)\cdot\mathscr{S}(g)$.

The converse is equivalent to showing that $\mathscr{S}g=\mathscr{F}g$ for any bounded $\mathscr{G}$-measurable function g and any splitting field $\mathscr{S}$. But $E(\mathscr{S}g)^2=E\mathscr{S}(g)\cdot\mathscr{F}(g)\leqq E(\mathscr{F}g)^2$, and the equality $\mathscr{S}g=\mathscr{F}g$ will follow from the converse of Schwarz's inequality if we prove $E(\mathscr{S}g)^2=E(\mathscr{F}g)^2$. This, however, follows from (2.1) by $E(\mathscr{S}g)^2=E\mathscr{S}(\mathscr{F}g\cdot g)=E\mathscr{F}(g)\cdot g=E(\mathscr{F}g)^2$.

The elementary properties of splitting fields proved by McKEAN [8] are now immediate consequences of Lemma 2.1. In particular, we have

Corollary 2.2. *(i)* $\mathscr{F}$ *is a splitting field*
(ii) $\mathscr{S}_0$ *is the minimal splitting field*
(iii) $\mathscr{F} \cap \mathscr{G} \subset \mathscr{S}_0$.

Simplifications occur when the fields $\mathscr{F}$ and $\mathscr{G}$ are generated by Gaussian systems. These are described in the following known lemma, which we include here for completeness.

Lemma 2.3. *Suppose* $\mathscr{F} = \sigma\{X_\phi: \phi \in \Phi\}$ *and* $\mathscr{G} = \sigma\{X_\gamma: \gamma \in \Gamma\}$ *where* $\{X_\alpha: \alpha \in \Phi \cup \Gamma\}$ *is a Gaussian family. Then*

$$\mathscr{S}_0 = \sigma\{\mathscr{F} X_\gamma: \gamma \in \Gamma\}.$$

Proof. Since polynomials in the variables $\{X_\gamma: \gamma \in \Gamma\}$ are dense in $L^2(\Omega, \mathscr{G}, P)$, it is enough to show for each monomial $X_{\gamma_1}^{k_1} \dots X_{\gamma_n}^{k_n}$ that $\mathscr{F}(X_{\gamma_1}^{k_1} \dots X_{\gamma_n}^{k_n})$ is measurable over $\sigma\{\mathscr{F} X_\gamma: \gamma \in \Gamma\}$. To this end we write $X_{\gamma_i} = Y_i + Z_i$ where $Y_i = \mathscr{F} X_{\gamma_i}$. Then the Z_i's are independent of $\mathscr{F}$, and we observe that $\mathscr{F}\left(\prod_1^n X_{\gamma_i}^{k_i}\right) = \mathscr{F}\left(\prod_1^n (Y_i + Z_i)^{k_i}\right)$ may be written as a sum of terms of the form $\mathscr{F}(\prod Y_i^{a_i} \cdot \prod Z_i^{b_i}) = \prod Y_i^{a_i} \mathscr{F}(\prod Z_i^{b_i})$. Since the Z_i's are independent of $\mathscr{F}$ we see that $\mathscr{F}(\prod Z_i^{b_i}) = E(\prod Z_i^{b_i})$ is a constant and the result follows.

We turn now to our definition of the Markov property. As formulated here the concept is perhaps too broad, and we must introduce further assumptions later to obtain deeper implications. The present generality has the advantage, however, of yielding the easy characterization in Theorem 3.3 of the next section.

Let $\{X_t: t \in T\}$ be a real or complex valued process whose time domain T is a smooth open set in some Euclidian space R_n. Everything that follows could easily be done when T is a smooth manifold. Because the changes necessary are easy to make we will not pursue the added generality. Let $D_- \subset T$ be an open set whose boundary in T is a smooth $n-1$ dimensional surface Γ in T. We write $\bar{D}_-$ for the closure $D_- \cup \Gamma$ of D_- and D_+ for the complement in T of $\bar{D}_-$.

With these conventions we introduce the following fields of events:

$$\begin{aligned}
&\text{the ``past''} &&= \Sigma(D_-) \equiv \sigma\{X_t: t \in D_-\} \\
&\text{the ``future''} &&= \Sigma(D_+) \equiv \sigma\{X_t: t \in D_+\} \\
&\text{the ``present''} &&= \Sigma(\Gamma) \equiv \bigcap\{\Sigma(0): 0 \supset \Gamma, 0 \text{ is open}\} \\
& &&\Sigma(\Gamma_-) = \bigcap\{\Sigma(D_- \cap 0): 0 \supset \Gamma, 0 \text{ is open}\} \\
& &&\Sigma(\Gamma_+) = \bigcap\{\Sigma(D_+ \cap 0): 0 \supset \Gamma, 0 \text{ is open}\}.
\end{aligned}$$

We then define the Markov property.

Definition 2.4. $\{X_t: t \in T\}$ is called Markovian if for each open set $D_- \subset T$ with smooth boundary Γ we have

$$\Sigma(\Gamma_+) = \Sigma(\Gamma_-) = \Sigma(\Gamma), \tag{2.3}$$

and

(2.4) $\Sigma(\Gamma)$ is the minimal splitting field of $\Sigma(D_-)$ and $\Sigma(D_+)$.

Condition (2.3) is of a technical nature and says we see the same things in the very near future as in the very near past. Thus $\Sigma(\Gamma)\subset\Sigma(D_+)\cap\Sigma(D_-)$ and the definition requires that the splitting field of $\Sigma(D_-)$ and $\Sigma(D_+)$ is as small as possible and involves only boundary data.

The analogy with classical definitions of a Markov process is clear. In one dimension, however, our definition allows many processes that are not classically Markovian. For example, the integral of a Wiener process is Markovian by our definition but not classically so. In Hida's terminology [5] it is a 2-ple Markov process.

For Gaussian processes Lemma 2.3 gives a useful translation of this definition into Hilbert space language. Each of the previously introduced fields is associated with a closed subspace of $L^2(\Omega, P)$. Namely, for the open set D_- we introduce:

$$H(D_-)=\text{closed linear span of } \{X_t\colon t\in\bar{D}_-\},$$

and the boundary space

$$H(\Gamma_-)=\bigcap\{H(D_-\cap 0)\colon 0\supset\Gamma, 0 \text{ is open}\}.$$

Analogously we define $H(D_+)$, $H(\Gamma_+)$ and $H(\Gamma)$. We will write H for $H(T)$.

When applied to variables in H, the conditional expectation operator $\Sigma(D_-)$ is simply the orthogonal projection onto the space $H(D_-)$, and Lemma 2.3 implies that the minimal splitting field of $\Sigma(D_-)$ and $\Sigma(D_+)$ is generated by the variables $\{\Sigma(D_-)X\colon X\in H(D_+)\}$. In this language the definition of the Markov property becomes:

Definition 2.4′. A Gaussian process $\{X_t\}$ is Markovian if for each open set $D_-\subset T$ with smooth boundary Γ,

(2.5) $$H(\Gamma_+)=H(\Gamma_-)=H(\Gamma)$$

and

(M.1) The projection of $H(D_+)$ onto $H(D_-)$ is $H(\Gamma)$.

Because of its later importance we introduce the special notation of $H_0(D_\pm)$ for the orthogonal complement of $H(\Gamma_\pm)$ in $H(D_\pm)$. Assuming condition (2.5) is satisfied, then each of the following two statements is equivalent to the Markov property for $\{X_t\}$.

(M.2) The orthogonal complement of $H(D_+)$ in H is $H_0(D_-)$.

(M.3) $$H=H_0(D_-)\oplus H(\Gamma)\oplus H_0(D_+).$$

3. Local Reproducing Kernel Spaces and the Markov Property for Gaussian Processes

In order to obtain convenient function space representations of the spaces $H(D)$ we introduce the formalism of reproducing kernel Hilbert spaces. For

Markov processes these spaces are easily characterized; when further assumptions are introduced in Part II we will see that the inner products have the form $\langle u, v\rangle = \sum_{\alpha,\beta} \int v_{\alpha\beta} D^\alpha u \overline{D^\beta v}$. This leads us to a class of Dirichlet problems (we have tried to use a suggestive notation).

Let $\{X_t\}$ be a Gaussian process with covariance function $R(t, s) = EX_t\overline{X}_s$. For an open set $D \subset T$ we introduce the function space $\mathscr{H}(D)$ which is isomorphic to $H(D)$; $\mathscr{H}(D) = \{u(s) = EX\overline{X}_s\colon s \in D \text{ and } X \in H(D)\}$ with the inner product

$$\langle u_1, u_2\rangle = EX_1\overline{X}_2 \quad \text{where} \quad u_i(s) = EX_i\overline{X(s)}, \qquad \text{for } i = 1, 2.$$

The map $X \to EX\overline{X}_s$ is easily seen to be an isometry of $H(D)$ onto $\mathscr{H}(D)$. Instead of $\mathscr{H}(T)$ we will usually write $\mathscr{H}$, and we list without proof the following well-known elementary properties of $\mathscr{H}(D)$ (see [2]):

(3.1) $\mathscr{H}(D)$ is spanned by the functions $R(t, \cdot)$ for $t \in D$.

(3.2) For each u in $\mathscr{H}(D)$, $u(t) = \langle u, R(t, \cdot)\rangle$; that is, R is a reproducing kernel for $\mathscr{H}(D)$.

The functions $EX\overline{X}_s$ in $\mathscr{H}(D)$ were defined only for s in D, but $H(D) \subset H$ so that it makes sense for s to vary over T. Each u in $\mathscr{H}(D)$ thus extends to a function $\tilde{u}(s)$ in $\mathscr{H}$ with $\|u\|_{\mathscr{H}(D)} = \|u\|_{\mathscr{H}}$, and we may consider $\mathscr{H}(D)$ as a subspace of $\mathscr{H}$. Also note that the restriction of a function u in $\mathscr{H}$ to D defines a function in $\mathscr{H}(D)$ which may be interpreted as the projection of u onto $\mathscr{H}(D)$. We summarize these remarks.

Each function in $\mathscr{H}(D)$ has a unique norm preserving extension to a function in $\mathscr{H}$. The projection of a function u in $\mathscr{H}$ onto $\mathscr{H}(D)$ may be interpreted either as its restriction to D or as the unique element in $\mathscr{H}$ of minimal norm which agrees with u on D.

If D_- and D_+ are a complementary pair of open sets with smooth boundary Γ, we introduce spaces $\mathscr{H}(\Gamma_\pm)$ which are isomorphic to the boundary spaces $H(\Gamma_\pm)$,

$$\mathscr{H}(\Gamma_\pm) = \{u_s = EX\overline{X}_s\colon s \in D_\pm \text{ and } X \in H(\Gamma_\pm)\}.$$

If $H(\Gamma_+) = H(\Gamma_-)$ we may identify $\mathscr{H}(\Gamma_+)$ with $\mathscr{H}(\Gamma_-)$ and write $\mathscr{H}(\Gamma) = \mathscr{H}(\Gamma_+) = \mathscr{H}(\Gamma_-)$. This makes sense because $\mathscr{H}(D_+)$ and $\mathscr{H}(D_-)$ are both contained in $\mathscr{H}$. Thus $\mathscr{H}(\Gamma)$ is a subspace both of $\mathscr{H}(D_+)$ and of $\mathscr{H}(D_-)$.

The orthogonal complement of $\mathscr{H}(\Gamma)$ in $\mathscr{H}(D_+)$ (resp. $\mathscr{H}(D_-)$) is denoted as $\mathscr{H}_0(D_+)$ (resp. $\mathscr{H}_0(D_-)$). From the definition of $H(\Gamma)$ we see that $\mathscr{H}_0(D_+)$ may be described as the closure in $\mathscr{H}(D_+)$ of

$$\{u \in \mathscr{H}(D_+)\colon u \text{ vanishes near } \Gamma\}. \tag{3.3}$$

Intuitively, $\mathscr{H}_0(D_+)$ consists of functions with zero boundary data, while $\mathscr{H}(\Gamma) \subset \mathscr{H}(D_+)$ consists of "harmonic" functions on D_+.

For reference we translate our earlier characterizations of the Markov property as follows.

Proposition 3.1. *A Gaussian process $\{X_t\colon t \in T\}$ satisfying the condition*

$$\mathscr{H}(\Gamma_+) = \mathscr{H}(\Gamma_-) = \mathscr{H}(\Gamma) \tag{3.4}$$

is Markovian if and only if one of the three following equivalent conditions is satisfied:

(M.1)′ *The projection of* $\mathscr{H}(D_+)$ *onto* $\mathscr{H}(D_-)$ *is* $\mathscr{H}(\Gamma)$

(M.2)′ $$\mathscr{H}_0(D_-)=\mathscr{H}\ominus\mathscr{H}(D_+)$$

(M.3)′ $$\mathscr{H}=\mathscr{H}_0(D_-)\oplus\mathscr{H}(\Gamma)\oplus\mathscr{H}_0(D_+).$$

The next definition is introduced to characterize the kernel spaces associated with Markov processes. $\mathscr{L}$ is a linear function space on T with an inner product.

Definition 3.2. $\mathscr{L}$ is said to be local if for each pair of complementary open sets D_- and D_+ with smooth boundary Γ, conditions (L.1) and (L.2) are satisfied.*

(L.1) If u_+ and u_- are in $\mathscr{L}$ with the support of $u_\pm$ in $\bar{D}_\pm$, then $\langle u_+, u_-\rangle=0$.

(L.2) If $u=u_+ + u_-$ is in $\mathscr{L}$ with the support of u_+ in D_+ and support of u_- in $\bar{D}_-$, then u_+ and u_- are in $\mathscr{L}$.

Theorem 3.3. *A Gaussian process* $\{X_t\colon t\in T\}$ *which satisfies* $\mathscr{H}(\Gamma_-)=\mathscr{H}(\Gamma_+)=\mathscr{H}(\Gamma)$ *is Markovian if and only if the space* $\mathscr{H}=\mathscr{H}(T)$ *is local.*

Proof. Suppose $\{X_t\}$ is Markovian. We will first show that (L.1) is satisfied. Thus let $u_\pm\in\mathscr{H}$ and assume the support of $u_\pm\subset\bar{D}_\pm$. Then $\langle u_+, R(t,\cdot)\rangle=u_+(t)=0$ for $t\in D_-$. Since the functions $\{R(t,\cdot)\colon t\in\bar{D}_-\}$ span $\mathscr{H}(D_-)$ we conclude that u_+ is orthogonal to $\mathscr{H}(D_-)$ and by (M.2′) that $u_+\in\mathscr{H}_0(D_+)\subset\mathscr{H}(D_+)$. The same reasoning shows that u_- is orthogonal to $\mathscr{H}(D_+)$. Thus $\langle u_+, u_-\rangle=0$ and (L.1) is satisfied.

Turning to (L.2), let $u=u_+ + u_-\in\mathscr{H}$ with the support of u_+ in D_+ and the support of u_- in $\bar{D}_-$. Write u_0 for the projection of u onto $\mathscr{H}(D_+)$. Then u_0 vanishes near Γ since u does. Thus $u_0\in\mathscr{H}_0(D_+)$, and applying (M.2′) we see that $u_0\in\mathscr{H}\ominus\mathscr{H}(D_-)$; that is, u_0 vanishes on D_-. But u_0 agrees with u on D_+ so that $u_+=u_0$ is in $\mathscr{H}$ and (L.2) is satisfied.

To prove the converse we will show that if $\mathscr{H}$ is local then (M.2′) holds; that is $\mathscr{H}\ominus\mathscr{H}(D_+)=\mathscr{H}_0(D_-)$. We begin with the inclusion $\mathscr{H}\ominus\mathscr{H}(D_+)\subset\mathscr{H}(D_-)$. Let P be the projection onto $\mathscr{H}(D_-)$ and $u\in\mathscr{H}\ominus\mathscr{H}(D_+)$. We know that $u-Pu=0$ on $\bar{D}_-$ and that the support of u is in $\bar{D}_-$. Thus by (L.1), $\|(1-P)u\|^2=\langle u-Pu, u\rangle=0$ and hence $u=Pu$ and $u\in\mathscr{H}(D_-)$.

To show $u\in\mathscr{H}_0(D_-)$ it is now sufficient to show that u is orthogonal to $\mathscr{H}(\Gamma)$. This is clear since $\mathscr{H}(\Gamma)\subset\mathscr{H}(D_+)$ and by assumption u is orthogonal to $\mathscr{H}(D_+)$.

The proof of the identity $\mathscr{H}\ominus\mathscr{H}(D_+)=\mathscr{H}_0(D_-)$ will be complete if we prove $\mathscr{H}(D_+)\cap\mathscr{H}_0(D_-)=\{0\}$, and for this it suffices to prove that any u in $\mathscr{H}(D_+)\cap\mathscr{H}_0(D_-)$ vanishes on D_-. Moreover, since such a u must be orthogonal to $\mathscr{H}(\Gamma)$ we can find a sequence 0_n of open neighborhoods of Γ for which $\lim u_n=0$, where u_n is the projection of u onto $\mathscr{H}(D_+\cap 0_n)$. The problem thus reduces to showing that $v_n=u-u_n$ vanishes on D_-.

* Conditions (L.1) and (L.2) are satisfied by such examples as $L^2(R_1)$ and the Sobolev spaces H_m^2. Condition (L.2) is introduced to eliminate such examples as the space of even L^2 functions.

But v_n vanishes on $\bar{D}_+ \cap 0_n \supset \Gamma$ so that by (L.2) we may write $v_n = v_+ + v_-$ where v_- vanishes on D_+, v_+ has support in D_+ and $v_\pm \in \mathscr{H}$. By (L.1), $\langle v_+, v_- \rangle = 0$ and $\|v_n\|^2 = \|v_+\|^2 + \|v_-\|^2$. On the other hand, $\|v_n\| \leqq \|v_+\|$ since $v_n \in \mathscr{H}(D_+)$ and v_+ agrees with v_n on D_+. Hence $\|v_-\| = 0$ and $v_n = v_+$ vanishes on D_-.

II. Markov Processes of Finite Order

4. Smooth Local Spaces of Finite Order

We now show that if the function space $\mathscr{H}$ is local and rich enough then its inner product is given by a Dirichlet form. This result follows from PEETRE'S [10] characterization of differential operators. For the sake of completeness we include PEETRE'S proof here.

First we introduce some notation. For a smooth open set $T \subset R_n$, $C_0^\infty(T)$ is the set of infinitely differential functions with compact support contained in T. We will write C_0^∞ instead of $C_0^\infty(R_n)$. If $\alpha = (\alpha_1, \ldots, \alpha_n)$ is a multi-index where α_i's are non-negative integers, we write $|\alpha| = \alpha_1 + \cdots + \alpha_n$. D^α denotes the differential operator $D^\alpha = \frac{d^{\alpha_1}}{d x_1^{\alpha_1}} \cdots \frac{d^{\alpha_n}}{d x_1^{\alpha_n}}$, and for the generic point $x = (x_1, \ldots, x_n)$ in R_n, x^α will denote the monomial $x_1^{\alpha_1} \ldots x_n^{\alpha_n}$. Let Δ denote the Laplacian

$$\Delta = \frac{d^2}{d x_1^2} + \cdots + \frac{d^2}{d x_n^2}.$$

For each $N \geqq 0$ we introduce an inner product $\langle \cdot, \cdot \rangle_N$ on $C_0^\infty(T)$; namely

$$\langle f, g \rangle_N = \int_T \sum_{|\alpha| \leqq N} D^\alpha f \, \overline{D^\alpha g}.$$

$H_0^N(T)$ denotes the completion of $C_0^\infty(T)$ with respect to the norm $\|f\|_N^2 = \langle f, f \rangle_N$. Clearly $H_0^N(T) \supset H_0^{N+1}(T)$. Functions in $H_0^N(T)$ have strong L^2 derivatives of all orders up to N; the Fourier representation easily shows that for $|\alpha| < N - n/2$ there is a constant $C(\alpha)$ such that the Sobolev like inequality

$$\sup_{t \in R_n} |D^\alpha f(t)| \leqq C(\alpha) \|f\|_N \tag{4.1}$$

holds for all f in C_0^∞, and hence by continuity this holds for all f in $H_0^N(T)$.

Convergence in $C_0^\infty(T)$ is defined as follows: $\{f_n\}$ converges to f if there is a bounded open set U with $\bar{U} \subset T$ for which $f_n \in C_0^\infty(U)$ for each n and such that f_n converges to f in $H_0^N(U)$ for all $N \geqq 0$.

The inequality (4.1) shows that $C_0^\infty(T)$ is complete.

We will call a bilinear form

$$B(f, g) = \sum_{\alpha, \beta} \int_T a_{\alpha\beta}(t) \, D^\alpha f(t) \, \overline{D^\beta g(t)}$$

a Dirichlet form if the coefficients $a_{\alpha\beta}(t)$ are locally square summable and if on each compact set all but a finite number of the $a_{\alpha\beta}$ vanish. Such a form is well-defined and continuous on $C_0^\infty(T)$. Moreover, if B is symmetric and positive then $C_0^\infty(T)$ together with the inner product $B(f, g)$ is a local space.

The following theorem states that any reasonable positive local form on $C_0^\infty(T)$ is continuous and that if a local form is continuous then it agrees with some Dirichlet form. The proof of parts (*ii*) and (*iii*) are adopted from PEETRE [10].

Theorem 4.1. (*i*) *A non-negative form* $\langle\cdot,\cdot\rangle$ *on* $C_0^\infty(T)$, *which satisfies the honesty condition: if* $u_n \to 0$ *in* $C_0^\infty(T)$ *and* $\langle u_n - u_m, u_n - u_m\rangle \to 0$ *as* $n, m \to \infty$, *then* $\langle u_n, u_n\rangle \to 0$, *is continuous.*

(*ii*) *If* $\langle\cdot,\cdot\rangle$ *is a non-negative local form on* $C_0^\infty(T)$, *there exists a discrete set* $\Delta \subset T$ *such that* $\langle\cdot,\cdot\rangle$ *is continuous on* $C_0^\infty(T-\Delta)$.

(*iii*) *If* $\langle\cdot,\cdot\rangle$ *is a continuous local form on* $C_0^\infty(T)$, *then* $\langle\cdot,\cdot\rangle$ *is a Dirichlet form.*

Proof of (i). For any relatively compact open subset Ω of T, the space $H_0^\infty(\Omega) \equiv \bigcap N_0^N(\Omega)$ is complete with respect to the quasi-norm

$$\|f\| = \sum_{n\geq 0} 2^{-N} \cdot \|f\|_N (1+\|f\|_N)^{-1}.$$

Since $\langle\cdot,\cdot\rangle$ is everywhere defined on $H_0^\infty(\Omega)$ we may define another quasi-norm $|f| = \|f\| + \langle f,f\rangle^{\frac{1}{2}}$ on $H_0^\infty(\Omega)$.

It now suffices to show that the two quasi-norms $\|\cdot\|$ and $|\cdot|$ are equivalent. To see this let $\{f_n\}$ be a Cauchy sequence with respect to $|\cdot|$. Then $\{f_n\}$ is Cauchy with respect to $\|\cdot\|$ and thus has a $\|\cdot\|$ limit f in $H_0^\infty(\Omega)$. This may then be used to determine a linear map S from the $|\cdot|$ completion of $H_0^\infty(\Omega)$ into $H_0^\infty(\Omega)$. S is easily seen to be continuous and onto, and therefore by the open mapping theorem S is open. Thus if S^{-1} exists it must be continuous, and this would show that $\|\cdot\|$ and $|\cdot|$ are equivalent. But S^{-1} will exist if and only if S is one to one or, what is the same thing, if and only if ker $S=\{0\}$. This, however, is precisely the content of our honesty condition and the result follows.

Proof of (ii). We will call a point $x \in T$ singular if the form $\langle\cdot,\cdot\rangle$ is unbounded on $C_0^\infty(0)$ for each neighborhood 0 of x; that is, if there are functions f_k in $C_0(0)$ with $\|f_k\|_k \leq 1$ and $\langle f_k, f_k\rangle \uparrow \infty$. We claim that any relatively compact open set Ω with $\bar{\Omega} \subset T$ contains only a finite number of singular points.

If this were not so we could find open sets $0_k \subset \Omega$ with disjoint closures and functions f_k in $C_0^\infty(0_k)$ which satisfy $\|f_k\|_k \leq 1/k^2$ and $\langle f_k, f_k\rangle \geq 1$. Then $f = \sum f_k$ is in $C_0^\infty(T)$ and because $\langle\cdot,\cdot\rangle$ is local we would have for each N,

$$\langle f, f\rangle \geq \left\langle \sum_1^N f_k, \sum_1^N f_k \right\rangle \geq N,$$

which is impossible.

Now let Δ be the discret set of singular points and let Ω be a relatively compact open set with $\bar{\Omega} \subset T-\Delta$. An elementary partition of unity argument easily yields the existence of an integer N and a constant C for which the inequality $\langle f,f\rangle \leq C\|f\|_N^2$ holds for all f in $C_0^\infty(\Omega)$. Thus $\langle\cdot,\cdot\rangle$ is continuous on $C_0^\infty(\Omega)$ and hence on $C_0^\infty(T-\Delta)$.

Proof of (iii). For each bounded open set U with $\bar{U} \subset T$ there is an N' for which

(4.2) $$|\langle f, g\rangle| \leq \|f\|_{N'} \|g\|_{N'}; \qquad \text{for } f, g \in C_0^{N'}(U).$$

Let $N=N'+k$ where $k>n/2$. The inclusion i: $H_0^N(U)\to H_0^{N'}(U)$ is known to be Hilbert-Schmidt so that if $\{f_j\}\subset C_0^\infty(U)$ is an orthonormal basis in $H_0^N(U)$ we have $\sum \|f_j\|_{N'}^2<\infty$.

On finite sums $\sum b_{ij} f_i(s) f_j(t)$ we define the functional

$$F\left(\sum b_{ij} f_i(s) f_j(t)\right)=\sum b_{ij}\langle f_i, \bar{f}_j\rangle . \tag{4.3}$$

But $\langle f_{ij}, \bar{f}_j\rangle \leqq \|f_i\|_N \|f_j\|_{N'}$ so that $\sum |\langle f_i, \bar{f}_j\rangle|^2<\infty$, and we see that F is continuous on a dense subspace of $H_0^{2N}(U\times U)$. The functional F thus determines a distribution $F\in C_0^\infty(U\times U)'$ for which

$$F(f(s)\, g(t))=\langle f, \bar{g}\rangle . \tag{4.4}$$

Because $\langle\cdot, \cdot\rangle$ is local the support of F is contained in the diagonal of $U\times U$, and using (4.4) we conclude that there are finitely many measures $\mu_{\alpha\beta}$ on U which give

$$\langle f, \bar{g}\rangle=\sum \int_U \mu_{\alpha\beta}(dt)\, D^\alpha f\, D^\beta g .$$

The measures $\mu_{\alpha\beta}$ may be written in the form $(1-\Delta)^m a'_{\alpha\beta}$ where the $a'_{\alpha\beta}$ are continuous functions. Integrating by parts and changing g to $\bar{g}$, we obtain the desired expression

$$\langle f, g\rangle=\sum \int_U a_{\alpha\beta}(t)\, D^\alpha f(t)\, \overline{D^\beta g(t)} .$$

Because U was arbitrary the proof is complete.

Note that the set Δ of singular points for a non-negative local form on $C_0^\infty(R_n)$ which commutes with translations must be closed under translations. Since Δ is discrete by part (*ii*) we see that Δ is empty.

Corollary 4.2. *A non-negative local form on $C_0^\infty(R_n)$ which commutes with translations is a Dirichlet form.*

5. Markov Processes of Finite Order

It is reasonable to expect that if an inner product $\langle\cdot, \cdot\rangle$ on $C_0^\infty(T)$ is given by a well-behaved Dirichlet form, then there should be naturally associated with $\langle\cdot, \cdot\rangle$ a Gaussian Markov process. In this section we will investigate these problems when the form $\langle\cdot, \cdot\rangle$ is elliptic. In this case we can obtain rather complete results, and the processes that we obtain are the analogues of the Markov processes of order p discussed at the end of Section 1.

A general answer to the question of when the completion of $(C_0^\infty(T), \langle\cdot, \cdot\rangle)$ is the reproducing kernel space associated with some Gaussian process is contained in the following known

Proposition 5.1. *Let $\mathscr{H}$ be the completion of $(C_0^\infty(T), \langle\cdot, \cdot\rangle)$. Then $\mathscr{H}$ is a reproducing kernel space if and only if for each $t\in T$ the functional $u\to u(t)$ is continuous with respect to the norm $\langle u, u\rangle^{1/2}$ on $C_0^\infty(T)$.*

Proof. If $\mathscr{H}$ has a reproducing kernel $R(t, s)$, then $u\to u(t)=\langle u, R(t, \cdot)\rangle$ is continuous.

Conversely if $u \to u(t)$ is continuous and if u_n is any sequence which converges in $\mathscr{H}$, we see that $u_n(t)$ converges for each t in T. Thus $\mathscr{H}(T)$ is a function space. Moreover, since $u \to u(t)$ is continuous we know that for each t there is a function $r_t(s)$ in $\mathscr{H}(T)$ such that $u(t) = \langle u, r_t \rangle$. Setting $R(t, s) = r_t(s)$, we see that R is a reproducing kernel for $\mathscr{H}$ and the proof is complete.

Note that $R(t, s)$ is non-negative definite and thus is the correlation function of some Gaussian process $\{X_t\colon t \in T\}$.

Now let

$$\langle u, v \rangle = \sum_{|\alpha|, |\beta| \leqq p} \int_T a_{\alpha\beta}(t) D^\alpha u(t) \overline{D^\beta v(t)} \tag{5.1}$$

be a positive symmetric Dirichlet form on $C_0^\infty(T)$. We will assume that the coefficients $a_{\alpha\beta}$ are bounded, uniformly continuous and infinitely differentiable. The form $\langle \cdot, \cdot \rangle$ is called *uniformly strongly elliptic* if there is a constant $C > 0$ such that for all $x \in R_n$ and $t \in T$ we have

$$\Big|\operatorname{Re} \sum_{|\alpha|, |\beta| = p} a_{\alpha\beta}(t) X^{\alpha+\beta}\Big| \geqq C |X|^{2p}. \tag{5.2}$$

If the form $\langle \cdot, \cdot \rangle$ is uniformly strongly elliptic then it follows from Garding's inequality (see [1], p. 78) that for each $\lambda > 0$ the two norms $\langle u, u \rangle^{1/2} + \lambda \|u\|_0$ and $\|u\|_p$ are equivalent on $C_0^\infty(T)$, where $\|\ \|_0$ is the L^2 norm and $\|\ \|_p$ is the pth order Sobolev norm. Thus if the norm $\|\ \|_0$ is continuous with respect to the inner product $\langle \cdot, \cdot \rangle$ we may conclude that the space $\mathscr{H}$ coincides as a function space with the space $H_0^p(T)$ and that the two norms are equivalent.

Theorem 5.2. *Let $\langle \cdot, \cdot \rangle$ be a uniformly strongly elliptic inner product of order p on $C_0^\infty(T)$ and let $\mathscr{H}$ be the completion of $C_0^\infty(T)$ with respect to this inner product. If the norm $\|u\| = \langle u, u \rangle^{1/2}$ is equivalent to the norm $\|\ \|_p$, then $\mathscr{H}$ is local. If moreover $2p > n$, then $u \to u(t)$ is continuous, and hence $\mathscr{H}$ is a reproducing kernel space.*

Proof. We have noted that $\mathscr{H}$ contains the same functions as $H_0^p(T)$, and thus for any pair $u_\pm$ in $\mathscr{H}$ we have

$$\langle u_+, u_- \rangle = \sum_{|\alpha|, |\beta| \leqq p} \int_T a_{\alpha\beta}(t) D^\alpha u_+(t) \overline{D^\beta u_-(t)};$$

and these integrals converge absolutely.

It follows that if Γ is a smooth surface which separates T into open sets $D_\pm$, then $\langle u_+, u_- \rangle = 0$ for any two functions u_+ and u_- in $\mathscr{H}$ with the support of $u_\pm$ in $D_\pm$. (L.1) is thus satisfied.

(L.2) follows from the observation that if $u = u_+ + u_-$ is in $\mathscr{H}$ and the support of u_+ is in D_+ while the support of u_- is in $\bar{D}_-$, then $u_+ \in H_0^p(D_+)$.

The last statement follows from inequality (4.1) with $\alpha = 0$.

Now let $(\mathscr{H}, \langle \cdot, \cdot \rangle)$ be as in Theorem 5.2 with $2p > n$ and let $R(t, s)$ be the associated reproducing kernel. For technical reasons we need to know that $R(t, s)$ is jointly continuous or, what is the same, that $t \to R_t$ is continuous from T to $\mathscr{H}$. This follows from the general theory of fundamental solutions for elliptic operators, but it also admits an elementary Hilbert space proof which we now give.

On $\mathscr{H}$ the inner products $\langle \cdot, \cdot \rangle$ and

$$(u, v)_p = \sum_{|\alpha| \leqq p} \int_T D^\alpha u(t) \overline{D^\beta v(t)}$$

are equivalent. Thus there is a reproducing kernel $\rho(t, s)$ in $(\mathscr{H}, (\cdot, \cdot)_p)$ and a bounded operator B on $\mathscr{H}$ for which

$$(u, v)_p = \langle u, Bv \rangle \quad \text{for all } u \text{ and } v \text{ in } \mathscr{H}.$$

The identity

$$u(t) = (u, \rho_t)_p = \langle u, R_t \rangle \quad \text{for all } u \text{ in } \mathscr{H}$$

implies that $R_t = B\rho_t$, and since B is continuous it suffices to show $t \to \rho_t$ is continuous.

But $\rho_t = P\tilde{\rho}_t$ where P: $H_0^p(R_n) \to H_0^p(T)$ is the projection and $\tilde{\rho}(t, s)$ is the reproducing kernel in $H_0^p(R_n)$. The inequality (4.1) shows $H_0^p(R_n)$ is populated with continuous functions $u(t) = (u, \tilde{\rho}_t)_p$. This implies $t \to \tilde{\rho}_t$ is weakly continuous. It is thus strongly continuous if $t \to \|\tilde{\rho}_t\|$ is continuous. But $\|\tilde{\rho}_t\|$ is constant since $(\cdot, \cdot)_p$ is translation invariant on $H_0^p(R_n)$.

Actually we can prove considerably more with these methods. By the translation invariance of $(\cdot, \cdot)_p$ it follows that $\tilde{\rho}(t, s)$ is a function $\tilde{\rho}(t-s)$ of the difference $t-s$, and elementary manipulations show that

$$\tilde{\rho}(t) = c \int_{R_n} e^{it \cdot x} \Big(\sum_{|\alpha| \leqq p} x^{2\alpha} \Big)^{-1}$$

and, moreover, that $\tilde{\rho}(t)$ is Lipschitz continuous. It then follows from

$$\|R_t - R_s\|^2 = \|BP\tilde{\rho}_t - BP\tilde{\rho}_s\|^2 \leqq \|B\|^2 \, \|\tilde{\rho}_t - \tilde{\rho}_s\|_p^2$$

that $R(t, s)$ is Lipschitz.

Standard arguments (see e.g. [12]) show that any centered Gaussian process $\{X_t \colon t \in T\}$ with a Lipschitz covariance function R may be modified to have continuous sample paths. We will assume this has been done and will investigate the boundary spaces $H(\Gamma)$ associated with a smooth $p-1$ dimensional surface Γ in T.

The first result says that in an appropriate weak sense $\{X_t\}$ has $p-1$ normal derivatives on Γ.

To formulate this we let $d\sigma$ be the surface measure on Γ and let $\dot{s}$ be a continuous choice of the unit normal vector to Γ at the point s on Γ. For each function f in $L^2(\Gamma)$ with compact support we introduce the function

$$F(h) = \int_\Gamma f(s) X_{s+h\dot{s}} \, d\sigma \tag{5.3}$$

of the real variable h.

Lemma 5.3. *As an H valued function $F(h)$ is $p-1$ times continuously differentiable.*

Proof. Let $X \in H$ and let $u(s) = EX\overline{X}_s$. Then $u \in \mathscr{H}$ and hence is in $H_0^p(T)$. It now follows from the Imbedding Theorem of Sobolev ([13], p. 69) that for $|\beta| \leqq p-1$, $D^\beta u(s) \in L^2_{\text{loc}}(\Gamma)$ and that as a function of h, with values in $L^2_{\text{loc}}(\Gamma)$,

$(D^\beta u)(s+h\dot{s})$ is continuous. In particular, for $f \in L^2(\Gamma)$ with compact support,

$$\int_\Gamma \bar{f}(s)\, u(s+h\dot{s})\, d\sigma$$

has $p-1$ continuous derivatives. But

$$\int_\Gamma \bar{f}(s)\, u(s+h\dot{s})\, d\sigma = EX \int_\Gamma \bar{f}(s)\, \bar{X}_{s+h\dot{s}}\, d\sigma,$$

and because $X \in H$ was arbitrary we conclude that $\int_\Gamma f(s) X_{s+h\dot{s}}\, d\sigma$ has $p-1$ continuous weak derivatives. But the continuity of weak derivatives implies that they are strong derivatives and the result follows.

Following McKEAN in [8] we now define a p-th order Markov process. Suppose that for all f the function $F(h)$ is for small h, $p-1$ times continuously differentiable in measure. We then introduce the "differential" σ-field

(5.4) $$\Sigma_p(\Gamma) = \sigma\{F^{(k)}_{(0)} : 0 \leqq k < p, f \in L^2(\Gamma) \text{ has compact support}\}.$$

If Γ splits T into D_+ and D_- then it is clear that $\Sigma_p(\Gamma) \subset \Sigma(\Gamma_+) \cap \Sigma(\Gamma_-)$. If, in addition, for all such Γ, $\Sigma_p(\Gamma)$ is the minimal splitting field of $\Sigma(D_-)$ and $\Sigma(D_+)$ we will call $\{X_t\}$ *Markovian of order p.*

If $\{X_t\}$ is Gaussian we will also introduce the Hilbert spaces

(5.5) $$H_p(\Gamma) = \text{span}\{F^{(k)}_{(0)} : 0 \leqq k < p, f \in L^2(\Gamma) \text{ has compact support}\}.$$

In this case we see that a Gaussian process $\{X_t : t \in T\}$ is *Markovian of order p* iff for each f in $L^2(\Gamma)$ with compact support, $F(h)$ has $p-1$ continuous derivatives and the projection of $H(D_+)$ onto $H(D_-)$ is $H_p(\Gamma)$.

Theorem 5.4. *Let $\{X_t : t \in T\}$ be a Gaussian process and let $\mathscr{H}$ be the associated reproducing kernel space. Suppose that $\mathscr{H}$ contains $C_0^\infty(T)$ as a dense subspace, that the inner product $\langle \cdot, \cdot \rangle$ on $\mathscr{H}$ is given by a uniformly strongly elliptic Dirichlet form of degree $p > n/2$, and that the norm on $\mathscr{H}$ is equivalent to $\| \ \|_p$. Then $\{X_t\}$ is Markovian of order p.*

Proof. By combining Theorem 5.2 and Lemma 5.3 it only remains to identify $H_p(\Gamma)$ with $H(\Gamma)$. But $H_p(\Gamma) \subset H(\Gamma)$, and we will show that $H_p(\Gamma) = H(\Gamma)$ by showing that the orthogonal complement of $H_p(\Gamma)$ in $H(\Gamma)$ is $\{0\}$. Thus we let $X \in H(\Gamma) \ominus H_p(\Gamma)$ and set $u(s) = EX\bar{X}_s$. Then $u \in \mathscr{H}(\Gamma)$ and because $\{X_t\}$ is Markovian we know that $C_0^\infty(D_-) \subset \mathscr{H}_0(D_-)$. Thus, for any $\phi \in C_0(D_-)$ we have $0 = \langle u, \phi \rangle$.

Now let $f \in L^2(\Gamma)$ have compact support and observe that for $k \leqq p-1$:

(5.6) $$\int_\Gamma \overline{f(s)} \frac{\partial^k}{\partial \eta^k} u(s)\, d\sigma = EX \left[\frac{d^k}{dh^k} \int_\Gamma \overline{f(s)}\, \bar{X}_{s+h\dot{s}}\, d\sigma \right]_{h=0}$$

where $\frac{\partial^k}{\partial \eta^k} u(s)$ is the k-th normal derivative of u at $s \in \Gamma$. But by assumption the right side equals zero and we conclude that

(5.7) $$\frac{\partial^k}{\partial \eta^k} u = 0 \text{ a.e. on } \Gamma \quad \text{for } k = 0, \dots, p-1.$$

It is now a simple approximation argument to show that because of the conditions (5.1) the restriction of u to D_- is in $H_0^p(D_-)$ and thus is the limit in $\mathscr{H}(T)$ of functions ϕ_n in $C_0^\infty(D_-)$.

We have already seen, however, that $\langle u, \phi\rangle = 0$ for each $\phi \in C_0^\infty(D_-)$, and we conclude that $u(s) \equiv 0$ on D_-. By the same argument $u(s) \equiv 0$ on D_+, and we see that u is the zero function and that $X \equiv 0$.

6. The Spaces $\mathscr{H}(D_\pm)$, $\mathscr{H}_0(D_\pm)$ and $\mathscr{H}(\Gamma)$

The prediction problem for $\{X_t\colon t \in T\}$ when $\langle\cdot,\cdot\rangle$ is uniformly strongly elliptic of order $p > n/2$ is given in the next section. As a first step we give a new characterization of the spaces $\mathscr{H}(D_\pm)$, $\mathscr{H}_0(D_\pm)$ and $\mathscr{H}(\Gamma)$.

The key idea, which goes back to McKean [8] and Molchan [9], is to introduce the operator

$$Au = \sum_{|\alpha|, |\beta| \leq p} (-1)^{|\alpha|} (D^\alpha a_{\alpha\beta} D^\beta u) \tag{6.1}$$

associated with the inner product

$$\langle u, v\rangle = \sum_{|\alpha|, |\beta| \leq p} \int_T a_{\alpha\beta}(t)\, D^\alpha u(t)\, \overline{D^\beta u(t)} \tag{6.2}$$

on $\mathscr{H}$. If the form $\langle\cdot,\cdot\rangle$ is uniformly strongly elliptic and satisfies the conditions of Theorem 5.4, the operator A is also uniformly strongly elliptic and formally self-adjoint.

The various subspaces $\mathscr{H}(D_+)$, $\mathscr{H}(D_-)$ and $\mathscr{H}(\Gamma)$ of $\mathscr{H}(T)$ are easily described in terms of A.

Recall that $\mathscr{H}(D_+)$ may be identified as the orthogonal complement in $\mathscr{H}(T)$ of $\mathscr{H}_0(D_-)$ and that the space $\mathscr{H}_0(D_-)$ may be identified as the closure in $H_0^p(T)$ of the space $C_0^\infty(D_-)$. Thus if $u \in \mathscr{H}(D_+)$ and $\phi \in C_0^\infty(D_-)$ we have $0 = \langle u, \phi\rangle$. Integrating by parts we see that

$$0 = \int_{D_-} u(t)\, \overline{A\phi(t)} \qquad \text{for all } \phi \in C_0^\infty(D_-).$$

But A is formally self-adjoint so that on D_- u is a weak solution of $Au = 0$. By the regularity theorems for elliptic operators ([1], p. 131) it follows that u is actually infinitely differentiable and is a classical solution of $Au = 0$ on D_-.

Conversely if $u \in \mathscr{H}$ and satisfies $Au = 0$ on D_- we see that $\langle u, \phi\rangle = 0$ for each $\phi \in C_0^\infty(D_-)$ and hence that $u \in \mathscr{H}(D_+)$. We thus have

Proposition 6.1. (*i*) *A function u defined on T is in the space $\mathscr{H}(D_\pm)$ if and only if $u \in H_0^p(T)$ and satisfies*

$$Au(t) = 0 \quad \textit{for } t \in D_\mp.$$

(*ii*) *A function u in $\mathscr{H}$ is in $\mathscr{H}(\Gamma) = \mathscr{H}(D_+) \cap \mathscr{H}(D_-)$ if and only if*

$$Au(t) = 0 \quad \textit{for } t \in D_+ \cup D_-.$$

The question of when a function $u(t)$ defined on D_- is the restriction of a function in $\mathscr{H}$ to D_- is answered by the following

Proposition 6.2. *A function $u(t)$ defined on D_- is the restriction to D_- of some function in $\mathcal{H}$ if and only if it has an extension $\tilde{u}(t)$ in $H_0^p(T)$ which satisfies*

$$A\tilde{u}(t)=0 \quad \text{for } t \text{ in } D_+ .$$

The restriction of $\tilde{u}(t)$ to D_+ is the unique solution in $\mathcal{H}(D_+)$ of the exterior Dirichlet problem

$$A\tilde{u}(t)=0 \quad \text{for } t\in D_+$$

with the same boundary data as u on Γ.

Proof. The only point which requires comment is the uniqueness of $\tilde{u}$, and for this it suffices to show that if $u\in\mathcal{H}$, u has zero boundary data on Γ, and $Au(t)\equiv 0$ on D_+ then $u(t)\equiv 0$ on D_+.

Let v be the projection of u onto $\mathcal{H}(D_+)$. Then $v(t)=EX\overline{X}_t\in\mathcal{H}(\Gamma)$ and has zero boundary data. Thus if $f\in L^2(\Gamma)$ has compact support then

$$0=\int_\Gamma \bar{f}(s)\frac{\partial^k}{\partial\eta^k}u(s)\,d\sigma=EX\left[\frac{d^k}{dh^k}\int_\Gamma \overline{f(s)}\,\overline{X}_{s+h\,\dot{s}}\,d\sigma\right]_{h=0}$$

for $k=0,\dots,p-1$, and we conclude that $EX\overline{F}^k_{(0)}=0$. Since $X\in H(\Gamma)$ and the $F^h(0)$ span $H(\Gamma)$, we conclude that $X=0$. Thus $v\equiv 0$ and since $u(t)$ agrees with $v(t)$ on D_+ we must have $u(t)\equiv 0$ on D_+.

Note. It follows from the Calderón extension theorem ([1], p. 171) and Proposition 6.2 that if D_- is bounded with $\overline{D}_-\subset T$ and if u is in the Sobolev space $H^p(D_-)$, then u extends to an element in $\mathcal{H}$.

7. Solution of the Prediction Problem

The prediction problem for $\{X_t\}$ is most easily discussed in terms of the Dirichlet problem for A. To this end we let $D_-\subset T$ be a bounded open set with smooth compact boundary Γ. Let D_+ be the complementary open set in T. For t in D_+, let $h_t(\cdot)$ be the projection of $R(t,\cdot)$ onto $\mathcal{H}(D_-)$. Then $h_t(s)$ agrees with $R(t,s)$ for s in D_- and satisfies $Ah_t(s)=0$ for s in D_+. Thus $R(t,\cdot)-h_t(\cdot)$ vanishes on D_- and hence belongs to $\mathcal{H}_0(D_+)$. Moreover, the restriction of $h_t(s)$ to each of the sets $D_-\cup\Gamma$ and $D_+\cup\Gamma$ is smooth.

Now let $u\in\mathcal{H}(D_-)$. Then

$$0=\langle u, R(t,\cdot)-h_t(\cdot)\rangle$$

and since $u(t)=\langle u, R(t,\cdot)\rangle$ we see that

$$u(t)=\langle u, h_t\rangle \quad \text{for } t\in D_+ . \tag{7.1}$$

Approximating u by a sequence $u_n\in\mathcal{H}$ with compact support we have $u(t)=\lim\langle h_t, u_n\rangle$ or

$$u(t)=\lim_{n\to\infty}\sum_{|\alpha|,|\beta|\leq p}\int_T a_{\alpha\beta}(s)\,D^\alpha u_n(s)\,\overline{D^\beta h_t(s)}\,. \tag{7.2}$$

Integrating by parts to bring the derivatives onto h_t and using the fact that $Ah_t(s)=0$ for $s\notin\Gamma$, we may pass to the limit to obtain a formula

$$u(t)=\int_\Gamma \sum_{j=0}^{p-1} b_j(t,s)\frac{\partial^i u}{\partial n_j}(s)\,d\sigma\,,\qquad t\in D_+ \tag{7.3}$$

where the functions $b_j(t, s)$ are C^∞ functions of s on Γ.

Formula (7.3) can now be used to solve the least squares prediction problem as was illustrated by McKean [8]. That is, we can use (7.3) to calculate the projection of X_t, $t\in D_+$ onto the space $H(D_-)$.

Set

$$\hat{X}(t)=\int_\Gamma \sum_{j=0}^{p-1} \overline{b_j}(t,s)\frac{\partial^j X}{\partial \eta_j}(s)\,d\sigma \tag{7.4}$$

where the derivatives $\dfrac{\partial^j X}{\partial \eta_j}$ are to be interpreted in the sense

$$\int_\Gamma f(s)\frac{\partial^j X}{\partial \eta^j}(s)\,d\Gamma=\frac{d^j}{dh^j}\int f(s)\,X_{s+h\dot{s}}\,d\sigma\,. \tag{7.5}$$

We claim that $\hat{X}(t)$ is the projection of X_t onto $H(D_-)$. To see this, we only have to show that for $\tau\in D_-$,

$$EX_\tau\overline{X}_t=EX_\tau\overline{\hat{X}}_t\,.$$

But for $\tau\in D_-$, $EX_\tau\overline{X}_t=R(\tau, t)$ is in $\mathscr{H}(D_-)$ and hence satisfies

$$\begin{aligned} R(\tau,t)&=\int_\Gamma \sum_{j=0}^{p-1} b_j(t,s)\frac{\partial^j}{\partial\eta^j}R(\tau,s)\,d\sigma \\ &=\overline{\sum_{j=0}^{p-1} EX_\tau\int_\Gamma \overline{b}_j(t,s)\frac{\partial^i}{\partial\eta^j}X_s\,d\sigma}=EX_\tau\overline{\hat{X}}_t\,. \end{aligned}\tag{7.6}$$

Since $\hat{X}_t\in H(D_-)$ it follows that $\hat{X}_t$ is the projection of X_t onto $H(D_-)$.

For completeness we formulate the

Proposition 7.1. *Assume the Gaussian process $\{X_t\colon\ t\in T\}$ is Markovian of order p and satisfies the conditions of Theorem 5.4. Let $\Gamma\subset T$ be a smooth compact surface which separates T into complementary open sets D_- and D_+ where D_- is bounded. Then there exist smooth functions $b_j(t, s)$, $t\in D_+$ and $s\in\Gamma$ so that for each smooth function u defined in a neighborhood of Γ,*

$$\tilde{u}(t)=\int_\Gamma \sum_{j=0}^{p-1} b_j(t,s)\frac{\partial^j u(s)}{\partial\eta^j}\,d\sigma\,,\qquad t\in D_+ \tag{7.7}$$

is the unique solution in $\mathscr{H}(D_+)$ of the exterior Dirichlet problem

$$A\tilde{u}(t)=0\,,\qquad t\in D_+ \tag{7.8}$$

with

$$\frac{\partial^j u}{\partial\eta^j}(s)=\frac{\partial^j u}{\partial\eta^j}(s)\,,\qquad s\in\Gamma\,.$$

For $t \in D_+$ *the least squares prediction of* X_t *for given* $\{X_s\colon\ s \in D_-\}$ *is then*

$$X_t = \int_\Gamma \sum_{j=0}^{p-1} \overline{b_j(t,s)} \frac{\partial^j X(s)}{\partial \eta^j} d\sigma. \tag{7.9}$$

Note that just as the problem of predicting X_t for $t \in D_+$ when we are given $\{X_s\colon\ s \in D_-\}$ is related to exterior Dirichlet problem for the operator A, the interpolation problem of predicting X_s for s in D_- when we are given $\{X_t\colon\ t \in D_+\}$ is related to the standard interior Dirichlet problem for A.

III. Stationary Gaussian Markov Processes on R_n

A process $\{X_t\colon\ t \in R_n\}$ is called stationary if for each choice of $t_1, \dots, t_k$ in R_n the distribution of $X_{t_1+t}, \dots, X_{t_k+t}$ is independent of $t \in R_n$. For stationary Gaussian processes it is possible to obtain much more detail than the general results in Part I.

The mean of a stationary process is constant and may be normalized with $EX_t \equiv 0$. The covariance function $EX_t\overline{X}_s$ is then a function $\rho(t-s)$ of the difference $t-s$. Being a covariance function, ρ is positive definite and by Bochner's theorem it is the Fourier transform of a unique positive finite measure $\Delta(dx)$ on R_n; that is,

$$\rho(t) = \int_{R_n} e^{it\cdot x} \Delta(dx).$$

In the next three sections we prove some elementary results which relate the Markovian character of $\{X_t\}$ to analytic properties of the spectral measure Δ. Section 8 is a general discussion of the Fourier representation. Section 9 is a further study of Δ under the assumption that $\{X_t\}$ has the Markov property for half-spaces. In Section 10 we prove that if $\{X_t\}$ is Markovian of degree p then Δ^{-1} is an elliptic polynomial of degree $2p$.

8. The Spectral Representation of $\mathscr{H}(T)$

The identity $EX_t\overline{X}_s = \int e^{it\cdot x} e^{-is\cdot x} \Delta(dx)$ implies that the map $X_t \to e^{it\cdot x}$ determines a linear isometry from the space H onto $L^2(R_n, \Delta)$. Thus each element X of H is associated with a unique element f of $L^2(R_n, \Delta)$. If $u(s) = EX\overline{X}_s$ is the corresponding element of $\mathscr{H}$ we have

$$u(s) = \int f(x) e^{-is\cdot x} \Delta(dx). \tag{8.1}$$

Hence

Proposition (8.1). $\mathscr{H}$ *consists of all functions of the form*

$$u(s) = \int f(x) e^{-is\cdot x} \Delta(dx)$$

where $f \in L^2(R_n, \Delta)$. *The map* $u(s) \to f(x)$ *is an isometry of onto* $L^2(R_n, \Delta)$, *and thus*

$$\langle u, u \rangle = \int |f(x)|^2 \Delta(dx).$$

We next establish a decomposition of $\{X_t\}$ analogous to the classical decomposition of a stationary Gaussian process into deterministic and regular parts.

We introduce the spaces

$$\mathscr{H}_0 = \text{closure of } u\ \{u \in \mathscr{H}:\ u \text{ has compact support}\}$$

$$\mathscr{H}_\infty = \mathscr{H} \ominus \mathscr{H}_0.$$

The space $\mathscr{H}_\infty$ corresponds to the tail field $\Sigma_\infty = \bigcap \sigma\{X_t:\ |t| > N\}$, and $\mathscr{H}_0$ corresponds to the largest field independent of Σ_∞.

Proposition (8.2). *Let $\Delta(dx) = \Delta_c(x)\,dx + \Delta_s(dx)$ be the Lebesgue decomposition of Δ into absolutely continuous and singular parts. Then:*

(i) If $\mathscr{H}_0 \neq \{0\}$, then

$$\mathscr{H}_0 = \{u(s) = \textstyle\int f(x)\, e^{-s\cdot x}\, \Delta_c(x)\, dx : f \in L^2(R_n, \Delta_c)\} \tag{8.2}$$

and

$$\mathscr{H}_\infty = \{u(s) = \textstyle\int f(x)\, e^{-is\cdot x}\, \Delta_s(dx) : f \in L^2(R_n, \Delta_s)\}. \tag{8.3}$$

Thus the splitting $\mathscr{H} = \mathscr{H}_0 \oplus \mathscr{H}_\infty$ corresponds to the splitting $L^2(R_n, \Delta) = L^2(R_n, \Delta_c) \oplus L^2(R_n, \Delta_s)$.

(ii) $\mathscr{H}_0 \neq \{0\}$ if and only if for some L^2 function $u(t) \neq 0$ with compact support

$$\int \frac{|F(x)|^2}{\Delta_c(x)}\, dx < \infty \tag{8.4}$$

where

$$F(x) = \frac{1}{(2\pi)^{n/2}} \int e^{it\cdot x} u(t)\, dt. \tag{8.5}$$

Proof. From the Fourier representation of $\mathscr{H}$ it is clear that the translation operator $u(s) \to u_t(s) \equiv u(t+s)$ is unitary on $\mathscr{H}$. Moreover, since it leaves the space of functions with compact support invariant, the space $\mathscr{H}_0$ must be invariant under translations.

Assuming $\mathscr{H}_0 \neq \{0\}$, let $u(s) = \int f(x)\, e^{-is\cdot x} \Delta(dx)$ have compact support. We now show that a function $v(s) = \int g(x)\, e^{-is\cdot x} \Delta(dx)$ is in $\mathscr{H}_0$ if and only if $g(x)$ vanishes almost everywhere with respect to Lebesgue measure dx.

If $v \in \mathscr{H}_\infty$, then v is orthogonal to $u_t(s)$ for all t in R_n. That is,

$$0 = \langle u_t(s), v(s)\rangle = \int f(x)\, e^{-it\cdot x}\, \overline{g(x)}\, \Delta(dx),$$

and this is equivalent to saying that $g(x) = 0$ a.e. with respect to the measure $|f(x)|\, \Delta(dx)$. But $u(s) = \int e^{is\cdot x} f(x)\, \Delta(dx)$ has compact support, so by the easy half of the Paley-Wiener theorem the measure $f(x)\, \Delta(dx)$ equals $F(x)\,dx$ where

$$F(x) = \frac{1}{(2\pi)^{n/2}} \int e^{it\cdot x} u(t)\, dx$$

is an entire function of exponential type. In particular, $F(x)$ vanishes at most on a set of Lebesgue measure zero, and hence $|f(x)|\ \Delta(dx)$ is equivalent to Lebesgue measure dx. Since $u \in \mathscr{H}_0$ was an arbitrary function with compact support, the proof of (i) is complete.

To prove (*ii*) let $u(s)$ be as before and observe that

$$\infty > \langle u, u\rangle = \int |f(x)|^2 \Delta_c(x)\, dx = \int \frac{|F(x)|^2}{\Delta_c(x)}\, dx .$$

Conversely, if $u \in L^2$ has compact support and $F(x)$ defined by (8.5) satisfies $\int \frac{|F(x)|^2}{\Delta_c(x)}\, dx < \infty$, then

$$f(x) \equiv \frac{F(x)}{\Delta_c(x)} \in L^2(R_n, \Delta_c) \quad \text{and} \quad u(s) = \int f(x)\, e^{-is\cdot x} \Delta_c(x)\, dx$$

is in $\mathscr{H}_0$. This completes the proof.

Our splitting $\mathscr{H} = \mathscr{H}_0 \oplus \mathscr{H}_\infty$ corresponds to a splitting of the space $H = H_0 \oplus H_\infty$, where

$$H_\infty = \bigcap_{N \geq 0} H(\{t : |t| > N\}), \tag{8.6a}$$

$$H_0 = H \ominus H_\infty . \tag{8.6b}$$

Let Y_t be the projection of X_t onto H_0 and let Z_t be the projection of X_t onto H_∞. Then $\{Y_t\}$ and $\{Z_t\}$ are independent stationary Gaussian processes and

$$X_t = Y_t + Z_t . \tag{8.7}$$

The correlation functions of Y_t and Z_t are respectively

$$\int e^{it\cdot x} \Delta_c(x)\, dx \quad \text{and} \quad \int e^{it\cdot x} \Delta_s(dx) .$$

Using a superscript notation such as H^y and H^y_∞ to denote the obvious Hilbert spaces defined for the processes $\{Y_t\}$ and $\{Z_t\}$, we have the following elementary relations:

$$H^x_0 = H^y = H^y_0, \qquad H^y_\infty = \{0\} \tag{8.8}$$

$$H^x_\infty = H^z = H^z_\infty, \qquad H^z_0 = \{0\} . \tag{8.9}$$

From our view point $\{Z_t\}$ is rather uninteresting, and we will assume in all that follows that $Z_t \equiv 0$ or what is the same that $H^x = H^x_0$. Thus the measure $\Delta(dx)$ is assumed absolutely continuous and we will speak of the spectral density $\Delta(x)$.

The spectral density of a Markov process is very smooth, and we will show in [11] that $\Delta^{-1}(x)$ must be an entire function of minimal exponential type. For the present, however, we will content ourselves with the following proposition. Note that $\{X_t\}$ is not assumed to be Markovian.

Proposition 8.3. *Suppose $\mathscr{H}_0 = \mathscr{H}$ and that for some bounded open set D, $\mathscr{H}(D) \cap \mathscr{H}_0 \neq \{0\}$. Then $\Delta(x)$ agrees a.e. with the ratio of two entire functions of finite exponential type.*

Proof. Let $u(s) = \int f(x)\, e^{-is\cdot x} \Delta(x)\, dx$ be in $\mathscr{H}(D)$ and have compact support. Then $f(x)\,\Delta(x)$ agrees a.e. with an entire function of exponential type. Because D is bounded we have for s in D and for large t, $u_t(s) = u(t+s) = 0$. That is, u_t

is orthogonal to $\mathscr{H}(D)$ and because $u \in \mathscr{H}(D)$ we have

$$0 = \langle u_t, u \rangle = \int f(x) e^{-it \cdot x} \bar{f}(x) \Delta(x) \, dx .$$

This implies that $f\bar{f}\Delta$ agrees a.e. with an entire function of exponential type so that $\Delta = (f\Delta \cdot \bar{f}\Delta)/(f\bar{f}\Delta)$ agrees a.e. with the ratio of two entire functions of exponential type.

9. The Markov Property for Half-Spaces

The easiest approach for obtaining information about Δ from Markovian conditions imposed on the process $\{X_t\}$ is to restrict considerations to the special case when the domains D_- and D_+ are half-spaces. This leads to a decomposition of the spaces $\mathscr{H}(D_-)$ and $\mathscr{H}(D_+)$ which reduces considerations to the one dimensional case. We now discuss this decomposition. An application of the techniques is made in the next section.

As usual $\{X_t\colon t \in R_n\}$ is a stationary Gaussian process with spectral density Δ. We treat R_n as the direct sum $R_1 \oplus R_{n-1}$ and write $t = (\sigma, \tau)$ with $\sigma \in R_1$ and $\tau \in R_{n-1}$ for an element t of R_n. For any real number r we write

$$D^r_- = \{t \in R_n\colon \sigma < r\} \tag{9.1}$$

$$D^r_+ = \{t \in R_n\colon \sigma > r\} . \tag{9.2}$$

The spectral density $\Delta(x)$ is defined on the dual of $R_1 \oplus R_{n-1}$, and we will write $x = (\xi, \eta)$ with $\xi \in R_1$ and $\eta \in R_{n-1}$. Δ is to be considered as a function $\Delta(\xi, \eta)$ of two variables. The covariance $R(t)$ takes the form

$$R(t) = R(\sigma, \tau) = \iint e^{i(\sigma \cdot \xi + \tau \cdot \eta)} \Delta(\xi, \eta) \, d\xi \, d\eta . \tag{9.3}$$

Corresponding to the spaces $\mathscr{H}(D^s_\pm)$ we introduce the appropriate subspaces of $L^2(R_n, \Delta)$. Namely

$$Z(D^r_\pm) = \text{closed span } \{e^{it \cdot x}\colon t \in D^r_\pm\} . \tag{9.4}$$

Then $u(s) \in \mathscr{H}(D^r_\pm)$ iff $u(s) = \int f(x) e^{-is \cdot x} \Delta(x) dx$ where $f \in Z(D^r_\pm)$.

By Fubini's theorem we may change Δ on a set of measure zero so that for all $\eta \in R_{n-1}$ the non-negative function $\Delta_\eta(\cdot) \equiv \Delta(\cdot, \eta)$ is in $L^1(R_1)$. Thus Δ_η is a permissible spectral density on R_1, and we may introduce the space $L^2(R_1, \Delta_\eta)$ and the subspaces of $L^2(R_1, \Delta_\eta)$

$$Z^{r+}_\eta \equiv \text{closed span } \{e^{i\sigma \cdot \xi}\colon \sigma > r\} \tag{9.5}$$

$$Z^{r-}_\eta \equiv \text{closed span } \{e^{i\sigma \cdot \xi}\colon \sigma < r\} . \tag{9.6}$$

The previously mentioned decomposition is based on the following proposition which shows how the spaces $Z(D^r_\pm)$ are built from the $Z^{r\pm}_\eta$. The proof we give is by no means the shortest possible but is instructive.

Proposition 9.1*. *A function f in $L^2(R_n, \Delta)$ is in $Z(D^r_\pm)$ if and only if $f(\cdot, \eta)$ is in $Z^{r\pm}$ for a.a. η.*

* In the language of direct integrals $L^2(R_n, \Delta) = \int \oplus L^2(R_1, \Delta_\eta) d\eta$; this proposition says that $Z(D^r_\pm) = \int \oplus Z^{r\pm}_\eta \, d\eta$.

Proof. We present the proof for $Z(D^r_-)$. Let $\mathscr{Z}^r$ denote the subspace of $L^2(R_n, \Delta)$ of functions f satisfying $f(\cdot, \eta)$ Z^{r-} for a.a. η. We must show $Z(D^r_-)=\mathscr{Z}^r$.

First we claim $\mathscr{Z}^r$ is closed. In fact, if f_k is a convergent subsequence in $\mathscr{Z}^r$ with $\lim f_k=f$ we have Fubini's theorem that

$$\int d\eta\{\int |f_k(\xi,\eta)-f(\xi,\eta)|^2 \Delta(\xi,\eta)\, d\xi\}=\|f_k-f\|^2 \tag{9.7}$$

and this converges to zero as $k\to\infty$. Choosing a subsequence $k'\to\infty$ so that $\int |f_{k'}(\xi,\eta)-f(\xi,\eta)|^2\Delta(\xi,\eta)\,d\xi\to 0$ for a.a. η we see that for a.a. η, $f(\cdot,\eta)$ is the limit in $L^2(R_1, \Delta_n)$ of the functions $f_{k'}(\cdot,\eta)\in Z^{r-}_\eta$. But Z^{r-}_η is closed so $f(\cdot,\eta)\in Z^{r-}_\eta$ and $f\in\mathscr{Z}^r$, and $\mathscr{Z}^r$ is closed.

The inclusion $Z(D^r_-)\subset\mathscr{Z}^r$ is now obvious since each f in $Z^r(D^r_-)$ is a limit of finite sums of the $e^{it\cdot x}$ where $t\in D^r_-$, and each such sum is in $\mathscr{Z}^r$.

To prove the inclusion $\mathscr{Z}^r\subset Z(D^r_-)$ we observe that each function of the form $\sum_1^N e^{i\sigma_k\xi}h_k(\eta)$ where $\sigma_k\leqq r$ and $\int |h_k(\eta)|^2\Delta(\xi,\eta)\,d\xi\,d\eta$ is in $Z(D^r_-)$. We now show that each $f\in\mathscr{Z}^r$ can be approximated by such functions and the result will follow.

For this we let $\{g_k(\xi)\}$ be an enumeration of the countable set of finite sums $\sum a_k e^{i\sigma_k\xi}$ where the $\sigma_k\leqq r$ and both the a_k and σ_k are rational. The set $\{g_k(\xi)\}$ is dense in each of the spaces Z^{r-}. For fixed $f\in\mathscr{Z}^r$ we define the approximating sequence

$$f_k(\xi,\eta)=\sum_1^k g_j(\xi)\,1_{B_{k,j}}(\eta) \tag{9.8}$$

where $1_{B_{k,j}}$ is the indicator function of the measurable set $B_{k,j}\subset R_{n-1}$, which is defined by the condition: $\eta\in B_{k,j}$ if and only if j is the smallest integer satisfying

$$\int |f(\xi,\eta)-g_j(\xi,\eta)|^2\Delta(\xi,\eta)\,d\xi=\min_{1\leqq i\leqq k}\{\int |f(\xi,\eta)-g_i(\xi,\eta)|^2\Delta(\xi,\eta)\,d\xi\}.$$

Then for each fixed η,

$$\int |f_k(\xi,\eta)-f(\xi,\eta)|^2\Delta(\xi,\eta)\,d\xi=\min_{1\leqq j\leqq k}\{\int |g_j(\xi,\eta)-f(\xi,\eta)|^2\Delta(\xi,\eta)\,d\xi \tag{9.9}$$

decreases to zero as $k\to\infty$. By the monotone convergence theorem we conclude that

$$\|f_k-f\|^2=\int d\eta\{\int |f_k(\xi,\eta)-f(\xi,\eta)|^2\Delta(\xi,\eta)\,d\xi\}\to 0. \tag{9.10}$$

Note that if f is an arbitrary element of $L^2(R_n, \Delta)$ and the sequence f_k is defined by (9.8), then f_k converges to the projection of f on $Z(D^r_-)$ while for almost all η, $f_k(\cdot,\eta)$ converges to the projection of $f(\cdot,\eta)$ on Z^{r-}_η.

The usefulness of this proposition in studying the Markov property is that it reduces the study of the projection of $\mathscr{H}(D^r_+)$ onto $\mathscr{H}(D^r_-)$ to the study of the projections of Z^{r+}_η onto Z^{r-}_η; and for fixed η this is a one variable problem.

An example of the results obtainable by this technique is the following easily proved

Proposition 9.2. *Suppose $\{X_t\colon t\in R_n\}$ is a stationary Gaussian process with spectral density $\Delta(x)$. Suppose also that $\{X_t\}$ is Markovian. Then for a.a. η, the*

projection of Z_η^{r+} *onto* Z_η^{r-} *is contained in*

$$\bigcap_{\varepsilon>0} \operatorname{span}\{e^{i\sigma\xi}: 0 \leqq r-\sigma < \varepsilon\}.$$

Assuming that $\mathscr{H}_\infty = \{0\}$ it then follows from results of LEVINSON & MCKEAN [7] that for a.a. η, $\Delta^{-1}(\xi, \eta)$ agrees a.e. with an entire function of minimal exponential type.

10. Stationary Markov Processes of Finite Order

We have seen in earlier sections that if $\{X_t: t \in R_n\}$ is a Gaussian Markov process, then under appropriate additional assumptions it follows that the inner product in $\mathscr{H}$ is given by a Dirichlet form. When the form is of order p, elliptic and satisfies other technical conditions, we were able to prove that $\{X_t\}$ was Markovian of order p.

In this section we assume that $\{X_t: t \in R_n\}$ is stationary and Markovian of order p. In light of the previous results a reasonable conjecture is that the inner product in $\mathscr{H}$ comes from a constant coefficient elliptic Dirichlet form. We prove the Fourier transform version of this conjecture.

Theorem 10.1. *Let* $\{X_t: t \in R_n\}$ *be a stationary Gaussian process with spectral density* $\Delta(x)$. *If* $\{X_t\}$ *is Markovian of order p then* $\Delta^{-1}(x)$ *is an elliptic polynomial* of degree* $2p$.

Proof. For $n=1$ this is well-known and was mentioned in the comment at the end of Section 1. In fact, $R(t)$ is the Green's function of the operator $\sum_0^{2p} a_\alpha \frac{d^\alpha}{dt^\alpha}$ if and only if $R(t) = \int e^{itx} p^{-1}(x)\, dx$ where

$$p(x) = \sum_0^{2p} a_\alpha (-ix)^\alpha.$$

We can also formulate the proposition for $n=1$ in terms of the space $L^2(R_1, \Delta)$. If $X_t^{(p-1)}$ exists then $x^{p-1} \in L^2(R_1, \Delta)$. Thus $\Delta^{-1}(x)$ is a polynomial if degree $2p$ if and only if x^{p-1} is in $L^2(R_1, \Delta)$ and the range of the projection of $\{e^{itx}: t \geqq 0\}$ onto $\{e^{itx}: t \leqq 0\}$ in $L^2(R_1, \Delta)$ is $\{1, x, \ldots, x^{p-1}\}$.

Turning to the case $n>1$ we know from Proposition 8.3 that $\Delta^{-1}(x) = P(x)\, Q^{-1}(x)$ where $P(x)$ and $Q(x)$ are entire functions. We can thus find some open set $U \subset R_n$ on which $Q(x)$ is non-zero. If we can show that the restriction of $P(x)\, Q^{-1}(x)$ to U is a polynomial it will then follow by analytic continuation that $P(x)\, Q^{-1}(x)$ is a polynomial.

Moreover, since $\Delta^{-1} = PQ^{-1}$ is analytic on U, we see that $\Delta^{-1}(x)$ must be a polynomial on U if we can show that in each of the separate variables $x_1, \ldots, x_n$, $\Delta^{-1}(x) = \Delta^{-1}(x_1, \ldots, x_n)$ is a polynomial. The proof of this will follow from the case $n=1$ and the reduction discussed in Section 9.

With this in mind we again write $t = (\tau, \sigma)$ for the general point t in $R_n = R_1 \oplus R_{n-1}$ and $x = (\xi, \eta)$ for the general point in the dual Euclidian space.

* The polynomial $\sum_{|\alpha| \leqq m} a_\alpha x^\alpha$ is elliptic if there is a constant $c>0$ with $c|x|^m \leqq |\sum_{|\alpha|=m} a_\alpha x^\alpha|$.

Without loss of generality we may assume U has the form

(10.1) $$U=\{(\xi,\eta)\in R_1\oplus R_{n-1}: |\xi|<\varepsilon; |\eta|<\varepsilon\}.$$

We must show for fixed η with $|\eta|<\varepsilon$ that as a function of ξ, $\Delta^{-1}(\xi,\eta)$ is a polynomial of degree $2p$ for $|\xi|<\varepsilon$.

Again we set

(10.2) $$D_-=\{t\in R_n: \sigma<0\}, \quad D_+=\{t\in R_n: \sigma>0\}$$
$$\Gamma=\{t\in R_n: \sigma=0\}.$$

If $f\in L^2(\Gamma)$ has compact support we set

$$F(\sigma)=\int_{R_{n-1}} f(\tau)\, X_{(\sigma,\tau)}\, d\tau .$$

Under the map $X_t\to e^{it\cdot x}$ from H onto $L^2(R_n,\Delta)$ the function $F(\sigma)$ corresponds to $e^{i\sigma\xi}\hat{f}(\eta)$ where

$$\hat{f}(\eta)=\int_\Gamma e^{i\tau\cdot\eta} f(\tau)\, d\tau,$$

and taking derivatives we have the correspondences

(10.3) $$F(0)\leftrightarrow\hat{f}(\eta), \quad F^{(k)}_{(0)}\leftrightarrow(i\xi)^k\,\hat{f}\eta) \quad \text{for } 1\leqq k\leqq p-1.$$

Moreover, $F^{(p-1)}_{(0)}$ exists if and only if

(10.4) $$\int_{R_n} |\xi^{p-1}\hat{f}(\eta)|^2\, \Delta(\xi,\eta)\, d\xi\, d\eta<\infty .$$

Thus if $\{X_t\}$ is Markovian of order p we conclude that (10.4) holds for each f in $L^2(R_{n-1})$ with compact support and that in $L^2(R_n,\Delta)$ the range of the projection of $\{e^{itx}: \sigma\geqq 0\}$ onto $\{e^{itx}: \sigma\leqq 0\}$ is contained in the closure of

(10.5) $$\{\xi^k\hat{f}(\eta): 0\leqq k<p \text{ and } f\in L^2(\Gamma) \text{ has compact support}\}.$$

But if (10.4) holds for all permissible f we have

10.6) $$\int_{R_1} |\xi^{p-1}|^2\, \Delta(\xi,\eta)\, d\xi<\infty \quad \text{for a.a. } \eta .$$

Proposition 9.3. now shows that for a.a. η, the projection of $\{e^{i\sigma\xi}: \sigma\geqq 0\}$ onto $\{e^{i\sigma\xi}: \sigma\leqq 0\}$ in $L^2(R_1,\Delta_\eta)$ is contained in $\{1,\xi,\ldots,\xi^{p-1}\}$. Thus for a.a. η, $\Delta^{-1}(\xi,\eta)$ is a polynomial of degree $2p$ in ξ. Since Δ^{-1} is continuous on U it follows that $\Delta^{-1}(\xi,\eta)$ is a polynomial of degree $\leqq 2p$ in ξ; and that for almost all η it is of degree $2p$. From the previous comments we see that $\Delta^{-1}(s)$ is a polynomial.

That $\Delta^{-1}(x)$ is an elliptic polynomial of degree $2p$ follows simply from the fact that any polynomial which is not elliptic of degree $2p$ may be reduced by an orthogonal change of coordinates to a polynomial which is of degree $k\neq 2p$ in the variable x_1. This possibility contradicts the earlier discussion, and we conclude Δ^{-1} is elliptic of order $2p$. The proof is complete.

If $\Delta^{-1}(x)\geqq 0$ is an elliptic polynomial of degree $2p$ one should expect that the associated stationary Gaussian process is Markovian of order p. If $\Delta^{-1}(x)$ has no zeros this is true and upon taking Fourier transforms follows immediately from

Theorem 5.4. However, when $\Delta^{-1}(x)$ has zeros the norm in $\mathscr{H}$ is not equivalent to the norm $\| \|_p$ and Theorem 5.4 is not applicable. In this case I have been unable to prove the general result although I believe it to be true.

The basic problem consists of showing that the associated Dirichlet problem for A is well-posed in $\mathscr{H}$ for an unbounded region D_- with non-compact boundary Γ. For a general Δ^{-1} with zeros this seems very difficult. However, if we restrict considerations to the case when Γ is compact there is no real problem. For functions u in $\mathscr{H}$ with a fixed compact support the norm in $\mathscr{H}$ is equivalent to the norm $\| \|_p$, and the solution of the prediction problem given in Section 8 is valid. We thus have

Proposition 10.2. *Let $\Delta(x)$ be the spectral density of a stationary Gaussian process $\{X_t\colon t\in R_n\}$ and suppose that $\Delta^{-1}(x)$ is an elliptic polynomial of degree $2p$.*

(i) If $\Delta^{-1}(x)$ has no zeros then $\{X_t\}$ is Markovian of order p.

(ii) If Δ^{-1} has zeros the Markov property of order p remains valid for $\{X_t\}$ when formulated with respect to compact surfaces Γ.

This research was partially supported by Army Research Office Grant ARO-D-1005.

References

1. Agmon, S., Lectures on Elliptic Boundary Value Problems. Princeton: Von Nostrand 1965.
2. Aronszajn, N., Theory of reproducing kernels. Trans. Amer. Math. Soc. **68**, 337–404 (1950).
3. Dolph, C. L., & M. A. Woodbury, On the relation between Green's functions and covariances of certain stochastic processes and its application to unbiased linear prediction. Trans. Amer. Math. Soc. **72**, 519–550 (1952).
4. Doob, J. L., The elementary Gaussian processes. Annals. of Math. Stat. **15**, 229–282 (1944).
5. Hida, T., Canonical representations of Gaussian processes and their applications. Mem. Coll Sci., U. of Kyoto, ser. A, Math. **33**, 109 155 (1960).
6. Hunt, G. A., Martingales et Processus de Markov. Paris: Dunod 1966.
7. Levinson, N., & H. P. McKean, Jr., Weighted Trigonometrical approximation on R^1 with application to the germ field of a stationary Gaussian noise. Acta Math. **112**, 99–143 (1964).
8. McKean, H. P., Jr., Brownian motion with a several dimensional time. Theory Prob. Applications **8**, 335–354 (1963).
9. Molchan, G. M., On some problems concerning Brownian motion in Lévy's sense. Theory Prob. Applications **12**, 682–690 (1967).
10. Peetre, J., Rectification a l'article «Une caractérisation abstraite des opérateurs différentiels». Math. Scand. **8**, 116–120 (1960).
11. Pitt, L., Some problems in the spectral theory of stationary processes on R_n (to appear).
12. Sirao, T., On the continuity of Brownian motion with a multidimensional parameter. Nagoya Math. Jour. **16**, 135–156 (1960).
13. Sobolev, S. L., Applications of Functional Analysis in Mathematical Physics. [Trans. from the Russian by F. E. Browder]. Amer. Math. Soc. Providence (1963).

Department of Mathematics
University of Virginia
Charlottesville 22901

(Received May 11, 1971)

Part V

APPLICATIONS TO SPLINE FITTING

Editor's Comments on Papers 21 Through 30

RKHS methods have been successfully applied to a wide variety of problems in the field of optimal approximation, which includes interpolation and smoothing by spline functions in one or more dimensions (curve and surface fitting). Paper 21 by Weinert surveys the one-dimensional case, emphasizing the RKHS formulation, connections with least-squares estimation, and the development of recursive spline algorithms. Papers 22 through 25 present details of results covered by the survey. Optimality properties of splines are discussed in Paper 22 by de Boor and Lynch, who also give an explicit formula for the reproducing kernel in the polynomial case. Kimeldorf and Wahba, in Paper 23, derive formulas for a certain kind of spline function and discuss one type of congruence relation between spline fitting and stochastic estimation. Vector-valued interpolating splines are treated in Paper 24 by Sidhu and Weinert, who show how to use system-theoretic ideas to recursively compute splines with a modified Kalman filter. In Paper 25, Weinert et al. present the corresponding results for scalar-valued smoothing splines. Other recent results in the one-dimensional case can be found in papers by Wahba [1], Craven and Wahba [2], Peele and Kimeldorf [3], de Figueiredo [4], Weinert et al. [5], and Sidhu and Weinert [6].

The multidimensional case is surveyed by Schumaker [7]. RKHS methods have been used by Mansfield [8, 9] for tensor product spaces and by Nielson [10] for gridded data, but the most direct generalization of the one-dimensional case was carried out by Duchon [11], based on some preliminary results of Atteia [12]. Duchon's surface splines are called "thin plate" splines since they approximate the equilibrium position of a thin plate deflected at scattered points. A more constructive approach to this topic is presented by Meinguet in Paper 26 (see also [13]), which also discusses algorithmic aspects. For an application of thin-plate splines to meteorological problems, see Wahba and Wendelberger [14]. In Paper 27, Wahba extends the thin-plate concept to splines on the sphere.

One important statistical application of spline fitting is in probability density estimation. Boneva et al. [15] developed the "histospline" density estimate, which is simply the derivative of the cubic spline interpolating the sample cumulative distribution function. In remarks at the end of the Boneva et al. paper, Parzen places the histospline in a RKHS framework. In Paper 28, Wahba studies the statistical properties of a variant of the histospline. In Paper 29, de Montricher et al. take a different approach by estimating the density directly from the data. Their estimate is still a spline function, however, whose form depends on the particular RKHS chosen for the problem. For further results, see Wahba [16], Tapia and Thompson [17], and Leonard [18]. The related problem of spectral density estimation using splines is treated from a RKHS point of view in Cogburn and Davis [19] and Wahba and Wold [20].

While splines are solutions to minimum norm problems in Sobolev-type spaces, one can formulate similar approximation problems in other types of RKHS, for example, spaces of bandlimited functions. Yao does just this in Paper 30, in which he studies interpolation, sampling expansions, and truncation error bounds in spaces of functions whose Fourier, Hankel, sine, or cosine transforms vanish outside a finite interval. For further work of this nature, see Campbell [21], Higgins [22], and Melkman [23].

Other optimal approximation problems to which RKHS methods have been applied include solution of linear operator equations [24, 25, 26, 27], solution of two-point boundary value problems [28], and simulation of nonlinear systems [29].

REFERENCES

1. G. Wahba, Improper Priors, Spline Smoothing, and the Problem of Guarding Against Model Errors in Regression, *R. Stat. Soc. J.* **40B:**364–372 (1978).
2. P. Craven and G. Wahba, Smoothing Noisy Data with Spline Functions, *Numer. Math.* **31:**377–403 (1979).
3. L. Peele and G. Kimeldorf, Time Series Prediction Functions Based on Imprecise Observations, *Ann. Stat.* **7:**801–811 (1979).
4. R. J. P. de Figueiredo, LM-g Splines, *J. Approx. Theory* **19:**332–360 (1977).
5. H. L. Weinert, U. B. Desai, and G. S. Sidhu, ARMA Splines, System Inverses, and Least-Squares Estimates, *Soc. Ind. Appl. Math. J. Control Optimiz.* **17:**525–536 (1979).
6. G. S. Sidhu and H. L. Weinert, Dynamical Recursive Algorithms for Lg-Spline Interpolation of EHB Data, *Appl. Math. Comput.* **5:**157–185 (1979).
7. L. L. Schumaker, Fitting Surfaces to Scattered Data, in *Approximation Theory II*, G. G. Lorentz, C. K. Chui, and L. L. Schumaker, eds., Academic Press, New York, 1976, pp. 203–268.
8. L. E. Mansfield, On the Optimal Approximation of Linear Functionals in Spaces of Bivariate Functions, *Soc. Ind. Appl. Math. J. Numer. Anal.* **8:**115–126 (1971).
9. L. E. Mansfield, Optimal Approximation and Error Bounds in Spaces of Bivariate Functions, *J. Approx. Theory* **5:**77–96 (1972).
10. G. M. Nielson, Multivariate Smoothing and Interpolating Splines, *Soc. Ind. Appl. Math. J. Numer. Anal.* **11:**435–446 (1974).
11. J. Duchon, Splines Minimizing Rotation-Invariant Semi-Norms in Sobolev Spaces, in *Constructive Theory of Functions of Several Variables,* W. Schempp and K. Zeller, eds., Springer Verlag, Berlin, 1977, pp. 85–100.
12. M. Atteia, Spline Functions and Aronszajn-Bergman Reproducing Kernels (in French), *Revue Française Inf. Rech. Operat.* **4:**31–43 (1970).
13. J. Meinguet, An Intrinsic Approach to Multivariate Spline Interpolation at Arbitrary Points, in *Polynomial and Spline Approximation,* B. N. Sahney, ed., Reidel, Dordrecht, 1979, pp. 163–190.
14. G. Wahba and J. Wendelberger, Some New Mathematical Methods for Variational Objective Analysis Using Splines and Cross Validation, *Monthly Weather Rev.* **108:**1122–1143 (1980).
15. L. I. Boneva, D. Kendall, and I. Stefanov, Spline Transformations: Three New Diagnostic Aids for the Statistical Data-Analyst, *R. Stat. Soc. J.* **33B:**1–70 (1971).

16. G. Wahba, Optimal Smoothing of Density Estimates, in *Classification and Clustering,* J. VanRyzin, ed., Academic Press, New York, 1977, pp. 423–458.
17. R. A. Tapia and J. R. Thompson, *Nonparametric Probability Density Estimation,* Johns Hopkins University Press, Baltimore, 1978, 176p.
18. T. Leonard, Density Estimation, Stochastic Processes, and Prior Information, *R. Stat. Soc. J.* **40B:**113–146 (1978).
19. R. Cogburn and H. T. Davis, Periodic Splines and Spectral Estimation, *Ann. Stat.* **2:**1108–1126 (1974).
20. G. Wahba and S. Wold, Periodic Splines for Spectral Density Estimation: The Use of Cross Validation for Determining the Degree of Smoothing, *Commun. Stat.* **4:**125–141 (1975).
21. L. L. Campbell, Series Expansions for Random Processes, in *Probability and Information Theory,* M. Behara, K. Krickeberg and J. Wolfowitz, eds., Springer-Verlag, Berlin, 1969, pp. 77–95.
22. J. R. Higgins, A Sampling Theorem for Irregularly Spaced Sample Points, *IEEE Trans. Inform. Theory* **IT-22:**621–622 (1976).
23. A. A. Melkman, n-Widths and Optimal Interpolation of Time- and Band-Limited Functions, in *Optimal Estimation in Approximation Theory,* C. A. Micchelli and T. J. Rivlin, eds., Plenum, New York, 1977, pp. 55–68.
24. M. Z. Nashed and G. Wahba, Generalized Inverses in Reproducing Kernel Hilbert Spaces: An Approach to Regularization of Linear Operator Equations, *Soc. Ind. Appl. Math. J. Math. Anal.* **5:**974–987 (1974).
25. J. W. Hilgers, On the Equivalence of Regularization and Certain Reproducing Kernel Hilbert Space Approaches for Solving First Kind Problems, *Soc. Ind. Appl. Math. J. Numer. Anal.* **13:**172–184 (1976); **15:**1301 (1978).
26. G. Wahba, On the Optimal Choice of Nodes in the Collocation-Projection Method for Solving Linear Operator Equations, *J. Approx. Theory* **16:**175–186 (1976).
27. G. Wahba, Practical Approximate Solutions to Linear Operator Equations When the Data are Noisy, *Soc. Ind. Appl. Math. J. Numer. Anal.* **14:**651–667 (1977).
28. M. L. Athavale and G. Wahba, Determination of an Optimal Mesh for a Collocation-Projection Method for Solving Two-Point Boundary Value Problems, *J. Approx. Theory* **25:**38–49 (1979).
29. R. J. P. de Figueiredo and T. A. W. Dwyer, A Best Approximation Framework and Implementation for Simulation of Large-Scale Nonlinear Systems, *IEEE Trans. Circuits and Syst.* **CAS-27:**1005–1014 (1980).

21

STATISTICAL METHODS IN OPTIMAL CURVE FITTING

Howard L. Weinert

Department of Electrical Engineering
The Johns Hopkins University
Baltimore, Maryland

Key Words and Phrases: splines; reproducing kernels; interpolation; smoothing; least-squares estimation.

ABSTRACT

Many optimal curve fitting and approximation problems have the same structure as certain estimation problems involving random processes. This structural correspondence has many useful consequences for curve fitting problems, including recursive algorithms and computable error bounds. The basic facts of this correspondence are reviewed and some new results on error bounds and optimal sampling are presented.

1. INTRODUCTION

Consider the basic problem of approximating $f(t)$ given (possibly inaccurate) measurements of $\lambda_1 f, \lambda_2 f, \ldots, \lambda_N f$, where $\{\lambda_j\}_1^N$ are linear functionals. The approximation can be carried out by fitting a curve $s(\cdot)$ to the data, and then approximating $f(t)$ by $s(t)$. If the data are error-free, it makes sense to require $\lambda_j s=\lambda_j f$, in which case we have an interpolation problem. If the data contain errors, we have a smoothing problem. In both cases, the data values will be denoted by $\{r_j\}_1^N$.

Before anything can be said about the choice of $s(\cdot)$, something must be said about the underlying function $f(\cdot)$. I will assume that $f(\cdot)$ belongs to the following Hilbert space (see Remark 1):

$$H = \{g(t),\ t \varepsilon\ [0,T]:\ g^{(n)} \text{ exists a.e. and } \int_0^T [g^{(n)}]^2 < \infty\}. \quad (1)$$

Here, $g^{(n)}$ denotes the nth derivative of g, and n is a fixed integer satisfying $1 \le n \le N$. Since dependent or unbounded functionals provide no useful information about $f(\cdot)$, I will also assume that the measurement functionals $\{\lambda_j\}_1^N$ are linearly independent and bounded on H. For future reference, a constant-coefficient (see Remark 2) differential operator L is defined as

$$L = \frac{d^n}{dt^n} + a_{n-1}\frac{d^{n-1}}{dt^{n-1}} + \cdots + a_1 \frac{d}{dt} + a_0 . \quad (2)$$

Now if the measurements are error-free, they localize $f(\cdot)$ to a linear variety U in H, where

$$U = \{g\ \varepsilon\ H:\ \lambda_j g = r_j\ ,\ 1 \le j \le N\} . \quad (3)$$

As the interpolating function $s_0(\cdot)$ choose a function in U that satisfies

$$\int_0^T (Ls_0)^2 = \min_{g \varepsilon U} \int_0^T (Lg)^2 . \quad (4)$$

If the measurements contain errors, the smoothing function $s_1(\cdot)$ is chosen as a solution to the optimization problem

$$\underset{g \varepsilon H}{\text{minimize}} \ \{ \int_0^T (Lg)^2 + \sum_{j=1}^{N} \rho_j^{-1} (r_j - \lambda_j g)^2 \}. \tag{5}$$

Here, the weights $\{\rho_j\}_1^N$ are strictly positive real numbers.

Both problems (4) and (5) have unique solutions if $\{\lambda_j\}_1^n$ are linearly independent on the null space of L. This condition will be assumed in all that follows. The solution to (4) is called the <u>Lg-interpolating spline</u>. The solution to (5) is called the <u>Lg-smoothing spline</u>. We can see that as all the weights ρ_j approach zero, the spline smoothing problem (5) reduces to the spline interpolation problem (4).

Because of their optimality properties, splines are superior to ordinary polynomials in curve fitting. At the same time, the functional form of a spline can be almost as simple as a polynomial (see Remark 2). Splines have been applied to many problems in numerical analysis, statistics, and engineering. For further information on spline theory, history, and applications, see Greville (1969); Schoenberg (1969); Schultz (1973); Jerome and Schumaker (1969); de Montricher, Tapia, and Thompson (1975); Weinert and Kailath (1976); de Figueiredo and Netravali (1976); Caprihan (1975); Bellman and Roth (1971); Larsen, Crawford, and Smith (1977).

<u>Remark 1</u>. Functions in H can be differentiated at least n times without producing a delta function. Similar curve fitting problems can be formulated in other types of Hilbert spaces; for example, the space of bandlimited functions as in Yao (1967).

Remark 2. All the results discussed in this paper extend in the obvious way to variable-coefficient differential operators provided the coefficients are sufficiently differentiable. As discussed later on, the operator L determines the functional form of the spline.

2. INTERPOLATING AND SMOOTHING SPLINES AS PROJECTIONS

Subsequent results require that the interpolation and smoothing problems (4) and (5) be reformulated as minimum-norm problems with inner product constraints, after which they can be solved using the Projection Theorem. The interpolation problem will be considered first (see Remark 3).

The space H is a reproducing kernel Hilbert space (see Aronszajn, 1950) with norm

$$\|g\|^2 = \sum_{j=1}^{n} (\lambda_j g)^2 + \int_0^T (Lg)^2 \;, \; g \in H \tag{6}$$

and reproducing kernel

$$K_0(t,\tau) = \sum_{j=1}^{n} z_j(t) z_j(\tau) + \int_0^T G(t,\xi) G(\tau,\xi) d\xi , t,\tau \in [0,T] \tag{7}$$

where $\{z_j(\cdot)\}_1^n$ is the basis for the null space of L that is dual to $\{\lambda_j\}_1^n$ and $G(\cdot,\cdot)$ is the Greens function for L that satisfies $\lambda_j G(\cdot,\tau)=0$ for $1\leq j\leq n$ and $\tau\in[0,T]$. In terms of the inner product determined by (6), we have the reproducing property

$$\langle g(\cdot), K_0(\cdot,t)\rangle = g(t) \;, \; g \in H \;, \; t \in [0,T] \tag{8}$$

and thus equation (3) can be restated as

$$U = \{g \in H: \langle g(\cdot), h_j(\cdot)\rangle = r_j , \; 1\leq j\leq N\} \tag{9}$$

where

$$h_j(t) = \lambda_j K_0(\cdot,t) \;, \; t \in [0,T]. \tag{10}$$

Since the first term in (6) is fixed for all $g \in U$, it is clear from (4) that the interpolating spline $s_0(\cdot)$ is the function in U that satisfies

$$\|s_0\|^2 = \min_{g \in U} \|g\|^2 . \tag{11}$$

In the light of (9) and (11), we see that $s_0(\cdot)$ is the projection of any function in U onto the span of $\{h_j(\cdot)\}_1^N$.

In order to get a similar formulation of the spline smoothing problem, we must work in the following space H^+ of order pairs:

$$H^+ = \{(g(\cdot),\underset{\sim}{\theta}): g \in H \text{ and } \underset{\sim}{\theta} = [\theta_1,\theta_2,\ldots, \theta_N]' \varepsilon E^N\}. \tag{12}$$

It can be shown (see Remark 4) that H^+ is a Hilbert space with norm

$$\|(g,\underset{\sim}{\theta})\|^2 = \int_0^T (Lg)^2 + \sum_{j=1}^N \rho_j^{-1} \theta_j^2 + \sum_{j=1}^n (\lambda_j g+\theta_j)^2. \tag{13}$$

Now define two index sets by

$$I = [0,T] \tag{14a}$$

$$J = \{1,2,\ldots, N\} \tag{14b}$$

and let

$$K_1(t,\tau)= \sum_{j=1}^n (1+\rho_j)z_j(t)z_j(\tau)+\int_0^T G(t,\xi)G(\tau,\xi)d\xi, t,\tau\varepsilon I. \tag{15}$$

Also let $\underset{\sim}{e}_j$, $j \in J$, be the jth unit N-vector, let $\underset{\sim}{z}(\cdot)$ be the N-vector

$$\underset{\sim}{z}(\cdot) = [z_1(\cdot), z_2(\cdot),\ldots, z_n(\cdot),0,\ldots, 0]' \tag{16}$$

and let $\underset{\sim}{Q}$ be the diagonal $N\times N$ matrix

$$\underset{\sim}{Q} = \text{diag } (\rho_1,\rho_2,\ldots, \rho_N) . \tag{17}$$

Then H^+ with norm given by (13) has the reproducing kernel

$$K_t^+ = \begin{cases} (K_1(\cdot,t),\ -\underset{\sim}{Q}\ \underset{\sim}{z}(t)), & t \in I \\ (-\underset{\sim}{z}'(\cdot)\underset{\sim}{Q}\ \underset{\sim}{e}_t,\ \underset{\sim}{Q}\ \underset{\sim}{e}_t), & t \in J\ . \end{cases} \tag{18}$$

In terms of the inner product determined by (13), the reproducing property of H^+ is

$$<(g(\cdot),\underset{\sim}{\theta}),\ K_t^+> = \begin{cases} g(t), & t \in I \\ \theta_t, & t \in J \end{cases},\quad (g(\cdot),\underset{\sim}{\theta}) \in H^+\ . \tag{19}$$

Now since $\{r_j\}_1^N$ are fixed, it is clear that the optimization problem in (5) is equivalent to minimizing $\|(g,\underset{\sim}{\theta})\|^2$ subject to the constraints $\lambda_j g+\theta_j = r_j$, $j \in J$. These constraints can be written as inner product constraints:

$$<(g,\underset{\sim}{\theta}),\ (d_j,\underset{\sim}{\omega}_j)> = \lambda_j g+\theta_j = r_j,\ j \in J \tag{20}$$

where

$$d_j(t) = \lambda_j K_1(\cdot,t) - \underset{\sim}{e}_j'\underset{\sim}{Q}\ \underset{\sim}{z}(t),\ t \in I\ ,\ j \in J \tag{21}$$

$$\underset{\sim}{\omega}_j = \underset{\sim}{Q}(\underset{\sim}{e}_j - \lambda_j\ \underset{\sim}{z}),\ j \in J\ . \tag{22}$$

Let

$$U^+ = \{(g,\underset{\sim}{\theta}) \in H^+ : <(g,\underset{\sim}{\theta}),(d_j,\underset{\sim}{\omega}_j)> = r_j,\ j \in J\}\ . \tag{23}$$

Then if $(\hat{g},\underset{\sim}{\hat{\theta}})$ solves

$$\underset{(g,\underset{\sim}{\theta})\in U^+}{\text{minimize}}\ \|(g,\underset{\sim}{\theta})\|^2\ . \tag{24}$$

we can see that the smoothing spline $s_1(t) = \hat{g}(t)$, $t \in I$. Therefore, $s_1(\cdot)$ is the component in H of the projection of any element of U^+ onto the span of $\{(d_j,\underset{\sim}{\omega}_j)\}_1^N$.

Remark 3. Nothing in the interpolation problem (4) forces a particular choice of norm and reproducing kernel for H. However, the choice in (6) and (7) seems to be the most convenient one for developing recursive spline algorithms and computable error bounds. This particular framework for interpolation was developed, to varying degrees, by Golomb and Weinberger (1959), de Boor and Lynch (1966), Weinert and Kailath (1974).

Remark 4. Details of this new approach to the smoothing problem can be found in Weinert, Byrd, and Sidhu (1977). It is clear that spline smoothing is nothing more than spline interpolation in an augmented space. Note that because of the cross-term in (13), H^+ is not the usual product of H and E^N; hence the more complicated reproducing kernel and reproducing property. For other approaches to spline smoothing, see Lyche and Schumaker (1973), Reinsch (1971), de Figueiredo and Caprihan (1977), Kimeldorf and Wahba (1970a, 1970b, 1971), Wahba and Wold (1975). Anselone and Laurent (1968) characterize Ls_1 as a component of the solution to a projection problem in $LH \times E^N$.

3. SPLINES AND STOCHASTIC ESTIMATES

In this section the stochastic estimation problems that are structurally equivalent to spline interpolation and smoothing are described (see Remark 5). Let $\{y_0(t), t \in I\}$ be a zero-mean random process with covariance function $K_0(\cdot,\cdot)$. It can be shown (see Parzen, 1961) that H is congruent (isometrically isomorphic) to the space of zero-mean random variables spanned by $\{y_0(t), t \in I\}$. As a result, if $\hat{y}_0(t)$ is the linear least-squares estimate of $y_0(t)$ given random variables $\{\lambda_j y_0\}_1^N$, and if $\hat{\hat{y}}_0(t)$ is the sample value of

$\hat{y}_0(t)$ obtained by setting $\lambda_j y_0 = r_j$, $j \in J$, then

$$s_0(t) = \hat{\hat{y}}_0(t), \ t \in I \ . \tag{25}$$

For the spline smoothing case, let $\{y_1(t), \ t \in I\}$ be a zero-mean random process with covariance function $K_1(\cdot,\cdot)$, and let $\underset{\sim}{v} = [v_1, v_2, \ldots, v_N]'$ be a zero-mean random vector such that

$$E[y_1(t)\underset{\sim}{v}] = -\underset{\sim}{Q}\,\underset{\sim}{z}(t), \ t \in I \tag{26}$$

$$E[\underset{\sim}{v}\,\underset{\sim}{v}'] = \underset{\sim}{Q} \ . \tag{27}$$

Now H^+ is congruent to the space of zero-mean random variables spanned by $\{y_1(t), \ t \in I; \ v_j, \ j \in J\}$. If $\hat{y}_1(t)$ is the linear least-squares estimate of $y_1(t)$ given noisy data $\{\lambda_j y_1 + v_j\}_1^N$, and if $\hat{\hat{y}}_1(t)$ is the sample value of $\hat{y}_1(t)$ obtained by setting $\lambda_j y_1 + v_j = r_j$, $j \in J$, then

$$s_1(t) = \hat{\hat{y}}_1(t), \ t \in I \ . \tag{28}$$

Thus the interpolating spline $s_0(\cdot)$ and smoothing spline $s_1(\cdot)$ are sample functions of the random processes $\hat{y}_0(\cdot)$ and $\hat{y}_1(\cdot)$, respectively. As a result recursive algorithms for $\hat{y}_0(\cdot)$ and $\hat{y}_1(\cdot)$ serve as recursive algorithms (see Remark 6) for $s_0(\cdot)$ and $s_1(\cdot)$. Of course, $\hat{y}_0(\cdot)$ and $\hat{y}_1(\cdot)$ can be found using just the covariance information given above. However, computation of the covariances $K_0(\cdot,\cdot)$ and $K_1(\cdot,\cdot)$ can be avoided by using the dynamical linear models that generate the processes $y_0(\cdot)$ and $y_1(\cdot)$ in response to white noise.

The dynamics of these models are immediately determined by the coefficients of the operator L , without intermediate computations. Before giving the models, we will restrict attention to a broad class of constraint functionals $\{\lambda_j\}_1^N$ called extended Hermite-Birkhoff functionals, which have the form

$$\lambda_j g = \sum_{k=1}^{n} \alpha_{jk} g^{(k-1)}(t_j) \ , \ j \in J \tag{29}$$

where $0 \le t_1 \le t_2 \le \cdots \le t_N \le T$ and the $\{\alpha_{jk}\}$ are known real numbers. For future reference, let

$$\underset{\sim}{c}_j = [\alpha_{j1}, \alpha_{j2}, \ldots, \alpha_{jn}] \ , \ j \in J \ . \tag{30}$$

For the spline interpolation case, it can be shown that if $u(\cdot)$ is a zero-mean, unit intensity white noise process, $y_0(\cdot)$ is generated by the model

$$\frac{d}{dt} \underset{\sim}{x}_0(t) = \underset{\sim}{A}\, \underset{\sim}{x}_0(t) + \underset{\sim}{b}\, u(t) \tag{31a}$$

$$y_0(t) = \underset{\sim}{c}\, \underset{\sim}{x}_0(t) \tag{31b}$$

where

$$\underset{\sim}{b} = [0 \ . \ . \ . \ 0 \ 1]' \tag{32}$$

$$\underset{\sim}{c} = [1 \ 0 \ . \ . \ . \ 0] \tag{33}$$

$$\underset{\sim}{A} = \left[\begin{array}{c|c} \underset{\sim}{0} & \underset{\sim}{I} \\ \hline -a_0 & -a_1 \ \ldots \ -a_{n-1} \end{array}\right] \tag{34}$$

and $\underset{\sim}{x}_0(\cdot)$ is the n-dimensional state vector. All that remains is to specify the unique initial conditions that guarantee that $y_0(\cdot)$ has covariance $K_0(\cdot,\cdot)$. The most convenient initialization point is t_n . Let $\underset{\sim}{\phi}(\cdot,\cdot)$ be the fundamental matrix for $\underset{\sim}{A}$; i.e.,

$$\underset{\sim}{\phi}(t,\tau) = \exp \underset{\sim}{A}(t-\tau) \ . \tag{35}$$

If $\underset{\sim}{M}$ is the $n \times n$ matrix with i^{th} row $\underset{\sim}{M}_i'$:

$$\underset{\sim}{M}_i' = \underset{\sim}{c}_i\, \underset{\sim}{\phi}(t_i, t_n) \tag{36}$$

and if $\underset{\sim}{\Delta}(\cdot)$ is the $n\times n$ matrix with i^{th} row $\underset{\sim}{\Delta}_i'(\cdot)$:

$$\underset{\sim}{\Delta}_i'(t) = \begin{cases} \underset{\sim}{c}_i \underset{\sim}{\phi}(t_i,t), & t \in [t_i,t_n] \\ \underset{\sim}{0}, & \text{otherwise} \end{cases} \tag{37}$$

then the initial conditions for (31) are

$$E[\underset{\sim}{x}_0(t_n)] = \underset{\sim}{0}$$

$$E[\underset{\sim}{x}_0(t_n)\underset{\sim}{x}_0'(t_n)] = \underset{\sim}{M}^{-1}[\underset{\sim}{I} + \int_{t_1}^{t_n} \underset{\sim}{\Delta}(\tau)\underset{\sim}{b}\underset{\sim}{b}'\underset{\sim}{\Delta}'(\tau)d\tau]\underset{\sim}{M}^{-T} \tag{38}$$

$$E[\underset{\sim}{x}_0(t_n)\; u(t)] = \begin{cases} \underset{\sim}{M}^{-1}\; \underset{\sim}{\Delta}(t)\underset{\sim}{b}, & t \in [t_1,t_n] \\ \underset{\sim}{0}, & \text{otherwise .} \end{cases} \tag{39}$$

For the spline smoothing case, the model for $y_1(\cdot)$ is identical to (31)-(39) except that the state vector, now denoted by $x_1(\cdot)$, has initial covariance

$$E[\underset{\sim}{x}_1(t_n)\underset{\sim}{x}_1'(t_n)] = \underset{\sim}{M}^{-1}[\underset{\sim}{I} + \underset{\sim}{Q}_n + \int_{t_1}^{t_n} \Delta(\tau)\underset{\sim}{b}\underset{\sim}{b}'\underset{\sim}{\Delta}'(\tau)d\tau]\underset{\sim}{M}^{-T} \tag{40}$$

where $\underset{\sim}{Q}_n = \text{diag}(\rho_1,\rho_2,\ldots,\rho_n)$. In addition, the correlation between $y_1(\cdot)$ and $\underset{\sim}{v}$ expressed by (26) can be accounted for by the alternate relations

$$E[\underset{\sim}{x}_1(t_n)\underset{\sim}{v}'] = -\underset{\sim}{M}^{-1}[\underset{\sim}{Q}_n|\underset{\sim}{0}] \tag{41}$$

$$E[u(t)\underset{\sim}{v}'] = \underset{\sim}{0}\;,\; t \in I\;. \tag{42}$$

<u>Remark 5</u>. The stochastic correspondence (25) for interpolation was proved by Weinert and Kailath (1974) and, in a different way, by Weinert and Sidhu (1978). The correspondence (28) for

smoothing was proved by Weinert, Byrd, and Sidhu (1977). Different choices of norm and reproducing kernel for H and H^+ lead to different stochastic correspondences. In fact, our early work in this area was motivated by that of Kimeldorf and Wahba (1970a, 1970b, 1971) who did make a different choice. Their stochastic correspondences for interpolation and smoothing involve random processes with unknown mean values, and do not seem to be as useful for developing recursive spline algorithms. Note that every reproducing kernel is a covariance function and vice-versa.

Remark 6. With a recursive algorithm, the spline can be computed using one data point at a time. Recursive algorithms are therefore useful in real-time applications and those in which data storage is a limiting factor. A recursive interpolation algorithm based on the stochastic realization (31)-(39) was developed by Weinert and Sidhu (1978). See also Sidhu and Weinert (1977). Similar results for smoothing were established by Weinert, Byrd, and Sidhu (1977). Proofs of (31)-(42) are also given in those papers. The generalization to vector-valued splines is treated by Sidhu and Weinert (1978). The uniqueness assumption stated in section 1 guarantees the invertibility of the matrix $\tilde{M}$. This uniqueness condition is equivalent to a strong type of observability for the model (31); see Weinert and Sidhu (1977) for details.

Remark 7. Note that in going from the estimation problem corresponding to interpolation to that corresponding to smoothing, one does not simply add noise to the previous observations. The signal covariance $K_0(\cdot,\cdot)$ must be changed to $K_1(\cdot,\cdot)$. Then, noise that is correlated with the signal is added. The change in signal covariance is easily accomplished here by changing the covariance matrix of the initial state of the realization of $K_0(\cdot,\cdot)$. An attempt to generalize the results of Weinert and Sidhu (1978) to the smoothing case was made by de Figueiredo and Caprihan (1977). However, I believe that the stochastic correspondence and resulting recursive algorithm reported there are incorrect, though a valid nonrecursive formula for smoothing splines is given.

4. ERROR BOUNDS FOR SPLINE INTERPOLATION AND SMOOTHING

Returning to the case of general constraint functionals $\{\lambda_j\}_1^N$, we will first examine what, if anything, can be said about the interpolation error $|f(t)-s_0(t)|$ when all that is known about $f(\cdot)$ is that $f \varepsilon H$. Much work has been done on upper-bounding this error (see e.g., Golomb and Weinberger, 1959 and Schultz, 1973) and all these bounds have the general form

$$|f(t)-s_0(t)| \leq dq(f) \tag{43}$$

for a fixed $t \varepsilon I$, where d is a known constant and $q(f)$ is some nonlinear functional of $f(\cdot)$. Obviously, in order to evaluate this bound, one must have additional a priori information about $f(\cdot)$ in the form of $q(f)$. In most cases, this information is either completely lacking or is too imprecise to give a tight bound. The best one can do in such cases is bound a normalized version of the interpolation error. In fact, using the Schwartz inequality, the reproducing property (8), and the stochastic correspondence discussed in section 3, it can be shown that

$$\max_{f \varepsilon H} \frac{|f(t)-s_0(t)|}{\| f \|} = e_0(t) \tag{44}$$

where $e_0^2(t)$ is the minimum mean-square error in estimating $y_0(t)$ from $\{\lambda_j y_0\}_1^N$; i.e.,

$$e_0^2(t) = E[(y_0(t)-\hat{y}_0(t))^2] . \tag{45}$$

Note that $e_0(t)$ can be computed once L is chosen, without additional a priori information about $f(\cdot)$. In fact, recursions for computing $e_0(t)$ can easily be added to the spline interpolation algorithm of Weinert and Sidhu (1978), thus avoiding explicit use of the covariance $K_0(\cdot,\cdot)$. Note also that $e_0(t)$ is independent of the spline interpolation data $\{r_j\}_1^N$ and can thus be computed before any measurements are made. As discussed in the next section, this fact permits the development of optimal sampling schemes. The corresponding result for spline smoothing is

$$\max_{(\underset{\sim}{f,\theta})\in H^+} \frac{|f(t)-s_1(t)|}{\|\underset{\sim}{(f,\theta)}\|} = e_1(t)$$

where

$$e_1^2(t) = E[(y_1(t)-\hat{y}_1(t))^2].$$

The above remarks on computing $e_0(t)$ apply also to $e_1(t)$.

Remark 8. Larkin (1972) took a different approach to the problem of lack of a priori information. He imposed a Gaussian probability distribution on H , effectively converting every element of H into a random process. His interpolation bounds are then of the form

$$|f(t)-s_0(t)| \le m(\alpha) \text{ with probability } \alpha \tag{46}$$

where t is fixed and $m(\alpha)$ depends on α, N, and $\{\lambda_j f\}_1^N$.

The problem with this approach is that there is no natural probability distribution associated with H . A nonlinear constraint is thus being added to the interpolation problem, equivalent to that of guessing some $q(f)$ in bounds of the type (43). As a result, $m(\alpha)$ has no relation to the actual interpolation error. Sacks and Ylvisaker (1970) obtained results analogous to (44)-(45) for the problem of optimal quadrature (approximate integration). Wahba (1969) gives partially computable error bounds for some general interpolation and smoothing problems.

5. OPTIMAL SAMPLING FOR SPLINE INTERPOLATION

In this section we shall restrict attention to the problem of interpolating from samples $\{f(t_j)\}_1^N$. Sampling functionals are of course a special case of (29). It is known that the interpolating spline $s_0(\cdot)$ satisfies the following conditions:

$$L^a L\, s_0(t) = 0\ ,\ t_j<t<t_{j+1}\ ,\ 1\leq j\leq N-1 \tag{47a}$$

$$L\, s_0(t) = 0\ ,\ 0<t<t_1 \text{ and } t_N<t<T \tag{47b}$$

where L^a is the formal adjoint of L . In addition $s(\cdot)$ has 2n-2 continuous derivatives on $[0,T]$. Obviously, L determines the functional form of the spline.

The following experimental design problem is of interest here. Suppose we can choose N points in $[0,T]$ at which to sample an unknown function $f \in H$, after which the samples will be interpolated with the spline $s_0(\cdot)$. If the spline is to give the best approximation to $f(\cdot)$ over the entire interval $[0,T]$, it makes sense in light of the bound (44) to choose the sampling times $\{t_j\}_1^N$ to minimize $\int_0^T e_0^2(t)dt$. This is a difficult nonlinear optimization problem, and results for an arbitrary L are not yet available, but if $L = \frac{d}{dt}$, in which case the spline is piecewise linear, then it is easy to show that

$$\int_0^T e_0^2(t)\,dt = \frac{1}{2} t_1^2 + \frac{1}{2} (T-t_N)^2 + \frac{1}{6} \sum_{j=1}^{N-1} (t_{j+1}-t_j)^2 .$$

As a result, the optimal sample times are

$$t_1^* = \frac{T}{3N-1} , \quad t_{j+1}^* = t_j^* + \frac{3T}{3N-1} , \quad 1 \le j \le N-1 . \tag{48}$$

Remark 9. Results on optimal sampling for other types of approximation problems, including approximate integration, can be found in Karlin (1969); Karlin, Micchelli, Pinkus, and Schoenberg (1976); Sacks and Ylvisaker (1970); Wahba (1971,1974,1976); Larkin (1970); Richter-Dyn (1971); Samaniego (1976); Bojanov (1977); Bojanov and Chernogorov (1977); Hajek and Kimeldorf (1974). The result (48) appears to be new.

ACKNOWLEDGEMENTS

I wish to acknowledge the valuable collaboration of T. Kailath, G. S. Sidhu, and R. H. Byrd at various stages of the research described in this paper. I am also indebted to the referee for valuable suggestions.

BIBLIOGRAPHY

Anselone, P. M. & Laurent, P. J. (1968). A general method for the construction of interpolating or smoothing spline functions. Num. Math. 12, 66-82.

Aronszajn, N. (1950). Theory of reproducing kernels. Trans. Amer. Math. Soc. 68, 337-404.

Bellman, R. & Roth, R. S. (1971). The use of splines with unknown end points in the identification of systems. J. Math. Anal. Appl. 34, 26-33.

Bojanov, B. D. (1977). Characterization and existence of optimal quadrature formulas for a class of differentiable functions. Soviet Math. Dokl. 18, 215-18.

Bojanov, B. D. & Chernogorov, V. G. (1977). An optimal interpolation formula. J. Approx. Theory 20, 264-74.

Caprihan, A. (1975). Finite-duration digital filter design by use of cubic splines. IEEE Trans. Circuits and Systems 22, 204-07.

Craven, P. & Wahba, G. (1977). Smoothing noisy data with spline functions: Estimating the correct degree of smoothing by the method of generalized cross-validation. Tech. Report 445, Dept. of Statistics, Univ. of Wisconsin, Madison.

DeBoor, C. & Lynch, R. (1966). On splines and their minimum properties. J. Math. Mech. 15, 953-69.

De Figueiredo, R. J. P. & Caprihan, A. (1977). An algorithm for the construction of the generalized smoothing spline with application to system identification. Proc. Conf. Inf. Sci. Systems. Baltimore: Barton Press, 494-500.

De Figueiredo, R. J. P. & Netravali, A. N. (1976). On a class of minimum energy controls related to spline functions. IEEE Trans. Auto. Control 21, 725-27.

De Montricher, G. F., Tapia, R. A., & Thompson, J. R. (1975). Nonparametric maximum likelihood estimation of probability densities by penalty function methods. Ann. Statist. 3, 1329-48.

Golomb, M. & Weinberger, H. (1959). Optimal approximation and error bounds. On Numerical Approximation (R. Langer, Ed.). Madison: Univ. Wisconsin Press, 117-90.

Greville, T. N. E., Ed. (1969). Theory and Applications of Spline Functions. New York: Academic Press.

Hajek, J. & Kimeldorf, G. (1974). Regression designs in autoregressive stochastic processes. Ann. Statist. 2, 520-27.

Jerome, J. & Schumaker, L. L. (1969). On Lg-splines. J. Approx. Theory 2, 29-49.

Karlin, S. (1969). The fundamental theorem of algebra for monosplines satisfying certain boundary conditions and applications to optimal quadrature formulas. Approximations with Special Emphasis on Spline Functions (I. J. Schoenberg, Ed.). New York: Academic Press, 467-84.

Karlin, S., Micchelli, C. A., Pinkus, A., & Schoenberg, I. J. (1976). Studies in Spline Functions and Approximation Theory. New York: Academic Press.

Kimeldorf, G. & Wahba, G. (1970a). A correspondence between Bayesian estimation of stochastic processes and smoothing by splines. Ann. Math. Statist. 41, 495-502.

Kimeldorf, G. & Wahba, G. (1970b). Spline functions and stochastic processes. Sankhya 132, 173-80.

Kimeldorf, G. & Wahba, G. (1971). Some results on Tchebycheffian spline functions. J. Math. Anal. Appl. 33, 82-95.

Larkin, F. M. (1970). Optimal approximation in Hilbert spaces with reproducing kernel functions. Math. Comput. 24, 911-21.

Larkin, F. M. (1972). Gaussian measure in Hilbert space and applications in numerical analysis. Rocky Mt. J. Math. 2, 379-421.

Larsen, R. D., Crawford, E. F., & Smith, P. W. (1977). Reduced spline representations for EEG signals. Proc. IEEE. 65, 804-07.

Lyche, T. & Schumaker, L. L. (1973). Computation of smoothing and interpolating natural splines via local bases. SIAM J. Numer. Anal. 10, 1027-38.

Parzen, E. (1961). An approach to time series analysis. Ann. Math. Statist. 32, 951-89.

Reinsch, C. (1971). Smoothing by spline functions, II. Num. Math. 16, 451-54.

Richter-Dyn, N. (1971). Minimal interpolation and approximation in Hilbert spaces. SIAM J. Numer. Anal. 8, 583-97.

Sacks, J. & Ylvisaker, D. (1970). Statistical designs and integral approximations. Proc. 12th Biennial Sem. Canadian Math. Congress (R. Pyke, Ed.). Montreal: Canadian Math. Congress, 115-36.

Samaniego, F. J. (1976). The optimal sampling design for estimating the integral of a process with stationary independent increments. IEEE Trans. Inf. Theory 22, 375-76.

Schoenberg, I. J., Ed. (1969). Approximations with Special Emphasis on Spline Functions. New York: Academic Press.

Schultz, M. H. (1973). Spline Analysis. Englewood Cliffs, NJ: Prentice-Hall, Inc.

Sidhu, G. S. & Weinert, H. L. (1977). Dynamical recursive algorithms for Lg-spline interpolation of EHB data. Submitted for publication.

Sidhu, G. S. & Weinert, H. L. (1978). Vector-valued Lg-splines. Submitted for publication.

Wahba, G. (1969). On the numerical solution of Fredholm integral equations of the first kind. Tech. Report 217, Dept. of Statistics, Univ. of Wisconsin, Madison.

Wahba, G. (1971). On the regression design problem of Sacks and Ylvisaker. Ann. Math. Statist. 42, 1035-53.

Wahba, G. (1974). Regression design for some equivalence classes of kernels. Ann. Statist. 2, 925-34.

Wahba, G. (1976). On the optimal choice of nodes in the collocation-projection method for solving linear operator equations. J. Approx. Theory 16, 175-86.

Wahba, G. & Wold, S. (1975). A completely automatic French curve: Fitting spline functions by cross validation. Commun. Statist. 4, 1-17.

Weinert, H. L., Byrd, R. H., & Sidhu, G. S. (1977). A stochastic framework for recursive computation of spline functions - Part II: Smoothing splines. Submitted for publication.

Weinert, H. L. & Kailath, T. (1974). Stochastic interpretations and recursive algorithms for spline functions. Ann. Statist. 2, 787-94.

Weinert, H. L. & Kailath, T. (1976). A spline-theoretic approach to minimum energy control. IEEE Trans. Auto. Control 21, 391-93.

Weinert, H. L. & Sidhu, G. S. (1977). On uniqueness conditions for optimal curve fitting. J. Optimization Theory Appl. 23, 211-16.

Weinert, H. L. & Sidhu, G. S. (1978). A stochastic framework for recursive computation of spline functions - Part I: Interpolating splines. IEEE Trans. Inf. Theory 24, 45-50.

Yao, K. (1967). Applications of reproducing kernel Hilbert spaces - bandlimited signal models. Inf. and Control 11, 429-44.

Received March, 1978.

22

Reprinted from *J. Math. Mech.* **15:**953–969 (1966)

On Splines and Their Minimum Properties

CARL DE BOOR & ROBERT E. LYNCH[1]

Communicated by G. BIRKHOFF

0. Introduction. It is the purpose of this note to show that the several minimum properties of odd degree polynomial spline functions [4, 18] all derive from the fact that spline functions are representers of appropriate bounded linear functionals in an appropriate Hilbert space. (These results were first announced in *Notices, Amer. Math. Soc.*, **11** (1964) 681.) In particular, spline interpolation is a process of best approximation, i.e., of orthogonal projection, in this Hilbert space. This observation leads to a generalization of the notion of spline function. The fact that such generalized spline functions retain all the minimum properties of the polynomial splines, follows from familiar facts about orthogonal projections in Hilbert space.

1. Polynomial splines and their minimum properties. A *polynomial spline function*, $s(x)$, of degree $m \geqq 0$, having the $n \geqq 1$ *joints* $x_1 < x_2 < \cdots < x_n$, is by definition a real valued function of class $C^{(m-1)}(-\infty, \infty)$, which reduces to a polynomial of degree at most m in each of the $n + 1$ intervals $(-\infty, x_1)$, $(x_1, x_2), \cdots, (x_n, +\infty)$. The most general such function is given by

$$s(x) = \sum_{i=0}^{m} \alpha_i x^i + \sum_{j=1}^{n} \beta_j (x - x_j)_+^m ,$$

where α_i, $i = 0, \cdots, m$, and β_j, $j = 1, \cdots, n$, are real numbers and

$$(x)_+^m = \begin{cases} x^m, & x \geqq 0, \\ 0, & x < 0. \end{cases}$$

Specifically, let $m = 2k - 1$, and $n \geqq k \geqq 1$, and let S_0 denote the family of polynomial spline functions of odd degree m with joints $x_1, \cdots, x_n$, which reduce to polynomials of degree at most $k - 1$ in each of the two intervals $(-\infty, x_1)$ and (x_n, ∞). Equivalently, S_0 consists of all polynomial spline functions $s(x)$ of degree m with joints $x_1, \cdots, x_n$ which satisfy

[1] The work of R. E. Lynch was supported in part by the National Science Foundation through Grant GP-217 and by the Army Research Office (Durham) through Grant DA-ARO(D)-31-124-G388, at The University of Texas.

$$\begin{aligned}(1.1)\qquad & s^{(j)}(x_1) = s^{(j)}(x_n) = 0, \qquad j = k, \cdots, 2k-2,\\ & s^{(2k-1)}(x) \equiv 0, \quad \text{all} \quad x \notin [x_1, x_n].\end{aligned}$$

Hence, for $n = k$, S_0 consists just of the set $\{\pi_{k-1}\}$ of polynomials of degree at most $k - 1$. Let $[a, b]$ be a finite interval containing all the joints $x_1, \cdots, x_n$ and consider S_0 as a subset of the class of functions [19]

$$(1.2)\qquad F^{(k)}[a, b] = \{f(x) \mid f \in C^{(k-1)}[a, b],\ f^{(k-1)} \text{ absolutely continuous}, \ f^{(k)} \in L^2[a, b]\}.$$

The elements of S_0 have the following properties [4], [18]:

Interpolation property: Given $f \in F^{(k)}[a, b]$, there exists a unique element $s(x) \in S_0$ satisfying

$$s(x_i) = f(x_i), \qquad i = 1, \cdots, n.$$

Denote this unique element by Pf.

First minimum property: If $f \in F^{(k)}[a, b]$ and $s \in S_0$, then

$$\int_a^b [f^{(k)}(x) - s^{(k)}(x)]^2\, dx \geqq \int_a^b [f^{(k)}(x) - (Pf)^{(k)}(x)]^2\, dx,$$

with equality if and only if $s = Pf + \pi_{k-1}$.

Second minimum property: If $f \in F^{(k)}[a, b]$, then

$$\int_a^b [f^{(k)}(x)]^2\, dx \geqq \int_a^b [(Pf)^{(k)}(x)]^2\, dx,$$

with equality if and only if $f = Pf$.

A third minimum property concerns the linear approximation of a linear functional L on $F^{(k)}[a, b]$ of the form

$$Lf = \sum_{i=0}^{k-1} \int_a^b f^{(i)}(y)\, d\mu_i(y),$$

where the $\mu_i(x)$ are functions of bounded variation. For later reference, we denote by $\mathcal{L}^{(k)}$ the set of all such linear functionals.

Third minimum property: For all $L \in \mathcal{L}^{(k)}$, and all $f \in F^{(k)}[a, b]$, the best approximation, L^*f, in the sense of Sard [15] (see below in Section 2) to Lf by an expression of the form $\sum_{i=1}^{n} \alpha_i f(x_i)$ is given by operating with L on Pf; for short, $L^* = LP$. This approximation is exact for all $f \in S_0$.

2. Representers in Hilbert space and their minimum properties. In order to relate these minimum properties of the elements of S_0 to "familiar facts about orthogonal projections in Hilbert space," we need the following facts:

Theorem 2.1. The linear space $F^{(k)}[a, b]$ is a Hilbert space with respect to the inner product

$$(2.1)\quad (f, g) = \sum_{i=1}^{k} f(x_i)g(x_i) + \int_a^b f^{(k)}(y)g^{(k)}(y)\,dy, \quad \text{all} \quad f, g \in F^{(k)}[a, b].$$

This Hilbert space possesses a reproducing kernel [1], [3], $K(x, y)$, *which is*

$$(2.2)\quad \begin{aligned} K(x, y) = \sum_{i=1}^{k} c_i(x)c_i(y) + (-1)^k\Big\{&(x - y)_+^{2k-1} \\ &+ \sum_{i=1}^{k}\sum_{j=1}^{k} (x_i - x_j)_+^{2k-1}c_i(x)c_j(y) \\ &- \sum_{i=1}^{k} [(x - x_i)_+^{2k-1}c_i(y) + (x_i - y)_+^{2k-1}c_i(x)]\Big\}\Big/(2k - 1)!, \end{aligned}$$

where

$$c_i(x) = \prod_{j=1, j\neq i}^{k} (x - x_j)/(x_i - x_j), \qquad i = 1, \cdots, k.$$

For all $L \in \mathfrak{L}^{(k)}$, *for all* $f \in F^{(k)}[a, b]$,

$$(2.3)\quad \begin{aligned} Lf = &\sum_{i=1}^{k} L(c_i)f(x_i) \\ &+ \frac{1}{(k-1)!}\int_a^b L_{(x)}\Big((x - y)_+^{k-1} - \sum_{i=1}^{k} c_i(x)(x_i - y)_+^{k-1}\Big)f^{(k)}(y)\,dy. \end{aligned}$$

Here and below, the subscript (x) [or (y)] indicates [5] that an operation is to be performed on a function of x [or y] for fixed y [or fixed x].

This theorem is a special case of Theorem 3.1 and Lemma 3.1 proved below in Section 3.

Corollary. *The linear functionals* L_i, $i = 1, \cdots, n$, *given by*

$$(2.4)\quad L_i f = f(x_i), \qquad i = 1, \cdots, n, \quad \text{all} \quad f \in F^{(k)}[a, b],$$

are bounded; their representers [3] *span* S_0.

Proof. A linear functional L on a real Hilbert space H with reproducing kernel $K(x, y)$ is bounded if and only if the function $\phi(x) = L_{(y)}K(x, y)$ is in H, and in that case, $\phi(x)$ is its representer, *i.e.*,

$$Lf = (f, \phi), \quad \text{all} \quad f \in H.$$

Since the linear functionals L_i, $i = 1, \cdots, n$, are linearly independent over $F^{(k)}[a, b]$, the corollary follows from the fact that, by (2.2), $K(x, x_i) \in S_0$, $i = 1, \cdots, n$.

Let now, more generally, $L_1, \cdots, L_n$ be n linearly independent bounded linear functionals over a real Hilbert space H, and let $S = \langle\phi_1, \cdots, \phi_n\rangle$ be the n-dimensional subspace of H spanned by the representers $\phi_1, \cdots, \phi_n$ of the L_i. Given $f \in H$, an element $g \in H$ is said to *interpolate* f *with respect to* $L_1, \cdots, L_n$ if

$$L_i g = L_i f, \qquad i = 1, \cdots, n.$$

Let P_S denote the orthogonal projection from H onto S, *i.e.*, for $f \in H$, $P_S f$ is the unique best approximation to f by an element in S with respect to the norm in H. Then P_S satisfies [7]

$$(P_S f, h) = (P_S f, P_S h) = (f, P_S h), \quad \text{all} \quad f, h \in H, \tag{2.5}$$

and

$$||f||^2 = ||f - P_S f||^2 + ||P_S f||^2, \quad \text{all} \quad f \in H. \tag{2.6}$$

By setting $h = \phi_i$ in (2.5), it follows that

$$L_i(P_S f) = L_i(f), \quad \text{all} \quad f \in H, \qquad i = 1, \cdots, n. \tag{2.7}$$

This proves

Lemma 2.1. Given $f \in H$, $P_S f$ is the unique element in S which interpolates f with respect to $L_1, \cdots, L_n$.

For illustration, if H is $F^{(k)}[a, b]$ with inner product (2.1), and the L_i are given by (2.4), Lemma 2.1 states that the spline function $Pf \in S_0$, which, by definition, interpolates f at the points $x_1, \cdots, x_n$, is also the best approximation to f with respect to the norm

$$||g||^2 = \sum_{i=1}^{k} (g(x_i))^2 + \int_a^b [g^{(k)}(y)]^2 \, dy,$$

thus implying the first minimum property in Section 1.

An element $f \in H$ is in S if and only if $f = P_S f$. This may be put somewhat differently. Given $f \in H$, let $W_f = \{h \mid h \in H, L_i(h) = L_i(f) \text{ for } i = 1, \cdots, n\}$. Then, it follows from Lemma 2.1 that

$$P_S h = P_S f, \quad \text{all} \quad h \in W_f. \tag{2.8}$$

Hence one has from (2.6) that

$$||P_S f|| \leqq ||h||, \text{ all } h \in W_f, \text{ with equality if and only if } h = P_S f; \tag{2.9}$$

and we have proved the following

Lemma 2.2. An element $f \in H$ is in S if and only if f is the element of minimal norm in the set W_f of all elements of H that interpolate f with respect to $L_1, \cdots, L_n$.

Specifically, if H is $F^{(k)}[a, b]$ with inner product (2.1) and the L_i are given by (2.4), then for all $h \in W_f$,

$$\sum_{i=1}^{k} (h(x_i))^2 = \sum_{i=1}^{k} (f(x_i))^2,$$

and the second minimum property of polynomial splines follows.

Returning once more to the general situation, let L be a bounded linear functional on H with representer ϕ. We wish to approximate L by a linear functional of the form $\sum_{i=1}^{n} \alpha_i L_i$ in such a way that the norm

$$\|R\| = \sup_{\|f\| \leq 1} |Rf| \tag{2.10}$$

of the error functional $R = L - \sum_{i=1}^{n} \alpha_i L_i$ be minimized. Since $\|R\| = \|\phi - \sum \alpha_i \phi_i\|$, we have

Lemma 2.3. *The element $\bar{\phi} = P_S\phi$ is the representer of the unique best approximation $\bar{L}$ to L by a bounded linear functional of the form $\sum_{i=1}^{n} \alpha_i L_i$ with respect to the norm* (2.10).

In particular, one has by (2.5) that

$$\bar{L}(f) = (f, P_S\phi) = (P_S f, \phi) = L(P_S f), \quad \text{all} \quad f \in H. \tag{2.11}$$

Corollary 1. *Given $f \in H$, the value of the best approximation to L at f is equal to the value of L at the best approximation to f.*

It is this property which makes $\bar{L}$ so attractive for use in computational work: any computational scheme which solves the interpolation problem also solves the problem of computing the best approximation to *any* bounded linear functional.

For $f \in S$, $P_S f = f$; therefore, by (2.11), $\bar{L}(f) = L(f)$, for $f \in S$.

Corollary 2. *The approximation $\bar{L}$ to L is exact for $f \in S$.*

Once again, let H be, in particular, $F^{(k)}[a, b]$ with inner product (2.1), and let the L_i be given by (2.4). Then Corollaries 1 and 2 of Lemma 2.3 imply the third minimum property of polynomial splines. To see this, we need to recall the definition of L^*, the best approximation to L in the sense of Sard [15], [18] by a linear combination of the L_i.

Since $n \geq k$, it is possible to choose numbers $\alpha_1, \cdots, \alpha_n$ so that $Rf = 0$ whenever $f \in \{\pi_{k-1}\}$. The set

$$\mathfrak{M} = \left\{ L' = \sum_{i=1}^{k} \alpha_i L_i \mid Lf = L'f \quad \text{for} \quad f \in \{\pi_{k-1}\} \right\}$$

is therefore not empty. By Peano's theorem [3], [11], or by (2.3) in Theorem 2.1, we can write Rf as

$$Rf = (L - L')f = \int_a^b K(t) f^{(k)}(t)\, dt, \quad \text{all} \quad L' \in \mathfrak{M},$$

where

$$K(t) = R_{(x)}(x - t)_+^{k-1}/(k - 1)!,$$

provided that $L \in \mathfrak{L}^{(k)}$. With this, L^* is defined as the unique element in $\mathfrak{M}$ which minimizes

$$\int_a^b (K(t))^2\, dt.$$

But this is just the demand that L^* minimize $\|L - L'\|$ over all $L' \in \mathfrak{M}$. It is now immediate that $\bar{L}$ and L^* agree. First, the assumption that $L \in \mathcal{L}^{(k)}$ is sufficient (though not necessary) to insure that $L_{(y)}K(x, y) \in F^{(k)}[a, b]$, so that $\bar{L}$ is defined. Also, by Corollary 2 of Lemma 2.3, $\bar{L} \in \mathfrak{M}$. Hence, since $\bar{L}$ minimizes $\|L - L'\|$ over all $L' = \sum \alpha_i L_i$, we have $\bar{L} = L^*$, and the third minimum property follows.

Finally, we mention some estimates for the error $\bar{R}f = Lf - \bar{L}f$. Since

$$Lf - \bar{L}f = (\phi - \bar{\phi}, f) = (\phi, (I - P_S)f),$$

use of Schwarz's inequality gives

$$(2.12)\quad |Lf - \bar{L}f| \leqq \|\phi - \bar{\phi}\|\,\|f\|, \quad\text{and}\quad |Lf - \bar{L}f| \leqq \|\phi\|\,\|f - P_S f\|.$$

But since, by (2.5), $((I - P_S)\phi, f) = ((I - P_S)\phi, (I - P_S)f)$, we have the better estimate

$$(2.13)\qquad |Lf - \bar{L}f| \leqq \|\phi - \bar{\phi}\|\,\|f - P_S f\|.$$

Hence if $\|f\| \leqq r$, which implies $\|f - P_S f\| \leqq (r^2 - \|P_S f\|^2)^{1/2}$, then

$$(2.14)\quad \bar{L}f - \|\phi - \bar{\phi}\|\,(r^2 - \|P_S f\|^2)^{1/2} \leqq Lf \leqq \bar{L}f + \|\phi - \bar{\phi}\|\,(r^2 - \|P_S f\|^2)^{1/2}.$$

The importance of the fact that this estimate depends only on the bound r and the numbers $L_i(f)$, $i = 1, \cdots, n$, and is optimal with respect to this information, is rightfully stressed in [5].

3. The Hilbert space $F^{(k)}[a, b]$. The linear space $F^{(k)}[a, b]$ can be made into a Hilbert space in various ways, thus providing various classes of functions which, due to the fact that they are representers of suitable linear functionals, have all the minimum properties of polynomial splines.

Specifically, let M be a k-th order ordinary linear differential operator in normal form,

$$(3.1)\qquad M = (d^k/dx^k) + \sum_{i=0}^{k-1} a_i(x)(d^i/dx^i),$$

and let $L_1, \cdots, L_k$ be k linear functionals. Under suitable conditions on the $a_i(x)$ and the L_i,

$$(3.2)\quad (e, f) = \sum_{i=1}^{k} L_i(e)L_i(f) + \int_a^b (Me)(y)(Mf)(y)\,dy, \quad \text{all} \quad e, f \in F^{(k)}[a, b],$$

is an inner product defined on $F^{(k)}[a, b]$, which makes $F^{(k)}[a, b]$ into a Hilbert space with reproducing kernel.

This is proved in the following theorem, which provides facts necessary to define and describe generalized splines and their minimum properties.

Theorem 3.1. Let M be any k-th order ordinary linear differential operator in normal form, (3.1), *where $k \geqq 1$ and $a_i \in C[a, b]$, $i = 0, \cdots, k - 1$. Let $\mathfrak{N}(M)$ denote the k-dimensional linear subspace of all functions f in $C^{(k)}[a, b]$ for which $Mf = 0$. Let $L_1, \cdots, L_k$ be any set of k linear functionals in $\mathcal{L}^{(k)}$, which are* *linearly independent over $\mathfrak{N}(M)$. Then $F^{(k)}[a, b]$ is a Hilbert space with respect to the inner product* (3.2), *and has a reproducing kernel. This reproducing kernel, K, is given by*

$$K(x, y) = \sum_{i=1}^{k} c_i(x)c_i(y) + \int_a^b G(x, t)G(y, t)\, dt, \qquad x, y \in [a, b], \tag{3.3}$$

where $c_1, \cdots, c_k$ is the dual basis to $L_1, \cdots, L_k$ in $\mathfrak{N}(M)$, and $G(x, y)$ is the Green's function for the differential equation $(Mf)(x) = e(x)$ with $L_i(f) = 0$, $i = 1, \cdots, k$.

Proof. The assumptions on the coefficients of the differential operator M are sufficient to insure [2, p. 117] that $\mathfrak{N}(M)$ is indeed of dimension k. Moreover, for each $x \in [a, b]$ and each $f \in F^{(k)}[a, b]$ there exists a unique element $s \in \mathfrak{N}(M)$ such that

$$s^{(j)}(x) = f^{(j)}(x), \qquad j = 0, \cdots, k-1.$$

Denote this element by $Q^{(x)}f$ and define a function h on $[a, b] \times [a, b]$ by

$$h(x, y) = Q^{(y)}{}_{(x)}(x - y)^{k-1}/(k-1)!. \tag{3.4}$$

Then

$$M_{(x)}h(x, y) = 0, \quad \text{all} \quad x, y \in [a, b], \tag{3.5}$$

$$\partial^j h(x, y)/\partial x^j|_{x=y} = \delta_{j,k-1}, \qquad j = 0, \cdots, k-1, \tag{3.5'}$$

and [2, p. 89] for all $f \in C^{(k)}[a, b]$,

$$f(x) = (Q^{(a)}f)(x) + \int_a^x h(x, y)(Mf)(y)\, dy, \quad \text{all} \quad x \in [a, b]. \tag{3.6}$$

Thus, the function

$$g(x, y) = \begin{cases} h(x, y), & x \geqq y, \\ 0, & x < y, \end{cases} \tag{3.7}$$

is just the Green's function for the initial value problem

$$Mf = e, \quad f^{(j)}(a) = 0, \qquad j = 0, \cdots, k-1.$$

For the proof of Theorem 3.1, we need to know that (3.6) holds not only for $f \in C^{(k)}[a, b]$, but also for all $f \in F^{(k)}[a, b]$.

Lemma 3.1. *Under the hypotheses of Theorem* 3.1, *the identity*

$$f(x) = (Q^{(a)}f)(x) + \int_a^b g(x, y)(Mf)(y)\, dy \tag{3.8}$$

is valid for all $f \in F^{(k)}[a, b]$ and all $x \in [a, b]$. More generally, for any linear functional $L \in \mathfrak{L}^{(k)}$,

$$Lf = L(Q^{(a)}f) + \int_a^b L_{(x)}g(x, y)(Mf)(y)\, dy, \quad \text{all} \quad f \in F^{(k)}[a, b]. \tag{3.9}$$

We were unable to find a reference for this lemma. We defer its proof to the Appendix and continue with the proof of Theorem 3.1.

By hypothesis, the linear functionals L_i, $i = 1, \cdots, k$, form a maximal linearly independent set over $\mathfrak{N}(M)$. There exist, therefore, k functions, $c_i \in \mathfrak{N}(M)$, $i = 1, \cdots, k$, such that

$$L_i c_j = \delta_{ij}, \qquad i, j = 1, \cdots, k.$$

With these, define a projection operator P from $F^{(k)}[a, b]$ onto $\mathfrak{N}(M)$ by

$$(3.10) \qquad (Pf)(x) = \sum_{i=1}^{k} (L_i f) c_i(x), \quad \text{all} \quad f \in F^{(k)}[a, b].$$

Then, in particular, $Pf = f$ for $f \in \mathfrak{N}(M)$, and, by Lemma 3.1,

$$(Pf)(x) = (P(Q^{(a)}f))(x) + P \int_a^b g(x, y)(Mf)(y)\, dy$$

$$= (Q^{(a)}f)(x) + \int_a^b P_{(x)} g(x, y)(Mf)(y)\, dy, \quad \text{all} \quad f \in F^{(k)}[a, b],$$

where, by Lemma 3.1, the interchange of integration and the operator P is justified by the assumption that the L_i are in $\mathfrak{L}^{(k)}$. Therefore, with the definition

$$(3.11) \qquad G(x, y) = (I - P)_{(x)} g(x, y), \qquad x, y \in [a, b],$$

it follows that

$$(3.12) \quad f(x) = (Pf)(x) + \int_a^b G(x, y)(Mf)(y)\, dy, \qquad x \in [a, b], \quad \text{all} \quad f \in F^{(k)}[a, b].$$

Hence, if for some $e \in L^2[a, b]$ and some $f_0 \in \mathfrak{N}(M)$,

$$(3.13) \qquad f(x) = f_0(x) + \int_a^b G(x, y) e(y)\, dy,$$

then

$$f \in F^{(k)}[a, b], \quad L_i f = L_i f_0, \quad i = 1, \cdots, k, \quad \text{and} \quad Mf = e \ \text{a.e.},$$

which also shows that G is the Green's function for $Mf = e$, $L_i f = 0$, $i = 1, \cdots, k$. With these facts established, all assertions of the theorem follow.

First, one checks that (3.2) indeed defines an inner product, *i.e.*, a symmetric positive definite bilinear form on $F^{(k)}[a, b]$.

Secondly, $F^{(k)}[a, b]$ is complete with respect to the norm $\|f\| = (f, f)^{1/2}$. For if $\{f_j\}_1^\infty$ is a Cauchy sequence in $F^{(k)}[a, b]$ with respect to this norm, then $\{L_i f_j\}_{j=1}^\infty$ is a Cauchy sequence of real numbers, so $\gamma_i = \lim_{j\to\infty} (L_i f_j)$ exists, $i = 1, \cdots, k$; furthermore, $\{Mf_j\}_{j=1}^\infty$ is a Cauchy sequence in $L^2[a, b]$ which, therefore, converges in L^2 to some $e \in L^2[a, b]$. But then,

$$f(x) = \sum_{i=1}^{k} \gamma_i c_i(x) + \int_a^b G(x, y) e(y)\, dy$$

is the limit point of the sequence $\{f_j\}_1^\infty$ in $F^{(k)}[a, b]$.

Finally, let K be the function on $[a, b] \times [a, b]$ defined by (3.3). Then, by the remarks following (3.12), $K(x, y)$, as a function of x, is in $F^{(k)}[a, b]$ for each $y \in [a, b]$, and

$$(3.14) \qquad \begin{aligned} M_{(y)}K(x, y) &= G(x, y), \qquad x, y \in [a, b], \\ L_{i(y)}K(x, y) &= c_i(x), \qquad i = 1, \cdots, k, x \in [a, b]. \end{aligned}$$

Therefore, using (3.12) once more, we get

$$\begin{aligned} f(x) &= \sum_{i=1}^{k} c_i(x)(L_i f) + \int_a^b G(x, y)(Mf)(y)\, dy \\ &= \sum_{i=1}^{k} L_{i(y)}K(x, y)L_i f + \int_a^b M_{(y)}K(x, y)(Mf)(y)\, dy \\ &= (K(x, y), f(y))_{(y)}, \quad x \in [a, b], \quad \text{all} \quad f \in F^{(k)}[a, b]. \end{aligned}$$

But this shows that K is the reproducing kernel for $F^{(k)}[a, b]$, which concludes the proof.

Remark. Lemma 3.1 has as a consequence many classical integral representations for the error of formulas for interpolation, quadrature, and numerical differentiation, *e.g.* those of Peano [11], [12], Radon [13], Remes [14], Milne [10], Golomb and Weinberger [5], Weinberger [23], and others. Explicitly, if $n + 1$ linear functionals $L_0, \cdots, L_n$ from $\mathfrak{L}^{(k)}$ are given, and if the error functional, $R = L_0 - \sum_{i=1}^{n} \alpha_i L_i$, vanishes on the null space $\mathfrak{N}(M)$ of the ordinary k-th order linear differential operator M, (3.1), then

$$(3.15) \qquad Rf = \int_a^b R_{(x)}g(x, y)(Mf)(y)\, dy, \quad \text{all} \quad f \in F^{(k)}[a, b],$$

where $g(x, y)$ is the Green's function associated with the initial value problem

$$(Mf)(x) = e(x), \quad f^{(j)}(a) = 0, \quad j = 0, \cdots, k - 1.$$

By Schwarz's inequality, this results in the error estimate

$$(3.16) \qquad |Rf|^2 \leqq \int_a^b [R_{(x)}g(x, y)]^2\, dy \int_a^b [(Mf)(y)]^2\, dy,$$

which corresponds to (2.12) and supplies the motivation for Sard's proposal to choose the approximation $\sum_{i=1}^{n} \alpha_i L_i$ to L_0 in such a way that $\int_a^b [R_{(x)}g(x, y)]^2\, dy$ be minimized.

4. Polynomial spline functions. As an illustration for the use of Theorem 3.1 and Lemmas 2.1–2.3 and their corollaries, we consider once more polynomial splines. Accordingly, we choose $M = (d^k/dx^k)$, so that $\mathfrak{N}(M) = \{\pi_{k-1}\}$. In this case, $g(x, y)$ in (3.7) is

$$(4.1) \qquad g(x, y) = (x - y)_+^{2k-1}/(2k - 1)!$$

so that the reproducing kernel (see (3.3) and (3.14)) is indeed given by (2.2), provided $L_i f = f(x_i)$, $i = 1, \cdots, k$. By the corollary to Theorem 2.1, the space S_0 is spanned by the representers of the linear functionals $L_i, \cdots, L_n$, given by $L_i f = f(x_i)$, $i = 1, \cdots, n$.

As noted above, because of the condition (1.1), the elements of S_0 restricted to the interval $[x_1, x_n]$ do not constitute all polynomial spline functions of degree $2k - 1$ with joints $x_2, \cdots, x_{n-1}$ in that interval. But the subspace S_1 spanned by the representers $\phi_1, \cdots, \phi_m$ of $L_1, \cdots, L_m$ does have this property in case the L_i are given by

$$L_i f = f^{(i-1)}(x_1), \qquad i = 1, \cdots, k,$$
$$L_{k-1+i} f = f(x_i), \qquad i = 2, \cdots, n-1,$$
$$L_{m+1-i} f = f^{(i-1)}(x_n), \qquad i = 1, \cdots, k,$$

where, as before, $a \leqq x_1 < \cdots < x_n \leqq b$, while only $n \geqq 2$. So $m = 2k + n - 2$.

We omit the straightforward verification of this fact and note only that with this choice for the L_i, Lemma 2.1 implies the remark following Lemma 2 of [4] as well as Theorem 2 of [4]; Lemma 2.2 implies Lemma 2′ of [4].

The first two minimum properties of polynomial splines were first pointed out explicitly* by Walsh, Ahlberg, and Nilson [21] in the special case $k = 2$ of cubic *periodic* splines. In the remainder of this section, we derive the three minimum properties for periodic polynomial splines of degree $2k - 1$. The statement of the first two will be somewhat more general than as given in [22].

The linear functionals,

$$L_1 f = f(a), \quad L_i f = f^{(i-2)}(b-) - f^{(i-2)}(a+), \qquad i = 2, \cdots, k,$$

are linearly independent over $\{\pi_{k-1}\}$, and are in $\mathfrak{L}^{(k)}$. Their dual basis $c_1, \cdots, c_n$ in $\{\pi_{n-1}\}$ can be computed recursively by

$$c_1(x) \equiv 1, \quad c_i(x) = d_i(x) - d_i(a), \qquad i = 2, \cdots, k,$$

where $d_1(x) \equiv 1/(b - a)$,

$$d_{i+1}(x) = \int_a^x d_i(y)\,dy - \int_a^b \int_a^x d_i(y)\,dy\,dx, \qquad i = 1, \cdots, k-1.$$

$F^{(k)}[a, b]$ is, therefore, a Hilbert space with respect to the inner product (3.2) (with $M = d^k/dx^k$), and has a reproducing kernel, $K(x, y)$, given by

$$K(x, y) = \sum_{i=1}^{k} c_i(x)c_i(y) + \gamma(x, y),$$

$$\begin{aligned}\gamma(x, y) = (-1)^k\Big\{&(x - y)_+^{2k-1}/(2k-1)! \\ &- \sum_{i=2}^{k} [(b - y)^{2k-i+1}c_i(x) - (x - a)^{2k-i+1}c_i(y)]/(2k - i + 1)! \\ &- \sum_{i=2}^{k}\sum_{j=2}^{k} (b - a)^{2k-i-j+3}c_i(x)c_j(y)/(2k - i - j + 3)!\Big\}.\end{aligned} \tag{4.2}$$

* All the minimum properties as listed in Section 1 are contained implicitly already in [5].

Let $L_{k+1}f = f^{(k-1)}(b-) - f^{(k-1)}(a+)$, and set $c_{k+1}(x) = L_{k+1(y)}K(x, y)$. The closed subspace

$$F_p^{(k)}[a, b] = \{f \varepsilon F^{(k)}[a, b] \mid L_i f = 0, \quad i = 2, \cdots, k+1\}$$

of $F^{(k)}[a, b]$ is also a Hilbert space with respect to (3.2), and one checks that $K_p(x, y)$, given by

$$K_p(x, y) = K(x, y) - \sum_{i=2}^{k+1} c_i(x)c_i(y)$$
$$= 1 - c_{k+1}(x)c_{k+1}(y) + \gamma(x, y), \tag{4.3}$$

is its reproducing kernel. On $F_p^{(k)}[a, b]$, (3.2) simplifies to

$$(e, f) = e(a)f(a) + \int_a^b e^{(k)}(y)f^{(k)}(y)\, dy, \quad \text{all} \quad e, f \varepsilon F_p^{(k)}[a, b]. \tag{4.4}$$

Let $a = x_1 < x_2 < \cdots < x_n = b$, $n \geqq 2$. Then, the linear functionals $N_1, \cdots, N_{n-1}$, given by $N_i f = f(x_i)$, $i = 1, \cdots, n-1$, are linearly independent and bounded over $F_p^{(k)}[a, b]$. Their representers, $K_p(x, x_i)$, $i = 1, \cdots, n-1$, span therefore an $(n-1)$-dimensional subspace, S_2, of $F_p^{(k)}[a, b]$. By (4.2) and (4.3), S_2 consists of piecewise polynomial functions in $C^{(2k-2)}[a, b]$ of degree $2n-1$ with joints at $x_1, \cdots, x_n$. Also, along with any other element of $F_p^{(k)}[a, b]$, any $s \varepsilon S_2$ satisfies

$$s^{(j)}(a+) = s^{(j)}(b-), \qquad j = 0, \cdots, k-1.$$

But, in fact, if $d \varepsilon [a, b]$ and $\phi_d(x) = K_p(x, d)$, then also

$$\phi_d^{(j)}(a+) = \phi_d^{(j)}(b-), \qquad j = k, \cdots, 2k-2, \tag{4.5}$$

which implies that S_2 is the set of periodic polynomial splines with period $(b-a)$ of degree $2k-1$ with joints $x_1, \cdots, x_{n-1}$ (considered as functions defined on $[a, b]$ only).

To verify (4.5), we observe that, in the notation of Theorem 3.1 (see (3.10), (3.11)),

$$P_{(y)}G(x, y) = P_{(y)}(I-P)_{(x)}g(x, y) = (I-P)_{(x)}P_{(y)}g(x, y)$$
$$= \sum_{i=1}^{k} (I-P)_{(x)}(L_{i(y)}g(x, y))c_i(y). \tag{4.6}$$

Since in this particular case, $g(x, y) = (x-y)_+^{k-1}/(k-1)!$, we have

$$L_{1(y)}g(x, y) \equiv 0, \quad L_{i(y)}g(x, y) = -(x-a)^{k-i+1}/(k-i+1)!, \quad i = 2, \cdots, k,$$

hence

$$L_{i(y)}g(x, y) \varepsilon \{\pi_{k-1}\}, \qquad i = 1, \cdots, k,$$

so that the coefficient of $c_i(y)$ in (4.6) vanishes identically for each $i = 1, \cdots k$. Hence $P_{(y)}G(x, y) \equiv 0$ on $[a, b] \times [a, b]$. But as $(\partial^k/\partial x^k)K(x, y) \equiv G(x, y)$, this implies that $\psi_d^{(j)}(a+) = \psi_d^{(j)}(b-)$, $j = k, \cdots, 2k-2$, where $\psi_d(x) = K(x, d)$. Hence, as $\phi_d^{(k)}(x) \equiv \psi_d^{(k)}(x) +$ constant, while

$$\phi_d^{(j)}(x) \equiv \psi_d^{(j)}(x), \qquad j = k+1, \cdots, 2k-2,$$

equation (4.5) follows.

We proved

Theorem 4.1. *Let S_2 be the set of polynomial spline functions of degree $2k - 1$ with joints $x_1, \cdots, x_{n-1}$, where $a = x_1 < \cdots < x_n = b$, which are periodic with period $(b - a)$. For $i = 1, \cdots, n-1$, let $\phi_i(x)$ be the representer in $F_p^{(k)}[a, b]$ with respect to (4.4) of the linear functional N_i, given by $N_i f = f(x_i)$. Then S_2, considered as a subset of $F_p^{(k)}[a, b]$, is spanned by $\phi_1, \cdots, \phi_{n-1}$.*

Corollary. *For all $f \in F_p^{(k)}[a, b]$,* (i) *there exists a unique element $s_f \in S_2$ satisfying $s_f(x_i) = f(x_i)$, $i = 1, \cdots, n-1$;* (ii) *for all $s \in S_2$,*

$$\int_a^b [f^{(k)}(x) - s^{(k)}(x)]^2\, dx \geqq \int_a^b [f^{(k)}(x) - s_f^{(k)}(x)]^2\, dx$$

with equality if and only if $s(x) - s_f(x) \equiv$ constant; (iii)

$$\int_a^b [f^{(k)}(x)]^2\, dx \geqq \int_a^b [s_f^{(k)}(x)]^2\, dx$$

with equality if and only if $f = s_f$; (iv) *if L is any linear functional in $\mathfrak{L}^{(k)}$, and $\bar{L}$ is that element of the set*

$$\left\{N = \sum_{i=1}^{n-1} \alpha_i N_i \mid (L - N)(1) = 0, \quad \alpha_i \text{ real}, \quad i = 1, \cdots, n-1\right\}$$

which minimizes

$$\int_a^b [(L - N)_{(y)}(x - y)_+^{k-1}/(k-1)!]^2\, dx,$$

then $\bar{L}f = Ls_f$.

5. General spline functions. We have proved Theorem 3.1 in all its generality since, practical considerations aside, there seems to be nothing inherently special about the use of the seminorm

$$\int_a^b [f^{(k)}(x)]^2\, dx \tag{5.1}$$

as compared with the seminorm

$$\int_a^b [Mf(x)]^2\, dx, \tag{5.2}$$

where M is the k-th order ordinary differential operator (3.1). Moreover, there is no mathematical reason to single out the point functionals

(5.3) $$L_i(f) = f(x_i), \qquad i = 1, \cdots, n,$$

from the more general set of bounded linear functionals over the Hilbert space $F^{(k)}[a, b]$. It would, therefore, seem acceptable to define a general set of spline functions S on $[a, b]$ belonging to the differential operator M and the set $L_1, \cdots, L_n$ of linearly independent bounded linear functionals over $F^{(k)}[a, b]$ as the n-dimensional subspace S spanned by functions $\phi_1, \cdots, \phi_n$ which are the representers of the linear functionals $L_1, \cdots, L_n$ with respect to the inner product (3.2). (M. Atteia in a note to appear in Comptes Rendus Acad. Sci. Paris has gone even farther and given a definition of "spline function" entirely in the setting of abstract Hilbert space.) According to Section 2, these splines retain all the (appropriately worded) minimum porperties of polynomial splines.

But in order not to dilute the notion of spline function too much, we prefer to follow Greville's definition of general spline function [6].

Definition. Let S be an m-dimensional subspace of $C^{(m)}[a, b]$, where $[a, b]$ is a finite interval and let $a \leqq x_1 < \cdots < x_n \leqq b$. A real valued function f on $[a, b]$ is a *spline function* with respect to S of order m with joints $x_1, \cdots, x_n$, provided that $f \in C^{(m-2)}[a, b]$ and f coincides in each interval $(a, x_1), (x_1, x_2), \cdots, (x_n, b)$ with some element in S.

If the coefficients a_i of the differential operator M in (3.1) are such that $a_i \in C^{(i)}[a, b]$, $i = 0, \cdots, k - 1$, (*i.e.*, if the coefficients are smoother than required in Theorem 3.1), then M possesses an adjoint differential operator M^* given by

(5.4) $$M^* = (-1)^k(d^k/dx^k) + \sum_{i=0}^{k-1} (-1)^i(d^i a_i(x)/dx^i),$$

which, after carrying out the differentiation and rearranging the terms, can be written as

(5.4′) $$M^* = (-1)^k\left[(d^k/dx^k) + \sum_{i=0}^{k-1} b_i(x)(d^i/dx^i)\right],$$

for appropriate $b_i \in C^{(i)}[a, b]$. As we now show, in this case the representers of the point functionals (5.3) in the Hilbert space $F^{(k)}[a, b]$ with inner product (3.2) are spline functions of order $2k$ with respect to the null space $\mathfrak{N}(M^*M)$ of M^*M.

Let $u_1, \cdots, u_k$ be a basis in $\mathfrak{N}(M)$. The function $h(x, y)$ which satisfies (3.5)–(3.5′) can be written as

$$h(x, y) = \sum_{i=1}^{k} v_i(y)u_i(x),$$

where the coefficients v_i are the solution of the system

$$\sum_{i=1}^{k} v_i(y)\, d^j u/dx^j = \delta_{j,k-1} \quad \text{for} \quad x = y \quad \text{and} \quad j = 0, \cdots, k - 1.$$

Consequently [8, p. 78], the functions v_i form a basis for $\mathfrak{N}(M^*)$. Hence as a function of y, $h(x, y) \varepsilon \mathfrak{N}(M^*)$. It then follows that as a function of y, $G(x, y)$ in (3.11) satisfies $M^*_{(y)}G(x, y) = 0$, $y \neq x$, But then, because of (3.14),

$$M^*_{(y)}M_{(y)}K(x_i, y) = 0, \qquad y \neq x_i, \quad j = 1, \cdots, n,$$

which shows that in the intervals (a, x_1), (x_1, x_2), $\cdots$, (x_n, b),

$$K(x_i, y) \varepsilon \mathfrak{N}(M^*M), \qquad j = 1, \cdots, n.$$

Since as a function of y, $G(x, y) \varepsilon C^{k-2}$, it follows from (3.14) that as a function of y, $K(x, y) \varepsilon C^{2k-2}$. Finally, since $K(x, y) \equiv K(y, x)$, the representers ϕ_i of the point functionals (5.3) are, $\phi_i(x) = K(x_i, x)$, which completes the demonstration.

[*Editor's Note:* It is necessary to assume that the coefficients are smooth enough to make the proof work.]

Theorem 5.1. *Let M be any k-th order ordinary linear differential operator in normal form, (3.1), where $k \geqq 1$, $a_i \varepsilon C^i[a, b]$, $i = 0, \cdots, k$, and $[a, b]$ is a finite interval. Let $a \leqq x_1 < x_2 < \cdots < x_n \leqq b$, with $n \geqq k$ and assume that the first k of the n linear functionals $L_1, \cdots, L_n$ given by (5.3) are linearly independent over $\mathfrak{N}(M)$. Then the space S spanned by the representers ϕ_i of the L_i with respect to the inner product (3.2) consists of all spline functions with respect to $\mathfrak{N}(M^*M)$ of order $2k$ with joints $x_1, \cdots, x_n$. For all $f \varepsilon F^{(k)}[a, b]$, there exists a unique element $s \varepsilon S$ denoted by Pf which interpolates f with respect to $L_1, \cdots, L_n$. For all $f \varepsilon F^{(k)}[a, b]$ and all $h \varepsilon S$,*

$$\int_a^b [(Mf)(y) - (M(Pf))(y)]^2\, dy \leqq \int_a^b [(Mf)(y) - (Mh)(y)]^2\, dy$$

with equality if and only if $h - Pf \varepsilon \mathfrak{N}(M)$. For all $f \varepsilon F^{(k)}[a, b]$ and all $h \varepsilon F^{(k)}[a, b]$ interpolating f with respect to $L_1, \cdots, L_n$,

$$\int_a^b [M(Pf)(y)]^2\, dy \leqq \int_a^b [(Mh)(y)]^2\, dy$$

with equality if and only if $h = Pf$. For all $L \varepsilon \mathfrak{L}^{(k)}$, Sard's best formula $L^* = \sum_{i=1}^n \alpha_i L_i$ for L (with respect to M) satisfies $L^*f = LPf$, for all $f \varepsilon F^{(k)}[a, b]$.

APPENDIX

Proof of Lemma 3.1. The assumptions on the differential operator M are sufficient [8, pp. 72–75] to insure that the function h defined in (3.4) has all partial derivatives with respect to x up to and including the k-th, and that these partial derivatives are continuous on $[a, b] \times [a, b]$ in x and y jointly. Therefore, there exists a constant N such that

$$|\partial^j h(x, y)/\partial x^j| \leqq N, \quad \text{all} \quad x, y \varepsilon [a, b], \quad j = 0, \cdots, k.$$

Furthermore, if $e \varepsilon C[a, b]$, then, using Leibniz's rule [9] and (3.5′), one computes

$$\frac{d^j}{dx^j} \int_a^x h(x, y)e(y)\, dy = \delta_{k,j}e(x) + \int_a^x \frac{\partial^j}{\partial x^j} h(x, y)e(y)\, dy, \qquad j = 1, \cdots, k.$$

Let $e \in L^2[a, b]$, then there exists a sequence $\{e_i\}_1^\infty$, $e_i \in C[a, b]$, such that as $i \to \infty$,

$$||e - e_i||_2 = \left[\int_a^b |e(y) - e_i(y)|^2\, dy\right]^{1/2} \to 0.$$

Set

$$f_i(x) = \int_a^x h(x, y)e_i(y)\, dy, \qquad i = 1, 2, \cdots,$$

and

$$s_j(x) = \delta_{kj}e(x) + \int_a^x \frac{\partial^j}{\partial x^j} h(x, y)e(y)\, dy, \qquad j = 0, \cdots k.$$

Note that s_k is defined only almost everywhere. We have for all $x \in [a, b]$, $j = 0, \cdots, k$, and $i = 1, 2, \cdots,$

$$\begin{aligned}\left|\int_a^x \frac{\partial^j}{\partial x^j} h(x, y)e(y)\, dy - \int_a^x \frac{\partial^j}{\partial x^j} h(x, y)e_i(y)\, dy\right| &\leqq \int_a^x \left|\frac{\partial^j}{\partial x^j} h(x, y)\right| |e(y) - e_i(y)|\, dy \\ &\leqq \left[\int_a^x \left|\frac{\partial^j}{\partial x^j} h(x, y)\right|^2 dy\right]^{1/2} ||e - e_i||_2 \\ &\leqq (b - a)^{1/2}N\, ||e - e_i||_2 .\end{aligned}$$

Therefore, $f_i^{(j)} \to s_j$ uniformly on $[a, b]$, for $j = 0, \cdots, k - 1$, and $f_i^{(k)} \to s_k$ in the L^2 norm. Hence setting $f = s_0$, we have $f^{(j)} = s_j$, $j = 0, \cdots, k$, where for $j = k$, the equality again holds only almost everywhere. In particular, $f^{(j)}(a) = 0$, $j = 0, \cdots, k - 1$, and

$$\begin{aligned}(Mf)(x) &= \sum_{i=0}^{k-1} a_i(x)f^{(i)}(x) + f^{(k)}(x) \\ &= \int_a^x M_{(x)}h(x, y)e(y)\, dy + e(x) = e(x).\end{aligned}$$

Therefore, if $f \in F^{(k)}[a, b]$, then

$$f(x) - (Q^{(a)}f)(x) = \int_a^x h(x, y)M(f - Q^{(a)}f)(y)\, dy = \int_a^x h(x, y)(Mf)(y)\, dy,$$

which proves the first part of Lemma 3.1.

Let L be the linear functional on $F^{(k)}[a, b]$ defined by

$$Lf = \int_a^b f^{(j)}(x)\, d\mu(x),$$

where μ is a function of bounded variation and $0 \leqq j \leqq k - 1$. If

$$f(x) = \int_a^x h(x, y)e(y)\, dy,$$

where $e \in L^2[a, b]$, then by the above and by Fubini's Theorem,

$$Lf = \int_a^b \int_a^x \frac{\partial^i}{\partial x^i} h(x, y)e(y)\, dy\, d\mu(x)$$

$$= \int_a^x \int_a^b \frac{\partial^i}{\partial x^i} h(x, y)\, d\mu(x)e(y)\, dy$$

$$= \int_a^x L_{(x)}h(x, y)e(y)\, dy,$$

which implies the second part of Lemma 3.1.

References

[1] Aronszajn, N., Theory of reproducing kernels, *Trans. Amer. Math. Soc.*, **68** (1950) 337–404.

[2] Birkhoff, G., & Rota, G-C., *Ordinary Differential Equations*, Ginn and Company, Boston, 1962.

[3] Davis, P. J., *Interpolation and Approximation*, Blaisdell Publishing Co., 1963.

[4] de Boor, C., Best approximation properties of spline functions of odd degree, *J. Math. Mech.*, **12** (1963) 747–750.

[5] Golomb, M., & Weinberger, H. F., Optimal approximation and error bounds, *Proc. Symp. on Numerical Approximation*, R. E. Langer (ed.), 117–190, Univ. Wisc. Press, 1959.

[6] Greville, T. N. E., Interpolation by generalized spline functions, *MRC Tech. Summary Report* No. 476 (1964).

[7] Halmos, P. R., *Introduction to Hilbert Space, 2nd ed.*, Chelsea Publication Co., New York, 1957.

[8] Kamke, E., *Differentialgleichungen, Lösungsmethoden und Lösungen I*, 5. Auflage, Akademische Verlagsgesellschaft Geest und Partig K.-G., Leipzig, 1956.

[9] Kaplan, W., *Advanced Calculus*, Addison-Wesley, Reading, Mass., 1952.

[10] Milne, W. E., The remainder in linear methods of approximation, *J. Research, U. S. Nat. Bur. Stand.*, **43** (1949) 501–511.

[11] Peano, G., Resto nelle formule di quadratura espresso con un integrale definito, *Atti Della Reale Accademia Dei Lincei, Rendiconti* (**5**), **22** (1913) 562–569.

[12] Peano, G., Residuo in formulas de quadratura, *Mathesis* (**4**), **34** (1914) 5–10.

[13] Radon, J., Restausdrücke bei Interpolations-und Quadraturformeln durch bestimmte Integrale, *Monatsh. Math. Phys.*, **42** (1935) 389–396.

[14] Remes, E. J., Sur les termes complementaires de certaines formules d'analyse approximative, *Dokl. Akad. Nauk SSSR*, **26** (1940) 129–133.

[15] Sard, A., *Linear Approximation*, Mathematical Surveys, No. 9, Amer. Math. Soc., 1963.

[16] Schoenberg, I. J., On interpolation by spline functions and its minimal properties, *Proc. Conf. on Approximation at Oberwolfach*, Birkhäuser, Basel, 1964.

[17] Schoenberg, I. J. Spline interpolation and best quadrature formulae, *Bull. Amer. Math. Soc.*, **70** (1964) 143–148.

[18] Schoenberg, I. J., On best approximations of linear operators, *Proc. Koninkl. Nederl. Akad. Wetensch. Series A*, **67**, No. 2 and *Indag. Math.*, **26** (1964), No. 2, 155–163.

[19] SCHOENBERG, I. J., Spline interpolation and the higher derivatives, *Proc. Nat. Acad. Sci.*, 51 (1964) 24–28.
[20] SCHOENBERG, I. J., On trigonometric spline interpolation, *J. Math. Mech.*, 13 (1964) 795–825.
[21] WALSH, J. L., AHLBERG, J. H., & NILSON, E. N., Best approximation properties of the spline fit, *J. Math. Mech.*, 11 (1962) 225–234.
[22] WALSH, J. L., AHLBERG, J. H., & NILSON, E. N., Best approximation and convergence properties of higher-order spline approximations, *J. Math. Mech.*, 14 (1965) 231–243.
[23] WEINBERGER, H. F., Optimal approximation for functions prescribed at equally spaced points, *J. Research, U. S. Nat. Bur. Stand.*, 65B (1961) 99–104.

2637 Whitewood
Ann Arbor, Michigan
and
Computation Center
University of Texas
Austin, Texas

ERRATA

Page 955, line 11 from the bottom should read: "kernel $K(x, y)$ is bounded only if the function. . ."

Page 961, equation (4.1) should read: "$g(x,y) = (x - y)_+^{(k-1)}/(k-1)!$"

Page 962, line 7 from the bottom should read: "$d_{i+1}(x) = \int_a^x d_i(y)\,dy - [1/(b-a)] \int_a^b \int_a^x d_i(y)\,dy\,dx,\ i = 1, \ldots, k-1.$"

Page 965, line 2 from the bottom should read: "$\sum_{i=1}^{k} v_i(y)\, d^j u_i/dx^j = \delta_{j,\,k-1}. \ldots$"

Page 968, line 4 should read: "$= \int_a^b \left[\int_y^b \frac{\partial^j}{\partial x^j} h(x,y)\, d\mu(x)\right] e(y)\, dy$"

Page 968, line 5 should read: "$= \int_a^b L_{(x)}\, g(x,y)e(y)\, dy$"

23

Reprinted from *J. Math. Anal. Appl.* **33**:82–95 (1971)

Some Results on Tchebycheffian Spline Functions*

George Kimeldorf

Department of Statistics, Florida State University, Tallahassee, Florida 32306

AND

Grace Wahba

Department of Statistics, University of Wisconsin, Madison, Wisconsin 53706

Submitted by L. A. Zadeh

This report derives explicit solutions to problems involving Tchebycheffian spline functions. We use a reproducing kernel Hilbert space which depends on the smoothness criterion, but not on the form of the data, to solve explicitly Hermite-Birkhoff interpolation and smoothing problems. Sard's best approximation to linear functionals and smoothing with respect to linear inequality constraints are also discussed. Some of the results are used to show that spline interpolation and smoothing is equivalent to prediction and filtering on realizations of certain stochastic processes.

1. Introduction

Suppose we are given a closed interval $I = [a, b]$, a set $\{y_i\}$ of n constants, and a set $\{t_i\}$ of n distinct constants in (a, b). Consider the class of functions

$$\mathscr{H} = \{u : D^{m-1}u \text{ is absolutely continuous and } Lu \in \mathscr{L}_2(I)\} \qquad (1.1)$$

where L is an m-th order linear differential operator. Of all functions $u \in \mathscr{H}$ satisfying

$$u(t_i) = y_i \qquad i = 1, 2,..., n, \qquad (1.2)$$

we seek one, say $\hat{u}$, which minimizes $\int_a^b (Lu)^2$. It is now a classical result that when $L = D^m$ and $n \geqslant m$, then a solution $\hat{u}$ exists and is unique. The function $\hat{u}$ is called the $(2m - 1)$-th degree natural polynomial spline of interpolation to the data $\{(t_i, y_i)\}$.

* Sponsored by the Mathematics Research Center, United States Army, Madison, Wisconsin, under Contract No.: DA-31-124-ARO-D-462.

Recently (see, for example [3, 7]) spline problems have been placed in the context of an abstract Hilbert space. In this paper, the class $\mathscr{H}$ is made into a Hilbert space in which the linear functionals F_i defined by

$$F_i u = u(t_i) \tag{1.3}$$

are continuous, and hence have representers ψ_i so that (1.2) can be written

$$\langle \psi_i , u \rangle = y_i \qquad i = 1, 2,..., n. \tag{1.4}$$

Moreover, there exists a subspace $\mathscr{H}_1$ of $\mathscr{H}$ such that $\int (Lu)^2 = \| P_1 u \|^2$ for all $u \in \mathscr{H}$ where P_1 is the projection of $\mathscr{H}$ on $\mathscr{H}_1$.

There are several important advantages of placing spline problems in an abstract context. One is that existence and uniqueness proofs become consequences of straightforward results in the geometry of Hilbert space. A second advantage is the facility with which the results can be extended to more general operators than D^m and to more general linear functionals. An important disadvantage of an abstract approach is the nonconstructive nature of the proofs, for in general there is no way to construct the representer of a given linear functional.

De Boor and Lynch [4] suggested the use of reproducing kernel Hilbert spaces in problems involving splines. If the linear functional $N : f \to f(t)$ is continuous for each $t \in I$, then it is well known that there exists a reproducing kernel which, if known, allows one to construct the representer of any bounded linear functional. In this paper the reproducing kernel structure is combined with known results on total positivity to provide a unified approach to a variety of problems involving Tchebycheffian spline functions.

In Section 2, we exhibit the explicit reproducing kernel structure for $\mathscr{H}$. Section 3 uses the results of Section 2 to provide an explicit solution to the following generalized Hermite-Birkhoff problem: Given a set $\{N_i\}$ of continuous linear functionals on $\mathscr{H}$ and a set $\{y_i\}$ of scalars, find a function $\hat{u} \in \mathscr{H}$ which minimizes $\int_a^b (Lu)^2$ subject to the constraints $N_i u = y_i$. Section 4 discusses approximations to linear functionals.

Sections 5 and 6 consider variations of the interpolation problem. In Section 5 we treat the following smoothing problem: Rather than constrain the function according to $N_i u = y_i$ we insist only that $N_i u$ be "near" y_i. More precisely, we seek a function $\hat{u} \in \mathscr{H}$ which minimizes

$$\int_a^b (Lu)^2 + \sum (N_i u - y_i)\, s_{ij}(N_j u - y_j)$$

where $S = [s_{ij}]$ is a positive definite matrix.

Section 6 extends a problem recently solved by Ritter [11] of replacing the equations $N_i u = y_i$ by inequalities of the form

$$y_i \leqslant N_i u \leqslant z_i \tag{1.5}$$

where the y_i and z_i are prescribed. We seek a function $\hat{u} \in \mathscr{H}$ which minimizes $\int_a^b (Lu)^2$ subject to (1.5). The reproducing kernel structure is used to show that this problem is reducible to a standard quadratic programming problem.

In Section 7 we describe a stochastic process for which spline interpolation and smoothing are equivalent to minimum variance unbiased linear prediction and smoothing.

2. The Reproducing Kernel Structure

Without loss of generality we restrict our attention to the interval [0, 1]. Let $\{a_i\}$ be a set of m strictly positive functions such that $a_i \in C^i$ and without loss of generality we take $a_i(0) = 1$. Define the m-th order differential operator L by

$$L = D \frac{1}{a_1} D \frac{1}{a_2} \cdots D \frac{1}{a_m} \tag{2.1}$$

where D is the differentiation operator. Define operators M_i by

$$M_0 = I, \qquad M_i = D \frac{1}{a_{m+1-i}} M_{i-1}, \qquad i = 1, 2,..., m-1 \tag{2.2}$$

where I is the identity operator. For $i = 1, 2,..., m$, let the function ω_i be defined by

$$\begin{aligned} \omega_1 &= a_m \\ \omega_2(t) &= a_m(t) \int_0^t a_m \quad t_{m-1})\, dt_{m-1} \\ &\vdots \\ \omega_m(t) &= a_m(t) \int_0^t a_{m-1}(t_{m-1}) \int_0^{t_{m-1}} a_{m-2}(t_{m-2}) \cdots \int_0^{t_2} a_1(t_1)\, dt_1 \cdots dt_{m-1} \end{aligned} \tag{2.3}$$

so that we have

$$(M_i \omega_{j+1})(0) = \delta_{i,j}, \qquad i, j = 0, 1,..., m-1 \tag{2.4}$$

where δ is the Kronecker delta. It is well known ([9] p. 379) that the set $\{\omega_i\}$ is an extended Tchebycheff system.

Let

$$\mathscr{H}_0 = \{u : D^{m-1}u \text{ is absolutely continuous and } Lu \equiv 0\}.$$

Hence $\mathscr{H}_0$ is an m-dimensional vector space spanned by the ω_i. From (2.2) and (2.4) we have that $\mathscr{H}_0$ is a Hilbert space with inner product

$$\langle u, v\rangle_0 = \sum_{i=0}^{m-1} [(M_i u)(0)]\,[(M_i v)(0)] \tag{2.5}$$

and orthonormal basis $\{\omega_i\}$. Let

$$\mathscr{H}_1 = \{u : D^{m-1}u \text{ is absolutely continuous, } Lu \in \mathscr{L}_2 \text{ and } (M_i u)(0) = 0,\ i = 0, 1,\ldots, m-1\}.$$

Clearly $\mathscr{H}_1$ is a Hilbert space with inner product

$$\langle u, v\rangle_1 = \int_0^1 (Lu)(Lv). \tag{2.6}$$

Let

$$\mathscr{H} = \{u : D^{m-1}u \text{ is absolutely continuous and } Lu \in \mathscr{L}_2\}$$

so that $\mathscr{H} = \mathscr{H}_0 \oplus \mathscr{H}_1$ is a Hilbert space with inner product

$$\langle u, v\rangle = \langle u, v\rangle_0 + \langle u, v\rangle_1 . \tag{2.7}$$

We now proceed to construct a reproducing kernel for $\mathscr{H}$; that is, we construct a function K defined on $[0, 1] \otimes [0, 1]$ such that for every fixed $s_0 \in [0, 1]$ we have: (1) $K(s_0, \cdot) \in \mathscr{H}$, and (2) $u \in \mathscr{H}$ if, and only if,

$$u(s_0) = \langle K(s_0, \cdot), u\rangle.$$

For an account of reproducing kernel Hilbert spaces, the reader is referred to Aronszajn [2].

Let K_1 be the symmetric positive definite function defined by

$$K_1(s, t) = a_m(s)\, a_m(t) \int_0^s \int_0^t \int_0^{s_{m-1}} \int_0^{t_{m-1}} \cdots \int_0^{s_2} \int_0^{t_2} \left[\prod_{j=1}^{m-1} a_j(s_j)\, a_j(t_j)\right] \min(s_1, t_1)\, ds_1\, dt_1 \cdots ds_{m-1}\, dt_{m-1} \tag{2.8}$$

and $G(s_0, \cdot) = LK_1(s_0, \cdot)$. Clearly, for fixed s_0, $K_1(s_0, \cdot) \in \mathscr{H}_1$. Moreover, it is known ([8] chap. 10) that G is the Green's function for the differential equation $Lu = b \in \mathscr{L}_2$ with boundary conditions $(M_i u)(0) = 0$, $i = 0, 1,..., m-1$. Hence $u \in \mathscr{H}_1$ if, and only if, for any fixed $s_0 \in [0, 1]$

$$\begin{aligned} u(s_0) &= \int_0^1 G(s_0, t)\,[(Lu)(t)]\,dt \\ &= \int_0^1 [LK(s_0, t)]\,[Lu(t)]\,dt \\ &= \langle K_1(s_0, \cdot), u\rangle. \end{aligned}$$

Therefore, K_1 is the reproducing kernel for $\mathscr{H}_1$ and

$$K_1(s, t) = \int^1 G(s, r)\, G(t, r)\, dr.$$

Letting

$$K_0(s, t) = \sum_{i=1}^{m} \omega_i(s)\, \omega_i(t), \tag{2.9}$$

we verify readily that K_0 is the reproducing kernel for $\mathscr{H}_0$, and hence the function

$$K = K_0 + K_1 \tag{2.10}$$

is the reproducing kernel for $\mathscr{H}$.

The proof of the following lemma is an elementary consequence of the reproducing kernel structure.

LEMMA 2.1. *Let $\mathscr{H}$ be a Hilbert space with reproducing kernel Q. If N is a linear functional on $\mathscr{H}$ and ψ the function defined by $\psi(s_0) = NQ(s_0, \cdot)$, then N is continuous if, and only if, $\psi \in \mathscr{H}$, in which case $Nu = \langle \psi, u\rangle$ for all $u \in \mathscr{H}$ (i.e. ψ is the representer of N).*

The following lemma follows from the preceding.

LEMMA 2.2. *Let P_0 and P_1 be the projection operators in $\mathscr{H}$ onto $\mathscr{H}_0$ and $\mathscr{H}_1$ respectively. Then if N is a continuous linear functional with representer ψ, we have*

$$(P_0\psi)(s_0) = \sum_{i=1}^{m} \omega_i(s_0)\,(N\omega_i),$$

and

$$(P_1\psi)(s_0) = NK_1(s_0, \cdot).$$

We remark that the linear functional N defined by $Nu = u^{(j)}(t_0)$ for $0 < t_0 < 1$ and $j = 0, 1,..., m - 1$ is continuous on $\mathscr{H}$. This fact, whose proof follows from Lemma 2.1, will be used in the examples in the sequel.

3. The Generalized Hermite-Birkhoff Problem

We are now in a position to solve a generalized Hermite-Birkhoff problem. In order to apply the results of Section 2 we need the following lemma.

LEMMA 3.1. *Let $\mathscr{H} = \mathscr{H}_0 \oplus \mathscr{H}_1$ be the direct sum of an m-dimensional Hilbert space $\mathscr{H}_0$ with basis $\{\omega_1, \omega_2, ..., \omega_m\}$ and any Hilbert space $\mathscr{H}_1$. Let P_0 and P_1 be the projections onto $\mathscr{H}_0$ and $\mathscr{H}_1$ respectively and let $\psi_1, \psi_2, ..., \psi_n$ be $n \geq m$ elements of $\mathscr{H}$ such that*

(i) *The set $\{P_0\psi_i : i = 1, 2,..., n\}$ spans $\mathscr{H}_0$, and*

(ii) *The set $\{P_1\psi_i : i = 1, 2,..., n\}$ is linearly independent.*

If $\mathbf{y}' = (y_1, y_2, ..., y_n)$ is any n-tuple of scalars, then the unique element $\hat{u} \in \mathscr{H}$ which minimizes $\langle P_1u, P_1u\rangle$ subject to the constraints

$$\langle \psi_i, u\rangle = y_i, \qquad i = 1, 2,..., n \tag{3.1}$$

is

$$\hat{u} = \boldsymbol{\omega}'(T\Sigma^{-1}T')^{-1} T\Sigma^{-1}\mathbf{y} + \boldsymbol{\xi}'\Sigma^{-1}[I - T'(T\Sigma^{-1}T')^{-1} T\Sigma^{-1}]\,\mathbf{y} \tag{3.2}$$

where T is the $m \times n$ matrix $[\langle\omega_i, \psi_j\rangle]$, Σ is the $n \times n$ matrix $[\langle P_1\psi_i, P_1\psi_j\rangle]$, $\boldsymbol{\omega} = (\omega_1, \omega_2, ..., \omega_m)'$, and $\boldsymbol{\xi} = (P_1\psi_1, P_1\psi_2, ..., P_1\psi_n)'$.

Proof. By assumption, T and Σ are of full rank. We can write $\hat{u} = \boldsymbol{\omega}'\boldsymbol{\alpha} + \boldsymbol{\xi}'\boldsymbol{\beta} + x$ where $\boldsymbol{\alpha}'$ and $\boldsymbol{\beta}'$ are some (as yet undetermined) m-tuple and n-tuple respectively of scalars, $x \in \mathscr{H}_1$, and $\langle x, \xi_i\rangle = 0$. Clearly we must take $x = 0$. The set of constraints $\langle\psi_1, \hat{u}\rangle = y_i$ is equivalent to $T'\boldsymbol{\alpha} + \Sigma\boldsymbol{\beta} = \mathbf{y}$ or

$$\boldsymbol{\beta} = \Sigma^{-1}(\mathbf{y} - T'\boldsymbol{\alpha}). \tag{3.3}$$

We have

$$\langle P_1\hat{u}, P_1\hat{u}\rangle = \boldsymbol{\beta}'\Sigma\boldsymbol{\beta}. \tag{3.4}$$

If (3.3) is substituted into (3.4), it is trivial to verify that (3.4) is minimized if and only if

$$\boldsymbol{\alpha} = (T\Sigma^{-1}T')^{-1} T\Sigma^{-1}\mathbf{y}$$

whence

$$\boldsymbol{\beta} = \Sigma^{-1}[I - T'(T\Sigma^{-1}T')^{-1} T\Sigma^{-1}]\,\mathbf{y}.$$

The explicit solution to the generalized Hermite-Birkhoff problem is stated in the following theorem.

THEOREM 3.1. *Let L be an m-th order differential operator of the form* (2.1), *let $\mathscr{H}$ be the Hilbert space defined in Section 2, let $\{\omega_i : i = 1, 2,..., m\}$ be as defined by* (2.3), *and let K_1 be defined by* (2.8). *Suppose $\{N_i : i = 1, 2,..., n\}$ is a set of $n \geqslant m$ continuous linear functionals on $\mathscr{H}$ such that*

(i) *The rank of the $m \times n$ matrix $T = [N_j\omega_i]$ is m, and*

(ii) *The n functions ξ_i defined for fixed s_0 by $\xi_i(s_0) = N_iK_1(s_0, \cdot)$ are linearly independent.*

If $\mathbf{y} = (y_1, y_2,..., y_n)'$ is any vector of scalars, then the unique function $\hat{u} \in \mathscr{H}$ which minimizes $\int_0^1 (Lu)^2$ subject to the constraints

$$N_iu = y_i, \qquad i = 1, 2,..., n \tag{3.5}$$

is given by (3.2) *where $\boldsymbol{\omega} = (\omega_1, \omega_2,..., \omega_m)'$, $\boldsymbol{\xi} = (\xi_1, \xi_2,..., \xi_n)'$ and Σ is the $n \times n$ matrix $[N_i\xi_j] = [\langle\xi_i, \xi_j\rangle]$.*

The proof of Theorem 3.1 follows directly from Lemma 3.1 by letting ψ_i be the representer of N_i and using the results of Section 2.

EXAMPLE 3.1. Let N_i be $N_i : f \to f(t_i)$ where the $t_i \in (0, 1)$ are distinct. Condition (i) of the theorem is satisfied because the set $\{\omega_j\}$ forms a Tchebycheff system and condition (ii) is satisfied because K_1 is positive definite on $(0, 1) \otimes (0, 1)$. The unique function $\hat{u} \in \mathscr{H}$ which minimizes $\int_0^1 (Lu)^2$ subject to the constraints $u(t_i) = y_i$ is called the natural L-spline of interpolation to the points $\{(t_i, y_i)\}$.

EXAMPLE 3.2. Let N_i be defined by $N_iu = u^{(m_i)}(t_i)$ where $m_i < m$ and $0 < t_i < 1$. It can be shown that K_1 is a Green's function of the type considered by Karlin ([8] chap. 10, Secs. 7, 8), and hence that condition (ii) of the theorem is satisfied. In general, however, condition (i) will fail. In the case $L = D^m$, condition (i) is equivalent to the condition of Schoenberg [15] that the problem be m-poised. The concept of m-poisedness is also studied by Ferguson [5].

4. BEST APPROXIMATION OF CONTINUOUS LINEAR FUNCTIONALS

Let us adopt the notation of Theorem 3.1 and suppose $\{N_i : i = 1, 2,..., n\}$ satisfies the hypotheses of the theorem. If N is a given continuous linear functional on $\mathscr{H}$ we desire an approximation $\hat{N}$ to N of the form

$$\hat{N} = \sum_{i=1}^{n} c_iN_i \tag{4.1}$$

where the c_i are constants such that

$$\hat{N}u = Nu \qquad \text{if} \qquad u \in \mathscr{H}_0 . \tag{4.2}$$

Equation (4.2) implies that the representer, say v, of $\hat{N} - N$ belongs to $\mathscr{H}_1$, and hence

$$(\hat{N} - N)u = \langle v, u\rangle = \int (Lv)(Lu). \tag{4.3}$$

Among all approximations $\hat{N}$ of the form (4.1) subject to the constraint (4.2), we seek one for which $\int (Lv)^2$ is minimized. This functional $\hat{N}$ is called the best approximation to N in the sense of Sard [12]. We shall prove that $\hat{N}$ exists, is unique, and satisfies

$$\hat{N}u = N\hat{u} \tag{4.4}$$

for all $u \in \mathscr{H}$ satisfying (3.5) where $\hat{u}$ is as in Theorem 3.1. We first need the following lemma.

LEMMA 4.1. *Adopting the notation and hypotheses of Lemma 3.1, we let* ψ *be any fixed element of* $\mathscr{H}$. *Then there exists a unique element* $\hat{\psi} \in \mathscr{H}$ *of the form* $\hat{\psi} = \sum c_i \psi_i$ *which minimizes* $\| P_1(\hat{\psi} - \psi)\|^2$ *subject to the constraint* $\| P_0(\hat{\psi} - \psi)\|^2 = 0$. *Furthermore, if* $u \in \mathscr{H}$ *satisfies* (3.1), *then*

$$\langle u, \hat{\psi}\rangle = \langle \hat{u}, \psi\rangle. \tag{4.5}$$

Proof. Writing $\hat{\psi} = \sum c_i \psi_i$, we seek a vector $\mathbf{c} = (c_1, ..., c_n)'$ of scalars which minimizes $\mathbf{c}'\Sigma\mathbf{c} - 2\mathbf{w}'\mathbf{c}$ subject to the constraint $\mathbf{v} = T\mathbf{c}$ where $\mathbf{v}$ is the vector $[\langle \omega_j, \psi\rangle]$ and $\mathbf{w}$ is the vector $[\langle P_1\psi_j, \psi\rangle]$. To verify the unique solution

$$\mathbf{c}' = \mathbf{v}'(T\Sigma^{-1}T')^{-1}\, T\Sigma^{-1} + \mathbf{w}\Sigma^{-1}[I - T'(T\Sigma^{-1}T')^{-1}\, T\Sigma^{-1}], \tag{4.6}$$

we let $\mathbf{e}$ be any vector such that $\mathbf{v} = T(\mathbf{c} + \mathbf{e})$. Hence we have $T\mathbf{e} = \mathbf{0}$ so that

$$(\mathbf{c}' + \mathbf{e}')\, \Sigma(\mathbf{c} + \mathbf{e}) - 2\mathbf{w}'(\mathbf{c} + \mathbf{e}) - \mathbf{c}'\Sigma\mathbf{c} + 2\mathbf{w}'\mathbf{c} = \mathbf{e}'\Sigma\mathbf{e},$$

which is non-negative, and which is zero if, and only if, $\mathbf{e} = \mathbf{0}$. Equation (4.5) can be verified directly using (3.2) and (4.6).

To prove (4.4) and the existence and uniqueness of $\hat{N}$, we use Lemma 4.1 and let N have representer ψ, N_i have representer ψ_i, and $\hat{N}$ have representer $\hat{\psi}$.

Equation (4.4) states that the "best" approximation to a functional N operating on a function u is the functional N operating on the spline which "interpolates" u. Schoenberg [14] proved (4.4) in the case when $L = D^m$

and $N_i u = u(t_i)$. Karlin and Ziegler [10] demonstrated (4.4) for L of the form (2.1) with $N_i u = u^{(m_i)}(t_i)$ for $0 \leqslant m_i \leqslant m - 1$, while Jerome and Schumaker [7] considered general continuous linear functionals with L of the form $L = \Sigma b_j D^j$ for $b_j \in C^j$.

5. The Generalized Smoothing Problem

Lemma 5.1. *Let $\mathscr{H} = \mathscr{H}_0 \oplus \mathscr{H}_1$ be the direct sum of an m-dimensional Hilbert space $\mathscr{H}_0$ with basis $\{\omega_1, \omega_2, ..., \omega_m\}$ and any Hilbert space $\mathscr{H}_1$. Let P_0 and P_1 be the projections onto $\mathscr{H}_0$ and $\mathscr{H}_1$ respectively. Suppose $S = [s_{ij}]$ is a $p \times p$ positive definite matrix and $\phi_1, \phi_2, ..., \phi_p$ are $p \geqslant m$ elements of $\mathscr{H}$ such that*

(i) *The set $\{P_0\phi_i : i = 1, 2, ..., p\}$ spans $\mathscr{H}_0$, and*

(ii) *The set $\{P_1\phi_i : i = 1, 2, ..., p\}$ is linearly independent.*

If $\mathbf{z}' = (z_1, z_2, ..., z_p)$ is any p-tuple of scalars, then the unique element $\hat{u} \in \mathscr{H}$ which minimizes

$$\sum\sum (\langle \phi_i, u\rangle - z_i)\, s_{ij}(\langle \phi_j, u\rangle - z_j) + \langle P_1 u, P_1 u\rangle \tag{5.1}$$

is

$$\hat{u} = \boldsymbol{\omega}'(UM^{-1}U')^{-1}\, UM^{-1}\mathbf{z} + \boldsymbol{\eta}'M^{-1}[I - U'(UM^{-1}U')^{-1}\, UM^{-1}]\,\mathbf{z} \tag{5.2}$$

where U is the $m \times p$ matrix $[\langle \omega_i, \phi_j\rangle]$, M is the $p \times p$ matrix

$$[\langle P_1\phi_i, P_1\phi_j\rangle] + S^{-1}, \quad \boldsymbol{\eta} = (P_1\phi_1, ..., P_1\phi_p)' \quad \textit{and} \quad \boldsymbol{\omega} = (\omega_1, \omega_2, ..., \omega_m)'.$$

Proof. We can write $\hat{u} = \boldsymbol{\omega}'\boldsymbol{\alpha} + \boldsymbol{\eta}'\boldsymbol{\beta} + x$ where $\boldsymbol{\alpha}'$ and $\boldsymbol{\beta}'$ are some (as yet undetermined) m-tuple and p-tuple respectively of scalars, $x \in \mathscr{H}_1$, and $\langle x, \eta_i\rangle = 0$. Clearly we must take $x = 0$. The quantity (5.1) to be minimized is

$$(U'\boldsymbol{\alpha} + \Sigma\boldsymbol{\beta} - \mathbf{z})'\, S(U'\boldsymbol{\alpha} + \Sigma\boldsymbol{\beta} - \mathbf{z}) + \boldsymbol{\beta}'\Sigma\boldsymbol{\beta},$$

which is minimized if, and only if,

$$\boldsymbol{\alpha} = (UM^{-1}U')^{-1}\, UM^{-1}\mathbf{z} \qquad \text{and} \qquad \boldsymbol{\beta} = -\, M^{-1}(U'\boldsymbol{\alpha} - \mathbf{z}).$$

The explicit solution to the generalized smoothing problem provided by Section 2 and Lemma 5.1 is stated in the following theorem:

Theorem 5.1. *Let L be an m-*th *order differential operator of the form* (2.1), *$\mathscr{H}$ be the Hilbert space of Section 2, $\{\omega_i : i = 1, 2, ..., m\}$ be as defined by* (2.3), *and let K_1 be defined by* (2.8). *Suppose $\{R_i : i = 1, 2, ..., p\}$ is a set of $p \geqslant m$ continuous linear functionals on $\mathscr{H}$ such that*

(i) *The rank of the* $m \times p$ *matrix* $U = [R_j \omega_i]$ *is* m, *and*

(ii) *The* p *functions* η_i *defined by* $\eta_i(s_0) = R_i K_1(s_0, \cdot)$ *are linearly independent.*

If $\mathbf{z} = (z_1, ..., z_p)'$ *is any vector of scalars and* $S = [s_{ij}]$ *is a* $p \times p$ *positive define matrix, then the unique element* $\hat{u} \in \mathscr{H}$ *which minimizes*

$$\sum\sum (R_i u - z_i)\, s_{ij}(R_j u - z_j) + \int_0^1 (Lu)^2 \tag{5.3}$$

is given by (5.2) *where* $\boldsymbol{\omega} = (\omega_1, \omega_2, ..., \omega_m)'$, $\boldsymbol{\eta} = (\eta_1, \eta_2, ..., \eta_p)'$, *and* $M = [R_i \eta_j] + S^{-1}$.

We remark that the function $\hat{u}$ of Theorem 5.1 is a spline function in the sense that it solves a generalized Hermite-Birkhoff interpolation problem. In fact, if we let $z_i = R_i \hat{u}$, then it is clear that $u = \hat{u}$ is the unique function in $\mathscr{H}$ which minimizes $\int (Lu)^2$ subject to the constraints $R_i u = z_i$.

EXAMPLE 5.1. Let R_i be given by $R_i : u \to u(t_i)$ where the $t_i \in (0, 1)$ are distinct, and let $L = D^m$. Then $\hat{u}$ is a polynomial spline function. This result, where S is a diagonal matrix, was announced by Schoenberg [13] in 1964. An analogous result with L a general linear differential operator was announced by Greville and Schoenberg [6] in 1965. Abstract existence and uniqueness theorems for smoothing spline functions were proved by Anselone and Laurent [1] in 1967.

Theorem 3.1 presents a solution to a minimization problem in which the values of certain linear functionals are constrained, while in Theorem 5.1, the values of the functionals are not constrained, but appear in the term to be minimized. We state the following theorem, whose proof is analogous to that of the preceding results, of which Theorems 3.1 and 5.1 are special cases.

THEOREM 5.2. *In the notation of Theorems 3.1 and 5.1, suppose* $\{N_i : i = 1, 2, ..., n\}$ *and* $\{R_i : i = 1, 2, ..., p\}$ *are sets of continuous linear functionals on* $\mathscr{H}$ *such that*

(i) *The rank of the* $m \times (p + n)$ *matrix* $V = [T; U]$ *is* m, *and*

(ii) *The set* $\{\xi_i\} \cup \{\eta_j\}$ *of* $n + p$ *functions is linearly independent.*

If $\mathbf{w} = (y_1, y_2, ..., y_n; z_1, z_2, ..., z_p)'$ *is any vector of scalars and* $S = [s_{ij}]$ *is any positive definite* $p \times p$ *matrix, then the unique function* $\hat{u} \in \mathscr{H}$ *which minimizes*

$$\sum\sum (R_i u - z_i)\, s_{ij}(R_j u - z_j) + \int_0^1 (Lu)^2$$

subject to the constraints

$$N_i u = y_i \qquad i = 1, 2, \ldots, n$$

is

$$\hat{u} = \boldsymbol{\omega}(VM^{-1}V')^{-1}\, VM^{-1}\mathbf{w} + \begin{bmatrix}\boldsymbol{\xi}\\ \boldsymbol{\eta}\end{bmatrix}' M^{-1}[I - V(VM^{-1}V')\, VM^{-1}]\,\mathbf{w}$$

where M *is the* $(n + p) \times (n + p)$ *partitioned matrix*

$$M = \begin{bmatrix}\Sigma & \Lambda\\ \Lambda' & \Gamma + S^{-1}\end{bmatrix}$$

in which $\Sigma = [N_i \xi_j]$, $\Lambda = [N_i \eta_j]$, *and* $\Gamma = [R_i \eta_j]$.

6. Linear Inequality Constraints

Klaus Ritter [11] recently proposed replacing equations (3.5) by inequalities. He showed the resulting minimization problem to be reducible to a standard problem in quadratic programming. In particular, we have the following result.

Theorem 6.1. *Let* L, $\mathscr{H}$, $\boldsymbol{\omega}$, K, T, Σ, *and* $\boldsymbol{\xi}$ *be defined as in Theorem 3.1. Let* $\{N_i\}$ *be a set of* n *continuous linear functionals on* $\mathscr{H}$, *and let* $\mathbf{y}_j = (y_{j1}, \ldots, y_{jn})'$ *be* n*-tuples of scalars* ($j = 1, 2$). *Then the function* $\hat{u} = \hat{\boldsymbol{\alpha}}'\boldsymbol{\omega} + \hat{\boldsymbol{\beta}}'\boldsymbol{\xi}$ *minimizes* $\int (Lu)^2$ *subject to the constraints*

$$y_{1i} \leqslant N_i u \leqslant y_{2i}, \qquad i = 1, 2, \ldots, n \tag{6.1}$$

if and only if, $\boldsymbol{\alpha} = \hat{\boldsymbol{\alpha}}$ *and* $\boldsymbol{\beta} = \hat{\boldsymbol{\beta}}$ *minimizes* $\boldsymbol{\beta}'\Sigma\boldsymbol{\beta}$ *subject to the constraints* $\mathbf{y}_1 \leqslant T'\boldsymbol{\alpha} + \Sigma\boldsymbol{\beta} \leqslant \mathbf{y}_2$.

The proof of Theorem 6.1 follows immediately from the following lemma, whose proof is elementary.

Lemma 6.1. *Let* $\mathscr{H}$, $\mathscr{H}_0$, $\mathscr{H}_1$, $\boldsymbol{\omega}$, P_0, *and* P_1 *be as in Lemma 3.1. Let* $\{\psi_i\}$ *be* n *elements of* $\mathscr{H}$, $\boldsymbol{\xi} = (P_1\psi_1, P_1\psi_2, \ldots, P_1\psi_n)'$, *and let* $\mathbf{y}_j = (y_{j1}, y_{j2}, \ldots, y_{jn})'$ *be* n*-tuples of scalars for* $j = 1, 2$. *Then the element* $\hat{u} = \boldsymbol{\alpha}'\boldsymbol{\omega} + \boldsymbol{\beta}'\boldsymbol{\xi} \in \mathscr{H}$ *minimizes* $\langle P_1 u, P_1 u\rangle$ *subject to the constraints*

$$y_{1i} \leqslant \langle \psi_i, u\rangle \leqslant y_{2i}, \qquad i = 1, 2, \ldots, n \tag{6.2}$$

if and only if, $\boldsymbol{\alpha} = \hat{\boldsymbol{\alpha}}$ *and* $\boldsymbol{\beta} = \hat{\boldsymbol{\beta}}$ *minimizes* $\boldsymbol{\beta}'\Sigma\boldsymbol{\beta}$ *subject to the constraints* $\mathbf{y}_1 \leqslant T'\boldsymbol{\alpha} + \Sigma\boldsymbol{\beta} \leqslant \mathbf{y}_2$.

7. Spline Interpolation as Statistical Prediction and Filtering

Let $Y(t)$, $0 \leqslant t \leqslant 1$, be the stochastic process defined by

$$Y(t) = \sum_{i=1}^{m} \theta_i \omega_i(t) + X(t), \qquad 0 \leqslant t \leqslant 1 \tag{7.1}$$

where $\{\theta_i\}_{i=1}^m$ are random variables independent of $X(t)$, $0 \leqslant t \leqslant 1$, $\{\omega_i(t)\}_{i=1}^m$ are given by (2.3), and $X(t)$, $0 \leqslant t \leqslant 1$ is the zero mean Gaussian stochastic process with covariance given by

$$EX(s)\, X(t) = K_1(s, t), \tag{7.2}$$

$K_1(s, t)$ being given by (2.8). $X(t)$ has a representation as

$$X(t) = a_m(t) \int_0^t a_{m-1}(t_{m-1})\, dt_{m-1} \int_0^{t_{m-1}} a_{m-2}(t_{m-2})\, dt_{m-2} \cdots \int_0^{t_2} a_1(t_1)\, dW(t_1) \tag{7.3}$$

where $W(u)$, $0 \leqslant u \leqslant 1$, is the Wiener process. Hence, we may say that $Y(t)$ formally satisfies the stochastic differential equation

$$LY(t) = \frac{dW(t)}{dt}$$

with the (random) boundary conditions $M_i Y(0) = \theta_{i+1}$, $i = 0, 1, 2, \ldots, m - 1$. $dW(t)/dt$ is commonly referred to as "white noise". Let now $\{\theta_i\}_{i=1}^m$ be independent normal random variables with mean zero and variance 1. We may define the Hilbert space $\mathscr{G}$ spanned by the family of random variables $\{y(t), 0 \leqslant t \leqslant 1\}$ as all finite linear combinations of random variables of the form

$$\rho = \sum_\nu c_\nu Y(t)_\nu$$

plus the closure of this linear manifold under the norm induced by the inner product

$$\langle \rho_1, \rho_2 \rangle = E\rho_1 \rho_2 . \tag{7.4}$$

(We have $EY(s)\, Y(t) = K(s, t)$, where K is given by (2.10).)

$\mathscr{G}$ is isomorphic to $\mathscr{H}$ under the correspondence induced by

$$Y(s_0) \leftrightarrow K(s_0, \cdot), \qquad s_0 \in [0, 1]. \tag{7.5}$$

Let $\rho_0, \rho_1, \ldots, \rho_n$ be n random variables in $\mathscr{G}$ and let $u, \psi_1, \psi_2, \ldots, \psi_n$ be the $n + 1$ elements in $\mathscr{H}$ which correspond to $\rho_0, \rho_1, \ldots, \rho_n$ under correspondence (7.5). We shall call an estimate $\tilde{\rho}_0$ of ρ_0 unbiased with respect to $\theta = (\theta_1, \theta_2, \ldots \theta_m)'$ if

$$E(\tilde{\rho}_0 \mid \theta) = E(\rho_0 \mid \theta). \tag{7.6}$$

Let $\hat{\rho}_0$ be the minimum variance unbiased estimate of ρ_0, based on data $\{\rho_i = y_i\}$, $i = 1, 2, \ldots, M$, that is, $\hat{\rho}_0$ is that linear combination $\sum_{i=1}^n d_i \rho_i$ which minimizes

$$E\left(\hat{\rho}_0 - \sum_{i=1}^n d_i \rho_i\right)^2 \tag{7.7}$$

subject to

$$E(\hat{\rho}_0 - \rho_0 \mid \theta) = 0. \tag{7.8}$$

This last condition can be shown to be equivalent to

$$\sum_{i=1}^n \theta_i (\hat{N}_0 - N_0)\, w_i(\cdot) \equiv 0 \tag{7.9}$$

where $\hat{N}_0$ and N_0 are the continuous linear functionals (i.e. elements in the dual of $\mathscr{H}$) whose representers $\hat{u}$ and u correspond to the random variables $\hat{\rho}_0$ and ρ_0 under the correspondence (7.5). As a consequence of Lemma 4.1, and the isomorphism, we have

THEOREM 7.1. *Let Σ, the $n \times n$ matrix $[E(\rho_i \rho_j \mid \theta)]$, and T, the $m \times n$ matrix $[E\theta_i \rho_j]$, be of full rank. Then the minimum variance unbiased estimate $\hat{\rho}_0$ for ρ_0, based on data $\{\rho_i = y_i\}_{i=1}^n$ is given by*

$$\hat{\rho}_0 = N_0 \hat{u}$$

where $\hat{u}$ is given by (3.2) *and N_0 is the continuous linear functional whose representer corresponds to ρ_0.*

EXAMPLE 7.1. Let $\rho_i = Y(t_i)$ for $t_i \in (0, 1)$ distinct. Then the minimum variance unbiased estimate $\hat{Y}(t)$ for $Y(t)$ based on data $\{Y(t_i) = y_i\}_{i=1}^n$ is, considered as a function of t, the natural L-spline of interpolation to the points $\{(t_i, y_i)\}$.

The generalized smoothing problem of Section 5 also has a statistical interpretation. We cite as a theorem only one form, which displays the smoothing problem as equivalent to the extraction of signal from noise.

THEOREM 7.2. *Let*

$$Z(t) = Y(t) + \epsilon(t)$$

where $\epsilon(t)$ is a Gaussian noise process independent of $Z(t)$ with $E\epsilon(s)\,\epsilon(t) = A(s, t)$. Let $\hat{Y}(t)$ be the minimum variance unbiased estimate of $Y(t)$ based on data $Z(t_i) = z_i$, $i = 1, 2,..., p$. Then $\hat{Y}(t)$ is given by

$$Y(t) = \langle K(t, \cdot), \hat{u} \rangle = u(t)$$

where $\hat{u}$ is given by (5.2) with $\phi_i(\cdot) = K(t_i, \cdot)$, $i = 1, 2,..., p$ and S^{-1} is the $p \times p$ matrix $[A(t_i, t_j)]$.

References

1. P. M. Anselone and P. J. Laurent, A general method for the construction of interpolating or smoothing spline functions, MRC Technical Summary Report #834, 1967, Mathematics Research Center, Univ. of Wisconsin.
2. N. Aronszajn, Theory of reproducing kernels, *Trans. Amer. Math. Soc.* **68** (1950), 337–404.
3. M. Attéia, Generalisation de la définition et des propriétés des "spline-fonctions," *C. R. Acad. Sci. Paris* **260** (1965), 3550–3553.
4. C. deBoor and R. E. Lynch, On splines and their minimum properties, *J. Math. Mech.* **15** (1966), 953–969.
5. D. Ferguson, The question of uniqueness for G. D. Birkhoff interpolation problems, *J. Approximation Theory* **2** (1969), 1–28.
6. T. N. E. Greville and I. J. Schoenberg, Smoothing by generalized spline functions (Abstract), *SIAM Rev.* **7** (1965), 617.
7. J. W. Jerome and L. L. Schumaker, On Lg-splines, *J. Approximation Theory* **2** (1969), 29–49.
8. S. Karlin, "Total Positivity," Vol. I, Stanford University Press, 1968.
9. S. Karlin and W. J. Studden, "Tchebycheff Systems: with Applications in Analysis and Statistics," Interscience, 1966.
10. S. Karlin and Z. Ziegler, Chebyshevian spline functions, *SIAM J. Numer. Anal.* **3** (1966), 514–543.
11. K. Ritter, Splines and quadratic programming, Conference on Approximations, University of Wisconsin, 1969.
12. A. Sard, "Linear Approximation," Mathematical Surveys No. 9, American Mathematical Society, Providence, 1963.
13. I. J. Schoenberg, Spline functions and the problem of graduation, *Proc. Nat. Acad. Sci. U.S.A.* **52** (1964), 947–950.
14. I. J. Schoenberg, On best approximations of linear operators, *Indag. Math.* **26** (1964), 155–163.
15. I. J. Schoenberg, On the Ahlberg-Nilson extension of spline interpolation: The *g*-splines and their optimal properties, *J. Math. Anal. Appl.* **21** (1968), 207–231.

24

Reprinted from *J. Math. Anal. Appl.* **70:**505–529 (1979)

Vector-Valued Lg-Splines
I. Interpolating Splines

GURSHARAN S. SIDHU

Instituto de Investigaciones en Matemáticas Aplicadas y en Sistemas, Universidad Nacional Autónoma de México, México 20, D.F.

AND

HOWARD L. WEINERT

Department of Electrical Engineering, The Johns Hopkins University, Baltimore, Maryland 21218

Submitted by T. T. Soong

The theory of Lg-splines developed by Jerome and Schumaker is extended to the vector-valued (multivariate) case. The extension is described in the framework of a reproducing-kernel Hilbert space which among other things allows the establishment of a congruent least-squares estimation problem for a vector-valued lumped random process. The results include a dynamic recursive algorithm for vector-valued Lg-splines with EHB data and a useful structural characterization theorem for such splines. Some results on computable approximation error bounds are also included.

I. INTRODUCTION

We shall be concerned in this paper with an extension of the Lg-splines first discussed in a 1969 paper [1] by Jerome and Schumaker. Their theory was developed in the space H_k of functions f, on the interval $W = [0, T]$, whose kth derivative $f^{(k)}$ exists a.e. and is square-integrable on W; i.e.

$$H_k = \left\{ f \text{ on } [0, T]: f^{(k)} \text{ exists a.e. and } \int_0^T \{f^{(k)}\}^2 < \infty \right\}. \tag{1.1}$$

Thus if we define the differential operator $L = D^k + \sum_{j=0}^{k-1} a_j(t)\, D^j$, where $D = d/dt$ and $a_j \in C^j(W)$, it is clear that $f \in H_k$ if and only if $\int_0^T \{Lf\}^2 < \infty$. Now, corresponding to real numbers $\{r_j, 1 \leqslant j \leqslant N\}$ and linear functionals

$\{\lambda_j, 1 \leqslant j \leqslant N\}$ on H_k Jerome and Schumaker defined an *Lg-spline interpolating* $\{r_j, 1 \leqslant j \leqslant N\}$ *with respect to* $\{\lambda_j, 1 \leqslant j \leqslant N\}$ *as an* $s \in H_k$ *such that*

$$s \in U_r = \{f \in H_k: \lambda_j f = r_j, 1 \leqslant j \leqslant N\} \tag{1.2a}$$

$$\int_0^T (Ls)^2 \leqslant \int_0^T (Lf)^2 \qquad \text{all } f \in U_r. \tag{1.2b}$$

Thus s is the function satisfying the interpolation constraints (1.2a) that is smoothest in the sense of (1.2b). Freedom in choosing L implies a consequent freedom of the choice of the smoothness criterion (1.2b).

Jerome and Schumaker have presented a rather complete development of these splines from a function analytic point of view. Their results include existence and uniqueness conditions (so-called poisedness of (1.2) with respect to L) and an elegant structural theorem. Their uniqueness condition has subsequently been given an operationally useful systems-theoretic interpretation in our work [2].

Lg-splines and their specialized relatives have been found to have a good deal of interesting structure—hence a large body of related theoretical work has been developed. On the other hand, their smoothness and interpolation properties have prompted considerable applications interest with consequent activity on efficient computational algorithms.

From the viewpoint of applications it has become clear that various extensions of Jerome and Schumaker's *Lg*-splines are required:

(A) *Splines on* R^n. These are of interest in various applications, for instance, those related to picture processing ($n = 2$) such as reconstruction from samples.

(B) *Vector-valued (Multivariate) splines.* Often one wishes to reconstruct a number of functions ($p > 1$) from their sample values. Since the samples of one function can convey information about the others, it is not adequate to interpolate the samples of each function *independently* of those of the other functions. A *simultaneous* interpolation would be more appropriate. This paper is concerned with the resolution of this problem.

(C) ARMA *splines.* It has been noted [3]–[6] that *Lg*-splines provide a solution of the following optimal control problem in $\mathscr{L}_2(W)$:

Determine the minimum energy input function u on $[0, T]$ *such that the output y of the linear system* $Ly = u$ *satisfies the constraints* (1.3)
$$\lambda_j y = r_j, \qquad 1 \leqslant j \leqslant N.$$

Obviously, since $u = Ly$, the optimum solution is $u_* = Ls$ where s is the *Lg*-spline of (1.2). Vector-valued splines would provide an extension of this correspondence to the case of systems with vector-valued inputs and outputs.

However, a severe restriction in (1.3) is the form of the linear systems the model $Ly = u$ corresponds to the so-called *autoregressive* (or *numerator—free*) *systems*. General linear differential systems have two operators L and M associated with them—roughly speaking, they are of the type $Ly = Mu$ and are said to be of *autoregressive moving-average* (ARMA) *type*. This clearly motivates the concept of an ARMA spline. We have already reported such results [5] with related work being given also in [6].

In a subsequent paper we shall discuss the case of vector-valued ARMA splines which brings with it a host of new difficulties.

II. Vector-Valued Lg-Splines—Definition, Existence and Uniqueness

For a given integer $p \geqslant 1$, and fixed nonnegative integers $n_1, n_2, \ldots, n_p$, let H be the space of $p \times 1$ vector-valued functions[1] $f = (f_1, f_2, \ldots, f_p)'$ on W such that $f_j \in H_{n_j}$, $1 \leqslant j \leqslant p$; that is to say[2]

$$H = H_{n_1} \times H_{n_2} \times \cdots \times H_{n_p}, \tag{2.1}$$

where each H_{n_j} is defined as in (1.1). Thus for each j, $f_j^{(n_j)} \in \mathscr{L}_2(W)$, the space of square-integrable functions on $[0, T]$.

Let L be a $p \times p$ matrix of ordinary differential operators, with its ijth entry L_{ij} of the form

$$L_{ij} = \sum_{l=0}^{n_j} a_{ij,l}(t)\, D^l, \qquad D = d/dt, \tag{2.2}$$

where we assume (see Appendix 1) that for each j

$$\begin{aligned} a_{ij,n_j} &= 1 \quad \text{if} \quad i = j, \\ &= 0 \quad \text{if} \quad i \neq j. \end{aligned} \tag{2.3}$$

Clearly now

$$f \in H \quad \text{if and only if} \quad Lf \in \mathscr{L}_2{}^p(W), \tag{2.4}$$

where $\mathscr{L}_2{}^p(W)$ is the p-fold Cartesian product of $\mathscr{L}_2(W)$.

Now if $\{\lambda_i\}_1^N$ are independent linear functionals on H, then for any real numbers $\{r_j\}_1^N$ one has:

[1] Primes (′) denote matrix transpose.

[2] Note that one could work in the space of functions mapping $\{1, 2, \ldots, p\} \times W$ into the reals. This is in fact implicit in particular in Section V.

DEFINITION 2.1. An $s \in U_r \subset H$ is a vector-valued Lg-spline interpolating $\{r_j, 1 \leqslant j \leqslant N\}$ with respect to $\{\lambda_j, 1 \leqslant j \leqslant N\}$ if

$$\int_0^T (Ls)'(Ls) \leqslant \int_0^T (Lf)'(Lf), \qquad \text{all } f \in U_r \tag{2.5a}$$

where

$$U_r = \{f \in H: \lambda_j f = r_j, 1 \leqslant j \leqslant N\}. \quad \blacksquare \tag{2.5b}$$

H can be made a Hilbert space under a variety of norms (one is given in Section III). First, let us obtain the natural generalization to the vector-valued case of the existence and uniqueness results of [1]:

THEOREM 2.2. (a) *Provided the* $\{\lambda_j, 1 \leqslant j \leqslant N\}$ *are continuous functionals on* H, *the vector-valued spline of definition* 2.1 *always exists.*

(b) *It is unique if and only if either of the following equivalent conditions holds:*

(i) $N_L \cap U_0 = \{0\}$ *where*

$$N_L = \{f \in H: Lf = 0\} \qquad \text{and} \qquad U_0 = \{f \in H: \lambda_j f = 0, 1 \leqslant j \leqslant N\}; \tag{2.6}$$

(ii) $N \geqslant n = n_1 + n_2 + \cdots + n_p$, *and among the* $\{\lambda_j, 1 \leqslant j \leqslant N\}$ *there are* n *functionals linearly independent on* N_L. (2.7)

Proof. (a) Proceeding as in [1] we can establish that LU_r, the image of U_r under L, is a closed linear variety in $\mathscr{L}_2^p(W)$ and hence contains a minimum $\mathscr{L}_2^p$-norm vector u_*. By the definition of $LU_r = \{Lf: f \in U_r\}$ there must then *exist* an $s \in U_r$ such that $Ls = u_*$. This s is a solution of (2.5).

(b) Let s be a solution of (2.5). Then, so is $s + g$ for any $g \in N_L \cap U_0$. Hence $N_L \cap U_0 = \{0\}$ is a necessary condition for uniqueness. It is also sufficient for, as we show below, the difference $g = s_{(1)} - s_{(2)}$ of any two solutions of (2.5) is in $N_L \cap U_0$. First, since λ_j is linear and $\lambda_j s_{(1)} = \lambda_j s_{(2)} = r_j$ we have $s_{(1)} - s_{(2)} \in U_0$. Also, from the proof of part (a), $Ls_{(1)} = Ls_{(2)} = u_*$ and thence $s_{(1)} - s_{(2)} \in N_L$.

To establish (ii) note that $u_* \in LU_r$ is unique. Hence the determination of s involves solving the system $Ls = u_*$ of p differential equations subject to boundary conditions $\lambda_j s = r_j$, $1 \leqslant j \leqslant N$. From the discussion of appendix I it is clear that the solution of these de's is unique if and only if $n = n_1 + n_2 + \cdots + n_p$ of the λ_j are linearly independent on N_L. $\blacksquare$

Henceforth we shall assume the conditions for existence and uniqueness, namely that the λ_j are continuous on H, that $N \geqslant n$, and that, after possible rearrangement,

$$\{\lambda_j, 1 \leqslant j \leqslant n\} \text{ are linearly independent on } N_L. \tag{2.8}$$

III. H as a Hilbert Space

The integer

$$n = \sum_{j=1}^{p} n_j \tag{3.1}$$

has a special significance:

Lemma 3.1. *N_L is a subspace of H with dimension equal to n.* ∎

This is proved in Appendix I using a state-space representation of $Lf = 0$.

As a consequence of Lemma 3.1, N_L has a basis $\{z_j, 1 \leqslant j \leqslant n\}$ chosen to be dual to the $\{\lambda_j, 1 \leqslant j \leqslant n\}$. Thus for $1 \leqslant i \leqslant n$, $1 \leqslant j \leqslant n$

$$\begin{aligned} Lz_j = 0, \qquad \lambda_i z_j = \delta_{ij} &= 1 \quad \text{for } i = j \\ &= 0 \quad \text{for } i \neq j. \end{aligned} \tag{3.2}$$

Next, let $G(\cdot, \cdot)$ be the Green's function of L relative to $\{\lambda_j, 1 \leqslant j \leqslant n\}$, i.e. G is the $p \times p$ matrix of functions on $W \times W$ given, for each $t \in W$, by

$$LG(\cdot, t) = I_p \delta(\cdot - t), \qquad \lambda_j G(\cdot, t) = \mathbf{0}', \qquad 1 \leqslant j \leqslant n \tag{3.3}$$

where I_p is the $p \times p$ identity matrix, and $\mathbf{0}$ a $p \times 1$ column of zeros. In (3.3) λ_j operates on the columns of $G(\cdot, t)$.

Now, any $f \in H$ can be written in the form

$$f(\cdot) = \sum_{j=1}^{n} (\lambda_j f)\, z_j(\cdot) + \int_0^T G(\cdot, t)\{Lf(t)\}\, dt \tag{3.4}$$

which prompts the inner product for H

$$\langle f, g \rangle_H = \sum_{j=1}^{n} (\lambda_j f)(\lambda_j g) + \int_0^T \{Lf(t)\}' \{Lg(t)\}\, dt \tag{3.5}$$

with the corresponding norm

$$\| f \|_H^2 = \sum_{j=1}^{n} (\lambda_j f)^2 + \int_0^T \{Lf(t)\}' \{Lf(t)\}\, dt. \tag{3.6}$$

Note that the first term in (3.4) belongs to N_L and is orthogonal to the second term. Thus (3.4) implies an orthogonal decomposition of H:

$$H = N_L \oplus H_1, \qquad N_L \perp H_1 \tag{3.7}$$

It is easily seen that N_L and H_1 are congruent to $\mathbb{R}^n$ and $\mathscr{L}_2{}^p(W)$ respectively.

Thus, since $\mathbb{R}^n$ and $\mathscr{L}_2^p(W)$ are complete so are N_L and H_1. We can then assert the following for their direct sum H:

LEMMA 3.2. *H equipped with the inner product* (3.5) *is a Hilbert space.*

Note that $\sum_{j=1}^n (\lambda_j f)^2$ is constant on U_r; thus definition 2.1 can be put in the minimum-norm form:

THEOREM 3.3. *The multivariate Lg-spline s interpolating $\{r_j, 1 \leq j \leq N\}$ is the unique minimum norm element of the linear variety U_r of H.* ∎

As a consequence, one can use the projection theorem [7, p. 64] to establish that

$$s = \mathscr{P}^{\mathscr{S}_N} g \qquad \text{any} \qquad g \in U_r, \tag{3.9}$$

where $\mathscr{P}^{\mathscr{S}_N}$ denotes projection onto $\mathscr{S}_N$, the orthogonal complement of U_0. It is easy to show that

$$\mathscr{S}_N = U_0^{\perp} = \operatorname{span}\{h_j, 1 \leq j \leq N\}, \tag{3.10}$$

where h_j are the representers of the λ_j; i.e.

$$h_j \in H \qquad \text{is such that} \qquad \lambda_j f = \langle f, h_j \rangle_H \quad \text{all} \quad f \in H. \tag{3.11}$$

Solving the normal equations corresponding to (3.9) one has

$$s(\cdot) = \mathbf{h}'(\cdot)\, \mathbf{R}^{-1} \mathbf{r} \tag{3.12}$$

where $\mathbf{h}'(\cdot) = (h_1(\cdot), h_2(\cdot),..., h_N(\cdot))$, $\mathbf{r}' = (r_1, r_2, ..., r_N)$, and $\mathbf{R}$ is the $N \times N$ matrix with ijth entry $\langle h_i, h_j \rangle_H$. The solution (3.12) is in fact not very useful in practice and is include here primarily for completeness.

IV. REPRODUCING KERNEL FOR H

Let K be the $p \times p$ matrix-valued function on $W \times W$ defined by

$$K(t, \tau) = \sum_{j=1}^{n} z_j(t)\, z_j'(\tau) + \int_0^T G(t, \xi)\, G'(\tau, \xi)\, d\xi \tag{4.1}$$

and let K_j^r and K_l^c denote respectively the jth row and lth column of K. The function $K(\cdot, \cdot)$ is of central importance in this paper. We start by establishing some of its key properties. Proofs are given only for the nonobvious ones.

PROPERTY 4.1. *Symmetry:*

$$K(t, \tau) = K'(\tau, t), \qquad all \qquad t, \tau \in W. \tag{4.2a}$$

Thus

$$K_j^c(t, \tau) = [K_j^r(\tau, t)]'. \tag{4.2b}$$

PROPERTY 4.2. *Inclusion of K in H:*

From (4.1) and (3.4) is is evident that, for each *fixed* $t \in W$,

$$K_j^c(\cdot, t) = [K_j^r(t, \cdot)]' \in H, \qquad 1 \leqslant j \leqslant p, \tag{4.3a}$$

Thus for the ijth entry K_{ij} of K we have

$$K_{ij}(\cdot, t) \in H_{n_i}; \qquad K_{ij}(t, \cdot) \in H_{n_j}. \tag{4.3b}$$

PROPERTY 4.3. *Reproducing Property:*

From (4.1), (3.2)–(3.3), *for each* $t \in W$,

$$LK(\cdot, t) = G'(t, \cdot), \tag{4.4a}$$

$$\lambda_j K(\cdot, t) = z_j'(t), \qquad 1 \leqslant j \leqslant n. \tag{4.4b}$$

Hence, using (3.5) it is readily verified that, for every $f = (f_1, f_2, ..., f_p)' \in H$, and *for each* $t \in W$,

$$\langle K_j^c(\cdot, t), f(\cdot)\rangle_H = \langle K_j^r(t, \cdot), f(\cdot)\rangle_H = f_j(t), \qquad 1 \leqslant j \leqslant p. \tag{4.5}$$

This is known as the *reproducing property* [8] *of K.*

H is said to be a *reproducing kernel Hilbert space* (RKHS) with K as its *reproducing kernel* (RK).

PROPERTY 4.4. *Nonnegative Definiteness of K:*

Using (4.5) it is easily shown that for any finite integer $m \geqslant 1$ and for each choice of $\tau_1, \tau_2, ..., \tau_m$ in W and real $p \times 1$ vectors $a_1, a_2, ..., a_m$:

$$\sum_{i=1}^{m} \sum_{j=1}^{m} a_i' K(\tau_i, \tau_j)\, a_j = \left\| \sum_{i=1}^{m} K(\cdot, \tau_i)\, a_i \right\|_H^2 \geqslant 0. \tag{4.6}$$

Thus K is a matrix-valued nonnegative definite function on $W \times W$ in the sense of Mercer [9].

PROPERTY 4.5. *The* RK *and the Representers of Continuous Linear Functionals:*

The representer h_i of λ_i is given by

$$h_i'(t) = \lambda_i K(\cdot, t) = [\lambda_i K_1^c(\cdot, t), ..., \lambda_i K_p^c(\cdot, t)], \tag{4.7}$$

and

$$\langle h_i, h_j \rangle_H = \lambda_{j(\tau)}[\lambda_{i(t)} K(t, \tau)]'. \tag{4.8}$$

Here the parenthetical subscripts on λ_i and λ_j show the independent variables with respect to which they operate.

Proof of (4.7)–(4.8). Any $f = (f_1, f_2, ..., f_p)' \in H$ can be decomposed in H in the form $f = \sum_{j=1}^{p} f_{[j]}$ where $f_{[j]} = (0, 0, ..., 0, f_j, 0, ..., 0)'$.

The linearity of λ_i then implies

$$\lambda_i f = \sum_{j=1}^{p} \lambda_i f_{[j]}. \tag{4.9}$$

Clearly thus λ_i induces continuous linear functionals $\lambda_{ij}: H_{n_j} \to \mathbb{R}$, and (4.9) can be written as:

$$\lambda_i f = \sum_{j=1}^{p} \lambda_{ij} f_j. \tag{4.10}$$

Thus using (4.5)

$$\begin{aligned} \lambda_{i(t)} f(t) &= \sum_{j=1}^{p} \lambda_{ij(t)} \langle K_j^c(\cdot, t), f(\cdot) \rangle_H \\ &= \left\langle \sum_{j=1}^{p} \lambda_{ij(t)} K_j^c(\cdot, t), f(\cdot) \right\rangle_H \end{aligned} \tag{4.11}$$

which implies that (note (4.2))

$$\begin{aligned} h_i'(\cdot) &= \left[\sum_{j=1}^{p} \lambda_{ij(t)} K_{1j}(\cdot, t), ..., \sum_{j=1}^{p} \lambda_{ij(t)} K_{pj}(\cdot, \tau) \right] \\ &= \left[\sum_{j=1}^{p} \lambda_{ij(t)} K_{j1}(t, \cdot), ..., \sum_{j=1}^{p} \lambda_{ij(t)} K_{jp}(t, \cdot) \right]. \end{aligned}$$

In view of (4.10) this gives us (4.7). Equation (4.8) is a ready consequence of (4.7) for, $\langle h_i, h_j \rangle_H = \lambda_j h_i(\cdot) = \lambda_j[\lambda_{i(\tau)} K(\tau, \cdot)]'$.

For future use we note that (4.7), (4.4b) and (3.2) imply

$$z_j = h_j, \qquad \langle h_i, h_j \rangle_H = \delta_{ij}, \qquad 1 \leqslant i, \ j \leqslant n. \tag{4.12}$$

PROPERTY 4.6. *The* RK *spans* H:

As a consequence of (4.5), $\langle f(\cdot), K_j^c(\cdot, t) \rangle_H = 0$ for each j and every $t \in W$ if and only if $f = 0$. Thus we have

$$\begin{aligned} H &= \operatorname{span}\{K_j^c(\cdot, t), 1 \leqslant j \leqslant p, t \in W\} \\ &= \operatorname{span}\{K_j^r(t, \cdot), 1 \leqslant j \leqslant p, t \in W\}. \end{aligned} \tag{4.13}$$

It can be shown furthermore that the RK is unique. Thus clearly K completely determines H.

The following theorem is of central importance:

THEOREM 4.1. *For each* $t \in W$

$$s'(t) = \mathscr{V}\mathscr{P}^{\mathscr{S}_N}K(\cdot, t), \tag{4.14}$$

where $\mathscr{V}: \mathscr{S}_N \to \mathbb{R}$ *through* $\mathscr{V}h_j = r_j$, $1 \leq j \leq N$, *and* $\mathscr{V}$ *linear.*

Proof. Let $s(\cdot) = [s_1(\cdot), s_2(\cdot),..., s_p(\cdot)]'$. Using (4.5) and the selfadjointness of $\mathscr{P}^{\mathscr{S}_N}$ (3.9) can be rewritten

$$s_j(t) = \langle \mathscr{P}^{\mathscr{S}_N}g(\cdot), K_j^c(\cdot, t)\rangle_H = \langle g(\cdot), \mathscr{P}^{\mathscr{S}_N}K_j^c(\cdot, t)\rangle_H$$

which upon expressing $\mathscr{P}^{\mathscr{S}_N}K_j^c(\cdot, t) = \sum_{i=1}^N \beta_{ji}(t)\, h_i(\cdot)$ gives

$$s_j(t) = \sum_{i=1}^N \beta_{ji}(t) \langle g(\cdot), h_i(\cdot)\rangle_H = \sum_{i=1}^N \beta_{ji}(t)\, r_i$$

since $g \in U_r$. ∎

One could go directly from Theorem 4.1 to the algorithm of Section VI using an extension of our approach of [10]. Instead we choose to proceed via the intermediate step of introducing a space Y of random variables that is isometrically isomorphic to H, and then obtain the algorithm in Y. This is done since the type of algorithm obtained is much better known in the related context of linear least-squares estimation for lumped random processes.

V. SPLINES AND STOCHASTIC ESTIMATION

From properties 4.2 and 4.4, K is a symmetric nonnegative definite function on $W \times W$. Thus there exists [11] a zero-mean $p \times 1$ vector-valued process $\{y(t), t \in W\}$ with covariance function K:

$$y(t) = (y_1(t), y_2(t),..., y_p(t))'; \qquad E\{y_j(t)\} = 0 \tag{5.1}$$

$$E\{y(t)\, y'(\tau)\} = K(t, \tau); \qquad t, \tau \in W. \tag{5.2}$$

Ley Y be the complete vector space of finite variance, zero-mean random variables spanned by the random variables $\{y_j(t); 1 \leq j \leq p, t \in W\}$ and with the usual inner product $\langle a, b\rangle_Y = E\{ab\}$. We use E to denote probabilistic expectation with respect to the probability law of the process $y(\cdot)$.

From (4.3), (4.5) and (5.2) it is clear that

$$\langle K_i^r(t, \cdot), K_j^r(\tau, \cdot)\rangle_H = K_{ij}(t, \tau) = \langle y_i(t), y_j(\tau)\rangle_Y \tag{5.3}$$

Since $\{K_i^r(t, \cdot); 1 \leqslant i \leqslant p, t \in W\}$ and $\{y_i(t); 1 \leqslant i \leqslant p, t \in W\}$ respectively span H and Y, (5.3) implies the following extension of a theorem first stated by Loeve [12, p. 408]:

THEOREM 5.1. *The space H with* RK $K(\cdot, \cdot)$ *and the space Y of random variables defined above are isometrically isomorphic (congruent). Furthermore $f \in H$ corresponds to $z \in Y$, denoted $f \sim z$, if and only if*

$$f_j(t) = E\{zy_j(t)\}. \quad \blacksquare \tag{5.4}$$

It is readily noted that under this congruence

$$K_j^c(\cdot, t) \quad \text{or} \quad K_j^r(t, \cdot) \sim y_j(t), \qquad 1 \leqslant j \leqslant p, \tag{5.5}$$

$$h_j(\cdot) = [\lambda_{j(t)} K(t, \cdot)]' \sim \lambda_{j(t)} y(t), \qquad 1 \leqslant j \leqslant N. \tag{5.6}$$

Thus if $\mathscr{D}_N = \text{span}\{\lambda_j y, 1 \leqslant j \leqslant N\}$ then

$$\mathscr{S}_N \sim \mathscr{D}_N . \tag{5.7}$$

Consequently

$$\mathscr{P}^{\mathscr{S}_N} K(t, \cdot) \sim \mathscr{P}^{\mathscr{S}_N} y(t), \qquad \text{each} \qquad t \subset W \tag{5.8}$$

where $\mathscr{P}^{\mathscr{D}_N} y(t) = (\mathscr{P}^{\mathscr{D}_N} y_1(t), \mathscr{P}^{\mathscr{D}_N} y_2(t),..., \mathscr{P}^{\mathscr{D}_N} y_p(t))$. The random vector $\mathscr{P}^{\mathscr{D}_N} y(t)$ is known as *the linear least-squares estimate (llse) of $y(t)$ given $\{\lambda_j y$, $1 \leqslant j \leqslant N\}$* since it minimizes, for $1 \leqslant j \leqslant p$, $E(y_j(t) - \rho_j)^2$ over all $\rho_j \in \mathscr{D}_N$. If we denote by $\hat{y}(t)$ the sample value of $\mathscr{P}^{\mathscr{D}_N} y(t)$ corresponding to $\{\lambda_j y = r_j$, $1 \leqslant j \leqslant N\}$ then we have the following generalization of the univariate results of [13]–[17].

THEOREM 5.2.

$$s(t) = \hat{y}(t). \quad \blacksquare \tag{5.9}$$

It is natural thus to seek algorithms for computing $\hat{y}(\cdot)$ and hence $s(\cdot)$. This is greatly facilitated by the fact that $y(\cdot)$ is a lumped process—it has a known finite-dimensional linear model as summarized in the following theorem:

THEOREM 5.3. *The random process $y(\cdot)$ of* (5.1) *is such that*

$$Ly = u \tag{5.10}$$

where $u(\cdot)$ is a $p \times 1$ vector-valued, zero mean, white process, i.e.,

$$Eu(t) = 0; \qquad Eu(t)\, u'(\tau) = I_p \delta(t - \tau). \quad \blacksquare \tag{5.11}$$

The proof is quite easy to obtain, albeit somewhat informally, since $y_j(t) \sim K_j^r(t, \cdot)$ implies that $L_{(t)}K(t, \cdot) = G'(\cdot, t) \sim L_{(t)} y(t)$. Writing $L_{(t)} y(t) = (u_1(t), u_2(t),..., u_p(t))'$ one has $u_j(t) \sim G_j^c(\cdot, t)$, and hence from (3.5) it is readily verified that

$$Eu_i(t)\, u_j(\tau) = \langle G_i^c(\cdot, t), G_j^c(\cdot, \tau)\rangle_H = \delta_{ij}\delta(t - \tau).$$

Now, in view of Appendix I we have a ready corollary that $y(\cdot)$ has a finite-dimensional state model:

COROLLARY 5.4. *The process $y(\cdot)$ is the output of the state model*

$$y(t) = Cx(t); \qquad \frac{d}{dt}\, x(t) = A(t)\, x(t) + Bu(t) \tag{5.12a}$$

where $u(\cdot)$ is as in (5.11) *and $A(\cdot)$, B, C are $n \times n$, $n \times p$ and $p \times n$ matrices given by*

$$B = \text{block diag}[[0,..., 0, 1]', n_j \times 1] \tag{5.12b}$$

$$C = \text{block diag}[[1, 0,..., 0], 1 \times n_j] \tag{5.12c}$$

$$A(\cdot) = \text{block}[A_{ij}, n_i \times n_j] \tag{5.12d}$$

$$A_{ii}(\cdot) = \left[\begin{array}{c|ccc} \mathbf{0} & & I_{n_i-1} & \\ \hline -a_{ii,0} & -a_{ii,1} & \cdots & -a_{ii,n_i-1} \end{array}\right], \qquad n_i \times n_i, \tag{5.12e}$$

$$A_{ij}(\cdot) = \left[\begin{array}{cccc} & & \mathbf{0} & \\ \hline -a_{ij,0} & -a_{ij,1} & \cdots & -a_{ij,n_j-1} \end{array}\right], \qquad n_i \times n_j, \quad i \neq j \tag{5.12f}$$

and $x(\cdot)$ is $n \times 1$ given by

$$x' = [y_1, y_1^{(1)},..., y^{(n_1-1)} \mid y_2, y_2^{(1)},..., y^{(n_2-1)} \mid \cdots \mid y_p, y_p^{(1)},..., y_p^{(n_p-1)}]. \tag{5.12g}$$

The model in (5.12) is not complete in that it does not allow us to completely obtain the covariance function of $y(\cdot)$. What is lacking is a set of boundary conditions for the differential system (5.12a). Although we can go further without additional restriction, the most fruitful results accrue for a very broad special class of λ_j considered in the next section.

VI. LG-SPLINES WITH EHB DATA

We now restrict attention to λ_j, $1 \leqslant j \leqslant N$, of the form

$$\lambda_j f = \sum_{i=1}^{p} \sum_{l=1}^{n_i} \alpha_{ij,l} f_i^{(l-1)}(t_j), \qquad t_j \in W, \tag{6.1}$$

where t_j and $\alpha_{ij,l}$, $1 \leqslant i \leqslant p$, $1 \leqslant l \leqslant n_i$ are known real numbers. Such functionals are a natural generalization to our setting of the so-called *extended Hermite–Birkhoff* (EHB) *functionals* [1]. With this restriction we shall refer to s as an *Lg-spline with* EHB *data.*

We assume that the *knots* t_j are ordered thus:

$$0 \leqslant t_1 \leqslant t_2 \leqslant \cdots \leqslant t_N \leqslant T. \tag{6.2}$$

If we define $1 \times n$ matrices, for $1 \leqslant j \leqslant N$,

$$c_j = (c_{j1}, c_{j2}, \dots, c_{jp}); \qquad c_{ji} = (\alpha_{ji,1}, \alpha_{ji,2}, \dots, \alpha_{ji,n_i}), \tag{6.3}$$

then

$$\lambda_j f = \sum_{j=1}^{p} c_{ji}(f_i(t_j), f_i^{(1)}(t_j), \dots, f_i^{(n_i-1)}(t_j))'. \tag{6.4}$$

Thus in view of (5.12g) the random variables $\lambda_j y$ are given by

$$\lambda_j y = c_j x(t_j), \tag{6.5}$$

where $x(\cdot)$ is the random vector defined in (5.12). ∎

Now the missing boundary conditions of (5.12) can be obtained:

THEOREM 6.1. *When the λ_j are of* EHB *type the boundary conditions of model* (5.12) *are given by:*

$$x(t_n) = \mathcal{O}^{-1} \left\{ \Lambda + \int_{t_1}^{t_n} \Delta(\xi)\, Bu(\xi)\, d\xi \right\}, \tag{6.6}$$

$$\Pi_n = E\{x(t_n)\, x'(t_n)\} = \mathcal{O}^{-1}\{I_n + Q\}\, \mathcal{O}^{-T}, \tag{6.7}$$

$$\begin{aligned} E\{x(t_n)\, u'(\xi)\} &= 0, && \xi < t_1 \ \text{or}\ \xi > t_n, \\ &= \mathcal{O}^{-1}\Delta(\xi)\, B, && t_1 \leqslant \xi \leqslant t_n, \end{aligned} \tag{6.8}$$

where

$$\Lambda' = (\lambda_1 y, \lambda_2 y, \dots, \lambda_n y), \tag{6.9}$$

$$\mathcal{O} = \begin{bmatrix} c_1\phi(t_1, t_n) \\ \vdots \\ c_{n-1}\phi(t_{n-1}, t_n) \\ c_n \end{bmatrix}, \qquad n \times n \text{ matrix}, \tag{6.10}$$

$$Q = \int_{t_1}^{t_n} \Delta(\xi)\, BB'\Delta'(\xi)\, d\xi, \qquad n \times n \text{ matrix}, \tag{6.11}$$

and $\Delta(\xi)$ is the $n \times n$ matrix with ith *row*

$$\begin{aligned} \Delta_i(\xi) &= c_i\phi(t_i, \xi), && t_i \leqslant \xi \leqslant t_n, \\ &= 0, && \text{otherwise}. \end{aligned} \tag{6.12}$$

Here ϕ is the state transition matrix of $A(\cdot)$, i.e.

$$\frac{\partial}{\partial t}\phi(t, \tau) = A(t)\,\phi(t, \tau), \qquad \phi(\tau, \tau) = I_n . \tag{6.13}$$

Proof. First, $y_j(t) \sim K_j{}^r(t, \cdot)$ implies that $z_i(\cdot) \sim \lambda_i y$, $1 \leqslant i \leqslant n$ (note (4.12), (5.6)). Thus (4.2) and Theorem 5.1 imply that for $1 \leqslant i, j \leqslant n$, $E\{(\lambda_i y)(\lambda_j y)\} = \delta_{ij}$ and hence

$$E\Lambda\Lambda' = I_n . \tag{6.14}$$

Also, $N_L \perp H_1$ implies that $\int_0^T G(\cdot, \xi) f(\xi)\, d\xi \perp z_j(\cdot)$, $1 \leqslant j \leqslant n$, for any $f \in \mathscr{L}_2{}^p(W)$. Thus in the congruent space Y we must have

$$\int_0^T f'(\xi)\, u(\xi)\, d\xi \perp \lambda_j y, \qquad 1 \leqslant j \leqslant n. \tag{6.15}$$

Using the definition (6.13) of ϕ it is easily seen that (5.12) has an integral form:

$$y(t) = Cx(t); \qquad x(t) = \phi(t, t_n)\, x(t_n) + \int_{t_n}^{t} \phi(t, \xi)\, Bu(\xi)\, d\xi. \tag{6.16}$$

Hence using (6.5)

$$\Lambda = \mathcal{O}x(t_n) - \int_{t_1}^{t_n} \Delta(\xi)\, Bu(\xi)\, d\xi, \tag{6.17}$$

of which (6.6) is an obvious consequence provided $\mathcal{O}$ is nonsingular. It is easily shown by using our technique of [2] that $\mathcal{O}$ is nonsingular if and only if condition (2.8) holds.

Now, in view of (6.4)–(6.15) we obtain from (6.17)

$$\mathcal{O}\Pi_n\mathcal{O}' = I + Q, \tag{6.18}$$

$$\begin{aligned} \mathcal{O}E\{x(t_n)\, u'(\xi)\} &= 0 + \int_{t_1}^{t_n} \Delta(\tau)\, BE\{u(\tau)\, u'(\xi)\}\, d\tau, \\ &= 0, \qquad \text{if} \quad \xi \notin [t_1, t_n] \\ &= \Delta(\xi)\, B, \qquad \text{if} \quad t_1 \leqslant \xi \leqslant t_n . \end{aligned} \tag{6.19}$$

Now (6.7)–(6.8) follow from (6.18)–(6.19). ∎

Having obtained a complete model for $\{y(t), t \in W\}$ we can now give a recursive algorithm for computing the sample function $\hat{y}(t)$ of its linear least-squares estimate $\mathscr{P}^{\mathscr{D}_N} y(t)$ given observations $\{\lambda_j y, 1 \leqslant j \leqslant N\}$. This, in view of Theorem 5.2 is also an algorithm for $s(\cdot)$.

VI.1. *Dynamical Recursive Algorithm*

$$s(t) = C\hat{x}(t \mid N), \qquad t \in W \tag{6.20}$$

where the $n \times 1$ vector function $\hat{x}(\cdot \mid N)$ is computed through the following recursive procedure consisting of algebraic updates at knots and differential equations (with determined end conditions) between adjacent knots.

Step 1. Initialization

(i) Compute Q_{j+1} and $M_{j+1}(t_{j+1})$ recursively for $1 \leqslant j \leqslant n-1$ through

$$M_1(t_1) = c_1; \qquad Q_1 = \mathbf{0} \quad (n \times n \text{ matrix}); \tag{6.21}$$

$$\Pi_j(t_j) = \mathbf{0} \qquad (n \times n \text{ matrix}), \qquad 1 \leqslant j \leqslant n-1 \tag{6.22}$$

$$\frac{d}{dt} M_j(t) = -M_j(t)\, A(t), \tag{6.23}$$

$$\frac{d}{dt} \Pi_j(t) = A(t)\, \Pi_j(t) + \Pi_j(t)\, A'(t) + BB', \qquad t_j \leqslant t \leqslant t_{j+1} \tag{6.24}$$

$$M'_{j+1}(t_{j+1}) = [M'_j(t_{j+1}) \mid c'_{j+1}) \tag{6.25}$$

$$Q_{j+1} = Q_j + \left[\begin{array}{c|c} M_j(t_{j+1}) \cdot \Pi_j(t_{j+1}) \cdot M'_j(t_{j+1}) & \mathbf{0}' \\ \hline \mathbf{0} & \mathbf{0} \end{array}\right] \tag{6.26}$$

(ii) Let $\mathcal{O} = M_n(t_n)$, $Q = Q_n$, and compute

$$x_0 = \mathcal{O}^{-1}(r_1, r_2, ..., r_n)'; \qquad \Pi_0 = \mathcal{O}^{-1} Q \mathcal{O}^{-T} \tag{6.27}$$

Step 2. Forward pass for $t \geqslant t_n$:

(iii) Set

$$\hat{x}(t_n \mid n) = x_0, \qquad P(t_n \mid n) = \Pi_0 . \tag{6.28}$$

(iv) Recursively compute and store $\{e_{j+1}, R^{\epsilon}_{j+1}, K'_{j+1}, n \leqslant j \leqslant N-1\}$ through

$$e_{j+1} = r_{j+1} - c_{j+1}\hat{x}(t_{j+1} \mid j); \qquad R^{\epsilon}_{j+1} = c_{j+1} P(t_{j+1} \mid j)\, c'_{j+1}, \tag{6.29}$$

$$K'_{j+1} = P(t_{j+1} \mid j)\, c'_{j+1}; \qquad (n \times 1 \text{ vector}), \tag{6.30}$$

where $\hat{x}(t_{j+1} \mid j)$, $P(t_{j+1} \mid j)$ are computed for $n \leqslant j \leqslant N-1$ through:

$$\frac{d}{dt} \hat{x}(t \mid j) = A(t)\, \hat{x}(t \mid j) \tag{6.31}$$

$$\frac{d}{dt} P(t \mid j) = A(t)\, P(t \mid j) + P(t \mid j)\, A'(t) + BB' \qquad t_j \leqslant t \leqslant t_{j+1} \tag{6.32}$$

$$\hat{x}(t_{j+1} \mid j+1) = \hat{x}(t_{j+1} \mid j) + K_{j+1}(R_{j+1}^{\epsilon})^{-1} e_{j+1} \tag{6.33}$$

$$P(t_{j+1} \mid j+1) = P(t_{j+1} \mid j) - K_{j+1}(R_{j+1}^{\epsilon})^{-1} K'_{j+1} \tag{6.34}$$

(v) Compute and store $\hat{x}(t_N \mid N)$ using (6.33).

Step 3. Computation of $\hat{x}(t \mid N)$:

(vi) Compute $\hat{x}(t \mid N)$ by integrating, starting at $t = t_N$ with the value of $\hat{x}(t_N \mid N)$ from step 2(v), the equations

$$\frac{d}{dt} \hat{x}(t \mid N) = A(t)\, \hat{x}(t \mid N) + BB'\mu(t \mid N) \tag{6.35}$$

where $\mu(t \mid N)$ is a piecewise continuous $n \times 1$ vector-valued function given by:

$$\begin{aligned} \mu(t \mid N) &= 0, & t < t_1 \quad \text{or} \quad t > t_N \\ &= \mu_j(t \mid N), & t_{j-1} < t < t_j\,, \quad 2 \leqslant j \leqslant N. \end{aligned} \tag{6.36}$$

(vii) The pieces $\mu_j(\cdot \mid N)$ are computed recursively, for $2 \leqslant j \leqslant N$, through

$$\mu_{N+1}(t_N \mid N) = 0 \tag{6.37}$$

$$\begin{aligned} \mu_j(t_j \mid N) &= \mu_{j+1}(t_j \mid N) + c_j'(R_j^{\epsilon})^{-1}\{e_j - K_j'\mu_{j+1}(t_j \mid N)\}, & j > n \\ &= \mu_{j+1}(t_j \mid N) - c_j'\beta_j\,, & j \leqslant n \end{aligned} \tag{6.38}$$

where

$$(\beta_1\,, \beta_2\,, \ldots, \beta_n)' = \mathcal{O}^{-T}\mu_{n+1}(t_n \mid N); \tag{6.39}$$

$$-\frac{d}{dt}\mu_j(t \mid N) = A'(t)\, \mu_j(t \mid N), \qquad t_{j-1} \leqslant t \leqslant t_j\,. \tag{6.40}$$

Remark 6.1. *Coincident Knots:*

If knots are coincident, i.e. $t_j = t_{j+1}$, then the integration of differential equations between t_j and t_{j+1} is trivialized; to wit, equations (6.23), (6.24), (6.31), (6.32), (6.41) are replaced by the identities:

$$M_j(t_{j+1}) = M_j(t_j), \tag{6.23$'$}$$

$$\Pi_j(t_{j+1}) = \Pi_j(t_{j+1}), \tag{6.24$'$}$$

$$\hat{x}(t_{j+1} \mid j) = \hat{x}(t_j \mid j), \tag{6.31$'$}$$

$$P(t_{j+1} \mid j) = P(t_j \mid j), \tag{6.32$'$}$$

$$\mu_j(t_{j-1} \mid N) = \mu_j(t_j \mid N). \tag{6.41$'$}$$

Remark 6.2. Equation (6.34) has an alternate form:

$$P(t_{j+1} \mid j+1) = \{I - K_{j+1}(R^{\epsilon}_{j+1})^{-1} c'_{j+1}\} P(t_{j+1} \mid j) \{I - K_{j+1}(R^{\epsilon}_{j+1})^{-1} c'_{j+1}\}'. \tag{6.34'}$$

The choice between (6.34) and (6.34′) is a tradeoff between the lower number of computations needed in (6.34) and the better numerical properties of (6.34′). For further remarks on this aspect see [10].

The proof of the algorithm is postponed till Section VIII; let us instead examine a structural theorem that follows from the algorithm.

VII. A Structural Theorem for the EHB-Data Case

In view of Theorem 5.2 it is clear that $s(\cdot)$ is an Lg-spline interpolating $\{r_j\}_1^N$ with respect to $\{\lambda_j\}_1^N$ *if and only if* $s(\cdot) = \hat{y}(\cdot)$, the sample function of the 1.1.s.e. of $y(\cdot)$ corresponding to observations $\lambda_j y = r_j$, $1 \leqslant j \leqslant N$. Now in turn the construction of the previous section shows that $\hat{y}(\cdot)$ is this 1.1.s.e. if and only if it is the solution of the algorithm (6.20–6.40). This argument is central to establishing the structural theorem of this section.

First, from (6.35), (6.20) and (5.12b–g) it is immediate that if $s = (s_1, s_2, \ldots, s_p)'$ then

$$\hat{x}(\cdot \mid N) = (s_1, s_1^{(1)}, \ldots, s_1^{(n_1-1)} \mid \cdots \mid s_p, s_p^{(1)}, \ldots, s_p^{(n_p-1)})' \tag{7.1}$$

and that

$$Ls(t) = w(t) = B'\mu(t \mid N). \tag{7.2}$$

The $p \times 1$ vector-valued function $w(\cdot)$ in turn has the state model (use (6.40)):

$$w(t) = B'\mu(t \mid N), \qquad -\frac{d}{dt}\mu(t \mid N) = A'(t)\,\mu(t \mid N); \qquad t_{j-1} < t < t_j. \tag{7.3}$$

Now, using the defining relations (5.12) for $A(\cdot)$ and B, it is quite easy to show that (7.3) is equivalent to

$$L^* w(t) = 0, \qquad t_{j-1} < t < t_j, \quad 2 \leqslant j \leqslant N, \tag{7.4}$$

where L^* is a $p \times p$ matrix whose ijth entry L^*_{ij} is the ordinary differential operator given by:

$$L^*_{ij} f = \sum_{k=0}^{n_i} (-1)^k D^k\{a_{ji,k} f\}, \qquad 1 \leqslant i, \ j \leqslant p, \tag{7.5}$$

Furthermore, straightforward calculations show that (6.40) implies that

$$\mu(t \mid N) = (\Theta_{1,0}s(t), \Theta_{1,1}s(t), \ldots, \Theta_{1,n_1-1}s(t), \ldots, \Theta_{p,0}s(t), \Theta_{p,1}s(t), \ldots, \Theta_{p,n_p-1}s(t))' \tag{7.6}$$

where $\Theta_{i,j}$, are $1 \times p$ rows of ordinary differential operators defined for $1 \leqslant i \leqslant p$, $0 \leqslant j \leqslant n_i - 1$ by

$$\Theta_{i,j}s(t) = [\theta_{i,j}(1), \theta_{i,j}(2), \ldots, \theta_{i,j}(p)]\, Ls(t), \tag{7.7a}$$

$$\theta_{i,j}(k) f = \sum_{l=0}^{n_i-1-j} (-1)^l D^l\{a_{ki,j+1+l}f\}. \tag{7.7b}$$

A knot $\tau = t_{j+1}$ shall be said to have *order* $l(\tau)$ if

$$t_j < t_{j+1} = t_{j+2} = \cdots = t_{j+l(\tau)} < t_{j+l(\tau)+1}; \qquad \tau = t_{j+1}. \tag{7.8}$$

Let

$$[f]_\tau = f(\tau+) - f(\tau-). \tag{7.9}$$

Then from (6.38) we have:

$$[\mu(\tau \mid N)]_\tau = \begin{bmatrix} [\Theta_{1,0}s]_\tau \\ \vdots \\ [\Theta_{p,n_p-1}s]_\tau \end{bmatrix} = [c'_{j+1}, c'_{j+2}, \ldots, c'_{j+l(\tau)}] \begin{bmatrix} \alpha_1 \\ \alpha_2 \\ \vdots \\ \alpha_{l(\tau)} \end{bmatrix} = \mathbf{M}\boldsymbol{\alpha} \tag{7.10}$$

for suitable real numbers α_j.

The linear independence of $\{\lambda_k, j + 1 \leqslant k \leqslant j + l(\tau)\}$ implies that $l(\tau) \leqslant n$ and that the $n \times l(\tau)$ matrix $\mathbf{M}$ has full rank. Thus $\mathbf{M}$ can be extended to a full rank $n \times n$ matrix $\mathbf{M}_+ = [\mathbf{M} \mid \Gamma]$, which together with the $n \times 1$ vector $\boldsymbol{\alpha}_+ = (\boldsymbol{\alpha}' \mid \mathbf{0})'$ allows us to rewrite (7.10) as

$$[\mu(\tau \mid N)]_\tau = \mathbf{M}_+\boldsymbol{\alpha}_+$$

or that

$$\mathbf{M}_+^{-1}[\mu(\tau \mid N)]_\tau = [\alpha_1, \alpha_2, \ldots, \alpha_{l(\tau)}, 0, \ldots, 0]'. \tag{7.11}$$

Let η_{ij} denote the ijth entry of $\mathbf{M}_+^{-1}$ and define row operators, for $1 \leqslant j \leqslant n$,

$$R_j^{(\tau)} = \sum_{i=1}^{p} \sum_{l=1}^{n_i} \eta_{j,N_i+l} \cdot \Theta_{i,l-1}; \qquad N_i = \sum_{k=1}^{i-1} n_k. \tag{7.12}$$

Then (7.22) is equivalent to

$$\begin{aligned} [R_j^{(\tau)}s]_\tau &= \alpha_j, && 1 \leqslant j \leqslant l(\tau) \\ &= 0, && l(\tau) + 1 \leqslant j \leqslant n. \end{aligned} \tag{7.13}$$

In view of the remarks at the beginning of the section (note that $\mu(t \mid N) = 0$

for $t < t_1$ or $t > t_N$) we have thus established the following generalization of Theorem 3.6 of [1]:

THEOREM 7.1. *$s \in H$ is an Lg-spline interpolating data $\{r_j, 1 \leqslant j \leqslant N\}$ with respect to* EHB $\{\lambda_j, 1 \leqslant j \leqslant N\}$ *on* $[0, T]$ *if and only if*

(a) $L^*Ls(t) = 0$, *when t is not a knot,*

(b) $Ls(t) = 0$, for $t < t_1$ *or* $t > t_N$,

(c) $\lambda_i s = r_i$, $1 \leqslant i \leqslant N$,

(d) $[R_i^{(\tau)}s]_\tau = 0$, *for* $l(\tau) + 1 \leqslant i \leqslant n$ *if τ is a knot of order* $l(\tau)$. ∎

Practical exploitation of Jerome and Schumaker's theorem has been investigated by various authors (see for example [21–22]).

VIII. Proof of Algorithm (6.20–6.40)

The basic idea is to exploit Theorem 5.2, and the known lumped model (5.12), (6.6–6.13) of the process $y(\cdot)$, to compute the sample function $\hat{y}(t)$ of the l.l.s.e. $\mathscr{P}^{\mathscr{D}_N}y(t)$ corresponding to $\lambda_j y = r_j$, $1 \leqslant j \leqslant N$.

Equations (6.21)–(6.26) are an algorithmic form of (6.6)–(6.13). It relies on the semigroup property of $\phi(\cdot, \cdot)$, namely that

$$\phi(t, \xi)\,\phi(\xi, \tau) = \phi(t, \tau) \qquad \text{all } t, \xi, \tau. \tag{8.1}$$

Hence $\phi(t, \xi)\,\phi(\xi, t) = \phi(t, t) = I_n$ and thus using (6.13):

$$\frac{\partial}{\partial t}\phi(\xi, t) = -\phi(\xi, t)\,A(t). \tag{8.2}$$

Now let $M_j(\cdot)$, $1 \leqslant j \leqslant n-1$, be $j \times n$ matrix-valued functions with ith row $c_i\phi(t_i, \cdot)$. Clearly, then $\mathcal{O} = M_n(t_n)$ and (6.21) and (6.25) are obvious consequences. Furthermore using (8.1), $M_j(t) = M_j(t_j)\,\phi(t_j, t)$ which in view of (7.2) gives us (6.23).

Next, let us define the $n \times n$ matrices, $1 \leqslant j \leqslant n-1$,

$$V_j(t) = [\Delta_1'(t) \mid \cdots \mid \Delta_j'(t) \mid \mathbf{0}]', \tag{8.3}$$

$$\Pi_j(t) = \int_{t_j}^{t} \phi(t, \xi)\,BB'\phi'(t, \xi)\,d\xi, \qquad t_j \leqslant t \leqslant t_{j-1}. \tag{8.4}$$

Thus (6.22) and (6.24) follow readily. Also, an account of (8.1)

$$V_j(t) = \left[\frac{M_j(t_{j+1})}{\mathbf{0}}\right]\phi(t_{j+1}, t)$$

and hence from (6.11)–(6.12)

$$Q = \sum_{i=1}^{n-1} \int_{t_i}^{t_{i+1}} V_i(\xi)\, BB' V_i'(\xi)\, d\xi = \sum_{i=1}^{n-1} \left[\begin{array}{c|c} M_i(t_{i+1})\, \Pi_i(t_{i+1})\, M_i'(t_{i+1}) & \mathbf{0} \\ \hline \mathbf{0} & 0 \end{array} \right]. \tag{8.5}$$

Clearly $Q = Q_n$ and the Q_j defined below obey the recursion (6.26):

$$Q_j = \sum_{i=1}^{j-1} \left[\begin{array}{c|c} M_i(t_{i+1})\, \Pi_i(t_{i+1})\, M_i'(t_{i+1}) & \mathbf{0} \\ \hline \mathbf{0} & 0 \end{array} \right]. \tag{8.6}$$

For $k \geqslant 1$ let $\mathscr{D}_k = \operatorname{span}\{\lambda_1 y, \ldots, \lambda_k y\}$ and $\mathscr{P}_k$ denote the projection operation onto $\mathscr{D}_k$. First (5.12a) and the linearity of $\mathscr{P}_k$ give $\mathscr{P}_N y(t) = C\mathscr{P}_N x(t)$. Thus (6.20) follows with the notation

$$\hat{x}(t \mid k) = \text{sample value of } \mathscr{P}_k x(t) \text{ corresponding to}$$
$$\lambda_j y = r_j, \qquad 1 \leqslant j \leqslant k. \tag{8.7}$$

We shall also use the following

$$\mathscr{E}_k x(t) \triangleq x(t) - \mathscr{P}_k x(t), \tag{8.8}$$

$$\Pi(t) \triangleq Ex(t)\, x'(t), \qquad \Sigma(t \mid k) \triangleq E[\mathscr{P}_k x(t)]\, [\mathscr{P}_k x(t)]' \tag{8.9}$$

$$P(t \mid k) \triangleq E[\mathscr{E}_k x(t)]\, [\mathscr{E}_k x(t)]'. \tag{8.10}$$

It is clear that $\mathscr{E}_k x(t) \perp \mathscr{P}_k x(t)$ and thus that

$$\Pi(t) = \Sigma(t \mid k) + P(t \mid k). \tag{8.11}$$

As a consequence of (6.15), $\mathscr{P}_n \int_{t_1}^{t_n} \Delta(\xi)\, Bu(\xi)\, d\xi = 0$ and hence from (6.6) $\mathscr{P}_n x(t_n) = \mathcal{O}^{-1}\Lambda$ which readily gives $\hat{x}(t_n \mid n) = \mathcal{O}^{-1}(r_1, r_2, \ldots, r_n)'$ and $\Sigma(t_n \mid n) = \mathcal{O}^{-1}\mathcal{O}^{-T}$. Thus $P(t_n \mid n) = \Pi(t_n) - \mathcal{O}^{-1}\mathcal{O}^{-T}$. In view of (8.11) one then has (6.27–6.28).

The basic idea of the rest of the algorithm is to compute an orthogonal basis $\{\epsilon_j, 1 \leqslant j \leqslant N\}$ for $\mathscr{D}_N$ and then calculate $\mathscr{P}_N x(t)$ as a linear combination of this new basis. In the context of stochastic provesses this Gram–Schmidt procedure has been known as the *Innovations method* [18]. Define

$$\epsilon_j = \lambda_j y - \mathscr{P}_{j-1}(\lambda_j y), \qquad 1 \leqslant j \leqslant N. \tag{8.12}$$

In view of (4.12) and (5.6), for $1 \leqslant j \leqslant n$, $\epsilon_j = \lambda_j y$; and from (6.5), for $j \geqslant n+1$

$$\epsilon_j = \lambda_j y - c_j \mathscr{P}_{j-1} x(t_j) = c_j \mathscr{E}_{j-1} x(t_j). \tag{8.13}$$

Thus if e_j denote the sample values of the ϵ_j we have

$$\begin{aligned} e_j &= r_j , & 1 \leqslant j \leqslant n \\ &= r_j - c_j\hat{x}(t_j \mid j-1), & n+1 \leqslant j \leqslant N; \\ R_j^{\epsilon} = E\epsilon_j^2 &= 1, & 1 \leqslant j \leqslant n \\ &= c_j P(t_j \mid j-1)\, c_j' , & j \geqslant n+1 \end{aligned} \tag{8.14}$$

as in (6.29). Now from (8.12) it is clear that

$$E\epsilon_i\epsilon_j = R_i^{\epsilon}\delta_{ij}; \qquad \mathscr{D}_k = \operatorname{span}\{\epsilon_1, \epsilon_2, \ldots, \epsilon_k\} \tag{8.15}$$

and thus that for any random variable $w \in Y$,

$$\mathscr{P}_k w = \sum_{j=1}^{k} E\{w\epsilon_j\}\,(R_j^{\epsilon})^{-1}\,\epsilon_j = \mathscr{P}_{k-1}w + E\{w\epsilon_k\}\,(R_k)^{-1}\,\epsilon_k; \qquad k \geqslant 1, \tag{8.16}$$

a recursive relation of great importance below.

First (8.16) with $w = x(t_{j+1})$ and $k = j+1$ gives

$$\mathscr{P}_{j+1}x(t_{j+1}) = \mathscr{P}_j x(t_{j+1}) + E\{x(t_{j+1})\,\epsilon_{j+1}\}\,(R_{j+1}^{\epsilon})^{-1}\,\epsilon_{j+1} \tag{8.17}$$

from which we have (6.33) with

$$K_{j+1}' = E\{x(t_{j+1})\,\epsilon_{j+1}\} \tag{8.18}$$

Equation (8.17) in turn, upon using (8.10) and (8.13), gives us (6.30). Also, since $\epsilon_{j+1} \perp \mathscr{D}_j$ it is clear that we have

$$\Sigma(t_{j+1} \mid j+1) = \Sigma(t_{j+1} \mid j) + K_{j+1}(R_{j+1}^{\epsilon})^{-1}\,K_{j+1}' . \tag{8.19}$$

This together with (8.11) gives (6.34). Next, from (5.12) and (6.13) one has

$$x(t) = \phi(t, t_j)\,x(t_j) + \int_{t_j}^{t} \phi(t, \xi)\,Bu(\xi)\,d\xi \tag{8.20}$$

which, for $t \geqslant t_j \geqslant t_n$, gives (recall that $Eu(\xi)\,x(t_n) = 0,\ \xi \geqslant t_n$)

$$P_j x(t) = \phi(t, t_j)\,P_j x(t_j) \tag{8.21}$$

of which (6.31) is a ready consequence. Also we then have $\Sigma(t \mid j) = \phi(t, t_j)\,\Sigma(t_j \mid j)\,\phi'(t, t_j)$ and from (8.20),

$$\Pi(t) = \phi(t, t_j)\,\Pi(t_j)\,\phi'(t, t_j) + \int_{t_1}^{t} \phi(t, \xi)\,BB'\phi'(t, \xi)\,d\xi,$$

the two of which together with (8.11) give (6.32).

Furthermore from (8.20)–(8.21)

$$\mathscr{E}_j x(t) = \phi(t, t_j)\, \mathscr{E}_j x(t_j) + \int_{t_j}^{t} \phi(t, \xi)\, Bu(\xi)\, d\xi.$$

Set $t = t_{j+1}$ and for $\mathscr{P}_j x(t_j)$ substitute from (8.17). Thus

$$\mathscr{E}_j x(t_{j+1}) = \Psi_{j+1,j} \mathscr{E}_{j-1} x(t_j) + \int_{t_j}^{t_{j+1}} \phi(t_{j+1}, \xi)\, Bu(\xi)\, d\xi \tag{8.22}$$

where

$$\Psi_{j+1,j} = \phi(t_{j+1}, t_j)\, \{I_p - K_j (R_j^{\epsilon})^{-1} c_j\}. \tag{8.23}$$

Next for $t \geqslant t_N$

$$x(t) = \phi(t, t_N)\, x(t_N) + \int_{t_N}^{t} \phi(t, \xi)\, Bu(\xi)\, d\xi.$$

Thus in view of (6.8)

$$\mathscr{P}_N x(t) = \phi(t, t_N)\, \mathscr{P}_N x(t_N) \qquad \text{for } t \geqslant t_N$$

or $\hat{x}(t \mid N) = \phi(t, t_N)\, \hat{x}(t_N \mid N)$ which gives (6.35) for the case of $t \geqslant t_N$. The case of $t \leqslant t_1$ is similar.

Let $t_n \leqslant t \leqslant t_N$. Suppose now that $t_{j-1} \leqslant t \leqslant t_j$ then

$$x(t) = \phi(t, t_{j-1})\, x(t_{j-1}) + \int_{t_{j-1}}^{t} \phi(t, \xi)\, Bu(\xi)\, d\xi$$

or

$$\mathscr{P}_N x(t) = \phi(t, t_{j-1})\, \mathscr{P}_N x(t_{j-1}) + \mathscr{P}_N \int_{t_{j-1}}^{t} \phi(t, \xi)\, Bu(\xi)\, d\xi. \tag{8.24}$$

But it is clear that since $E\{u(\xi)\, x'(\tau)\} = 0$ for $\xi > \tau$ we have $E\{u(\xi)\, \epsilon_l\} = 0$ for $l \leqslant j - 1$. Thus

$$\mathscr{P}_N \int_{t_{j-1}}^{t} \phi(t, \xi)\, Bu(\xi)\, d\xi = \sum_{l=j}^{N} \int_{t_{j-1}}^{t} \phi(t, \xi)\, BE\{u(\xi)\, \epsilon_l\}\, (R_l^{\epsilon})^{-1}\, \epsilon_l\, d\xi. \tag{8.25}$$

But using (8.13):

$$E\{u(\xi)\, \epsilon_l\} = E\{u(\xi)\, \mathscr{E}_{l-1} x'(t_l)\}\, c_l'. \tag{8.26}$$

which upon repeatedly using (8.22) gives

$$E\{u(\xi)\, \epsilon_l\} = \left[E\{u(\xi)\, \mathscr{E}_{j-2} x'(t_{j-1})\}\, \Psi_{j,j-1}' + \int_{t_{j-1}}^{t_j} E\{u(\xi)\, u'(\theta)\}\, B' \phi'(t_j, \theta)\, d\theta\right] \Psi_{l,j}' c_l'$$

$$= B' \phi'(t_j, \xi)\, \Psi_{l,j}' c_l'; \qquad l \geqslant j; \quad t_{j-1} < \xi < t_j. \tag{8.27}$$

Thus from (8.25)–(8.27)

$$\hat{x}(t \mid N) = \phi(t, t_{j-1})\, \hat{x}(t_{j-1} \mid N) + \int_{t_{j-1}}^{t} \phi(t, \xi)\, BB'\phi'(t_j, \xi)\, d\xi \cdot \sum_{l=j}^{N} \Psi'_{l,j} c'_l (R_l^\epsilon)^{-1} e_l . \tag{8.28}$$

which gives

$$\frac{d}{dt} \hat{x}(t \mid N) = A(t)\, \hat{x}(t \mid N) + BB' \left\{ \phi'(t_j, t) \sum_{l=j}^{N} \Psi'_{l,j} c'_l (R_l^\epsilon)^{-1} e_l \right\} .$$

This in turn gives (6.35) with

$$\mu(t \mid N) = \mu_j(t \mid N) = \phi'(t_j, t) \sum_{l=j}^{N} \Psi'_{l,j} c'_l (R_l^\epsilon)^{-1} e_l; \qquad t_{j-1} \leqslant t \leqslant t_j . \tag{8.29}$$

Equations (6.37), (6.40) are then immediate. Note, also that

$$\mu_j(t_j \mid N) = \sum_{l=j+1}^{N} \Psi'_{l,j} c'_l (R_l^\epsilon)^{-1} e_l + c'_j (R_j^\epsilon)^{-1} \epsilon_j$$

which upon using (8.29) gives (6.39a).

The case of $t_1 < t < t_n$ is a little more difficult. First of all as in (8.28), now

$$\hat{x}(t \mid N) = \phi(t, t_n)\, \hat{x}(t_n \mid N) + \sum_{l=n+1}^{N} \int_{t_n}^{t} \phi(t, \xi)\, BE\{u(\xi)\, \epsilon_l\}\, (R_l^\epsilon)^{-1} \epsilon_l\, d\xi. \tag{8.30}$$

But using (8.13) and (8.22) [note also the whiteness of $u(\cdot)$, (5.11)]

$$E\{u(\xi)\, \epsilon_l\} = E\{u(\xi)\, x'(t_{n+1})\}\, \Psi'_{l,n+1} c'_l , \qquad l \geqslant n+1.$$

However $x(t_{n+1}) = \phi(t_{n+1}, t_n)\, x(t_n) + \int_{t_n}^{t_{n+1}} \phi(t_{n+1}, \xi)\, Bu(\xi)\, d\xi$. Thus (use (6.8))

$$\begin{aligned} E\{u(\xi)\, \epsilon_l\} &= E\{u(\xi)\, x'(t_n)\}\, \phi'(t_{n+1}, t_n)\, \Psi'_{l,n+1} c'_l \\ &= B'\Delta'(\xi)\, \mathcal{O}^{-T} \phi'(t_{n+1}, t_n)\, \Psi'_{l,n+1} c'_l , \qquad l \geqslant n+1. \end{aligned} \tag{8.31}$$

Thus from (8.29)–(8.31):

$$\frac{d}{dt} \hat{x}(t \mid N) = A(t)\, \hat{x}(t \mid N) + BB'\Delta'(t)\, \mathcal{O}^{-T} \mu_{n+1}(t_n \mid N)$$

which is (6.35) with

$$\mu(t \mid N) = \Delta'(t)\, \mathcal{O}^{-T} \mu_{n+1}(t_n \mid N), \qquad t_1 \leqslant t \leqslant t_n .$$

Then (6.38b)–(6.39) are easily shown using (6.10) and (6.12).

IX. Concluding Remarks

This paper has been aimed at extending logically and completely the existent theory of Lg-splines to the situation of simultaneous interpolation of several functions. In later papers in the series we shall examine the same extension for smoothing splines and ARMA splines. In particular, the latter brings with it an increasing complexity as far as the various aspects considered in Appendix I are concerned.

APPENDIX I: The Structure of $Ly = u$

We have defined in (2.2) the $p \times p$ matrix $L(D)$ of ordinary differential operators $L_{ij}(D)$ subject to the restriction (2.3).

Given a $p \times p$ matrix $M(D)$ of polynomials let p_j, $1 \leqslant j \leqslant p$, denote the degree of the highest degree term in column j of $M(D)$. Now, let us define Γ_M as the matrix obtained as follows. Place in each position ij of Γ_M the coefficient of D^{p_j} in the ijth entry of $M(D)$. Matrix $M(D)$ is said to be *column proper* [20] if Γ_M is nonsingular. Now, restriction (2.3) ensures that for $L(D)$, we have $p_j = n_j$, $1 \leqslant j \leqslant p$, and that Γ_L is a $p \times p$ identity matrix and hence that $L(D)$ is column proper. This is central to the rest of the development.

Using various, by now well-known, results on the canonical forms of linear systems (see for example [20]) it is easily shown that the system of p differential equations

$$Ly = u \tag{A.1}$$

with $y \in H$ and $u \in \mathscr{L}_2{}^p$ has the state-space representation of equations (5.12). This can be verified by direct expansion of (5.12). Now we establish Lemma 3.1.

Proof of Lemma 3.1. Note that $y \in N_L$ if and only if $Ly = 0$. In terms of (5.12), $N_L = \{y \in H: y(t) = Cx(t), dx(t)/dt = A(t)\, x(t)\}$. It is easily seen then that, picking any $t_0 \in W$, $y \in N_L$ can be written as $y(t) = C\phi(t, t_0)\, x_0$ where $x_0 = x(t_0)$ is an arbitrary $n \times 1$ real vector. In other words

$$N_L = \text{span}\{\text{columns of } C\phi(\cdot, t_0)\}. \tag{A.2}$$

However we can show that these n columns are independent in H and hence that $\dim N_L = n$. To do so it is enough to show that $y(\cdot) = C\phi(\cdot, t_0)\, x_0 = 0$ implies $x_0 = 0$. But using (5.12), with $u(\cdot) \equiv 0$, it is easily shown that $y(\cdot) \equiv 0$ implies $\{y_j^{(k)}(t_0) = 0, 1 \leqslant j \leqslant p, 0 \leqslant k \leqslant n_j - 1\}$ and thus in view of (5.12g) that $x_0 = x_0(t_0) = 0$, which completes the proof. ∎

Strictly speaking, restriction (2.3) can be eliminated. All that is needed is that $L(D)$ be a column proper polynomial matrix. Then Lemma 3.1 and all the subsequent results continue to hold. However, $A(\cdot)$, B, and C become more complex in structure. One can then, in fact, reduce the equations $L(D)\,y = u$, by forming suitable linear combinations, to a new form $\bar{L}(D)\,y = \bar{u} = \bar{M}u$, where $\bar{M}$ is an invertible $p \times p$ matrix of constants such that $\bar{L}(D) = \bar{M}L(D)$ satisfies the condition $\Gamma_{\bar{L}} = I_p$; then $\bar{L}(D)\,y = \bar{M}u$ gives rise to suitable $A(\cdot)$, B, C matrices with minor changes in B and C.

A more difficult situation arises when $L(D)$ is not column proper. Then a suitable row operation procedure [20] is needed to reduce it to a column proper form $\bar{L}(D)$. But then the column degrees of $\bar{L}(D)$ obey $p_j \leqslant n_j$, and consequently $n = \dim N_L \leqslant \sum_{j=1}^{p} n_j$; with this modification the rest of the results continue to hold, modulo certain changes of detail.

References

1. J. W. Jerome and L. L. Schumaker, On Lg-splines, *J. Approximation Theory* **2** (1969), 29–49.
2. H. L. Weinert and G. S. Sidhu, On uniqueness conditions for optimal curve fitting, *J. Optimization Theory Appl.* **2** (October 1977), 211–216.
3. A. N. Netravali and R. J. P. de Figueiredo, On a class of minimum energy controls related to spline functions, *IEEE Trans. Automatic Control.* *AC*-**21** (1976), 725–727.
4. H. L. Weinert and T. Kailath, A spline-theoretic approach to minimum energy control, *IEEE Trans. Automatic Control* *AC*-**21** (1976), 391–393.
5. H. L. Weinert and G. S. Sidhu, "A Spline Theoretic Approach to Minimum Energy Control—Part II: Systems with Numerator Dynamics and ARMA Splines," Tech. Rpt. 75-11, Dept. of Electrical Engineering, The Johns Hopkins University, Baltimore, Maryland, 1975.
6. R. J. P. de Figueiredo, "LMg-Splines," Report 75-10, Dept. of Electrical Engineering, Rice University, Houston, Texas, 1975.
7. D. G. Luenberger, "Optimization by Vector Space Methods," Wiley, New York, 1969.
8. N. Aronszajn, Theory of reproducing kernels, *Trans. Amer. Math. Soc.* **68** (1950), 337–404.
9. J. Mercer, Functions of positive and negative type and their connection with the theory of integral equations, *Philos. Trans. Roy. Soc. London Ser. A* **209** (1909), 415–446.
10. G. S. Sidhu and H. L. Weinert, Dynamic recursive algorithms for Lg-spline interpolation of EHB data, *Appl. Math. and Computation*, in press.
11. E. Parzen, "Time Series Analysis Papers," p. 267, Holden-Day, San Francisco, 1967.
12. M. Loève, Fonctions aleàtoires du second ordre, *in* "Processus stochastiques et mouvement Brownien" (P. Lévy, Ed.), 2nd ed., pp. 367–420, Gauthier-Villars, Paris, 1965.
13. G. Kimeldorf and G. Wahba, Spline functions and stochastic processes, *Sankhyā Ser. A* **132** (1970), 173–180.
14. G. Kimeldorf and G. Wahba, A correspondence between Bayesian estimation on stochastic processes and smoothing by splines, *Ann. Math. Statist.* **41** (1970), 495–502.
15. H. L. Weinert, "A Reproducing Kernel Hibert Space Approach to Spline Problems with Applications in Estimation and Control," Ph.D. thesis, Dept. of Electrical Engineering, Stanford University, Stanford, Calif., 1972.

16. H. L. WEINERT AND T. KAILATH, Stochastic interpretations and recursive algorithms for spline functions, *Ann. of Statist.* **2** (1974), 787–794.
17. H. L. WEINERT AND G. S. SIDHU, A stochastic framework for recursive computation of spline functions—Part I: Interpolating splines, *IEEE Trans. Information Theory,* **IT-24** (1978), 45–50.
18. T. KAILATH, The innovations approach to detection and estimation theory, *Proc. IEEE* **58** (May 1970), 680–695.
19. H. L. WEINERT AND G. S. SIDHU, "Stochastic Error Analysis of Spline Interpolation," JHU-EE Report 77-4, Dept. of Electrical Engineering, The Johns Hopkins University, Baltimore, Maryland, 1977.
20. W. A. WOLOVICH, "Linear Multivariable Systems," Springer-Verlag, New York/Berlin, 1974.
21. C. CARASSO AND P. J. LAURENT, On the numerical construction and practical use of interpolating spline functions, "Proc. IFIP Congress Information Processing 68 (Edinburgh 1968)," Vol. 1, pp. 86–89, *MR* **40**, No. 8219.
22. M. J. MUNTEANU AND L. L. SCHUMAKER, On a method of Carasso and Laurent for constructing interpolating splines, *Math. Comp.* **27**, No. 122 (April 1973), 317–325.

25

Reprinted from *J. Optimization Theory Appl.* **30:**255–268 (1980)

A Stochastic Framework for Recursive Computation of Spline Functions: Part II, Smoothing Splines

H. L. WEINERT,[1] R. H. BYRD,[2] AND G. S. SIDHU[3]

Communicated by D. G. Luenberger

Abstract. Many of the optimal curve-fitting problems arising in approximation theory have the same structure as certain estimation problems involving random processes. We develop this structural correspondence for the problem of smoothing inaccurate data with splines and show that the smoothing spline is a sample function of a certain linear least-squares estimate. Estimation techniques are then used to derive a recursive algorithm for spline smoothing.

Key Words. Spline smoothing, reproducing kernels, least-squares estimation, recursive algorithms.

1. Introduction

The problem that we address in this paper is that of reconstructing a function from a finite collection of inaccurate measurements. The reconstruction problem can be approached in many ways. We could, for example, specify probability distributions for the quantities of interest and then apply an optimal estimation algorithm. However, in many engineering and numerical analysis settings, we may be unwilling or unable to apply probabilistic modeling techniques directly, either because the quantities of interest are patently nonrandom or because the relevant statistics are unknown. In these situations, spline smoothing has proved a useful alternative. In order to apply this technique, we assume that the unknown

[1] Assistant Professor, Department of Electrical Engineering, The Johns Hopkins University, Baltimore, Maryland.
[2] Assistant Professor, Department of Mathematical Sciences, The Johns Hopkins University, Baltimore, Maryland.
[3] Professor Investigador, Instituto de Investigaciones en Matematicas Aplicadas y en Sistemas, Universidad Nacional Autonoma de Mexico, Mexico City, Mexico.

function is in the Hilbert space H, where

$$H=\left\{f(t),\ t\in[0,T]\colon f^{(n)} \text{ exists a.e. and } \int_0^T [f^{(n)}]^2<\infty\right\}, \tag{1}$$

$f^{(n)}$ is the nth derivative of f, and n is a fixed, strictly positive integer. Let L be the following differential operator with domain H:

$$L=(d^n/dt^n)+a_{n-1}(t)(d^{n-1}/dt^{n-1})+\cdots+a_1(t)(d/dt)+a_0(t), \tag{2}$$

where $a_j(\cdot)$ has j continuous derivatives on $[0, T]$. Suppose that $\{\lambda_j\}_1^N$, $N\geq n$, are linear measurement functionals on H that are continuous and linearly independent. Let $\{\rho_j\}_1^N$ be strictly positive weights, and let $\{r_j\}_1^N$ denote the measurement values. The smoothing spline is defined as follows.

Definition 1.1. A function $s(\cdot)\in H$ is an Lg-spline smoothing $\{r_j\}_1^N$ with respect to $\{\lambda_j\}_1^N$ and $\{\rho_j\}_1^N$ if it solves the following optimization problem:

$$\underset{f\in H}{\text{minimize}}\left\{\int_0^T (Lf)^2+\sum_{j=1}^N \rho_j^{-1}(r_j-\lambda_j f)^2\right\}. \tag{3}$$

The meaing of the cost functional (3) becomes clear if one considers two limiting cases. First, as all the weights approach zero, the problem in (3) reduces to that of minimizing just the integral term, subject to the constraints

$$\lambda_j f=r_j, \qquad 1\leq j\leq N.$$

The solution to this limiting problem is just the Lg-spline *interpolating* $\{r_j\}_1^N$ with respect to $\{\lambda_j\}_1^N$. On the other hand, as all the weights approach infinity, the solution to (3) approaches the function in the null space of L that gives a least-squares fit to the data $\{r_j\}_1^N$. Thus, the smoothing spline provides a compromise between fidelity to the data and smoothness of the reconstruction.

Note that we are making no direct probabilistic assumptions here. Instead, we must choose an operator L, which determines the functional form of the smoothing spline, and weights $\{\rho_j\}_1^N$, which reflect our degree of confidence in the data $\{r_j\}_1^N$. As we shall see, however, these choices determine the statistics of a signal-in-noise linear least-squares estimation problem such that the smoothing spline is a sample function of the optimal estimate. This result, which is developed using reproducing kernel Hilbert space (RKHS) methods, permits us to use estimation techniques to recursively compute smoothing splines. Although numerous authors (Refs. 1–9) have obtained formulas and characterizations for smoothing splines of varying degrees of generality, all the existing algorithms are nonrecursive. A recursive algorithm is needed for real-time applications and those in which data storage capabilities are limited. In the course of our development, we also present new results on the structure of the spline-smoothing problem. A preliminary version of this work was presented in Ref. 10.

This paper is a sequel to Ref. 11, which deals with the limiting case of spline interpolation.

2. Construction of RKHS

We first note that a solution to problem (3) always exists and is unique if $\{\lambda_j\}_1^n$ are linearly independent on the null space of L, that is, if $Lf = 0$ and $\lambda_j f = 0$, $1 \le j \le n$, imply $f \equiv 0$. We will assume in all that follows that this condition holds (see Refs. 3 and 12 for more details).

Let $\{z_i(\cdot)\}_1^n$ be the basis for the null space of L that is dual to $\{\lambda_j\}_1^n$; i.e.,

$$Lz_j = 0, \qquad \lambda_i z_j = \begin{cases} 1, & i = j, \\ 0, & i \neq j, \end{cases} \qquad 1 \le (i, j) \le n, \tag{4}$$

and let $G(\cdot, \cdot)$ be the Green's function for L satisfying

$$LG(\cdot, \tau) = \delta(\tau - \cdot), \qquad \lambda_j G(\cdot, \tau) = 0, \qquad 1 \le j \le n, \qquad \tau \in [0, T]. \tag{5}$$

Then, any $f \in H$ can be uniquely decomposed as

$$f(t) = \sum_{j=1}^{n} (\lambda_j f) z_j(t) + \int_0^T G(t, \tau)[Lf(\tau)]\, d\tau. \tag{6}$$

Now, consider the space H^+ of ordered pairs:

$$H^+ = \{(f(\cdot), \theta) \colon f(\cdot) \in H \text{ and } \theta = [\theta_1, \theta_2, \dots, \theta_N]' \in E^N\}. \tag{7}$$

It can easily be shown using (6) that H^+ is a Hilbert space with inner product

$$\langle (f, \theta), (g, \omega) \rangle = \int_0^T (Lf)(Lg) + \sum_{j=1}^{N} \rho_j^{-1} \theta_j \omega_j + \sum_{j=1}^{n} (\lambda_j f + \theta_j)(\lambda_j g + \omega_j) \tag{8}$$

and associated norm

$$\|(f, \theta)\|^2 = \int_0^T (Lf)^2 + \sum_{j=1}^{N} \rho_j^{-1} \theta_j^2 + \sum_{j=1}^{n} (\lambda_j f + \theta_j)^2. \tag{9}$$

Now, let I denote the interval $[0, T]$, and let J denote the finite index set $\{1, 2, \dots, N\}$:

$$I = [0, T], \qquad J = \{1, 2, \dots, N\}. \tag{10}$$

Also, let

$$K(t,\tau)=\sum_{j=1}^{n}(1+\rho_j)z_j(t)z_j(\tau)+\int_0^T G(t,\xi)G(\tau,\xi)\,d\xi, \qquad t,\tau\in I. \tag{11}$$

It is clear from (4)–(6) that

$$LK(\cdot,\tau)=G(\tau,\cdot),\ \lambda_j K(\cdot,\tau)=(1+\rho_j)z_j(\tau), \qquad 1\le j\le n, \tag{12}$$

and that

$$K(\cdot,\tau)\in H \qquad \text{for each } \tau\in I. \tag{13}$$

Next, let e_j, $j\in J$, be the jth unit N-vector, let $z'(\cdot)$ be the N-vector

$$z'(\cdot)=[z_1(\cdot),z_2(\cdot),\ldots,z_n(\cdot),0,\ldots,0], \tag{14}$$

and let Q be the diagonal $N\times N$ matrix

$$Q=\operatorname{diag}(\rho_1,\rho_2,\ldots,\rho_N). \tag{15}$$

Now define K_t^+ as follows:

$$K_t^+=\begin{cases}(K(\cdot,t),-Qz(t)), & t\in I,\\ (-z'(\cdot)Qe_t,\ Qe_t), & t\in J.\end{cases} \tag{16}$$

We claim that K_t^+ is the reproducing kernel for the Hilbert space H^+ with inner product (8). To verify this, we must check that K_t^+ satisfies the following two conditions (see Ref. 13):

$$K_t^+\in H^+, \qquad \text{for each } t\in I \text{ or } t\in J, \tag{17a}$$

$$\langle (f(\cdot),\theta),K_t^+\rangle=\begin{cases}f(t) & t\in I,\\ \theta_t, & t\in J,\end{cases} \qquad \text{for all } (f,\theta)\in H^+. \tag{17b}$$

Eq. (17a) follows from (4), (7), (13). Eq. (17b) can be verified by substituting (16) into (8). The details are in the Appendix. Because of the cross term in the norm (9), H^+ is not the usual product of H and E^N; therefore, the RK of (16) and the reproducing property (17b) are more complicated.

3. Spline Smoothing as a Constrained Minimum-Norm Problem

Using the RKHS structure developed in the last section, spline smoothing can be reformulated as a constrained minimum-norm problem in H^+. Examination of (9) shows that the problem

$$\begin{aligned}&\text{minimize } \|(f,\theta)\|^2,\\ &\text{subject to } \lambda_j f+\theta_j=r_j, \qquad j\in J,\end{aligned} \tag{18}$$

is equivalent to (3), since the $\{r_j\}$ are fixed. In order to apply the projection theorem to (18), we must express the constraints as inner product constraints in H^+. What we are looking for is the element $h_j^+ \in H^+$ such that

$$\langle (f, \theta), h_j^+ \rangle = \lambda_j f + \theta_j, \qquad j \in J. \tag{19a}$$

In other words, h_j^+ is the representer in H^+ of the functional λ_j^+, where

$$\lambda_j^+(f, \theta) = \lambda_j f + \theta_j. \tag{19b}$$

Thus, if

$$h_j^+ = (h_j(\cdot), \omega_j), \tag{20}$$

we have via the reproducing property

$$h_j(t) = \lambda_j^+ K_t^+ = \lambda_j K(\cdot, t) - e_j' Q z(t), \qquad t \in I, \qquad j \in J, \tag{21}$$

$$\omega_j = [\lambda_j^+ K_1^+ \cdots \lambda_j^+ K_N^+]' = Q(e_j - \lambda_j z), \qquad j \in J. \tag{22}$$

Eqs. (21)–(22) can be checked by substituting into (19a). The following theorem has now been established.

Theorem 3.1. Let $(\hat{f}, \hat{\theta})$ solve the following problem:

$$\underset{(f,\theta)\in U}{\text{minimize}} \|(f, \theta)\|^2,$$

$$U = \{(f, \theta) \in H^+ : \langle (f, \theta), h_j^+ \rangle = r_j, j \in J\}.$$

Then, the smoothing spline is $s(t) = \hat{f}(t)$, $t \in I$.

The following two corollaries follow immediately from the projection theorem.

Corollary 3.1. Let

$$S = \text{span}\{h_j^+\}_1^N.$$

Then, $(s(\cdot), \hat{\theta})$ is the projection onto S of any element in U.

Corollary 3.2. The smoothing spline is given by

$$s(t) = h'(t) R^{-1} r, \qquad t \in I, \tag{23}$$

where

$$h'(t) = [h_1(t), h_2(t), \ldots, h_N(t)],$$

$$r' = [r_1, r_2, \ldots, r_N],$$

$$[R]_{ij} = \langle h_i^+, h_j^+ \rangle.$$

Theorem 3.1 shows that spline *smoothing* is nothing more than *interpolation* in the augmented space H^+. This theorem has been established without additional assumptions and without any modifications to the original problem of Definition 1.1. The useful reproducing kernel structure comes free of charge. Note that the smoothing spline could be computed recursively via (23) by orthogonalizing $\{h_j^+\}_1^N$. However, this requires the explicit computation of $\{h_j^+\}_1^N$, something which can be avoided by the approach given below.

4. Smoothing Splines and Least-Squares Estimates

Since the function $K(\cdot, \cdot)$ of (11) is symmetric and nonnegative definite, it is the covariance function of some zero-mean random process $\{y(t), t \in I\}$. Thus,

$$E[y(t)] = 0, \qquad E[y(t)y(\tau)] = K(t, \tau), \qquad t, \tau \in I. \tag{24}$$

Also, let

$$v' = [v_1, v_2, \ldots, v_N]$$

be a zero-mean random vector with

$$E[y(t)v] = -Qz(t), \qquad t \in I, \tag{25}$$

$$E[vv'] = Q. \tag{26}$$

Let Y be the Hilbert space of zero-mean random variables spanned by $\{y(t), t \in I; v_j, j \in J\}$, with inner product

$$\langle a, b \rangle_Y = E[ab], \qquad a, b \in Y. \tag{27}$$

By a slight extension of a classical result (see Ref. 14), we have the following theorem.

Theorem 4.1. There is a congruence map (inner product-preserving isomorphism) between Y and H^+; $(f(\cdot), \theta) \in H^+$ and $w \in Y$ are images under the congruence, written $(f(\cdot), \theta) \sim w$, iff

$$f(\cdot) = E[wy(\cdot)] \quad \text{and} \quad \theta = E[wv].$$

Thus, for example,

$$h_j^+ \sim \lambda_j y + v_j, \qquad j \in J. \tag{28}$$

Now, let $(f(\cdot), \theta)$ be any element in U, and let $\gamma \in Y$ be its image under the congruence map. Since inner products are preserved,

$$E[\gamma(\lambda_j y + v_j)] = \langle \gamma, \lambda_j y + v_j \rangle_Y = \langle (f, \theta), h_j^+ \rangle = r_j, \qquad j \in J. \tag{29}$$

Let $\hat{\gamma}$ be the linear least-squares estimate (l.l.s.e.) of γ, given $\{\lambda_j y + v_j\}_1^N$. Then, $\hat{\gamma}$ is the projection of γ onto

$$D = \text{span}\{\lambda_j y + v_j\}_1^N.$$

In light of Corollary 3.1 and (28), we have

$$(s(\cdot), \hat{\theta}) \sim \hat{\gamma}, \tag{30}$$

which implies, using Theorem 4.1,

$$s(\cdot) = E[\hat{\gamma} y(\cdot)]. \tag{31}$$

Now, let $\hat{y}(t)$ be the l.l.s.e. of $y(t)$, given $\{\lambda_j y + v_j\}_1^N$. We can write

$$\hat{y}(t) = \sum_{j=1}^{N} \beta_j(t)(\lambda_j y + v_j). \tag{32}$$

Since projection operators are self adjoint, we have from (29), (31)–(32):

$$s(t) = \langle \hat{\gamma}, y(t) \rangle_Y = \langle \gamma, \hat{y}(t) \rangle_Y = \sum_{j=1}^{N} \beta_j(t) \langle \gamma, \lambda_j y + v_j \rangle_Y = \sum_{j=1}^{N} \beta_j(t) r_j. \tag{33}$$

We have now established the following important result relating smoothing splines and stochastic estimates.

Theorem 4.2. If $\hat{\hat{y}}(t)$ is the sample value of $\hat{y}(t)$ obtained by setting

$$\lambda_j y + v_j = r_j, \qquad j \in J,$$

then (32)–(33) imply that

$$s(t) = \hat{\hat{y}}(t), \qquad t \in I. \tag{34}$$

Kimeldorf and Wahba (Ref. 4) earlier obtained a different correspondence between spline smoothing and stochastic estimation. They showed that the smoothing spline is a sample function of a certain minimum-variance unbiased linear estimate of a random process with unknown mean value. This more complicated correspondence does not appear to lead to useful algorithms.

Our correspondence, then, involves signal-in-noise estimation where the signal and noise are correlated with each other and where the relevant statistics are determined by L and $\{\rho_j\}_1^N$; see (24)–(26). The smoothing spline can then be computed recursively by developing a recursive algorithm for $\hat{y}$ and then replacing $\lambda_j y + v_j$ with r_j, $j \in J$. As in the limiting case of spline interpolation, we can avoid computation of $K(\cdot, \cdot)$ and $z(\cdot)$ by developing a dynamical model that generates $y(\cdot)$. We first note that, in light of (5) and (12), $Ly(\cdot)$ is a zero-mean, unit intensity, white-noise proces . As a result, we can write the following model for $y(\cdot)$:

$$(d/dt)x(t) = A(t)x(t) + bu(t), \tag{35a}$$

$$y(t) = cx(t), \tag{35b}$$

where

$$x'(\cdot) = [y(\cdot), y^{(1)}(\cdot), \ldots, y^{(n-1)}(\cdot)], \tag{35c}$$

$$b' = [0, \ldots, 0, 1], \tag{35d}$$

$$c = [1, 0, \ldots, 0], \tag{35e}$$

$$A(\cdot) = \left[\begin{array}{c|c} 0 & I \\ \hline -a_0(\cdot) & -a_1(\cdot) \cdots -a_{n-1}(\cdot) \end{array}\right], \tag{35f}$$

and $u(\cdot)$ is zero-mean, unit intensity, white noise. Since the entries in the last row of $A(\cdot)$ are the coefficients of L, the model is immediately parametrized once L is chosen. All that remains is to specify the (unique) initial conditions that guarantee that $y(\cdot)$ has covariance $K(\cdot,\cdot)$. Before doing this, we shall restrict attention to a rather broad class of constraint functionals $\{\lambda_j\}$, called extended Hermite–Birkhoff (EHB) functionals, which have the form

$$\lambda_j f = \sum_{k=1}^{n} \alpha_{jk} f^{(k-1)}(t_j), \qquad j \in J, \tag{36}$$

where

$$0 \le t_1 < t_2 < \cdots < t_N \le T$$

and the $\{\alpha_{jk}\}$ are known real numbers. Note that, if

$$c_j = [\alpha_{j1}, \alpha_{j2}, \ldots, \alpha_{jn}], \qquad j \in J, \tag{37}$$

then

$$\lambda_j y = c_j x(t_j). \tag{38}$$

The most convenient initialization point is t_n, so we can write

$$\lambda_j y = c_j \phi(t_j, t_n) x(t_n) + \int_{t_n}^{t_j} c_j \phi(t_j, \tau) b u(\tau)\, d\tau, \tag{39}$$

where $\phi(\cdot,\cdot)$ is the fundamental matrix of $A(\cdot)$. Writing the first n equations in (39) in vector form and solving for $x(t_n)$, we have

$$x(t_n) = M^{-1}\eta + M^{-1} \int_{t_1}^{t_n} \Delta(\tau) b u(\tau)\, d\tau; \tag{40}$$

here,

$$\eta' = [\lambda_1 y, \lambda_2 y, \ldots, \lambda_n y]; \tag{41}$$

M is the $n \times n$ matrix with i^{th} row M_i':

$$M_i' = c_i \phi(t_i, t_n); \tag{42}$$

and $\Delta(\cdot)$ is the $n \times n$ matrix with i^{th} row $\Delta_i'(\cdot)$:

$$\Delta_i'(\tau) = \begin{cases} c_i \phi(t_i, \tau), & \tau \in [t_i, t_n], \\ 0, & \text{otherwise.} \end{cases} \tag{43}$$

The uniqueness assumption stated at the beginning of Section 2 guarantees the invertibility of M. Now, from (4), (5), (12), (24), (35a), (35b),

$$E[\eta u(t)] = 0, \qquad t \in I, \tag{44}$$

$$E[\eta\eta'] = I + Q_n, \tag{45}$$

where Q_n is the upper left $n \times n$ portion of Q. Thus,

$$E[x(t_n)x'(t_n)] = M^{-1}\left[I + Q_n + \int_{t_1}^{t_n} \Delta(\tau) bb' \Delta'(\tau)\, d\tau\right] M^{-T}, \tag{46}$$

$$E[x(t_n)u(t)] = \begin{cases} M^{-1}\Delta(t) b, & t \in [t_1, t_n], \\ 0, & \text{otherwise.} \end{cases} \tag{47}$$

These two initial conditions complete the specification of the model (35) for $y(\cdot)$. In terms of this model, (25) can be replaced by

$$E[x(t_n)v'] = -M^{-1}[Q_n \,\vdots\, 0], \tag{48}$$

$$E[u(t)v'] = 0, \qquad t \in I. \tag{49}$$

5. Recursive Algorithm for Lg-Spline Smoothing

A recursive algorithm for the Lg-spline $s(\cdot)$ smoothing data $\{r_j\}_1^N$ with respect to EHB functionals $\{\lambda_j\}_1^N$ and weights $\{\rho_j\}_1^N$ is given below. The derivation proceeds along the same lines as in the interpolation case in Ref. 11 and will be omitted.

Step 1. Starting with

$$x_{n/n} = M^{-1} r_0, \qquad \text{where } r_0' = [r_1, r_2, \ldots, r_n], \tag{50}$$

$$P_{n/n} = M^{-1}\left[Q_n + \int_{t_1}^{t_n} \Delta(\tau) bb' \Delta'(\tau)\, d\tau\right] M^{-T}, \tag{51}$$

compute the following quantities for $j = n+1, n+2, \ldots, N$:

$$\epsilon_j = r_j - c_j x_{j/j-1}, \tag{52a}$$

$$R_j = c_j P_{j/j-1} c_j' + \rho_j \text{ (a scalar)}, \tag{52b}$$

$$K_j = P_{j/j-1} c_j', \tag{52c}$$

$$x_{j/j-1} = \phi(t_j, t_{j-1}) x_{j-1/j-1}, \tag{52d}$$

$$x_{j/j} = x_{j/j-1} + K_j R_j^{-1} \epsilon_j, \tag{52e}$$

$$P_{j/j-1} = \phi(t_j, t_{j-1}) P_{j-1/j-1} \phi'(t_j, t_{j-1}) + \int_{t_{j-1}}^{t_j} \phi(t_j, \tau) bb' \phi'(t_j, \tau)\, d\tau, \tag{52f}$$

$$P_{j/j} = [I - K_j R_j^{-1} c_j] P_{j/j-1} [I - K_j R_j^{-1} c_j]'. \tag{52g}$$

Step 2. For $t \geq t_N$,

$$s(t) = c\phi(t, t_N) x_{N/N}. \tag{53}$$

Step 3. For $t_{j-1} \leq t \leq t_j$ and $j = N, N-1, \ldots, n+1$,

$$s(t) = cx(t/N), \tag{54}$$

where

$$x(t/N) = \phi(t, t_j) x(t_j/N) + \left[\int_{t_j}^{t} \phi(t, \tau) bb' \phi'(t_j, \tau)\, d\tau\right] \mu_j, \tag{55a}$$

$$x(t_N/N) = x_{N/N}, \tag{55b}$$

$$\mu_j = \psi_{j+1}' \mu_{j+1} + c_j' R_j^{-1} \epsilon_j, \qquad \mu_N = c_N' R_N^{-1} \epsilon_N, \tag{56}$$

$$\psi_{j+1} = \phi(t_{j+1}, t_j)[I - K_j R_j^{-1} c_j]. \tag{57}$$

Step 4. For $t_1 \leq t \leq t_n$,

$$s(t) = cx(t/N), \tag{58}$$

where

$$x(t/N) = \phi(t, t_n) x(t_n/N) + \left[\int_{t_n}^{t} \phi(t, \tau) bb' \Delta'(\tau)\, d\tau\right] M^{-T} \phi'(t_{n+1}, t_n) \mu_{n+1}. \tag{59}$$

Step 5. For $t \leq t_1$,

$$s(t) = c\phi(t, t_1) x(t_1/N). \tag{60}$$

The only difference between the above smoothing algorithm and the interpolation algorithm in Ref. 11 is the presence of Q_n in (51) and ρ_j in

(52b). As discussed in Section 1, in the limiting case in which all the weights are zero, these extra terms disappear, and smoothing reduces to interpolation.

6. Example

To illustrate the concepts discussed in this paper, we will derive the simplest type of smoothing spline: the piecewise linear spline smoothing two samples. In this case,

$$L=(d/dt), \qquad n=1, \qquad N=2,$$

and

$$\lambda_j f = f(t_j), \qquad j=1,2.$$

The uniqueness condition is obviously satisfied. The quantities needed for the recursive algorithm of Section 5 are

$$A=0, \qquad b=1, \qquad c=1, \qquad c_1=c_2=1, \qquad \phi(\cdot,\cdot)=1,$$
$$M=1, \qquad \Delta(\cdot)=0 \text{ a.e.}$$

The initial conditions for Step 1 are therefore

$$x_{1/1}=r_1, \qquad P_{1/1}=\rho_1;$$

thus,

$$x_{2/1}=r_1, \qquad P_{2/1}=\rho_1+t_2-t_1=K_2,$$
$$R_2=\rho_1+\rho_2+t_2-t_1, \qquad \epsilon_2=r_2-r_1,$$
$$x_{2/2}=r_1+(r_2-r_1)(\rho_1+t_2-t_1)/(\rho_1+\rho_2+t_2-t_1).$$

Step 2 yields

$$s(t)=[\rho_2 r_1+(\rho_1+t_2-t_1)r_2]/(\rho_1+\rho_2+t_2-t_1), \qquad t\geq t_2.$$

From Step 3, for $t_1\leq t\leq t_2$, we have

$$\mu_2=(r_2-r_1)/(\rho_1+\rho_2+t_2-t_1),$$
$$x(t/2)=x_{2/2}+(t-t_2)\mu_2;$$

thus,

$$s(t)=[(\rho_2+t_2-t)r_1+(\rho_1+t-t_1)r_2]/(\rho_1+\rho_2+t_2-t_1),$$
$$t_1\leq t\leq t_2.$$

Step 4 is inoperative; and, since

$$x(t_1/2)=x_{2/2}+(t_1-t_2)\mu_2,$$

Step 5 gives

$$s(t) = [(\rho_2 + t_2 - t_1)r_1 + \rho_1 r_2]/(\rho_1 + \rho_2 + t_2 - t_1), \qquad t \le t_1.$$

Note that $s(\cdot)$ is continuous and that, for all t, $s(t)$ lies between r_1 and r_2. In fact, if ρ_2 is finite and $\rho_1 \to \infty$,

$$s(t) \to r_2.$$

If instead ρ_1 is finite and $\rho_2 \to \infty$,

$$s(t) \to r_1.$$

If $\rho_1 = \rho_2 = \rho$ and $\rho \to \infty$,

$$s(t) \to (r_1 + r_2)/2.$$

If both weights are zero, $s(t)$ becomes the piecewise linear spline interpolating r_1 and r_2 at t_1 and t_2. As a point of information, in this example

$$z'(t) = [1, 0],$$

$$K(t, \tau) = \begin{cases} 1 + \rho_1 + \min(t, \tau) - t_1, & (t, \tau) \ge t_1, \\ 1 + \rho_1 + t_1 - \max(t, \tau), & (t, \tau) \le t_1, \\ 1 + \rho_1, & \text{otherwise}, \end{cases}$$

$$h_1(t) = 1, \qquad \omega_1' = [0, 0], \qquad \omega_2' = [-\rho_1, \rho_2],$$

$$h_2(t) = \begin{cases} 1 + \rho_1 + \min(t, t_2) - t_1, & t \ge t_1, \\ 1 + \rho_1, & t \le t_1, \end{cases}$$

$$\langle h_1^+, h_j^+ \rangle = 1, \qquad j = 1, 2,$$

$$\langle h_2^+, h_2^+ \rangle = 1 + \rho_1 + \rho_2 + t_2 - t_1.$$

7. Conclusions

A new correspondence between spline smoothing and linear least-squares estimation has been used to develop a recursive algorithm for Lg-splines smoothing extended Hermite–Birkhoff data. This correspondence can also be used to derive a least upper bound for the spline smoothing error in terms of the mean-square error of the estimation problem (see Ref. 15). In addition, our results have been extended to spline problems associated with estimation of autoregressive-moving average processes (Ref. 16) and to vector-valued splines (Ref. 17).

8. Appendix

We present herewith the details of the verification of the reproducing property (17b). Let (f, θ) be any element of H^+. Then, using (16) and (8), we have for $t \in I$:

$$\langle (f, \theta), K_t^+ \rangle = \langle (f, \theta), (K(\cdot, t), -Qz(t)) \rangle$$

$$= \int_0^T [Lf(\cdot)][LK(\cdot, t)] + \sum_{j=1}^N \rho_j^{-1} \theta_j(-e_j'Qz(t))$$

$$+ \sum_{j=1}^n (\lambda_j f + \theta_j)(\lambda_j K(\cdot, t) - e_j'Qz(t)).$$

Using (12), (14)–(15), we get

$$\langle (f, \theta), K_t^+ \rangle = \int_0^T G(t, \tau)[Lf(\tau)]\, d\tau - \sum_{j=1}^n \rho_j^{-1} \theta_j \rho_j z_j(t)$$

$$+ \sum_{j=1}^n (\lambda_j f + \theta_j)((1 + \rho_j) z_j(t) - \rho_j z_j(t))$$

$$= \int_0^T G(t, \tau)[Lf(\tau)]\, d\tau - \sum_{j=1}^n \theta_j z_j(t)$$

$$+ \sum_{j=1}^n (\lambda_j f) z_j(t) + \sum_{j=1}^n \theta_j z_j(t) = f(t).$$

The last equation follows from (6). Now, let $t \in J$, and use (16), (8), (14), (4), (15). We have

$$\langle (f, \theta), K_t^+ \rangle = \langle (f, \theta), (-z'(\cdot)Qe_t, Qe_t) \rangle$$

$$= \sum_{j=1}^N \rho_j^{-1} \theta_j (e_j' Q e_t) + \sum_{j=1}^n (\lambda_j f + \theta_j)(-e_j' Q e_t + e_j' Q e_t)$$

$$= \sum_{j=1}^N \rho_j^{-1} \theta_j \rho_j \delta_{jt} = \theta_t.$$

References

1. WHITTAKER, E. T., *On a New Method of Graduation*, Proceedings of the Edinburgh Mathematical Society, Vol. 41, pp. 63–75, 1923.
2. SCHOENBERG, I. J., *Spline Functions and the Problem of Graduation*, Proceedings of the National Academy of Sciences, Vol. 52, pp. 947–950, 1964.

3. ANSELONE, P. M., and LAURENT, P. J., *A General Method for the Construction of Interpolating or Smoothing Spline Functions*, Numerische Mathematik, Vol. 12, pp. 66–82, 1968.
4. KIMELDORF, G., and WAHBA, G., *Some Results on Tchebycheffian Spline Functions*, Journal of Mathematical Analysis and Applications, Vol. 33, pp. 82–95, 1971.
5. REINSCH, C., *Smoothing by Spline Functions, II*, Numerische Mathematik, Vol. 16, pp. 451–454, 1971.
6. LYCHE, T., and SCHUMAKER, L. L., *Computation of Smoothing and Interpolating Natural Splines Via Local Bases*, SIAM Journal on Numerical Analysis, Vol. 10, pp. 1027–1038, 1973.
7. MUNTEANU, M. J., *Generalized Smoothing Spline Functions for Operators*, SIAM Journal on Numerical Analysis, Vol. 10, pp. 28–34, 1973.
8. DEFIGUEIREDO, R. J. P., and CAPRIHAN, A., *An Algorithm for the Construction of the Generalized Smoothing Spline*, Proceedings of the Johns Hopkins Conference on Information Sciences and Systems, pp. 494–500, 1977.
9. WAHBA, G., and WOLD, S., *A Completely Automatic French Curve: Fitting Spline Functions by Cross Validation*, Communications in Statistics, Vol. 4, pp. 1–17, 1975.
10. WEINERT, H. L., BYRD, R. H., and SIDHU, G. S., *Estimation Techniques for Recursive Smoothing of Deterministic Data*, Proceedings of the Johns Hopkins Conference on Information Sciences and Systems, pp. 54–57, Baltimore, Maryland, 1977.
11. WEINERT, H. L., and SIDHU, G. S., *A Stochastic Framework for Recursive Computation of Spline Functions: Part I, Interpolating Splines*, IEEE Transactions on Information Theory, Vol. 24, pp. 45–50, 1978.
12. WEINERT, H. L., and SIDHU, G. S., *On Uniqueness Conditions for Optimal Curve Fitting*, Journal of Optimization Theory and Applications, Vol. 23, pp. 211–216, 1977.
13. ARONSZAJN, N., *Theory of Reproducing Kernels*, Transactions of the American Mathematical Society, Vol. 68, pp. 337–404. 1950.
14. PARZEN, E., *An Approach to Time Series Analysis*, Annals of Mathematical Statistics, Vol. 32, pp. 951–989, 1961.
15. WEINERT, H. L., *Statistical Methods in Optimal Curve Fitting*, Communications in Statistics, Vol. B7, pp. 417–435, 1978.
16. WEINERT, H. L., DESAI, U. B., and SIDHU, G. S., *ARMA Splines, System Inverses, and Least-Squares Estimates*, SIAM Journal on Control and Optimization, Vol. 17, pp. 525–536, 1979.
17. SIDHU, G. S., and WEINERT, H. L., *Vector-Valued Lg-Splines*, Journal of Mathematical Analysis and Applications, Vol. 70, pp. 505–529, 1979.

26

Reprinted from *J. Appl. Math. Phys.* **30**:292–304 (1979)

Multivariate Interpolation at Arbitrary Points Made Simple

By Jean Meinguet, Institut de Mathématique pure et appliquée, Université Catholique de Louvain, B-1348 Louvain-la-Neuve, Belgium

Dedicated to Professor E. Stiefel

1. An Optimal Interpolation Problem

For simplicity, all functions and vector spaces considered in this paper are real. The classical problem of minimizing the quadratic functional

$$|v|_m^2 := \int_a^b |v^{(m)}(t)|^2\, dt \qquad (1)$$

under the interpolatory constraints

$$v(a_i) = \alpha_i, \quad 1 \le i \le N, \qquad (2)$$

with $N \ge m \ge 1$, the a_i being any prescribed distinct points of $[a, b]$ and the α_i any prescribed real scalars, has for natural setting the vector space

$$\mathscr{H}^m[a, b] := \{v \in C^{m-1}[a, b] : v^{(m-1)} \text{ absolutely continuous, } v^{(m)} \in L_2(a, b)\}, \qquad (3)$$

or equivalently, the class of functions whose distributional derivatives up to the mth one are in $L_2(a, b)$. Equipped with the Sobolev seminorm $|\cdot|_m$, $\mathscr{H}^m[a, b]$ is a semi-Hilbert function space, the linear variety defined by (2) being accordingly closed and Hausdorff. The existence of a unique solution u of the above problem, and the fundamental characterization of u as an odd degree polynomial spline, readily follow from the orthogonal projection theorem; a further analysis of this *close association between spline interpolation and orthogonal projection* leads to the so-called first integral relation (see e.g. [1], p. 155), from which various important *intrinsic properties* (such as, for example, the minimum norm property and the best approximation property) can be obtained easily.

As analyzed in recent papers (see e.g. [6], [11]), and motivated in detail in [11] (especially in relation with a constructive theory of functions of several variables), a proper abstract setting for a natural n-dimensional generalization of this optimal interpolation problem and results is provided by the (generalized) *Beppo Levi space* $X \equiv BL^m(\mathscr{R}^n)$ of order m over $\mathscr{R}^n$ (m and n are integers ≥ 1, to be regarded throughout as given). Defined as

$$X = \{v \in \mathscr{D}' : \partial^\alpha v \in L_2 \text{ for } |\alpha| = m\}, \qquad (4)$$

where $\alpha \coloneqq (\alpha_1, \ldots, \alpha_n) \in N^n$ and $|\alpha| \coloneqq \alpha_1 + \cdots + \alpha_n$, X is thus simply the vector space of all the (Schwartz) distributions (i.e., continuous linear functionals on the vector space $\mathscr{D}$ of infinitely differentiable functions with compact support in $\mathscr{R}^n$, provided with the canonical Schwartz topology) for which all the partial derivatives (in the distributional sense) of (total) order m are square integrable in $\mathscr{R}^n$. X is naturally equipped with the semi-inner product $(\cdot,\cdot)_m$ corresponding to the *rotation invariant* seminorm

$$|v|_m \coloneqq \left\{ \sum_{i_1,\ldots,i_m=1}^{n} \int_{\mathscr{R}^n} |\partial_{i_1\ldots i_m} v(x)|^2 \, dx \right\}^{1/2}, \tag{5}$$

where every partial derivative $\partial_{i_1\ldots i_m} v \equiv \partial^m v/\partial x_{i_1} \cdots \partial x_{i_m}$ is to be interpreted in the distributional sense; the kernel of $|\cdot|_m$ is known to be simply the vector space $\mathscr{P} \equiv \mathscr{P}_{m-1}$ of dimension

$$M = \binom{n+m-1}{n} \tag{6}$$

of all polynomials over $\mathscr{R}^n$ of (total) degree $\leq m - 1$. It should be noted that $|v|_m^2$ may be physically interpreted (at least if $m = n = 2$ and under some simplifying assumptions) as the *bending energy of a thin plate of infinite extent*, v denoting the deflection normal to the rest position (supposed of course to be plane); this is naturally suggested by the classical interpretation of (1) (if $m = 2$) as the potential energy of a statically deflected thin beam (which indeed is proportional to the integral of the square of the curvature of the elastica of the beam). As they define the equilibrium positions of a thin plate of infinite extent that deforms in bending only (under deflections specified at a number of independent points), the solutions of optimal interpolation problems can be appropriately termed *surface splines* (which in fact emphasizes their intrinsic multivariate structure). Originally introduced by engineers (for interpolating wing deflections and computing slopes for aeroelastic calculations, see [9]), this ingenious device proves most interesting to analyze mathematically; in this connection, various deep results have been obtained recently by Duchon (see [5, 6, 7]); for a deliberately more constructive approach, see [11], where indeed a prominent role is played by *representation formulas* in function and distribution spaces, these complementary results being obtained by resorting to such basic mathematical tools as convolutions and Fourier transforms of distributions.

Beppo Levi spaces have many interesting properties, which are partly reminiscent of those of the widely known *Sobolev spaces* (on $\mathscr{R}^n$). The following result is the only one, however, to be directly relevant here.

Theorem 1. *Suppose*

$$m > n/2. \tag{7}$$

Then the seminormed space $X \equiv BL^m(\mathscr{R}^n)$, *defined by* (4) *and* (5), *is a semi-Hilbert function space of continuous functions on* $\mathscr{R}^n$, *all the evaluation linear functionals with finite support in* $\mathscr{R}^n$ *that annihilate* $\mathscr{P} \equiv \mathscr{P}_{m-1}$ *being accordingly bounded.*

Presented in [6] as a straightforward consequence of an 'iterated' version of a theorem essentially due to Krylov (concerning distributions whose every partial derivative of order 1 is a function), this theorem has been given a constructive proof in [11].

In view of this basic result, the optimal (or surface spline) interpolation problem we want to analyze hereafter, namely Problem (P), can be formulated as follows. Let there be given:

A finite set $A = (a_i)_{i \in I}$ of distinct points of $\mathscr{R}^n$ containing a $\mathscr{P}$-*unisolvent* subset, by which we mean a set $B = (a_j)_{j \in J}$ of M points of A, M being defined by (6), such that there exists a unique $p \in \mathscr{P}$ satisfying the interpolating conditions

$$p(a_j) = \alpha_j, \quad \forall j \in J, \tag{8}$$

for any prescribed real scalars α_j, $\forall j \in J$.

A set of real scalars $(\alpha_i)_{i \in I}$, or equivalently provided that $m > n/2$, the linear variety:

$$V := \{v \in X: v(a_i) = \alpha_i, \forall i \in I\}; \tag{9}$$

whenever $\alpha_i := f(a_i)$, $\forall i \in I$, where f denotes a function defined (at least) on A, V can be interpreted as the set of X-*interpolants* of f on A.

Then we have the following definition of *Problem* (P): *Find* $u \in V$ *such that*

$$|u|_m = \inf_{v \in V} |v|_m, \tag{10}$$

it being understood once and for all that $m > n/2$.

2. Representation Formulas in Beppo Levi Spaces

By virtue of the $\mathscr{P}$-unisolvence of the subset B of the given set A of interpolation points in $\mathscr{R}^n$, there exists in $\mathscr{P} \equiv \mathscr{P}_{m-1}$ a unique basis $(p_j)_{j \in J}$ that is dual to the set of shifted Dirac measures $(\delta_{(a_j)})_{j \in J}$ (in the sense that $p_i(a_j) = \delta_{ij}$, $\forall i, j \in J$, where δ_{ij} is the Kronecker symbol). For every $v \in X$, and always *provided that* $m > n/2$, the (uniquely defined) $\mathscr{P}$-*interpolant* Pv of v on B is accordingly given by the formula (of Lagrange type)

$$Pv := \sum_{j \in J} v(a_j) p_j; \tag{11}$$

owing to this definition, the mapping $P: X \to X$ is a *linear projector* of X with range $\mathscr{P} \equiv \mathscr{P}_{m-1}$ and kernel

$$X_0 := \{v \in X: v(a_j) = 0, \forall j \in J\}, \tag{12}$$

so that

$$X = \mathscr{P}_{m-1} \oplus X_0; \tag{13}$$

equipped with the seminorm $|\cdot|_m$, X_0 is a Hilbert space (it is indeed complete, like X to which it is isometrically isomorphic, while trivially Hausdorff), the *direct sum decomposition* (13) being then topological.

As a matter of fact, in view of Theorem 1, X_0 (for $m > n/2$) is more precisely a *Hilbert function space* (of continuous functions on $\mathscr{R}^n$), which means that, for each fixed $x \in \mathscr{R}^n$, the evaluation linear functional $\delta_{(x)}$ on X_0 is *bounded*; hence the following *representation formula in the Hilbert space X_0 for $m > n/2$*:

$$v(x) \equiv \langle \delta_{(x)}, v \rangle = (K_x, v)_m, \quad \forall v \in X_0, \forall x \in \mathscr{R}^n, \tag{14}$$

where $K_x \in X_0$ denotes the (necessarily unique) *Fréchet–Riesz representer* of $\delta_{(x)}$ and $\langle \cdot, \cdot \rangle$ is the *duality bracket* between dual topological vector spaces. But the image of $\mathscr{D}$ under the linear projector $I - P$ (where I denotes the identity mapping of X) is contained in X_0, so that (14) is satisfied for all functions $\varphi - P\varphi$, $\varphi \in \mathscr{D}$, therefore, by definition of the partial differentiation for distributions, we may write

$$(K_x, \varphi - P\varphi)_m \equiv (K_x, \varphi)_m = \langle (-1)^m \Delta^m K_x, \varphi \rangle, \quad \forall \varphi \in \mathscr{D}, \tag{15a}$$

where

$$\Delta^m \equiv \sum_{i_1, \ldots, i_m = 1}^{n} (\partial_{i_1 \ldots i_m})^2 \tag{16}$$

is the (distributional) mth *iterated Laplacian*; in view of the elementary identity

$$(\varphi - P\varphi)(x) = \langle \delta_{(x)} - \sum_{j \in J} p_j(x) \delta_{(a_j)}, \varphi \rangle, \quad \forall \varphi \in \mathscr{D}, \tag{15b}$$

it finally follows from (14) and (15a) that K_x is a solution of the (distributional) partial differential equation

$$(-1)^m \Delta^m K_x = \delta_{(x)} - \sum_{j \in J} p_j(x) \delta_{(a_j)}, \quad \forall x \in \mathscr{R}^n, \tag{17}$$

in $X_0 \subset \mathscr{D}'$.

Now it is quite easy to find in $\mathscr{D}'$ a particular solution of (17). It is indeed well known that a *fundamental solution* of Δ^m in $\mathscr{R}^n$ (i.e., a solution $E \in \mathscr{D}'$ of the distributional equation $\Delta^m E = \delta$) is the rotation invariant function on the complement of the origin in $\mathscr{R}^n$ defined by the following formulas:

$$E(y) \coloneqq \begin{cases} cr^{2m-n} \ln r, & \text{if } 2m \geq n \text{ and } n \text{ is even}, \\ dr^{2m-n}, & \text{otherwise}, \end{cases} \tag{18a}$$

where $r(y) \equiv |y|$ denotes as usual the radial coordinate (or Euclidean norm) of the point $y \in \mathscr{R}^n$ and

$$c := \frac{(-1)^{n/2+1}}{2^{2m-1}\pi^{n/2}(m-1)!\,(m-n/2)!}, \tag{18b}$$

$$d := \frac{(-1)^m \Gamma(n/2 - m)}{2^{2m}\pi^{n/2}(m-1)!}. \tag{18c}$$

On the basis of the superposition principle of linear operators, we readily get a particular distribution satisfying (17), viz.,

$$H_x(y) := (-1)^m\left[E(x-y) - \sum_{j\in J} p_j(x)E(a_j - y)\right], \quad \forall x, y \in \mathscr{R}^n, \tag{19}$$

from the transforms of E by the translations defined by the shifted Dirac measures on the right-hand side of (17).

It turns out that $H_x \in X$ if $m > n/2$. As justified in every detail in [11], Theorem 1 can indeed be regarded as a rather simple corollary of this non-trivial result, which can be proved directly (see Theorem 2 in [11]) or, more elegantly, by exploiting basic properties of the Fourier transformation of distributions (see [6] and [11]). On the other hand, it is well known that the only solutions in X of the mth iterated Laplace equation are the polynomials of degree $\leq m - 1$. Hence it finally follows that, if $m > n/2$, there exists a unique $K_x \in X_0$ satisfying (17), namely the (continuous) function:

$$\begin{aligned} K_x(y) &:= (I - P)H_x(y) \\ &\equiv (-1)^m\Big\{E(x-y) - \sum_{i\in J} p_i(x)E(a_i - y) - \sum_{j\in J} p_j(y)E(x - a_j) \\ &\qquad + \sum_{i\in J}\sum_{j\in J} p_i(x)p_j(y)E(a_i - a_j)\Big\}; \end{aligned} \tag{20}$$

needless to say, the fact that K_x involves no functions more complicated than logarithms, and is easily coded, is most welcome in the matter of application of the concrete representation formula (14). Under assumption (7), the set $\{K_x \colon \forall x \in \mathscr{R}^n\}$ is the so-called *reproducing kernel* (or *kernel function*) of the Hilbert function space X_0; it can be regarded equivalently as the real-valued continuous function $(x, y) \mapsto K(x, y) \equiv K_x(y)$ on $\mathscr{R}^n \times \mathscr{R}^n$. It should be noted in passing that, for $n = 1$, (18a, c) reduces to the so-called *symmetric* fundamental solution of the (univariate) distributional equation $E^{(2m)} = 0$, viz.,

$$E(y) := (1/2)|y|^{2m-1}/(2m-1)! \equiv [y_+^{2m-1} - y^{2m-1}/2]/(2m-1)!, \tag{21}$$

the closed form expression (originally given by Golomb and Weinberger in [8], see pp. 151–153) for the reproducing kernel of the space $(I - P)\mathscr{H}^m[a, b]$ equipped with the Sobolev norm $|\cdot|_m$ being then immediately obtained from (20); surprisingly enough, the present multivariate approach need not be more complicated than the univariate one.

The paramount importance of reproducing kernels in approximation theory and in numerical analysis, especially when their computational evaluation is easy, stems from their *reproducing property* (expressed here by (14)). Among other significant corollaries (such topics are excellently surveyed in [3], pp. 316–326 and [12], pp. 82–102), the following are relevant here:

According to the basic formula

$$K(x, y) = (K_x, K_y)_m, \quad \forall x, y \in \mathscr{R}^n, \tag{22}$$

which is obvious from (14), the real number $K(x, y)$ can be interpreted as the inner product in X_0 of the Riesz representers of $\delta_{(x)}$ and $\delta_{(y)}$; it thus follows that $K(x, y)$ is a *symmetric* function of the two variables x and y, which is *positive definite* in the sense that, for every $N \geq 1$ and every set of points $b_1, \ldots, b_N \in \mathscr{R}^n$, the $N \times N$ matrix whose (i, j) element is $K(b_i, b_j)$ for $i, j \in [1, N]$ is positive semidefinite (it is indeed simply the *Gram matrix* of the sequence (K_{b_i}), $1 \leq i \leq N$).

The Riesz representer l of *any* bounded linear functional L on X_0 is explicitly given by

$$l(x) = \langle L, K_x \rangle, \quad \forall x \in \mathscr{R}^n, \tag{23a}$$

and its norm is such that

$$\|L\| \equiv |l|_m = [\langle L_x, \langle L_y, K(x, y)\rangle\rangle]^{1/2} \tag{23b}$$

where the subscripts indicate explicitly the variable to consider when applying the concerned L (the other variable is held fixed).

Hence the following far-reaching result, which indeed may be regarded as a *practical multivariate extension* (to the Sobolev-like space X) *of the classical Peano kernel theorem* (relative to $C^m[a, b]$).

Theorem 2. *Let L denote any bounded linear functional on $X \equiv BL^m(\mathscr{R}^n)$ equipped with the seminorm $|\cdot|_m$. Then, provided that $m > n/2$, the representation formula*

$$\langle L, v \rangle = (l, v)_m, \quad \forall v \in X, \tag{24a}$$

where

$$l(x) \equiv (-1)^m \langle L_y, E(x - y)\rangle (\text{mod } \mathscr{P}_{m-1}), \quad \forall x \in \mathscr{R}^n, \tag{24b}$$

is valid, the associated Peano-like kernel being accordingly the (distributional) mth total derivative $D^m l \in (L_2)^M$. Moreover, the error coefficient C in the sharp appraisal

$$|\langle L, v \rangle| \leq C |v|_m, \quad \forall v \in X, \tag{25a}$$

is given explicitly by

$$C \equiv \|L\| := [(-1)^m \langle L_x, \langle L_y, E(x - y)\rangle\rangle]^{1/2}. \tag{25b}$$

It is thus surprisingly easy to determine the best possible constant in *inequalities of Bramble–Hilbert type on* $\mathscr{R}^n$.

3. Some Intrinsic Properties of Surface Splines

Let us proceed to the *solution of Problem* (P). In view of the direct sum decomposition (13), the restriction of the associated projector $I - P$ to the linear variety $V \subset X$ defined by (9) is clearly an injection. Therefore, finding $u \in V$ such that (10) holds amounts strictly to finding an element w of minimal norm $|\cdot|_m$ in the image of V under $I - P$, which is the linear variety

$$W \coloneqq \{v \in X_0 : v(a_k) = \alpha'_k, \forall k \in K\} \tag{26a}$$

where

$$K \equiv I - J \coloneqq \{k \in N : 1 \le k \le N \equiv \text{Card}(K)\} \tag{26b}$$

for definiteness, and

$$\alpha'_k \coloneqq \alpha_k - \sum_{j \in J} \alpha_j p_j(a_k), \quad \forall k \in K. \tag{26c}$$

As a finite intersection of translated kernels in X_0 of linearly independent bounded linear functionals $\delta_{(a_k)}$, $1 \le k \le N$, W is a non-empty closed convex subset of the Hilbert function space X_0. By virtue of the *orthogonal projection theorem*, there exists in W a unique element w of minimal norm, which is characterized by the orthogonality property:

$$(w, v)_m = 0, \quad \forall v \in W_0, \tag{27a}$$

where

$$W_0 \coloneqq \{v \in X_0 : v(a_k) = 0, \forall k \in K\}. \tag{27b}$$

Since W_0 can be defined equivalently, owing to (14), as the set of all $v \in X_0$ that are orthogonal to K_{a_k} for all $k \in K$, this optimal function w must belong to the span of $\{K_{a_k} : 1 \le k \le N\}$ and is thus necessarily of the form

$$w = \sum_{k=1}^{N} \gamma_k K_{a_k}; \tag{28a}$$

expressing now that w belongs to W as defined by (26), we get, for determining the (real) coefficients γ_k, the Cramer system of linear equations

$$\sum_{j=1}^{N} K(a_i, a_j)\gamma_j = \alpha'_i, \quad 1 \le i \le N, \tag{28b}$$

whose coefficient matrix is symmetric and positive definite (as Gram matrix of a sequence of linearly independent elements of X_0). The latter property implies that w, and accordingly the unique solution u of Problem (P) given by

$$u = w + \sum_{j \in J} \alpha_j p_j, \tag{29}$$

depend continuously on the data $(\alpha_i)_{i \in I}$, *Problem* (P) *being consequently well-posed.*

A significant consequence of the foregoing is the simple *geometric characterization:*

$$u = V_0^{\perp} \cap V \tag{30}$$

where V, the translation defined by (9) of

$$V_0 \coloneqq \{v \in X: v(a_i) = 0, \forall i \in I\} \equiv W_0, \tag{31}$$

is the linear variety of all *X-interpolants* (on A) to the *prescribed data* $(\alpha_i)_{i \in I}$ and $V_0^{\perp}$, the orthogonal complement in X of V_0, is the vector subspace of all *optimal X-interpolants* (on A) to *arbitrary data*. Owing to (30), the mapping Q that assigns to each $v \in X$ its optimal X-interpolant on A is nothing else but the *orthogonal projector of X onto* $V_0^{\perp}$; exploiting this remarkable *association between surface spline interpolation and orthogonal projection* reveals a number of intrinsic properties of optimal X-interpolants, which of course are strongly reminiscent of the classical minimum properties (fully described, for example, in [1] and [4]) of (odd degree) polynomial splines. Among these properties, let us single out here:

The *minimum norm property*: if $v \in X$, then

$$|Qv|_m^2 \leq |v|_m^2, \tag{32}$$

with equality iff $v = Qv$.

The *best approximation property*: if $v \in X$ and $w \in V_0^{\perp}$, then

$$|v - Qv|_m^2 \leq |v - w|_m^2, \tag{33}$$

with equality iff $w \equiv Qv (\text{mod}\, \mathscr{P})$.

A simple proof, of Holladay type (see e.g. [1], pp. 76–77), immediately follows from the so-called *first integral relation* (or generalized Pythagoras' Theorem), viz.,

$$|v|_m^2 = |v - Qv|_m^2 + |Qv|_m^2, \quad \forall v \in X, \tag{34}$$

which itself can be regarded as a direct consequence of the so-called *fundamental identities*. Used in [1] as a cornerstone for an intrinsic theory of univariate splines, the latter result can be extended as follows to surface splines $w \in V_0^{\perp}$.

Theorem 3. *Any* $w \in V_0^{\perp}$ *has a unique representation of the form*

$$w(y) = (-1)^m \sum_{i \in I} \gamma_i E(a_i - y) + q(y), \quad \forall y \in \mathscr{R}^n, \tag{35a}$$

where $q \in P$ and the coefficients γ_i are real scalars satisfying the relations

$$\sum_{i \in I} \gamma_i p(a_i) = 0, \quad \forall p \in \mathscr{P}_{m-1}. \tag{35b}$$

The corresponding fundamental identity is

$$|v - w|_m^2 = |v|_m^2 - |w|_m^2 - 2 \sum_{i \in I} \gamma_i [v(a_i) - w(a_i)] \tag{36a}$$

or equivalently,

$$(v, w)_m = \sum_{i \in I} \gamma_i v(a_i), \quad \forall v \in X, \forall w \in V_0^{\perp}. \tag{36b}$$

The proof follows essentially from (29), where w is defined by (28a) and each of the K_{a_k} is obtained from (20), and from the representation formula (14). It is most remarkable that, for $n = 1$, the definition (35a, b) of *surface splines* (relative to the set A of given interpolation points) reduces strictly to the classical definition of *natural splines* (relative to the set A of given *knots* in $\mathscr{R}$). The verification is left to the reader, as also the extension to surface splines of other classical results such as the famous *Schoenberg theorem* (and the related *hypercircle inequality*, see e.g. [10], Section 3), which essentially asserts that the *best approximation in the sense of Sard* (relative to the prescribed set A) to *any* bounded linear functional on $BL^m(\mathscr{R}^n)$ for $m > n/2$ is characterized by the condition to be exact on $V_0^{\perp}$.

4. Algorithmic Aspects of Surface Spline Interpolation

Throughout this section, every possible dependence on the number $N := \text{Card}(K)$ of interpolation points in the set $A - B$ is emphasized explicitly by appending N in superscript position. As for the superscript T, it is for transpose.

In view of (28a, b), the unique element w^N of minimal norm $|\cdot|_m$ in the linear variety W^N defined by (26a, b, c) can be obtained by eliminating the N-vector $c^N := (\gamma_1^N, \ldots, \gamma_N^N)^T$ of (real) unknowns from the pair of matrix equations

$$G^N c^N = b^N, \quad \text{with } b^N := (\alpha_1', \ldots, \alpha_N')^T, \tag{37a}$$

$$(R^N)^T c^N = w^N, \quad \text{with } R^N := (K_{a_1}, \ldots, K_{a_N})^T, \tag{37b}$$

$G^N \equiv G(K_{a_1}, \ldots, K_{a_N})$ denoting the $N \times N$ *Gram matrix* of the sequence $(K_{a_1}, \ldots, K_{a_N})$ with respect to $(\cdot, \cdot)_m$ and being consequently a *positive definite symmetric* real matrix. It follows that w^N is given by

$$w^N = (R^N)^T (G^N)^{-1} b^N \tag{38a}$$

or equivalently, using determinants, by

$$w^N = -\det \begin{bmatrix} G^N & b^N \\ (R^N)^T & 0 \end{bmatrix} \Big/ \det(G^N); \tag{38b}$$

as to the remainder (or error) associated with this approximation of elements of W^N by w^N, it may be expressed similarly, in determinantal form, as

$$v - w^N = \det \begin{bmatrix} G^N & b^N \\ (R^N)^T & v \end{bmatrix} \Big/ \det (G^N), \quad \forall v \in W^N. \tag{39}$$

Taking into account the fact that $(w^N, v - w^N)_m = 0$, $\forall v \in W^N$, we readily get

$$\inf_{v \in W^N} |v|_m^2 \equiv |w^N|_m^2 = -\det \begin{bmatrix} G^N & b^N \\ (b^N)^T & 0 \end{bmatrix} \Big/ \det (G^N) \tag{40}$$

and

$$|v - w^N|_m^2 = \det [G(K_{a_1}, \ldots, K_{a_N}, v)]/\det [G(K_{a_1}, \ldots, K_{a_N})], \quad \forall v \in W^N. \tag{41}$$

As regards the *algorithmic interpretation* of equations (38a, b), there exist of course various possibilities, according to the type of representation for w^N that proves more suitable for practical use. In particular, whenever the interpolation points a_i (for $i \in I^N \subset I$) may be regarded as essentially fixed data, the solution $l^N \equiv (\lambda_1^N, \ldots, \lambda_N^N)$ of the matrix equation

$$G^N l^N = R^N \tag{42}$$

is probably worth determining once and for all; indeed, (38a) reduces then simply to the *biorthonormal expansion of Lagrange type*

$$w^N = (l^N)^T b^N \equiv \sum_{j=1}^{N} \alpha_j' \lambda_j^N, \tag{43}$$

which is characterized by the set of relationships

$$\langle \delta_{(a_j)}, \lambda_i^N \rangle \equiv (K_{a_j}, \lambda_i^N)_m = \delta_{ij}, \quad \forall i, j \in K^N \equiv [1, N] \subset K, \tag{44}$$

expressing the biorthonormality of the prescribed evaluation linear functionals $\delta_{(a_j)}$ on X_0 and the linear combinations λ_i^N defined by (42) of the associated functions $K_{a_j} \in X_0$. As emphasized in [3] (see pp. 35–39), various classical problems can be solved conveniently in this way, among others: the general (univariate) Hermite interpolation problem, its special cases (pointwise interpolation, osculatory interpolation, two point Taylor interpolation, ...) and its multivariate extensions by tensor product techniques. On the other hand, for all-round exploratory work, interpolation formulas of Lagrange type are rather unsuitable, for *lack of flexibility* when passing from w^N to w^{N+1} by adjunction of a new point a_{N+1} to the already used interpolation points a_i, $1 \leq i \leq N$.

This shortcoming is adequately covered by the *biorthonormal expansion of Newton type* to be described now, which indeed is constructed so as to have the so-called *permanence property* (this is characteristic of Fourier series and other orthonormal or biorthonormal expansions): the transition from w^N to w^{N+1} necessitates merely the addition of one more term to the expansion of w^N, all the terms obtained formerly remaining thus unchanged. The price to be paid for the convenience of the permanence property is simply that the prescribed linear functionals $\delta_{(a_j)}$ (for $j \in K$) must first be combined in a certain way before the use of the respective generalized Newton representation formula can be invoked. As exemplified now in connection with w^N, a general setting in which this type of biorthonormality and permanence can be obtained is provided by the theory of matrices and determinants. It turns out that the determinant quotient in (38b) is nothing else but the last pivot in the so-called *LDU (or triangular) factorization* of the bordered matrix of order $N + 1$ in the numerator, which indeed can be expressed uniquely as such a product of three matrices (L^{N+1} being unit lower triangular, D^{N+1} diagonal, U^{N+1} unit upper triangular) since none of the leading principal minors $\det(G^j)$, $1 \le j \le N$, vanishes (all of them are even positive). This is readily verified by exploiting the well-known Binet–Cauchy theorem of corresponding minors, which yields more generally the expression as a determinantal ratio of each element of the matrices L^{N+1}, D^{N+1} and U^{N+1}; specifically, for $1 \le j \le N$, the jth element $\tilde{K}_{a_j}$ of the last row of L^{N+1} is given by

$$\tilde{K}_{a_j} = \det\begin{bmatrix} G^{j-1} & R^{j-1}(a_j) \\ (R^{j-1})^T & K_{a_j} \end{bmatrix} \Big/ \det(G^j), \tag{45}$$

the jth element $\tilde{d}_j$ of the diagonal of D^{N+1} by

$$\tilde{d}_j = \det(G^j)/\det(G^{j-1}), \tag{46}$$

and the jth element $\tilde{\alpha}'_j$ of the last column of U^{N+1} by

$$\tilde{\alpha}'_j \equiv \langle \tilde{\delta}_{(a_j)}, w^N \rangle, \tag{47a}$$

with

$$\tilde{\delta}_{(a_j)} = \det\begin{bmatrix} G^{j-1} & \delta^{j-1} \\ (R^{j-1}(a_j))^T & \delta_{(a_j)} \end{bmatrix} \Big/ \det(G^j) \tag{47b}$$

and $\delta^{j-1} := (\delta_{(a_1)}, \ldots, \delta_{(a_{j-1})})^T$. Since the element in the bottom right-hand corner of D^{N+1} (resp. L^{N+1} and U^{N+1}) is $-w^N$ (resp. 1), we immediately obtain the interpolation formula

$$w^N = \sum_{j=1}^{N} \tilde{K}_{a_j} \tilde{d}_j \tilde{\alpha}'_j \tag{48}$$

by equating to zero the $(N + 1, N + 1)$-element in the product $(LDU)^{N+1}$; this expansion is biorthonormal in the sense that

$$\langle \tilde{\delta}_{(a_j)}, \tilde{d}_i \tilde{K}_{a_i} \rangle \equiv (\tilde{K}_{a_j}, \tilde{d}_i \tilde{K}_{a_i})_m = \delta_{ij}, \quad \forall i, j \ge 1; \tag{49}$$

having manifestly the required permanence property, the representation (48) of w^N is of Newton type. A number of interesting examples of biorthogonal systems of this kind are known, among others: the classical Newton interpolation formula and its confluent forms, the Euler–Maclaurin expansion, the Fourier expansions (see [3], pp. 46–49). Although the *results* (45, 46, 47, 48) of the biorthonormalization of the functions $K_{a_1}, K_{a_2}, \ldots$ against the functionals $\delta_{(a_1)}, \delta_{(a_2)}, \ldots$ are classical (see e.g. [3], pp. 41–45, where they are proved inductively), the purely *constructive approach* outlined above does not seem to be widely known in spite of its fundamental simplicity; it also suggests a most interesting interpretation of the expansion (48) we will mention in passing: (48), with the explicit definitions (45, 46, 47), can be regarded as the *Schweinsian expansion* (see [2], p. 109 and [13], p. 369) of the negative of the determinant quotient in (38b), the jth term in (48) giving simply the contribution to w^N of the jth *exchange-step* required by the LDU factorization; this interpretation is briefly mentioned in [2] (see p. 110, ex. 3), but only in connection with the classical Newton formula.

Since G^N is a positive definite symmetric (real) matrix, its LDU factorization is actually a LDL^T factorization where D^N is a positive diagonal matrix, so that there exists a unique lower triangular (real) matrix L^{*N} with positive diagonal elements for which

$$G^N = (L^{*N})(L^{*N})^T, \tag{50}$$

this variant being known as the *Cholesky decomposition* of G^N. Equation (38a) can thus be rewritten in the form

$$w^N = (R^{*N})^T b^{*N} \equiv \sum_{j=1}^{N} \alpha_j'^* K_{a_j}^*, \tag{51}$$

where $R^{*N} := (K_{a_1}^*, \ldots, K_{a_N}^*)$ and $b^{*N} := (\alpha_1'^*, \ldots, \alpha_N'^*)$, the solutions of the triangular system

$$L^{*N}(R^{*N} \mid b^{*N}) = (R^N \mid b^N), \tag{52}$$

are readily obtained by *back-substitution*. It should be noted that $(L^{*N})^T$ is nothing else but the upper triangular matrix constructed while orthonormalizing with respect to $(\cdot,\cdot)_m$ by the *Gram–Schmidt algorithm* the sequence of components K_{a_k} of the N-covector $(R^N)^T$, so that

$$\langle \delta_{(a_j)}^*, K_{a_i}^* \rangle \equiv (K_{a_j}^*, K_{a_i}^*)_m = \delta_{ij}, \quad \forall i, j \geq 1, \tag{53}$$

determinantal expressions for $K_{a_j}^*$, $\alpha_j'^*$, $\delta_{(a_j)}^*$ following directly from the comparison of (51) with (48) in view of (45, 46, 47). It should be realized (see also [12], p. 87) that $(\tilde{d}_j \tilde{K}_{a_j})$ is in actual fact the Riesz representer of the functional $\delta_{(a_j)}$ for the closed vector subspace of all functions of X_0 vanishing at the points $a_1, \ldots, a_{j-1}$.

In conclusion, the whole solution process of Problem (P) can be described in a *recursive* form, the standard algorithms to be used (Cholesky factorization and back-substitution) being indeed *numerically stable* and *very economical* as regards the number of arithmetic operations (see e.g. [14], p. 122 and [15], p. 27). Moreover, the computation can be easily organized so that each element of L^{*N}, b^{*N} is to be determined primarily as an inner product; the well known technique of accumulating inner products in double precision for additional accuracy can thus be used throughout, if necessary.

References

[1] J. H. Ahlberg, E. N. Nilson, and J. L. Walsh, *The Theory of Splines and Their Applications*, Academic Press Inc., New York (1967).
[2] A. C. Aitken, *Determinants and Matrices*, Oliver and Boyd Ltd., Edinburgh (1956).
[3] P. J. Davis, *Interpolation and Approximation*, Blaisdell Publ. Co., New York (1963).
[4] C. de Boor and R. E. Lynch, *On Splines and their Minimum Properties*, J. Math. Mech. *15*, 953–969 (1966).
[5] J. Duchon, *Fonctions-spline à énergie invariante par rotation*, Rapport de recherche n° 27, Université de Grenoble (1976).
[6] J. Duchon, *Interpolation des fonctions de deux variables suivant le principe de la flexion des plaques minces*, R.A.I.R.O. Analyse numérique *10*, 5–12 (1976).
[7] J. Duchon, 'Splines Minimizing Rotation—Invariant Semi-norms in Sobolev Spaces', in *Constructive Theory of Functions of Several Variables, Oberwolfach 1976*, W. Schempp and K. Zeller, ed., pp. 85–100, Springer-Verlag, Berlin–Heidelberg (1977).
[8] M. Golomb and H. F. Weinberger, 'Optimal Approximation and Error Bounds', in *On Numerical Approximation*, R. E. Langer, ed., pp. 117–190, The University of Wisconsin Press, Madison (1959).
[9] R. L. Harder and R. N. Desmarais, *Interpolation Using Surface Splines*, J. Aircraft *9*, 189–191 (1972).
[10] J. Meinguet, *Optimal Approximation and Error Bounds in Seminormed Spaces*, Num. Math. *10*, 370–388 (1967).
[11] J. Meinguet, 'An Intrinsic Approach to Multivariate Spline Interpolation at Arbitrary Points'. To appear in *Proc. NATO Advanced Study Institute on Polynomial and Spline Approximation, Calgary 1978*, B. N. Sahney, ed., 29 pp., D. Reidel Publ. Co., Dordrecht (1979).
[12] H. S. Shapiro, *Topics in Approximation Theory*, Springer-Verlag, Berlin–Heidelberg (1971).
[13] H. W. Turnbull, *The Theory of Determinants, Matrices, and Invariants*, Dover Publ., Inc., New York (1960).
[14] B. Wendroff, *Theoretical Numerical Analysis*, Academic Press Inc., New York (1966).
[15] J. H. Wilkinson and C. Reinsch, *Linear Algebra*, Springer-Verlag, Berlin–Heidelberg (1971).

Abstract

The concrete method of 'surface spline interpolation' is closely connected with the classical problem of minimizing a Sobolev seminorm under interpolatory constraints; the intrinsic structure of surface splines is accordingly that of a multivariate extension of natural splines. The proper abstract setting is a Hilbert function space whose reproducing kernel involves no functions more complicated than logarithms and is easily coded. Convenient representation formulas are given, as also a practical multivariate extension of the Peano kernel theorem. Owing to the numerical stability of Cholesky factorization of positive definite symmetric matrices, the whole construction process of a surface spline can be described as a recursive algorithm, the data relative to the various interpolation points being exploited in sequence.

Résumé

La méthode concrète d'interpolation par surfaces-spline est étroitement liée au problème classique de la minimisation d'une semi-norme de Soboleff sous des contraintes d'interpolation; la structure intrinsèque des surfaces-spline est dès lors celle d'une extension multivariée des fonctions-spline naturelles. Le cadre abstrait adéquat est un espace fonctionnel hilbertien dont le noyau reproduisant ne fait pas intervenir de fonctions plus compliquées que des logarithmes et est aisé à programmer. Des formules commodes de représentation sont données, ainsi qu'une extension multivariée d'intérêt pratique du théorème du noyau de Peano. Grâce à la stabilité numérique de la factorisation de Cholesky des matrices symétriques définies positives, la construction d'une surface-spline peut se faire en exploitant point après point les données d'interpolation.

(Received: September 25, 1978)

27

SPLINE INTERPOLATION AND SMOOTHING ON THE SPHERE*

GRACE WAHBA†

Abstract. We extend the notion of periodic polynomial splines on the circle and thin plate splines on Euclidean d-space to splines on the sphere which are invariant under arbitrary rotations of the coordinate system. We solve the following problem: Find $u \in \mathcal{H}_m(S)$, a suitably defined reproducing kernel (Sobolev) space on the sphere S to, A) minimize $J_m(u)$ subject to $u(P_i) = z_i$, $i = 1, 2, \cdots, n$, and B) minimize

$$\frac{1}{n}\sum_{i=1}^{n}(u(P_i)-z_i)^2+\lambda J_m(u),$$

where

$$J_m(u)=\int_0^{2\pi}\int_0^{\pi}(\Delta^{m/2}u(\theta,\phi))^2\sin\theta\,d\theta\,d\phi, \qquad m \text{ even}$$

$$=\int_0^{2\pi}\int_0^{\pi}\left\{\frac{(\Delta^{(m-1)/2}u)_\phi^2}{\sin^2\theta}+(\Delta^{(m-1)/2}u)_\theta^2\right\}\sin\theta\,d\theta\,d\phi, \qquad m \text{ odd}.$$

Here Δ is the Laplace–Beltrami operator on the sphere and $J_m(u)$ is the natural analogue on the sphere, of the quadratic functional $\int_0^{2\pi}(u^{(m)}(\theta))^2\,d\theta$ on the circle, which appears in the definition of periodic polynomial splines. $J_m(u)$ may also be considered to be the analogue of

$$\sum_{i=0}^{m}\binom{m}{i}\int_{-\infty}^{\infty}\int_{-\infty}^{\infty}\left(\frac{\partial^m u}{\partial x^i\,\partial y^{m-i}}\right)^2 dx\,dy$$

appearing in the definition of thin plate splines on the plane. The solution splines are obtained in the form of infinite series, which do not appear to be convenient for certain kinds of computation. We then replace J_m in A) and B) by a quadratic functional Q_m which is topologically equivalent to J_m on $\mathcal{H}_m(S)$ and obtain closed form solutions to the modified problems which are suitable for numerical calculation, thus providing practical pseudo-spline solutions to interpolation and smoothing problems on the sphere. Convergence rates of the splines and pseudo-splines will be the same. A number of results established or conjectured for polynomial and thin plate splines can be extended to the splines and pseudo-splines constructed here.

1. Introduction. This work is motivated by the following problem. The 500 millibar height (the height above sea level at which the pressure is 500 millibars) is measured (with error) at a large number n of weather stations distributed around the world. It is desired to find a smooth function $u = u(\theta, \phi)$ defined on the surface of the earth (θ = latitude, ϕ = longitude) which is an estimate of the 500 millibar height at position (θ, ϕ). There are many ways that this can be done. In this paper we develop what appears to be the natural generalization to the sphere of periodic interpolating and smoothing splines on the circle (see Golomb [12], Wahba [27]) and thin plate splines on Euclidean d-space (see Duchon [6], Meinguet [18], Wahba [28]).

To obtain a periodic interpolating or smoothing spline on the circle C one seeks the solution to one of the problems: Find $u \in \mathcal{H}_m(C)$ to minimize

A) $\quad J_m(u)$ subject to $u(t_i) = z_i, \quad i = 1, 2, \cdots, n$

or

B) $$\frac{1}{n}\sum_{i=1}^{n}(u(t_i)-z_i)^2+\lambda J_m(u).$$

* Received by the editors November 29, 1979. This research was supported by the U.S. Army Research Office under contract DAAG29-77-0209.

† Department of Statistics, University of Wisconsin, Madison, Wisconsin 53706.

Here

$$J_m(u)=\int_0^{2\pi}(u^{(m)}(t))^2\,dt, \tag{1.1}$$

$t_i\in[0,2\pi]$ and $\mathscr{H}_m(C)=\{u: u, u', \cdots, u^{(m-1)}$ abs. cont., $u^{(m)}\in\mathscr{L}_2[0,2\pi]$, $u^{(j)}(0)=u^{(j)}(2\pi)$, $j=0,1,\cdots,m-1\}$. To find a thin plate interpolating or smoothing spline on Euclidean d-space E^d, one finds $u\in\mathscr{H}_m(E^d)$ to minimize A) or B) above, where now $t_i=(x_{1i}, x_{2i},\cdots,x_{di})\in E^d$ and

$$J_m(u)=\sum_{i_1,i_2,\cdots,i_{m-1}}^{d}\int_{E^d}\left(\frac{\partial^m u}{\partial x_{i_1}\,\partial x_{i_2},\cdots,\partial x_{i_m}}\right)^2 dx_1\,dx_2\cdots,dx_d. \tag{1.2}$$

$\mathscr{H}_m(E^d)$is defined in Meinguet [18]. To obtain a thin plate interpolating or smoothing spline it is necessary that $2m-d>0$, since otherwise the evaluation functionals $u\to u(t_i)$ will not be bounded in $\mathscr{H}_m(E^d)$ and thus will not have representers which are used in the construction of the solution.

Duchon has called the solutions to problems involving J_m in Euclidean d-space thin plate splines, because, in two dimensions with $m=2$,

$$J_2(u)=\int_{-\infty}^{\infty}\int_{-\infty}^{\infty}\left[\left(\frac{\partial^2 u}{\partial x_1^2}\right)^2+2\left(\frac{\partial^2 u}{\partial x_1\,\partial x_2}\right)^2+\left(\frac{\partial^2 u}{\partial x_2^2}\right)^2\right]dx_1\,dx_2$$

is the bending energy of a thin plate. Interpolating and smoothing thin plate splines have been computed in a number of examples by Franke, Utreras, Wahba, Wahba and Wendelberger, and Wendelberger for data given in the form of an analytic function which is evaluated by computer at $t_1, t_2,\cdots,t_n$ [9], for function data with simulated errors [26], [28], [30] and for measured 500 millibar height data [31], with very satisfying results. Fisher and Jerome in a classic early paper [8] answered some important questions concerning interpolation problems on Ω a bounded set in R^d associated with general elliptic operators.

For the analysis of meteorological data, we would like to be able to compute smoothing splines on the sphere. To motivate the definition of J_m for the sphere, we first take a look at the Sobolev spaces $\mathscr{H}_m(C)$ of periodic functions on the circle. $\mathscr{H}_m(C)$ is the collection of square integrable functions u on $[0,2\pi]$ which satisfy

$$a_0^2+\sum_{\nu=1}^{\infty}\nu^{2m}a_\nu^2+\sum_{\nu=1}^{\infty}\nu^{2m}b_\nu^2<\infty, \tag{1.3}$$

where

$$a_\nu=\frac{1}{\sqrt{\pi}}\int_0^{2\pi}\cos\nu\theta u(\theta)\,d\theta,\qquad \nu=0,1,\cdots,$$

$$b_\nu=\frac{1}{\sqrt{\pi}}\int_0^{2\pi}\sin\nu\theta u(\theta)\,d\theta,\qquad \nu=1,2,\cdots.$$

We have

$$J_m(u)\doteq\int_0^{2\pi}(u^{(m)}(\theta))^2\,d\theta=\sum_{\nu=1}^{\infty}\nu^{2m}a_\nu^2+\sum_{\nu=1}^{\infty}\nu^{2m}b_\nu^2 \tag{1.4}$$

for $u\in\mathscr{H}_m(C)$. $\mathscr{H}_m(C)$ is thus a space of (periodic, square integrable) functions whose Fourier coefficients $\{a_\nu, b_\nu\}$ decay sufficiently fast to satisfy (1.3). The functions $\{\cos\nu\theta, \sin\nu\theta\}$ are the (periodic) eigenfunctions of the operator $D^{2m}(D^{2m}u=u^{(2m)})$

which appears when $J_m(u)$ of (1.1) is integrated by parts and u is sufficiently smooth and periodic:

$$J_m(u)=\int_0^{2\pi} u\cdot D^{2m}u\,d\theta.$$

If one formally integrates (1.2) by parts, and u is sufficiently smooth and decreases to 0 at infinity, then one obtains

$$J_m(u)=(-1)^m\int\cdots\int_{E^d} u\cdot\tilde{\Delta}^m u\,dx_1\cdots dx_d,$$

where $\tilde{\Delta}u$ is the Laplacian,

$$\tilde{\Delta}u=\frac{\partial^2 u}{\partial x_1^2}+\frac{\partial^2 u}{\partial x_2^2}+\cdots+\frac{\partial^2 u}{\partial x_d^2}.$$

The analogue of $\tilde{\Delta}$ on the sphere is the Laplace–Beltrami operator defined by

$$\Delta u=\frac{1}{\sin^2\theta}u_{\phi\phi}+\frac{1}{\sin\theta}(\sin\theta u_\theta)_\theta,$$

where $\theta\in[0,\pi]$ is latitude and $\phi\in[0,2\pi]$ is longitude. This is the restriction of the Laplacian in 3-space to the surface of the sphere; see Courant and Hilbert [3, Chapt. V, VII], and Whittaker and Watson [32]. The role of the eigenfunctions $\{(1/\sqrt{\pi})\cos\nu\theta,(1/\sqrt{\pi})\sin\nu\theta\}$ in $\mathcal{H}_m(C)$ is played in $\mathcal{H}_m(S)$, (S is the sphere) by the normalized spherical harmonics $\{Y_\nu^k(\theta,\phi)\}_{\nu=0}^{\infty}{}_{k=-\nu}^{\nu}$ (defined in § 2), which are the (periodic) eigenfunctions of the Laplace–Beltrami operator Δ^m, and the role of the eigenvalues $\{\nu^{2m},\nu^{2m}\}_{\nu=1}^\infty$ of D^{2m} is played by the eigenvalues of Δ^m. Δ^m has the single square integrable periodic eigenfunction $Y_0^0(\theta,\phi)=1$, corresponding to the eigenvalue 0. We now define $\mathcal{H}_m(S)$ as the space of square integrable functions u on S with

(1.5) $$|u_{00}|<\infty,\qquad \sum_{\nu=1}^{\infty}\sum_{k=-\nu}^{\nu}\frac{u_{\nu k}^2}{\lambda_{\nu k}}<\infty,$$

where

(1.6) $$u_{\nu k}=\int_S Y_\nu^k(P)u(P)\,dP$$

and $\{\lambda_{\nu k}^{-1}\}$, $(\lambda_{\nu k}^{-1}=[\nu(\nu+1)]^k)$ are the eigenvalues of Δ^m corresponding to $\{Y_\nu^k\}$. $\mathcal{H}_m(S)$ is thus a space of square integrable functions whose Fourier Bessel coefficients with respect to the spherical harmonics decay sufficiently fast to satisfy (1.5). Let J_m be defined by

(1.7)
$$J_m(u)=\int_0^{2\pi}\int_0^{\pi}(\Delta^{m/2}u)^2\sin\theta\,d\theta\,d\phi,\qquad m\text{ even},$$
$$=\int_0^{2\pi}\int_0^{\pi}\frac{(\Delta^{(m-1)/2}u)_\phi^2}{\sin^2\theta}+(\Delta^{(m-1)/2}u)_\theta^2\sin\theta\,d\theta\,d\phi,\qquad m\text{ odd}.$$

It is not hard to show that, for $u\in\mathcal{H}_m(S)$,

(1.8) $$J_m(u)=\sum_{\nu=1}^{\infty}\sum_{k=-\nu}^{\nu}\frac{u_{\nu k}^2}{\lambda_{\nu k}}.$$

A number of results which are known or conjectured for polynomial splines on the circle and thin plate splines on E^d will carry over to the thin plate splines and pseudo-splines on the sphere. They include optimality properties of the generalized cross-validation estimate of λ and m [4], [25], convergence rates for smoothing splines with noisy data, properties of associated orthogonal series density estimates, and interpretation of interpolating and smoothing splines as Bayes estimates when u is modeled as the solution to the stochastic differential equation $\Delta^{m/2} u =$ "white noise." Details and further references may be found in Wahba [29]. The corresponding splines when u is modeled as a general stationary autoregressive moving average process on S are also given in Wahba [29], as well as possible models encompassing nonstationarity (anisotropy). The reader interested in meteorological applications may be interested in consulting Stanford [24], where ensembles of $\{u^2_{\nu k}\}$ defined in (1.6) have been computed from measured satellite radiance data and are suggestive of an appropriate choice of m in certain meteorological applications. The results here also show that variational techniques for meteorological data analysis similar to these pioneered by Sasaki [21] and others can be carried out on the sphere; see also Wahba and Wendelberger [30]. Part of the importance of the present work is in its potential applicability to important meteorological problems, some of which are mentioned near the end of § 2.

We seek $u \in \mathcal{H}_m(S)$ to minimize

A) $\quad J_m(u)$ subject to $u(P_i) = z_i, \qquad i = 1, 2, \cdots, n,$

B) $$\frac{1}{n} \sum_{i=1}^{n} (u(P_i) - z_i)^2 + \lambda J_m(u),$$

where $P_i \in S$, and J_m is defined by (1.7).

We cannot solve these problems for $m = 1$ for the same reason they cannot be solved in E^d for $2m - d \leqq 0$, that is, because the evaluation functions are not continuous in $\mathcal{H}_1(S)$, that is, $\mathcal{H}_1(S)$ is not a reproducing kernel space. However, for $m = 2, 3, \cdots$ we will give the explicit solution to those two problems, which we will call thin plate splines on the sphere. It is actually not hard to obtain the solutions, since we can construct a reproducing kernel for $\mathcal{H}_m(S)$, with $J_m(\cdot)$ as a seminorm, from the well-known eigenfunctions and eigenvalues of the Laplace–Beltrami operator. Given the reproducing kernel [2], the solutions to such problems are well known, and in fact problems A) and B) can be solved with $u(P_i)$ replaced by $L_i u$, where L_i is any continuous linear functional on $\mathcal{H}_m(S)$. See, for example, Kimeldorf and Wahba [17] and references cited there.

Unfortunately we only know the aforementioned reproducing kernels in the form of infinite series. It appears that no closed form expression exists which is convenient for computational purposes. Wendelberger [31] has computed the reproducing kernels given below for m from 2 to 10 by evaluating the infinite series, and it is likely that satisfactory computational procedures for interpolation and smoothing splines on the sphere can be developed based on the infinite series. However, for general continuous linear functionals it may be important to have a reproducing kernel in closed form. Furthermore, to compute certain functionals of the solution, for example, derivatives, it may be important to have a closed form solution. For this reason we suggest replacing $J_m(\cdot)$ by another quadratic functional $Q_m(\cdot)$ which is topologically equivalent to $J_m(\cdot)$ in the sense that there exist α and β, $0 < \alpha < \beta < \infty$ such that

$$\alpha J_m(u) \leqq Q_m(u) \leqq \beta J_m(u), \quad \text{all } u \in \mathcal{H}_m(S).$$

We give the reproducing kernel associated with Q_m in closed form. It involves only logarithms and powers of monomials of sines and cosines, and appears quite suitable for the numerical computation of the solutions of A) and B) and related problems with J_m replaced by Q_m. We will call the resulting interpolating and smoothing functions thin plate pseudo-splines on the sphere. Convergence rates for the thin plate pseudo-splines will be the same as those for the thin plate splines on the sphere because of the topological equivalence of J_m and Q_m.

In § 2 we derive the thin plate spline solutions to problems A) and B), and in § 3 we obtain the thin plate pseudo-spline solutions, where J_m is replaced by Q_m.

We remark that the development of § 2 can no doubt be generalized to establish splines associated with the Laplace–Beltrami operator on compact Riemannian manifolds other than the circle and the sphere; see Gine [10], Hannan [14], Yaglom [33] and Schoenberg [23]. However this is not pursued further.

2. Spherical harmonics and the solution to problems A) and B) on the sphere. The spherical harmonics $\{U_\nu^k(\theta, \phi)\}$ are defined by

$$\begin{aligned} U_\nu^k(\theta,\phi) &= \cos k\phi P_\nu^k(\cos\theta), \qquad k=1,2,\cdots,\nu \\ &= \sin k\phi P_\nu^k(\cos\theta), \qquad k=-1,-2,\cdots,-\nu \\ &= P_\nu(\cos\theta), \qquad k=0, \quad \nu=0,1,2,\cdots, \end{aligned}$$

where $P_\nu^k(z)$ are the Legendre functions of the kth order,

$$P_\nu^k(z) = (1-z^2)^{k/2}\left(\frac{d^k}{dz^k}\right)P_\nu(z),$$

and $P_\nu(z)$ is the νth Legendre polynomial. Recursion formulas for generating the P_ν^k may be found in Abramowitz and Stegun [1].

It is well known that the $\{U_\nu^k, k=-\nu,\cdots,\nu, \nu=0,1,\cdots\}$ form an $\mathcal{L}_2(S)$-complete set of eigenfunctions of the Laplace–Beltrami operator of (1.4) satisfying

$$\Delta U_\nu^k = -\nu(\nu+1)U_\nu^k, \qquad k=-\nu,\cdots,\nu, \quad \nu=0,1,\cdots.$$

See Courant and Hilbert [3], Sansone [22]. Let

$$Y_\nu^0 = \sqrt{\frac{2\nu+1}{4\pi}}\,U_\nu^0, \qquad \nu=0,1,\cdots,$$

$$Y_\nu^k = 2\sqrt{\frac{2\nu+1}{4\pi}\frac{(\nu-k)!}{(\nu+k)!}}\,U_\nu^k, \qquad k=1,2,\cdots, \quad \nu=1,2,\cdots.$$

Then (Sansone [22, p. 264, 268])

$$\int_S (Y_\nu^k(P))^2\,dP = 1,$$

and we have the addition formula

$$\sum_{k=-\nu}^{\nu} Y_\nu^k(P)Y_\nu^k(P') = \frac{2\nu+1}{4\pi}P_\nu(\cos\gamma(P,P')),$$

where $\gamma(P, P')$ is the angle between P and P'. The $\{Y_\nu^k\}$ form an orthonormal basis for $\mathcal{L}_2(S)$. Jones [16] has used a finite set of spherical harmonics to estimate 500 millibar heights by regression methods. The spherical harmonics are also utilized in several numerical weather prediction models [11].

Let $\mathcal{H}_m^0(S)$ be the subset of $\mathcal{L}_2(S)$ with an expansion of the form

$$u(P) \sim \sum_{\nu=1}^{\infty} \sum_{k=-\nu}^{\nu} u_{\nu k} Y_\nu^k(P), \tag{2.1}$$

where

$$u_{\nu k} = \int_S u(P) Y_\nu^k(P)\, dP,$$

satisfying

$$\sum_{\nu=1}^{\infty} \sum_{k=-\nu}^{\nu} \frac{u_{\nu k}^2}{\lambda_{\nu k}} < \infty,$$

where

$$\lambda_{\nu k} = [\nu(\nu+1)]^{-m}. \tag{2.2}$$

Functions in $\mathcal{H}_m^0(S)$ satisfy

$$\int_S u(P)\, dP = 0,$$

since the 0, 0th term $Y_0^0 \equiv 1$ has been omitted from the expansion (2.1).

$\mathcal{H}_m^0(S)$ is clearly a Hilbert space with the norm defined by

$$\|u\|_m^2 = \sum_{\nu=1}^{\infty} \sum_{k=-\nu}^{\nu} \frac{u_{\nu k}^2}{\lambda_{\nu k}} \tag{2.3}$$

for any $m \geqq 0$. For $m > 1$, define $K(P, P')$, $(P, P') \in S \times S$ by

$$\begin{aligned} K(P, P') = K_m(P, P') &= \sum_{\nu=1}^{\infty} \sum_{k=-\nu}^{\nu} \lambda_{\nu k} Y_\nu^k(P) Y_\nu^k(P') \\ &\equiv \frac{1}{4\pi} \sum_{\nu=1}^{\infty} \frac{2\nu+1}{\nu^m(\nu+1)^m} P_\nu(\cos \gamma(P, P')). \end{aligned} \tag{2.4}$$

Since $|P_\nu(z)| \leqq 1$ for $|z| \leqq 1$ (Sansone [22, p. 187]), the series converges uniformly for any $m > 1$ and $K(P, P')$ is a well-defined positive definite function on $S \times S$ with

$$\langle K(P, \cdot), K(P', \cdot)\rangle_m = K(P, P'),$$

where $\langle \cdot, \cdot \rangle_m$ is the inner product induced by (2.3). Furthermore, it is easily verified that, for m an integer >1,

$$\int_S K(P, R) \Delta_{(R)}^m K(P', R)\, dR = K(P, P'),$$

where $\Delta_{(R)}^m$ means the operator Δ^m applied to the variable R. This follows since $\Delta^m Y_\nu^k \equiv \lambda_{\nu k}^{-1} Y_\nu^k$. Thus, for m an integer >1, $K(\cdot, \cdot)$ reproduces under the inner product induced by the norm $J_m^{1/2}(\cdot)$, and

$$J_m(u) = \sum_{\nu=1}^{\infty} \sum_{k=-\nu}^{\nu} \frac{u_{\nu k}^2}{\lambda_{\nu k}},$$

with

$$u_{\nu k} = \int_S u(P) Y_\nu^k(P)\, dP \quad \text{for any } u \in \mathcal{H}_m^0.$$

$\mathcal{H}_m^0(S)$ is therefore the reproducing kernel Hilbert space $(r\,k\,h\,s)$ with reproducing kernel $(rk)K(\cdot, \cdot)$.

The space $\mathcal{H}_m(S)$ in which one wants to solve problems A) and B) is

$$\mathcal{H}_m(S) = \mathcal{H}_m^0(S) \oplus \{1\},$$

where $\{1\}$ is the one-dimensional space of constant functions. $\mathcal{H}_m^0(S)$ and $\{1\}$ will be orthogonal subspaces in $\mathcal{H}_m(S)$ if we endow $\mathcal{H}_m(S)$ with the norm defined by

$$\|u\|^2 = J_m(u) + \frac{1}{4\pi}\left(\int_S u(P)\,dP\right)^2.$$

The following theorem is an immediate consequence of these facts and Kimeldorf and Wahba [17, Lemmas 3.1, 5.1].

THEOREM 1. *The solutions* $u_{n,m}$ *and* $u_{n,m,\lambda}$ *to problems* A) *and* B) *on the sphere are given by*

$$u_{n,m,\lambda}(P) = \sum_{i=1}^{n} c_i K(P, P_i) + d, \tag{2.5}$$

where $\mathbf{c} = (c_i, \cdots, c_n)'$ *and* d *are given by*

$$\mathbf{c} = (K_n + n\lambda I)^{-1}[I - T(T'(K_n + n\lambda I)^{-1}T)^{-1}T'(K_n + n\lambda I)^{-1}]\mathbf{z}, \tag{2.6}$$

$$d = (T'(K_n + n\lambda I)^{-1}T)^{-1}T'(K_n + n\lambda I)^{-1}\mathbf{z}, \tag{2.7}$$

where K_n *is the* $n \times n$ *matrix with* j, kth *entry* $(K_n)_{ij}$ *given by*

$$(K_n)_{ij} = K(P_i, P_j), \tag{2.8}$$

$$T = (1, \cdots, 1)' \tag{2.9}$$

and

$$\mathbf{z} = (z_1, \cdots, z_n)'.$$

Also

$$u_{n,m} \equiv u_{n,m,0}.$$

The continuous linear functionals $L_i u = u(P_i)$ may be replaced in the problem statements by any set of n linearly independent continuous linear functionals on $\mathcal{H}_m(S)$ which are not all identically 0 on $\{1\}$. Then, as is usual in *rk* theory, to obtain the solution one replaces $K(P, P_i)$ in (2.5) by $L_i K(P, \cdot)$, $K(P_i, P_j)$ in (2.8), by $L_{i(P)} L_{j(P')} K(P, P')$, and the ith component of T in (2.9) by $L_i(1)$. (See Kimeldorf and Wahba [17].) One example of useful L_i is $L_i u = \int_{S_i} u(P)\,dP$; i.e., the data functionals are regional averages (see Dyn and Wahba [5]). Furthermore, if $z_i = L_i u + \varepsilon_i$, where u is fixed, unknown function in $\mathcal{H}_m(S)$ and the $\{\varepsilon_i\}$ can be modeled as i.i.d. $\mathcal{N}(0, \sigma^2)$ random variables, then (provided λ is chosen properly; see [4], [30]) an estimate of Lu for L any continuous linear functional on $\mathcal{H}_m(S)$ is provided by $Lu_{m,n,\lambda}$. $Lu = \sin\theta u_\theta(P)$ and $Lu = u_\phi(P)$ are continuous linear functionals on $\mathcal{H}_m(S)$ for $m \geqq 3$. Therefore, this provides a technique for estimating meteorological properties of interest involving the derivatives of u, for example, the geostrophic wind; see [30]. Other potential applications are to the estimation of budgets (Johnson and Downey [15]), and the geostrophic vorticity (Haltiner and Martin [13]).

We remark that the equations (2.6) and (2.7) for $\mathbf{c}$ and d can be readily verified to be equivalent to

$$(K_n + n\lambda I)\mathbf{c} + dT = \mathbf{z}, \tag{2.10}$$

$$T'\mathbf{c} = 0. \tag{2.11}$$

If we assume K_n and T are given, then (2.10) and (2.11) lend themselves more readily to numerical solution than the computation of (2.6) and (2.7). See Paihua Montes [19], Wahba [28], Wendelberger [31].

In order to have a closed form expression for $K(P_i)$ it is necessary to sum the series

$$k_m(z)=\sum_{\nu=1}^{\infty}\frac{2\nu+1}{\nu^m(\nu+1)^m}P_\nu(z). \tag{2.12}$$

A closed form expression for $m=1$ ($z\neq 0$) can be obtained but does not interest us here.

To attempt to sum (2.12) for $m=2$, we note that

$$\frac{2\nu+1}{\nu^2(\nu+1)^2}\equiv\frac{1}{\nu^2}-\frac{1}{(\nu+1)^2}\equiv\int_0^1 \log h\left(1-\frac{1}{h}\right)h^\nu\,dh,\qquad \nu=1,2,\cdots. \tag{2.13}$$

Using the generating formula for Legendre polynomials (Sansone [22, p. 169]),

$$\sum_{\nu=1}^{\infty}h^\nu P_\nu(z)=(1-2hz+h^2)^{-1/2}-1,\qquad -1<h<1, \tag{2.14}$$

gives

$$k_2(z)=\int_0^1 \log h\left(1-\frac{1}{h}\right)\left(\frac{1}{\sqrt{1-2hz+h^2}}-1\right)dh. \tag{2.15}$$

Repeated attempts to integrate this by parts using formulas for indefinite integrals involving expressions of the form $\sqrt{1-2zh+h^2}$, and related integrals to be found in Pierce and Foster [20] and Dwight [7], led us to terms with a closed form expression plus a term involving Dwight [7, formula 731.1] whose right-hand side is an infinite series. This exercise, plus a helpful conversation with R. Askey who suggested that the sum could be reduced to a dilogarithm, convinced us that no readily computable closed form expression was to be found. For this reason we seek to change the problem slightly so that readily computable interpolating and smoothing formulas can be obtained. We do this in the next section.

3. Thin plate pseudo-splines on the sphere. We seek a norm $Q_m^{1/2}(u)$ on $\mathscr{H}_m^0(S)$ which is topologically equivalent to $J_m^{1/2}(u)$ on $\mathscr{H}_m^0(S)$ and for which the reproducing kernel can be obtained in closed form convenient for computation.

Define

$$Q_m(u)=\sum_{\nu=1}^{\infty}\sum_{k=-\nu}^{\nu}\frac{u_{\nu k}^2}{\xi_{\nu k}},\qquad u_{\nu k}=\int_S u(P)Y_\nu^k(P)\,dP,$$

where

$$\xi_{\nu k}=\left[\left(\nu+\frac{1}{2}\right)(\nu+1)(\nu+2)\cdots(\nu+2m-1)\right]^{-1}. \tag{3.1}$$

Since

$$\frac{1}{m^{2m}\xi_{\nu k}}\leqq\frac{1}{\lambda_{\nu k}}\leqq\frac{1}{\xi_{\nu k}},\qquad \nu=1,2,\cdots,\quad k=-\nu,\cdots,\nu,\quad m=2,3,\cdots,$$

we have

$$\frac{1}{m^{2m}}Q_m(u)\leqq J_m(u)\leqq Q_m(u),\qquad u\in\mathscr{H}_m^0(S),$$

and, thus the norms $J_m^{1/2}(\cdot)$ and $Q_m^{1/2}(\cdot)$ are topologically equivalent on $\mathcal{H}_m^0(S)$. The reproducing kernel $R(P, P')$ for $\mathcal{H}_m^0(S)$ with norm $Q_m^{1/2}(\cdot)$ is then

$$\begin{aligned} R(P, P') = R_m(P, P') &= \sum_{\nu=1}^{\infty} \sum_{k=-\nu}^{\nu} \xi_{\nu k} Y_\nu^k(P) Y_\nu^k(P') \\ &= \frac{1}{2\pi} \sum_{\nu=1}^{\infty} \frac{1}{(\nu+1)(\nu+2)\cdots(\nu+2m-1)} P_\nu(\cos\gamma(P, P')). \end{aligned} \tag{3.2}$$

A closed form expression can be obtained for $R(P, P')$ as follows. Since

$$\frac{1}{r!}\int_0^1 (1-h)^r h^\nu \, dh \equiv \frac{1}{(\nu+1)\cdots(\nu+r+1)}, \qquad r = 0, 1, 2, \cdots,$$

then by using the generating function (2.14) for the Legendre polynomials we have

$$\begin{aligned} R(P, P') &= \frac{1}{2\pi} \sum_{\nu=1}^{\infty} \frac{1}{(\nu+1)(\nu+2)\cdots(\nu+2m-1)} P_\nu(z) \\ &= \frac{1}{2\pi}\left[\frac{1}{(2m-2)!} q_{2m-2}(z) - \frac{1}{(2m-1)!}\right], \end{aligned} \tag{3.3}$$

where

$$z = \cos\gamma(P, P')$$

and

$$q_m(z) = \int_0^1 (1-h)^m (1-2hz+h^2)^{-1/2} \, dh, \qquad m = 0, 1, \cdots. \tag{3.4}$$

Formulas for $\int h^m (1-2hz+h^2)^{-1/2} \, dh$, $m = 0, 1, 2$ and recursion formulas for general m in terms of the formulas for $m-1$ and $m-2$ can be found in Pierce and Foster [20, pp. 165, 174, 177, 196]. q_m was obtained by hand for $m = 0, 1, 2$ and 3. In the middle of this dull exercise P. Bjornstad observed that the MACSYMA program at MIT, which could be called from the computer science department at Stanford where this exercise was taking place, could be used to evaluate $q_m(z)$ recursively. He kindly wrote such a program and the results appear in Table 1. Thus, for example, $R(P, P')$ for $m = 2$ involves q_2 and, from the table,

$$q[2] = \frac{A(12W^2 - 4W) - 6CW + 6W + 1}{2},$$

giving

$$q_2(z) = \frac{1}{2}\left\{\ln\left(1 + \sqrt{\frac{2}{1-z}}\right)\left[12\left(\frac{1-z}{2}\right)^2 - 4\frac{(1-z)}{2}\right] - 12\left(\frac{1-z}{2}\right)^{3/2} + 6\left(\frac{1-z}{2}\right) + 1\right\}.$$

Note that $q[0]$ which appears in the $m = 1$ case does not lead to a proper *rk* since $q_0(1)$ is not finite. However, a proper *rk* exists for any $m > 1$, and the table can be used to define q_{2m-2} for $m = \frac{3}{2}, 2, \frac{5}{2}, \cdots, 6$.

We collect these results in

THEOREM 2. *The solutions $u_{n,m}$ and $u_{n,m,\lambda}$ to the problems: Find $u \in \mathcal{H}_m(S)$ to*

A′) *minimize $Q_m(u)$ subject to $u(P_i) = z_i$, $i = 1, 2, \cdots, n$,*

B′) *minimize* $\frac{1}{n}\sum_{i=1}^{n} (u(P_i) - z_i)^2 + \lambda Q_m(u)$, *are given by*

$$\hat{u}_{n,m} = \hat{u}_{n,m0}, \qquad \hat{u}_{n,m,\lambda}(P) = \sum_{i=1}^{n} c_i R(P, P_i) + d,$$

where $R(P, P')$ *is defined by* (3.3) *and* (3.4) *and* $\mathbf{c}$ *and* d *are determined by*

$$(R_n + n\lambda I)\mathbf{c} - dT = \mathbf{z}, \qquad T'\mathbf{c} = 0,$$

where R_n *is the* $n \times n$ *matrix with* j, kth *entry* $R(P_i, P_j)$ *and* $T = (1, \cdots, 1)'$.

TABLE 1

$$q_m(z) = \int_0^1 (1-h)^m (1-2hz+h^2)^{-1/2}\, dh, \qquad m = 0, 1, \cdots, 10,$$

Key. $q[m] = q_m[z]$, $A = \ln(1 + 1/\sqrt{W})$, $C = 2\sqrt{W}), W = (1-z)/2$

$q[0]$	A
$q[1]$	$2AW - C + 1$
$q[2]$	$(A(12W^2 - 4W) - 6CW + 6W + 1)/2$
$q[3]$	$(A(60W^3 - 36W^2) + 30W^2 + C(8W - 30W^2) - 3W + 1)/3$
$q[4]$	$(A(840W^4 - 720W^3 + 72W^2) + 420W^3 + C(220W^2 - 420W^3) - 150W^2 - 4W + 3)/12$
$q[5]$	$(A(7560W^5 - 8400W^4 + 1800W^3) + 3780W^4 + C(-3780W^4 + 2940W^3 - 256W^2)$ $-2310W^3 + 60W^2 - 5W + 6)/30$
$q[6]$	$(A(27{,}720W^6 - 37{,}800W^5 + 12{,}600W^4 - 600W^3) + 13{,}860W^5$ $+ C(-13{,}860W^5 + 14{,}280W^4 - 2772W^2) - 11970W^4 + 1470W^3 + 15W^2 - 3W + 5)/30$
$q[7]$	$(A(360{,}360W^7 + 582{,}120W^6 + 264{,}600W^5 + 29{,}400W^4) + 180{,}180W^6$ $+ C(-180{,}180W^6 + 231{,}000W^5 - 71{,}316W^4 + 3072W^3)$ $-200{,}970W^5 + 46{,}830W^4 - 525W^3 + 21W^2 - 7W + 15)/105$
$q[8]$	$(A(10{,}810{,}800W^8 - 20{,}180{,}160W^7 + 11{,}642{,}400W^6 + 2{,}116{,}800W^5 + 58{,}800W^4)$ $+ 5{,}405{,}400W^7 + C(-5{,}405{,}400W^7 + 8{,}288{,}280W^6 - 3{,}538{,}920W^6 + 363{,}816W^4)$ $-7{,}387{,}380W^6 + 2{,}577{,}960W^5 - 159{,}810W^4 - 840W^3 + 84W^2 - 40W + 105)/840$
$q[9]$	$(A(61{,}261{,}200W^9 - 129{,}729{,}600W^8 + 90{,}810{,}720W^7 - 23{,}284{,}800W^6 + 1{,}587{,}600W^5)$ $+ 30{,}630{,}600W^8 + C(-30{,}630{,}600W^8 + 54{,}654{,}600W^7 - 29{,}909{,}880W^6$ $+ 5{,}104{,}440W^5 - 131{,}072W^4) - 49{,}549{,}500W^7 + 23{,}183{,}160W^6$ $-2{,}903{,}670W^5 + 17{,}640W^4 - 420W^3 + 72W^2 - 45W + 140)/1{,}260$
$q[10]$	$(A(232{,}792{,}560W^{10} - 551{,}350{,}800W^9 + 454{,}053{,}600W^8 - 151{,}351{,}200W^7 + 17{,}463{,}600W^6$ $- 317{,}520W^5) + 116{,}396{,}280W^9 + C(-116{,}396{,}280W^9 + 236{,}876{,}640W^8$ $- 158{,}414{,}256W^7 + 38{,}507{,}040W^6 - 2{,}462{,}680W^5) - 217{,}477{,}260W^8$ $+ 127{,}987{,}860W^7 - 24{,}954{,}930W^6 + 930{,}006W^5 + 2{,}940W^4$ $- 180W^3 + 45W^2 - 35W + 126)/1{,}260$

Of course the remarks following Theorem 1 concerning general continuous linear functionals and computing procedures apply here also.

4. Acknowledgments. We would like to thank P. Bjornstad for writing the computer program which generated Table 1, G. Golub for his hospitality at the Stanford University Numerical Analysis Group where some of this work was done and M. Ghil for helpful discussions concerning related stochastic differential equations.

REFERENCES

[1] M. ABRAMOWITZ AND I. STEGUN, *Handbook of Mathematical Functions*, National Bureau of Standards, Applied Mathematics Series 55, 1964.

[2] N. ARONSZAJN, *Theory of reproducing kernels*, Trans. Amer. Math. Soc., 68, (1950), pp. 337–303.

[3] R. COURANT AND D. HILBERT, *Methods of Mathematical Physics, vol.* 1, Interscience, New York, 1953.

[4] P. CRAVEN AND G. WAHBA, *Smoothing noisy data with spline functions: estimating the correct degree of smoothing by the method of generalized cross-validation*, Numer. Math., 31 (1979), pp. 377–403.

[5] N. DYN AND G. WAHBA. On the estimation of functions of several variables from aggregated data. University of Wisconsin-Madison, Mathematics Research Center, TSR 1974, July 1979.

[6] J. DUCHON, Interpolation des fonctions de deux variables suivante le principe de la flexion des plaques minces. R.A.I.R.O. Analyse Numerique, 10 (1976), pp. 5–12.

[7] H. B. DWIGHT, *Tables of Integrals and Other Mathematical Data*, Macmillan, New York, 1947.

[8] S. D. FISHER AND J. W. JEROME, *Elliptic variational problems in L^2 and L^∞*, Indiana Univ. Math. J., 23 (1974), pp. 685–698.

[9] R. FRANKE, *A critical comparison of some methods for interpolation of scattered data*, Naval Postgraduate School Report NPS-53-79-003, 1979.

[10] E. GINÉ, *Invariant tests for uniformity on compact Riemannian manifolds based on Sobolev norms*, Ann. Statist., 36 (1975), pp. 1243–1266.

[11] Global Atmospheric Research Programme (GARP), *Modelling for the first GARP global experiment*, GARP Publications, Series No. 14, World Meteorological Organization, Geneva, Switzerland.

[12] M. GOLOMB, *Approximation by periodic spline interpolants on uniform meshes*, J. Approx. Theory, 1 (1968), pp. 26–65.

[13] G. J. HALTINER AND F. L. MARTIN, *Dynamic and Physical Meteorology*, McGraw-Hill, New York, 1979, p. 358.

[14] E. J. HANNAN, *Group representations and applied probability*, J. Appl. Probab., 2 (1965), pp. 1–68.

[15] D. R. JOHNSON AND W. K. DOWNEY, *The absolute angular momentum budget of an extratropical cyclone, Quasi Langrangian diagnostics* 3, Monthly Weather Review, 104 (1976), pp. 3–14.

[16] R. H. JONES, *Stochastic process on a sphere as applied to meterological* 500*mb forecasts*, in Time Series Analysis, Murray Rosenblatt, ed., John Wiley, New York, 1963.

[17] G. KIMELDORF AND G. WAHBA, *Some results on Tchebycheffian spline functions*, J. Math. Anal. Applic., 33, (1971), pp. 82–95.

[18] J. MEINGUET, *Multivariate interpolation at arbitrary points made simple*, ZAMP, to appear.

[19] L. PAIHUA MONTES, *Quelque methodes numeriques pour le calcul de fonctions splines a une et plusiers variables*, Thesis, Université Scientifique et Medicale de Grenoble, 1978.

[20] B. O. PIERCE AND R. M. FOSTER, *A Short Table of Integrals*, Grun and Company, 1956.

[21] Y. SASAKI, *A theoretical interpretation of anisotropically weighted smoothing on the basis of numerical variational analysis*, Monthly Weather Review, 99 (1971), pp. 698–707.

[22] G. SANSONE, *Orthogonal Functions*, revised English edition, Interscience, New York, 1959.

[23] I. J. SCHOENBERG, *Positive definite functions on spheres*, Duke Univ. Math. Journal, 9 (1942), pp. 96–108.

[24] J. L. STANFORD, *Latitudinal-wavenumber power spectra of stratospheric temperature fluctuations*, J. Atmospheric Sci., 36 (1979), pp. 921–931.

[25] F. UTERAS DIAZ, *Quelques resultats d'optimalite pour la methode de validation croisée*, No. 301 mathématiques appliquées, Université Scientifique et Medicale de Grenoble, 1978.

[26] F. UTRERAS DIAZ, *Cross validation techniques for smoothing spline functions in one or two variables*, in Smoothing Techniques for Curve Estimation, T. Gasser and M. Rosenblatt, eds., Lecture Notes in Mathematics 757, Springer-Verlag, New York, 1979, pp. 196–231.

[27] G. WAHBA, *Smoothing noisy data by spline functions*, Numer. Math., 24 (1975), pp. 383–393.

[28] ———, *How to smooth curves and surfaces with splines and cross-validation*, in Proceedings of the 24th Conference on the Design of Experiments in Army Research Development and Testing, ARO Report 79-2, U.S. Army Research Office, Research Triangle Park, North Carolina, 1979, also University of Wisconsin, Madison, Department of Statistics Technical Report 555, 1979.

[29] ———, *Spline interpolation and smoothing on the sphere*, University of Wisconsin, Madison, Department of Statistics Technical Report 584, October, 1979.

[30] G. Wahba and J. Wendelberger, *Some new mathematical methods for variational objective analysis using splines and cross-validation*, Monthly Weather Review, 108 (1980), pp. 1122–1143.

[31] J. Wendelberger, Thesis, to appear.

[32] E. T. Whittaker and G. N. Watson, *A course of modern analysis*, Cambridge University Press, Cambridge, 1958.

[33] A. M. Yaglom, *Second order homogeneous random fields*, in Fourth Berkeley Symposium in Probability and Statistics, 2,, J. Neyman, ed., University of California Press, pp. 593–622.

ERRATA

Table 1, q [6], line 2 should read: ". . . -2772 W^3. . ."

Table 1, q [7], line 1 should read: "(A(360,360W^7 — 582,120W^6 + 264,600W^5 — 29,400W^4) + 180,180W^6"

Table 1, q [8], line 1 should read: ". . . -2,116,800W^5. . ."

Table 1, q [8], line 2 should read: ". . .-3,538,920W^5. . ."

28

Reprinted from *Ann. Stat.* **3**:30–48 (1975)

INTERPOLATING SPLINE METHODS FOR DENSITY ESTIMATION I. EQUI-SPACED KNOTS[1]

BY GRACE WAHBA

University of Wisconsin

Statistical properties of a variant of the histospline density estimate introduced by Boneva-Kendall-Stefanov are obtained. The estimate we study is formed for x in a finite interval, $x \in [a, b] = [0, 1]$ say, by letting $\hat{F}_n(x)$, $x \in [0, 1]$ be the unique cubic spline of interpolation to the sample cumulative distribution function $F_n(x)$ at equi-spaced points $x = jh$, $j = 0, 1, \cdots, l + 1$, $(l + 1)h = 1$, which satisfies specified boundary conditions $\hat{F}_n'(0) = a$, $\hat{F}_n'(1) = b$. The density estimate $\hat{f}_n(x)$ is then $\hat{f}_n(x) = d/dx\, \hat{F}_n(x)$. It is shown how to estimate a and b. A formula for the optimum h is given. Suppose f has its support on $[0, 1]$ and $f^{(m)} \in \mathscr{L}_p[0, 1]$. Then, for $m = 1, 2, 3$ and certain values of p, it is shown that

$$E(\hat{f}_n(x) - f(x))^2 = O(n^{-(2m-2/p)/(2m+1-2/p)})\,.$$

Bounds for the constant covered by the "O" are given. An extension to the $\mathscr{L}_p$ case of known convergence properties of the derivative of an interpolating spline is found, as part of the proofs.

1. Introduction. In this paper we study the convergence properties of a histospline density estimate of the type introduced by Boneva, Kendall and Stefanov (BKS) [3] and discussed by Schoenberg [16], [17]. Although BKS considered the estimation of densities supported on the entire real line as well as on a finite interval, we consider here only densities supported on a finite interval, say, [0, 1].

Let $W_p^{(m)}$ be the Sobolev space of functions

$$\{f: f^{(\nu)} \text{ abs. cont.}, \nu = 0, 1, \cdots, m - 1, f^{(m)} \in \mathscr{L}_p[0, 1]\}\,.$$

Let $h > 0$ satisfy $1/h = l + 1$, where l is a positive integer. Let h_j be the fraction of independent observations from some density f, falling between jh and $(j + 1)h$, $j = 0, 1, \cdots, l$. As a density estimate $\hat{f}$, BKS seek the unique function in the space $W_2^{(1)}$ which minimizes

$$\int_0^1 (g'(x))^2\, dx \tag{1.1}$$

subject only to

$$\int_{jh}^{(j+1)h} g(x)\, dx = h_j\,, \qquad j = 0, 1, \cdots, l\,. \tag{1.2}$$

Let $\hat{F}$ be the unique function in the space $W_2^{(2)}$ which minimizes

$$\int_0^1 (G''(x))^2\, dx \tag{1.3}$$

Received April 1973; revised June 1974.

[1] This report was supported by the Air Force Office of Scientific Research under Grant No. AFOSR 72-2363.

Key words and phrases. Spline, histospline, density estimate, optimal convergence rates.

subject to

$$G(0) = 0 \tag{1.4}$$
$$G(ih) = \textstyle\sum_{j=0}^{i-1} h_j \,, \qquad i = 1, 2, \cdots, l+1 \,.$$

Clearly, $\hat{F}$ satisfies $\hat{F}' = \hat{f}$, where $\hat{f}$ is the solution associated with the problem of (1.1) and (1.2). Therefore this BKS histospline density estimate is the derivative of the so-called natural cubic spline of interpolation $\hat{F}$ to the sample cdf F_n at the points ih, $i = 0, 1, \cdots, l+1$. See Schoenberg [15], Section 3 for a discussion of the natural cubic splines of interpolation. The spline $\hat{F}$ is a cubic polynomial in each interval $[jh, (j+1)h]$, $j = 0, 1, \cdots, l$, uniquely characterized on [0, 1] by (1.4) and the conditions $\hat{F}$, $\hat{F}'$, $\hat{F}''$ continuous and $F''(0) = F''(1) = 0$. The drawback to using the natural cubic spline is that maximum possible convergence rates in the cases $m = 2$ and $m = 3$ defined below will not obtain in a neighborhood of the boundaries unless F also satisfies $F''(0) = F''(1) = 0$. (See [9].) Another histospline was considered by BKS and subsequently Schoenberg [16], [17]. It is the derivative of the solution $\hat{F}$ to the problem: Find the unique function in the space $W_2^{(2)}$ which minimizes (1.3) subject to (1.4) and the additional conditions

$$G'(0) = 0 \tag{1.5}$$
$$G'(1) = 0 \,.$$

The histospline we study is a variation of this. We replace (1.5) by

$$G'(0) = \hat{a}_1 \tag{1.6}$$
$$G'(1) = \hat{b}_1$$

where $\hat{a}_1$ and $\hat{b}_1$ are estimates of $f(0)$ and $f(1)$ formed from the sample cdf in a manner to be described. Thus, if f has its support on [0, 1] we let $\hat{F}_n$ be the solution to the problem: Find the unique function in the space $W_2^{(2)}$ which minimizes (1.3) subject to

$$G(jh) = F_n(jh) \,, \qquad j = 0, 1, \cdots, l+1 \,,$$
$$G'(0) = \hat{a}_1$$
$$G'(1) = \hat{b}_1$$

and our density estimate $\hat{f}_n(x)$ is

$$\hat{f}_n(x) = \frac{d}{dx} \hat{F}_n(x) \,. \tag{1.7}$$

This estimate can be less smooth than the first (unconstrained) BKS estimate. Of course, we may replace $\hat{a}_1$ and $\hat{b}_1$ by 0 if it is known that $f(0) = 0$, $f(1) = 0$.

As do BKS, ([3], pages 34–35), we consider that h is a parameter to be chosen. Our criterion is minimum mean square error at a point. Our theorems provide results on the optimum choice of h.

Let $W_p^{(m)}(M)$ be given by

$$W_p^{(m)}(M) = \{f: f \in W_p^{(m)}, \|f^{(m)}\|_p \leqq M\}$$

where

$$\|f^{(m)}\|_p = [\int |f^{(m)}(\xi)|^p \, d\xi]^{1/p}, \qquad p \geqq 1$$
$$\|f^{(m)}\|_\infty = \sup_\xi |f^{(m)}(\xi)| .$$

It is shown in Wahba [21], based on a result of Farrell [5], that if $\hat{f}_n(x)$, $n = 1, 2, \cdots$ is any sequence of estimates of the true density f at the point x, and ε is any positive number, that if

$$\sup_{f \in W_p^{(m)}(M)} E(\hat{f}_n(x) - f(x))^2 = b_n n^{-(2m-2/(p+\varepsilon))/(2m+1-2/(p+\varepsilon))}$$

then there exists $D_0 > 0$ such that $b_n \geqq D_0$ for infinitely many n. Thus, the best possible mean square convergence rate uniform over $W_p^{(m)}(M)$ is not better than $n^{-(2m-2/(p+\varepsilon))/(2m+1-2/(p+\varepsilon))}$ for arbitrarily small ε. It is known for various density estimates that the rate $n^{-(2m-2/p)/(2m+1-2/p)}$ is achieved, that is,

$$(1.8) \qquad \sup_{f \in W_p^{(m)}(M)} E(\hat{f}_n(x) - f(x))^2 \leqq Dn^{-(2m-2/p)/(2m+1-2/p)} ,$$

where D depends on m, p, M, the method (and, possibly, bounds on f). See, for example [21], where the Parzen kernel type estimate ([12]), the Kronmal–Tarter orthogonal series estimate ([11]) and the polynomial algorithm for density estimation ([20]) are studied. The ordinary histogram with optimally chosen "bin" size also satisfies (1.8) with $m = 1$.

It is the purpose (and main theorem) of this note to prove that the histospline density estimate of (1.6) with optimally chosen h shares the rate of convergence property (1.8) of these other estimates. The result (1.8) is proved for $m = 1, 2$ and 3, and several sets of values for p. An upper bound for D is given. In particular, for $m = 1$, $p = 2$, $h_{\text{opt}} \sim cn^{-\frac{1}{4}}$. Recent work indicates that the $m = 1$, $p = 2$ result is true for the BKS histospline of (1.1) and (1.2). (R. Kuhn personal communication.)

In order to achieve the rate (1.8) for higher m, it is necessary (and doubtless, sufficient) to use higher degree splines. A proof of (1.8) for $m > 3$ is not forthcoming at this time, however, due to the complexity of the formula for higher degree splines.

2. Explicit expressions for the histospline estimate, and outline of proof of the main theorem. Our development of an explicit formula for an interpolating spline will be slightly unorthodox, for the purpose of easing the proofs of the main theorem. The reader may consult [2], [6], [7], [15] and the bibliography [14] for additional background on splines.

We endow $W_2^{(2)}$ with the inner product

$$(2.1) \qquad \langle F, G \rangle = F(0)G(0) + F'(0)G'(0) + \int_0^1 F''(x)G''(x)\,dx .$$

$W_2^{(2)}$ is then a reproducing kernel Hilbert space (RKHS) with the reproducing kernel

$$
\begin{aligned}
Q(s,t) &= 1 + st + \textstyle\int_0^{\min(s,t)} (s-u)(t-u)\,du \\
&= 1 + st + \left(\frac{ts^2}{2} - \frac{s^3}{6}\right), \qquad s < t \\
&= 1 + st + \left(\frac{st^2}{2} - \frac{t^3}{6}\right), \qquad s > t\,.
\end{aligned} \tag{2.2}
$$

Denote this Hilbert space $\mathscr{H}_Q$, with norm $\|\cdot\|_Q$. The true cdf F is always assumed to be in $\mathscr{H}_Q$, that is, $f = F' \in W_2^{(1)}$.

Let Q_t be the function defined on [0, 1] by

$$Q_t(s) = Q(s,t)\,, \qquad s, t \in [0,1]\,, \tag{2.3}$$

and let Q_t' be the function defined on [0, 1] by

$$
\begin{aligned}
Q_t'(s) &= \frac{d}{du} Q(s,u)\Big|_{u=t}, \qquad s, t \in [0,1] \\
&= s + \frac{s^2}{2}, \qquad s \leqq t \\
&= s + st - \frac{t^2}{2}, \qquad s \geqq t\,.
\end{aligned} \tag{2.4}
$$

By the properties of RKHS (see, e.g. [10]), $Q_t, Q_{t'} \in \mathscr{H}_Q$ for each t, and

$$
\begin{aligned}
\langle G, Q_t\rangle &= G(t)\,, \qquad G \in \mathscr{H}_Q,\ t \in [0,1] \\
\langle G, Q_t'\rangle &= G'(t)\,.
\end{aligned} \tag{2.5}
$$

Consider the solution to the problem:
Find $G \in \mathscr{H}_Q$ to min $\|G\|_Q$ subject to

$$
\begin{aligned}
G'(0) &= \langle G, Q_0'\rangle = a_1 \\
G(0) &= \langle G, Q_0\rangle = a_0 \\
G(s_i) &= \langle G, Q_{s_i}\rangle = y_i\,, \qquad i = 1, 2, \cdots, l \\
G(1) &= \langle G, Q_1\rangle = b_0 \\
G'(1) &= \langle G, Q_1'\rangle = b_1
\end{aligned} \tag{2.6}
$$

where $0 < s_1 < \cdots < s_l < 1$.
Denoting $\bar{s} = (s_1, s_2, \cdots, s_l)$, $\bar{y} = (y_1, y_2, \cdots, y_l)$, $\bar{a} = (a_1, a_0)$, $\bar{b} = (b_0, b_1)$, let $S(x) = S(x; \bar{s}; \bar{a}, \bar{y}, \bar{b})$ be the solution to this problem. Then, by observing that $S \in \mathscr{S}_l(\bar{s})$ defined by

$$\mathscr{S}_l(\bar{s}) = \text{span}\,\{Q_0', Q_{s_i}, i = 0, 1, \cdots, l+1, Q_1'\} \tag{2.7}$$

where $s_0 = 0$, $s_{l+1} = 1$, it may be established that

$$
\begin{aligned}
&S(x; \bar{s}; \bar{a}, \bar{y}, \bar{b}) \\
&\quad = (Q_0'(x), Q_0(x), Q_{s_1}(x), \cdots, Q_{s_l}(x), Q_1(x), Q_1'(x))Q_{l+4}^{-1}(\bar{a}; \bar{y}; \bar{b})'
\end{aligned} \tag{2.8}
$$

where Q_{l+4} is the $(l+4) \times (l+4)$ Grammian matrix of the basis for $\mathscr{S}_l(\bar{s})$.

Q_{l+4} is of full rank (see for example [19]) and the entries may be found from (2.5). By observing the nature of the inner product in $\mathscr{H}_Q$, it is easily seen that $S(x; \bar{s}; \bar{a}, \bar{y}, \bar{b})$ is also the solution to: Find $G \in \mathscr{H}_Q$ to

$$\min \int_0^1 (G''(x))^2 \, dx$$

subject to (2.6). The solution to this problem is well known ([15]) to be the unique cubic spline satisfying (2.6). It may easily be checked from (2.3), (2.4) and (2.8) that S has the characteristic properties of a cubic spline, viz. S is a polynomial of degree less than or equal to three in each interval $[s_i, s_{i+1}]$, $i = 0, 1, \cdots, l$, and S, S' and S'' are continuous.

The density estimate $\hat{f}_n(x)$ that we study is thus given by

$$\hat{f}_n(x) = \frac{d}{dx} \hat{F}_n(x) \,, \qquad (2.9)$$

$$\hat{F}_n(x) = S(x; \bar{s}_h; \hat{a}, \bar{F}_n, \hat{b})$$

with

$$\bar{s}_h = (h, 2h, \cdots, lh) \,, \qquad (l+1)h = 1$$

$$\hat{a} = (\hat{a}_1, 0)$$

$$\bar{F}_n = (F_n(h), F_n(2h), \cdots, F_n(lh))$$

$$\hat{b} = (1, \hat{b}_1) \,.$$

Equation (2.8) is not the computationally best method for computing $\hat{F}_n$, because Q_{l+4} is ill-conditioned for large l; however, computing routines for $S(x)$ and $S'(x)$ are commonly available. See, for example, [1]. The estimates $\hat{a}_1$ and $\hat{b}_1$ depend on m, $(= 1, 2,$ or $3)$ and are defined as follows: Let $l_{0,\nu}(x)$ be the polynomial of degree m satisfying

$$(2.10) \qquad \begin{aligned} l_{0,\nu}(x) &= 1 \,, & x &= \nu h \,, \\ &= 0 \,, & x &= jh \,, \quad j \neq \nu \,, \quad j = 0, 1, \cdots, m \,. \end{aligned}$$

and let $l_{1,\nu}(x)$ be the polynomial of degree m satisfying

$$(2.11) \qquad \begin{aligned} l_{1,\nu}(x) &= 1 \,, & x &= 1 - \nu h \\ &= 0 \,, & x &= 1 - jh \,, \quad j \neq \nu \,, \quad j = 0, 1, \cdots, m \,. \end{aligned}$$

Let

$$(2.12) \qquad \hat{a}_1 = \hat{F}_n'(0) = \hat{f}_n(0) = \frac{d}{dx} \sum_{\nu=0}^m l_{0,\nu}(x) \Big|_{x=0} F_n(\nu h)$$

$$(2.13) \qquad \hat{b}_1 = \hat{F}_n'(1) = \hat{f}_n(1) = \frac{d}{dx} \sum_{\nu=0}^m l_{1,\nu}(x) \Big|_{x=1} F_n(1 - \nu h) \,.$$

$\hat{a}_1$ is the derivative at 0 of the mth degree polynomial interpolating the sample cdf at $0, h, \cdots, mh$, and similarly for $\hat{b}_1$. It follows from (2.8) that $S(x; \bar{s}; \bar{a}, \bar{y}, \bar{b})$ is linear in the entries of $\bar{a}$, $\bar{y}$, and $\bar{b}$, that is

$$S(x; \bar{s}; \bar{a} + \bar{\varepsilon}_a, \bar{y} + \bar{\varepsilon}, \bar{b} + \bar{\varepsilon}_b) = S(x; \bar{s}; \bar{a}, \bar{y}, \bar{b}) + S(x; \bar{s}; \bar{\varepsilon}_a, \bar{\varepsilon}, \bar{\varepsilon}_b)$$

where $\bar{\varepsilon}_a$, $\bar{\varepsilon}$ and $\bar{\varepsilon}_b$ are 2-, l- and 2-vectors, respectively. $(d/dx)S(x; \bar{s}, \bar{a}, \bar{y}, \bar{b})$ also has this linearity property.

Let F be the true cdf, and let $\tilde{F}$ be the cubic spline of interpolation to F, with knots jh, $j = 1, 2, \cdots, l$, and matching F and F' at the boundaries, that is

$$\tilde{F}(x) = S(x; \bar{s}_h; \bar{F}_a, \bar{F}_h, \bar{F}_b)$$

where

$$\bar{F}_a = (F'(0), 0)$$
$$\bar{F}_h = (F(h), F(2h), \cdots, F(lh))$$
$$\bar{F}_b = (1, F'(1)) .$$

(Note, by the nature of the minimization problem (2.6), that $\tilde{F}$ is the projection of F onto $\mathscr{S}_l(\bar{s})$.)

We may write

$$f(x) - \hat{f}_n(x) = \frac{d}{dx}(F(x) - \hat{F}_n(x))$$

(2.14) $$= \frac{d}{dx}(F(x) - \tilde{F}(x)) + \frac{d}{dx}(\tilde{F}(x) - \hat{F}_n(x))$$

(2.15) $$= \frac{d}{dx}(F(x) - \tilde{F}(x)) + \frac{d}{dx}H_n(x) ,$$

where

(2.16) $$H_n(x) = S(x; \bar{s}_h; \bar{\varepsilon}_a, \bar{\varepsilon}, \bar{\varepsilon}_b)$$

and

(2.17) $$\begin{aligned} \bar{\varepsilon}_a &= (\varepsilon_0', 0) , & \varepsilon_0' &= F'(0) - \hat{a}_1 , & & \\ \bar{\varepsilon} &= (\varepsilon_1, \varepsilon_2, \cdots, \varepsilon_l) , & \varepsilon_j &= F(jh) - F_n(jh) , & j &= 1, 2, \cdots, l , \\ \bar{\varepsilon}_b &= (0, \varepsilon_{l+1}') , & \varepsilon_{l+1}' &= F'(1) - \hat{b}_1 . & & \end{aligned}$$

The first term on the right of (2.15), which we shall call the bias term, is non-random and depends only on how well F can be approximated by an interpolating cubic spline. The second, or variance term is a (linear) function of the random variables ε_0', ε_{l+1}' and ε_i, $i = 1, 2, \cdots, l$.

Then, as usual,

(2.18) $$E(f(x) - \hat{f}_n(x))^2 \leqq 2\left(\frac{d}{dx}(F(x) - \tilde{F}(x))\right)^2 + 2E\left(\frac{d}{dx}H_n(x)\right)^2 .$$

Bounds on the absolute bias, $|d/dx(F(x) - \tilde{F}(x))|$ appear in the approximation theory literature in various forms, for equally spaced, as well as arbitrarily spaced knots. If $F^{(m+1)} \in \mathscr{L}_2[0, 1]$, then it is known ([18], Theorems 5.1, 5.2 and 5.3) that, for $m = 1, 2, 3$,

(2.19) $$\sup_x \left|\frac{d}{dx}(F(x) - \tilde{F}(x))\right| \leqq K_2(m)\|F^{(m+1)}\|_2 h^{m-\frac{1}{2}}$$

where $\|\cdot\|_p$ is the $\mathscr{L}_p$ norm, and $K_2(m)$ is a constant depending on m. Generalizations of (2.19) to arbitrary m are given when $\tilde{F}$ is replaced by an interpolating spline of higher degree. For $F^{(m+1)} \in \mathscr{L}_\infty[0, 1]$, $m = 3$, ([9]) gives

$$\sup_x \left|\frac{d}{dx}(F(x) - \tilde{F}(x))\right| \leqq K_\infty(m)\|F^{(m+1)}\|_\infty h^m , \tag{2.20}$$

and [9] is easily extendable to $m = 1, 2$. (For earlier results, see [13].) Some information about generalizations of (2.20) up to, but not beyond $m = 5$ are known ([4]). We would like to have the result

$$F^{(m+1)} \in \mathscr{L}_p \Rightarrow \sup_x \left|\frac{d}{dx}(F(x) - \tilde{F}(x))\right| \leqq K_p(m)\|F^{(m+1)}\|_p h^{m-1/p} , \quad p \geqq 1 , \tag{2.21}$$

or, equivalently, $f \in W_p^{(m)} \Rightarrow$

$$\sup_x \left(\frac{d}{dx}(F(x) - \tilde{F}(x))\right)^2 \leqq \frac{A'}{2}\|f^{(m)}\|_p^2 h^{2m-2/p} , \qquad p \geqq 1 \tag{2.22}$$

where $A' = A'(m, p)$. We are not aware of such results for $p \neq 2$ or ∞. We provide a proof of (2.22) good for $m = 2, 3$, $1 \leqq p \leqq 2$. In the proof, the dependency on the knots $\{s_i\}_{i=1}^l$ is retained so that the results may be used in a sequel paper where the knots are determined by the order statistics. Combining these results will give us a bound on the bias for

$$\begin{aligned} m &= 1 , & p &= 2 ,\ \infty \\ m &= 2 , & 1 \leqq p &\leqq 2 ,\ \infty \\ m &= 3 , & 1 \leqq p &\leqq 2 ,\ \infty . \end{aligned}$$

The establishment of bounds on the variance term is tedious for cubic splines, and we are unable to do it for higher degree splines. It will be shown that

$$E\left(\frac{d}{dx}H_n(x)\right)^2 \leqq \frac{B}{2}\frac{1}{nh} + \frac{A''}{2}\|f^{(m)}\|_p^2 h^{2m-2/p} \tag{2.23}$$

where A'', B are constants to be given. Then we will have

$$\sup_{f \in W_p^{(m)}(M)} E(f(x) - \hat{f}_n(x))^2 \leqq AM^2h^{2m-2/p} + B\frac{1}{nh} \tag{2.24}$$

where $A = A' + A''$.
The right-hand side of (2.24) is minimized by taking $h = k_n/n$ with

$$k_n = \left[\frac{1}{(2m - 2/p)}\frac{B}{M^2A}\right]^{1/(2m+1-2/p)} \cdot n^{(2m-2/p)/(2m+1-2/p)} . \tag{2.25}$$

Then, we will have the main result, which is:

$$\sup_{f \in W_p^{(m)}(M)} E[f(x) - \hat{f}_n(x)]^2 \leqq Dn^{-(2m-2/p)/(2m+1-2/p)} \tag{2.26}$$

where

$$D = \frac{(2m + 1 - 2/p)}{(2m - 2/p)^{(2m-2/p)}}(M^2AB^{2m-2/p})^{1/(2m+1-2/p)} . \tag{2.27}$$

Details of these assertions are in the next section.

3. Proof of the main theorem.

3.1. *Bounds on the bias term.* The case $m = 3$, $p = \infty$ is covered by

PROPOSITION 1. *Let* $F^{(iv)} \in \mathscr{L}_\infty[0, 1]$. *Then*

$$\left(\frac{d}{dx}(F(x) - \tilde{F}(x))\right)^2 \leqq (\tfrac{3}{16})^2 \|F^{(iv)}\|_\infty^2 h^6 . \tag{3.1}$$

PROOF. This is Theorem 2 of [9]; the proof there may be extended from $F^{(iv)}$ continuous to $F^{(iv)} \in \mathscr{L}_\infty$.

The case $m = 1$ or 2 and $p = \infty$ is covered by

PROPOSITION 2. *Let* $F^{(iii)} \in \mathscr{L}_\infty[0, 1]$. *Then*

$$\left(\frac{d}{dx}(F(x) - \tilde{F}(x))\right)^2 \leqq (\tfrac{9}{4})^2 \|F^{(iii)}\|_\infty^2 h^4 . \tag{3.2}$$

Suppose only that $F^{(ii)} \in \mathscr{L}_\infty[0, 1]$. Then

$$\left(\frac{d}{dx}(F(x) - \tilde{F}(x))\right)^2 \leqq (\tfrac{9}{2})^2 \|F^{(ii)}\|_\infty^2 h^2 . \tag{3.3}$$

PROOF. This may be proved from the argument in [9] by following the proof of Theorem 2 in [9], and noting that, if $F^{(iii)} \in \mathscr{L}_\infty$, then r_i of [9], equation (8) is bounded by $3h|\sup_\xi F^{(iii)}(\xi)|$, if only $F^{(ii)} \in \mathscr{L}_\infty$, then r_i of [9], equation (8) is bounded by $6|\sup_\xi F^{(ii)}(\xi)|$.

The next series of lemmas result in a theorem which provides bounds on the bias for $m = 1$, $p = 2$, and $m = 2, 3$, $1 \leqq p \leqq 2$.

LEMMA 1.

$$\left(\frac{d}{dx}(F(x) - \tilde{F}(x))\right)^2 \leqq \|Q_x' - \tilde{Q}_x'\|_Q^2 \|F - \tilde{F}\|_Q^2 \tag{3.4}$$

where $\tilde{Q}_x'$ *is the projection of* Q_x' *in* $\mathscr{H}_Q$ *onto* $\mathscr{S}_l(\bar{s})$.

PROOF. Since $\tilde{F}$ is the projection of F onto $\mathscr{S}_l(\bar{s})$,

$$\left|\frac{d}{dx}(F(x) - \tilde{F}(x))\right| = |\langle Q_x', F - \tilde{F}\rangle| = |\langle Q_x' - \tilde{Q}_x', F - \tilde{F}\rangle| . \tag{3.5}$$

LEMMA 2.

$$\|Q_x' - \tilde{Q}_x'\|_Q^2 \leqq \tfrac{1}{3}h . \tag{3.6}$$

PROOF. See Appendix.

LEMMA 3. *Let* $F^{(iv)} \in \mathscr{L}_p[0, 1]$, $1 \leqq p \leqq 2$. *Then*

$$\|F - \tilde{F}\|_Q^2 \leqq \tfrac{1}{48} \|F^{(iv)}\|_p^2 h^{5-2/p} . \tag{3.7}$$

PROOF. See Appendix.

LEMMA 4. *Let* $F^{(iii)} \in \mathscr{L}_p[0, 1]$, $1 \leqq p \leqq 2$. *Then*

$$\|F - \tilde{F}\|_Q^2 \leqq \tfrac{1}{3} \|F^{(iii)}\|_p^2 h^{3-2/p} . \tag{3.8}$$

PROOF. See Appendix.

THEOREM 1. *Let $f^{(m)} \in \mathscr{L}_p$ for $m = 1$, $p = 2$, or $m = 2, 3$, $1 \leqq p \leqq 2$. Then*

$$\left(\frac{d}{dx}(F(x) - \tilde{F}(x))\right)^2 \leqq A\|f^{(m)}\|_p^2 h^{2m-2/p} \tag{3.9}$$

where

$$\begin{aligned} A &= \tfrac{1}{3}\,, & m &= 1\,, & p &= 2 \\ A &= \tfrac{1}{9}\,, & m &= 2\,, & 1 &\leqq p \leqq 2 \\ A &= \tfrac{1}{144}\,, & m &= 3\,, & 1 &\leqq p \leqq 2\,. \end{aligned}$$

PROOF. The result follows upon combining Lemmas 1, 2, 3 and 4, and noting, in the case $m = 1$, $p = 2$, that $\|F - \tilde{F}\|_Q^2 \leqq \|f^{(1)}\|_2^2$.

3.2. *Bounds on the variance term.* We seek a bound on

$$E\left[\frac{d}{dx}H_n(x)\right]^2$$

where

$$H_n(x) = S(x; \bar{s}_h, \bar{\varepsilon}_a, \bar{\varepsilon}, \bar{\varepsilon}_b) \tag{3.10}$$

and

$$\begin{aligned} \bar{s}_h &= (h, 2h, \cdots, lh)\,, & & \\ \bar{\varepsilon}_a &= (\varepsilon_0', \varepsilon_0)\,, & \varepsilon_0 &= 0\,, \quad \varepsilon_0' = F'(0) - \hat{a}_1\,, \\ \bar{\varepsilon} &= (\varepsilon_1, \varepsilon_2, \cdots, \varepsilon_l)\,, & \varepsilon_j &= \tilde{F}(jh) - \hat{F}_n(jh) = F(jh) - \hat{F}_n(jh)\,, \quad j = 1, 2, \cdots, \\ \bar{\varepsilon}_b &= (\varepsilon_{l+1}, \varepsilon'_{l+1})\,, & \varepsilon_{l+1} &= 0\,, \quad \varepsilon'_{l+1} = F'(1) - \hat{b}_1\,. \end{aligned}$$

Lemma 5 bounds the derivative of a cubic spline in terms of h and the data $\bar{\varepsilon}_a$, $\bar{\varepsilon}$, $\bar{\varepsilon}_b$.

LEMMA 5. *For $jh \leqq x < (j+1)h$, $j = 0, 1, \cdots, l$,*

$$\left|\frac{d}{dx}H_n(x)\right| \leqq 8\left\{\sum_{i=0}^{l} c_i \frac{|\psi_i|}{h} + \frac{1}{2^{j+1}}|\varepsilon_0'| + \frac{1}{2^{l+2-j}}|\varepsilon'_{l+1}|\right\} \tag{3.11}$$

where

$$\begin{aligned} \psi_i &= \varepsilon_{i+1} - \varepsilon_i = [F((j+1)h) - F(jh)] - [F_n((j+1)h) - F_n(jh)] \\ c_i &= \frac{1}{2^{|i-j|+1}} + \frac{1}{2^{|i+1-j|+1}}\,, \qquad i = 0, 1, \cdots, l,\ i \neq j \\ c_j &= \frac{1}{2} + \frac{1}{2^2} + \frac{1}{8}\,. \end{aligned}$$

PROOF. See Appendix.

Bounds on $E[d/dx\, H_n(x)]^2$ may now be found by bounding the random variables on the right of (3.11).

Now

$$\psi_j = \varepsilon_{j+1} - \varepsilon_j = [F((j+1)h) - F(jh)] - [F_n((j+1)h) - F_n(jh)]$$
$$= p_j - \frac{\#\text{ of observations between } jh \text{ and } (j+1)h}{n}$$

where

$$p_j = \textstyle\int_{jh}^{(j+1)h} f(\xi)\, d\xi < h\Lambda$$

and

$$\Lambda = \sup_\xi f(\xi)\,.$$

Thus $n(p_j - \psi_j)$ is binomial $B(n, p_j)$ and so

$$E(\psi_j)^2 = \frac{1}{n} p_j(1 - p_j) \leqq \frac{\Lambda h}{n}\,. \tag{3.12}$$

To complete the bound on the variance term, we need to know $E(\varepsilon_0')^2 \equiv E(F'(0) - \hat{a}_1)^2$ and $E(\varepsilon_{l+1}')^2 \equiv E(F'(1) - \hat{b}_1)^2$. The answer is given by

LEMMA 6. *Let $\hat{a}_1$ and $\hat{b}_1$ be given by* (2.12) *and* (2.13) *for $m = 1, 2, 3$. Then*

$$\left.\begin{matrix} E(F'(0) - \hat{a}_1)^2 \\ E(F'(1) - \hat{b}_1)^2 \end{matrix}\right\} \leqq \alpha \|f^{(m)}\|_p^2 h^{2m-2/p} + \beta \frac{\Lambda}{nh} \tag{3.13}$$

where

$$\alpha = 8m^{2m-2/p}/[(m-1)!]^2 \tag{3.14}$$
$$\beta = 2m^3(m+1)^2\Lambda\,.$$

PROOF. See Appendix.

Note that, if $x = 0$, or $x = 1$,

$$E(f(0) - \hat{f}_n(0))^2 = E(F'(0) - \hat{a}_1)^2$$
$$E(f(1) - \hat{f}_n(1))^2 = E(F'(1) - \hat{b}_1)^2$$

and the mean square error is given by the right-hand side of (3.13). For $x \neq 0, 1$, we combine Lemmas 5, 6 and (3.12) to obtain, for $jh \leqq x < (j+1)h$,

$$\begin{aligned} E\left(\frac{d}{dx} H_n(x)\right)^2 &\leqq 8^2 \left\{2 \textstyle\sum_{r,s=0}^{l} c_r c_s E \frac{|\psi_r \psi_s|}{h^2} + 4\left(\frac{1}{2^{2j+2}} + \frac{1}{2^{2l+4-2j}}\right)\right. \\ &\qquad \left.\times \left(\alpha\|f^{(m)}\|_p^2 h^{2m-2/p} + \beta\frac{\Lambda}{nh}\right)\right\} \\ &\leqq 8^2 \left\{2(3\tfrac{1}{8})^2 \frac{\Lambda}{nh} + 4\left(\frac{1}{2^{2(x/h)}} + \frac{1}{2^{2(1-x)/h}}\right)\right. \\ &\qquad \left.\times \left(\alpha\|f^{(m)}\|_p^2 h^{2m-2/p} + \beta\frac{\Lambda}{nh}\right)\right\}. \end{aligned} \tag{3.15}$$

Note that if x is bounded away from 0 and 1, then $[1/2^{2(x/h)} + 1/2^{2(1-x)/h}] \to 0$ rapidly as $h \to 0$.

3.3. *Final result.* Summarizing the results from (3.1), (3.2), (3.3), (3.9) and

(3.15) gives

$$E(f(x) - \hat{f}_n(x))^2 \leqq A\|f^{(m)}\|_p^2 h^{2m-2/p} + B\frac{1}{nh}$$

where

(3.16)
$$A = 2\left[A' + 64\alpha\left(\frac{4}{2^{2x/h}} + \frac{4}{2^{2(1-x)/h}}\right)\right]$$

(3.17)
$$B = 2\left[B' + 64\beta\left(\frac{4\Lambda}{2^{2x/h}} + \frac{4\Lambda}{2^{2(1-x)/h}}\right)\right]$$

and

$$\begin{aligned} A' &= (\tfrac{9}{2})^2 && m = 1, \quad p = \infty \\ &(\tfrac{9}{4})^2 && m = 2, \quad p = \infty \\ &(\tfrac{3}{16})^2 && m = 3, \quad p = \infty \\ &\tfrac{1}{3} && m = 1, \quad p = 2 \\ &\tfrac{1}{9} && m = 2, \quad 1 \leqq p \leqq 2 \\ &\tfrac{1}{144} && m = 3, \quad 1 \leqq p \leqq 2 \end{aligned}$$

$$B' = 2 \cdot (8 \cdot 3\tfrac{1}{8})^2$$

and α and β are given by (3.14). It can be shown easily that there exists $\Lambda = \Lambda(m, p, M) < \infty$ such that $\sup_{f \text{ a density} \in W_p^{(m)}(M)} \sup_\xi f(\xi) \leqq \Lambda$, so that our results are uniform over $W_p^{(m)}(M)$. This demonstration is omitted. We have proved

THEOREM 2. *Suppose f has its support on $[0, 1]$ and $f \in W_p^{(m)}(M)$ for one of the following cases*:

$$\begin{aligned} m &= 1, && p = 2, && p = \infty \\ m &= 2, && 1 \leqq p \leqq 2, && p = \infty \\ m &= 3, && 1 \leqq p \leqq 2, && p = \infty. \end{aligned}$$

Let F_n be the sample cdf *based on n independent observations from F, and let $\hat{F}_n(x)$ be the cubic spline of interpolation to F_n at the points jh, $j = 0, 1, \cdots, l + 1$; $(l + 1)h = 1$, which satisfies the boundary conditions $\hat{F}_n'(0) = \hat{a}_1$, $\hat{F}_n'(1) = \hat{b}_1$, where $\hat{a}_1$ and $\hat{b}_1$ are given by* (2.12) *and* (2.13). *Let $\hat{f}_n(x) = d/dx\, \hat{F}_n(x)$, and suppose h is chosen as $h = k_n/n$,*

$$k_n = \left[\frac{1}{(2m - 2/p)}\frac{B}{M^2 A}\right]^{1/(2m+1-2/p)} n^{(2m-2/p)/(2m+1-2/p)}$$

where A and B are given by (3.16) *and* (3.17). *Then*

(3.18)
$$\sup_{f \in W_p^{(m)}(M)} E[f(x) - \hat{f}_n(x)]^2 \leqq Dn^{-(2m-2/p)/(2m+1-2/p)}$$

with

(3.19)
$$D = \frac{(2m + 1 - 2/p)}{(2m - 2/p)^{(2m-2/p)}}(M^2 A B^{2m-2/p})^{1/(2m+1-2/p)}.$$

APPENDIX

This appendix contains the proofs of Lemmas 2–6. The proofs are carried out where the knots $\{s_i\}_{i=1}^l$ do not necessarily satisfy $s_{i+1} - s_i = h$, but only $0 = s_0 < s_1 < \cdots < s_l < s_{l+1} = 1$. The purpose of this generality is to allow the lemmas to be referenced for a later report which deals with the situation where the knots are determined by the order statistics. Let I_j be the interval $[s_j, s_{j+1}]$, for $j = 0, 1, \cdots, l$.

LEMMA 2. *Let $\tilde{Q}_x'$ be the projection of Q_x' onto $\mathscr{S}_l(\bar{s})$, and let $x \in I_j$. Then*

$$\|Q_x' - \tilde{Q}_x'\|_Q^2 \leqq \tfrac{1}{3}(s_{j+1} - s_j)\,, \qquad j = 0, 1, \cdots, l\,.$$

PROOF. For $x \in I_j$, define R_x' in $\mathscr{H}_Q$ by

$$R_x' = \frac{1}{(s_{j+1} - s_j)}(Q_{s_{j+1}} - Q_{s_j})\,.$$

Since $R_x' \in \mathscr{S}_l(\bar{s})$ and $\tilde{Q}_x'$ is the projection of Q_x' onto $\mathscr{S}_l(\bar{s})$

$$\|Q_x' - \tilde{Q}_x'\|_Q \leqq \|Q_x' - R_x'\|_Q\,. \tag{A1.1}$$

To compute the square of the right side of (A1.1), note from (2.4) that

$$Q_x'(0) = 0 \tag{A1.2}$$

$$\frac{d}{ds} Q_x'(s)\bigg|_{s=0} = 1 \tag{A1.3}$$

$$\begin{aligned} \frac{d^2}{ds^2} Q_x'(s) &= 1 & s < x \\ &= 0 & s > x\,. \end{aligned} \tag{A1.4}$$

After some calculations,

$$R_x'(0) = 0 \tag{A1.5}$$

$$\frac{d}{ds} R_x'(s)\bigg|_{s=0} = 1 \tag{A1.6}$$

$$\begin{aligned} \frac{d^2}{ds^2} R_x'(s) &= 1\,, & 0 \leqq s \leqq s_j \\ &= \frac{(s_{j+1} - s)}{(s_{j+1} - s_j)}\,, & s_j \leqq s \leqq s_{j+1} \\ &= 0\,, & s_{j+1} \leqq s \leqq 1\,. \end{aligned} \tag{A1.7}$$

$$\frac{d^2}{ds^2} Q_x'(s) - \frac{d^2}{ds^2} R_x'(s) = 0\,, \qquad \text{for} \quad s \notin I_j \tag{A1.8}$$

and

$$\begin{aligned} \|Q_x' - R_x'\|_Q^2 &= \frac{1}{(s_{j+1} - s_j)^2}\left[\textstyle\int_{s_j}^x (u - s_j)^2\, du + \int_x^{s_{j+1}} (s_{j+1} - u)^2\, du\right] \\ &\leqq \tfrac{1}{3}(s_{j+1} - s_j)\,. \end{aligned}$$

LEMMA 3. *Let $F \in \mathscr{H}_Q$ satisfy $F^{(\mathrm{iv})} = \rho \in \mathscr{L}_p[0, 1]$. Let $\tilde{F}$ be the projection of F onto $\mathscr{S}_l(\bar{s})$. Then*

$$\|F - \tilde{F}\|_Q^2 \leqq \tfrac{1}{48} \textstyle\sum_{j=0}^l (s_{j+1} - s_j)^{5-2/p}\left[\int_{s_j}^{s_{j+1}} |\rho(\xi)|^p\, d\xi\right]^{2/p}\,. \tag{A2.1}$$

If $s_{j+1} - s_j = h$, $j = 0, 1, \cdots, l$, *and* $1 \leqq p \leqq 2$,

$$\|F - \tilde{F}\|_Q^2 \leqq \tfrac{1}{48} h^{5-2p} \|F^{(\mathrm{iv})}\|_p^2 .$$

PROOF. First we show that $F^{(\mathrm{iv})} = \rho$ implies that

$$F(t) = \textstyle\int_0^1 Q(t, s)\rho(s)\, ds + c_1 Q_0(t) + c_2 Q_1(t) + c_3 Q_0'(t) + c_4 Q_1'(t) \tag{A2.2}$$

for some $\{c_i\}$. But $Q_0(t, s) = \int_0^1 (s - u)_+(t - u)_+\, du$ is the Green's function for the operator D^4, with boundary conditions

$$\begin{aligned} G^{(\nu)}(0) &= 0 , & \nu &= 0, 1 \\ G^{(\nu)}(1) &= 0 , & \nu &= 2, 3 . \end{aligned}$$

Thus, F always has a representation

$$F(t) = \textstyle\int_0^1 Q_0(t, s)\rho(s)\, ds + \sum_{i=0}^{3} d_i t^i . \tag{A2.3}$$

But

$$\textstyle\int_0^1 Q_0(t, s)\rho(s)\, ds = \int_0^1 Q(t, s)\rho(s)\, ds - \int_0^1 (1 + st)\rho(s)\, ds .$$

Since $Q_0(t)$, $Q_0'(t)$, $Q_1(t)$ and $Q_1'(t)$ span the same space as $\{1, t, t^2, t^3\}$, $\{c_i\}$ can always be found so that (A2.2) equals (A2.3).

Next, if v is any element in $\mathscr{H}_Q$ of the form

$$v = \textstyle\sum_{i=0}^{l+1} c_i Q_{s_i} + aQ_0' + bQ_1' .$$

Then, since $v \in \mathscr{S}_l(\bar{s})$,

$$\|F - \tilde{F}\|_Q \leqq \|F - v\|_Q . \tag{A2.4}$$

The proof now proceeds by finding an element $v \in \mathscr{S}_l(\bar{s})$ so that the right-hand side of (A2.4) is bounded by the right-hand side of (A2.1). For $x \in I_j$, define $R_x \in \mathscr{S}_l(\bar{s})$ by

$$R_x = \frac{(s_{j+1} - x)}{(s_{j+1} - s_j)} Q_{s_j} + \frac{(x - s_j)}{(s_{j+1} - s_j)} Q_{s_{j+1}} , \qquad j = 0, 1, \cdots, l .$$

Define $v \in \mathscr{S}_l(\bar{s})$ by

$$\begin{aligned} v &= \textstyle\int_0^1 R_x \rho(x)\, dx + c_1 Q_0 + c_2 Q_0' + c_3 Q_1 + c_4 Q_1' \\ &\equiv \textstyle\sum_{j=0}^{l} \left\{ Q_{s_j} \int_{s_j}^{s_{j+1}} \frac{(s_{j+1} - x)}{(s_{j+1} - s_j)} \rho(x)\, dx + Q_{s_{j+1}} \int_{s_j}^{s_{j+1}} \frac{(x - s_j)}{(s_{j+1} - s_j)} \rho(x)\, dx \right\} \\ &\quad + c_1 Q_0 + c_2 Q_0' + c_3 Q_1 + c_4 Q_1' . \end{aligned}$$

Now

$$F - v = \textstyle\int_0^1 (Q_x - R_x)\rho(x)\, dx$$

and, by the properties of the reproducing kernel, it can be shown that

$$\|F - v\|_Q^2 = \textstyle\int_0^1 \int_0^1 \rho(x)\rho(x')\langle Q_x - R_x, Q_{x'} - R_{x'}\rangle\, dx\, dx' .$$

Since

$$Q_x(0) - R_x(0) = 0$$

$$\frac{d}{ds}(Q_x(s) - R_x(s))\bigg|_{s=0} = 0$$

$$\frac{d^2}{ds^2}(Q_x(s) - R_x(s)) = (x-s)_+ - \frac{(x-s_j)(s_{j+1}-s)}{(s_{j+1}-s_j)}, \qquad x \in I_j;\ s \in I_j$$

$$= 0, \qquad x \in I_j,\ s \in I_k,\ k \neq j,$$

it follows that

$$\langle Q_x - R_x, Q_{x'} - R_{x'} \rangle = 0$$

if $x \in I_j$, $x' \in I_k$ with $j \neq k$. Thus

$$\|F - v\|_Q^2 \leqq \textstyle\sum_{j=0}^{l} \{\int_{s_j}^{s_{j+1}} |\rho(x)| \|Q_x - R_x\|_Q \, dx\}^2 .$$

Furthermore

$$\|Q_x - R_x\|_Q^2 = \int_{s_j}^{s_{j+1}} \left[(x-s)_+ - \frac{(x-s_j)_+}{(s_{j+1}-s_j)}(s_{j+1}-s)_+ \right]^2 ds$$

$$\leqq \tfrac{1}{48}(s_{j+1}-s_j)^3, \qquad x \in I_j,$$

so that

$$\|F - v\|_Q^2 \leqq \tfrac{1}{48} \textstyle\sum_{j=0}^{l} (s_{j+1}-s_j)^3 [\int_{s_j}^{s_{j+1}} |\rho(t)| \, dt]^2 .$$

For $1/p + 1/p' = 1$, a Hölder inequality gives

$$\int_{I_j} |\rho(t)| \, dt \leqq [\int_{I_j} dt]^{1/p'} [\int_{I_j} |\rho(t)|^p \, dt]^{1/p}$$

$$= (s_{j+1}-s_j)^{1-1/p} [\int_{I_j} |\rho(t)|^p \, dt]^{1/p} .$$

Thus,

$$\|F - v\|_Q^2 \leqq \tfrac{1}{48} \textstyle\sum_{j=0}^{l} (s_{j+1}-s_j)^{5-2/p} [\int_{I_j} |\rho(t)|^p \, dt]^{2/p} .$$

If $(s_{j+1} - s_j) = h$ and $1 \leqq p \leqq 2$, then

$$\|F - v\|_Q^2 \leqq \tfrac{1}{48} h^{5-2/p} [\int_0^1 |\rho(t)|^p \, dt]^{2/p} .$$

LEMMA 4. *Let $F \in \mathscr{H}_Q$ satisfy $F^{(\text{iii})} = \eta \in \mathscr{L}_p[0, 1]$. Let $\tilde{F}$ be the projection of F onto $\mathscr{S}_l(\bar{s})$. Then*

$$\|F - \tilde{F}\|_Q^2 \leqq \tfrac{1}{3} \textstyle\sum_{j=0}^{l} (s_{j+1}-s_j)^{3-2/p} [\int_{s_j}^{s_{j+1}} |\eta(\xi)|^p \, d\xi]^{2/p} .$$

If $(s_{j+1} - s_j) = h$, and $1 \leqq p \leqq 2$, then

$$\|F - \tilde{F}\|_Q^2 \leqq \tfrac{1}{3} h^{3-2/p} \|F^{(\text{iii})}\|_p^2 .$$

PROOF. As in the proof of Lemma 3, by the Green's function properties of $Q_x'(s)$ there exist c_1, c_2, c_3, c_4 such that

$$F(t) = \int_0^1 Q_x'(t)\eta(x) \, ds + c_1 Q_0(t) + c_2 Q_0'(t) + c_3 Q_1(t) + c_4 Q_1'(t) .$$

Let

$$v = \int_0^1 R_x'\eta(x) \, dx + c_1 Q_0 + c_2 Q_0' + c_3 Q_1 + c_4 Q_1'$$

where R_x' is defined as in the proof of Lemma 2,

$$R_x' = \frac{1}{(s_{j+1}-s_j)}(Q_{s_{j+1}} - Q_{s_j}) \qquad \text{for} \quad x \in I_j .$$

Then

$$||F - \tilde{F}||_Q^2 \leqq ||F - v||_Q^2 = \int_0^1 \int_0^1 \eta(x)\eta(x')\langle Q_x' - R_x', Q_{x'}' - R_{x'}'\rangle\, dx\, dx' .$$

Also, it can be shown that

$$\langle Q_x' - R_x', Q_{x'}' - R_{x'}'\rangle = 0 \qquad \text{if} \quad x \in I_j, \quad x' \in I_k, \quad j \neq k ;$$

so that

$$||F - v||_Q^2 \leqq \sum_{j=0}^{l} \{\int_{s_j}^{s_{j+1}} |\eta(x)|||Q_x' - R_x'||\, dx\}^2 .$$

By Lemma 2,

$$||Q_x' - R_x'||^2 \leqq \tfrac{1}{3}(s_{j+1} - s_j) \qquad \text{for } x \in I_j$$

from which the result follows as in Lemma 3.

LEMMA 5. *Let* $S(x) = S(x, \bar{s}; \bar{\varepsilon}_a, \bar{\varepsilon}, \bar{\varepsilon}_b)$ *be the cubic spline of interpolation defined by* (2.8) *with*

$$\begin{aligned} \bar{s} &= (s_1, s_2, \cdots, s_l) & & \\ \bar{\varepsilon}_a &= (\varepsilon_0', \varepsilon_0) , & \varepsilon_0 &= 0 \\ \bar{\varepsilon} &= (\varepsilon_1, \varepsilon_2, \cdots, \varepsilon_l) & & \\ \bar{\varepsilon}_b &= (\varepsilon_{l+1}, \varepsilon_{l+1}') , & \varepsilon_{l+1} &= 0 . \end{aligned}$$

Let

$$\begin{aligned} \Delta_i &= (s_{i+1} - s_i) , & i &= 0, 1, \cdots, l , \\ \psi_i &= (\varepsilon_{i+1} - \varepsilon_i) , & i &= 0, 1, \cdots, l . \end{aligned}$$

Then, for $x \in I_j$,

$$\left|\frac{d}{dx} S(x)\right| \leqq 8 \left\{\sum_{i=0}^{l} c_i \frac{\Delta_j}{\Delta_i} \frac{|\psi_i|}{\Delta_i} + \frac{1}{2^{j+1}} \frac{\Delta_j}{\Delta_0} |\varepsilon_0'| + \frac{1}{2^{l+2-j}} \frac{\Delta_j}{\Delta_l} |\varepsilon_{l+1}'|\right\} \tag{A4.1}$$

where

$$\begin{aligned} c_i &= \frac{1}{2^{|i-j|+1}} + \frac{1}{2^{|i+1-j|+1}} & i &= 0, 1, \cdots, l, i \neq j \\ &= \frac{1}{2} + \frac{1}{2^2} + \frac{1}{8} , & i &= j . \end{aligned}$$

PROOF. Define

$$R_x' = \frac{1}{\Delta_j} (Q_{s_{j+1}} - Q_{s_j}) \qquad \text{for} \quad x \in I_j .$$

Then

$$\frac{d}{dx} S(x) = \langle S, Q_x'\rangle_Q = \langle S, R_x'\rangle_Q + \langle S, Q_x' - R_x'\rangle .$$

Since $S(s_j) = \varepsilon_j$, $j = 0, 1, \cdots, l + 1$,

$$\langle S, R_x'\rangle = \frac{1}{\Delta_j} (\varepsilon_{j+1} - \varepsilon_j) = \frac{\psi_j}{\Delta_j} . \tag{A4.2}$$

Since the spline S is a cubic between the knots $0, s_1, \cdots, s_l, 1$, $d^2/dx^2\, S(x)$ is linear between the knots, and by the properties of cubic splines, continuous. Define κ_i by

$$\frac{d^2}{ds^2} S(s)\bigg|_{s=s_i} = \kappa_i .$$

Thus

$$\frac{d^2}{ds^2} S(s) = \frac{1}{\Delta_j} [\kappa_j(s_{j+1} - s) + \kappa_{j+1}(s - s_j)] \qquad \text{for} \quad s \in I_j .$$

Combining this with (A1.2)—(A1.8) gives

$$\langle S, Q_x' - R_x' \rangle_Q$$

$$= \frac{1}{\Delta_j^2} \int_{s_j}^{x} (s - s_j)[\kappa_j(s_{j+1} - s) + \kappa_{j+1}(s - s_j)]\, ds$$

$$+ \frac{1}{\Delta_j^2} \int_{x}^{s_{j+1}} (s_{j+1} - s)[\kappa_j(s_{j+1} - s) + \kappa_{j+1}(s - s_j)]\, ds , \qquad \text{for} \quad x \in I_j$$

and so

$$\text{(A4.3)} \qquad |\langle S, Q_x' - R_x' \rangle_Q| \leqq \Delta_j \max(|\kappa_j|, |\kappa_{j+1}|) , \qquad x \in I_j .$$

To proceed, we need to know the relationship between the κ_i and the data $\bar{s}$, $\bar{\varepsilon}_a$, $\bar{\varepsilon}$, $\bar{\varepsilon}_b$. By using a formula found in Kershaw, [9], equation (5), we may express this relationship for cubic splines. It is

$$(\kappa_0, \kappa_1, \cdots, \kappa_l, \kappa_{l+1}) = 6A^{-1}(\xi_0, \xi_1, \cdots, \xi_l, \xi_{l+1})$$

where A is the $(l + 2) \times (l + 2)$ matrix given by

$$A = \begin{bmatrix} 2 & 1 & & & & & \\ \alpha_1 & 2 & 1 - \alpha_1 & & & 0 & \\ & \alpha_2 & 2 & 1 - \alpha_2 & & & \\ & & & \cdot & & & \\ & & & & \cdot & & \\ & & & & & \cdot & \\ 0 & & & & \alpha_l & 2 & 1 - \alpha_l \\ & & & & & 1 & 2 \end{bmatrix}$$

where

$$\alpha_i = \frac{\Delta_{i-1}}{\Delta_i - \Delta_{i-1}} \qquad i - 1, 2, \cdots, l$$

and

$$\xi_0 = (\psi_0 - \Delta_0 \varepsilon_0')/\Delta_0^2$$

$$\xi_i = \left(\frac{\psi_i}{\Delta_i} - \frac{\psi_{i-1}}{\Delta_{i-1}}\right)\Big/(\Delta_i + \Delta_{i-1}) , \qquad i = 1, 2, \cdots, l$$

$$\xi_{l+1} = -(\psi_l - \Delta_l \varepsilon_{l+1}')/\Delta_l^2 .$$

For $i = 1, 2, \cdots, l$, ξ_i is the second divided difference of $S(x)$ at (s_{i-1}, s_i, s_{i+1}). We are now going to appeal to another result of Kershaw's, which gives bounds on the entries of A^{-1}. Let a^{rs}, $r, s = 0, 1, \cdots, l + 1$, be the r, sth entry of A^{-1}. According to [8],

$$|a^{rs}| \leqq \frac{4}{3} \frac{1}{2^{|r-s|+1}} , \qquad r, s = 0, 1, \cdots, l + 1 .$$

Therefore, since $\kappa_j = 6 \sum_{i=0}^{l+1} a^{ji}\xi_i$,

$$\text{(A4.4)} \qquad |\kappa_j| \leqq 6 \cdot \tfrac{4}{3} \cdot \sum_{i=0}^{l+1} \frac{1}{2^{|i-j|+1}} |\xi_i|$$

and, combining (A4.2), (A4.3), and (A4.4) gives

$$|\langle S, Q_x' - R_x'\rangle| \leqq \Delta_j \cdot 6 \cdot \tfrac{4}{3} \cdot \textstyle\sum_{i=0}^{l+1} \frac{1}{2^{|i-j|+1}} |\xi_i|, \qquad x \in I_j$$

and

$$|\langle S, Q_x'\rangle| \leqq \frac{|\psi_j|}{\Delta_j} + 8 \cdot \textstyle\sum_{i=1}^{l} \frac{1}{2^{|i-j|+1}} \left\{\frac{|\psi_i|}{\Delta_i} + \frac{|\psi_{i-1}|}{\Delta_{i-1}}\right\} \frac{\Delta_j}{\Delta_i + \Delta_{i-1}}$$

$$+ 8 \frac{1}{2^{j+1}} \frac{\Delta_j}{\Delta_0} \left\{\frac{|\psi_0|}{\Delta_0} + |\varepsilon_0'|\right\} + 8 \frac{1}{2^{l+2-j}} \frac{\Delta_j}{\Delta_l} \left\{\frac{|\psi_l|}{\Delta_l} + |\varepsilon_{l+1}'|\right\}$$

$$\leqq 8 \cdot \textstyle\sum_{i=0}^{l} c_i \frac{|\psi_i|}{\Delta_i} \frac{\Delta_j}{\Delta_i} + \frac{8}{2^{j+1}} \frac{\Delta_j}{\Delta_0} |\varepsilon_0'| + \frac{8}{2^{l+2-j}} \frac{\Delta_j}{\Delta_l} |\varepsilon_{l+1}'|$$

where

$$c_i = \frac{1}{2^{|i-j|+1}} + \frac{1}{2^{|i+1-j|+1}}, \qquad i = 0, 1, 2, \cdots, l,\ i \neq j$$

$$c_j = \frac{1}{2} + \frac{1}{2^2} + \frac{1}{8}.$$

We remark that the lack of a generalization of this lemma for higher degree splines is the stumbling block in generalizing the main theorem to higher m.

LEMMA 6. *Suppose* $f^{(m)} \in \mathscr{L}_p$ *on* $[0, 1]$. *Let* F_n *be the sample* cdf *for* n *independent observations from* f. *Let*

$$\hat{f}_n(0) = \frac{d}{dx} \textstyle\sum_{\nu=0}^{m} l_{0,\nu}(x)\Big|_{x=0} F_n(\nu h)$$

$$\hat{f}_n(1) = \frac{d}{dx} \textstyle\sum_{\nu=0}^{m} l_{1,\nu}(x)\Big|_{x=1} F_n(1 - \nu h)$$

where $l_{0,\nu}$ *and* $l_{1,\nu}$ *are the Lagrange polynomials defined in* (2.10) *and* (2.11). (*For* $m = 1$, $\hat{f}_n(0) = (1/n)F_n(h)$.) *Then*

$$\text{(A5.1)} \qquad \left.\begin{matrix} E(f(0) - \hat{f}_n(0))^2 \\ E(f(1) - \hat{f}_n(1))^2 \end{matrix}\right\} \leqq \frac{8m^{2m-2/p}}{[(m-1)!]^2} \|f^{(m)}\|_p^2 h^{2m-2/p} + 2m^3(m+1)^2 \frac{\Lambda}{nh}.$$

PROOF.

$$\text{(A5.2)} \qquad |f(0) - \hat{f}_n(0)| \leqq \left|f(0) - \frac{d}{dx} \textstyle\sum_{\nu=0}^{m} l_{0,\nu}(x)\Big|_{x=0} F(\nu h)\right|$$

$$+ \left|\frac{d}{dx} \textstyle\sum_{\nu=0}^{m} l_{0,\nu}(x)[F(\nu h) - F_n(\nu h)]\right|.$$

By combining Lemma 3.1 of [21], and Theorem 3 of [20], and noting that $\prod_{j=0}^{m} (0 - jh) = 0$ in equation (3.28) of [20], it can be shown that

$$\left|f(0) - \frac{d}{dx} \textstyle\sum_{\nu=0}^{m} l_{0,\nu}(x)\Big|_{x=0} F(\nu h)\right|^2$$

$$\text{(A5.3)} \qquad \leqq \left[\frac{2}{(m-1)!}\right]^2 \left[\textstyle\int_0^{mh} |f^{(m)}(\xi)|^p \, d\xi\right]^{2/p} (mh)^{m-1/p},$$

$$p \geqq 1, \ m = 1, 2, \cdots.$$

It can be verified, for $m = 1, 2, 3$, that

$$\left|\frac{d}{dx} l_{0,\nu}(x)\Big|_{x=0}\right| \leqq \frac{m}{h}. \tag{A5.4}$$

Now

$$F_n(\nu h) = \frac{\#\text{ observations in } [0, \nu h]}{n},$$

and hence $nF_n(\nu h)$ is binomial $B(n, \sum_{j=1}^{\nu} p_j)$, where $p_j = \int_{(j-1)h}^{jh} f(\xi)\,d\xi$, and hence

$$E[F(\nu h) - F_n(\nu h)]^2 = (\textstyle\sum_{j=1}^{\nu} p_j)(1 - \sum_{j=1}^{\nu} p_j)/n \leqq \dfrac{\Lambda m h}{n}. \tag{A5.5}$$

Putting together (A5.2) (A5.3) (A5.4) and (A5.5) gives

$$E(f(0) - \hat{f}_n(0))^2 \leqq \frac{8m^{2m-2/p}}{((m-1)!)^2} ||f^{(m)}||_p{}^2 h^{2m-2/p} + 2m^3(m+1)^2 \frac{\Lambda}{nh}.$$

The proof is carried out similarly for $x = 1$.

REFERENCES

[1] Academic Computing Center. *Approximation and Interpolation, Reference Manual for the* 1108. Chapter 6. MACC, Univ. of Wisconsin.

[2] AHLBERG, J. H., NILSON, E. N. and WALSH, J. L. (1967). *The Theory of Splines and their Applications.* Academic Press, New York.

[3] BONEVA, L., KENDALL, D. and STEFANOV, I. (1971). Spline transformations: Three new diagnostic aids for the statistical data analyst. *J. Roy. Statist. Soc.* **33** 1–70.

[4] DEBOOR, CARL (1968). On the convergence of odd degree spline interpolation. *J. Approx. Theor.* **1** 452–463.

[5] FARRELL, R. H. (1972). On best obtainable asymptotic rates of convergence in estimation of a density function at a point. *Ann. Math. Statist.* **43** 170–180.

[6] GREVILLE, T. N. E. (1969). Introduction to spline functions, in *Theory and Applications of Spline Functions* (T. N. E. Greville, ed.). Academic Press, New York. 1–35.

[7] GREVILLE, T. N. E. (1971). Another look at cubic spline interpolation of equidistant data. Univ. of Wisconsin MRC Technical Summary Report #1148, Madison.

[8] KERSHAW, D. (1970). Inequalities on the elements of the inverse of a certain tridiagonal matrix. *Math. Comp.* **24** 155–158.

[9] KERSHAW, D. (1971). A note on the convergence of interpolatory cubic splines. *SIAM J. Numer. Anal.* **8** 67–74.

[10] KIMELDORF, GEORGE, and WAHBA, GRACE (1971). Some results on Tchebycheffian spline functions, *J. Math. Anal. Appl.* **33** 82–95.

[11] KRONMAL, R. and TARTER, M. (1968). The estimation of probability densities and cumulatives by Fourier series methods. *J. Amer. Statist. Assoc.* **63** 925–952.

[12] PARZEN, E. (1962). On the estimation of a probability density function and mode. *Ann. Math. Statist.* **33** 1065–1076.

[13] SHARMA, A. and MEIR, A. (1966). Degree of approximation of spline interpolation. *J. Math. Mech.* **15** 759–767.

[14] VAN ROOY, P. L. J. and SCHURER, F. (1971). A bibliography on spline functions. Dept. of Mathematics, Technological Univ. Eindhoven, Netherlands, T. H.—Report 71-WSK-02.

[15] SCHOENBERG, I. J. (1967). On spline functions, in *Inequalities*, (O. Shisha, ed.). 255–291. Academic Press, New York.

[16] Schoenberg, I. J. (1972). Notes on spline functions II. On the smoothing of histograms. Univ. of Wisconsin MRC Technical Summary Report #1222, Madison.
[17] Schoenberg, I. J. (1972). Splines and histograms. Univ. of Wisconsin MRC Technical Summary Report #1273, Madison.
[18] Schultz, Martin (1970). Error bounds for polynomial spline interpolation. *Math. Comp.* **24** 507–515.
[19] Wahba, Grace (1971). A note on the regression design problem of Sacks and Ylvisaker. *Ann. Math. Statist.* **42** 1035–1053.
[20] Wahba, Grace (1971). A polynomial algorithm for density estimation. *Ann. Math. Statist.* **42** 1870–1886.
[21] Wahba, Grace (1975). Optimal convergence properties of variable knot kernel, and orthogonal series methods for density estimation. *Ann. Statist.* **3** 15–29.

Department of Statistics
University of Wisconsin
1210 West Dayton Street
Madison, Wisconsin 53706

Reprinted from *Ann. Stat.* **3:**1329–1348 (1975)

NONPARAMETRIC MAXIMUM LIKELIHOOD ESTIMATION OF PROBABILITY DENSITIES BY PENALTY FUNCTION METHODS[1]

BY G. F. DE MONTRICHER, R. A. TAPIA, AND J. R. THOMPSON

Princeton University and Rice University

Tne maximum likelihood estimate of a probability density function based on a random sample does not exist in the nonparametric case. For this reason and others based on heuristic Bayesian considerations Good and Gaskins suggested adding a penalty term to the likelihood. They proposed two penalty terms; however they did not establish existence or uniqueness of their maximum penalized likelihood estimates. Good and Gaskins also suggested an alternate approach for calculating the maximum penalized likelihood estimate which avoids the nonnegativity constraint on the estimate. In the present work the existence and uniqueness of both of Good's and Gaskins' maximum penalized likelihood estimates are rigorously demonstrated. Moreover, it is shown that one of these estimates is a positive exponential spline with knots only at the sample points and that in this case the alternate approach leads to the correct estimate; however in the other case the alternate approach leads to the wrong estimate. Finally, it is shown that a well-known class of reproducing kernel Hilbert spaces leads very naturally to maximum penalized likelihood estimates which are polynomial splines with knots at the sample points.

1. Introduction. Let Ω denote the interval (a, b). In this study we consider the problem of estimating the (unknown) probability density function $f \in L^1(\Omega)$ which gave rise to the random sample $x_1, \cdots, x_N \in \Omega$. The set Ω may be bounded or unbounded.

As usual define $L(v)$, the *likelihood* that $v \in L^1(\Omega)$ gave rise to the random sample $x_1, \cdots, x_N$ by

$$L(v) = \prod_{i=1}^{N} v(x_i)\,. \tag{1.1}$$

Let $H(\Omega)$ be a manifold in $L^1(\Omega)$ and consider the following optimization problem:

$$\text{maximize} \quad L(v)\,; \quad \text{subject to} \qquad v \in H(\Omega)\,, \qquad \int_\Omega v(t)\,dt = 1 \quad \text{and} \quad v(t) \geqq 0 \qquad \forall\, t \in \Omega\,. \tag{1.2}$$

Here dt denotes the Lebesgue measure on Ω. By a *maximum likelihood estimate* based on the random sample $x_1, \cdots, x_N$ (corresponding to the manifold $H(\Omega)$) we mean any solution of the constrained optimization problem (1.2). The

Received August 1974; revised March 1975.

[1] Invited paper presented to annual meeting of the Institute of Mathematical Statistics in Edmonton, Alberta, Canada on August 15, 1974. Research sponsored by the Office of Naval Research under Contract NR-042-283.

AMS 1970 *subject classifications.* Primary 62G05; Secondary 62E10.

Key words and phrases. Density estimation, maximum likelihood estimation, spline, monospline.

estimate is said to be *parametric* if $H(\Omega)$ is a finite dimensional manifold and *nonparametric* if $H(\Omega)$ is an infinite dimensional manifold.

In general a nonparametric maximum likelihood estimate does not exist. To see this observe that the nonexistent solution is idealized by a linear combination of Dirac delta spikes at the samples and gives a value of $+\infty$ to the likelihood functional. Hence, in any infinite dimensional manifold which has the property that it is possible to construct a sequence of functions which integrate to one, are nonnegative and converge pointwise to a Dirac delta spike, the likelihood will be unbounded and a maximum likelihood estimate will not exist. Morever most infinite dimensional manifolds in $L^1(\Omega)$ will have this property, e.g., the continuous functions, the differentiable functions, the infinitely differentiable functions and the polynomials. Of course we may consider pathological examples of infinite dimensional manifolds on which the likelihood is bounded. As an extreme case let $H(\Omega)$ be all continuous functions on Ω which vanish at the samples $x, \cdots, x_N$. Then the likelihood is identically zero on the infinite dimensional manifold $H(\Omega)$ and is therefore bounded. Furthermore any member of $H(\Omega)$ which is nonnegative and integrates to one is a maximum likelihood estimate for the random sample $x_1, \cdots, x_N$.

The fact that in general the nonparametric maximum likelihood estimate does not exist implies that, except in the extreme case where the parametric form of the unknown density function f is known a priori, the parametric maximum likelihood approach for large dimensional problems must necessarily lead to unsmooth estimates and a numerically ill-posed problem. This leaves the practitioner with the following dilemma: For small dimensional problems he has no flexibility and the solution will be greatly influenced by the choice of the manifold $H(\Omega)$; while for large dimensional problems the solution must necessarily approximate a linear combination of Dirac delta spikes, be unsmooth and create numerical problems.

The following example will illustrate many of these points. For a given positive integer n partition Ω into n half-open half-closed disjoint intervals $T_1, \cdots, T_n$ of equal length $h = (b - a)/n$. Let $I(T_i)$ denote the characteristic function of the interval T_i and let ν_i denote the number of samples in the interval T_i. The well-known histogram estimate for f based on the random sample $x_1, \cdots, x_N$ is given by

$$f^* = \sum_{i=1}^{n} \frac{\nu_i}{Nh} I(T_i) . \tag{1.3}$$

LEMMA 1.1. *The probability density estimate* (1.3) *is the maximum likelihood estimate for the random sample* $x_1, \cdots, x_N$ *corresponding to the n-dimensional manifold* $H(\Omega)$ *where* $H(\Omega)$ *is the linear span of the characteristic functions* $\{I(T_i): i = 1, \cdots, n\}$.

PROOF. A typical member ω of $H(\Omega)$ has the form

$$\omega = \sum_{i=1}^{n} y_i I(T_i) .$$

The nonnegativity constraint has the form $y_i \geqq 0$, $i = 1, \cdots, n$ and will not be active at the solution. To see this observe that if $\nu_i > 0$, then the optimal solution of problem (1.2) will have $y_i > 0$. Since the log is a concave function we may work with the log likelihood instead of the likelihood. Hence we are interested in determining $y_1, \cdots, y_n$ which maximizes

$$G(y_1, \cdots, y_n) = \sum_{i=1}^{n} \nu_i \log(y_i)$$

subject to the integral constraint $h \sum_{i=1}^{n} y_i = 1$. From the theory of Lagrange multipliers, see e.g., Fiacco–McCormick (1969) Chapter 2, we must have

$$\nu_i + \lambda y_i = 0\,, \qquad i = 1, \cdots, n \tag{1.4}$$
$$h \sum_{i=1}^{n} y_i = 1$$

for some scalar λ. It is not difficult to see that $y_i = \nu_i/(hN)$ gives the unique solution of (1.4) and that this solution satisfies the sufficiency conditions for a maximizer. This proves the lemma.

Notice that for a fixed sample as $n \to \infty$ the estimate f^* given by (1.3) has the property that $f^*(x_i) \to +\infty$, $i = 1, \cdots, N$ while $f^*(x) \to 0$ if $x \notin \{x_1, \cdots, x_N\}$. Hence for large n our maximum likelihood estimate is very unsmooth and unsatisfactory. Whether or not we obtain a reasonable estimate is completely dependent on the delicate and tricky art of choosing n, properly. It is also of interest to note that the numerical properties of (1.4) are very poor for large n.

For the reasons stated above and others based on heuristic Bayesian considerations Good and Gaskins (1971) suggested adding a penalty term to the likelihood which would penalize rough (unsmooth) estimates. They suggested two specific penalty terms; however, they left the reader in the unfortunate situation of not knowing whether their maximum penalized likelihood estimates exist. Good and Gaskins also suggested an alternate approach for constructing the maximum penalized likelihood estimate which avoids the nonnegativity constraint. Again they do not demonstrate that the original approach and the alternate approach give the same estimate or that the estimate obtained from the alternate approach exists.

In Section 2 we establish a general existence and uniqueness theory for a large class of maximum penalized likelihood estimates. This general theory is used to show that a well-known class of reproducing kernel Hilbert spaces (Sobolev spaces) lead quite naturally to maximum penalized likelihood estimates which are polynomial splines (monosplines in the terminology of Schoenberg (1968)) with knots at the sample points.

In Section 3 a rigorous demonstration, using the theory developed in Section 2, is given of the existence and uniqueness of one of Good's and Gaskins' maximum penalized likelihood estimates. It is also shown that this estimate is a positive exponential spline with knots only at the sample points and that the alternate approach suggested by Good and Gaskins gives the correct estimate.

In Section 4 using the theory developed in Section 2 a rigorous demonstration of the existence and uniqueness of Good's and Gaskins' other maximum penalized estimate is given. Moreover, it is also demonstrated that the estimate obtained from Good's and Gaskins' alternate approach is not the maximum penalized likelihood estimate.

Much of our analysis uses the notions of the Fréchet gradient, the Fréchet derivative and the second Fréchet derivative in an abstract Hilbert space. The reader not familiar with these notions is referred to Tapia (1971).

Many important statistical considerations, e.g., consistency, unbiasedness, convergence rate, choice of penalty term and numerical implementation are unanswered here. Instead, it is the purpose of the present work to set the maximum penalized likelihood approach on solid approximation theoretic ground. Hopefully, the insights gained from this study will lead to investigation of the classical statistical properties of the estimates considered.

Before moving on to Section 2 a few brief historical comments are in order. Rosenblatt (1956) performed the first analytical study of the theoretical properties of histograms. Parzen (1962) constructed a class of estimators which properly included the histogram estimators and examined the consistency properties of the estimators in this class. These results have been improved upon recently by Wahba (1971). Kimeldorf and Wahba (1970) introduced the application of spline techniques in contemporary statistics. Boneva, Kendall and Stefanov (1971) and Schoenberg (1972) examined the use of spline techniques for obtaining from histograms smooth estimates of a probability density function. It is of interest to us that essentially all previous authors seem either to ignore the nonnegativity constraint or to attempt handling it with the seemingly clever trick of working with a function whose square is to be used as the estimate of the probability density; however in many cases this approach tacitly ignores the nonnegativity constraint. More will be said about the use of this approach in Sections 3 and 4.

2. Maximum penalized likelihood estimators. Let $H(\Omega)$ be as in Section 1 and consider a functional $\Phi: H(\Omega) \to R$. Given the random sample $x_1, \cdots, x_N \in \Omega$ the Φ-*penalized likelihood* of $v \in H(\Omega)$ is defined by

$$\hat{L}(v) = \prod_{i=1}^{N} v(x_i) \exp(-\Phi(v)) . \tag{2.1}$$

Consider the constrained optimization problem:

$$\text{maximize} \quad \hat{L}(v); \quad \text{subject to} \tag{2.2}$$
$$v \in H(\Omega), \qquad \int v(t)\,dt = 1 \quad \text{and} \quad v(t) \geqq 0, \qquad \forall\, t \in \Omega .$$

The form of the penalized likelihood (2.1) is due to Good and Gaskins (1971). Their specific suggestions are analyzed in Sections 3 and 4.

Any solution to problem (2.2) is said to be a *maximum penalized likelihood estimate* based on the random sample $x_1, \cdots, x_N$ corresponding to the manifold $H(\Omega)$ and the penalty function Φ. The terms parametric and nonparametric have the same meaning in this context as they did in the context of Section 1. In the case when $H(\Omega)$ is a Hilbert space a very natural penalty function to use is $\Phi(v) = \|v\|^2$ where $\|\cdot\|$ denotes the norm on $H(\Omega)$. Consequently when $H(\Omega)$ is a Hilbert space and we refer to the penalized likelihood functional on $H(\Omega)$ or to the maximum penalized likelihood estimate corresponding to $H(\Omega)$ with no reference to the penalty functional Φ we are assuming that Φ is the square of the norm in $H(\Omega)$. The Hilbert space inner product will be denoted by $\langle \cdot, \cdot \rangle$ so that $\langle x, x\rangle = \|x\|^2$. When $H(\Omega)$ is a Hilbert space it is said to be a reproducing kernel Hilbert space if point evaluation is a continuous operation, i.e., $v_n \to v$ in $H(\Omega)$ implies $v_n(x) \to v(x)$ $\forall\, x \in \Omega$; see Goffman and Pedrick (1965).

For problem (2.2) to make sense we would like $H(\Omega)$ to have the property that for $x_1, \cdots, x_N \in \Omega$ there exists at least on $v \in H(\Omega)$ such that

$$\text{(2.3)} \qquad \int_\Omega v(t)\,dt = 1\,, \qquad v(t) \geqq 0 \quad \forall\, t \in \Omega \quad \text{and} \quad v(x_i) > 0 \qquad i = 1, \cdots, N\,.$$

PROPOSITION 2.1. *Suppose that $H(\Omega)$ is a reproducing kernel Hilbert space and D is a closed convex subset of $\{v \in H(\Omega) : v(x_i) \geqq 0\}$ with the property that D contains at least one function which is positive at the samples $x_1, \cdots, x_N$. Then the penalized likelihood functional* (2.1) *has a unique maximizer in D.*

PROOF. Since $H(\Omega)$ is a reproducing kernel Hilbert space we have $|v(x_i)| \leqq K_i\|v\|$ for $i = 1, \cdots, N$. It follows that

$$\text{(2.4)} \qquad |\hat{L}(v)| \leqq C_1\|v\|^N \exp(-\|v\|^2)\,.$$

The function $\theta(\lambda) = \lambda^N \exp(-\lambda)^2$ is bounded above by $(N/2)^{N/2} \exp(-N/2)$; hence $|\hat{L}(v)| \leqq C_2$. If $M = \sup\{\hat{L}(v) : v \in D\}$, then there exists $\{v_j\} \subset D$ such that $\hat{L}(v_j) \to M$. From our hypothesis $M > 0$. Notice that $\theta(\lambda) \to 0$ as $\lambda \to +\infty$. Hence from (2.4) $\|v_j\| \leqq C_3$ $\forall\, j$. The ball $\{v \in H(\Omega) : \|v\| \leqq C_3\}$ is weakly compact. Hence $\{v_j\}$ contains a weakly convergent subsequence which we also denote by $\{v_j\}$. Let v^* denote the weak limit of $\{v_j\}$. We have that $v_j(x_i) \to v^*(x_i)$ as $j \to \infty$ for each $i = 1, \cdots, N$. The norm is a continuous convex functional; hence weakly lower semicontinuous so that $\liminf \|v_j\| \geqq \|v^*\|$. It follows that

$$\text{(2.5)} \qquad \lim_j \prod_{i=1}^N v_j(x_i) \exp(-\|v_j\|^2) \leqq \prod_{i=1}^N v^*(x_i) \exp(-\|v^*\|^2)\,.$$

However the left-hand side of (2.5) is equal to M and the right-hand side is equal to $\hat{L}(v^*)$; so $M \leqq \hat{L}(v^*)$. Now since D is closed and convex it is weakly closed; hence $v^* \in D$. This establishes the existence of a maximizer.

Since $M > 0$, maximizing $\hat{L}$ over D is equivalent to maximizing $J = \log \hat{L}$ over D. A straightforward calculation gives the second Fréchet derivative of J as

$$\text{(2.6)} \qquad J''(v)(\mu, \eta) = -\sum_{i=1}^N \frac{\mu(x_i)\eta(x_i)}{v(x_i)^2} - 2\langle \mu, \eta\rangle\,.$$

Now since $J''(v)$ is negative definite J is strictly concave and can therefore have at most one maximizer on a convex set.

THEOREM 2.1. *Suppose $H(\Omega)$ is a reproducing kernel Hilbert space, integration over Ω is a continuous functional and there exists at least one $v \in H(\Omega)$ satisfying* (2.3) *Then the maximum penalized likelihood estimate corresponding to $H(\Omega)$ exists and is unique.*

PROOF. The proof follows from Proposition 2.1 since the constraints in (2.2) give a closed convex subset of $\{v \in H(\Omega) : v(x_i) \geqq 0, i = 1, \cdots, N\}$.

Suppose (a, b) is a finite interval. For each integer $s \geqq 1$ we let $H_0^s(a, b)$ denote the Sobolev space of functions defined on $[a, b]$ whose first $s - 1$ derivatives are absolutely continuous and vanish at a and at b and whose sth derivative is in $L^2(a, b)$. The inner product in $H_0^s(a, b)$ is defined by

$$\langle \mu, v \rangle = \int_a^b \mu^{(s)}(t) v^{(s)}(t)\, dt\,. \tag{2.7}$$

It is well known that the space $H_0^s(a, b)$ is a Hilbert space with the inner product given by (2.7).

LEMMA 2.1. *The space $H_0^s(a, b)$ is a reproducing kernel Hilbert space and integration over (a, b) is a continuous operation.*

PROOF. Suppose $u, u_n \in H_0^s(a, b)$ and $u_n \to u$ in $H_0^s(a, b)$. By the Cauchy-Schwarz inequality in $L^2(a, b)$ we have for $x \in [a, b]$

$$|u(x) - u_n(x)| = |\int_a^x (u'(t) - u_n'(t)\, dt| \leqq (b - a)\|u' - u_n'\|_{L^2(a,b)}\,; \tag{2.8}$$

hence point evaluation is a continuous operation. A straightforward integration by parts and the Cauchy-Schwarz inequality lead to

$$|\int_a^b [u(t) - u_n(t)]\, dt| \leqq \tfrac{1}{2}(b^2 - a^2)\|u' - u_n'\|_{L^2(a,b)}\,; \tag{2.9}$$

hence integration over (a, b) is a continuous operation.

THEOREM 2.2. *The maximum penalized likelihood estimate corresponding to the Hilbert space $H_0^s(a, b)$ exists, is unique and is a polynomial spline (monospline) of degree $2s$. Moreover, if the estimate is positive in the interior of an interval, then in this interval it is a polynomial spline (monospline) of degree $2s$ and of continuity class $2s - 2$ with knots exactly at the sample points.*

PROOF. The existence and uniqueness are a consequence of Lemma 2.1 and Theorem 2.1, since clearly there exists at least one $v \in H_0^s(a, b)$ satisfying (2.3).

When no confusion can arise we will delete the variable of integration in definite integrals. Consider an interval $I_+ = [\alpha, \beta] \subset [a, b]$. Let $I = \{t \in [a, b] : t \notin [\alpha, \beta]\}$. Define the two functionals J_+ and J_- on $H_0^s(a, b)$ by

$$J_+(v) = \textstyle\sum_i \log v(x_i) - \int_{I_+} [v^{(s)}]^2\,,$$

and

$$J_-(v) = \textstyle\sum_i \log v(x_i) - \int_{I_-} [v^{(s)}]^2\,,$$

where the summation in the first formula is taken over all i such that $x_i \in I_+$

and the summation in the second formula is taken over all i such that $x_i \in I_-$. It should be clear that

$$J(v) = J_+(v) + J_-(v)$$

where as before $J(v) = \log \hat{L}(v)$ and $\hat{L}$ is the penalized likelihood in $H_0^s(a, b)$. Let v_* denote the maximum penalized likelihood estimate for the samples $x_1, \cdots, x_N$. Suppose v_* is positive on the interval I_+. We claim that v_* restricted to this interval solves the following constrained optimization problem:

$$\text{maximize} \quad J_+(v)\,; \quad \text{subject to}$$

$$(2.10) \qquad v \in H_0^s(\alpha, \beta)\,, \quad v^{(m)}(\alpha) = v_*^{(m)}(\alpha)\,, \quad v^{(m)}(\beta) = v_*^{(m)}(\beta)\,, \quad m = 0, \cdots, s-1\,,$$

$$\textstyle\int_{I_+} v = \int_{I_+} v_* \quad \text{and} \quad v(t) \geqq 0\,, \qquad t \in I_+\,.$$

To see this observe that if v_+ satisfies the constraints of problem (2.10) and $J_+(v_*) < J_+(v_+)$, then the function v^* defined by

$$\begin{aligned} v^*(t) &= v_+(t)\,, \quad t \in I_+ \\ &= v_*(t)\,, \quad t \in I_- \end{aligned}$$

satisfies the constraints of problem (2.2) with $H_0^s(a, b)$ playing the role of $H(\Omega)$ and $J(v_*) = J_+(v_*) + J_-(v_*) < J_+(v_+) + J_-(v_+) = J(v^*)$, which in turn implies that $\hat{L}(v_*) < \hat{L}(v^*)$; however this contradicts the optimality of v^*. Now define the functional G on $H_0^s(\alpha, \beta)$ by

$$G(v) = J_+(v_* + v) \qquad \text{for} \quad v \in H_0^s(\alpha, \beta)\,.$$

Consider the constrained optimization problem

$$(2.11) \qquad \text{maximize} \quad G(v)\,; \quad \text{subject to}$$

$$v \in H_0^s(\alpha, \beta) \qquad \text{and} \qquad \textstyle\int_{I_+} v = 0\,.$$

If v satisfies the constraints of problem (2.11), then $v_* + tv$ satisfies the constraints of problem (2.10) for t sufficiently small, since v_* is positive in I_+. It follows that the zero function is the unique solution of problem (2.11). From the theory of Lagrange multipliers we therefore must have

$$(2.12) \qquad \nabla G(0) + \lambda v_0 = 0\,,$$

where λ is a real number, $\nabla G(0)$ is the Fréchet gradient of G at 0 and v_0 is the Fréchet gradient of the functional $v \to \int_{I_+} v$ in the space $H_0^s(\alpha, \beta)$. Clearly in this case v_0 is merely the Riesz representer of the functional $v \to \int_{I_+} v$. Specifically

$$\textstyle\int_{I_+} v_0^{(s)} v^{(s)} = \int_{I_+} v.$$

Integrating by parts in the distribution sense we see that $v_0^{(2s)} = 1$; hence v_0 is a polynomial of degree $2s$ in $[\alpha, \beta]$. A straightforward calculation shows that

$$(2.13) \qquad \nabla G(0) = \left(\sum_i \frac{v_i}{v_*(x_i)} - 2v_*\right)$$

where the summation is taken over i such that $x_i \in I_+$ and v_i is the Riesz representer of the functional $v \to v(x_i)$ in $H_0^s(\alpha, \beta)$, i.e.,

$$\int_{I_+} v_i^{(s)} v^{(s)} = v(x_i) .$$

As before integrating by parts in the distribution sense we see that $v_i^{(2s)} = \delta_i$ where δ_i is the Dirac mass at the point x_i. It follows that v_i is a polynomial spline of degree $2s - 1$ and of continuity class $2s - 2$ with a knot exactly at the sample point x_i. From (2.12) and (2.13) we have that v_* restricted to the interval $[\alpha, \beta]$ is a polynomial spline of degree $2s$ and of continuity class $2s - 2$ with knots exactly at the sample points in $[\alpha, \beta]$. A simple continuity argument takes care of the case when v_* is only positive on the interior of $[\alpha, \beta]$. Schoenberg (1968) defines a monospline to be the sum of a polynomial of degree $2s$ and a polynomial spline of degree $2s - 1$. This proves the theorem.

Before we analyze the Good and Gaskins estimates in Section 3 and Section 4 we must develop more background on Sobolev spaces. The reader desiring a more complete treatment is referred to Lions and Magenes (1968).

By the Sobolev space of order s on the real line we mean

$$H^s(-\infty, \infty) = \{\mu \in S' : (1 + \omega^2)^{s/2} F[\mu](\omega) \in L^2(-\infty, \infty)\} \tag{2.14}$$

where S' is the space of distributions with polynomial decrease at infinity and $F[\mu]$ denotes the Fourier transform of μ. The norm of $\mu \in H^s(-\infty, \infty)$ is given by

$$\|\mu\|_{H^s(-\infty,\infty)} = \|(1 + \omega^2)^{s/2} F[\mu](\omega)\|_{L^2(-\infty,\infty)} . \tag{2.15}$$

If s is an integer, then $\mu \in H^s(-\infty, \infty)$ if and only if $\mu, \mu^{(1)}, \cdots, \mu^{(s)} \in L^2(-\infty, \infty)$ and an equivalent norm is given by

$$[\textstyle\sum_{i=0}^{s} w_i \|\mu^{(i)}\|^2_{L_2(-\infty,\infty)}]^{\frac{1}{2}} \tag{2.16}$$

where $w_i \geqq 0$ and $w_0, w_s > 0$.

LEMMA 2.2. *The Sobolev space $H^s(-\infty, \infty)$ is a reproducing kernel Hilbert space if and only if $s > \frac{1}{2}$.*

PROOF. The dual of H^s is H^{-s}. A reproducing kernel Hilbert space is a space such that the Dirac distributions are in the dual; hence we want

$$(1 + \omega^2)^{-s/2} F(\delta_x) \in L^2(-\infty, \infty)$$

where δ_x is the Dirac distribution at the point x. Since the Fourier image of a Dirac mass if a constant we must have

$$(1 + \omega^2)^{-s/2} \in L^2(-\infty, \infty) .$$

This proves the lemma.

3. The first maximum penalized likelihood estimator of Good and Gaskins. Motivated by information theoretic considerations Good and Gaskins (1971) consider the maximum penalized likelihood estimate corresponding to the penalty function

$$\Phi_1(v) = \alpha \int_{-\infty}^{\infty} \frac{v'(t)^2}{v(t)} dt \quad (\alpha > 0).$$

They do not define the manifold $H(\Omega)$; but it is obvious from the constraints that must be satisfied and the fact that

$$\tfrac{1}{4}\Phi_1(v) = \alpha \int_{-\infty}^{\infty} \left(\frac{dv^{\frac{1}{2}}}{dt}\right)^2 dt$$

what the underlying manifold $H(\Omega)$ should be, namely $v^{\frac{1}{2}} \in H^1(-\infty, \infty)$. This leads us to analyzing the following constrained optimization problem:

$$\text{maximize} \quad L_1(v) = \prod_{i=1}^{N} v(x_i) \exp(-\Phi_1(v)); \quad \text{subject to}$$

(3.1) $$v^{\frac{1}{2}} \in H^1(-\infty, \infty), \qquad \int_{-\infty}^{\infty} v(t)\, dt = 1 \quad \text{and} \quad v(t) \geqq 0 \quad \forall t \in (-\infty, \infty).$$

In an effort to avoid the nonnegativity constraint in problem (3.1) Good and Gaskins considered working with the $v^{\frac{1}{2}}$ instead of v. Specifically if we let $u = v^{\frac{1}{2}}$, then restating problem (3.1) in terms of u we obtain

(3.2) $$\text{maximize} \quad \prod_{i=1}^{N} u(x_i)^2 \exp\left(-4\alpha \int_{-\infty}^{\infty} u'(t)^2\, dt\right); \quad \text{subject to} \quad u \in H^1(-\infty, \infty) \quad \text{and} \quad \int_{-\infty}^{\infty} u(t)^2\, dt = 1.$$

Problem (3.2) is solved for u^* and then $v^* = (u^*)^2$ is accepted as the solution to problem (3.1). This seemingly clever trick is somewhat standard in the literature. The following lemma tells us when this trick can be used.

LEMMA 3.1. *Let H be a subset of $L^2(\Omega)$ and J a functional defined on H. Consider Problem* I

$$\textit{maximize} \quad J(v^{\frac{1}{2}}); \quad \textit{subject to} \quad v^{\frac{1}{2}} \in H, \quad \int_\Omega v(t)\, dt = 1 \quad \textit{and} \quad v(t) \geqq 0 \quad \forall t \in \Omega$$

and Problem II

$$\textit{maximize} \quad J(u); \quad \textit{subject to} \quad u \in H \quad \textit{and} \quad \int_\Omega u(t)^2\, dt = 1.$$

Let u^* be a solution of Problem II. *Then $v^* = (u^*)^2$ solves Problem* I *if and only if $|u^*| \in H$ and $J(u^*) = J(|u^*|)$.*

PROOF. Observe that $v^* = |u^*|$ hence if $|u^*| \notin H$, then $J(v^*)^{\frac{1}{2}}$ is not defined and meaningless. While if $|u^*| \in H$ and $J(u^*) = J(|u^*|)$, then for any $v \geqq 0$ such that $v^{\frac{1}{2}} \in H$ we have

$$J(v^{\frac{1}{2}}) \leqq J(u^*) = J(|u^*|) = J(v^*)^{\frac{1}{2}}.$$

Now noticing that the proper constraints are always satisfied we have the lemma.

Two points are immediately of interest. The first being that the conditions of the lemma clearly hold when u^*, the solution to Problem II, is nonnegative. The second being that the space $H^1(-\infty, \infty)$ and the function $\hat{L}$ in problems (3.1) and (3.2) satisfy the conditions of the lemma. Hence Good's and

Gaskins' alternate approach gives the correct estimate in this case; however in their other case (analyzed in Section 4) this is unfortunately not true.

Problem (3.2) cannot possibly have a unique solution. To see this notice that if u^* is a solution, then so is $-u^*$. Adding the nonnegativity constraint to problem (3.2) and restating in the form obtained by taking the square root of the objective functional (since it is nonnegative) we arrive at the following constrained optimization problem:

$$\text{(3.3)} \qquad \text{maximum} \quad \hat{L}(v) = \textstyle\prod_{i=1}^{N} v(x_i) \exp(-\Phi(v)) \,; \quad \text{subject to}$$
$$v \in H^1(-\infty, \infty)\,, \qquad \textstyle\int_{-\infty}^{\infty} v(t)^2\,dt = 1 \quad \text{and} \quad v(t) \geqq 0\,, \quad \forall\, t \in (-\infty, \infty)$$

where

$$\Phi(v) = 2\alpha \textstyle\int_{-\infty}^{\infty} v'(t)^2\,dt$$

and α is given in problem (3.1).

PROPOSITION 3.1.

(i) *If v solves problem* (3.1), *then $v^{\frac{1}{2}}$ solves problem* (3.2) *and problem* (3.3).

(ii) *If u solves problem* (3.2), *then $|u|$ solves problem* (3.3) *and u^2 solves problem* (3.1).

(iii) *If v solves problem* (3.3), *then v solves problem* (3.2) *and v^2 solves problem* (3.1).

PROOF. The proof follows from Lemma 3.1 and the fact that if $v \geqq 0$, then

$$\Phi(v^{\frac{1}{2}}) = \tfrac{1}{2}\Phi_1(v)$$

and

$$\hat{L}_1(v) = \hat{L}(v^{\frac{1}{2}})^2\,.$$

COROLLARY 3.1. *If problem* (3.3) *has a unique solution, then problem* (3.1) *has a unique solution; and although problem* (3.2) *cannot have a unique solution, it will have solutions and the square of any of these solutions will give the unique solution of problem* (3.1).

The remainder of this section is dedicated to demonstrating that problem (3.3) has a unique solution which is a positive exponential spline with knots only at the sample points. The same will then be true of Good's and Gaskins' first maximum penalized estimate.

Along with problem (3.3) we will consider the constrained optimization problem obtained by only requiring nonnegativity at the sample points:

$$\text{maximize} \quad \hat{L}(v)\,; \quad \text{subject to}$$
$$\text{(3.4)} \qquad v \in H^1(-\infty, \infty)\,, \qquad \textstyle\int_{-\infty}^{\infty} v(t)^2\,dt = 1 \quad \text{and} \quad v(x_i) \geqq 0\,, \quad i = 1, \cdots, N\,.$$

Given $\lambda > 0$ and α in problem (3.3) we may also consider the constrained optimization problem:

(3.5)
$$\text{maximum} \quad \hat{L}_\lambda(v) = \prod_{i=1}^N v(x_i)\exp(-\Phi_\lambda(v)) ; \quad \text{subject to}$$
$$v \in H^1(-\infty, \infty) , \qquad \int_{-\infty}^{\infty} v(t)^2\,dt = 1 \quad \text{and} \quad v(x_i) \geqq 0 , \qquad i = 1, \cdots, N$$

where

$$\Phi_\lambda(v) = 2\alpha \int_{-\infty}^{\infty} v'(t)^2\,dt + \lambda \int_{-\infty}^{\infty} v(t)^2\,dt .$$

Our study of problem (3.5) will begin with the study of the following constrained optimization problem:

(3.6)
$$\text{maximize} \quad \hat{L}_\lambda(v) ; \quad \text{subject to}$$
$$v \in H^1(-\infty, \infty) \qquad \text{and} \qquad v(x_i) \geqq 0 , \qquad i = 1, \cdots, N$$

where $\hat{L}_\lambda$ is given by problem (3.5). Let $L^2 = L^2(-\infty, \infty)$.

PROPOSITION 3.2. *Problem* (3.6) *has a unique solution. Moreover if* v_λ *denotes this solution, then*

(i) v_λ *is a exponential spline with knots at the sample points* $x_1, \cdots, x_N$;
(ii) $v_\lambda(t) > 0$, $\forall\, t \in (-\infty, \infty)$; *and*
(iii) $\|v_\lambda\|_{L^2} \geqq (N/(4\lambda))^{\frac{1}{2}}$.

PROOF. From Lemma 2.2 $H^1(-\infty, \infty)$ is a reproducing kernel space. Also $\|v\|_\lambda{}^2 = \Phi_\lambda(v)$ gives a norm equivalent to the original norm on $H^1(-\infty, \infty)$. The existence of v_λ now follows from Proposition 2.1 with $D = \{v \in H^1(-\infty, \infty)$: $v(x_i) \geqq 0, i = 1, \cdots, N\}$. We will denote the Φ_λ inner product by $\langle\ ,\ \rangle_\lambda$. Let v_i be the representer in the Φ_λ inner product of the continuous linear functional given by point evaluation at the point x_i, $i = 1, \cdots, N$, i.e.

$$\langle v_i, \eta\rangle_\lambda = \eta(x_i) , \qquad \forall\, \eta \in H^1(-\infty, \infty) .$$

Equivalently

$$2\alpha \int_{-\infty}^{\infty} v_i'(t)\eta'(t)\,dt + \lambda \int_{-\infty}^{\infty} v_i(t)\eta(t)\,dt = \eta(x_i) , \qquad \forall\, \eta \in H^1(-\infty, \infty) .$$

Integrating by parts in the distribution sense gives

$$\int_{-\infty}^{\infty} [-2\alpha v_i''(t) + \lambda v_i(t)]\eta(t)\,dt = \eta(x_i) , \qquad \forall\, \eta \in H^1(-\infty, \infty) ;$$

hence

$$-2\alpha v_i'' + \lambda v_i = \delta_i , \qquad i = 1, \cdots, N \tag{3.7}$$

where $\delta_i(t) = \delta_0(t - x_i)$ and δ_0 denotes the Dirac distribution, i.e., $\int_{-\infty}^{\infty} \delta_0(t)\eta(t)\,dt = \eta(0)$. If we let v_0 be the solution of (3.7) for $i = 0$, then

$$v_0(t) = \frac{1}{2(2\alpha\lambda)^{\frac{1}{2}}} \exp((\lambda(2\alpha))^{\frac{1}{2}} t) , \quad t < 0$$
$$= \frac{1}{2(2\alpha\lambda)^{\frac{1}{2}}} \exp(-(\lambda/(2\alpha))^{\frac{1}{2}} t) , \quad t > 0$$

and $v_i(t) = v_0(t - x_i)$ for $i = 1, \cdots, N$. Since v_λ is the maximizer we have that $v_\lambda(x_i) > 0$, $i = 1, \cdots, N$ we necessarily have that the Fréchet derivative of $\hat{L}_\lambda$

at v_λ must be the zero functional; equivalently the gradient of $\hat{L}_\lambda$ or for that matter the gradient of $\log \hat{L}_\lambda$ must vanish at v_λ since $\hat{L}_\lambda$ and $\log \hat{L}_\lambda$ have the same maxima. A calculation similar to that used in the proof of Proposition 2.1 gives

$$\nabla_\lambda \log \hat{L}_\lambda(v) = 2v - \sum_{i=1}^{N} \frac{v_i}{v(x_i)} \tag{3.8}$$

where ∇_λ denotes the gradient. It follows from (3.8) that

$$v_\lambda = \tfrac{1}{2} \sum_{i=1}^{N} \frac{v_i}{v_\lambda(x_i)} . \tag{3.9}$$

Properties (i) and (ii) are now immediate. Since $\langle v_i, v_\lambda \rangle_\lambda = v_\lambda(x_i)$ from (3.9) we have

$$\|v_\lambda\|_\lambda^2 = N/2 . \tag{3.10}$$

A straightforward calculation shows that

$$v_i'(t) v_j'(t) \leqq \frac{\lambda}{2\alpha} v_i(t) v_j(t) , \qquad \text{for } i, j = 1, \cdots, N .$$

So

$$\begin{aligned} v_\lambda'(t)^2 &= \frac{1}{4}\left[\sum_i \left(\frac{v_i'(t)}{v_\lambda(x_i)}\right)^2 + \sum_{i,j} \frac{v_i'(t) v_j'(t)}{v_\lambda(x_i) v_\lambda(x_j)}\right] \\ &\leqq \frac{\lambda}{8\alpha}\left[\sum_i \left(\frac{v_i(t)}{v_\lambda(x_i)}\right)^2 + \sum_{i,j} \frac{v_i(t) v_j(t)}{v_\lambda(x_i) v_\lambda(x_j)}\right] = \frac{\lambda}{2\alpha} v_\lambda(t)^2 . \end{aligned}$$

Integrating in t gives

$$2\alpha \|v_\lambda'\|^2_{L^2(-\infty,\infty)} \leqq \lambda \|v_\lambda\|^2_{L^2(-\infty,\infty)} .$$

By definition of the Φ_λ-norm and (3.10) we have property (iii). This proves the proposition.

PROPOSITION 3.3. *Problem* (3.4) *has a unique solution.*

PROOF. Let $B = \{v \in H^1(-\infty, \infty) : \int_{-\infty}^{\infty} v(t)^2\, dt \leq 1$ and $v(x_i) \geqq 0, i = 1, \cdots, N\}$. Clearly B is closed and convex. If $\hat{L}_\lambda$ is given by (3.5), then by Proposition 2.1 the functional has a unique maximizer in B; say u_λ. Now by property (iii) of Proposition 3.2 if we choose $0 < \lambda < \frac{1}{4}$, then v_λ the unique solution of problem (3.6) will be such that $\|v_\lambda\|_{L^2(-\infty,\infty)}) > 1$. We will show that for this range of λ, $\|u_\lambda\|_{L^2(-\infty,\infty)} = 1$. Consider $v_\theta = \theta v_\lambda + (1 - \theta) u_\lambda$. We know that $\log \hat{L}_\lambda$ is a strictly concave functional (see the proof of Proposition 2.1). Moreover $\log \hat{L}_\lambda(v_\lambda) \geqq \log \hat{L}_\lambda(u_\lambda)$; hence $\log \hat{L}_\lambda(v_\theta) \geqq \log \hat{L}_\lambda(u_\lambda)$ for $0 < \theta < 1$. Now suppose $\|u_\lambda\|_{L^2(-\infty,\infty)} < 1$ and consider

$$g(\theta) = \|v_\theta\|_{L^2(-\infty,\infty)} .$$

We have $g(0) < 1$ and $g(1) > 1$. So for some $0 < \theta_0 < 1$, $g(\theta_0) = 1$ and $\log \hat{L}_\lambda(u_\lambda) \leq \log \hat{L}_\lambda(v_{\theta_0})$. This is a contradiction since u_λ is the unique maximizer of $\hat{L}_\lambda$ in B; hence $\|u_\lambda\|_{L^2(-\infty,\infty)} = 1$. This shows that u_λ is the unique solution of problem (3.5) for $0 < \lambda < \frac{1}{4}$. However, the term $\lambda \int_{-\infty}^{\infty} v(t)^2\,dt$ is constant over the constraint set in problems (3.4) and (3.5); hence problems (3.4) and (3.5) have the same solutions for any $\lambda > 0$. This proves the proposition since we have demonstrated that problem (3.3) has a unique solution for at least one λ.

PROPOSITION 3.4. *Problem* (3.3) *has a unique solution which is positive and an exponential spline with knots at the points* $x_1, \cdots, x_N$.

PROOF. If we can demonstrate that $\tilde{v}$ the unique solution of problem (3.4) has these properties we will be through. Let $G(v) = \log \hat{L}(v)$ where $\hat{L}$ is given in problem (3.3) and let

$$g(v) = \int_{-\infty}^{\infty} v(t)^2\,dt$$

for $v \in H^1(-\infty, \infty)$. Clearly $\tilde{v}(x_i) > 0$ for $i = 1, \cdots, N$; hance from the theory of Lagrange multipliers there exist λ such that $\tilde{v}$ satisfies the equations

$$(3.11) \qquad G'(v) - \lambda g'(v) = 0 \quad \text{and} \quad g(v) = 1\,.$$

Using $L^2(-\infty, \infty)$ gradients in the sense of distributions (3.11) is equivalent to

$$(3.12) \qquad -4\alpha v'' + 2\lambda v = \sum_{i=1}^{N} \frac{\delta_i}{v(x_i)} \quad \text{and} \quad g(v) = 1$$

where δ_i is the distribution such that $\int_{-\infty}^{\infty} v(t)\delta_i(t)\,dt = v(x_i)$, $i = 1, \cdots, N$. Since we have already established that problem (3.4) has a unique solution if follows that (3.12) must have a unique solution in $H^1(-\infty, \infty)$; namely $\tilde{v}$. If $\lambda \leqq 0$, then any solution of the first equation in (3.12) would be a sum of trigonometric functions and could not possibly satisfy the constraint $g(v) = 1$, i.e., cannot be contained in $L^2(-\infty, \infty)$. It follows that $\lambda > 0$. Now observe that

$$G - \lambda g = \log \hat{L}_\lambda$$

where $\hat{L}_\lambda$ is given by problem (3.5); hence if $\tilde{v}$ satisfies (3.11) (from the first equation alone) it must also be a solution of problem (3.6) for this λ and therefore has the desired properties according to Proposition 3.2. This proves the proposition.

PROPOSITION 3.5. *The first nonparametric maximum penalized likelihood estimate of Good and Gaskins exists and is unique*; *specifically the maximum penalized likelihood estimate corresponding to the penalty function*

$$\Phi(v) = \alpha \int_{-\infty}^{\infty} \frac{v'(t)^2}{v(t)}\,dt \quad (\alpha > 0)$$

and the manifold

$$H(\Omega) = \{v : v \geqq 0 \ \textit{and}\ v^{\frac{1}{2}} \in H^1(-\infty, \infty)\}$$

exists and is unique. Moreover the estimate is positive and an exponential spline with knots only at the sample points.

PROOF. The proof follows from Proposition 3.1 and Proposition 3.5.

4. The second maximum penalized likelihood estimator of Good and Gaskins. Consider the functional $\Phi : H^2(-\infty, \infty) \to R$ defined by

$$\Phi(v) = \alpha \int_{-\infty}^{\infty} v'(t)^2\, dt + \beta \int_{-\infty}^{\infty} v''(t)^2\, dt \tag{4.1}$$

for some $\alpha \geqq 0$ and $\beta > 0$. By a second maximum penalized likelihood estimate of Good and Gaskins we mean any solution of the following constrained optimization problem:

$$\begin{aligned} &\text{maximize} \quad \hat{L}_1(v) = \prod_{i=1}^{N} v(x_i) \exp(-\Phi(v^{\frac{1}{2}}))\,; \quad \text{subject to} \\ &v^{\frac{1}{2}} \in H^2(-\infty, \infty)\,, \quad \int_{-\infty}^{\infty} v(t)\, dt = 1 \quad \text{and} \quad v(t) \geqq 0 \quad \forall\, t \in (-\infty, \infty)\,. \end{aligned} \tag{4.2}$$

As in the first case (described in the previous section) Good and Gaskins suggest avoiding the nonnegativity constraint by calculating the solution of problem (4.2) from the following constrained optimization problem:

$$\begin{aligned} &\text{maximize} \quad \prod_{i=1}^{N} v(x_i)^2 \exp(-\Phi(v))\,; \quad \text{subject to} \\ &v \in H^2(-\infty, \infty) \quad \text{and} \quad \int_{-\infty}^{\infty} v(t)^2\, dt = 1 \end{aligned} \tag{4.3}$$

where Φ is given by (4.1).

Notice that if $u \in H^2(-\infty, \infty)$ is positive at some points and negative at other points, then $|u|$ is in general not a member of $H^2(-\infty, \infty)$, since in general its derivative is not continuous. It follows from Lemma 3.1 that the only way we can obtain solutions to problem (4.2) from solutions of problem (4.3) is for problem (4.3) to have solutions which are completely of one sign. However we will presently show that this is not the case.

Along with problem (4.3) we consider the constrained optimization problem:

$$\begin{aligned} &\text{maximize} \quad \hat{L}(v) = \prod_{i=1}^{N} v(x_i) \exp(-\tfrac{1}{2}\Phi(v))\,; \quad \text{subject to} \\ &v \in H^2(-\infty, \infty)\,, \quad \int_{-\infty}^{\infty} v(t)^2\, dt = 1 \quad \text{and} \quad v(x_i) \geqq 0\,, \\ &\qquad i = 1, \cdots, N\,. \end{aligned} \tag{4.4}$$

Problem (4.4) was obtained from problem (4.3) by taking the square root of the functional to be maximized (since it is nonnegative) and requiring nonnegativity at the sample points; hence the two problems only differ by the nonnegativity constraints at the sample points. This simple difference will allow us to establish uniqueness of the solution of problem (4.4); whereas problem (4.3) cannot have a unique solution. By Lemma 3.2 we can obtain the solution of problem (4.2) from the solution of problem (4.4) if and only if these solutions are nonnegative. Moreover we will presently demonstrate that the solutions of problem (4.4) are not necessarily nonnegative. It will then follow that we cannot obtain the second estimate by considering problem (4.4). If we naively use v_*^2, where v_* solves problem (4.4), as an estimate for the probability density function giving rise to the random sample $x_1, \cdots, x_N$, then clearly v_*^2 will be nonnegative and integrate to one and is therefore a probability density; however the estimate obtained in this manner will not in the strict sense of our definition be a maximum penalized likelihood estimate. For this reason we will refer to this latter estimate as the *pseudo maximum penalized likelihood estimate* of Good and Gaskins.

The next six propositions are needed to show that the second maximum penalized likelihood estimate and the pseudo maximum penalized likelihood estimate of Good and Gaskins exist, are unique and are distinct, and that Good's and Gaskins' alternate approach cannot be used to obtain their second maximum penalized likelihood estimate.

PROPOSITION 4.1. *The second maximum penalized likelihood estimate and the pseudo maximum likelihood estimate of Good and Gaskins are distinct.*

PROOF. We will show that it is possible for problem (4.4) to have solutions which are not nonnegative. Toward this end let $N = 1$, $x_1 = 0$, $\alpha = 0$, and $\beta = 2$. Let $G(v) = \log \hat{L}(v)$, i.e.,

$$G(v) = \log v(0) - \int_{-\infty}^{\infty} v''(t)^2\,dt$$

and let

$$g(v) = \int_{-\infty}^{\infty} v(t)^2\,dt\,.$$

As in the proof of Proposition 3.4 using the theory of distributions and the theory of Lagrange multipliers we see that the solutions of problem (4.4) in this case are exactly the solutions of

$$v^{(iv)} + \lambda v = \frac{\delta_1}{2v(0)} \qquad \text{and} \qquad g(v) = 1 \tag{4.5}$$

where δ_1 is defined in the proof of Proposition 3.4. If we let $\tilde{v}$ denote the Fourier transform of v, then taking the Fourier transform of the first expression in (4.5) gives

$$\tilde{v}(\omega) = [2v(0)(\lambda + 16\Pi^4\omega^4)]^{-1}\,.$$

Since $||\tilde{v}||_{L^2(-\infty,\infty)} = ||v||_{L^2(-\omega,\omega)} = 1$ we must have

$$\int_{-\infty}^{\infty} \frac{d\omega}{(\lambda + 16\Pi^4\omega^4)^2} = 4v(0)^2\,. \tag{4.6}$$

For the integral in (4.6) to exist we must have $\lambda > 0$. Now the inverse Fourier transform of $(\lambda + 16\Pi^4\omega^4)^{-1}$ is given by v where

$$\begin{aligned} v(t) &= \frac{e^{bt}}{8b^3}[\cos bt - \sin bt]\,, \quad t \leq 0 \\ &= \frac{e^{-bt}}{8b^3}(\cos bt + \sin bt]\,, \quad t > 0 \end{aligned} \tag{4.7}$$

with $b = \lambda^{\frac{1}{4}}/2^{\frac{1}{2}}$. From (4.7) $v(0) = (8b^3)^{-1}$ and from (4.6) $v(0)^2 = \frac{1}{4}\lambda^{-\frac{3}{4}}K$ where $K = ||(1 + 16\Pi^4\omega^4)^{-1}||_{L^2(-\infty,\infty)}$. Hence $\lambda^{\frac{3}{4}} = 2K$ and $b = 2^{\frac{1}{2}}K$ which is clearly not nonnegative. This proves the proposition.

In the above proof we assumed the existence of a solution. At that point we did not know that this was so; however our proof is valid since in Proposition 4.4 we will show that problem (4.4) always has a unique solution.

COROLLARY 4.1. *Problem* (4.3) *has solutions which are not nonnegative.*

PROOF. The solution constructed in the proof of Proposition 4.1 also solves problem (4.3). To see this observe that if v satisfies the constraints of problem (4.3), then so does $-v$ and the functional values are the same. Also either v or $-v$ will satisfy the constraints of problem (4.4) since in the present case there is only one sample.

COROLLARY 4.2. *Good's and Gaskins' alternate approach cannot be used to obtain their second maximum penalized likelihood estimate.*

Given $\lambda > 0$ consider the constrained optimization problem:

$$\text{maximize} \quad \hat{L}_\lambda(v) = \prod_{i=1}^{N} v(x_i) \exp(-\Phi_\lambda(v)) \; ; \quad \text{subject to}$$

$$(4.8) \qquad v \in H^2(-\infty, \infty) , \qquad \int_{-\infty}^{\infty} v(t)^2 \, dt = 1 \quad \text{and} \quad v(x_i) \geqq 0 , \quad i = 1, \cdots, N .$$

where

$$\Phi_\lambda(v) = \Phi(v) + \lambda \int_{-\infty}^{\infty} v(t)^2 \, dt$$

with $\Phi(v)$ given by (4.1).

As before we also consider the constrained optimization problem obtained by dropping the integral constraint:

$$(4.9) \qquad \text{maximize} \quad \hat{L}_\lambda(v) \; ; \quad \text{subject to}$$
$$v \in H^2(-\infty, \infty) \qquad \text{and} \qquad v(x_i) \geqq 0 , \qquad i = 1, \cdots, N .$$

PROPOSITION 4.2. *Problem* (4.9) *has a unique solution. Moreover if v_λ denotes this solution, then*

$$||v_\lambda||_{L^2(-\infty,\infty)} \to +\infty \qquad as \quad \lambda \to 0 .$$

PROOF. By Lemma 2.1 the Sobolev space $H^2(-\infty, \infty)$ is a reproducing kernel space. Moreover, if

$$||v||_\lambda^2 = \Phi_\lambda(v) ,$$

then an integration by parts gives

$$(4.10) \qquad \begin{aligned} ||v'||_{L^2}^2 &= |\langle v, v'' \rangle_{L^2}| \leqq ||v||_{L^2} ||v''||_{L^2} \\ &\leqq \tfrac{1}{2}[||v||_{L^2}^2 + ||v''||_{L^2}^2] \end{aligned}$$

where L^2 denotes $L^2(-\infty, \infty)$; hence $||\cdot||_\lambda$ is equivalent to the original norm on $H^2(-\infty, \infty)$. The existence and uniqueness of v_λ now follows from Proposition 2.1.

We must now show that $||v_\lambda||_{L^2} \to +\infty$ as $\lambda \to 0$. From the fundamental theorem of calculus we have

$$(4.11) \qquad \begin{aligned} v(x)^2 &= \int_{-\infty}^{x} \frac{dv(t)^2}{dt} \, dt = 2 \int_{-\infty}^{x} v(t)v'(t) \, dt \\ &\leqq 2||v||_{L^2} ||v'||_{L^2} . \end{aligned}$$

Also, $||v''||_{L^2} \leqq ||v||_\lambda/\beta^{\frac{1}{2}}$ so that from (4.10) and (4.11)

$$v(x)^2 \leqq 2||v||^{\frac{3}{2}}_{L^2}(||v||_\lambda/\beta^{\frac{1}{2}}) . \tag{4.12}$$

Evaluating (4.12) at x_i, taking logs (since $v(x_i) \geqq 0$) and summing over i gives

$$\sum_{i=1}^N \log v(x_i) \leqq \frac{N}{4} \log \left(\frac{4}{\beta^{\frac{1}{2}}} ||v||_\lambda\right) + \frac{3N}{4} \log (||v||_{L^2}) . \tag{4.13}$$

Hence from (4.13) we see that

$$\log \hat{L}_\lambda(v) \leqq \frac{3N}{4} \log (||v||_{L^2}) + \frac{N}{4} \log \left(\frac{4}{\beta^{\frac{1}{2}}} ||v||_\lambda\right) - ||v||_\lambda^2 . \tag{4.14}$$

In a manner exactly the same as that used to establish (3.10) we have that $||v_\lambda||_\lambda^2 = N/2$. Hence from (4.14) and the fact that $\log \hat{L}_\lambda(v) \leqq \log \hat{L}_\lambda(v_\lambda)$ we obtain

$$\log \hat{L}_\lambda(v) \leqq \frac{3N}{4} \log (||v_\lambda||_{L^2}) + \frac{N}{8} \log (8N/\beta) - \frac{N}{2} , \tag{4.15}$$

for any $v \in \{u \in H^2(-\infty, \infty) : u(x_i) \geqq 0, i = 1, \cdots, N\}$.

Let a and b be such that

$$a < \min_i (x_i) \quad \text{and} \quad \max_i (x_i) < b .$$

Given $\lambda > 0$ and ε and δ define the function θ_λ in the following piecewise fashion:

$$\begin{aligned} \theta_\lambda(t) &= \lambda^\varepsilon \exp(-(t-a)^2/2\sigma^2) && \text{for } t \in (-\infty, a) \\ &= \lambda^\varepsilon && \text{for } t \in [a, b] \\ &= \lambda^\varepsilon \exp(-(t-b)^2/2\sigma^2) && \text{for } t \in (b, +\infty) \end{aligned}$$

where $\sigma = \lambda^\delta$. Straightforward calculations can be used to show

$$\begin{aligned} \log \left(\textstyle\prod_{i=1}^N \theta_\lambda(x_i)\right) &= \varepsilon N \log (\lambda) , \\ ||\theta_\lambda||^2_{L^2} &= (b-a)\lambda^{2\varepsilon} + ((\Pi\lambda)^{\frac{1}{2}})^{2\varepsilon+\delta} \\ ||\theta_\lambda'||^2_{L^2} &= ((2\Pi\lambda)^{\frac{1}{2}})^{2\varepsilon-\delta} , \\ ||\theta_\lambda''||^2_{L^2} &= 2((2\Pi\lambda)^{\frac{1}{2}})^{2\varepsilon-3\delta} , \end{aligned}$$

and

$$||\theta_\lambda||_\lambda^2 = (b-a)\lambda^{2\varepsilon+1} + ((\Pi\lambda)^{\frac{1}{2}})^{2\varepsilon+\delta+1} + 4\alpha((2\Pi\lambda)^{\frac{1}{2}})^{2\varepsilon-\delta} + 2\beta((2\Pi\lambda)^{\frac{1}{2}})^{2\varepsilon-3\delta} . \tag{4.16}$$

If we want $||\theta_\lambda||_\lambda^2 \to 0$ as $\lambda \to 0$ it is sufficient to choose all exponents of λ in (4.16) positive. If we also want

$$\log \left(\textstyle\prod_{i=1}^N \theta_\lambda(x_i)\right) \to +\infty \quad \text{as} \quad \lambda \to 0$$

we should choose $\varepsilon < 0$. This leads to the inequalities

$$\begin{aligned} 2\varepsilon + 1 &> 0 \\ 2\varepsilon + \delta + 1 &> 0 \\ 2\varepsilon - \delta &> 0 \\ 2\varepsilon - 3\delta &> 0 \\ \varepsilon &< 0 . \end{aligned} \tag{4.17}$$

The system of inequalities (4.17) has solutions; specifically $\varepsilon = -\frac{1}{32}$ and $\delta = -\frac{1}{8}$ is one such solution. With this choice of ε and δ we see that $\log \hat{L}_\lambda(\theta_\lambda) \to +\infty$ as $\lambda \to 0$. It follows from (4.15) by choosing $v = \theta_\lambda$ that $\|v_\lambda\|_{L^2} \to +\infty$ as $\lambda \to 0$. This proves the proposition.

PROPOSITION 4.3. *Problem* (4.3) *has a unique solution.*

PROOF. By Proposition 4.2 there exists $\lambda > 0$ such that if v_λ is the unique solution of problem (4.9), then $\|v_\lambda\|_{L^2} > 1$. Now, if $B = \{v \in H^2(-\infty, \infty) : \int_{-\infty}^{\infty} v(t)^2\, dt \leqq 1 \text{ and } v(x_i) \geqq 0,\ i = 1, \cdots, N\}$, then B is closed and convex. The proof of the proposition is now exactly the same as the proof of Proposition 3.3.

PROPOSITION 4.4. *The pseudo maximum penalized likelihood estimate of Good and Gaskins exists and unique.*

PROOF. Since problems (4.4) and (4.8) have the same solutions the proposition follows from Proposition 4.3.

By the change of unknown function $v \to v^{\frac{1}{2}}$ we see that problem (4.2) is equivalent to the following constrained optimization problem:

$$\text{maximize} \quad \hat{L}(v) = \prod_{i=1}^{N} v(x_i) \exp(-\tfrac{1}{2}\Phi(v)) \; ; \quad \text{subject to}$$

$$(4.18) \qquad v \in H^2(-\infty, \infty)\,, \quad \int_{-\infty}^{\infty} v(t)^2\, dt = 1 \quad \text{and} \quad v(t) \geqq 0 \quad \forall\, t \in (-\infty, \infty)$$

where $\Phi(v)$ is given by (4.1).

In turn for $\lambda > 0$ problem (4.18) is equivalent to

$$\text{maximize} \quad \hat{L}_\lambda(v) \; ; \quad \text{subject to}$$

$$(4.19) \qquad v \in H^2(-\infty, \infty)\,, \quad \int_{-\infty}^{\infty} v(t)^2\, dt = 1 \quad \text{and} \quad v(t) \geqq 0 \quad \forall\, t \in (-\infty, \infty)$$

where $\hat{L}_\lambda$ is defined in problem (4.8).

As in the previous two cases we also consider the constrained optimization problem:

$$(4.20) \qquad \text{maximize} \quad \hat{L}_\lambda(v) \; ; \quad \text{subject to} \quad v \in H^2(-\infty, \infty) \quad \text{and} \quad v(t) \geqq 0 \quad \forall\, t \in (-\infty, \infty)$$

where $\hat{L}_\lambda(v)$ is defined in problem (4.8).

PROPOSITION 4.5. *Problem* (4.20) *has a unique solution. Moreover if* v_λ^+ *denotes this solution, then*

$$\|v_\lambda^+\|_{L^2} \to +\infty \qquad as \quad \lambda \to 0\,.$$

PROOF. The existence of v_λ^+ follows from Proposition 2.1 as in the proof of Proposition 4.2. Let us first show that

$$(4.21) \qquad \|v_\lambda^+\|_\lambda \leqq (N/2)^{\frac{1}{2}}\,.$$

From Lions (1968) we see that

$$(4.22) \qquad \hat{L}_\lambda'(v_\lambda^+)(\eta - v_\lambda^+) \leqq 0$$

for all nonnegative η in $H^2(-\infty, \infty)$. We have

$$\hat{L}_\lambda'(v)(\eta) = \sum_{i=1}^N \frac{\eta(x_i)}{v(x_i)} - 2\langle v, \eta \rangle_\lambda ;$$

hence

$$\hat{L}_\lambda'(v_\lambda^+)(v_\lambda^+) = N - 2\|v_\lambda\|_\lambda^2 . \tag{4.23}$$

Now choosing $\eta = 0$ in (4.22) and using (4.23) we arrive at (4.21). The functions θ_λ defined in the proof of Proposition 4.2 satisfy the constraints of this problem; hence

$$\log \hat{L}_\lambda(\theta_\lambda) \leqq \log \hat{L}_\lambda(v_\lambda^+) .$$

From (4.14) and (4.21) we have

$$\log \hat{L}_\lambda(\theta_\lambda) \leqq \frac{3N}{4} \log (\|v_\lambda^+\|_{L^2}) + \frac{N}{8} \log (8N/\beta) + \frac{N}{2} . \tag{4.24}$$

The proof now follows from (4.24) since $\log L_\lambda(\theta_\lambda) \to +\infty$ as $\lambda \to 0$.

PROPOSITION 4.6. *The second maximum penalized likelihood estimate of Good and Gaskins exists and is unique.*

PROOF. Using Proposition 4.5 the argument used to prove Proposition 4.3 shows that problem (4.19) has a unique solution which is also the unique solution of problem (4.18). This proves the proposition.

Acknowledgments. The authors would like to thank R. H. Byrd, B. F. Jones, P. E. Pfeiffer and W. A. Veech for helpful discussions. They also sincerely thank the referee and the editor for comments on the manuscript which led to a more readable paper.

REFERENCES

[1] BONEVA, L., KENDALL, D and STEFANOV, I. (1971). Spline transformations: Three new diagnostic aids for the statistical data-analyst. *J. Roy. Statist. Soc. Ser. B* **33** 1–77.

[2] FIACCO, A. V., and MCCORMICK, G. P. (1968). *Nonlinear Programming: Sequential Unconstrained Minimization Techniques.* Wiley, New York.

[3] GOFFMAN, C., and PEDRICK, G. (1965). *First Course in Functional Analysis.* Prentice-Hall, New Jersey.

[4] GOOD, I. J., and GASKINS, R. A. (1971). Nonparametric roughness penalties for probability densities. *Biometrika* **58** 255–277.

[5] KIMELDORF, G. S. and WAHBA, G. (1970). A correspondence between Bayesian estimation on stochastic processes and smoothing by splines. *Ann. Math. Statist.* **41** 495–502.

[6] LIONS, J. S. (1968). *Contrôle Optimal de Systèmes Gouvernés par des Équations aux Dérivées Partielles.* Dunod, Paris.

[7] LIONS, J. L. and MAGENES, E. (1968). *Problèmes aux Limites Non Homogènes et Applications,* **1** Dunod, Paris.

[8] PARZEN, E. (1962). On estimation of a probability density function and mode. *Ann. Math. Statist.* **33** 1065–1076.

[9] ROSENBLATT, M. (1956). Remarks on some nonparametric estimates of a density function. *Ann. Math. Statist.* **27** 832–837.

[10] SCHOENBERG, I. J. (1968). Monosplines and quadrature formulae. *Theory and Application of Spline Functions* (T. N. E. Greville, ed.). Academic, New York.

[11] SCHOENBERG, I. J. (1972). Splines and histograms, with an appendix by Carl de Boor. Mathematics Research Center Report 1273, University of Wisconsin, Madison.

[12] TAPIA, R. A. (1971). The differentiation and integration of nonlinear operators. *Nonlinear Functional Analysis and Applications* (Louis B. Rall, ed.). Academic, New York.

[13] WAHBA, G. (1971). A polynomial algorithm for density estimation. *Ann. Math. Statist.* **42** 1870–1886.

G. F. DE MONTRICHER
54 RUE DU TOUC
3100 TOULOUSE
FRANCE

R. A. TAPIA AND J. R. THOMPSON
DEPARTMENT OF MATHEMATICAL SCIENCES
RICE UNIVERSITY
HOUSTON, TEXAS 77001

30

Reprinted from *Inform. Control* **11**:429–444 (1967)

Applications of Reproducing Kernel Hilbert Spaces—Bandlimited Signal Models*

K. YAO

Department of Engineering, University of California, Los Angeles, California 90024

The finite energy Fourier-, Hankel-, sine-, and cosine-transformed bandlimited signals are specific realizations of the abstract reproducing kernel Hilbert space (RKHS). Basic properties of the abstract RKHS are applied to the detailed study of bandlimited signals. The relevancy of the reproducing kernel in extremum problems is discussed. New and known results in sampling expansions, minimum energy and non-uniform interpolations, and truncation error bounds are presented from a unified point of view of the RKHS. Some generalizations and extensions are stated.

LIST OF SYMBOLS

H, H_i	Reproducing Kernel Hilbert Spaces (RKHS)
H', H''	Subspaces of RKHS
x, f, f_0, g, h	Functions in RKHS
K, K_i	Reproducing kernels
Y, Ω	Spaces of functions
T	Index set (continuum; time)
s, t, t_i	Elements in T
$F(\omega)$	Function in the transformed domain
$M, M_i, E, C, A, c_i, c, m_i$	Constants
G_i, D_i, d_i	$n \times n$ determinants
$\mathbf{K}$	$n \times n$ matrix
λ, λ_j	Eigenvalues
$\boldsymbol{\theta}, \boldsymbol{\theta}_j, \hat{\mathbf{f}}$	$n \times 1$ column vectors
$\boldsymbol{\theta}^T, \boldsymbol{\theta}_j^T, \mathbf{f}^T$	$1 \times n$ row vector
$\theta_{j,i}$	i^{th} element of $\boldsymbol{\theta}_j$
ϕ_i	i^{th} element of a c.o.n. system

* This work was supported by NASA-JPL Contract No. 951733. Part of this work was done while the author was a NAS-NRC Postdoctoral Research Fellow, 1965–1966, at The University of California, Berkeley.

ψ_i	i^{th} sampling function
J_ν	Bessel function of the first kind and order ν, $\nu \geqq -\frac{1}{2}$
L	Bound linear functional
I, I'	Index set (integral)
$i_0(s)$	Nearest integer to s
$E_{I'}$	Truncation error

In communication and information theories, bandlimited (deterministic and random) signal models are used foı analysis and representations. These models are used because often they represent fairly well the actual signals encountered in practice. Furthermore, many mathematical properties can and have been derived from these models. This paper deals with some applications of reproducing kernel Hilbert space methods to bandlimited signal models.

The basic mathematical properties of the reproducing kernel Hilbert space (henceforth abbreviated as RKHS) were studied by Moore (1935), Bergman (1950), and Aronszajn (1950). Applications of RKHS methods to second-order stochastic processes were given by Loève (1948). RKHS methods have been found useful in time series, detection, filtering, and prediction problems. (Parzen, 1961, 1962 and Kailath, 1967).

In Section I, two equivalent definitions of the abstract RKHS are stated. Finite-energy, bandlimited signals associated with Fourier-, Hankel-, sine-, and cosine-transforms are shown to be specific realizations of the abstract RKHS.

In Section II, the relevancy of the reproducing kernel in some extremum problems is discussed. For a given single sampling instant, t, the minimum energy signal that satisfies an interpolation requirement at t and the signals with a given upper energy bound that maximize the square of the signals' value at t are easily obtained. When there are n distinct sampling instants, the solutions to these two problems are obtained from orthonormalization and the solution of a classical eigenvalue equation involving the reproducing kernel indexed by the sampling instants.

In Section III, various properties of sampling expansions in RKHS are discussed. A simple relationship between sampling expansion and complete orthonormal expansion is obtained. Sampling expansions for the four classes of bandlimited signals are stated. General as well as specific (Shannon sampling expansion) truncation error bounds are calculated. Finally, necessary and sufficient conditions are given for a finite expansion of the reproducing kernel indexed by the sampling instants to satisfy the minimum energy and interpolation propetries.

The results in Sections I, II, and III are often stated in the simplest. and not in the most general, terms. Various generalizations and extensions are discussed in Section IV.

I. PRELIMINARY

A Hilbert space is a complete infinite-dimensional inner-product space. The elements of this space can be functions defined on a set T. In particular, the abstract reproducing kernel Hilbert space (RKHS), H, is a Hilbert space of functions defined on a set T such that there exists a unique function, $K(s, t)$, defined on $T \times T$ with the following properties:

$$K(\cdot, t) \in H, \qquad \forall t \in T. \tag{1}$$

$$x(t) = (x, K(\cdot, t)), \qquad \forall t \in T, \quad \forall x \in H. \tag{2}$$

The function $K(s, t)$ is called the reproducing kernel of the abstract RKHS. (Aronszajn, 1950).

An equivalent definition of the abstract RKHS can also be given. The abstract proper functional Hilbert space, H, is a Hilbert space of functions defined on a set T such that the linear functional, $x(t)$, is bounded for every $x \in H$ and every $t \in T$. From the Riesz linear functional representation theorem, it is clear that the abstract reproducing kernel Hilbert space is equivalent to the abstract proper functional Hilbert space. For simplicity, we assume the scalars and the functions are all real-valued in the RKHS.

In signal analysis, the signals are often characterized in terms of some properties in the Fourier-transformed domain. The class of finite-energy, Fourier-transformed bandlimited signals can easily be shown to be a specific realization of the abstract RKHS. The reproducing kernel is given by the familiar sinc function.

THEOREM 1. *Let H_1 be the class of $L_2(-\infty, \infty)$ functions such that their Fourier-transforms*

$$F(\omega) = \underset{A\to\infty}{\text{l.i.m.}} \int_{-A}^{A} f(t)e^{-i\omega t}\,dt, \qquad f \in L_2(-A, A),$$

vanish almost everywhere outside of $(-\pi, \pi)$. Then H_1 is a reproducing kernel Hilbert space on $T_1 = (-\infty, \infty)$. The unique reproducing kernel

$K_1(s, t)$ is given by

$$K_1(s, t) = [\sin \pi(t - s)]/\pi(t - s). \tag{3}$$

Proof. From the inverse Fourier-transform, any $f \in H_1$ is given by

$$f(t) = \text{l.i.m.} \frac{1}{2\pi} \int_{-\pi}^{\pi} F(\omega) e^{i\omega t} \, d\omega, \qquad F \in L_2(-\pi, \pi).$$

If $F \in L_2(-\pi, \pi)$, then $F \in L_1(-\pi, \pi)$. Schwarz inequality shows

$$|f(t)| \leqq \frac{1}{2\pi} \left[\int_{-\pi}^{\pi} |F(\omega)|^2 \, d\omega \right]^{1/2} \left[\int_{-\pi}^{\pi} |e^{i\omega t}|^2 \, d\omega \right]^{1/2} < \infty,$$

for any $f \in H_1$ and any finite t. Thus, H_1 is a proper functional Hilbert space. Then H_1 is a specific realization of the abstract RKHS. The reproducing kernel given by Eq. (3) is obtained after applying the convolution theorem to the inverse Fourier-transform of the indicator function of $(-\pi, \pi)$.

In two-dimensional signal analysis, the signals are sometimes characterized in terms of some properties in the Bessel-transformed domain. The classes of finite-energy, Bessel-, sine-, and cosine-transformed band-limited signals can be shown to be specific realizations of the abstract RKHS in the same manner as that of Theorem 1.

COROLLARY 1. *Let H_2 be the class of $L_2(0, \infty)$ functions such that their Hankel-transforms of order ν, $\nu \geqq -\frac{1}{2}$,*

$$F(\omega) = \underset{A \to \infty}{\text{l.i.m.}} \int_0^A (\omega t)^{1/2} J_\nu(\omega t) f(t) \, dt, \qquad f \in L_2(0, A),$$

vanish almost everywhere outside of $(0, \pi)$. [The inverse Bessel-transform is given by

$$f(t) = \text{l.i.m.} \int_0^{\pi} (\omega t)^{1/2} J_\nu(\omega t) F(\omega) \, d\omega, \qquad t \in (0, \infty) F \in L_2(0, \pi) \Big].$$

Then H_2 is a reproducing kernel Hilbert space on $T_2 = (0, \infty)$. The unique reproducing kernel is given by

$$K_2(s, t) = \frac{\pi (ts)^{1/2}}{s^2 - t^2} \left(t J_\nu(s\pi) J_\nu'(t\pi) - s J_\nu(t\pi) J_\nu'(s\pi) \right).$$

COROLLARY 2. *Let H_3 be the class of $L_2(0, \infty)$ functions such that their* *sine-transforms*

$$F(\omega) = \underset{A\to\infty}{\text{l.i.m.}} \left(\frac{2}{\pi}\right)^{1/2} \int_0^A \sin \omega t f(t)\, dt, \qquad f \in L_2(0, \infty),$$

vanish almost everywhere outside of $(0, \pi)$. [*The inverse sine-transform is given by*

$$\left. f(t) = \text{l.i.m.} \left(\frac{2}{\pi}\right)^{1/2} \int_0^\pi \sin \omega t F(\omega)\, d\omega, t \in (0, \infty), F \in L_2(0, \pi) \right].$$

Then H_3 *is a reproducing kernel Hilbert space on* $T_3 = (0, \infty)$. *The unique reproducing kernel is given by*

$$K_3(s, t) = \frac{1}{\pi}\left(\frac{\sin \pi(t-s)}{(t-s)} - \frac{\sin \pi(t+s)}{(t+s)}\right).$$

COROLLARY 3. *Let* H_4 *be the class of* $L_2(0, \infty)$ *functions such that their cosine-transforms*

$$F(\omega) = \underset{A\to\infty}{\text{l.i.m.}} \left(\frac{2}{\pi}\right)^{1/2} \int_0^A \cos \omega t f(t)\, dt, \qquad f \in L_2(0, A),$$

vanish almost everywhere outside of $(0, \pi)$. [*The inverse cosine-transform is given by*

$$\left. f(t) = \text{l.i.m.} \left(\frac{2}{\pi}\right)^{1/2} \int_0^\pi \cos \omega t F(\omega)\, d\omega,\ t \in (0, \infty), F \in L_2(0, \pi) \right].$$

Then H_4 *is a reproducing kernel Hilbert space on* $T_4 = (0, \pi)$. *The unique reproducing kernel* $K_4(s, t)$ is given by

$$K_4(s, t) = \frac{1}{\pi}\left(\frac{\sin \pi(t-s)}{(t-s)} + \frac{\sin \pi(t+s)}{(t+s)}\right).$$

II. EXTREMUM PROBLEMS

In a given class of signals there are certain properties of the signals which are obtained from the solutions of extremum problems. If this class of signals is a RKHS, then the reproducing kernel plays an important role in these extremum problems. Suppose K is the reproducing kernel of a RKHS H, where H is a subspace of a Hilbert space Y. Then it turns out that the projection of any $y \in Y$ onto H is given by the inner product of y and K. (Aronszajn, 1950). For example, in H_1, the class of

Fourier bandlimited signals, maximum concentration properties of these signals over finite time intervals were obtained from self-adjoint compact integral equations where the kernel of the integral operator is given by the reproducing kernel $K_1(s, t)$. (Slepian and Pollak, 1961 and Landau and Pollak, 1961). In this section we shall consider two extremum problems in the abstract RKHS in which the sampling instants are specified. In these problems the reproducing kernel of the space again plays an important role

First consider a simple version of these two problems. Suppose t is a fixed point in the set T of the abstract RKHS H. What signal $f \in H$ with the specified value $f(t) = M$, where M is a real constant, has the smallest energy $\|f\|^2$? On the other hand, what signals $f \in H$ with energy $\|f\|^2 \leqq E$ have the maximum value for $f^2(t)$?

From Eq. (2) and the Schwarz inequality, it is clear that the solutions to both problems are the same. In the first case, the signal $f(s) = MK(s, t)/K(t, t)$ has the minimum energy $\|f\|^2 = M^2$ in the subspace of H with the constraint of $f(t) = M$. (Note that $K(t, t) = \|K\|^2 \neq 0$, unless K vanishes identically and thus H is void.) In the second case, $f(s) = \pm E^{1/2}K(s, t)/K(t, t)$ has the maximum value $f^2(t) = E$ in the subspace of H with the constraint of $\|f\|^2 \leqq E$.

Now consider the above two problems when there are n distinct, but arbitrarily-specified sampling instants, $t_i \in T$, $i = 1, \cdots, n$, and n real, finite, but arbitrarily-specified sampled values, M_i, $i = 1, \cdots, n$. Then the above two problems are generally not equivalent. Theorem 2 and Theorem 3 deal with these two problems.

In the first problem what signal $f \in H$ satisfying $f(t_i) = M_i$, $i = 1, \cdots, n$, (when t_i and M_i are specified), has the smallest energy $\|f\|^2$? The solution to a slightly extended version of this problem is given by Theorem 2 as the unique signal $f \in H$ which interpolates over a finite number of points and approximates any other specified signal $g \in H$ with minimum energy. When $g = 0$, Theorem 2 reduces to the first problem. The proof of Theorem 2 is based on the well-known Gram-Schmidt orthonormalization procedure so often used in interpolation theory. The proof of Theorem 2 is omitted since it is similar to Theorems 9.4.1 and 9.4.3 of Davis (1963).

THEOREM 2. *Consider the abstract reproducing kernel Hilbert space H and the reproducing kernel $K(s, t)$ defined on a set T. Let n be any finite positive integer, the sampling instants $\{t_1, \cdots, t_n\}$ be any set of finite distinct* *points in T, the sample values $\{M_1, \cdots, M_n\}$ be any set of real finite constants, and any $g \in H$. Denote H', the subspace of H under the interpolation constraints, to be*

$$H' = \{f \in H : f(t_i) = M_i, i = 1, \cdots, n\}.$$

Then

$$f_0(s) = g(s) + \sum_{i=1}^{n} d_i D_i(s)$$

is the unique element in H' that attains

$$\min_{f \in H'} \| f - g \|^2 = \| f_0 - g \|^2 = \sum_{i=1}^{n} d_i^2,$$

where for $i = 1, \cdots, n$,

$$d_i = \frac{1}{(G_{i-1} G_i)^{1/2}} \begin{vmatrix} K(t_1, t_1) & \cdots & K(t_n, t_1) \\ \vdots & & \\ K(t_1, t_{n-1}) & \cdots & K(t_n, t_{n-1}) \\ m_1 & \cdots & m_n \end{vmatrix},$$

$$m_i = M_i - g(t_i)$$

$$G_i = \det [K(t_j, t_k)]_{j,k=1,\cdots,i}, \qquad G_0 = 1$$

$$D_i(s) = \frac{1}{(G_{i-1} G_i)^{1/2}} \begin{vmatrix} K(t_1, t_1) & \cdots & K(t_n, t_1) \\ \vdots & & \\ K(t_1, t_{n-1}) & \cdots & K(t_n, t_{n-1}) \\ K(s, t_1) & \cdots & K(s, t_n) \end{vmatrix}.$$

In the second problem, for specified distinct sampling instants t_i, $i = 1, \cdots, n$, what signals $f \in H$ with energy $\| f \|^2 \leqq E$ yield the maximum value of $\sum_{i=1}^{n} f^2(t_i)$?

THEOREM 3. *Consider the abstract reproducing kernel Hilbert space H with the reproducing kernel $K(s, t)$ defined on a set T. Let n be any finite positive number and the sampling instants, $\{t_1 \cdots, t_n\}$, be any set of finite distinct points in T. Denote H'', the subspace of H under the energy constraint, to be*

$$H'' = \{f \in H : \| f \|^2 \leqq E\},$$

where E is a finite positive number. Then

$$f_0(s) = \pm\left[E/\lambda_n\left(\sum_{j=1}^{n} \theta_{n,j}^2\right)\right]^{1/2} \sum_{j=1}^{n} \theta_{n,j} K(s, t_j) \tag{4}$$

are elements in H'' that attain

$$\max_{f \in H''} \sum_{i=1}^{n} f^2(t_i) = \sum_{i=1}^{n} f_0^2(t_i) = \lambda_n E, \tag{5}$$

where λ_n is the largest eigenvalue and $\boldsymbol{\theta}_n$ is the corresponding eigenvector of the matrix equation

$$\begin{aligned} \mathbf{K}\boldsymbol{\theta} &= \lambda\boldsymbol{\theta}, \quad \textit{where} \\ \mathbf{K} &= [K(t_i, t_j)]_{i,j=1,\cdots,n}, \\ \boldsymbol{\theta}_i^T &= [\theta_{i,1}, \cdots, \theta_{i,n}], \qquad i = 1, \cdots, n. \end{aligned} \tag{6}$$

Proof. From the defining property of the reproducing kernel and the Schwarz inequality,

$$\begin{aligned} \left[\sum_{i=1}^{n} f^2(t_i)\right]^2 &= \left[\sum_{i=1}^{n} f(t_i), (f(s), K(s, t_i))\right]^2 \\ &= \left[\left(f(s), \sum_{i=1}^{n} f(t_i)K(s, t_i)\right)\right]^2 \\ &\leqq \|f\|^2 \left\|\sum_{i=1}^{n} f(t_i)K(s, t_i)\right\|^2 \\ &= \|f\|^2 \sum_{i=1}^{n}\sum_{j=1}^{n} f(t_i)f(t_j)K(t_i, t_j). \end{aligned} \tag{7}$$

In particular,

$$\frac{\sum_{i=1}^{n} f^2(t_i)}{\|f\|^2} \leqq \max_{\substack{f(t_i), i=1,\cdots,n \\ f \in H}} \frac{\sum_{i=1}^{n}\sum_{j=1}^{n} f(t_i)f(t_j)K(t_i, t_j)}{\sum_{i=1}^{n} f^2(t_i)} = \lambda_n, \tag{8}$$

where λ_n is a positive constant. Let $f_1(t_i)$, $i = 1, \cdots, n$, be such that

$$\sum_{i=1}^{n} f_1^2(t_i) = A \tag{9}$$

and

$$\sum_{i=1}^{n}\sum_{j=1}^{n} f_1(t_i)f_1(t_j)K(t_i, t_j) = A\lambda_n.$$

Then

$$\sum_{j=1}^{n} f_1(t_j)K(t_i, t_j) = \lambda_n f_1(t_i).$$

Let

$$h(s) = \sum_{j=1}^{n} f_1(t_j)K(s, t_j). \tag{10}$$

Then

$$\sum_{i=1}^{n} h^2(t_i) = A\lambda_n^2 \tag{11}$$

and

$$\| h \|^2 = A\lambda_n . \tag{12}$$

Thus, for $h(s)$ given by Eq. (10), the equality in Eq. (8) is attained. It is known that the maximum of the normalized quadratic form given by Eq. (8) is actually attained by the eigenvector, $\boldsymbol{\theta}_n$, corresponding to the largest eigenvalue, λ_n, of Eq. (6). Furthermore, the maximum of the normalized quadratic form is given by λ_n. (Courant and Hilbert, 1932). Thus, $f_1(t_i) = \theta_{n,i}$, $i = 1, \cdots, n$, and Eq. (4) follows from Eqs. (9), (10), and (11) while Eq. (5) follows from Eqs. (9), (10) and (12).

III. SAMPLING EXPANSIONS

In Section I, the classes of finite-energy, Fourier-, Hankel-, sine-, and cosine-transformed bandlimited signals were shown to be specific realizations of the abstract RKHS. For theoretical and practical reasons, sampling expansions may be of interest in these classes of signals. A class, Ω, of functions defined on a set T is said to possess a *sampling expansion* for a set of sampling instants $\{t_i \in T, i \in I\}$, if there exists a set of sampling functions, $\{\psi_i(s, t_i), i \in I\}$, such that

1. $\psi_i(s, t_i) \in \Omega, \qquad i \in I.$
2. $\psi_i(t_j, t_i) = \begin{cases} 1, & i = j \\ 0, & i \neq j. \end{cases}$
3. For any $f \in \Omega$, there is a uniformly convergent expansion given by $f(s) = \sum_{i \in I} f(t_i)\psi_i(s, t_i), \qquad s \in T.$

Theorem 4 states one simple relationship between complete orthonormal expansions and sampling expansions in RKHS.

THEOREM 4. *Consider the abstract reproducing kernel Hilbert space H with the reproducing kernel $K(s, t)$ defined on a set T. Let $\{\phi_i(s, t_i), t_i \in T, i \in I\}$ be a complete orthonormal system in H. If there are non-zero real con-*

stants c_i, such that

$$\phi_i(s, t_i) = c_i K(s, t_i), \qquad i \in I, \tag{13}$$

and

$$|K(t, t)| \leqq c < \infty, \qquad t \in T, \tag{14}$$

then the complete orthonormal expansion of any $f \in H$ given by

$$f(s) = \sum_{i \in I} a_i \phi_i(s, t_i), \qquad s \in T, \qquad a_i = (f, \phi_i), \tag{15}$$

is a sampling expansion.

Proof. From Eqs. (13) and (15),

$$a_i = (f, \phi_i) = c_i(f(s), K(s, t_i)) = c_i f(t_i).$$

Let $\psi_i(s, t_i) = c_i \phi_i(s, t_i)$. Then

$$c_i \phi_i(t_j, t_i) = c_i(\phi_i(s, t_i), K(s, t_j))$$

$$= \frac{c_i}{c_i} (\phi_i(s, t_i), \phi_j(s, t_j)) = \begin{cases} 1, & j = i \\ 0, & j \neq i. \end{cases}$$

Finally, in a RKHS, convergence in norm implies uniform convergence if $K(t, t)$ satisfies Eq. (14). (See Davis, 1963, Theorem 12.6.4).

In particular, the four sets of functions,

1. $\left\{\dfrac{\sin \pi(s - i)}{\pi(s - i)}, s \in T_1, -\infty < i < \infty\right\}$
2. $\left\{\dfrac{t_i(2s)^{1/2} J_\nu(\pi s)}{(t_i^2 - s^2)}, s \in T_2, \text{where } \{\pi t_i\} \text{ are the positive zeros of } J_\nu,\right.$ $\left.\nu \geqq -\dfrac{1}{2}, 1 \leqq i < \infty\right\}$
3. $\left\{\dfrac{2i}{(s + i)} \dfrac{\sin \pi(s - i)}{\pi(s - i)}, s \in T_3, 1 \leqq i < \infty\right\}$
4. $\left\{\dfrac{\sin \pi s}{\pi s}, \dfrac{2s}{(s + i)} \dfrac{\sin \pi(s - i)}{\pi(s - i)}, s \in T_4, 1 \leqq i < \infty\right\}$

are complete and orthonormal in H_j, where $j = 1, 2, 3, 4$, respectively. Since conditions required by Eqs. (13) and (14) in Theorem 4 are satisfied, we obtain the following four sampling expansions:

COROLLARY 4. *In RKHS H_1, any $f \in H_1$, possesses a sampling expansion given by*

$$f(s) = \sum_{i=-\infty}^{\infty} f(i) \frac{\sin \pi(s - i)}{\pi(s - i)}, \qquad -\infty < s < \infty. \tag{16}$$

COROLLARY 5. *In RKHS H_2, any $f \in H_2$, possesses a sampling expansion given by*

$$f(s) = \sum_{i=1}^{\infty} f(t_i) \frac{2(st_i)^{1/2}}{\pi J_{\nu+1}(\pi t_i)} \frac{J_\nu(\pi s)}{(t_i^2 - s^2)}, \qquad 0 < s < \infty, \qquad (17)$$

where $\{\pi t_i\}$ are the positive zeros of J_ν, $\nu \geqq -\frac{1}{2}$.

COROLLARY 6. *In RKHS H_3, any $f \in H_3$, possesses a sampling expansion given by*

$$f(s) = \sum_{i=1}^{\infty} f(i) \frac{2i}{(s+i)} \frac{\sin \pi(s-i)}{\pi(s-i)}, \qquad 0 < s < \infty.$$

COROLLARY 7. *In RKHS H_4, any $f \in H_4$, possesses a sampling expansion given by*

$$f(s) = f(0) \frac{\sin \pi s}{\pi s} + \sum_{i=1}^{\infty} f(i) \frac{2s}{(s+i)} \frac{\sin \pi(s-i)}{\pi(s-i)}, \qquad 0 < s < \infty.$$

The sampling expansion given by Eq. (16) is generally known as the Shannon sampling theorem and was derived by Whittaker, E. (1912). The sampling expansion given by Eq. (17) was first discussed by Whittaker, J. (1935). For other references and discussions see Yao and Thomas (1965).

In practice, we have only a finite number of terms in any sampling expansion. Theorem 5 gives an upper bound on the truncation error when a finite number of terms are used in place of all the terms in the sampling expansion of Theorem 4.

THEOREM 5. *Consider the abstract RKHS H and the sampling expansion*

$$f(s) = \sum_{i \in I} f(t_i)\psi_i(s, t_i), \qquad s \in T, \qquad f \in H$$

of Theorem 4. Let I' be a proper subset of I with some finite number of integers. The truncation error, $E_{I'}(s)$, is defined by

$$E_{I'}(s) = \sum_{i \in (I-I')} f(t_i)\psi_i(s, t_i).$$

Then

$$|E_{I'}(s)| \leqq [E - \sum_{i \in I'} c_i^2 f^2(t_i)]^{1/2} [\sum_{i \in (I-I')} c_i^2 K^2(s, t_i)]^{1/2}, \qquad f \in H'' \qquad (18)$$

is valid for any $f \in H''$, where $H'' = \{f \in H: \|f\|^2 \leqq E\}$. Furthermore, if $[E - \sum_{i \in I'} c_i^2 f^2(t_i)]$ is non-negative, there is a $f \in H''$ which attains the upper bound of $E_{I'}(s)$.

Proof. The hypercircle inequality of Golomb and Weinberger (1959) (See Davis, 1963, Theorem 9.4.7) states

$$|Lf - Lf_0|^2 \leqq [E - \|f_0\|^2][\sum_{i\epsilon(I-I')} (L\phi_i)^2], \tag{19}$$

where L is a bounded linear functional in a Hilbert space S, $f \epsilon S'' = \{g \epsilon S: \|g\|^2 \leqq E\}$, f_0 is an element of smallest norm satisfying

$$(f_0, \phi_i) = b_i, \qquad i \epsilon I',$$

$\{b_i, i \epsilon I'\}$ are fixed constants, and $\{\phi_i, i \epsilon I\}$ is a complete orthonormal system for S. Now, take $H = S$, $H'' = S''$, $b_i = c_i f(t_i)$,

$$f_0(s) = \sum_{i\epsilon I'} f(t_i)c_i\phi_i(s, t_i) = \sum_{i\epsilon I'} f(t_i)\psi_i(s, t_i),$$

$$f(s) = \sum_{i\epsilon I} f(t_i)\psi_i(s, t_i), \qquad f \epsilon H'',$$

and $L_s f = (f(u), K(u, s)) = f(s)$. Then Eq. (18) follows from Eq. (19). The attainment of the upper bound follows in the same manner as the attainment of the equality in the hypercircle inequality.

In the case of the Shannon sampling expansion, the result given in Theorem 5 can be used to calculate a simple upper truncation bound.

COROLLARY 8. *Consider the RKHS H_1 and the sampling expansion*

$$f(s) = \sum_{i=-\infty}^{\infty} f(i) \frac{\sin \pi(s-i)}{\pi(s-i)}, \qquad -\infty < s < \infty, \qquad f \epsilon H_1.$$

Let $I' = \{i_0(s) - M \leqq i \leqq i_0(s) + N\}$, where $i_0(s)$ is the nearest integer to time s, and M and N are positive integers. The truncation error, $E_{M,N}(s)$, is given by

$$E_{M,N}(s) = \sum_{i=-\infty, i_0(s)-M-1}^{i_0(s)+N+1} f(i) \frac{\sin \pi(s-i)}{\pi(s-i)}, \qquad -\infty < s < \infty, \qquad f \epsilon H_1.$$

For any $f \epsilon H'' = \{f \epsilon H_1 : \|f\|^2 \leqq E\}$, an upper bound of $E_{M,N}(s)$ is given by

$$|E_{M,N}(s)| < \begin{cases} \dfrac{E_0^{1/2}}{\pi}\left[\dfrac{1}{M} + \dfrac{2}{2N-1}\right]^{1/2}, & s \epsilon \left(i_0(s), i_0(s) + \dfrac{1}{2}\right] \\ \dfrac{E_0^{1/2}}{\pi}\left[\dfrac{2}{2M-1} + \dfrac{1}{N}\right]^{1/2}, & s \epsilon \left(i_0(s) - \dfrac{1}{2}, i_0(s)\right], \end{cases}$$

where

$$E_0 = \sum_{i=-\infty, i_0(s)+M-1}^{i_0(s)+N+1,\infty} f^2(i) < E < \infty .$$

Proof. From Theorem 5, in the case of the Shannon sampling expansion, $c_i = 1$ for all integers. Since $|\sin \pi(s - i)|$ is bounded by one and

$$\sum_{i=-\infty, i_0(s)-M-1}^{i_0(s)+N+1,\infty} \frac{1}{(s-i)^2} < \left[\frac{1}{M} + \frac{2}{2N-1}\right], \qquad s \in \left(i_0(s),\, i_0(s) + \frac{1}{2}\right],$$

$$\sum_{i=-\infty, i_0(s)-M-1)}^{i_0(s)+N+1,\infty} \frac{1}{(s-i)^2} < \left[\frac{2}{2M-1} + \frac{1}{N}\right], \qquad s \in \left(i_0(s) - \frac{1}{2},\, i_0(s)\right],$$

the upper bound of $E_{M,N}(s)$ follows immediately from Theorem 5.

Finally, we consider a sampling expansion problem in H_1 that relates various interpolation and minimum energy properties of Theorems 2, 3, 4, and 5. A finite Shannon sampling expansion given by

$$g(s) = C \sum_{i=1}^{n} f(t_i) \frac{\sin \pi(s - t_i)}{\pi(s - t_i)}, \tag{20}$$

where $C = 1$, $t_i = i$, $i = 1, \cdots, n$, $f \in H_1$, satisfies the following two properties:

$$\text{Interpolation Property} \quad g(t_i) = f(t_i),\ i = 1, \cdots, n, \tag{21}$$

$$\text{Minimum Energy Property} \quad \| g \|^2 = \min_{\substack{h \in h_1, \\ h(t_i) = f(t_i), \\ i=1,\cdots,n}} \| h \|^2. \tag{22}$$

In general, if the sampling instants, t_i, $i = 1, \cdots, n$, are distinct but arbitrary real points, the expansion given by Eq. (20) need not have the interpolation and minimum energy properties of Eqs. (21) and (22). A necessary and sufficient condition for the expansion given by Eq. (20) to satisfy the conditions of Eqs. (21), (22) is given in Theorem 6.

THEOREM 6. *A necessary and sufficient condition for a finite expansion given by Eq.* (20) *with arbitrary distinct real* $\{t_i, i = 1, \cdots, n\}$, *to satisfy the interpolation and minimum energy properties of Eqs.* (21), (22), *is that* $C = 1/\lambda_n$, $f(t_i) = \theta_{n,i}$ $i = 1, \cdots, n$, *where* λ_n *is the largest eigenvalue and* $\boldsymbol{\theta}_n$ *is the corresponding eigenvector of the matrix equation,*

$$\mathbf{K}\boldsymbol{\theta} = \lambda\boldsymbol{\theta},$$

$$\mathbf{K} = \left[\frac{\sin \pi(t_i - t_j)}{\pi(t_i - t_j)}\right]_{i,j=1,\cdots n}, \qquad \boldsymbol{\theta}^T = [\theta_1, \cdots, \theta_n].$$

Proof. Sufficiency: Let $f(t_i) = \theta_{n,i}$ and $C = 1/\lambda_n$. Then Eq. (19) becomes

$$g(t_j) = (1/\lambda_n) \sum_{i=1}^{n} \theta_{n,i} \frac{\sin \pi(t_j - t_i)}{\pi(t_j - t_i)} = \theta_{n,j} = f(t_j),$$

and Eq. (21) is satisfied. From Eq. (7), for any $h \in H_1$,

$$\left[\sum_{i=1}^{n} h^2(t_i)\right]^2 \leqq \| h \|^2 \left[\sum_{i=1}^{n} \sum_{j=1}^{n} h(t_i)h(t_j) \frac{\sin \pi(t_i - t_j)}{\pi(t_i - t_j)}\right].$$

Thus

$$\min_{\substack{h \in H_1, \\ h(t_i)=\theta_{n,i} \\ i=1,\cdots,n}} \| h \|^2 = (1/\lambda_n) \sum_{i=1}^{n} \theta_{n,i}^2. \tag{23}$$

By direct calculation

$$\| g \|^2 = (1/\lambda_n)^2 \sum_{i=1}^{n} \sum_{j=1}^{n} \theta_{n,i}\, \theta_{n,j} \frac{\sin \pi(t_i - t_j)}{\pi(t_i - t_j)} = (1/\lambda_n) \sum_{i=1}^{n} \theta_{n,i}^2. \tag{24}$$

Thus, Eqs. (23), (24) show that the property of Eq. (22) is satisfied by Eq. (20).

Necessity: Suppose Eq. (20) satisfies Eq. (21). That is,

$$f(t_j) = g(t_j) = C \sum_{i=1}^{n} f(t_i) \frac{\sin \pi(t_j - t_i)}{\pi(t_j - t_i)}, \qquad j = 1, \cdots, n.$$

Then

$$\mathbf{K}\hat{\mathbf{f}} = (1/C)\hat{\mathbf{f}}, \tag{25}$$

has eigenvalue $(1/C)$ and corresponding eigenvector $\hat{\mathbf{f}}$, where $\hat{\mathbf{f}}^T = [f(t_1), \cdots, f(t_n)]$. Suppose Eq. (20) satisfies Eq. (22). Then

$$\frac{\| g \|^2}{\sum_{i=1}^{n} f^2(t_i)} \min_{\substack{h \in H_1 \\ h(t_i)=f(t_i) \\ i=1,\cdots,n}} \frac{\| h \|^2}{\sum_{i=1}^{n} h^2(t_i)} = \frac{1}{\lambda_n}. \tag{26}$$

From Eqs. (23), (24), (25), and (26) then $C = 1/\lambda_n$ and $\hat{\mathbf{f}} = \boldsymbol{\theta}_n$.

In particular, if $t_i = i$, $i = 1, \cdots, n$, then $K(t_j - t_i) = \sin \pi(t_j - t_i)/\pi(t_j - t_i) = \delta_{ij}$ and $\lambda_n = 1 = C$. Then $\theta_{n,i}$, $i = 1, \cdots, n$, can take on any finite value. This conclusion clearly agrees with the previously known results.

IV. GENERALIZATIONS

Thus far, we have considered some of the more elementary aspects of reproducing kernel Hilbert spaces and bandlimited signals. It is clear that many generalizations are possible in several directions. Besides Fourier-, Hankel-, sine-, and cosine-transformed bandlimited signals, other unitary-transformed bandlimited signals can also be shown to be specific realizations of RKHS. Furthermore, if a class of bandlimited signals is a RKHS, then that class of bandpass signals is also a RKHS. The reproducing kernel in that case is obtained from the inverse transform of the indicator function of the support of the signals in the transformed domain.

If the number of sampling instants, n, is allowed to become infinite, then many of the above results are still valid. The proofs of some of these results then become considerably more involved. There are also analogous results in most of the above cases when constraints are imposed on the derivatives of the signals as well as the signals.

In conclusion, this paper has considered various new and known results in bandlimited signals and sampling expansions from the point of view of the RKHS. The actions of the reproducing kernel in extremum problems were emphasized. The RKHS approach simplified matters in some cases but offered a unified point of view in all cases.

Received: July 7, 1967

REFERENCES

Aronszajn, N. (1950), Theory of reproducing kernels. *Trans. Am. Math. Soc.* **68**, 337–404.

Bergman, S. (1950), "The Kernel Function and Conformal Mapping." Am. Math. Soc., New York.

Courant, R. and Hilbert, D. (1953), "Methods of Mathematical Physics." Wiley, New York.

Davis, P. (1963), "Interpolation and Approximation." Ginn, New York.

Kailath, T. (1967), "Statistical Detection Theory," unpublished.

Landau, H. and Pollak, H. (1961), Prolate spheroidal wave functions, Fourier analysis and uncertainty—II, *Bell System Tech. J.*, **40**, 65–84.

Loève, M. (1948), "Fonctions aléatoires de second ordre." Supplement to P. Levy, "Procès stochastiques et mouvement Brownien." Gauthier-Villars, Paris.

Moore, E. H. (1935), General analysis. *Mem. Am. Philosophical Soc.* Part I, 1935, Part II, 1939.

Parzen, E. (1961), An approach to time series analysis. *Ann. Math. Stat.*, **32**, 951–989.

Parzen, E. (1962), Extraction and detection problems and reproducing kernel Hilbert spaces. *SIAM. J. Control*, **1**, 35–62.

Shannon, C. E. (1948), A mathematical theory of communication. *Bell System Tech. J.*, **27**, 379–423, 623–656.

Slepian, D. and Pollak, H. (1961), Prolate spheroidal wave functions, Fourier analysis and uncertainty—I, *Bell System Tech. J.*, **40**, 43–64.

Whittaker, E. T. (1914), On the functions which are represented by the expansions of the interpolation theory. *Proc. Royal Soc., Edinburgh*, **35**, 181–194.

Whittaker, J. M. (1935), "Interpolatory Function Theory," Cambridge Univ. Press, London and New York.

Yao, K. and Thomas, J. B. (1965), On some representations and sampling expansions for bandlimited signals. Communication Lab., Dept. of Elec. Eng., Princeton Univ., Tech. Report No. 9.

AUTHOR CITATION INDEX

SUBJECT INDEX

About the Editor

HOWARD L. WEINERT is an associate professor of electrical engineering and computer science at The Johns Hopkins University. He was an undergraduate at Rice University and received the B.A. summa cum laude in 1967. He did his graduate work at Stanford University, earning the Ph.D. in 1972. He came to Hopkins in 1974 after two years at Systems Control, Inc.

Dr. Weinert has conducted research and taught courses in the areas of least-squares estimation, spline fitting, optimal control, linear system theory, and signal detection. He is currently developing parallel algorithms and fault-tolerant computational structures for statistical signal processing problems.

Dr. Weinert is a senior member of IEEE and a member of SIAM.

El Juego del Ángel